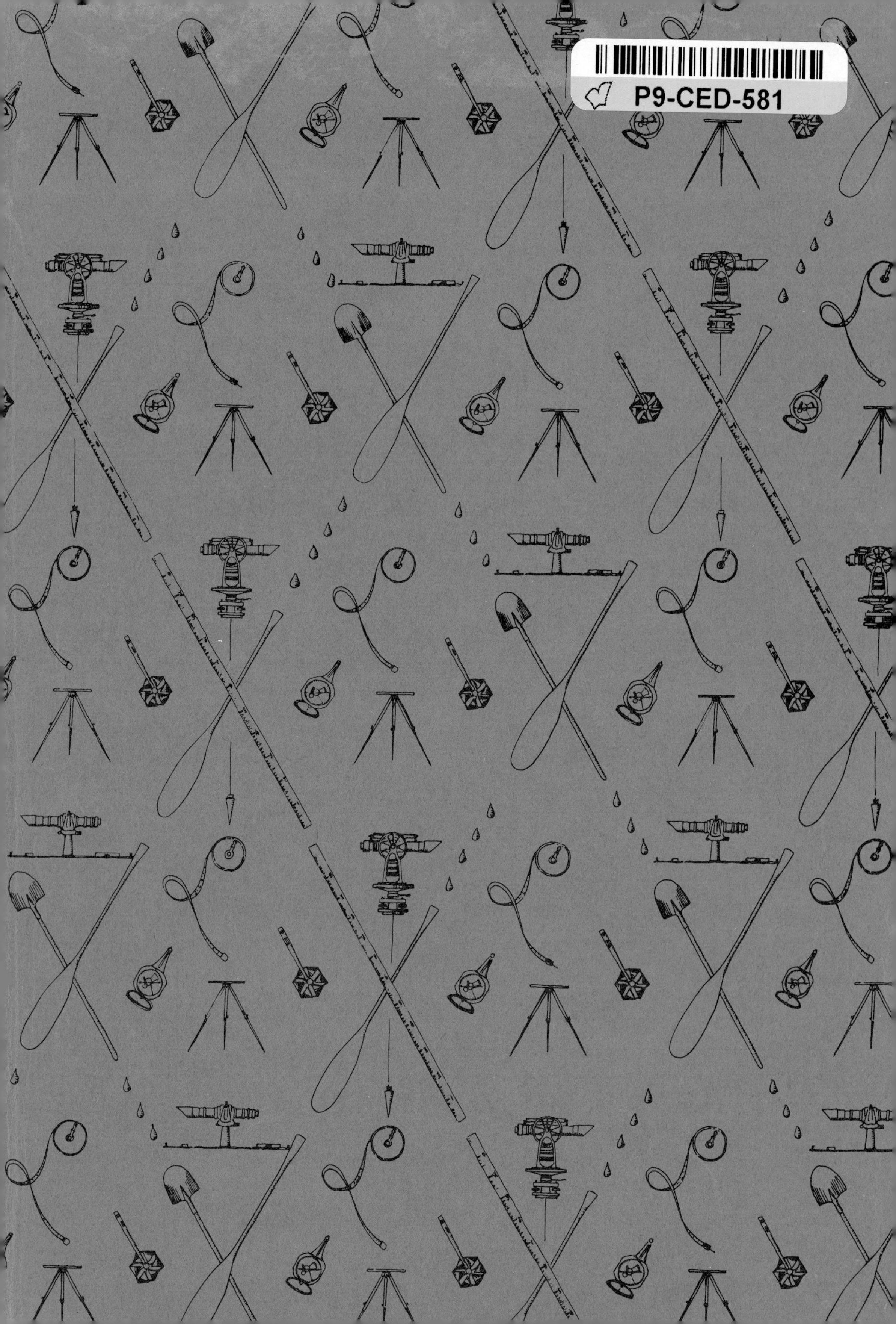

P9-CED-581

Water in Environmental Planning

WATER
in Environmental Planning

Thomas Dunne

University of Washington

Luna B. Leopold

University of California, Berkeley

W. H. Freeman and Company
San Francisco

The photograph on the dust jacket, taken near Williams, Colusa County, California, shows a valley typical of the Coast Range. The sinuous ephemeral stream is incised in the alluvium forming the valley flat. The dark areas in the swales and low spots are those producing saturated overland flow. Hortonian overland flow is probably the main runoff-producing process on the steep hillslopes.
Photograph on dust jacket is copyrighted by William A. Garnett.

Library of Congress Cataloging in Publication Data

Dunne, Thomas, 1943–
Water in environmental planning.

Includes bibliographies and indexes.
1. Environmental engineering. 2. Regional planning.
3. Hydrology. I. Leopold, Luna Bergere, 1915–
joint author. II. Title.
TD160.D85 333.9′1 78-8013
ISBN 0-7167-0079-4

Printed in the United States of America

3 4 5 6 7 8 9

To Bettylou and Barbara

A Note on Units

Although hydrologists and geomorphologists throughout the world work with metric units, the voluminous data for the United States are published primarily in English units. The use of metric units in the United States is increasing, but for many years to come the users of hydrologic and geomorphic information will have to deal with English units as well as the metric system. In this text, we have used metric units wherever it is convenient. We have not, however, converted all the diagrams drawn from other published works. This mixture of units may be a little inconvenient, but it is a situation to which we must all adjust.

Contents

13. Snow Hydrology 465

Part III. Geomorphology 491

14. Drainage Basins 493

15. Hillslope Processes 506

16. River Channels 590

17. Sediment Production and Transport 661

Preface

In this book we show how a knowledge of hydrology, fluvial geomorphology, and river quality is useful in planning. Many specialists in such disciplines as biology, engineering, forestry, geography, geology, hydrology, landscape architecture, and regional planning have taken up the challenge of applying their knowledge to the avoidance or solution of environmental problems. In developing nations, many scientists originally trained in a specialized discipline find themselves in the position of advising their governments on a wide range of scientific issues, including the assessment and management of natural resources and the maintenance of environmental quality during development. We refer to all these people as "planners," and we do not wish to restrict the title to those who have planning degrees or work in planning offices. Anyone who makes his or her knowledge available for the planning process is included in our term.

During the past few years we have taught interdisciplinary courses at the senior and graduate levels demonstrating the use of hydrology, fluvial geomorphology, and the study of river attributes in problems concerning the maintenance or reclamation of environmental quality. We and our students have been hampered by the lack of an appropriate textbook. Many excellent hydrology texts are available for the specialist, particularly in engineering. But many students and professionals need to understand the basis of hydrologic forecasting, hillslope stability, water quality, and other matters without knowing all the details of how to design major structures or treatment plants. Also, with the increasing use of nonstructural solutions to environmental problems and the increasing interest in avoiding environmental degradation in the first place, students are interested in applications of science that allow early recognition of problems and a wider range of alternative strategies than is covered in traditional texts.

Many planning problems require specialized engineering design not covered by this book, and we are not attempting to replace texts dealing with these topics, although we have included some material usually found in books on engineering hydrology. We recommend, however, that planners become involved in the analysis of processes and environments in ways that are now largely ignored by both planners and specialists. In many regions development is occurring so rapidly that specialist engineering services cannot keep up with the need for information. In other cases, individual developments are not thought to justify the cost of an environmental analysis. So, by default, no analysis is made and many problems arise. Particularly in the chapters on groundwater, runoff processes, flood hazard, hillslope and channel processes, and esthetics, we recommend observations that the planner could make to offset the lack of information provided by specialist agencies.

Furthermore, most texts in engineering hydrology deal with large areas or drainage basins of more than 100 square kilometers. But most areas of concern to environmental planners are small—a few square kilometers or a few hectares. For such small areas, hydrologic data are scarce. We present a variety of methods of estimating or measuring, in a simple manner, relevant factors on small areas.

Our first aim is to make planners *aware* of some of the environmental constraints placed upon development by natural processes. A great deal of damage has been done in the past simply because planning was nonexistent or because planners did not appreciate the interplay of processes that affects their schemes. We also hope that planners will learn to communicate more effectively with the specialists upon whom they must rely from time to time. Planners should understand, for example, what engineering hydrologists can and cannot do and how they go about solving a problem.

It has been our experience in both practice and teaching that most of the techniques used for design and for quantitative predictions of environmental effects can only be understood by making calculations. This may be unfamiliar or even distasteful to many planners, but we know of no other way.

Planners should be able to analyze and check the quantitative statements made in engineering reports, proposals, and environmental impact statements. Often, some simple quantitative field observations and calculations by the planner and an understanding of natural processes and environments will indicate whether there is a need to call in a specialist, or whether a certain course of action should be avoided altogether. We have therefore used a quantitative approach throughout the book. The level of mathematics used is covered long before a student leaves high school. Some of our students undergo traumas for a week or so as they rediscover logarithmic tables, dust off their old slide rules, invest in hand calculators, and confront graph paper again. Before long, however, they find that there is no real difficulty in handling symbols and numbers. The power of some of the simple techniques we introduce convinces them that a quantitative approach is valuable. Worked examples are provided for practice. We intend that this book be *used,* not simply read as a literature review

The instructor of a particular course may not wish to include all the material in this text. A course may be limited to a portion of the material and may incorporate local planning problems. On the basis of our experience, we suggest the following chapters of the text to form the basis of a one-quarter, five-credit course with a laboratory for students with various interests:

Landscape Architecture and Environmental Planning: Chapters 2, 6, 7, 9, 10, 11, 12, 15, 16, 18, 22

Physical Geography and Environmental Geology: Chapters 2, 4, 5, 6, 7, 8, 9, 10, 12, 13, 15, 16, 17, 18

Agricultural Management and Rural Planning: Chapters 2, 4, 5, 6, 7, 8, 9, 10, 11, 12, 15, 16, 18, 20, 21

Forestry: Chapters 2, 3, 4, 5, 6, 7, 8, 9, 10, 12, 13, 20, 21, 22

Engineering: Chapters 2, 4, 6, 7, 9, 10, 11, 12, 13, 15, 16, 17, 18

For use in the laboratory section of a course, Typical Problems appear at the end of most chapters. Some are worked out to illustrate a technique; others are not. The instructor will want to change many of them to incorporate local data and special problems. Other exercises might involve replacing some of the graphs with compilations for the region of interest to the class. Professional planners will certainly need to do this.

Of course, we would like all readers to take time to examine the entire book, because only then will they fully appreciate that we are trying to show how many planning problems involve a complex mixture of hydrology, geomorphology, and river quality. To understand and solve such problems requires an appreciation of the linkages between the three disciplines introduced in the text. We hope that material included here will lead to greater communication across the boundaries of the three fields.

We are indebted to several friends for their help during the writing of this book. Dr. A. Lincoln Washburn, then Director of the Quaternary Research Center at the University of Washington, was a source of constant encouragement and enthusiasm about the value of an interdisciplinary approach to earth science. Professor Don C. Erman gave valuable advice in the preparation of material on stream biota in Chapter 21. Professor William A. Garnett, artist in photography, provided the image used on the dust jacket. Our colleague and friend, Dr. William W. Emmett was a participant in the development of many of the techniques for field collection of data and methods of data analysis. Dr. David Western was especially helpful in the development of ideas and data for the examples from Kenya.

To companions beside rivers on several continents, we express our gratitude for ideas and assistance, especially to William W. Emmett, William E.

Dietrich, Herbert E. Skibitzke, Robert Myrick, David Western, George S. Ongweny, William B. Bull, Asher Schick, Denis St.-Onge, Bettylou Dunne, and Barbara B. Leopold.

May 1978

THOMAS DUNNE
LUNA B. LEOPOLD

Typical Problems

Symbols

The numbers in parentheses refer to the chapters in which the symbol is used in the manner described.

α	Albedo, reflectivity of a surface (4, 5, 13)
Δ	Slope of the curve relating saturation vapor pressure and temperature (4, 5)
Δ	Difference in (7, 10)
γ	Psychometric constant (4, 5)
θ	Soil moisture content (9)
θ_s	Soil moisture content at saturation (9)
μ	Dynamic viscosity of water (7)
μ_f	Dynamic viscosity of water at field temperature (7)
μ_s	Dynamic viscosity of water at 15°C (7)
ρ	Density of water (4, 5, 13)
σ	Stefan-Boltzmann constant (4, 5, 13)
Φ	Rate at which rainfall is absorbed by a drainage basin (10)
φ	Angle of internal friction (15)
A	Area of a lake or drainage basin (4, 10, 14)
A	Cross-sectional area of an aquifer or channel (7, 16)
A	Annual soil loss in the Universal Soil-Loss Equation (15)
A_s	Area of saturated soil in a drainage basin (9, 10)
AET	Actual evapotranspiration (5, 8)
AW	Available water content of a soil (5, 8)
AWC	Available water capacity of a soil (5, 8)

a Empirical constant (4, 13)
a Exponent in Thornthwaite equation (5)

B Basal area of obstacle (13)
BF Baseflow (9)
BOD Biochemical oxygen demand (20)
b Exponent of Q for width relation in hydraulic geometry (16)
bkf As a subscript refers to bankfull conditions (16, 18)

C Canopy interception (3)
C Cloudiness (4, 5, 13)
C Runoff coefficient in the Rational Formula (10)
C Cropping management factor in the Universal Soil-Loss Equation (15)
C Chezy resistance coefficient (16)
C_p Coefficient in formula for peak discharge of the synthetic unit hydrograph (10)
C_t Coefficient in formula for lag to peak of the synthetic unit hydrograph (10)
C_0, C_1, C_2 Coefficients in the Muskingum flood-routing method (10)
°C Degrees Celsius
CN Curve number (10)
COD Chemical oxygen demand (20)
c Empirical constant (16)
c_w Specific heat of water (4, 13)
cfs Cubic feet per second (16)

D Soil moisture deficit (8)
D Diameter of bed sediment (16)
D Dissolved oxygen deficit of a stream (20)
D_A Drainage area (16)
D_c Minimum dissolved oxygen deficit in a stream (20)
D_i Initial dissolved oxygen deficit of a stream-water and wastewater mixture (20)
D_R, D_r Durations of rainstorms for synthetic unit hydrograph (10)
DPS Direct precipitation on saturated area (9)
d Monthly fraction of annual hours of daylight (5)
d Depth of flow (7, 16, 17, 18)

E	Evaporation of intercepted water (3)
E	Kinetic energy of rainfall (15)
E_a	Component of evaporation or evapotranspiration due to mass transfer of water vapor (4, 5)
E_o	Evaporation from open water (4, 5)
E_s	Evaporation from snow (4)
E_t	Potential evapotranspiration (5)
ET	Evapotranspiration (3)
e	Base of natural logarithms (20)
e_a	Vapor pressure of the atmosphere at height a above a surface (4, 5, 13)
e_s	Vapor pressure of water, vegetation, or snow surface (4, 5, 13)
e_{sa}	Saturation vapor pressure of the atmosphere at temperature T_a (4, 5)
e_2	Vapor pressure at height of 2 meters (4)
F	Cumulative percentage frequency of a variable (2)
F	Fractional canopy cover of a forest (13)
°F	Degrees Fahrenheit
FC	Field capacity of a soil (6, 8)
f	Exponent of Q for depth relation in hydraulic geometry (16)
f	Darcy-Weisbach friction factor (16)
f_o	Initial infiltration rate (9)
f_c	Final constant infiltration rate (9)
$f(\ )$	Function of
G	Vegetative surface area (3)
G_c	Suspended sediment concentration (19)
GWR	Groundwater runoff (8)
GWS	Groundwater storage (8)
gh	Gauge height or stage (16)
H	Average grass height (3)
H	Net radiation expressed in terms of the equivalent depth of water evaporation (4, 5)
H	Height of a water surface (10)
H	Difference in elevation from a basin outlet to the most distant ridge (10)
h	Head in an aquifer (7)

h Height (10)

Δh Net change of water surface elevation (4)

I Total interception by canopy and litter (3)

I Rate or volume of inflow (4, 10)

I Thornthwaite annual heat index (5)

I Rainfall intensity in the Rational Runoff Formula (10)

I_o Solar radiation per day at outer edge of atmosphere (4)

I_t Index of antecedent precipitation on day t (10)

I_0 Index of antecedent precipitation at the beginning of a period (10)

IUH Instantaneous unit hydrograph (10)

K Sensible heat transfer to a lake or a vegetation surface expressed in terms of equivalent depth of water evaporation (4, 5)

K Seasonal crop factor in the Blaney-Criddle Formula (5)

K Coefficient of permeability, also called hydraulic conductivity (7)

K Ratio of computed flood peaks after urbanization to flood peaks before urbanization (10)

K Coefficient in the Muskingum flood-routing method (10)

K Soil erodibility factor in the Universal Soil-Loss Equation (15)

K Heat lost to atmosphere from a stream per degree of excess water temperature above the atmospheric temperature (19)

K_f Field coefficient of permeability at field water temperature (7)

K_s Standard coefficient of permeability at a water temperature of 15.6°C (7)

°K Degrees Kelvin (4, 5, 13)

k Monthly crop factor in the Blaney-Criddle Formula (5)

k Constant expressing the rate of reduction of basin wetness in the formula for antecedent precipitation index (10)

k Degree-day factor (or index) for snowmelt (13)

k Empirical constant (16)

k Deoxygenation constant or rate of exertion of biochemical oxygen demand (20)

L Litter interception (3)

L Latent heat of vaporization of water (4, 5)

L Length of catchment along the mainstream from the basin outlet to the most distant ridge (10, 14)

L Width of reservoir outlet or weir length (10)

L	Hillslope length factor in the Universal Soil-Loss Equation (15)
L	Distance downstream (19)
L	Initial biochemical oxygen demand of a mixture of stream water and wastewater (20)
L_c	Distance along mainstream from the outlet of a basin to the areal centroid of the catchment (10)
l	Distance (7)
l_c	Lag time between centroid of rainfall and centroid of runoff (9, 10)
l_p	Lag time between centroid of rainfall and peak of hydrograph (10)
M	Rate of release of meltwater from a snowpack (13)
M_e	Snowmelt due to latent heat flux (13)
M_h	Snowmelt due to sensible heat flux (13)
M_{lw}	Snowmelt due to net longwave radiation (13)
M_p	Snowmelt due to the flux of heat from rainfall (13)
M_s	Snowmelt due to solar radiation (13)
m	Rank of an event according to its size (2, 10, 12)
m	Exponent of Q in velocity relation in hydraulic geometry (16)
msl	Mean sea level
N	Long-term normal precipitation (2)
N	Maximum possible hours of sunshine (4)
N	Mass-transfer coefficient (4)
N	Number of days (10)
N	Normal stress (15)
n	Number of items in a record (2, 5, 10, 12)
n	Observed duration of bright sunshine (4)
n	Manning roughness coefficient (16)
n_w	Number of flow tubes in width w (7)
O	Rate or volume of outflow (4, 10)
OF	Overland flow (8)
P	Depth of rainfall (2, 3, 8, 13)
P	Erosion-control practice factor in the Universal Soil-Loss Equation (15)
PET	Potential evapotranspiration (5, 8)

PWP Permanent wilting point (5, 6)

p Probability (2, 10, 12)

p Atmospheric pressure (4, 5, 13)

p Piezometric pressure in a groundwater body (7)

p Pressure of water (15)

p_o Atmospheric pressure at sea level (13)

Q Discharge rate (7, 10, 12)

Q_{ave} Average annual discharge (16)

Q_{bkf} Discharge at bankfull (16)

Q_e Energy used for evaporation (4, 13)

Q_{et} Energy used for evapotranspiration (5)

Q_f Discharge of a given frequency (16)

Q_h Energy transferred as sensible heat (4, 5, 13)

Q_{lw} Net longwave radiation (4, 5, 13)

Q_{lwf} Net longwave radiation under a forest canopy (13)

Q_m Energy used to melt snow (13)

Q_n Net allwave radiation (4, 5)

Q_p Energy advected in rainfall (13)

Q_{pk} Peak discharge (10)

Q_{rs} Reflected solar radiation (4, 5, 13)

Q_s Incoming solar radiation or insolation (4, 5, 13)

Q_T Discharge of the T-year flood (10)

Q_v Net advected energy (4, 5)

Q_{ve} Energy advected out of a lake by evaporated water (4)

Q_θ Change of stored energy (5, 13)

q Probability that an event will occur within a specified time interval (2, 10)

q Discharge per unit width (7)

q Thermal quality of a snowpack, the ratio of the weight of ice to the total weight of a unit volume of the snowpack (13)

R Net rainfall entering the soil (3)

R Bowen's ratio (4)

R Volume of storm runoff (10)

R Rainfall erosivity index (15)

R Hydraulic radius (16)

RF Return flow (9)

RO	Runoff (8)
r	Reaeration constant (20)
S	Stemflow (3)
S	Net groundwater seepage (4)
S	Soil moisture surplus (8)
S	Slope of channel (10, 16)
S	Volume of storage (10)
S	Hillslope gradient factor in the Universal Soil-Loss Equation (15)
SM	Soil moisture (8)
$SSSF$	Subsurface stormflow (9)
s	Standard deviation (2)
s	Shear stress (15)
s	Gradient of a channel (16)
$S_{\overline{x}}$	Standard error of the mean (12)
T	Recurrence interval (2, 10, 12)
T	Throughfall (3)
T	Transmissibility of an aquifer (7)
T_a	Air temperature at height *a* above a surface (4, 5, 13)
T_a	Mean monthly air temperature (5)
T_b	Base temperature for calculations (4, 13)
T_b	Time base of a hydrograph or the duration of storm runoff (10)
T_{bi}	Time base of an instantaneous unit hydrograph (10)
T_d	Dewpoint temperature (13)
T_f	Temperature of a forest canopy (13)
T_L	Excess temperature of stream water over that of the atmosphere at distance *L* downstream (19)
T_m	Recurrence interval
T_o	Excess temperature of stream water over that of the atmosphere (19)
T_p	Time to peak, or duration of the rising limb of a hydrograph (10)
T_{pi}	Time to peak for an instantaneous unit hydrograph (10)
T_r	Duration of the recession limb of a hydrograph (10)
T_s	Surface temperature (4, 5, 13)
t	Time, number of days in a calculation period (10, 20)
t	Turbidity (19)
t_b	Time base of unit hydrograph (10)

t_c Time of concentration (10)

t_c Time of occurrence of low point on the oxygen sag curve (20)

t_p Lag to peak or lag time between centroid of rainfall and peak of the hydrograph (10)

t_{pR} Lag to peak of a unit hydrograph for a storm with duration D_R (10)

u Windspeed (4)

u Velocity (7, 16)

u_a, u_b Windspeed at height a or b above surface (4, 5, 13)

u_f Windspeed in a forest (13)

u_* Shear velocity (16)

V Volume of a lake (4)

v Velocity of flow (7, 16)

w Width (7, 19)

w Width of unit hydrograph (10)

wp Wetted perimeter (16)

w_{50}, w_{75} Width of a synthetic unit hydrograph at discharges respectively 50 and 75 percent of the peak value

x Distance along abscissa (7)

x Coefficient in the Muskingum flood-routing method (10)

x_m Annual maximum of recurrence interval T_m (10)

y Thickness of an aquifer (7)

Z Percentage of impervious area in an urban drainage basin (10)

z Elevation of a measurement point (7)

z_a Height above a surface at which measurements are made (4, 5, 13)

z_o Roughness length of a surface (4, 5, 13)

Water in Environmental Planning

I

Introduction

1

Six Field Examples

Water is central to many planning problems concerned with natural and altered environments. It is often a focus for interdisciplinary analysis and planning that brings together an engineering hydrologist and a plant ecologist, a forester and a sanitary engineer, or a geomorphologist and a specialist in urban design. The awareness of mutual concerns and the value of shared experience are growing. The central position of water in planning the avoidance or the rectification of environmental problems is leading specialists to discover new and interesting problems to which their talents can be applied.

We are excited about these developments and expect them to continue. We hope to communicate some of our excitement to the reader by outlining the potential for making a contribution to environmental planning based upon an understanding of hydrologic processes. To illustrate some of the potential, we will briefly outline six planning problems in which an understanding of hydrologic processes can make an important contribution. The reader should note that not all the work we mention was done by hydrologists, or even by people who considered themselves to be planners.

We are using the term "hydrology" here not only to involve the movement of water over and under the land surface, but also to include a variety of geomorphic, geochemical, and biologic processes that depend upon the storage and movement of water. These latter topics are usually discussed separately in courses on fluvial geomorphology, water chemistry, and aquatic biology. We try to bring them together in one text because they are related in many ways, and also because those who would like to participate in environmental planning should have at least an introductory knowledge of all three fields.

The Hydrologic Cycle

The case studies of environmental problems described in this chapter concern water. More particularly, they concern the ways in which the fluid is stored and transferred over and under the earth's surface. Some of the issues concern not only the water itself but its transport of sediment or dissolved materials, and its alteration of the earth's surface by erosion and deposition. Another theme running through the examples is that the activity of water is subject to natural fluctuations and to human modification.

It is useful for planners to have a conceptual framework within which to analyze environmental problems that have arisen or to anticipate the consequences of development. Such a framework is provided by the *hydrologic cycle.*

The hydrologic cycle, represented schematically in Figure 1-1, describes the ways in which water moves around the earth. During its endless circulation from ocean to atmosphere to earth and back to ocean, the water is stored temporarily in streams, lakes, the soil, or groundwater and becomes available for use.

In the cycle, solar energy evaporates water from the ocean. This water is carried by winds over the continents, and when atmospheric conditions are favorable, a portion of the water is precipitated, generally as rain or snow. If cold conditions prevail at the ground surface, snow will be stored there until enough energy is available for melting. The meltwater will thereafter follow the same pathways as rainwater. Let us consider, therefore, only the case of rain.

Before reaching the surface of the earth, most rain is caught by vegetation. Some of the water is stored upon leaf surfaces during wetting, and the remainder falls to the ground from leaves and branches, or runs down trunks and stems. A small amount of water never reaches the ground, but is evaporated back to the atmosphere from vegetation during and after the rainstorm. The process by which water is "short-circuited" back to the atmosphere in this way is known as *interception.*

Upon reaching the ground surface, a portion of the rain is absorbed by the soil. Rain that is not absorbed remains on the surface of the ground, fills small depressions, and eventually spills over and runs quickly downslope into streams as *overland flow,* which generates floods. The absorbed rainwater seeps into the soil by the process of *infiltration,* and is held there as *soil moisture* by capillary forces. If the soil moisture content is raised sufficiently, infiltrating water will displace older soil water, which may percolate laterally through the topsoil into streams as *subsurface storm runoff* or vertically to the *groundwater zone* where the pores of the soil or rock are completely filled with water. From this zone water moves slowly into streams, swamps, or lakes, providing surface runoff during dry weather.

Not all the infiltrated water reaches a stream, however. Some of it remains in the topsoil after rain and is returned to the atmosphere by *evaporation* from the soil surface, or by *transpiration* from the leaves of plants. Other water evaporates from streams, lakes, and swamps.

The concept of the hydrologic cycle can be extended to include the movement of sediment, chemicals, heat, and biota contained in water. Viewed in this way, the hydrologic cycle includes the aqueous phases of the cycles of sediment, gases and minerals, heat, and living matter. The unifying element in such a discussion is the storage and transport of water and its constituents. This extended definition of the hydrologic cycle enhances the value of the cycle as a framework for the analysis of many problems in planning and ecology.

Rocks exposed at the surface of the earth combine with water, various gases (especially oxygen and carbon dioxide), and organic acids by a set of geologic processes known as *weathering*. The products of these reactions are soil and chemical solutions. The latter flow through the soil and the groundwater zone to streams, determining the chemical properties of each water resource and its suitability for irrigation, washing, drinking, and other uses.

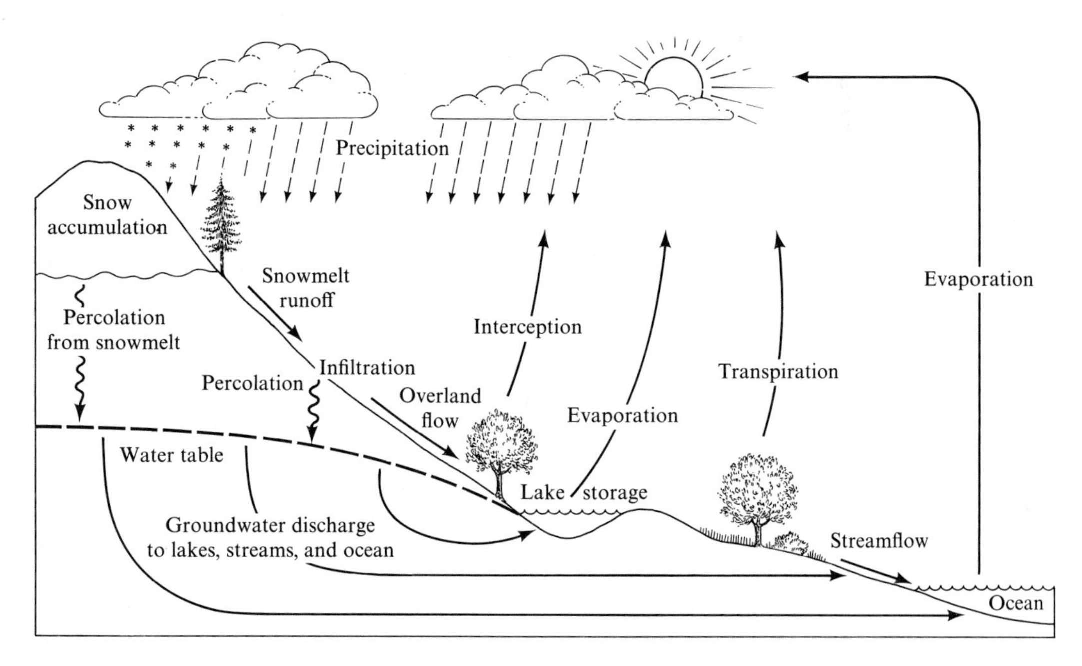

Figure 1-1 Schematic diagram of the hydrologic cycle.

The residual soil is eventually eroded by processes such as rainsplash, sheetwash, gullying, and landsliding, which carry the loosened particles into rivers through which they are transported to the sea. The transporting agencies may be intermittent, and the eroded material may be deposited temporarily on hillslopes, in channels or floodplains of rivers, or in lakes. It is by the erosion, transport, and deposition of weathered material that assemblages of landforms are produced. Soils and landscape, then, are products of weathering and the movement of sediment through the hydrologic cycle. Sediment and solutes are eventually returned to the sea and incorporated into new rocks to complete the cycling of geologic materials.

As water flows through the hydrologic cycle it exchanges heat with its environment. Most of the exchanges are natural and reflect the weather, but large concentrated quantities of heat are also added to streams and lakes when the water is used as a coolant in power plants and industrial processes. The discharge of waste heat, therefore, is a hydrologic problem, concerned with questions of water supply, the size and the shape of the receiving water body, and exchanges of energy between water bodies and the atmosphere.

Aquatic organisms from viruses and bacteria to game fish take advantage of the transport of nutrients in streams and lakes. Their survival and numbers are also partly a hydrologic problem. The maintenance of a healthy aquatic ecosystem is a goal of wise environmental planning and management and requires some knowledge of hydrologic processes. Aquatic organisms also provide a tool or an index with which the environmental conditions affecting a stream or lake can be measured. In Chapter 21, therefore, we review some principles governing the use of aquatic organisms as indicators of the general quality of a water resource for drinking, recreation, and other uses.

Finally, the hydrologic cycle is an appropriate framework for analyzing human modification of land and water resources. In each chapter of the book we will be concerned with this topic because people are major agents in the hydrologic cycle. They alter the land surface, manipulate the quantities of water in storage in various parts of the cycle, and radically change the concentrations of sediment, solutes, heat, and biota. A great many problems that confront a planner, therefore, can be analyzed by considering the paths that water takes, what the water is doing at various stages along each path, and how the quantity, pressure, chemistry, or any other characteristic of the water is altered by human action.

Amboseli National Park, Kenya

At the foot of Mount Kilimanjaro in southern Kenya, Amboseli National Park is centered on a dried-up Pleistocene lake bed (Figure 1-2). Water percolating downslope through lava and ash beds of the Kilimanjaro volcano saturates the lake sediments and emerges as springs in two central swamps

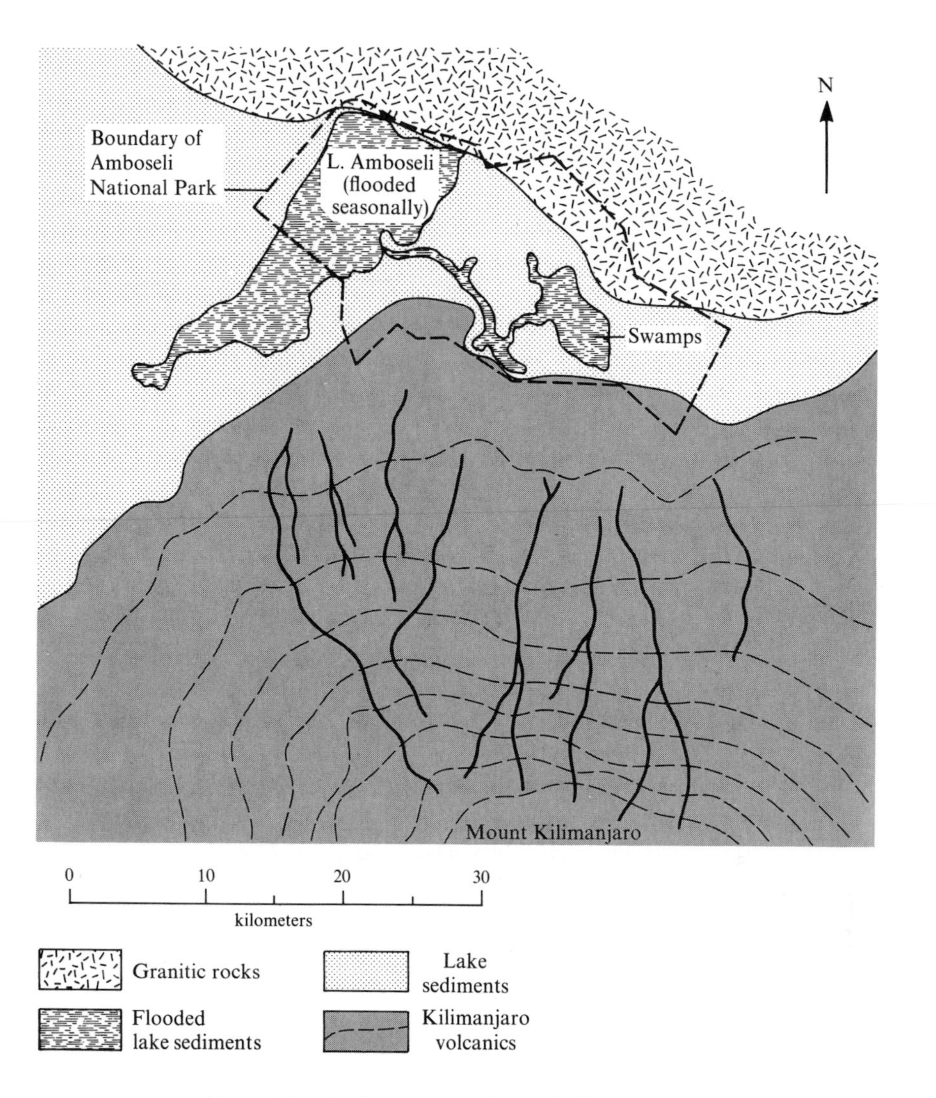

Figure 1-2 Geologic map of Amboseli National Park.

(Figure 1-3). This groundwater is very saline. The climate of the area is arid (350–400 mm of rainfall per annum) and supports dry savannah grassland and bush. Until the mid-1950's, however, the lake bed had an extensive cover of stately yellow-fever trees (*Acacia xanthophloea*) which provided a habitat for woodland game such as baboon, impala, and leopard. After this time the fever-tree woodlands declined rapidly, especially after 1961–1962, and were replaced by a sparse cover of salt-resistant short grass and a few bushes. The woodland game were replaced by plains game such as zebra, wildebeest, and kongoni.

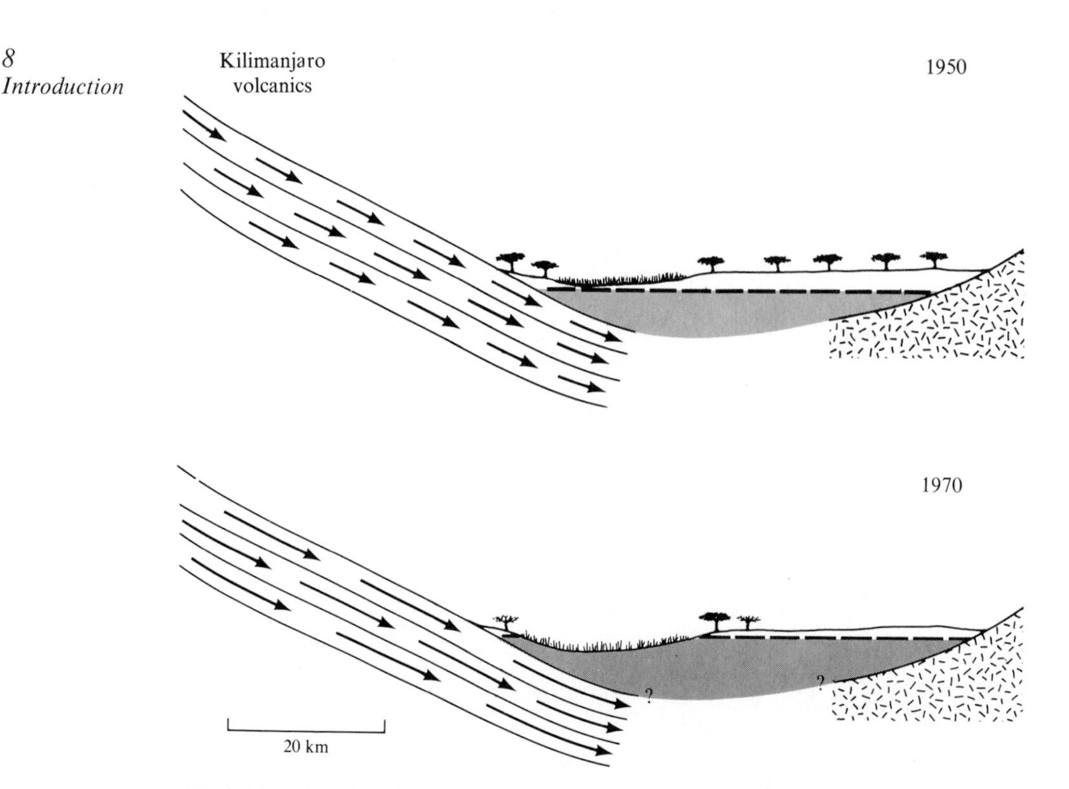

Figure 1-3 Alteration of the groundwater regime beneath the lake sediments of the Amboseli basin. In 1950 the inflow of groundwater from Mount Kilimanjaro was sufficient only to maintain a water table more than 4 m below the ground surface. The basin was covered by woodland and inhabited by a woodland fauna. After increased rainfall and groundwater inflow (indicated by the longer arrows in the lower diagram) during the 1960's, the highly saline groundwater rose into the root zone, killing the trees to leave a barren surface populated by a few salt-tolerant grasses and a fauna of plains game, around a central swamp that is kept fresh by a small inflow of shallow groundwater.

Many conservationists blamed this destruction of the vegetation upon the nearest "obvious" traditional culprits, namely Maasai herdsmen, and pressure developed to remove them. David Western, an ecologist, did not jump to the obvious conclusion. By careful analysis of the spatial distribution of dead and dying trees, the chemical characteristics of soils and vegetation, rainfall records, Maasai tradition, and the writings of early European explorers, he concluded that the drastic change of habitat in Amboseli was mainly caused by a shift in the hydrology of the region.

During 1961–1962 rainfall increased dramatically throughout East Africa, and 1961–1970 was much wetter than the average of the preceding 60 years. To a lesser extent 1951–1952 was also a period with rainfall higher than average. During the later wet period the groundwater beneath the Amboseli lake bed rose three to four meters because of increased rainfall on the slopes of Kilimanjaro. A smaller rise was recorded in the 1950's. Invasion of the root zone by highly saline groundwater (see Figure 1-3) introduced soluble salts into the upper 50 cm of soil, causing salinization and death of the trees by physiological drought. Salt-tolerant plant species replaced the woodlands.

Such changes have occurred in the past, the last being before 1890 during another period of heavy rainfall. Between these wet periods, the groundwater declines and the salts are leached from the root zone by percolating rainwater. During the past decade the water table has begun to decline slightly and fitfully, and there is some recolonization by less salt-tolerant plants.

An understanding of the Amboseli ecosystem, therefore, required an appreciation of changes in rainfall, groundwater, and the water budget of the park. Some knowledge of the role of hydrology is necessary for understanding the true cause of recent changes, and also for developing a long-term perspective on possible future changes that might affect the park. Further rise of the water table would threaten tourist facilities or complicate sewage disposal, road construction, and other developments.

Subsidence of Venice

In the fifth century AD the early inhabitants of Venice found refuge from the invading Huns on several low islands of silt and sand in a shallow lagoon in the Adriatic Sea. Venice grew into a great center of commerce and the arts, and though long past its prime it is still one of the world's loveliest cities. During the twentieth century the city has experienced increasingly severe flooding problems, which stem partly from natural causes and partly from changes wrought by industrial development. Storm tides now frequently inundate the famous Piazza San Marco and invade many of the surrounding buildings such as the Doges' Palace. Many of the world's finest buildings, sculptures, and paintings are threatened because Venice is sinking slowly into the sea.

The northern portion of the Adriatic is sinking slowly as the earth's crust buckles under the weight of several thousand meters of sediment brought by rivers from the Alps during the last few million years. The site of Venice and its neighboring mainland are interconnected parts of this recent geologic sequence of unconsolidated sand, silt, and clay (see Figures 1-4 and 1-5). These sediments contain groundwater that is exploited by pumping from wells in Venice itself and on the mainland.

In the early twentieth century the growth of a major industrial center at Marghera led to heavy groundwater withdrawals, which lowered the water pressure in the underlying sediments. This pressure supports a part of the weight of the overlying geologic materials, and when it is lowered the sediments undergo compaction and the ground surface subsides. Measured

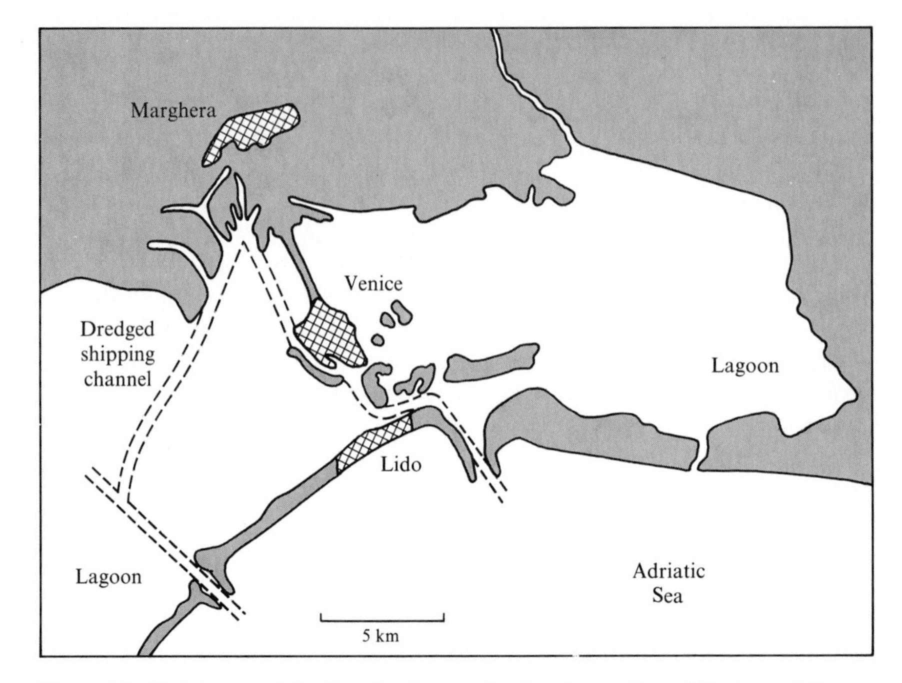

Figure 1-4 Sketch map of the Venetian lagoon showing the position of Venice and the industrial zone around Marghera.

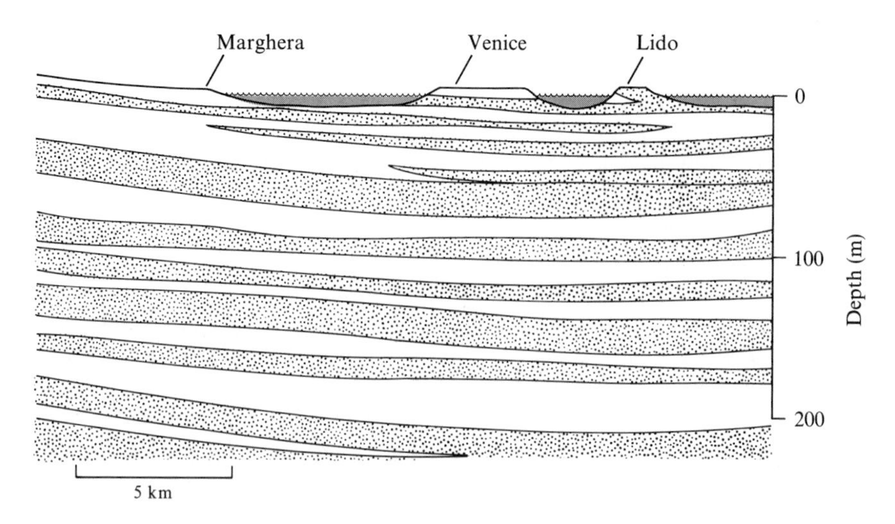

Figure 1-5 Schematic geologic section beneath the Venetian lagoon. The dotted area represents sand and gravel, and the unshaded area indicates clay and silt. (From Gambolati et al. 1974. Copyright 1974 by the American Association for the Advancement of Science.)

amounts of subsidence for the period 1952–1969 were about 10 cm in Venice and about 14 cm at the center of the Marghera well field. These small amounts of subsidence are critical in a setting like the Venetian Lagoon. In addition to threatening the architectural and art treasures, the small alteration of land elevation relative to the sea creates difficulties with storm drainage and the operation of sewerage systems. Partial filling of the lagoon and the dredging of deep shipping channels have compounded the problem by confining the high tides gathered by strong southerly winds in the Adriatic. The subsidence of Venice has been accelerated, therefore, at a time when tides have been allowed to penetrate farther and higher into the lagoon.

Interest arose in predicting the future impact of continued groundwater pumping. R. Allan Freeze, now at the University of British Columbia, was then working as a hydrogeologist at the IBM Research Center in Yorktown Heights, New York. He had developed a computerized mathematical model of groundwater flow, which allowed him to predict water pressures throughout a complicated geologic sequence under various rates of pumping. Together with other IBM research scientists (Gambolati et al. 1974), Freeze applied his model to the problem. With the computed water pressures the team could then predict the rate and extent of sediment compaction and subsidence at Venice.

Their results, in Figure 1-6, show the probable rates of subsidence under several possible future schedules of groundwater pumping. If the water

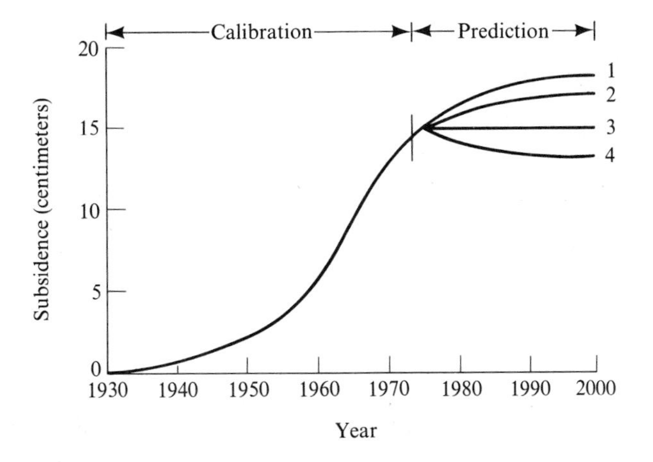

Figure 1-6 Calculated subsidence of the land surface at Venice. From 1930 to 1973 the calculation results were fitted to field observations to "calibrate" the method, which could then be used for prediction of subsidence during the next quarter century under various schedules of groundwater withdrawal (1–4) as described in the text. (From Gambolati et al. 1974. Copyright 1974 by the American Association for the Advancement of Science.)

withdrawals at both Venice and Marghera are held constant, as they have been from 1969 onward, about 3 cm of further subsidence could be expected in Venice by the end of this century (curve 1). Even such a small change is important at a site so close to mean sea level. Complete cessation of pumping in Venice alone would save about 1 cm of this subsidence (curve 2). Reduction of the Marghera consumption to 75 percent of its 1969 value and cessation of withdrawals in Venice would arrest further subsidence (curve 3). Closure of all wells would allow a slight recovery of elevation of perhaps 2 cm within the next 25 years as the groundwater is recharged and pressures rise again (curve 4).

The land will not rise to its former elevation, however, when the groundwater pressures recover. Most of the compaction of the sediments occurred in silty and clayey layers, which hardly expand at all once they have been compacted. Over 85 percent of the Venice subsidence is permanent.

This exercise demonstrates several ways in which hydrology can make a valuable contribution to planning. It allows prediction of the hydrologic impact of water-supply development. Second, the earth scientists were able to outline the effects of various management options. Together with a large team of scientists, engineers, and planners they have contributed to a complex strategy for saving Venice and the Venetian lagoon. The overall plan involves such actions as bringing a water supply to Marghera through a 20-km aqueduct so that groundwater pumping can be stopped altogether, protecting the lagoon from storm tides through the construction of large floating gates at the entrances to the lagoon, and improving storm drainage and sewage handling.

Most important of all, however, these scientists and others have demonstrated one example of the rather subtle processes by which a hydrologic disturbance can trigger environmental degradation at some distant place and time. They have reminded us that planners can profit best of all from an *awareness* of hydrologic processes, which will lead them to consider the environmental consequences of their actions at an early stage in the planning process. If this awareness of natural processes can be cultivated, society need not pay for careless development with its priceless treasures.

Snohomish Valley Floods

During November 1975 the Cascade Mountains of western Washington received heavy snowfall. Then in early December a large atmospheric disturbance moved slowly across the region and funneled warm, moist air from the Pacific Ocean over the snowpack. Heavy rainfall occurred for five days and was augmented by rapid snowmelt. The result was catastrophic flooding along every major river in western Washington, the damage costs being estimated roughly at $50 million.

The flooding was greatest on the Snohomish River, which drains a 4600-square-kilometer basin extending from the Cascade crest through the Puget

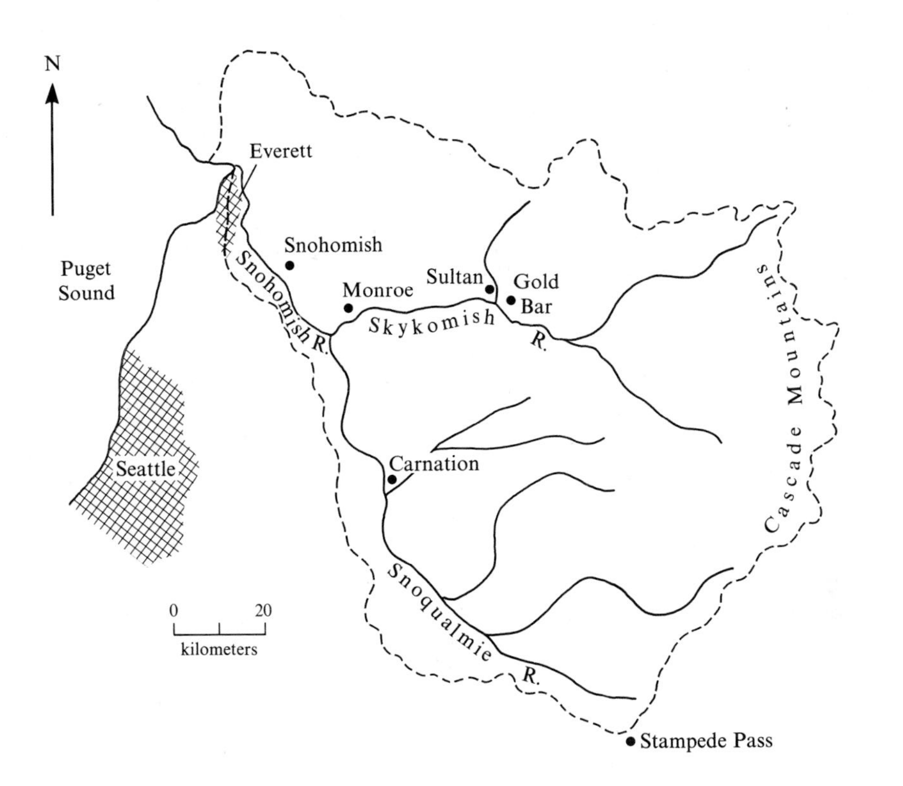

Figure 1-7 Map of the Snohomish River basin, Washington.

Lowland to a delta at the city of Everett (Figure 1-7). The story of this flood illustrates many of the physical, economic, and social processes the planner must understand and deal with.

The Snohomish River is formed by the junction of the Skykomish and Snoqualmie rivers, which flow from mountainous catchments that are clothed in coniferous forest. The catchment has been heavily logged and contains a dense network of haul roads and skids trails. The valley floors widen below the locations of Gold Bar and Carnation, and rich pasture and agricultural land along the rivers support a moderately dense rural population, which is served by several small towns.

It is not possible to state precisely the amount of water that the storm released over the Snohomish basin; the amount of rainfall and snowmelt varied with elevation. At sea level there was no snow cover and approximately 100 mm of rain fell in four days. Stampede Pass with an elevation of 1200 m, near the mean altitude of the basin, received over 400 mm of rain in four days, while the warm, moist air released more than 100 mm of meltwater from the thick snowpack (see the upper part of Figure 1-8).

Large volumes of runoff were generated by the storm and moved downstream as a series of flood waves, as indicated in Figure 1-8, which shows

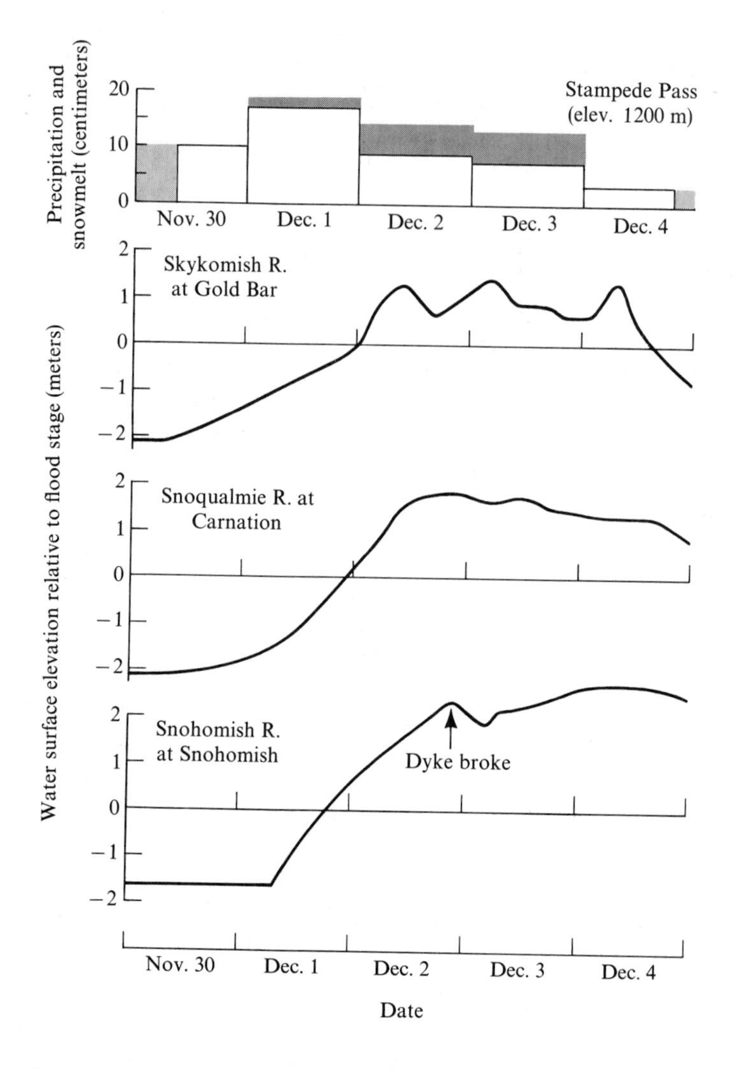

Figure 1-8 Records of precipitation, snowmelt, and river stage for locations shown in Figure 1-7 during the December flood of 1975 in the Snohomish basin. On the top graph, light gray indicates snow, dark gray indicates snowmelt in terms of the water released, and the unshaded portion represents rainfall. (Data from the National Weather Service and the U.S. Geological Survey.)

how the water surface elevation of the rivers varied through the flood at a station on each of the major tributaries and at another below their confluence. Both of the tributaries exceeded flood stage at midnight on December 1, while the main stem flooded several hours earlier. In the steep Skykomish valley, three separate flood waves can be seen in the record, their peaks reaching Gold Bar 20–30 hours after each burst of rainfall and melt. In the larger, less steep Snoqualmie basin the peaks did not respond as quickly, and they were largely attenuated into a single peak that crested slightly higher above flood stage than did those in the Skykomish basin. At Monroe, the Snohomish rose slowly to a single broad peak 2.5 m above flood stage.

The National Weather Service issued its first flood warning in the late morning of December 1, stating that the Snohomish River would rise one meter above flood stage at Snohomish early on December 2. The county Office of Emergency Services and rescue personnel were alerted. The earliest flooding occurred by 4:00 AM on December 2, when towns such as Sultan at the mouths of steep mountain tributaries became inundated to a depth of about half a meter.

As the flood wave gathered and moved downstream, the atmospheric disturbance slowed its westward movement. The storm lasted for a much longer time and produced more rain and snowmelt than had been expected. The National Weather Service had to keep revising its forecast of the crest upward as information on rainfall and snowmelt rates was telephoned or radioed in to the central office. By midday on December 2, for example, the forecast for the Snohomish crest at Snohomish Village had been revised to 2 m above flood stage and eventually reached 2.5 m. Similar revisions were necessary for the tributaries. Even though the flood prediction was not particularly accurate or fast, it was adequate to notify people that a very large flood was going to occur.

In the forested mountainous part of the catchment, tens of kilometers of haul road were gullied or blocked by landslides. Large quantities of sediment were carried from the roads and landslide scars, and their surfaces were covered with a slurry of silt and sand carrying a few 5- to 10-cm-diameter pebbles. Many culverts were blocked with tree debris, and the roadside ditches flooded over, scouring 50-cm-deep gullies across the road in less than an hour. Losses to the logging industry were very high.

Because several towns and many farms lie in the flood-prone area of the tributary valleys there was little that could be done apart from evacuating people and animals to higher ground. This was done, and although property losses and damage to municipal facilities such as sewage treatment plants were severe, they were not as catastrophic as in the lower valley.

Below Monroe, the rich farmlands were protected by flood-control dykes extending as far as the delta. The farmers felt secure behind these defenses, and they went to bed without making any evacuation plans for their large dairy herds. In some places, the dykes were designed to be overtopped by extraordinary floods, and a large pumping station had been installed to pump water from the flooded fields behind the dykes back into the river. Before the dykes and pumping station had been installed, floodwaters had remained on the fields for weeks. The system had operated successfully during minor floods over a 12-year period, but it now faced its first test in a large flood of the kind for which it was designed.

At 9:00 PM on December 2, the dyke immediately upstream of the pumping station failed; the station was undermined and fell into the river. The river level declined abruptly for a short time as water invaded the valley floor (see Figure 1-8). Over the farmlands, the water level rose one meter in two or three hours, isolating farmers from their cattle. The rise continued throught the night and next day. Fifteen hundred high-grade dairy cattle plus an undetermined number of horses and smaller animals were drowned in their barns. Feed, fertilizer, and machinery losses were also high, as were the costs of soaking and sediment damage in homes. The reaction of many victims was summarized by one farmer surveying his drowned herd: "We could have gotten out, but we didn't expect this." He had lived through many minor floods without suffering any great loss.

Further downstream the dykes were overtopped and breached at 40 places. More cattle were drowned in spite of hasty last-minute rescue attempts. Several thousand cattle died in Snohomish County in one day. Near the sea, high tides aggravated the problem by pouring salt water through the breached dykes. By December 4, 20,000 hectares of valley floor were under water (see Figure 1-9).

Property damage in the Snohomish basin alone was estimated to cost $20 million; almost none of it was covered by insurance. Another $30 mil-

Figure 1-9 Floodwaters invade the inhabited portion of the Snohomish valley floor. (U.S. Army Corps of Engineers, Seattle District.)

lion was estimated as the damage costs for other basins in the state. Miles of rural roads and bridges were destroyed, and the dyking and pumping system was in ruins.

The history of this flood has many ingredients from which a planner can learn. Considering first the physical aspects, between one-quarter and one-half of the water released by rain and snowmelt ran off as streamflow within a few days. This is a high proportion for a well-vegetated region such as the Cascades, but these large rain-on-melting-snow events are notorious throughout the northern United States for producing damaging floods. The peak rate of runoff at Monroe was 2900 m^3/sec, or 0.72 m^3/sec/km^2 of catchment. These figures again are high for such a large basin in the region. At Gold Bar the peak discharge on the Skykomish was 1.31 m^3/sec/km^2 from 1386 sq km, and at Carnation the Snoqualmie peaked at 0.95 m^3/sec/km^2 from 1562 sq km. The time delay between the occurrence of a burst of rainfall and the resulting peak varied from about half a day in the steep, mountain tributary above Sultan (drainage area 200 km^2) to more than two days at Snohomish. The main rivers exceeded flood stage within 36 hours of the onset of rain. These figures give an idea of the time available for warning and evacuation–if the flood warning is prompt and is taken seriously.

The average frequency of such a flood can be assessed from records of past floods. In spite of hasty but well-publicized claims that this was "the 100-year flood," the average frequency with which an event of this magnitude is to be expected is about 30 years at Gold Bar and Carnation, and about 20 years for the lower Snohomish near Monroe. This flood was the "largest in history" only in the sense that since the last, larger flood, more settlement had occurred in the flood-prone area and people thought they were secure, so they took few if any precautions.

The planner might also be interested in whether it was known beforehand which areas would be flooded by events with this frequency. Although detailed engineering surveys of the whole flood-prone area had not been completed, the inundated area was predictable, and it was known that many homes, barns, businesses, sewage treatment plants, and other facilities were in hazardous areas.

The planner might also become embroiled in the question of land use and floods. Many claims are made that widespread logging is responsible for rapid runoff. There are no data to support or to refute this argument for the Snohomish basin, but it is possible to make some rough judgments based on an understanding of how water gets into streams during floods in this area, and how flood waves move downstream. It is likely that the runoff from logged areas was slightly larger and faster than that from undisturbed forest, mainly because of the contribution from haul roads. By extrapolating from studies on small catchments in other regions, it is possible to judge that the effect would be moderately high from small, recently logged parcels of land. The effect would probably be reduced drastically downstream, however, as the storage function of the river channel and valley floor damped out differences between runoff from logged and unlogged areas. In large river basins, the dangerous floods are produced by large, slow-moving rainstorms that thoroughly soak the catchment and generate runoff from a large proportion of it. Over snowy regions, rain on melting snow is a particularly hazardous combination in this respect. Under these extreme conditions, differences of runoff between vegetation and land-use types tend to be relatively unimportant.

Over the preceding 15 years a major controversy had been occurring about flood control in the valley. Farmers, land developers, and the U.S. Army Corps of Engineers want to build a flood-control dam on a major tributary of the Snoqualmie River. It is thought that in addition to protecting farmland, the flood-control facilities could be used for water sports, and land values would be enhanced by an influx of recreation and homes. Their plan is opposed by a coalition of conservationists, kayakers, fishermen, some valley residents, and other groups. Immediately after the flood, local newspapers carried editorials and letters demanding more flood control. How should the planner assess the arguments? It is a sad fact that the state and federal agencies concerned with hydrology provided no technical guidance to the public when this clamor arose again, even though the proposal for the dam never claimed that it would make any significant reduction of flood levels in the lower Snohomish valley, where the bulk of the damage occurred. The idea was fostered in the public's mind that the damage could have been avoided if only the selfish conservationists had allowed a dam to be built.

After the flood many planners in the region were left again to ponder the wisdom of trying to control floods by means of dykes or dams and thereby encouraging people to settle in the supposed safety of the "protected" area. Even farmers who had been perceptive enough to evacuate their animals in earlier floods were lulled into a false sense of security by the knowledge that the government agencies were protecting them.

In Chapters 10 and 11 we discuss the prediction of floods and strategies of flood control. Most of the material deals with small areas of up to a few square kilometers because the readers for whom this book is intended will have to deal frequently with storm drainage problems on small catchments. But planners, and in fact all well-informed citizens concerned about environmental issues, also need to know something of how hydrologic predictions are made for flood warning or control in large basins. For this reason we have included in Chapters 10 and 11 brief descriptions of some techniques employed in large-scale flood problems. With this limited background, a planner or other concerned citizen should at least be able to ask some hard questions to examine the wisdom of flood-control schemes.

Landslides in Seattle

During the last glacial period (10,000 to 25,000 years ago) the lowland around Puget Sound in northwest Washington underwent a complex history of geologic change. A hilly landscape produced by earlier glacial and river erosion and deposition was inundated by a rising lake when a large glacier extended south along Puget Sound from British Columbia and blocked drainage to the sea. In this lake, streams from the flanking Olympic and Cascade mountains deposited fine silt and clay which is now called the Lawton Clay (see Figure 1-10). As the glacier moved up the sound, streams carried sands from its extending front and deposited them over the clay as the Esperance Sand. The ice then advanced over the fluvio-glacial deposits, carving them into hills and valleys, and plastering sandy and gravelly till over the landscape, as shown in the geologic cross section. After the ice retreated into Canada, the rising sea entered Puget Sound, and wave action steepened the edges of the hills into cliffs.

Around the Puget Lowland the juxtaposition of steep hillslopes and geologic contacts of permeable sands over impermeable silts and clays creates conditions that favor landsliding. As described in Chapter 15, water percolating through the permeable till and sand following heavy rain is impeded by the lower strata, and water pressure builds up, weakening the unconsolidated sediments to the point of collapse. The process is most common in the two situations indicated in Figure 1-10, which are also among the most highly prized residential sites offering magnificent views of the snow-covered Olympic and Cascade mountains.

Figure 1-11 shows a situation like that on the right-hand side of Figure 1-10. A steep cliff is developed in the Esperance Sand; its steepness is maintained by landsliding. Material slumps down onto the wide bench above the Lawton Clay and pushes older landslide debris slowly across the bench toward the sea. Because most of the bench has a gentle slope and fairly smooth topography, it is often not recognized as a zone of landslide hazard. The fallen blocks move slowly across it, however, and make the surface a highly unstable place for buildings, even beyond the reach of the initial failure. In Seattle's wet climate, slowly moving soil becomes thickly vegetated, masking old slide scars and fresh fallen debris, but after wet weather,

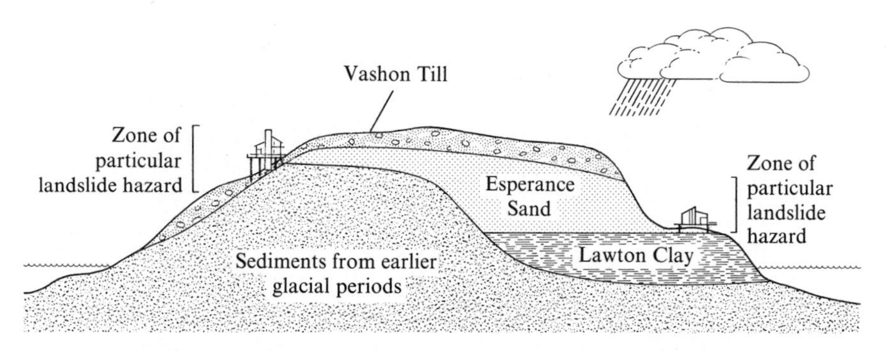

Figure 1-10 Schematic geologic cross section through a typical Seattle hill. The zones of particular landslide hazard are localized where steep hillsides intersect geologic contacts with a thin, permeable unit overlying an impermeable (dense, silty, or clayey) sediment. (Modified after Tubbs 1974.)

Figure 1-11 The cliffs and bench on which landslides occur at Alki Point, Seattle. Failures on the upper cliff undermine roads and houses and move the fallen material across the bench on which the trees are growing. The light-colored structureless material below the trees and above the banded, well-stratified Lawton Clay is old landslide debris on its way to a second landslide over the lower cliff.

when block movement is accelerated, cracks can be observed between the moving sections. Eventually the landslide debris reaches the seaward edge of the bench and creates a new set of landslides.

Thus there are three sites in which hillslope failures threaten buildings. At the top of the upper cliff the threat is undermining (see Figure 15-32). During the spring of 1974, a sidewalk running along the edge of the upper cliff in Figure 1-11 was undermined, and the large white house lost one-quarter of its garden. The failure has continued slowly since that time. The bench in the figure is covered by a dense growth of young maple trees and several feet of structureless light material are visible in the cliff immediately below them. This material is the Esperance Sand landslide debris in transit across the bench. In spite of its instability, the bench shown in the picture has been surveyed for a large residential complex, and some electric utility cables and lamp standards were actually installed before the ground motion accelerated in early 1974 and the building permit was withdrawn. Elsewhere in Seattle, this bench is a favorite building site and houses suffer direct damage from landslides, as well as the structural problems arising from the more subtle ground motions across the bench.

The third hazardous site, of course, is at the foot of the lower cliff, which receives the landslide debris and trees that have traversed the bench. In 1974 several houses were destroyed by this process, and some of the debris is visible in the photograph. There are no vacant lots visible, however, because the destroyed buildings were replaced within a year by new houses, apartment buildings, and a condominium.

In some parts of the Seattle region the landslide problem is aggravated by poor grading and drainage practices on residential lots. Artificial fills placed on impermeable substrates are often rendered unstable by the build-up of water pressures. Diversion of stormwater onto slopes has a similar effect. Undercutting of hillsides for roads and house lots triggers other slides.

Donald Tubbs, while a graduate student in geological sciences at the University of Washington, studied the spatial relationship of landslides to the pattern of geologic contacts. He was able to define and map zones with various degrees of landslide hazard. His map is a valuable guide to local planners, engineers, and homeowners. The timing of major episodes of landsliding was also related to precipitation patterns, and a rainfall index suitable for short-term warning of high landslide risk was developed by Tubbs from a statistical analysis of 30 years of landslide reports and weather records. The study has been widely publicized in Seattle and has been used in developing new grading ordinances, distributing building permits, and designing site plans for parks and residential developments.

In spite of a generally high level of awareness of the landslide hazard among long-term residents of Seattle, a considerable amount of building continues to occur in high-risk zones. A large residential development has recently been built on the topographic bench at the contact of the Esperance Sand and the Lawton Clay (see Figure 1-12). Excavation of old landslide

Figure 1-12 A new condominium built on the same landslide bench shown in Figure 1-11, but in a different part of the city. The steep, dark slope on the left is the upper cliff in sand; the graded area in the foreground is landslide debris, and the buildings themselves stand close to the sand–clay contact. Since the photograph was taken, shallow landslides down the front of the graded mound have piled debris against the building almost to the lowest window level.

debris at the site is now beginning to trigger renewed hillslope failure, which is piling wet debris against the uphill side of the buildings. Maintenance problems are increasing and a lawsuit has been filed by the landowner upslope for undermining his property. The site provides a useful field trip for students in landscape architecture and in geology because it is surrounded by every conceivable piece of hydrologic, geomorphic, and botanical evidence of hillslope instability, described in Chapter 15. In another part of Seattle, the city recently withdrew building permission after development had begun on a similar project on the same topographic bench on the same geologic contact. Throughout the city this well-defined zone of hazard continues to be used, but at considerable cost to the unwitting purchasers of endangered property.

Channel Changes on the Yakima River

In southwestern Washington, the Yakima River flows in a steep, divided channel through a set of gaps cut in bedrock and across intervening valleys floored with coarse gravels (see Figure 1-13). It drains 8000 sq km of the

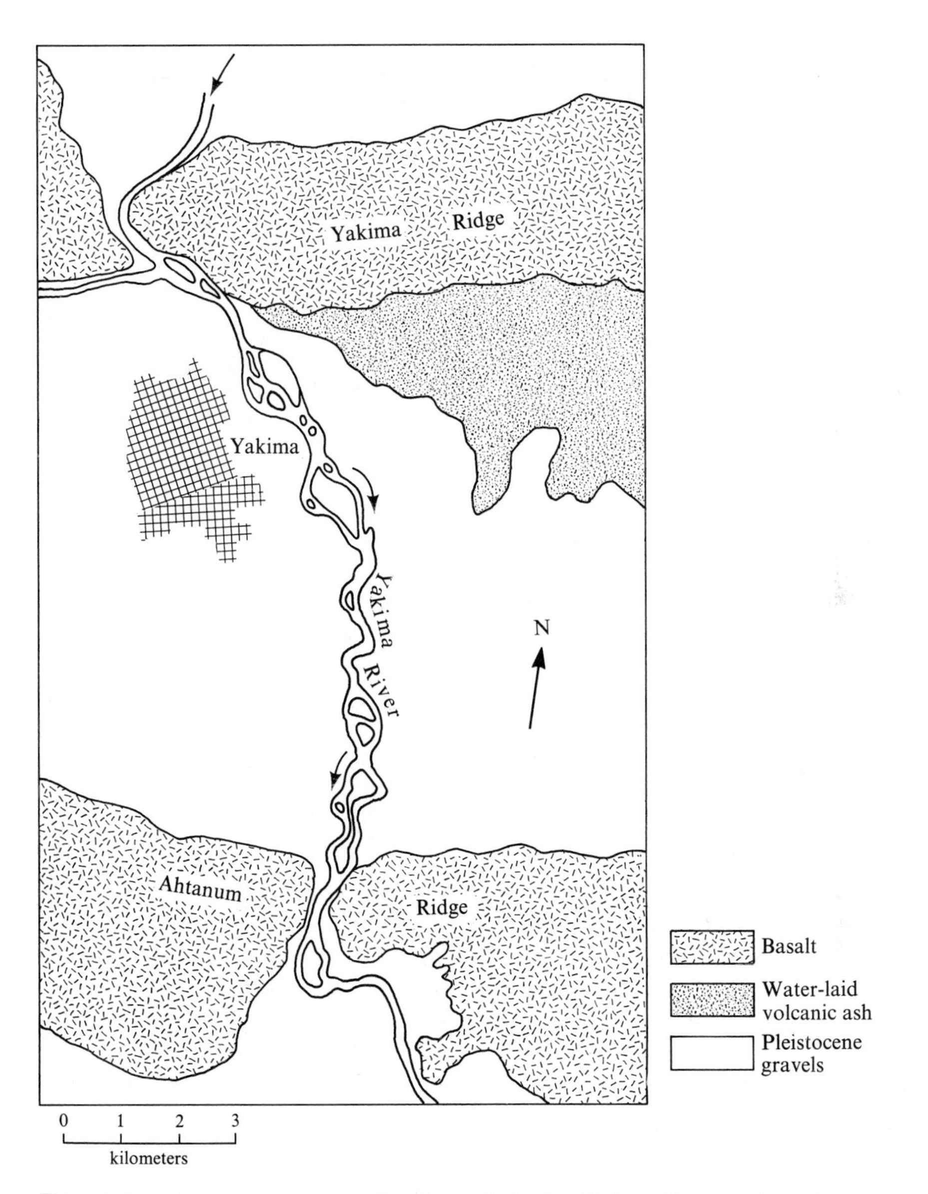

Figure 1-13 Geologic map showing that the south-flowing Yakima River cuts through two basalt ridges in narrow gaps and flows across an intervening broad, gravel-floored valley.

Cascade Mountains and its snow-fed discharge (the mean annual flow is 72 m^3/sec) is the basis of irrigated agriculture in the lower, arid portion of the basin. Near the town of Yakima a plan was proposed to convert the valley floor into a park 12 km long and 2 km wide. Because the park was to be intimately associated with the river, Grant R. Jones, the landscape architect and environmental planner responsible for the park, recognized the need for understanding why the river looks and acts as it does. He requested an account of the recent history of the river, projections of its probable future behavior, and the definition of various constraints and opportunities presented by the river.

During the past 12–15 million years the valley had a history of earth movements, deposition of volcanic lavas and ashes, and erosion by the river, which resulted in the alternation of parallel lava ridges and intervening gravel-floored valleys. During the Pleistocene epoch (10,000 to 2 million years ago), glaciers advanced into the upper part of the valley and the river was swollen with glacial meltwater and large quantities of gravelly sediment, which was deposited in the valleys. After the glaciers retreated, the river cut down about 3 m into its Pleistocene deposits, leaving a terrance of sand and gravel along both sides of the valley. This terrace now defines a natural corridor within which floodwaters are confined.

Examination of the geologic and topographic maps suggested that the valley has been tilted toward the south in recent geologic time and that this movement is continuing. During the last 15,000 years the river has cut down into the northern half of the park reach at a faster rate than to the south. In the northern section, also, the river has undergone less lateral shifting than the southern part of the reach, which has a more complicated channel with signs of recent deposition of channel bars and natural levees.

Divided channels of the kind found along the Yakima are called *braided* channels. They generally form within a narrow range of geologic and hydrologic conditions, and they operate in ways that are predictable within broad limits and that place severe constraints upon development along the valley floor. Braided channel patterns usually develop where flood discharges are high and fluctuate rapidly; where sediment transport rates along the stream bed are high; and where the channel gradient is steep, and the stream banks are formed in weak, noncohesive sand and gravel. The reach under study meets all but the second of these criteria. We will describe in Chapter 16 how these channels form and develop. At this point, it is sufficient to say only that many braided streams move rapidly across their valley floors and pose a threat to structures in the valley floor.

Aerial photographs of the valley (Figure 1-14) reveal many abandoned channels, indicating that the Yakima has swept across its whole valley in the recent geologic past. Four sets of aerial photographs were available from 1939 to 1973, and from them the successive positions of the river channels could be mapped. The maps showed the river channel to be very mobile. During the 34-year measurement period the average rate of lateral migration was 10 m/yr. Superimposed upon this gradual shifting, however, is a much more dangerous and less predictable type of movement in which the river overtops its bank and cuts an entirely new channel, perhaps 100 to 300 meters from its earlier course.

Channel diversion is a natural process along a large, fast river with weak bank materials, but in the vicinity of Yakima it has been accentuated by

Figure 1-14 Aerial photograph of the Yakima River valley. Earlier positions of the channel are visible as light-colored streaks of sand and gravel deposits. Strips of vegetation also outline some of the abandoned channels.

gravel mining in the valley floor. Deep gravel pits are excavated, and after abandonment they are isolated from the river by only a weak gravel dyke, which the river can easily breach. Figure 1-15 shows a case where gravel has been excavated from two pits for the construction of an interstate highway. In 1971 the river invaded the upper gravel pit, flowed out of the back of the excavation and reentered the main channel via a second pit 1800 m downstream. Overnight, the river was diverted up to 600 m to the west and is now cutting into an embankment under the highway.

Although it is not possible to predict such channel behavior with confidence, it was possible to produce a map of the "worst case" of channel migration during the next 10 to 25 years. This was done by examining the channel banks and valley floor for likely diversion sites. Some were obvious, and the path of the river after diversion could be easily foreseen; a few were more equivocal. After the probable diversions were mapped, recent directions and rates of gradual shifting were projected over the next 25 years, and the hazard zone was defined on this basis. Most of the predicted movement lies in the southern reach. The resulting map allowed the classification

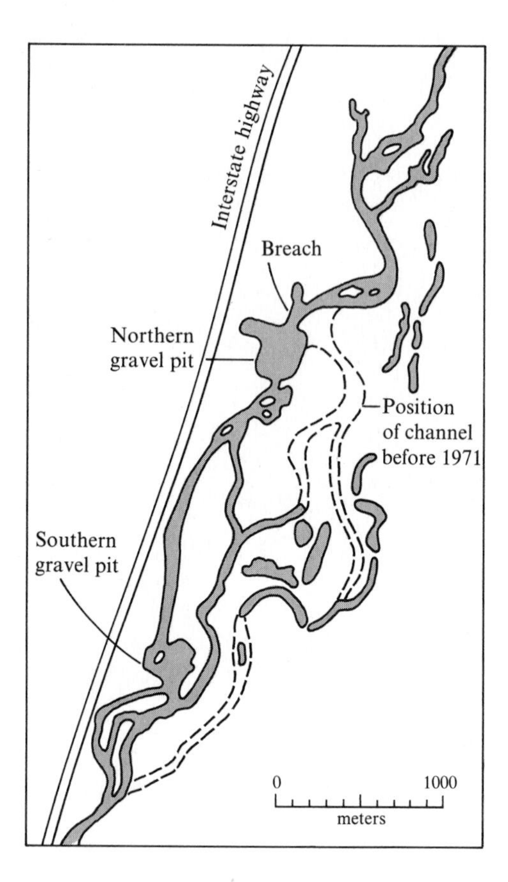

Figure 1-15 Map of a channel diversion that occurred in 1971. The river abandoned the channels shown by the dashed lines, breached the dyke protecting the northern gravel pit, and returned to the main channel through another pit to the south. The shaded area represents the 1975 distribution of water at average flow.

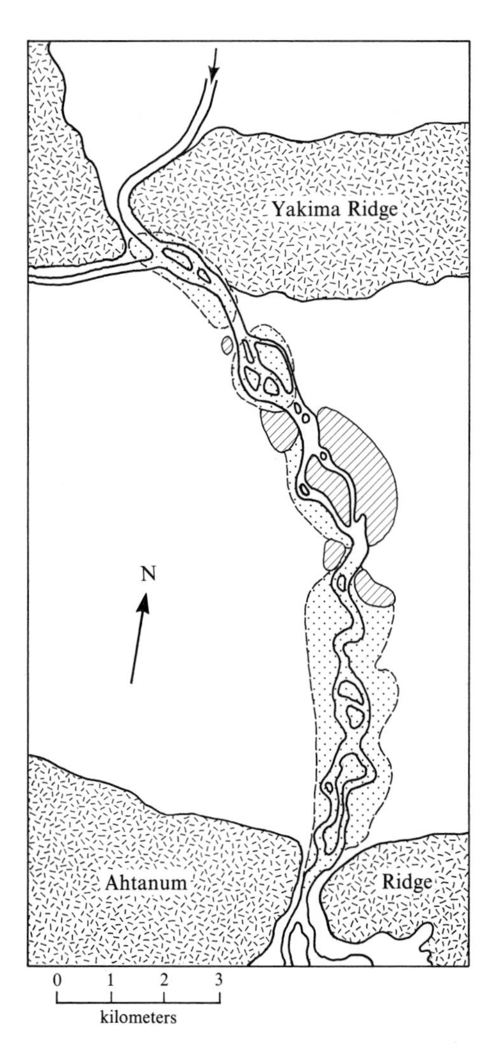

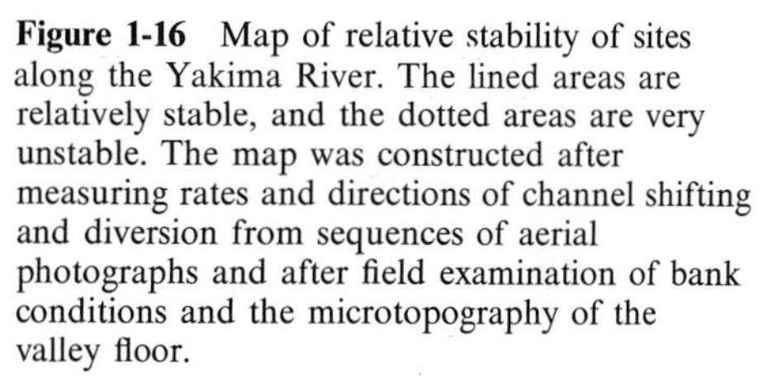
Figure 1-16 Map of relative stability of sites along the Yakima River. The lined areas are relatively stable, and the dotted areas are very unstable. The map was constructed after measuring rates and directions of channel shifting and diversion from sequences of aerial photographs and after field examination of bank conditions and the microtopography of the valley floor.

of the valley floor into stable and unstable zones as one basis for siting facilities. Careful field examination of the channel banks, some artificial dykes, and the microtopography of the valley floor made it possible to locate sites at which the river will probably break out of its channel and take a new course sometime this century (see Figure 1-16).

As another design constraint, the flood hazard was defined with the aid of an 80-year record of discharge on the river. The construction of several irrigation-supply dams has resulted in a slight reduction of flood hazard in the valley since 1933. Settlement has moved onto the valley floor encouraged by the flood control provided by the dams and by some dykes in the less active, northern part of the reach. There has not been a major flood since the settlement occurred, but the map of the flood hazard (see Figure 11-25) indicates that such an event would inundate homes. In 1975 a moderate flood (the "10-year" flood) caused minor damage.

Calculations of the amount of sediment carried by the river were made and checked against a few field observations. The river transports an average of 250,000 tonnes of sediment per year past this site, which is a low rate for a river of its size. Nevertheless, it is important that the sediment transport processes not be interfered with by any kind of engineering works. Obstruction of the flow would lead to rapid sedimentation and diversion of the channel. The quantity of sediment in motion is also important in the evolution of the river channel once it has been diverted into a gravel pit.

It was also possible to estimate the size of the largest rocks that could be moved by the river in large floods. A combination of hydraulic equations and field observation after a 10-year flood showed that the river could dislodge boulders up to 0.6 m in diameter. The dykes along the river, protected by rocks much smaller than this, have failed or have been disturbed in several places. Although the dykes can temporarily confine floodwaters, they provide little insurance against the movement of the channel itself, because the river is competent to move the rocks of which the dykes are built. Current proposals to extend the dyking system from the naturally more confined northern reach into the rapidly shifting southern part of the valley do not seem wise in view of this threat.

The planning team led by Grant R. Jones used the report on geomorphology and hydrology, and particularly the map of relative channel stability, as a basis for partitioning the valley floor into segments that could be used for a wildlife refuge, passive recreation, active recreation, and a visitor center. Some other land-use proposals for parts of the valley, such as the siting of a motel and trailer park were seen to be unwise in view of the flood hazard and recent rates and directions of channel migration. The proposed motel site has since been isolated from the interstate highway by a channel diversion.

Water Chemistry in Brandywine Creek

Brandywine Creek drains about 840 sq km of Piedmont plateau and Atlantic coastal plain into the Delaware River at Wilmington, Delaware. It is the principal source of water supply for the city, providing more than 100,000 m^3/day for domestic supply alone. The proximity of Wilmington and Philadelphia has made the catchment of the Brandywine attractive for residential development, and its population now exceeds 100,000.

As residential development spreads rapidly westward through the basin, problems are arising with the disposal of wastes into the river. Many residences are served only by septic-tank disposal systems, many of which function poorly because they are located on unsuitable sites. The wastes from some small residential developments are processed in small "package" treatment plants, while the main towns are served by larger sewage treatment plants. Each of these systems releases phosphorus, nitrogen, some oxygen-demanding compounds, and various other dissolved substances into the river. Small industrial concerns, such as paper mills, add their load of biodegradable wastes. Finally the agricultural lands of the basin shed sediment, fertilizers, and pesticides. Concern about water quality and the general esthetic degradation that usually results from haphazard development provoked local residents, with the assistance of a group of earth scientists, lawyers, and planners, to set up a Tri-County Conservancy for the Brandywine. One of the charges of this organization is to obtain information on the environmental impact of various land-use and waste disposal practices. The conservancy is also responsible for developing methods of predicting the effects of future land-use changes and for proposing management strategies to offset the degradation of land and water resources.

The first steps in developing any program that is supposed to lead to recommendations for planning must involve defining some primary concerns and then collecting field information about these problems. One of the subjects isolated for study was the role of land use on the phosphorus load of the Brandywine. As discussed in Chapter 20, relatively high concentrations of this element are commonly associated with the triggering of algal blooms and with various other changes in an aquatic ecosystem that are generally referred to as *eutrophication.*

Thomas H. Cahill, director of Environmental Programs for the conservancy, embarked upon a water sampling program at sites along the river that received water from forest, agricultural land, industrial plants, municipal sewage-treatment plants, and residential districts in which wastes are disposed of in septic systems. Water samples were collected during storms and periods of low flow and were analyzed for various kinds of phosphorus, nitrogen, oxidizable organic compounds, and other substances. Some examples of the resulting data are presented in Figures 1-17 and 1-18.

The spatial pattern of measurements shown in Figure 1-17 indicates the magnitude of the effects of municipal and industrial waste disposal. The upper tributaries of the basin draining forest and agricultural land were losing 0.02 kg/km^2/day at the time of sampling. Large inputs occur below the major population centers, particularly at Downington, where paper mill effluent augments the load. The association between phosphorus concentrations and biologic activity in the stream can be seen in Figure 1-18, which shows that as the phosphorus concentration increases downstream so do the concentrations of two indicators of the biomass of organisms in the stream.

Sampling through time at each station revealed relationships of phosphorus concentration to stream discharge that differed between periods of storm runoff and periods of dry-weather flow. Point discharges of phosphorus were diluted as stream discharge increased after storms. At stations not dominated by point discharges, however, phosphorus concentrations

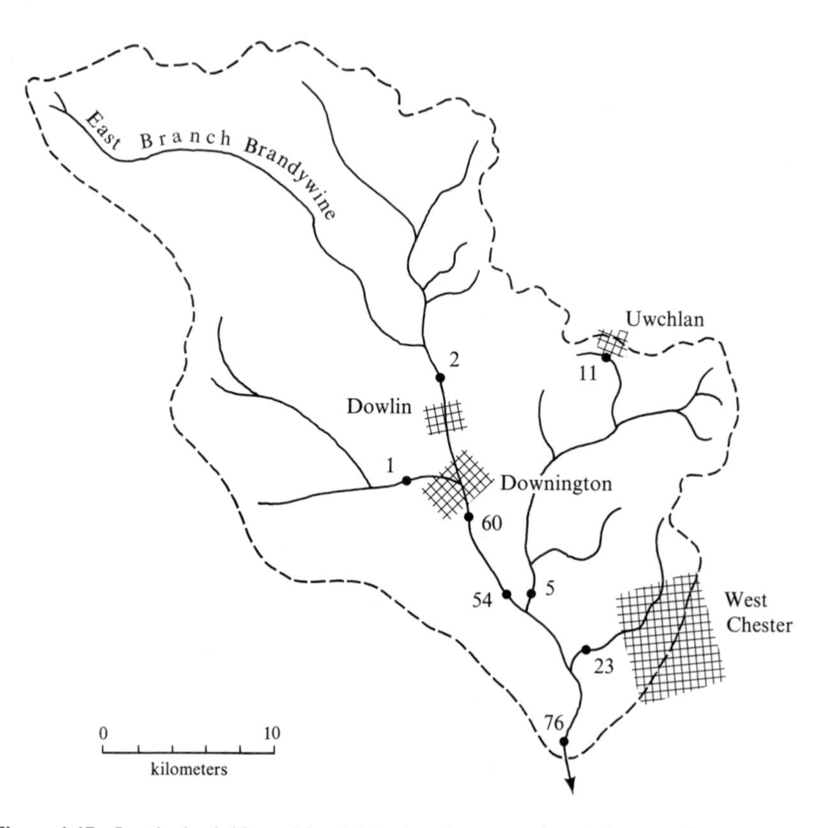

Figure 1-17 Load of soluble and insoluble phosphorus passing stations on the East Branch of the Brandywine Creek, August 13–14, 1973. Numbers are in kilograms of phosphorus per day. Hatched areas are towns and villages. (Modified after Cahill et al. 1975.)

increased dramatically as storm runoff occurred. Runoff from forested and cultivated hillslopes washed into the stream large amounts of phosphorus in solid organic form and adsorbed on soil particles.

The total annual transport of phosphorus by storm runoff was roughly equal to that occurring during dry weather. Enrichment of the stream, however, was much more dependent upon the non-storm point discharge of nutrients, a portion of which accumulated on the stream bed between high flows. While the transport of nutrients from the upper forested and cultivated reaches was significant during storm periods, it did not produce undesirable enrichment. The researchers calculated that only a small fraction of the nutrients applied to the agricultural land find their way into the river. Even the storm runoff from the urban areas of the catchment contributed more phosphorus than the agricultural land.

Studies of this type provide valuable information for planners, who should therefore be aware of such work. The field data can show the degree of effectiveness of various nutrient-reduction programs. They indicate the

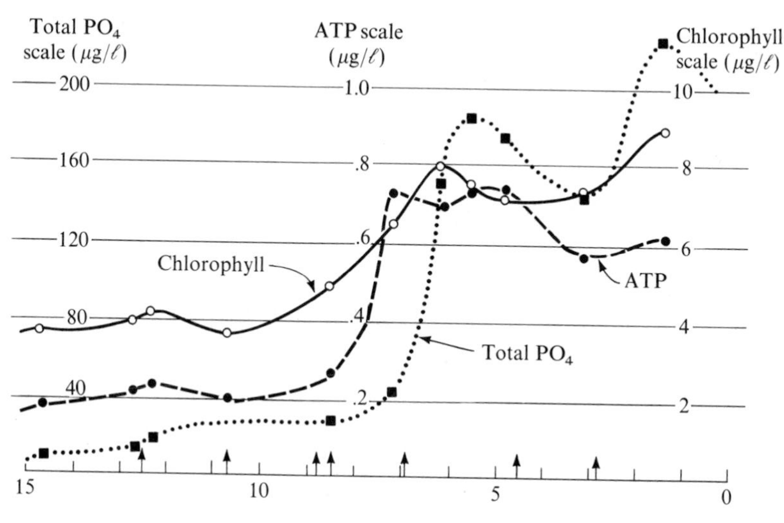

Figure 1-18 The general downstream increase of phosphate concentration in the East Branch of the Brandywine Creek is accompanied by increases in two indices of the biomass of aquatic organisms. Adenosine triphosphate (ATP) indicates the concentration of nonphotosynthetic biomass, while the chlorophyll content is a measure of the photosynthetic biomass. The arrows represent inputs of phosphorus at sites such as tributaries and sewage treatment plants along the stream. (After Cahill et al. 1975.)

relative importance of reducing phosphorus input from point sources, which are relatively easy to deal with, and from the disseminated "nonpoint" sources on fields and in woodlands. Similar studies of other pollutants are being carried out to compare the impact of various land-use practices on water quality. Emphasis is now being placed on the hydrologic processes by which the pollutants reach streams. Kunkle's (1970) investigation of bacterial contributions is one example of such studies which have considerable value in land-use planning for the conservation of water quality.

Readers now have a glimpse of the kinds of planning problems in which water is a central issue. We hope that this will help them through some of the longer chapters concerned with rather dry subjects such as precipitation analysis. But if planners want to make quantitative predictions concerning land and water management, they must become familiar with the analytical tools presented in the following text.

Bibliography

CAHILL, T. H., IMPERATO, P., NEBEL, P. K. AND VERHOFF, F. H. (1975) Phosphorus dynamics in a natural river system; pp. 44–56 in *Water–1974,* American Institute for Chemical Engineering, Symposium Series 145.

GAMBOLATI, G., GATTO, P. AND FREEZE, R. A. (1974) Predictive simulation of the subsidence of Venice; *Science,* vol. 183, pp. 849–851.

HYNES, H. B. N. (1960) *The biology of polluted waters;* Liverpool University Press, 202 pp.

HYNES, H. B. N. (1970) *The ecology of running waters;* University of Toronto Press, 555 pp.

JUDGE, J., MOLDVAN, A. AND BOSWELL, V. R. (1972) Venice fights for life; *National Geographic,* vol. 142, pp. 591–631.

KUNKLE, S. H. (1970) Concentrations and cycles of bacterial indicators in farm surface runoff; *Proceedings of the Cornell Agricultural Waste Management Conference,* Ithaca, NY, pp. 49–60.

TUBBS, D. W. (1974) Landslides in Seattle; *Information Circular 52,* Division of Geology and Earth Resources, Olympia, WA.

TUBBS, D. W. (1975) Causes, mechanisms, and prediction of landsliding in Seattle; Ph.D. dissertation, University of Washington, 87 pp.

WESTERN, D. AND VAN PRAET, C. (1973) Cyclical changes in the habitat and climate of an East African ecosystem; *Nature,* vol. 241, pp. 104–106.

II
Hydrology

2

Precipitation

Precipitation is the major factor controlling the hydrologic cycle of a region. Much of the ecology, geography, and land use of a region depends upon the functions of the hydrologic cycle, and therefore precipitation provides both constraints and opportunities in land and water management. The planner should be aware of how the various characteristics of precipitation are quantified and analyzed, how measurements are made, and in what forms useful precipitation data are published.

Precipitation has many characteristics that affect planning. The relative amounts of rain and snow, their seasonal timing, and the sizes and intensities of individual storms, for example, affect activities as different as city budgeting for snow removal or storm sewer design and seasonal runoff forecasting for hydroelectric power or irrigation. In this chapter we will discuss only rainfall because snow hydrology is treated in Chapter 13. For other planning problems dealing with regional water management or choice of crops one may have to focus on seasonal and annual totals of precipitation and their reliability. Still other planning situations require information on the magnitude and intensity of individual storms for the design of engineering structures. The variety of problems is great, and although the planner does not need to become expert in meteorology or in some of the more esoteric forms of analysis, he should be acquainted with the most frequently used analytical tools for studying and predicting precipitation.

In this chapter we also use the topic of precipitation as a vehicle for introducing a few elementary concepts about probability, which is fundamental to hydrologic prediction and design. It is particularly important that the planner grasp these simple concepts because he or she will be confronted by statements of probability in technical reports and proposals. It is important to appreciate the true meaning and limitations of such statements as well as to recognize the power of statistical reasoning. The introduction of these concepts may lengthen the chapter beyond what the reader considers to be appropriate for a treatment of rainfall. We ask you to be patient with us through this rather long, early chapter.

Measurement of Precipitation at a Point

Precipitation is measured in a *rain gauge,* which may be of the nonrecording or the recording type. Nonrecording rain gauges, illustrated in Figure 2-1, are simply cylinders that must be emptied at regular intervals to measure the catch. With them it is possible to gauge only the amount of precipitation falling during any interval (usually 1, 6, or 24 hours).

Recording gauges (sometimes called automatic or autographic) register not only the amount of precipitation but also its timing and intensity. Figures

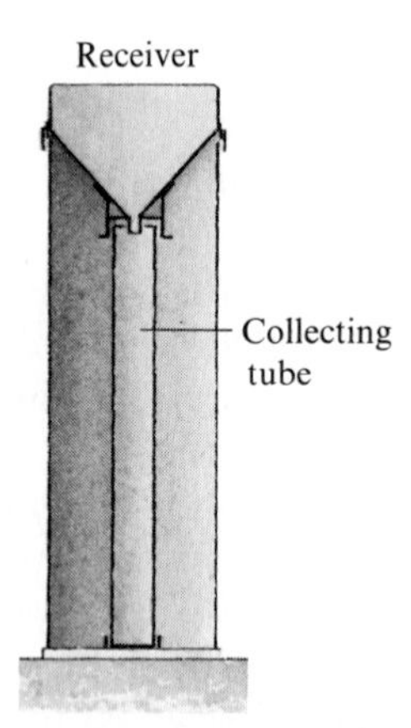

Figure 2-1 A nonrecording rain gauge. The conical collector, shaped to minimize splash, funnels water into a narrow measuring tube where it can be measured to the nearest 0.01 inch (0.25 mm) with a dip stick. The dimensions of the gauge vary among countries. In the United States a diameter of 8 inches (20 cm) and a height of 31 inches (79 cm) are standard. In exposed locations a wind shield surrounds the gauge to decrease turbulence around the orifice and allow a more representative catch. The gauge can be used for snow.

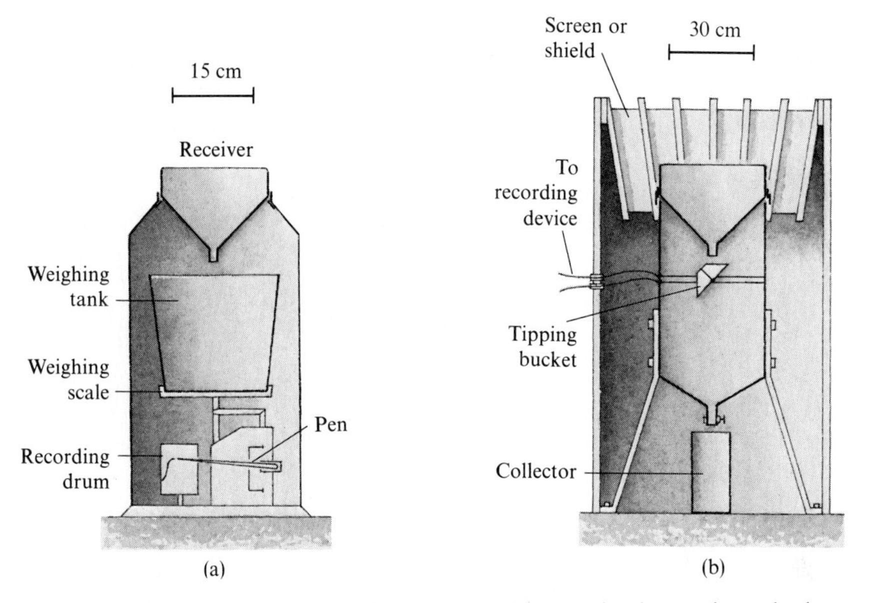

Figure 2-2 (a) Weighing rain gauge. The conical receiver funnels rainwater into a bucket standing on a weighing scale which is connected to a pen. The pen records the accumulation of water on a paper chart wrapped around a clock-driven recording drum. Can be used for snow. (b) Tipping bucket gauge. Two small triangular buckets each calibrated to hold 0.01 inch (0.25 mm) of rainwater are balanced on a knife edge. The collector funnels water into each of these buckets in turn. When one is full, it tips on its pivot, emptying its contents into the reservoir and activating an electrical switch that records the accumulation of each 0.01 inch of rain on a chart in a nearby shelter. Cannot be used for collecting snow.

2-2 and 2-3 illustrate the two most common types. The instruments are connected to a recording device such as a machine that punches holes in a paper tape or a pen that records the accumulation of rainfall on a paper chart wrapped around a clock-driven drum. For a more detailed discussion of precipitation measurement, the reader should consult reports by Kurtyka (1953) or the World Meteorological Organization (1965).

Measurement of Precipitation over an Area

Errors of precipitation measurement at a point are usually of the order of several percent for a single storm and range up to 30 percent for poorly exposed gauges in large storms with strong winds. Errors in the long-term average values are generally less. Much more important from the hydrologic forecasting and planning point of view are the uncertainties that arise in estimating the precipitation falling onto an area such as a drainage basin from measurements at only a few rain gauges in and around the area. The several methods of estimating the areal average precipitation on a drainage

Figure 2-3 A weighing-type, recording rain gauge installed in a catchment at Boco Mountain near Wolcott, Colorado. (Photograph by R. F. Hadley.)

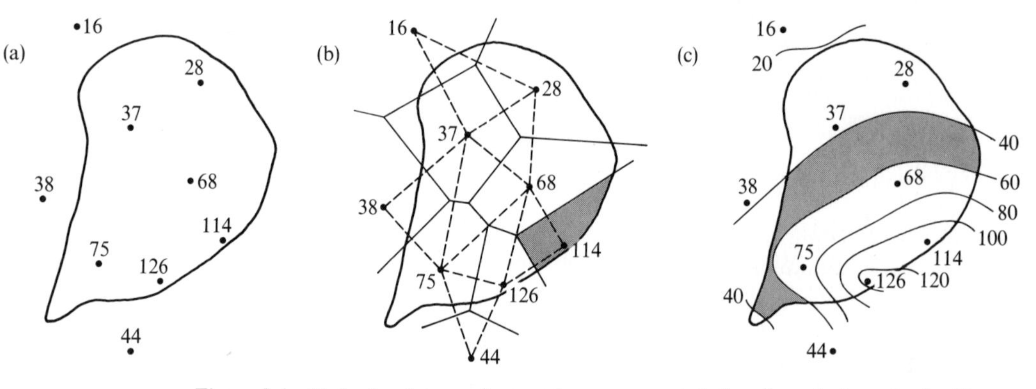

Figure 2-4 Methods of computing areal average precipitation for a drainage basin. The numbers represent storm totals measured at rain gauges, and the units are millimeters. (a) Basin outline and measured values. (b) Thiessen polygons. (c) Isohyets. See Typical Problem 2-1 for explanation of the figures and description of the methods.

basin are illustrated for a single storm in Figure 2-4 and in Typical Problem 2-1. An *arithmetic average* will suffice in regions of moderate relief and of general, cyclonic rainfall where precipitation gradients are not strong. The *Thiessen-weighted average* is a better measure where the distribution of gauges is not uniform and where precipitation gradients are strong. The *isohyetal method* can take into account strong precipitation gradients caused by topography or by thunderstorm cells. The last two methods, however, are time-consuming and yield answers close to the arithmetic average unless the spatial distribution of rain gauges is not uniform. The Thiessen and isohyetal methods are useful, however, for the portrayal and quantification of spatial patterns of precipitation, which are often more important than areal averages in water-resource management, ecology, and physical geography.

The precision with which the rainfall of an area can be estimated depends upon the density of the gauge network and the size and type of storm event. In desert regions experiencing localized thunderstorms, it is much more difficult to assess the average rainfall than in regions where precipitation originates mainly from general cyclonic storms. Sparse gauge networks tend to underestimate maximum amounts and intensities and if the record is short can grossly underestimate rainfall characteristics required for planning soil conservation practices, culverts, and other structures. It is often useful to supplement networks of standard gauges with small tube gauges or tin cans read after each storm by farmers and other unofficial observers. After severe storms it is also possible to obtain rainfall measurements from pails, cattle troughs, and other nonstandard containers in order to increase the precision of spatial patterns and average areal precipitation.

In mountainous regions, it is even more difficult to define spatial patterns of precipitation and to estimate areal averages. In areas of uniform meteorological conditions and exposure, precipitation may increase uniformly with altitude, as shown in Figure 2-5. Such a curve can be used to construct an isohyetal map or average value for an area. Elsewhere, the relationship may be much more complex, depending upon the season and type of storm, and the orientation, width, and exposure of the topographic barrier. Methods of defining the precipitation supply under such conditions are described by Dawdy and Langbein (1960), Rainbird (1967), Schermerhorn (1967), and Spreen (1947). The topic has great practical importance and requires considerable ingenuity because of the paucity of gauges in mountainous regions.

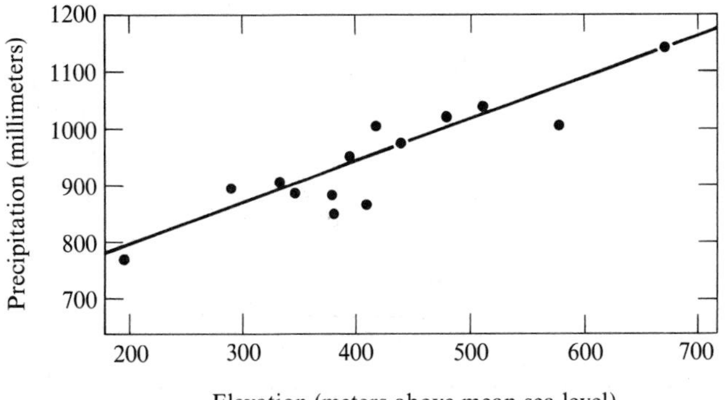

Figure 2-5 Relationship between mean annual precipitation and elevation, northeastern Vermont. (Data from R. L. Hendrick, U.S. Department of Agriculture, Danville, Vermont.)

Analysis of Rainfall Data

Different characteristics of rainfall are important to specialists in various fields, and therefore the number of ways of analyzing rainfall data are virtually unlimited. The methods chosen depend upon the nature of the available data and the purpose of the investigation. At most stations only daily totals of precipitation are measured. These totals may refer to several storms during a day or to a part of one storm that bridges two measuring periods. In this sense the records are a somewhat artificial and inaccurate description of precipitation. The only characteristics of rainfall that can be gleaned from such data are those pertaining to specified intervals of time, such as days, months, seasons, or years. A relatively small number of rainfall-measuring stations are equipped with continuously recording gauges, which yield data on the characteristics of individual storms such as timing and intensity as well as total amount. Some of the analytical techniques described below are applicable only to rainfall in specified periods of time, while others are relevant to the characteristics of individual rainstorms.

Another consideration in choosing appropriate methods of analysis is the particular planning problem for which a description of the rainfall regime is required. Ecologists and agronomists may be interested specifically in seasonal totals, the frequency of small amounts of rain, or the probability of droughts for their studies of crops or natural plant populations. The geomorphologist or the agricultural engineer studying soil erosion problems requires data on the frequency of severe, erosive rainfalls, while civil engineers may concentrate on the intensity, duration, and areal extent of the large, infrequent storms that test the design of their structures. The hydrologist is able to make statistical predictions of these rainfall characteristics. In some instances, it is necessary to devise entirely new ways of looking at raw precipitation data in order to answer a specific question relevant to water supply, forest management, plant survival, or landform development. This text introduces only a few examples of the most commonly used types of analysis. Other aspects of precipitation, such as snowfall and snow accumulation, and the description of rainfall energies are included in later chapters.

Estimating Missing Data

Gaps in a rainfall record are common for a variety of reasons. A gauge may have been installed after the period of interest or may have been closed down for some period. The gauge may not have functioned correctly or may not have been read regularly. Such gaps may be filled by estimates of precipitation based on other records. The simplest way of doing this is by regression and correlation of measurements from the station of interest with data from a nearby gauge when both were functioning. The gap in the record at the first station may then be closed by using the records of the second station and the calculated regression relationship. The dependability of the estimate can be improved by using data from several surrounding gauges to estimate data at a single gauge. A simpler technique involves estimating a missing rainfall record at some station, A, from that at three or four surrounding gauges by the following relationship:

$$P_A = \frac{1}{3}\left(\frac{N_A}{N_B} \cdot P_B + \frac{N_A}{N_C} \cdot P_C + \frac{N_A}{N_D} \cdot P_D\right) \qquad (2\text{-}1)$$

where P_B, P_C, and P_D are precipitation recorded at three surrounding stations during the gap at station A and N_A, N_B, N_C, N_D are the long-term normal precipitation at the four stations. They may be annual, seasonal, or monthly precipitation, depending upon which is more appropriate for the particular case.

Each of the methods described works well in widespread cyclonic rainfall over areas of fairly uniform topography. Problems are encountered in mountainous regions and in localized convective storms, when it may be useful to draw isohyetal maps to estimate missing data.

A special case of the estimation of missing data is the problem of calculating the monthly distribution of precipitation from seasonal totals collected in remote storage gauges in desert regions or mountains. This is done by prorating the seasonal total according to the time distribution of rainfall at a station with daily measurements. Errors in the procedure result from possible differences in monthly regime between the two stations.

Checking the Consistency of Precipitation Records

An earlier section considered errors in point measurement of precipitation. Relocation of a gauge that changed exposure, the growth of trees close to a gauge site, or the use of shields may alter gauge catch significantly. In order to maintain a homogeneous record, the precipitation record should be adjusted to eliminate the effects of such instrument changes. This can be done by using *double-mass curves,* an example of which is given in Figure 2-6.

To check the homogeneity of a station at location A, several nearby stations, known to have a homogeneous record, are selected and their annual totals are averaged for each year. The accumulated average is plotted against the accumulated annual total at station A, as shown in Figure 2-6. In this case, there is a break of slope in the relation, showing that station A accumulated rain more quickly relative to the other four stations after 1951. At this point, a search of records should be made for an indication of any change at station A occurring in 1951. If a change was recorded, or if there is good reason to believe that some change might have occurred, the post-1951 data at station A are adjusted to bring them into line with the pre-1951 relation. This is done by multiplying the rainfall of each year after 1951 by the ratio of the slopes of the two lines. In this case, the later totals are multiplied by 0.95/1.16.

A thorough check of the gauge history should be made to confirm any changes indicated by double-mass analysis. Other possibilities must also be taken into account. A station that shows a marked change of precipitation

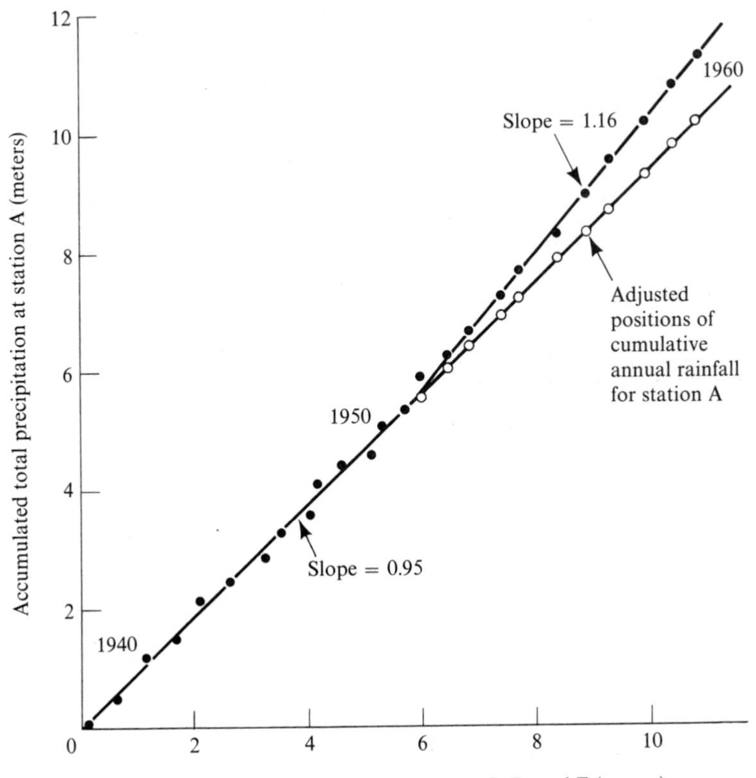

Figure 2-6 Double-mass analysis of annual rainfall for checking the homogeneity of the records at station A and for adjusting them to agree with the average recorded at four other stations, B–E.

over time may be downwind of a large growing industrial city. Industrial pollutants from the city might act as nuclei for raindrops and might be causing an increase of precipitation as the city grows. Other factors might also be causing a real change of precipitation, rather than simply a change of gauge catch.

Introductory Comments on Statistical Analysis of Hydrologic Data

A hydrologic record is a sample of events at the measuring station. This sample is used to make estimates of the population of all potential events at that point, and in particular of the events that may occur in the future. To make deductions about a population from a sample, the theory of *statistical estimation* is used. Several problems arise in applying simple statistical analysis to hydrologic data. For correct use of the technique, the sample data should be random, independent, and homogeneous. For some purposes they should also have a normal frequency distribution and be of equal variability. Double-mass analysis for checking the homogeneity of a record has already been introduced. In a hydrologic record the data are not strictly random, since they are restricted to a relatively short period of observation, and are usually so sparse that none can be discarded. The degree of independence of data varies with the phenomenon under study. Daily rainfall totals are not independent, as wet days tend to follow wet days; that is, wet and dry days tend to occur in runs. Annual rainfall totals are more independent, though they also tend to occur in runs of wet and dry years, a phenomenon known as *persistence.* This tendency increases the variability of rainfall totals and reduces the precision of statistical statements, so that a longer period of measurement is required to obtain a desired level of precision. This is also discussed in the section on water supply. In spite of these problems, simple statistical methods are used most frequently for all but the largest design problems in water-resource planning. For this reason, only simple statistical methods are used here.

A first step in statistical analysis of hydrologic data involves finding a *theoretical frequency distribution,* which approximates the *observed frequency distribution* of the sample record. In Figure 2-7 the bar graph shows the observed frequency of annual rainfall over a 138-year period at Baltimore, Maryland. The bar graph, or histogram, can be closely approximated by the symmetrical, bell-shaped curve superimposed upon the figure. The smooth curve is a theoretical frequency distribution known as the *normal distribution,* and it has many useful mathematical properties. Once such a theoretical distribution has been fitted to a sample record, the former can be used to estimate the population of all conceivable values of the hydrologic

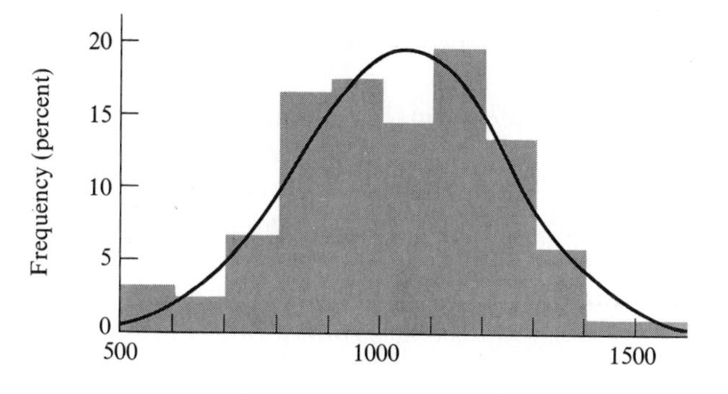

Figure 2-7 Frequency distribution of annual precipitation at Baltimore, Maryland, 1817–1954. The mean is 1041 mm and the standard deviation is 202 mm. (Data from the U.S. Weather Bureau.)

variable under study. It is possible, through the use of simple arithmetic, to fit such a curve to the observed distribution and also to calculate a variety of probability statements. Any elementary text, such as Freund and Williams (1958), gives instructions for this. Here, however, we will concentrate on graphical methods of fitting and prediction. We will concentrate first upon amounts of precipitation measured in fixed periods of time.

Analysis of Total Rainfall within Specific Measurement Periods

Annual totals of precipitation usually have a normal frequency distribution, and average annual precipitation at a place is expressed in terms of the *arithmetic mean.* The mean of our sample record is the best estimate we have of the true long-term mean of the population. At least as important as the mean annual rainfall is the variability of individual years about the mean, which is usually expressed by a statistic called the *standard deviation* (s), defined as:

$$s = \sqrt{\frac{\sum_{i=1}^{n} (x_i - \overline{x})^2}{n - 1}} \qquad (2\text{-}2)^*$$

where x_i is the precipitation of the ith year, $\overline{x}$ is the mean annual precipitation of the sample record, and n is the number of years of record.

Areas under the normal curve can be taken to represent probabilities of obtaining values of the variable under consideration. In Figure 2-8, for example, the shaded area to the left of 600 mm on the abscissa constitutes 1.2 percent of the area under the curve and indicates that 1.2 percent of all annual rainfalls will be less than 600 mm. In a normal distribution, 68 percent of all occurrences fall within one standard deviation above and below the mean, as indicated in Figure 2-8. Hence, in approximately two-thirds of all years, Baltimore will receive between 839, which is ($\overline{x} - s$), and 1243, which is ($\overline{x} + s$), mm.

Instead of computing the normal curve that fits a histogram of rainfall as shown in Figure 2-7, it is easier and more useful to work with the *cumulative frequency distribution* of the data. This is done in Figure 2-9, in which we have plotted measured values of annual precipitation against the percentage of all events less than or equal to these values. This cumulative frequency value is obtained by ranking the n annual precipitation values according to their size and by giving the smallest rainfall the rank $m = 1$. The largest

---

*For more rapid computation, Equation 2-2 can be manipulated to:

$$s = \sqrt{\frac{n \sum_{i=1}^{n} x_i^2 - \left(\sum_{i=1}^{n} x_i\right)^2}{n(n - 1)}}$$

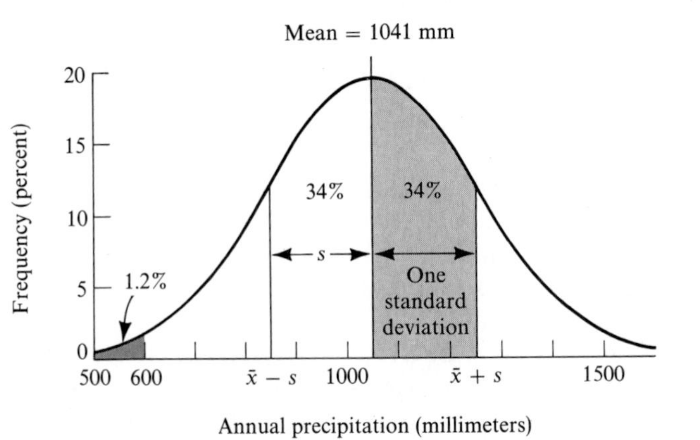

Figure 2-8 The relationship between probability and relative area under the normal curve. Within one standard deviation above and below the mean lie 68 percent of all occurrences represented by this frequency distribution. The data are annual rainfall totals from Baltimore with a mean of 1041 mm and a standard deviation of 202 mm.

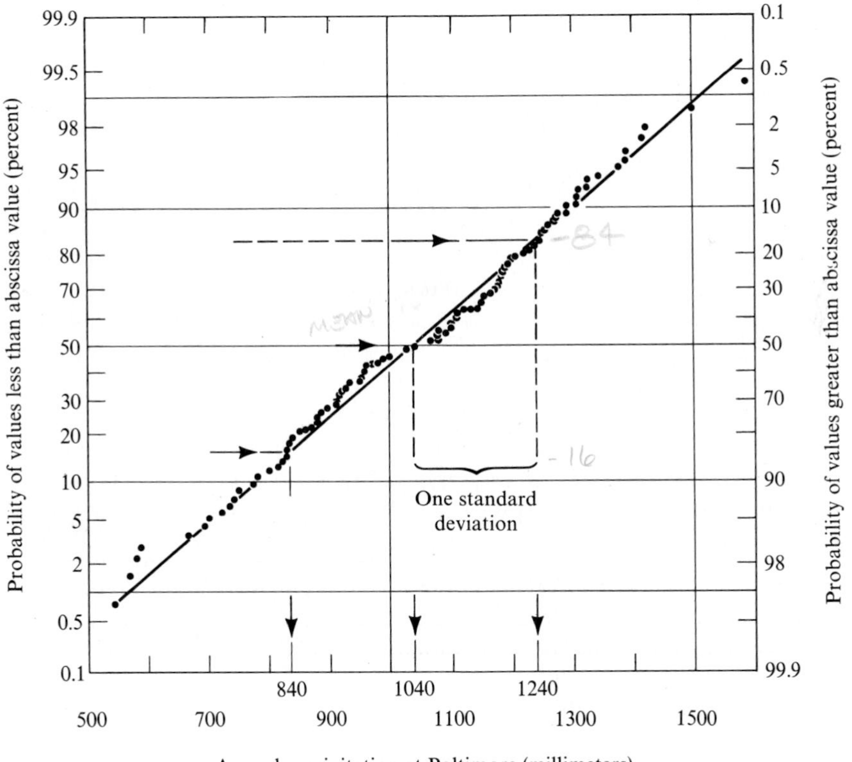

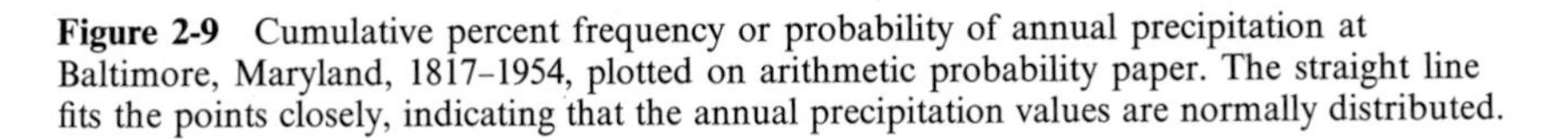

Figure 2-9 Cumulative percent frequency or probability of annual precipitation at Baltimore, Maryland, 1817–1954, plotted on arithmetic probability paper. The straight line fits the points closely, indicating that the annual precipitation values are normally distributed.

value has a rank $m = 138$ in the case of this 138-year period. The percentage of all events less than or equal to each rainfall value, then is given by the formula:

$$F_i = \frac{m}{n+1} \cdot 100\% \qquad \text{(2-3)*}$$

where F_i is the cumulative percentage frequency of the variable, or the percentage of years with a rainfall equal to or less than the particular annual rainfall having rank m. The percentage frequency of past events is taken as the probability (also in percent) of future events.

The cumulative frequency distribution of annual rainfall can be drawn by plotting the calculated values of F_i or of probability against the measured rainfall values, as shown in Figure 2-9. The values are plotted on a sheet of arithmetic probability paper, which has a vertical scale designed so that the cumulative frequency curve of a normal distribution plots as a straight line. Such a graph takes a few minutes to calculate and plot and can be used for a variety of analyses.

In Figure 2-9 the annual values of precipitation at Baltimore plot as a nearly straight line on arithmetic probability paper, indicating that the annual values are normally distributed. The mean annual precipitation can be read from the curve at the point representing 50 percent probability, which on the abscissa is 1040 mm. In a normal distribution, 50 percent of the values lie above and 50 percent lie below the mean. The standard deviation can be read by noting that 68 percent of all occurrences in a normal distribution lie within one standard deviation on either side of the mean, between probability values of 16 percent and 84 percent on the ordinate. Reading across from these probability values to the curve and down to the abscissa indicates values of annual precipitation which are one standard deviation away from the mean on either side. Thus one standard deviation is $1240 - 1040 = 200$ mm, or similarly, between the 16 and 50 percent probabilities, $1040 - 840 = 200$ mm.

The straight line smooths the irregularity of the observed distribution and can be used to read the probability of annual rainfalls less than any chosen value or greater than any value. There is a 4.7 percent chance, for example, that the rainfall of any year in Baltimore will be less than 700 mm. Conversely, there is a 95.3 percent chance that it will exceed this value. Such statements are valuable for the design of water supply and similar planning problems. Alternatively, if the planner is concerned about land drainage or some other problem associated with overabundant water supply, the curve can be used to read the annual total that will be exceeded 10, 25, etc., percent of the time. If upper and lower rainfall limits of crop tolerance are

*This formula is a very slight modification of the more common $F_i = m/n$. If the reader wishes to pursue the issue further, he should see Fair et al. 1966, p. 7-7, or Langbein (1960) for a fuller discussion of plotting positions. With a large sample, such as that shown in Figure 2-9, it is a little easier to group the data into classes before constructing the cumulative frequency distribution. This method is described in every elementary statistics textbook.

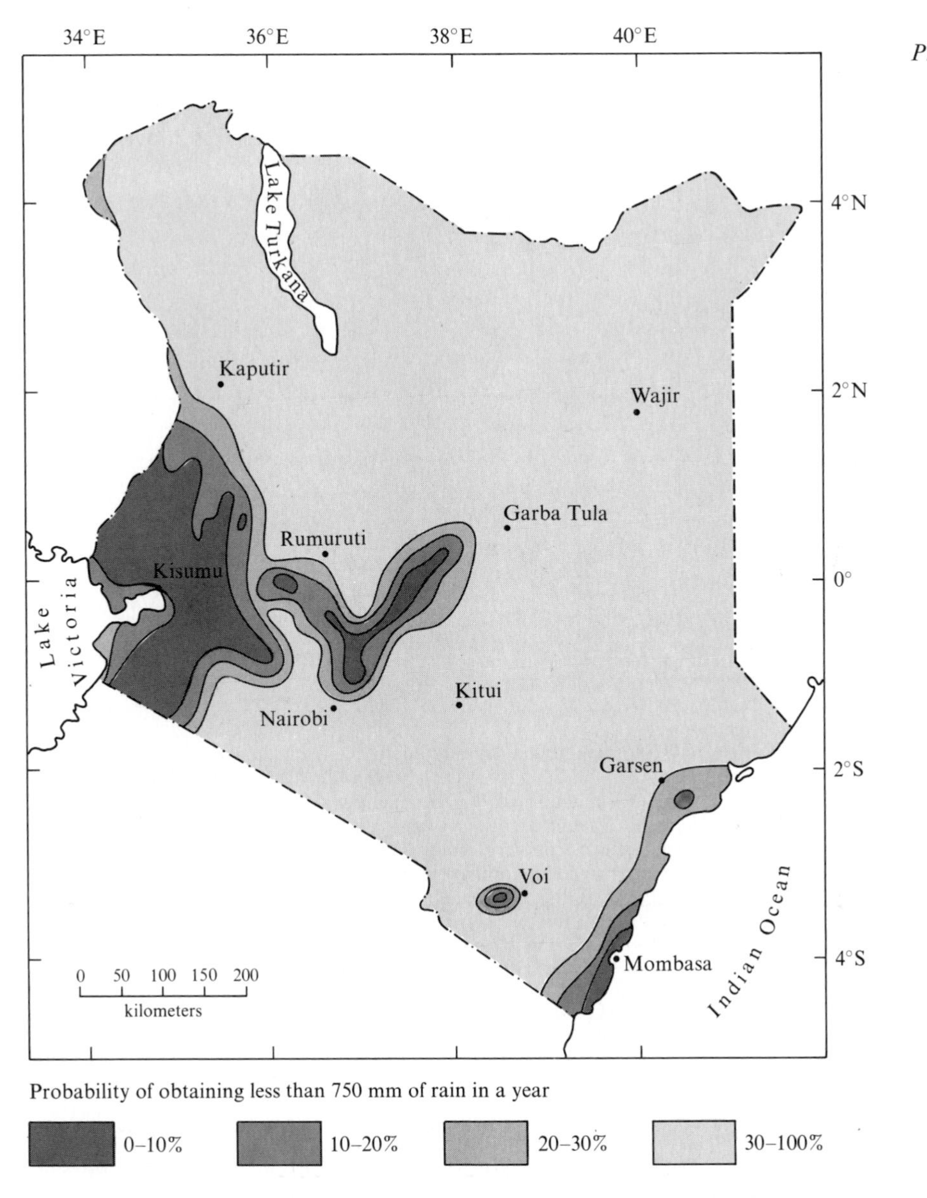

Figure 2-10 Map of drought probability in Kenya. (From the Atlas of Kenya.)

known, the probability of crop failure due to drought and waterlogging can be assessed. With probability curves from a number of stations, it is possible to map the chances of drought (see Figure 2-10), crop failure, or other problems associated with the supply of precipitation.

Fortunately, many hydrologic variables are normally distributed. Others can be made normal by transforming the original values into their loga-

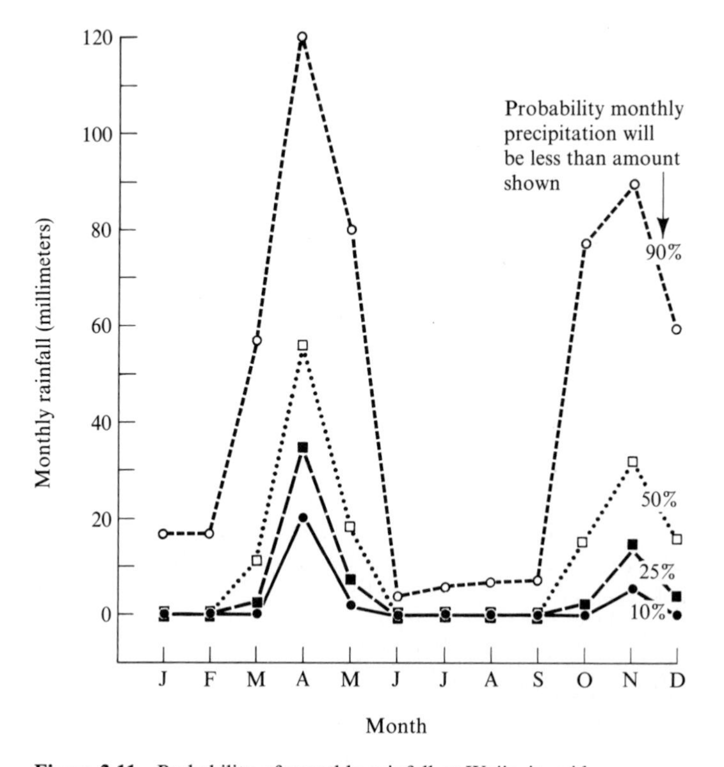

Figure 2-11 Probability of monthly rainfall at Wajir, in arid northeastern Kenya. The curves are labeled with values of probability that monthly precipitation will be less than the amount indicated on the ordinate. The values were obtained by transforming the measured monthly rainfall totals into their logarithms and plotting the cumulative frequency curve of the logarithms for each month on probability paper like that in Figure 2-9. Then, reading from the 10-, 25-, 50-, and 90-percent values on the probability scale across to the curve and down to the abscissa, it was possible to identify rainfall values that exceed 10, 25, 50, and 90 percent of all the values to be expected for that month. These monthly rainfall values were then transferred to the diagram above. The length of the sample record was 30 years.

rithms, square roots, or cube roots. Annual precipitation totals are usually distributed normally, except in arid regions, where it is often better to transform the values into their logarithms before fitting a normal distribution. Seasonal and monthly data should also be transformed in this way. They can then be used as shown in Figure 2-11 for a valuable description of the reliability of precipitation, for the planning of water supply or of probable demand for irrigation water or for understanding variations in seedling germination and crop yields. A further discussion of the use of frequency analysis applied to the planning of optimum choices of crops for single and double cropping practices is given by Manning (1956).

The Characteristics of Individual Storms

Information on the intensity of individual storms is used in calculations of storm runoff, which in turn are used for design purposes in the planning of storm drains, flood-control structures, culverts, and river bridges. Such information is also useful in calculating the probable hydrologic impact of major land-use changes, or the benefits to be gained by some land management practice such as floodplain zoning or the terracing of agricultural lands. Geomorphologists require information on storm intensities because most land-forming work is done during a few intense storms each year.

Recording rain gauges provide data on the time of beginning and end of individual rainstorms, as well as the timing and intensity of rainfall bursts during the storm. Figure 2-12 shows the record of rainfall accumulation on the chart of a weighing rain gauge. Table 2-1, the tabulated record of this event, indicates the maximum intensities for various periods of time. The analyst may have to abstract data from such charts, although an increasing number of data-collection agencies with the aid of computerized reading and reduction are now able to process information as fast as it is gathered. It is then often published as tables of intensity values for each period of every storm, or as tables of the maximum annual intensity values for various durations.

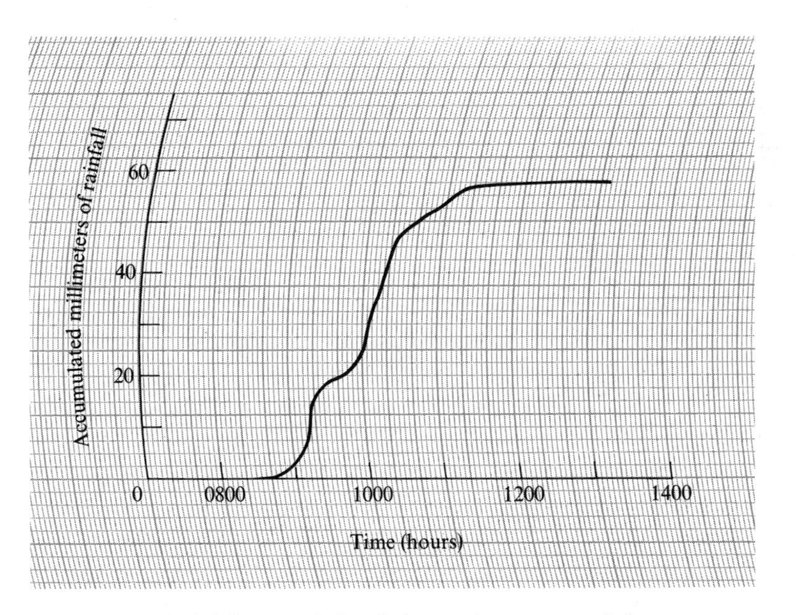

Figure 2-12 Rainfall accumulation during a rainstorm, recorded on a paper strip chart attached to the rotating drum of a weighing-type, recording rain gauge.

Table 2-1 Rainfall intensity-duration relationship during the storm shown in Figure 2-12. (a) Tabulation of the rainstorm. (b) Tabulation of maximum rainfall intensities for various durations.

(a)

TIME FROM BEGINNING OF STORM (MIN)	CUMULATIVE RAIN (MM)	TIME INTERVAL (MIN)	RAINFALL DURING INTERVAL (MM)
0	0	—	—
20	2.5	20	2.5
40	15.5	20	13.0
60	20.1	20	4.6
80	25.2	20	5.1
100	44.4	20	19.2
120	51.3	20	6.9
140	55.1	20	3.8
160	57.7	20	2.6

(b)

DURATION (MIN)	MAXIMUM RAINFALL DURING INTERVAL (MM)	MAXIMUM RAINFALL INTENSITY FOR DURATION (MM/HR)
20	19.2	57.6
40	26.1	39.2
60	31.2	31.2
100	48.8	29.3
120	52.6	26.3
160	57.7	21.6

Total Storm Rainfall

The most obvious characteristic of rainstorms is the total amount of precipitation received. As an indication of the most extreme events to which natural or man-made features of the landscape are likely to be exposed, Figure 2-13 shows the maximum recorded rainfall amounts during various time periods. Most of the stations that define the envelope curve are in tropical or subtropical areas, or are in areas subject to influx of moist tropical air masses. One interesting characteristic of this envelope curve is that it is continually being revised upward as the period of record lengthens, as stations are added to national networks, and as old records are found in remote areas. Figure 2-13 includes the envelope curve given by Foster in his 1948 compilation of intense rainfalls. The latest (1967) compilation by UNECAFE shows that the observed maximum for any duration is now thought to be about 1.5 times that accepted in 1948.

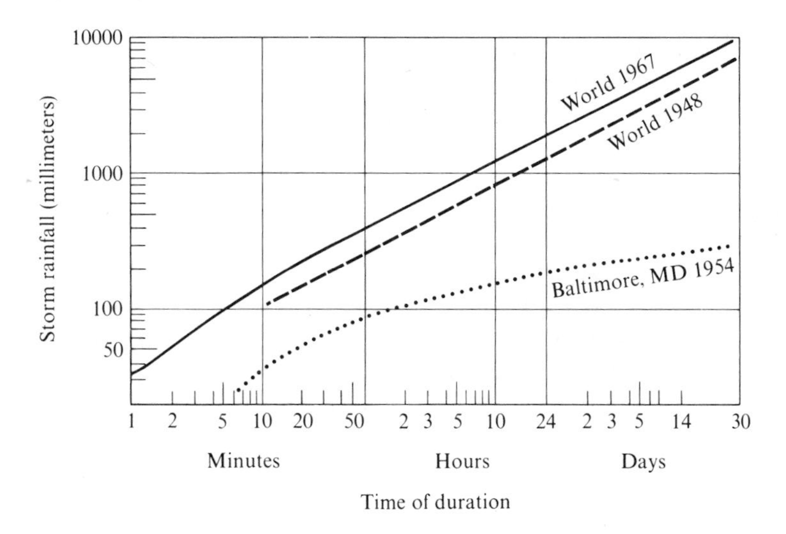

Figure 2-13 Envelope curves for maximum recorded rainfall for various periods of time. (Data from Jennings 1950; UNECAFE, in Todd 1970; U.S. Weather Bureau.)

Of more use for the solution of local problems is a graph of the recorded maxima within a uniform region. An envelope curve for the U.S. Weather Bureau first-order station at Baltimore, Maryland, is also included in Figure 2-13. Care should be taken, however, in using such data for design purposes. Most extreme rainfalls are associated with localized thunderstorms of restricted areal extent. National networks of first-order stations are not dense enough to sample such events adequately. Records at these stations tend to underestimate maximum point rainfalls, as indicated by records from cooperative and unofficial stations. As a general rule the maximum point rainfall of any short-period storm in the vicinity of a first-order station is fairly well represented by twice the station value. We will return to that point later in the chapter.

Intensity-Duration-Frequency Analysis of Point Rainfall

In engineering design, it is uneconomical to design structures to cope with extreme events in low-priority, low-investment situations. A designer takes a calculated risk and designs a structure that will accommodate the largest rainstorm that can be expected in some generally agreed-upon time interval. On the other hand, a designer of large, expensive structures may wish to know how good an estimate of future events a short record is. In this case, he or she will not trust past extremes. In geomorphology, very large rainfalls are responsible for catastrophic processes, which produce landforms, but generally the amount of work done in extreme storms is exceeded by the amount accomplished by more frequent storms of lesser magnitude. The geomorphologist, therefore, would like to know the magnitude and frequency of a complete spectrum of rainstorms in the area. After a disastrous storm producing severe erosion on agricultural land, channel realignment, and other damage, one of the economic considerations in deciding what to do in order to prevent such a future occurrence is to assess how frequently storms of similar magnitude are likely to occur. If such a storm is only likely to occur on the average once every 500 years, it may not be worth spending large sums of money to protect against such a recurrence. Soil conservationists, then, as well as civil engineers, geomorphologists, and other specialists interested in planning problems are interested in the frequency of storms of different intensities and durations. It is convenient to consider first intensity, duration, and frequency for a single station. Areal averages can be introduced later.

Experience indicates that very intense storms are of short duration and are rare (i.e., of low probability). Storms of long duration tend to be less intense, and extremely long storms supplying large amounts of rain are also rare. We need to quantify these inverse relationships between intensity, duration, and frequency.

The basic data used for intensity-duration-frequency analysis of point rainfall consist of the largest events of each year (e.g., the largest 5-minute rainfall of each year or the largest 6-hour rainfall of each year). This set is called the *annual-maximum series* and is a sample of the population of all 5-minute (or other duration) annual extremes at the measuring station. In order to generalize the experience of the relatively short sample record and possibly to extrapolate to the magnitude of rarer events, a theoretical frequency curve, known as the *extreme-value distribution,* is fitted to the observed frequency distribution of annual maxima (Gumbel 1945).

As with the normal curve it is possible to fit an extreme-value distribution to observed data by plotting the cumulative frequency distribution on special probability paper, which is often called "Gumbel paper" or "extreme-value paper." The plotting position formula in Equation 2-3 could be used for this. In design problems concerning storms, however, we are usually concerned with the probability of the occurrence of events *greater* than a certain magnitude. To obtain this *exceedence probability* the *largest* intensity value is given the ranking $m = 1$. An example of the extreme-value distribution is shown in Figure 2-14. Note, however, that the probability scale at the top of this graph is labeled in terms of decimal fractions rather than percentages. A probability of 0.10 in this case indicates that in any year there is a 10-percent chance that the rainfall value on the ordinate will be exceeded by the 24-hour annual maximum of that year. One has to become acquainted with using both expressions of chance interchangeably.

The frequency of hydrologic events can be discussed in terms of either probability or *recurrence interval* (sometimes called *return period*). Exceedence probability (p) refers to the chance that the annual-maximum event

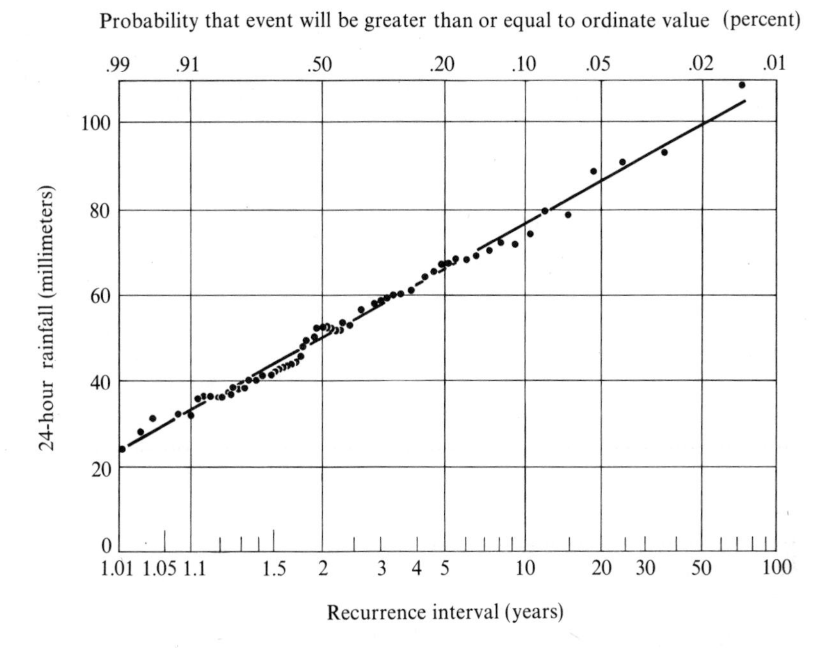

Figure 2-14 Rainfall frequency curve for 24-hour, annual maximum storms at Buffalo, New York, 1891–1961. (Data from U.S. Weather Bureau.)

of any year (e.g., 5-minute maximum rainfall) will equal or exceed some given value. We may read from Figure 2-14, for example, that there is a probability of 0.05 (5 percent) that in any year the annual-maximum, 24-hour rainfall at Buffalo, New York, will equal or exceed 88 mm. Recurrence interval (T) is the average interval (in years) between events equaling or exceeding a given magnitude. The rainstorm referred to above would be expected on the average five times in 100 years. Its average recurrence interval, therefore, would be 20 years. Such an event would be referred to as the 20-year, 24-hour rainstorm, and its return period can be calculated from

$$T = \frac{1}{p} = \frac{n+1}{m} \tag{2-4}$$

m = rank
1 = lowest
m = highest
n = # of events

The lower scale in Figure 2-14 is labeled in terms of recurrence interval, and the necessary calculations are listed in Typical Problem 2-5.

A line can be drawn by eye through the points in Figure 2-14 and used to estimate the return period of an event of any magnitude or the magnitude of an event of any frequency. The chances of gross error, of course, increase rapidly if a rare event is estimated by extrapolation from a short record. Quite often, however, this is necessary, but in such cases the estimate should be checked from all possible sources, and the great uncertainty should be taken into account in decision making.

Extrapolating curves on extreme-value probability paper can lead to gross error. One would prefer to use such plots for estimating only the size of an event with a return period no longer than the record. Unfortunately, one often has no choice. Throughout most of the world, records longer than 25 years are rare. Yet engineers, landscape architects, geomorphologists, and others often need to estimate the size of rainfalls of 50- to 100-year recurrence intervals. When doing so, they should recognize the dangers and try to obtain information on such events from a variety of sources to reach a consensus. In Chapter 10 we will discuss the application of statistical analysis to floods and will quantify the level of uncertainty inherent in such methods as the one described above.

The Gumbel extreme-value frequency distribution is not the only one that could be used for estimating large events. It is, however, the most popular and has received the widest application in various parts of the world. Hershfield and Kohler (1960) tested its applicability to rainfalls of 10-minute to 24-hour duration from 128 stations throughout the United States and found that the method yielded results of acceptable accuracy.

The concept of an average recurrence interval should not be taken to mean that the 20-year event, for example, will occur at equal 20-year intervals. If such an event occurs this year, the probability of it occurring next year remains 0.05. The recurrence interval refers only to the average spacing of events over a very large number of years. It is possible for the 20-year, 1-hour storm to occur in two consecutive years, or not to occur at all during 30 years or more. In using calculations of the average frequency of intense storms, therefore, one should also take note of the variation in their occurrence.

If there is a probability p that a rainstorm greater than or equal to X will occur within the next year, the probability that such a storm will *not* occur is $(1 - p)$. The probability that it will not occur for the next two years is $(1 - p)^2$, and the probability that no such event will occur for the next n years is $(1 - p)^n$. The chances that one such event (greater than or equal to X) *will* occur in the next n years is

$$q = 1 - (1 - p)^n \tag{2-5}$$

If, for example, we wish to calculate the probability of the 100-year, 1-hour storm occurring within the next 20 years,

$$\begin{aligned} q &= 1 - (1 - .01)^{20} \\ &= 1 - 0.82 \\ &= 0.18 \end{aligned}$$

Figure 2-15 shows the probability of various extreme events occurring within specified periods of time. These calculations demonstrate the large uncer-

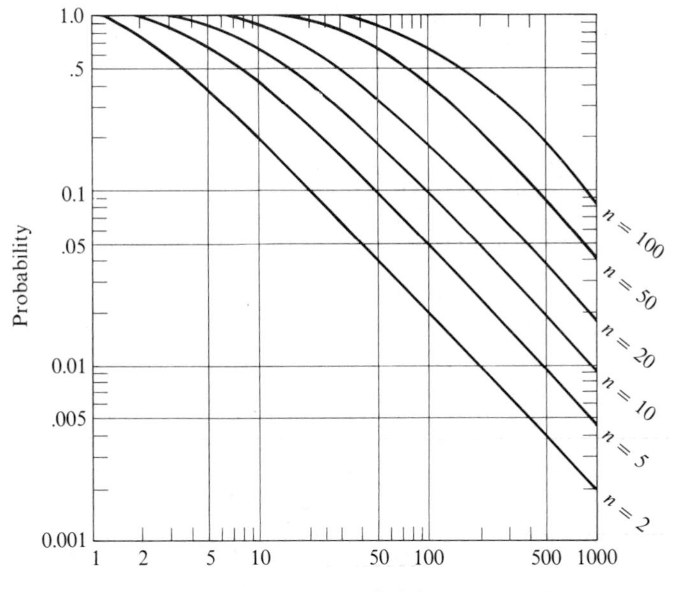

Figure 2-15 Probability of extreme events occurring within specified numbers of years. The abscissa is the average recurrence interval of the event; the ordinate is the probability of its recurrence within the period n years.

tainties inherent in statistical analysis of extreme hydrologic events, and should remind the planner that an extreme event, such as the 100-year storm, is not something that will necessarily occur at some time in the distant future.

The Partial-Duration Series

Use of the annual-maximum series ignores the second and third highest events of a year, even though they may be larger than the maxima of other years. To include these events, use is sometimes made of the *partial-duration series,* which consists of all events greater than some arbitrary base magnitude, usually the smallest number of the annual-maximum series. The partial-duration series does not fit the extreme-value distribution, and therefore cannot be extrapolated by the methods previously outlined. Instead, the return period and magnitude of the partial-duration series can be plotted on semi-logarithmic paper or double-logarithmic paper, using Equation 2-4 and a curve can be sketched by eye. This technique is particularly useful for estimating events of low recurrence interval from a short record.

Table 2-2 Empirical factors for converting rainstorm magnitudes calculated from the annual-maximum series to the partial-duration-series values for the same recurrence interval. Based on rainfall records from a sample of 50 widely scattered stations in the United States. (From U.S. Weather Bureau 1957.)

RECURRENCE INTERVAL (YR)	CONVERSION FACTOR
2	1.13
5	1.04
10	1.01
20	1.00

The difference between the probability estimated from the partial-duration series and the annual-maximum series is that the latter predicts the probability of an event occurring *as an annual maximum*. The partial-duration-series curve predicts the probability of an event occurring regardless of its place in the annual hierarchy. For return periods greater than 10 years, there is almost no difference between the partial-duration series and the annual-maximum series. There is a difference, however, between the probabilities calculated from the two series for small events. Table 2-2 gives empirical factors for converting the magnitude of the partial-duration series to that of the annual-maximum series. If, for example, the annual-maximum series suggests the 5-year, 6-hour rainfall to be 100 mm, the 5-year, 6-hour storm calculated from the partial-duration series would be 104 mm. In Chapter 10 the matter is discussed again in relation to river floods.

Construction of the Intensity-Duration-Frequency Regime at a Station

Studies of intensity-duration-frequency relations at a point often use both the partial-duration series and the annual-maximum series. The U.S. Weather Bureau (1955, 1957, and several other reports) surveyed intensity-frequency relationships for durations for 20 minutes to 24 hours. Events with recurrence intervals of 1 to 10 years were estimated from a freehand curve sketched through the partial-duration series plotted on double-logarithmic paper. Events with a recurrence interval of 20 years or more were estimated by extreme-value analysis of the annual-maximum series. For return periods between 10 and 20 years, the magnitudes of events were estimated by graphical smoothing. The resulting intensity-duration-frequency graphs are published for major weather stations (see Figure 2-16). It is useful to summarize the intensity-duration-frequency regime of a large number of stations by means of maps such as those in Figure 2-17. Interpolations for ungauged stations can be made from these maps, although

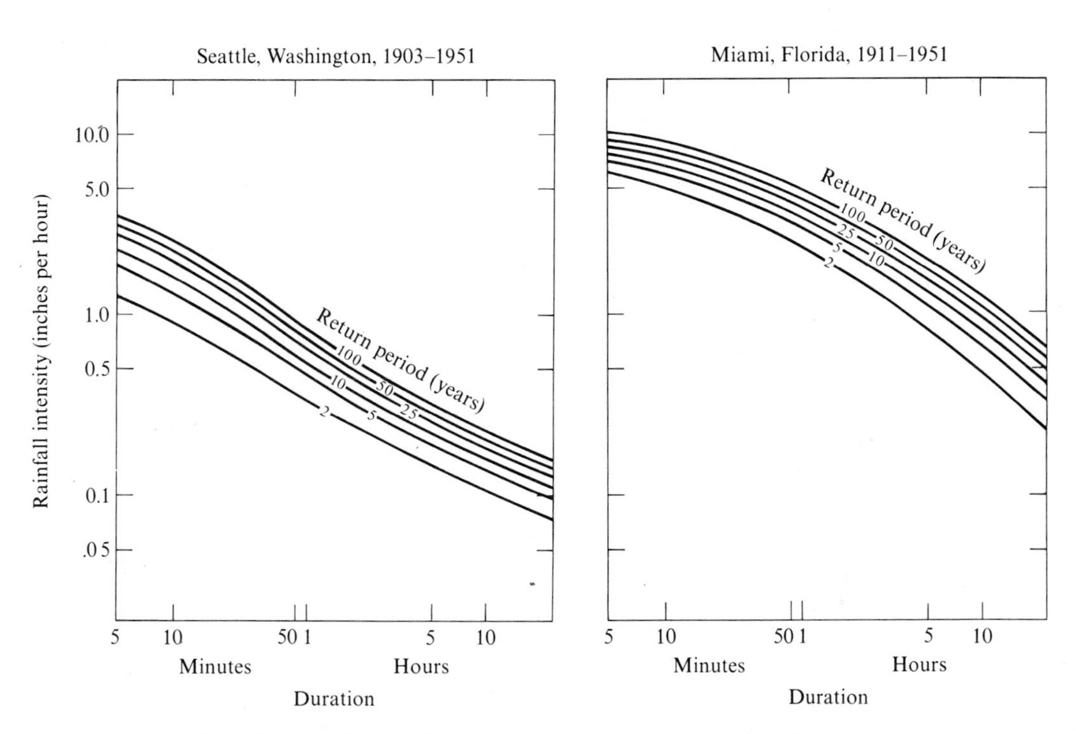

Figure 2-16 Intensity-duration-frequency curves for two stations. The differences between the two sets of curves reflect contrasts in the air-mass climatology of the cities. (From U.S. Weather Bureau.)

there are large uncertainties in doing this for areas of strong relief (U.S. Weather Bureau 1957). Interpolations for various storm durations and frequencies can be made from these maps, as illustrated in Typical Problem 2-6.

In many parts of the world there are few recording rain gauges whose records can be used for intensity-duration-frequency analysis. In these regions it is valuable to analyze the few records of intensity and to correlate the results with more commonly measured rainfall parameters. Rodda (1967), for example, correlated maximum 24-hour rainfalls for various recurrence intervals with mean annual rainfall at stations throughout the United Kingdom (see Figure 2-18). He could then estimate 24-hour storms at stations without recording gauges from a map of annual precipitation. Rodier and Auvray (1965) used the same kind of analysis to estimate 10-year, 24-hour design storms for runoff prediction over a vast area of West and Central Africa. Hershfield and Wilson (1958) pioneered such work and produced several generalizations that are useful in areas without dense recording networks. They suggested, for example, that the mean of the 1-hour and 24-hour rainfalls at a station is approximately equal to the 6-hour rainfall, and that the 100-year rainfall of a particular duration is roughly

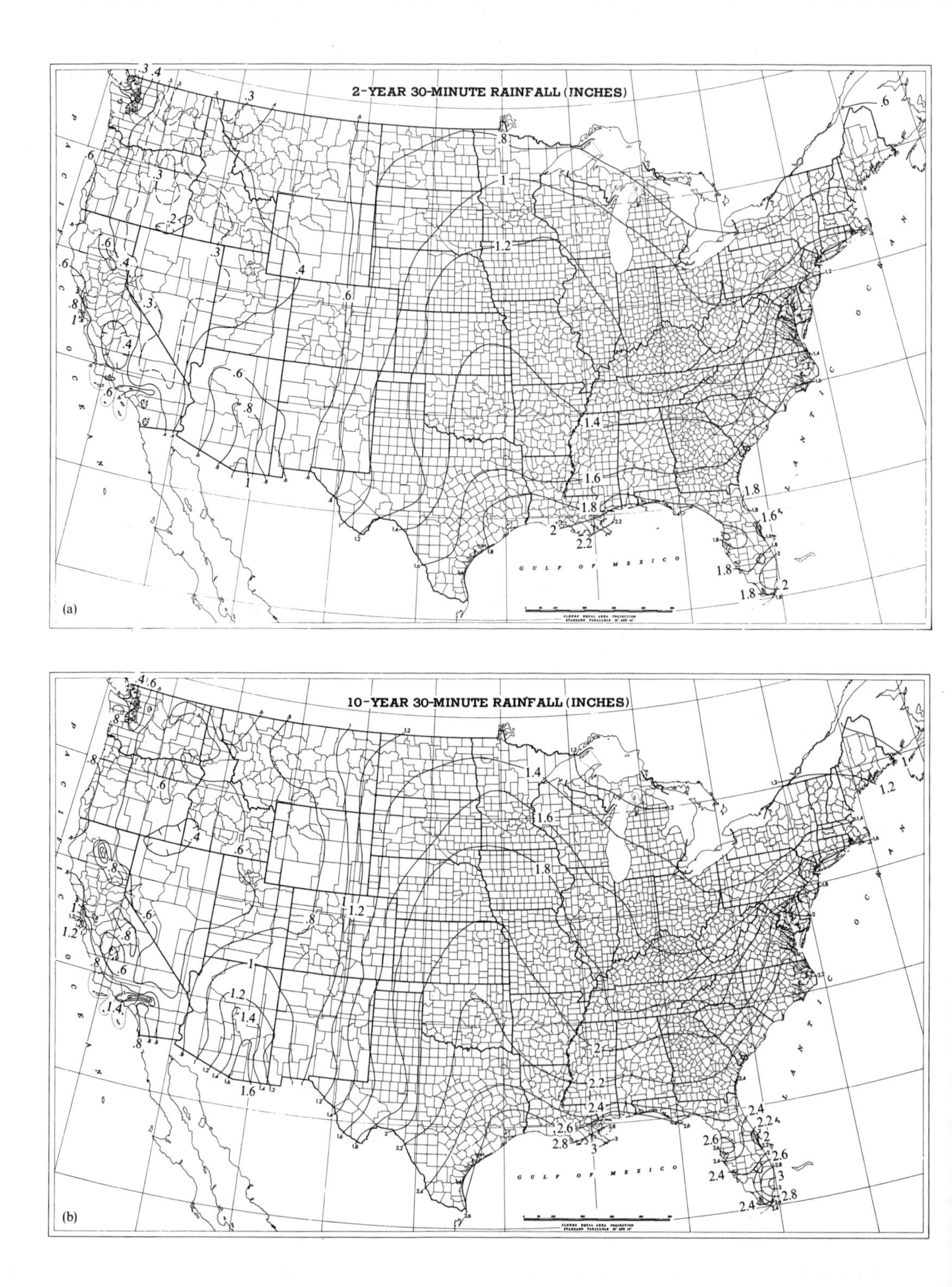
2-YEAR 30-MINUTE RAINFALL (INCHES)
(a)
GULF OF MEXICO
PACIFIC OCEAN
ATLANTIC OCEAN
10-YEAR 30-MINUTE RAINFALL (INCHES)
(b)

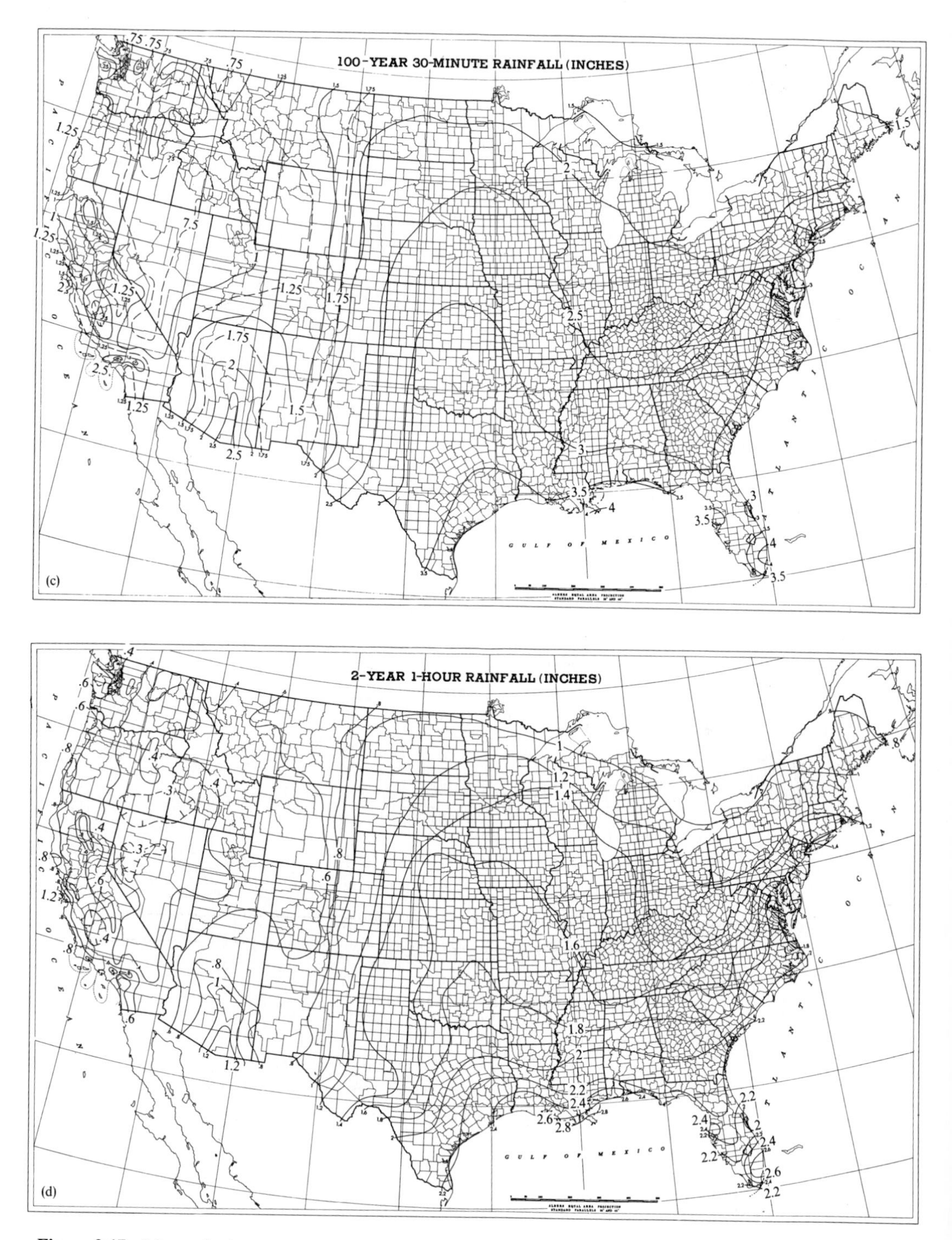

Figure 2-17 Maps of rainstorm amounts (inches) for varying durations and recurrence intervals in the United States. (a) 2-year, 30-minute rainfall, inches. (b) 10-year, 30-minute rainfall, inches. (c) 100-year, 30-minute rainfall, inches. (d) 2-year, 1-hour rainfall, inches.

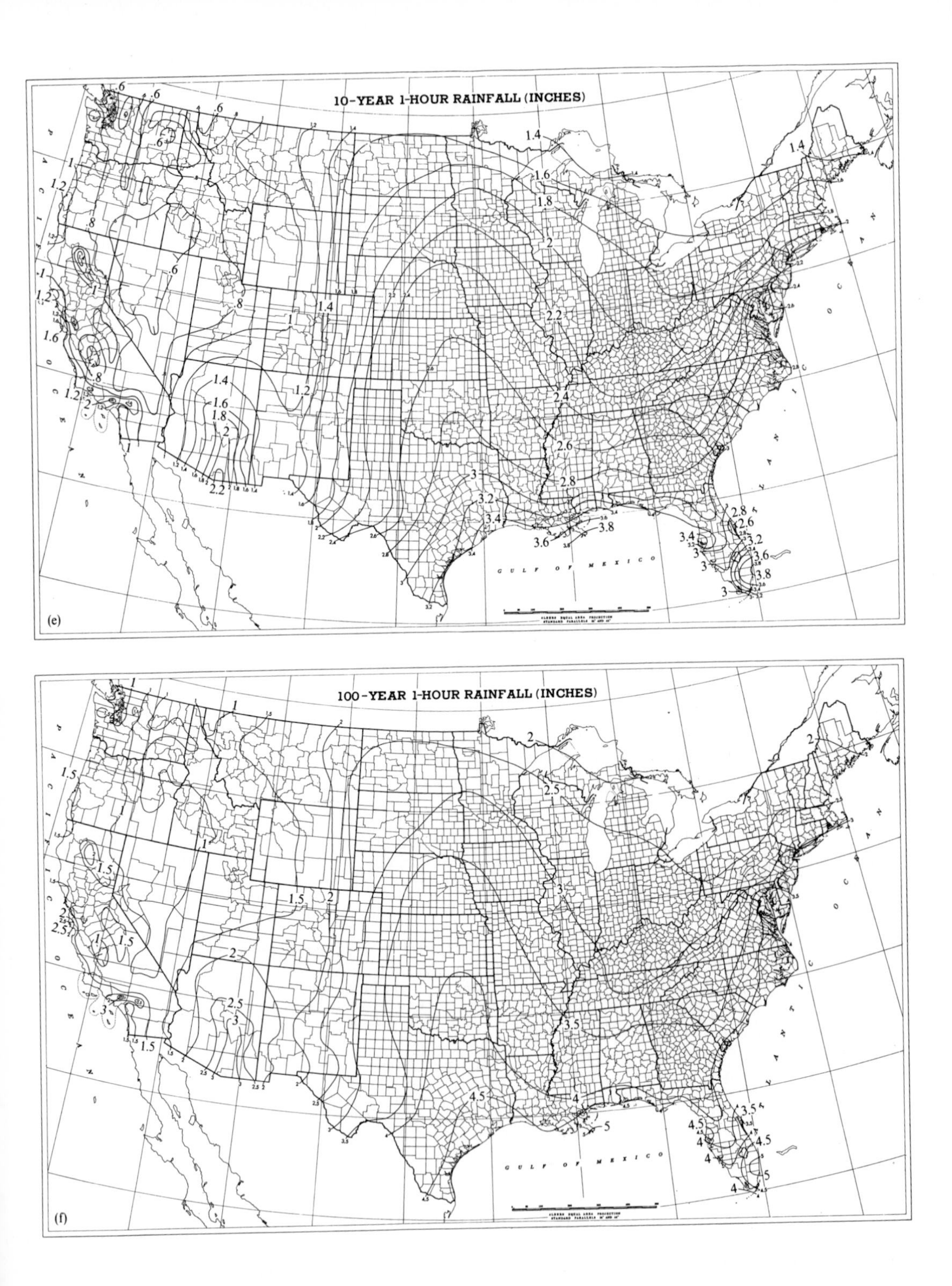

10-YEAR 1-HOUR RAINFALL (INCHES)
GULF OF MEXICO
(e)
100-YEAR 1-HOUR RAINFALL (INCHES)
GULF OF MEXICO
(f)

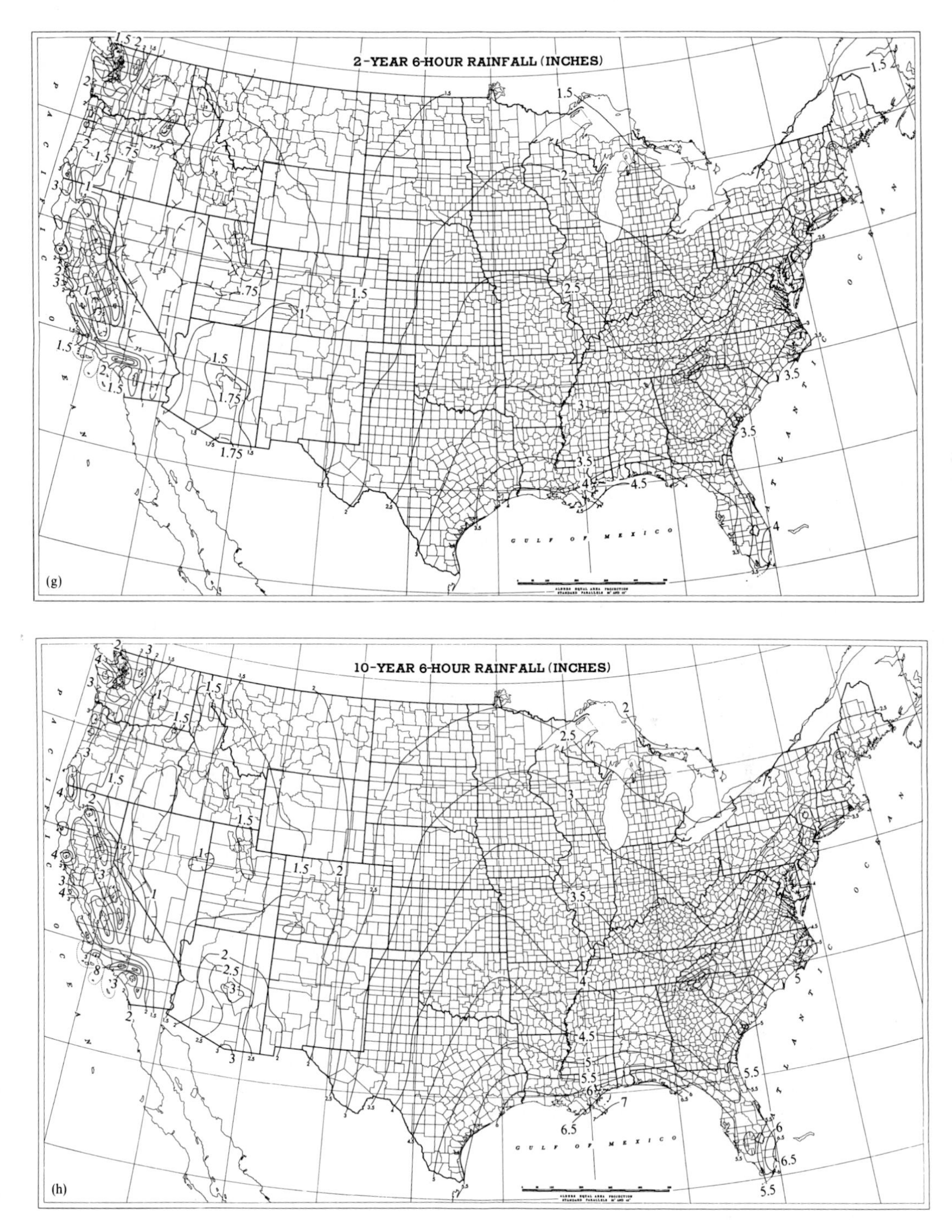

Figure 2-17, *continued* (e) 10-year, 1-hour rainfall, inches. (f) 100-year, 1-hour rainfall, inches. (g) 2-year, 6-hour rainfall, inches (h) 10-year. 6-hour rainfall, inches.

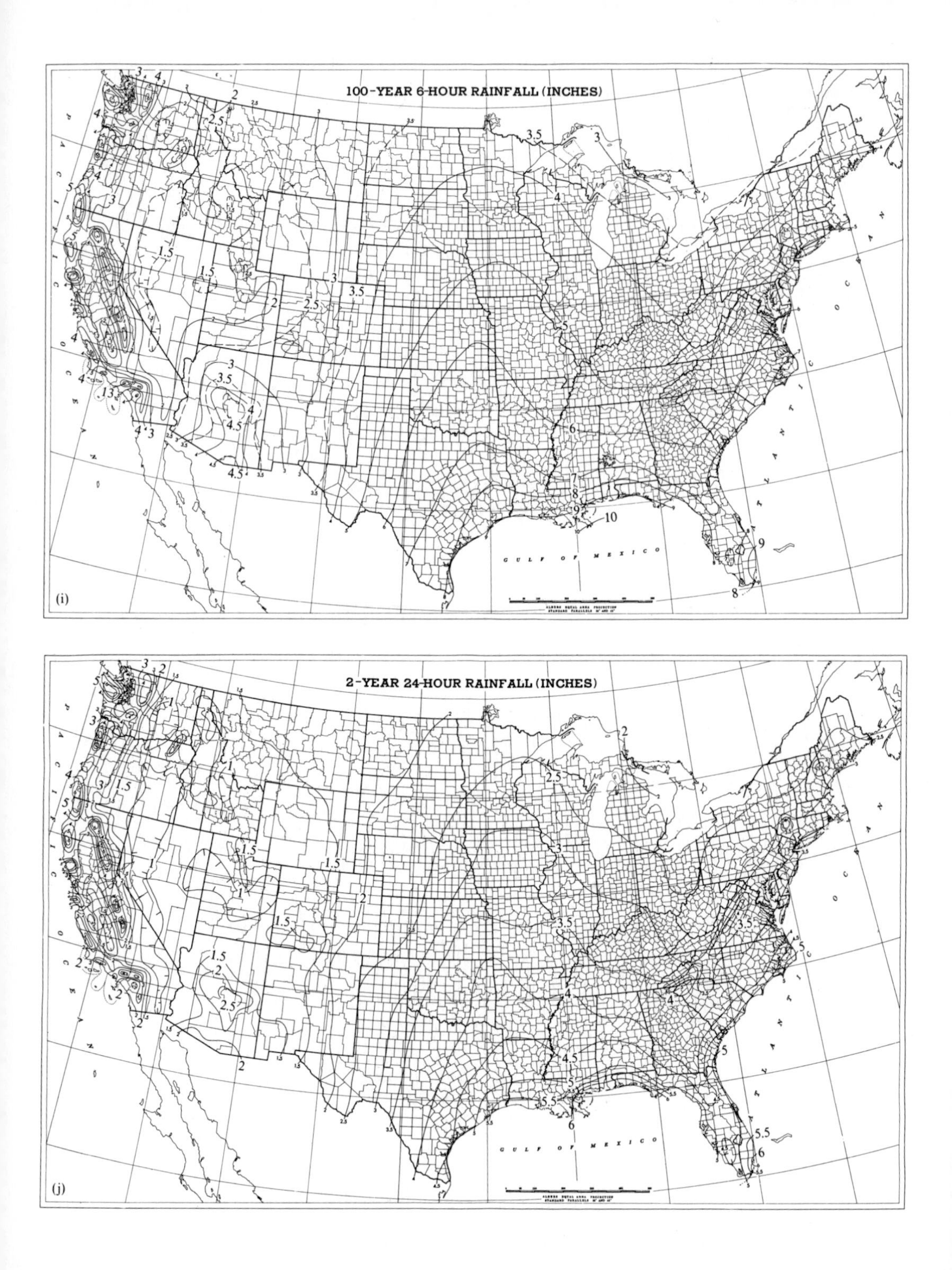

100-YEAR 6-HOUR RAINFALL (INCHES)
GULF OF MEXICO
(i)
2-YEAR 24-HOUR RAINFALL (INCHES)
GULF OF MEXICO
(j)

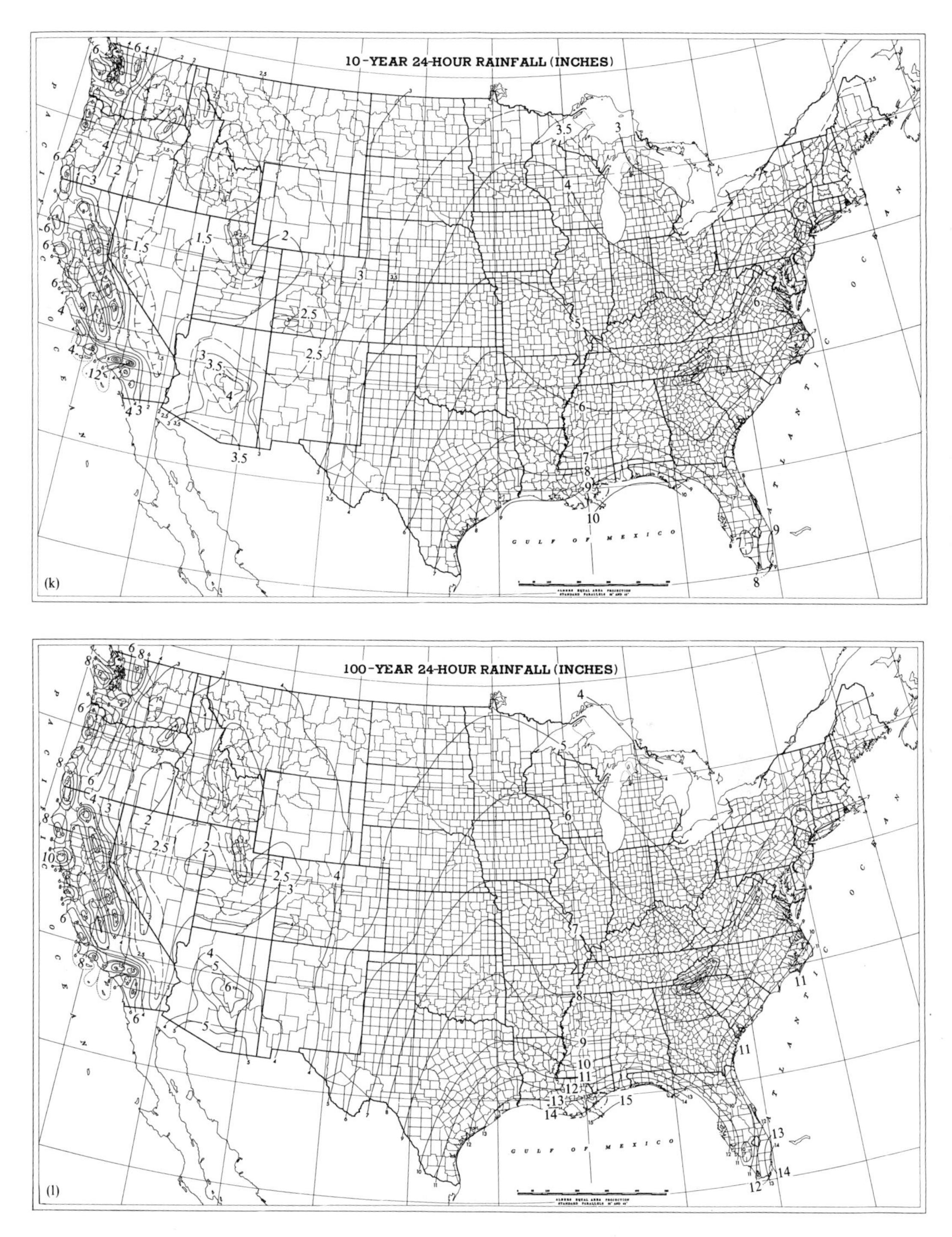

Figure 2-17, *continued* (i) 100-year, 6-hour rainfall, inches. (j) 2-year, 24-hour rainfall, inches. (k) 10-year, 24-hour rainfall, inches. (l) 100-year, 24-hour rainfall, inches. (From Hershfield 1961, data from U.S. Weather Bureau.)

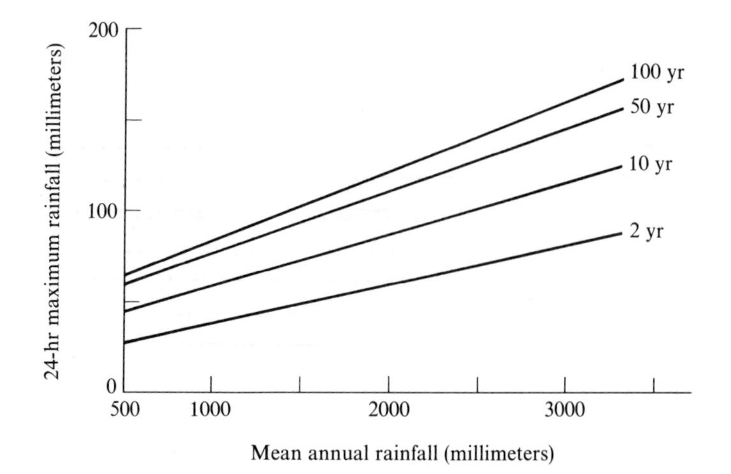

Figure 2-18 Relationship between mean annual rainfall and 24-hour maximum amounts for various recurrence intervals at stations in the British Isles. (From Rodda 1967.)

2.2 times as great as the 2-year storm of the same duration. Figure 2-19 portrays relationships developed by the U.S. Army Corps of Engineers for estimating intense rainfalls throughout the world. The map is used in Typical Problem 10-6. Refinement of these relationships and the development of new ones would be a valuable contribution of applied climatology not only to hydrology but to geomorphology, civil and environmental engineering, and planning.

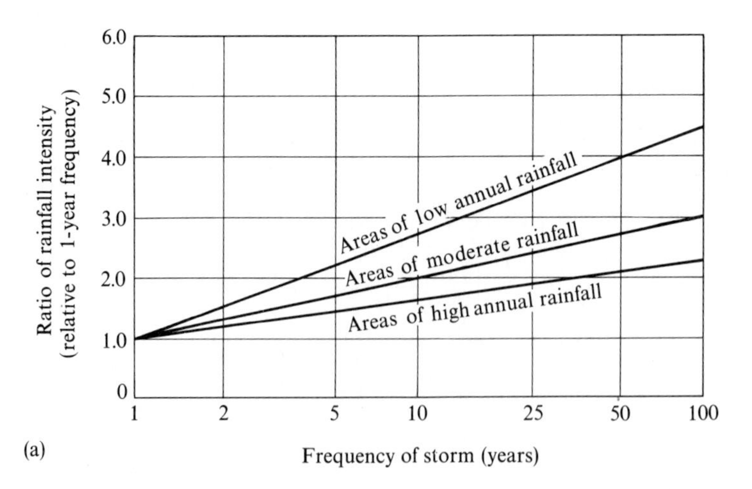

Figure 2-19 Rainfall intensity-frequency relationships for the world. (a) Relation between rainfall intensities of various frequencies. (b) Opposite: Isopluvial map of 2-year, 1-hour maximum precipitation (inches). (From U.S. Army Corps of Engineers 1968.)

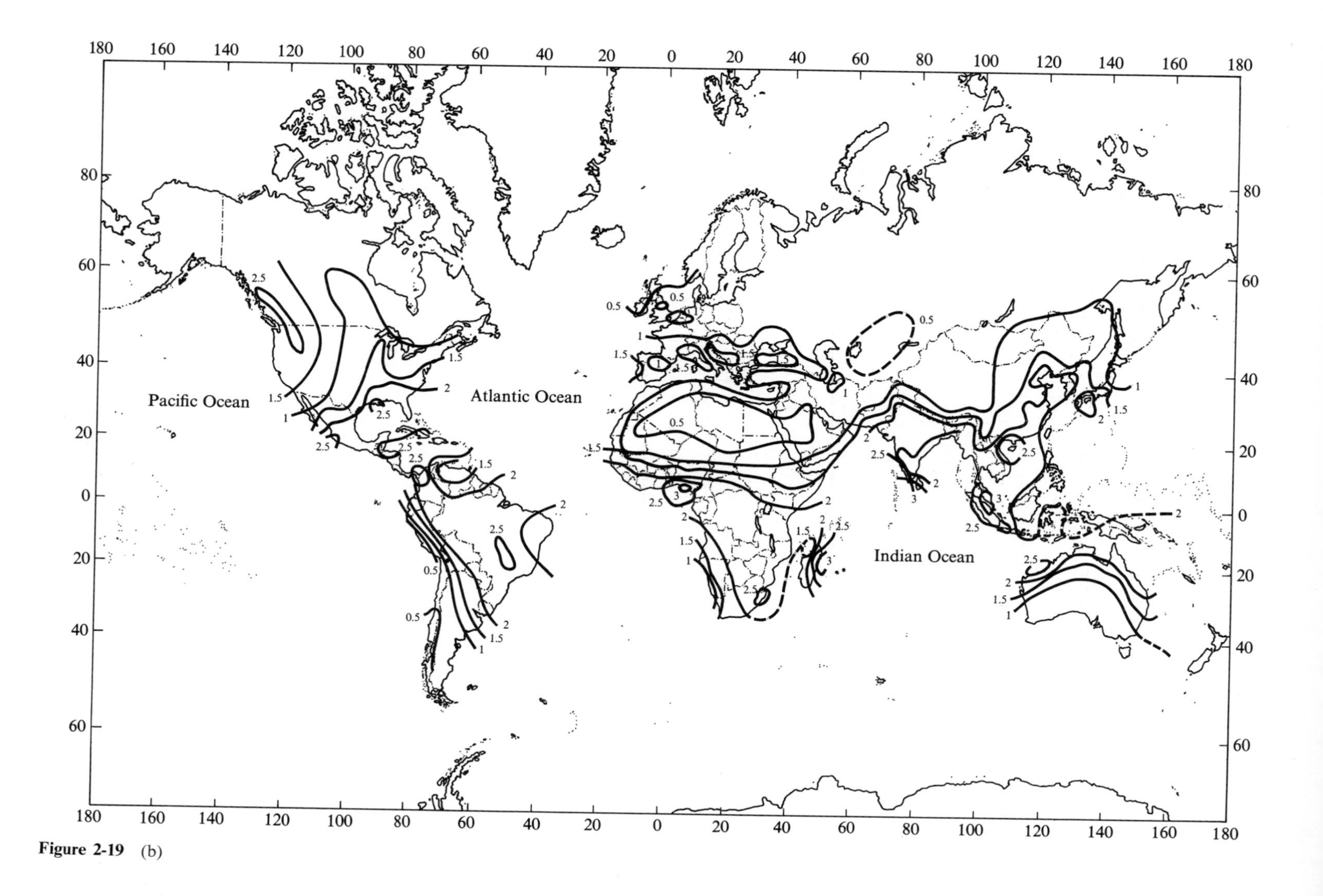

Figure 2-19 (b)

Seasonal Variation of Intensity-Duration-Frequency Regime

In many climatic regions there is a strong concentration of intense rainstorms in one season of the year. It is useful to know how frequently storms forming the annual-maximum or partial-duration series occur during each month. A heavy storm, for example, occurring on a dry soil with a dense vegetative cover during summer may cause less flooding and erosion than a smaller storm occurring on wet, bare, plowed soils during winter. Quantitative statements of seasonal regime, therefore, are useful in assessing the value of cropping sequences in soil conservation, in appreciating the probable seasonal variation in flood hazard and its relationship to land use, or in the planning and the management of small reservoirs for flood control and other purposes. The U.S. Weather Bureau (now called the National Weather Service) has published several regional summaries of the probability of occurrence of the 2-year, 5-year, 10-year, and other rainstorms in particular months. An example of the results is given in Figure 2-20, taken from a publication that describes the method for constructing such curves.

Figure 2-20 Seasonal variation of the probability of experiencing rainstorms of different recurrence intervals in a given month of the year. Isolines are labeled in percent probability. The graphs apply to the northeastern United States. (a) 1-hour duration. (b) 24-hour duration. (From U.S. Weather Bureau 1959.)

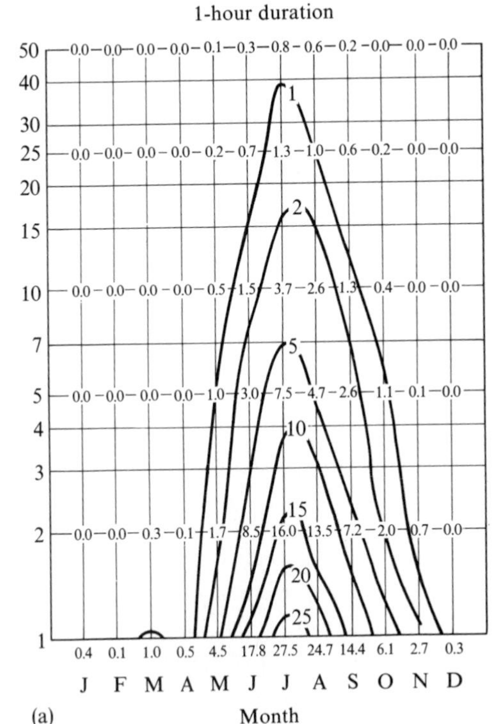

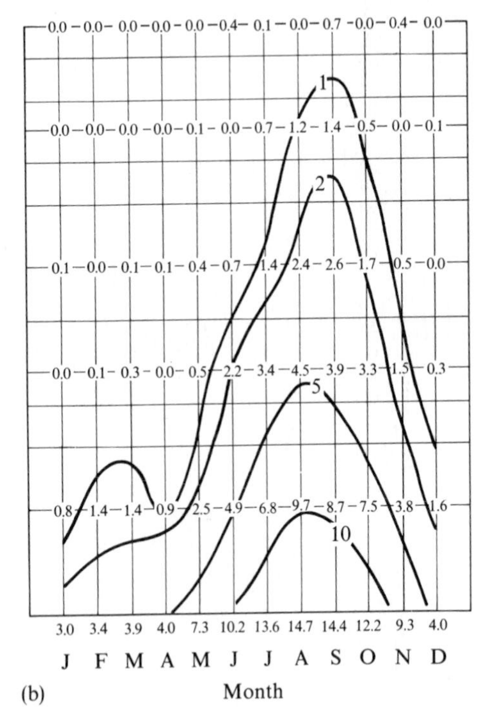

Temporal Distribution of Rainfall During a Storm

The time distribution of rainfall during a storm affects the time distribution of runoff from small watersheds (see Typical Problem 10-21). Very little information is available on the variation of intensity patterns in different areas of the world. The U.S. Soil Conservation Service publishes typical mass curves of rainfall for various storm durations in a few regions of the United States. An example is given in Table 2-3. If these data are not applicable to a particular problem of interest, the only alternative is to plot a sample of accumulation graphs for some representative storms from the region and average them.

Table 2-3 Cumulative time distributions of precipitation for various storm durations for the San Francisco Bay region. (From Rantz 1971.) See U.S. Soil Conservation Service (1972) for mass curves for other regions.

CUMULATIVE PERCENTAGE OF TIME	CUMULATIVE PERCENTAGE OF PRECIPITATION WITHIN STORMS OF INDICATED DURATION (D_r), IN HOURS					
	1	2	3	4	5	6
0	0	0	0	0	0	0
10	5	5	6	6	7	7
20	10	10	12	13	15	16
25	14	14	16	18	20	22
30	19	19	21	23	25	27
35	35	35	36	37	38	39
40	52	52	52	52	52	52
45	61	61	61	60	60	59
50	70	70	69	68	67	66
55	75	75	73	72	71	70
60	79	79	76	75	74	74
70	85	85	82	82	81	80
80	91	91	88	88	88	87
90	96	96	94	94	94	94
100	100	100	100	100	100	100

Spatial Characteristics of Storm Rainfall

A point measurement of rainfall is not representative of areas in excess of a few square miles. Average rainfall tends to vary inversely with the area covered, and it is necessary for many design purposes to quantify the depth-area relationship characteristic of a single storm or of all storms in a climatic region.

Depth-area curves are constructed by drawing isohyetal maps of total rainfall, measuring the area within each isohyet and computing the average precipitation within the enclosed area. Figure 2-21 shows some examples of such curves for convective storms in the southwestern United States. The lines representing individual storms are approximately parallel. As a result, rough estimates of total rainfall over an area can be obtained if rainfall at the center of the storm is known. Conversely, if the areal extent of the storm is known and data from a few gauges give a rough average rainfall, an estimate can be made of the precipitation at the storm center.

Dense gauge networks are required for definition of the area-depth relationships of intense storms. A review of results from dense gauge networks in Illinois showed that values of rainfall intensity derived from the U.S. Weather Bureau network of stations should be increased by ratios of 1.23, 1.19, and 1.10, respectively, for point estimates, 10-square-mile storms, and 500-square-mile storms.

If the time distribution of rainfall is of interest, a *depth-area-duration analysis* is performed. The technique for doing this is described by Gilman (1964), and an example of the results from one storm is presented in Figure 2-22.

The results presented so far refer to direct observations on individual storms. The hydrologist is also concerned with prediction in a statistical sense, as described in the section on intensity-duration-frequency analysis. Averaging point-rainfall values of a given frequency does not yield an areal average of the same frequency. It is not permissible to average the 10-year, 1-hour storm at five stations in a 100-square-mile basin to obtain the 10-year, 1-hour storm over the whole 100 square miles. Unfortunately, very little work has been done on statistical depth-area-duration relationships. The U.S.

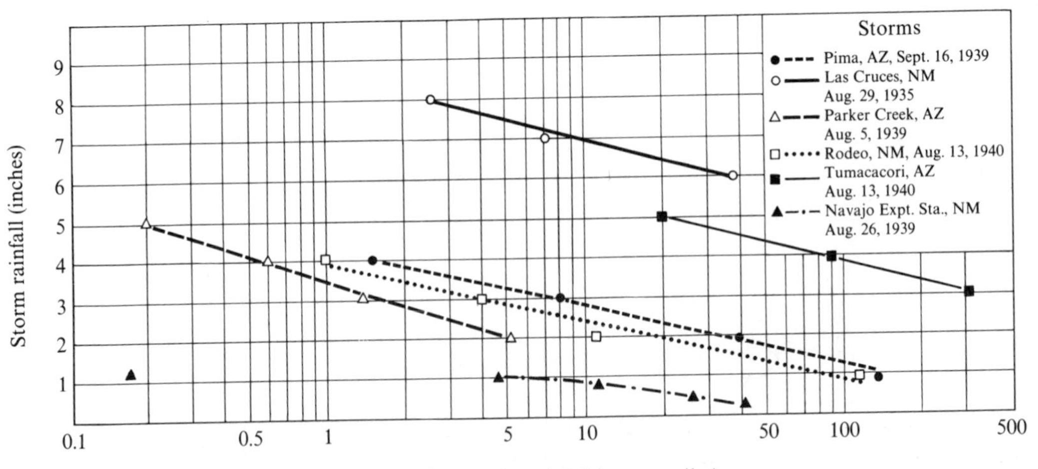

Figure 2-21 Depth-area curves for local summer storms, New Mexico and Arizona. (From Leopold 1942.)

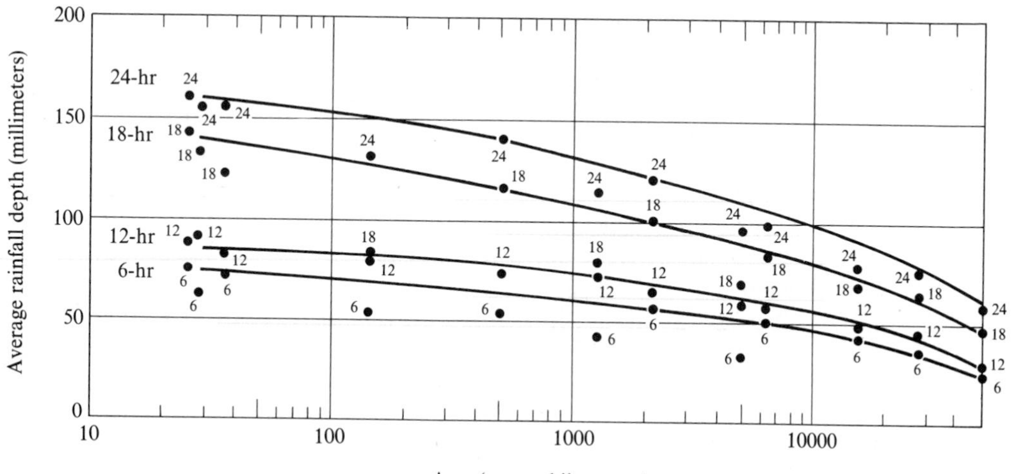

Figure 2-22 Maximum depth-area-duration curves for a storm. The points represent the maximum average depth over the designated area within the number of hours indicated at each symbol. (From Gilman, *Handbook of Applied Hydrology,* edited by Ven te Chow. Copyright © 1964 by McGraw-Hill, Inc. Used with permission of McGraw-Hill Book Company.)

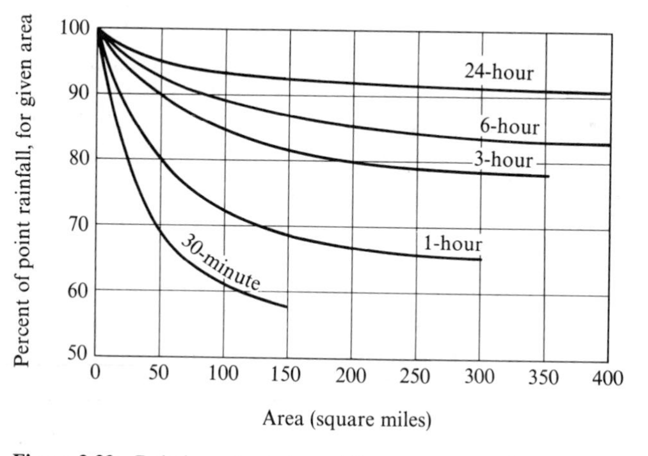

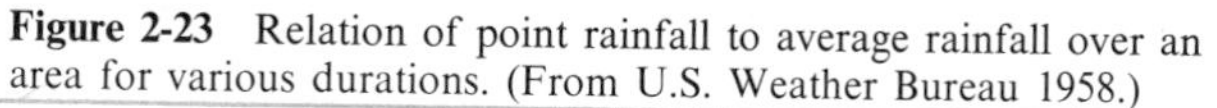
Figure 2-23 Relation of point rainfall to average rainfall over an area for various durations. (From U.S. Weather Bureau 1958.)

Weather Bureau (1957) conducted such a study for parts of the eastern United States, and found a relationship between point rainfall of a specified duration (regardless of its frequency) and the average rainfall over areas up to 400 square miles (see Figure 2-23). Once the magnitude of the rainfall of some frequency has been evaluated for one station, it can be multiplied by the appropriate value taken from the ordinate of this diagram to obtain the average rainfall for the area of interest.

Probable Maximum Precipitation

The failure of some engineering structures, such as dams above heavily populated areas, would cause great loss of life and much social and economic dislocation. In the design of such costly and important systems, therefore, it is usual to employ a much higher estimate of precipitation than that obtained from a frequency analysis of, say, the 100-year storm. The estimate is usually based upon meteorological judgment of the probable upper limit of precipitation that can occur given the atmospheric moisture content and the wind conditions in the region during major storms. The details of how such estimates are made are beyond the scope of this text but are illustrated by U.S. Weather Bureau (1956b, 1960) publications, which also present maps of the probable maximum precipitation for various durations and areas throughout the United States. The reports provide worked examples of the selection of probable maximum rainfalls for particular drainage basins. The technique and its application to the prediction of the maximum probable flood for the design of a spillway on a large dam in the Indus basin are described by Binnie and Mansell-Moulin (1966).

Long-Term Variations of Precipitation

In many hydrologic problems involving precipitation, there arises the question whether precipitation is changing. This question is difficult to answer in view of the great inter-annual variability, which tends to mask any trends in precipitation data. Also, there are few instrumental records of precipitation before 1900 and almost none before 1850 (Veryard 1963). The few long records, however, do show important trends over wide areas. Seoul, South Korea, for example, has a record of rainfall stretching back to the fifteenth century. Ten-year running means of annual totals shows variations from 750 mm to 1400 mm (Arakawa 1957). Much of the evidence for precipitation trends was reviewed in a UNESCO (1963) publication. Since that time Lamb (1966) and Winstanley (1973) have demonstrated major recent changes of precipitation that can be related to shifts of the global wind pattern. The significance of these changes is evident from the recent large-scale failure of crops, rangelands and water-supply systems, and the mass starvation in Sahelian Africa.

For an examination of changes extending back beyond the period of instrumental record, indirect evidence of climatic change must be used. Fritts (1969) has reviewed methods of analyzing tree rings for extending our knowledge of annual or seasonal rainfalls back several hundred years before the initiation of instrumental observations. Lamb (1972) has interpreted indirect evidence from many sources to construct climatic records for approximately 1000 years. Such long-term analyses of what can happen to climate are of great significance for the development of large water schemes, or for an understanding of environmental changes like those experienced in the Sahel region of Africa. By international agreement, the "average" climate of the world used to be defined as the average value of precipitation, temperature, and other variables for the period 1931–1960. As the period of observation was extended forward and backward in time, it became apparent that the period 1931–1960 had one of the most abnormal climatic regimes in the last several hundred years. The average climate of a place is now generally updated each decade to a new 30-year period, 1941–1970, 1951–1980, and so on. Many published records and atlas maps, however, are based upon data from 1931–1960, and they should be checked against older and more recent information wherever possible.

Climatic changes do not occur with sufficient regularity for prediction, and we do not yet understand their causes well enough to make deterministic predictions. In making decisions about the management of water and land, however, one must be aware both of the great inter-annual variability of rainfall and of the possibility of runs of wet or dry years that may cause important changes in runoff, sedimentation, or communities of vegetation or animals, or in the functioning of large water-resource developments. The possibility of such long-term variation should be considered in any plans involving systems that depend upon precipitation.

Other, more subtle changes of precipitation regime may also have great importance. During the growth of the city of Montreal, for example, there has been a decrease in the ratio of snow to rain associated with a slight increase of temperature in the downtown area. Leopold (1951) analyzed four long records of rainfall in New Mexico. There was no trend in annual totals, but the frequency of daily rainfall totals of 2.5–12.5 mm was low in the second half of the nineteenth century, compared with the period 1896–1939. Rains greater than 25 mm were relatively more frequent during 1849–1895 than in the first half of the twentieth century. Leopold reasoned that a reduction of vegetative cover (caused by low-moisture supply in gentle rains) and an increase of highly erosive intense rainfalls could have combined to trigger extensive erosion, which began around 1885 in the valleys of the Southwest. Howe et al. (1967) found that days with more than 63.5 mm of rain were more frequent in Mid-Wales in the period 1940–1964 than during 1911–1940, while annual totals showed almost no change. The change produced an increased frequency of flooding in the River Severn at Shrewsbury. There are many other examples of subtle changes of precipitation regime that have important consequences in hydrology, ecology, and land use. The subject of climatic fluctuation is discussed again with some examples of its effect on hydrology in Chapters 10 and 12.

Note on Sources of Precipitation Data

The major data-gathering agency for meteorology in the United States is the National Weather Service of the Department of Commerce. At each district office, tabulations of rainfall data are available from both recording and nonrecording stations. For selected stations these data are published regularly, and for a few stations, usually around large cities, summaries of the weather have been published. These summaries include intensity-duration-frequency curves and similar data that are useful in planning.

Anyone seeking data on precipitation should contact the local office of the National Weather Service. A useful list of available data on precipitation (and many other types of meteorological data) is available from the Superintendent of Documents, U.S. Government Printing Office. It is entitled "Selective Guide to Published Climatic Data Sources."

Other useful precipitation summaries have been compiled by various agencies, usually during some large-scale regional study of the climate or water resources of a region. For example, in 1968 the Pacific Northwest River Basins Commission compiled a three-volume study entitled "Climatological Handbook of the Columbia Basin States." Many similar compilations can be unearthed by searches in local libraries or by contacting state and federal agencies concerned with water resources or drainage design.

In other countries, meteorological departments or water resource agencies make similar compilations.

Bibliography

ARAKAWA, H. (1957) Climatic change as revealed by data from the Far East; *Weather,* vol. 12, pp. 46–51.

ATKINSON, B. W. (1970) The reality of the urban effect on precipitation: a case study; in *Urban climates,* World Meteorological Organization, Geneva.

BINNIE, G. M. AND MANSELL-MOULIN, M. (1966) The estimated probable maximum storm and flood on the Jhelum River: a tributary of the Indus; in *River flood hydrology,* The Institution of Civil Engineers, London, pp. 189–210.

CHAGNON, S. A. (1970) Recent studies of urban effects on precipitation in the United States; in *Urban climates,* World Meteorological Organization, Geneva.

DAWDY, D. R. AND LANGBEIN, W. B. (1960) Mapping mean annual precipitation; *International Association of Scientific Hydrology Bulletin,* vol. 19, no. 3, pp. 16–23.

FAIR, G. M., GEYER, J. C. AND OKUN, D. A. (1966) *Water and wastewater engineering;* vol. 1, *Water supply and wastewater removal,* John Wiley & Sons, New York.

FOSTER, E. E. (1948) *Rainfall and runoff;* Macmillan, New York.

FREUND, J. E. AND WILLIAMS, F. J. (1958) *Modern business statistics;* Prentice-Hall, Englewood Cliffs, NJ, 539 pp.

FRITTS, H. C. (1969) Tree-ring analysis: a tool for water resources research; *EOS, American Geophysical Union Transactions,* vol. 50, pp. 22–30.

GILMAN, C. S. (1964) Rainfall; Section 9 in *Handbook of applied hydrology* (ed. Ven te Chow), McGraw-Hill, New York.

GLOVER, J. (1957) The relationship between total seasonal rainfall and yield of maize in the Kenya Highlands; *Journal of Agricultural Science,* vol. 49, pp. 285–290.

GUMBEL, E. J. (1945) Floods estimated by the probability method; *Engineering News-Record,* vol. 134, pp. 833–837.

HERSHFIELD, D. M. (1961) Rainfall frequency atlas of the United States; *U.S. Weather Bureau Technical Paper 40.*

HERSHFIELD, D. M. (1962) A note on the variability of annual precipitation; *Journal of Applied Meteorology,* vol. 1, pp. 575–578.

HERSHFIELD, D. M. AND KOHLER, M. A. (1960) An empirical appraisal of the Gumbel extreme-value procedure; *Journal of Geophysical Research,* vol. 65, pp. 1737–1746.

HERSHFIELD, D. M. AND WILSON, W. T. (1958) Generalizing of rainfall intensity-frequency data; *Proceedings of the General Assembly of Toronto, International Association of Scientific Hydrology, Publication No. 43,* pp. 499–506.

HERSHFIELD, D. M. AND WILSON, W. T. (1960) A comparison of extreme rainfall depths from tropical and nontropical storms; *Journal of Geophysical Research,* vol. 65, pp. 959–982.

HOWE, G. M., SLAYMAKER, H. O. AND HARDING, D. M. (1967) Some aspects of the flood hydrology of the upper catchments of the Severn and the Wye; *Institute of British Geographers, Transactions,* vol. 41, pp. 35–58.

JENNINGS, A. H. (1950) World's greatest observed point rainfall; *Monthly Weather Review,* vol. 78, pp. 4–5.

KRAUS, E. B. (1954) Secular changes in the rainfall regime of southeastern Australia; *Quarterly Journal of the Royal Meteorological Society,* vol. 80, pp. 591–601.

KURTYKA, J. C. (1953) Precipitation measurement study; *Illinois State Water Survey: Report of Investigations,* no. 20.

LAMB, H. H. (1966) *The changing climate;* Methuen, London, 236 pp.

LAMB, H. H. (1972) *Climate: present, past and future;* Methuen, London, 596 pp.

LAMBOR, J. (1967) Frequency of intense rainfall for the territory of Poland; *Journal of Hydrology,* vol. 5, pp. 158–162.

LANGBEIN, W. B. (1960) Plotting positions in frequency analysis; *U.S. Geological Survey Water Supply Paper 1543-A,* pp. 48–50.

LEOPOLD, L. B. (1942) Areal extent of intense rainfalls, New Mexico and Arizona; *EOS, American Geophysical Union Transactions,* vol. 23, pp. 558–562.

LEOPOLD, L. B. (1951) Rainfall intensity: an aspect of climatic variation; *EOS, American Geophysical Union Transactions,* vol. 32, pp. 347–357.

MANNING, H. L. (1956) The statistical assessment of rainfall probability and its application in Uganda agriculture; *Proceedings of the Royal Society of London,* Series B, vol. 144, pp. 460–480.

MCGUINNESS, J. L. AND HARROLD, L. L. (1965) Role of storm surveys in small watershed research; *Water Resources Research,* vol. 1, pp. 219–222.

RAINBIRD, A. F. (1967) Methods of estimating areal average precipitation; *WMO/IHD Report No. 3,* Geneva.

REICH, B. M. (1963) Short-duration rainfall-intensity estimates and other design aids for regions of sparse data; *Journal of Hydrology,* vol. 1, pp. 3–28.

RODDA, J. C. (1967) A country-wide study of intense rainfall in the United Kingdom; *Journal of Hydrology,* vol. 5, pp. 58–69.

RODIER, J. A. AND AUVRAY, C. (1965) Flood computations in West Africa; *International Association of Scientific Hydrology, Proceedings of the Symposium of Budapest, Publication No. 66,* no. 1, pp. 12–38.

SCHERMERHORN, V. P. (1967) Relations between topography and annual precipitation in Western Oregon and Washington; *Water Resources Research,* vol. 3, pp. 707–712.

SPREEN, W. C. (1947) A determination of the effect of topography on precipitation; *EOS, American Geophysical Union Transactions,* vol. 28, pp. 285–290.

STIDD, C. K. (1943) Cube root normal precipitation distribution; *EOS, American Geophysical Union Transactions,* vol. 34, pp. 31–35.

TODD, D. K. (1970) *The water encyclopaedia;* Water Information Center, Port Washington, New York, 559 pp.

UNESCO (1963) Changes of climate; *Arid Zone Research,* no. 20, Paris.

U.S. ARMY CORPS OF ENGINEERS (1968) Planning and design of roads, air bases, and heliports in the theater of operations; *Technical Manual TM5-330.*

U.S. GEOLOGICAL SURVEY (1954) Water-loss investigations: Lake Hefner studies; *U.S. Geological Survey Professional Paper 270.*

U.S. WEATHER BUREAU (1945) Thunderstorm rainfall: Parts 1 and 2; *Hydrometeorological Report No. 5.*

U.S. WEATHER BUREAU (1946) Manual for depth-area-duration analyses of storm precipitation; *Technical Paper 1.*

U.S. WEATHER BUREAU (1955) Rainfall intensity-duration-frequency curves for selected stations in the United States, Alaska, Hawaiian Islands and Puerto Rico; *Technical Paper 25.*

U.S. WEATHER BUREAU (1956a) Rainfall intensities for local drainage design in the western United States; *Technical Paper 28.*

U.S. WEATHER BUREAU (1956b) Seasonal variation of the probable maximum precipitation east of the 105th meridian for areas from 10 to 1000 square miles and duration of 6, 12, 24 and 48 hours; *Hydrometeorological Report No. 33.*

U.S. WEATHER BUREAU (1957) Rainfall intensity-frequency regime: Part I, The Ohio Valley; *Technical Paper 29.*

U.S. WEATHER BUREAU (1959) Rainfall intensity-frequency regime: Part IV, Northeastern United States; *Technical Paper 29.*

U.S. WEATHER BUREAU (1960) Generalized estimates of probable maximum precipitation for the United States west of the 105th meridian for areas to 400 square miles and durations to 24 hours; *Technical Paper 38.*

U.S. WEATHER BUREAU (1961) Rainfall frequency atlas of the United States; *Technical Paper 40.*

VERYARD, R. G. (1963) A review of studies on climatic fluctuations during the period of the meteorological record; *Changes of climate,* UNESCO, Paris, pp. 3–16.

WINSTANLEY, D. (1973) Rainfall patterns and general circulation; *Science,* vol. 245, pp. 190–194.

WORLD METEOROLOGICAL ORGANIZATION (1971) Guide to hydrometeorological practices; *WMO Technical Publication No. 3,* Geneva.

2-1 Average Precipitation for an Area

Many of the hydrologic analyses concerning water supply and floods to be described in later chapters require an estimate of the average depth of precipitation over a drainage basin during a storm, a year, or some other period. Spatial averages of other hydrologic quantities measured at a point also need to be treated in this way. This typical problem, therefore, is concerned with calculation of the average precipitation over the drainage basin outlined in Figure 2-4 from the plotted measurements at nine rain gauges.

Solution

1. *Arithmetic average,* using only the gauges within the basin is 74.7 mm.

2. *Thiessen-weighted average* involves a weighting factor that is proportional to the fraction of the drainage basin area represented by each gauge. Neighboring gauges on the map are joined by dashed lines, and perpendicular bisectors (solid) are drawn with a set of drawing compasses for each line. The bisectors meet to form a polygon around each gauge. The area of each polygon within the drainage basin is measured with a planimeter and expressed as a decimal fraction of the total drainage basin area. The shaded area in Figure 2-4(b) is 0.08 of the total drainage area of the basin. This is used as the weighting factor for the measured rainfall of 114 mm in that polygon. The rainfall at each gauge is then multiplied by its appropriate fraction of the basin area, and the products for each gauge are added to form the Thiessen-weighted average for the whole basin, as in Table 2-4. The Thiessen-weighted average rainfall for the basin is 61.3 mm.

3. *Isohyetal average* is obtained by contouring precipitation values as shown in Figure 2-4(c). The area between two adjacent contours (isohyets) is then measured with a planimeter and expressed as a decimal fraction of the drainage basin area. The shaded area in Figure 2-4(c) is 0.28 of the total drainage basin area. The average precipitation for the area between two isohyets is the mean of the isohyetal values, and this mean is weighted by the fractional area between the contours. The area-weighted precipitation values are then

Table 2-4 Calculating the Thiessen-weighted average rainfall.

RAINFALL AT GAUGE (MM)	AREA OF POLYGON WITHIN DRAINAGE BASIN AS A FRACTION OF TOTAL BASIN AREA	RAINFALL WEIGHTED BY FRACTIONAL AREA (COL. 1 × COL. 2)
16	0.03	0.48
28	0.18	5.04
37	0.21	7.77
38	0.03	1.14
68	0.22	14.96
75	0.17	12.75
114	0.08	9.12
126	0.08	10.08
44	0	0
Total	1.00	61.34 mm

Table 2-5 Calculating the isohyetal average precipitation.

ISOHYETAL RANGE (MM)	AVERAGE RAINFALL BETWEEN ISOHYETS (MM)	PROPORTION OF DRAINAGE BASIN AREA BETWEEN THE ISOHYETS	AVERAGE RAINFALL WEIGHTED BY FRACTIONAL AREA (COL. 2 × COL. 3)
20–40	30	0.31	9.30
40–60	50	0.28	14.00
60–80	70	0.21	14.70
80–100	90	0.10	9.00
100–120	110	0.08	8.80
120–126	123	0.02	2.46
Total	—	1.00	58.26

summed to obtain the isohyetal average precipitation for the basin as in Table 2-5.

The isohyetal average for the basin is 58.3 mm. It is clear from these results that large differences can result in calculating the average rainfall of an area by different methods. The last two methods give a more accurate result if there are strong precipitation gradients and a non-uniform pattern of gauges. Otherwise, the arithmetic average will suffice and its use will save time.

In the frontispiece of the U.S. Geological Survey (1954) report on Lake Hefner, the reader will find another map and example on which to practice these techniques.

2-2 Statistical Analysis of Annual Precipitation

For many planning purposes concerned with water supply or agricultural development one must consider the chances of the precipitation in any year being less than certain critical values. Table 2-6 presents a record of annual rainfall at Mombasa Observatory for the period 1931–1960.

1. Calculate and plot on a sheet of arithmetic probability paper the cumulative frequency distribution of these values. Fit a straight line through them by eye. The line is the cumulative frequency curve of a normal distribution that fits the data.

2. From the line, estimate the mean annual rainfall.

3. From the line, estimate the standard deviation of annual rainfall.

4. Estimate the probability of obtaining less than 700 mm in any one year.

5. If a certain crop has been found by experiment to require between 1000 mm and 1500 mm to grow successfully, how many failures would you expect in 20 years at Mombasa?

Solution

1. The annual values are first of all ranked from the smallest (710 mm has a rank of $m = 1$) to the largest (1881 mm; $m = 30$). Equation 2-3 is then used to calculate the cumulative frequency of the years with rainfall less than each of the ranked values. Thus for the value 710 mm:

$$F_i = \frac{1}{30 + 1} \times 100 = 3.23\%$$

This probability value is plotted against 710 mm, and for the value with a rank of 2 (724 mm):

$$F_i = \frac{2}{31} \times 100 = 6.45\%$$

The process is repeated to produce Figure 2-24.

2. Reading across to the best-fit line from the 50-percent value on the ordinate gives a mean annual rainfall of 1220 mm, as shown.

3. Reading across to the line from the 16- and 84-percent values on the ordinate yields 920 mm and 1520 mm on the abscissa. Each of these values lies one standard deviation away from the

Table 2-6 Annual rainfall at Mombasa Observatory, 1931–1960.

YEAR	RAINFALL (MM)	YEAR	RAINFALL (MM)	YEAR	RAINFALL (MM)
1931	1306	1941	1285	1951	1529
1932	1345	1942	1087	1952	710
1933	1032	1943	939	1953	1552
1934	1580	1944	1296	1954	841
1935	1293	1945	1233	1955	846
1936	1497	1946	1263	1956	1082
1937	1469	1947	1881	1957	1215
1938	1392	1948	1221	1958	1017
1939	1037	1949	724	1959	1319
1940	1633	1950	915	1960	1077

mean. The standard deviation, therefore, is approximately 300 mm.

4. Reading up to the best-fit line from an abscissa value of 700 mm indicates a 4.5-percent chance of the rainfall in any one year falling below 700 mm.

5. Reading up to the curve from 1000 mm indicates a 24-percent chance that the annual rainfall will fall below 1000 mm. By the same method, there is an 83-percent chance of rainfall being below 1500 mm, and therefore there is a 17-percent chance of 1500 mm being exceeded. The chance, therefore, of the occurrence of either less than 1000 mm or more than 1500 mm is 24 + 17 = 41 percent. In 20 years, therefore, the crop should fail in 41 percent of the years, or eight times (approximately).

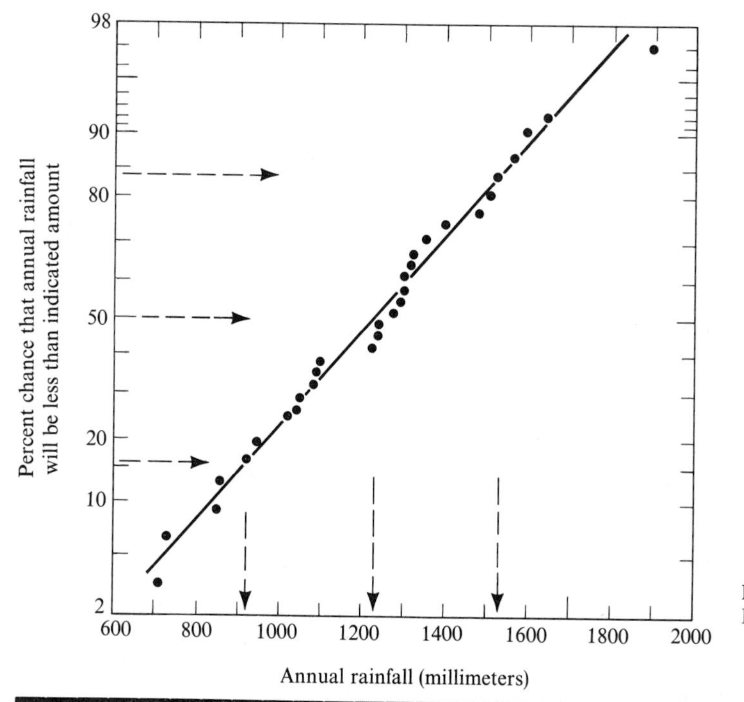

Figure 2-24 Rainfall at Mombasa Observatory.

2-3 Rapid Graphical Estimation of Rainfall Probability

The town of Moonshot, Oregon, is experiencing a rapid growth of population because of the recent relocation of an assembly plant for hand calculators near the town. To meet the increased demand for water, the town planners are considering construction of a dam in a small watershed in nearby mountains. A quick estimate of the variation of annual precipitation in the watershed will be needed for a forthcoming water-supply committee meeting with a local engineering consulting firm. Analysis of records from a U.S. Forest Service weather station in the watershed shows an average precipitation of 1270 mm and a standard deviation of 254 mm. Use normal probability paper and construct a graph to determine the drought precipitation that has a 10-percent chance of occurring in any year (i.e., which is exceeded 90 percent of the time).

Solution

The average annual precipitation (1270 mm) plots on the 50-percent probability line and the standard deviations (1016 and 1524 mm) plot on the 16- and 84-percent probability lines. There is a 10-percent chance that in any year a rainfall of 940 mm or less will occur. See Figure 2-25.

For another exercise, find the mean annual precipitation in an area of interest in the United States from an atlas of precipitation, determine the standard deviation using the map presented in Hershfield (1962), and construct a diagram to determine precipitation probabilities using normal probability paper. The ambitious student can then compare the results with values of probability obtained from the compilation of annual precipitation data for the same area by the technique shown in Typical Problem 2-2.

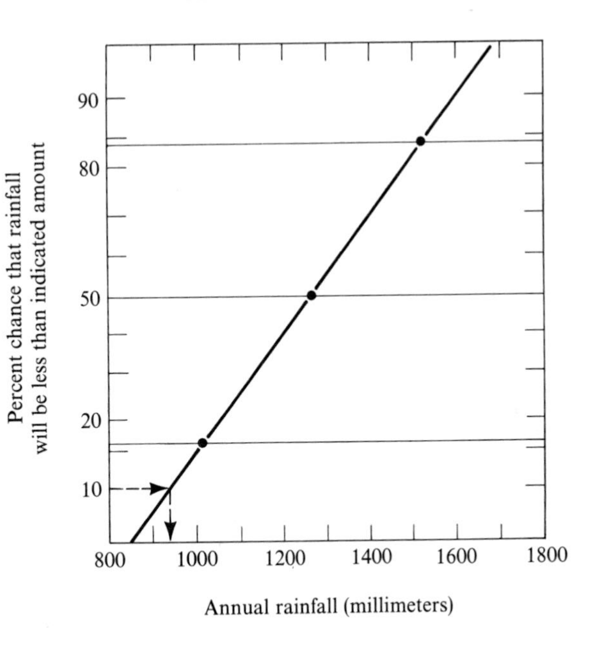

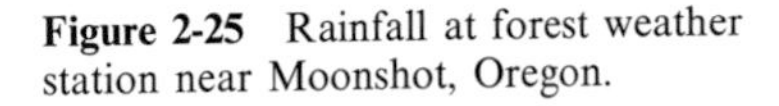
Figure 2-25 Rainfall at forest weather station near Moonshot, Oregon.

2-4 Rainfall Probabilities and Crop Yields

A planner interested in nutrition and rural development needs an estimate of the annual variability of maize yields in a district of a developing country. Crops are planted in the area in late March during the rainy season and are harvested in the October–November period.

For an 11-year period, field measurements have demonstrated a correlation between maize yields and April–August rainfall in two neighboring regions (Table 2-7). The areas have uniform soils and management but differing rainfalls. The data overlap in such a way that one is led to believe that the two sets of data can be combined. Other reasons confirm this (the data and this assertion are from a paper by Glover 1957).

Combining these two sets of data, obtain by graphical means a relationship between the seasonal rainfall and average maize yield. The graph will show that the yields are low during both dry and wet years.

Table 2-8 contains a set of April–August rainfalls for the period 1931–1960 for a station in the district of concern to the nutritionist. Plot the cumulative frequency curve of these data on arithmetic probability paper using Equation 2-3, and sketch a curve through the points. The curve will not be straight, but it can still be used to interpolate probabilities, though extrapolation would be risky. Recall, however, what we said at the end of this chapter about the use of the period 1931–1960 for characterizing the average climate of a place.

Table 2-7 Average maize yields and growing-season rainfall for two areas with uniform soil and management.

AREA A		AREA B	
MAIZE YIELD (KG/HA)	RAINFALL FOR THE PERIOD APRIL–AUGUST (MM)	MAIZE YIELD (KG/HA)	RAINFALL FOR THE PERIOD APRIL–AUGUST (MM)
560	48.5	405	111.5
920	49.5	900	100.0
1100	50.5	1160	91.5
1360	53.5	1610	89.0
1520	58.0	1520	91.0
1790	57.0	1610	86.0
1610	58.5	1920	86.5
1460	59.0	1790	81.0
1750	59.0	1720	74.5
2160	79.0	1920	74.0
2010	83.0	2080	72.5

Table 2-8 Growing season rainfalls for a station in the center of one district.

YEAR	APRIL–AUGUST RAINFALL (MM)	YEAR	APRIL–AUGUST RAINFALL (MM)	YEAR	APRIL–AUGUST RAINFALL (MM)
1931	72	1941	68	1951	107
1932	46	1942	68	1952	66
1933	54	1943	48	1953	58
1934	60	1944	53	1954	50
1935	94	1945	51	1955	82
1936	75	1946	54	1956	84
1937	73	1947	64	1957	63
1938	56	1948	99	1958	53
1939	52	1949	88	1959	67
1940	45	1950	64	1960	79

Using the two graphs, estimate the chance that in any year the maize yield will fall below 1000 kg/ha. This should be done by reading from the first graph the high and low rainfalls that correspond to a yield of 1000 kg/ha. Then from the second graph, read the probability of the seasonal rainfall being less than the low value and the probability of it exceeding the high value. Add these probabilities to obtain the answer.

Depending upon how you sketch the curves, the answer should be approximately 20 percent (one year in five).

2-5 Intensity-Duration-Frequency Analysis of Precipitation

For many design purposes to be introduced in Chapter 10 it is necessary to calculate the recurrence interval of different precipitation intensities for a range of durations. Using the annual maximum precipitation series for the period 1935–1961 at Buffalo, New York (Table 2-9), construct the

Table 2-9 Maximum annual precipitation at Buffalo, New York (mm).

YEAR	5 MIN	10 MIN	15 MIN	30 MIN	1 HR	2 HR	24 HR
1935	5	7	8	10	11	12	24
1936	6	9	12	15	16	16	67
1937	7	9	9	13	19	33	52
1938	7	10	14	22	24	26	52
1939	5	9	11	18	29	31	64
1940	6	8	11	15	20	26	56
1941	8	11	13	17	18	18	40
1942	7	12	16	21	23	30	56
1943	6	9	12	17	19	20	34
1944	10	17	21	24	44	58	93
1945	6	10	13	17	18	19	89
1946	7	13	18	26	41	42	69
1947	8	13	14	18	15	18	53
1948	10	17	20	29	37	47	54
1949	12	19	23	33	50	59	67
1950	12	17	21	21	22	22	52
1951	9	13	14	20	21	23	42
1952	8	9	10	12	15	15	36
1953	14	25	39	42	45	58	58
1954	8	13	16	17	18	18	59
1955	7	13	15	19	34	38	79
1956	9	14	15	22	27	36	52
1957	5	8	10	14	18	20	61
1958	6	11	14	19	20	25	37
1959	9	13	18	24	29	37	68
1960	10	10	11	14	16	17	36
1961	7	13	16	26	39	45	54

intensity-duration-frequency curves for durations from 5 minutes to 24 hours and for recurrence intervals 1.1 years to 50 years.

Solution

1. Rank the precipitation values from the largest to the smallest. For a storm of 5-minute duration, the largest value is 14 mm ($m = 1$) and the smallest is 4 mm ($m = 27$).

2. Use Equation 2-4 to compute the recurrence interval for each value (Table 2-10). The smallest value for the 5-minute duration precipitation has a recurrence interval of

$$T = \frac{n+1}{m} = \frac{28}{27} = 1.04 \text{ years}$$

and the largest value has a recurrence interval of

$$T = \frac{28}{1} = 28 \text{ years}$$

3. Plot results on Gumbel extreme-value paper, (the graph paper can be constructed by tracing Figure 2-14). The curve for the 5-minute duration is shown in Figure 2-26; the reader should add curves for other durations.

Table 2-10 Calculating the recurrence intervals for the 5-minute-duration precipitation.

YEAR	AMOUNT (MM)	RANK (m)	RECURRENCE INTERVAL $T = \frac{n+1}{m}$	YEAR	AMOUNT (MM)	RANK (m)	RECURRENCE INTERVAL $T = \frac{n+1}{m}$
1935	5	25	1.12	1949	12	2	14.0
1936	6	20	1.4	1950	12	3	9.33
1937	7	14	2.0	1951	9	7	4.0
1938	7	15	1.87	1952	8	12	2.33
1939	5	26	1.08	1953	14	1	28.0
1940	6	21	1.33	1954	8	13	2.15
1941	8	10	2.80	1955	7	18	1.56
1942	7	16	1.75	1956	9	8	3.50
1943	6	72	1.27	1957	5	27	1.04
1944	10	4	7.0	1958	6	24	1.17
1945	6	23	1.22	1959	9	9	3.11
1946	7	17	1.65	1960	10	6	4.67
1947	8	11	2.55	1961	7	19	1.47
1948	10	5	5.6				

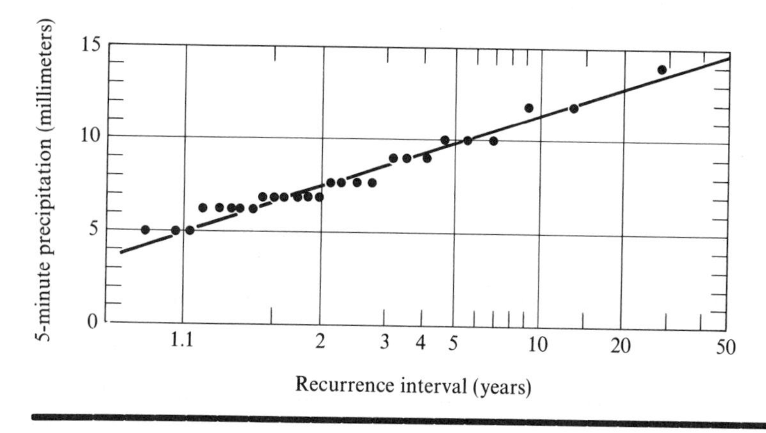

Figure 2-26 Frequency of 5-minute precipitation at Buffalo, New York.

2-6 Interpolation from Rainfall-Intensity-Duration-Frequency Graphs

For many of the runoff calculations described in Chapter 10 it is necessary to know the rainfall intensity for some duration or recurrence interval other than those shown in Figure 2-17. Other values must be interpolated from the maps.

Using Figure 2-17 find the intensity of the 25-year rainfall with a duration of 3.5 hours for Chicago.

Solution

From Figure 2-17 plot on a sheet of double-logarithmic paper the rainfall value for each duration and frequency for the location of Chicago as shown in Figure 2-27. Curves are sketched between the three points representing each storm duration, as shown in the diagram. From these curves a rainstorm with any recurrence interval can be interpolated for the four durations shown. The 25-year, 6-hour storm, for example, is 3.3 inches (intensity = 3.3/6 = 0.55 in/hr).

If the planner requires the rainfall of some other duration, values can be read from the four curves for any recurrence interval. For example, the four values for a 25-year recurrence interval are 1.93, 2.24, 3.38, and 4.50. These can be plotted in Figure 2-28. The 25-year storm with a duration of 3.5 hours can then be read from this graph in Figure 2-28. The 25-year storm with a duration of 3.5 hours can then be read from this graph as 3 inches (rainfall intensity = 3/3.5 = 0.86 in/hr).

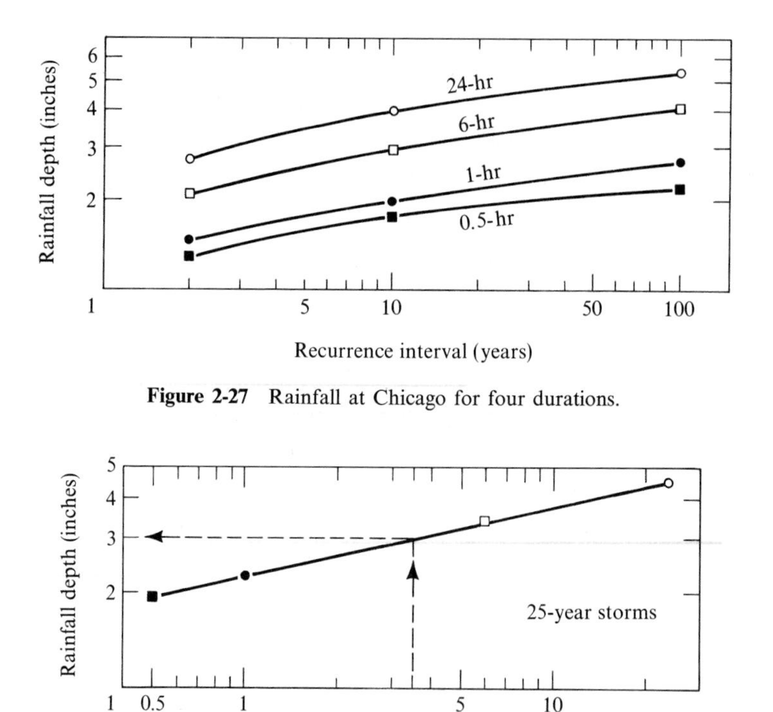

Figure 2-27 Rainfall at Chicago for four durations.

Figure 2-28 Rainfall at Chicago for 25-year storms.

3

Interception

Definitions and Process

Not all the precipitation discussed in Chapter 2 reaches the soil to supply plants or generate flood runoff. A part is caught by vegetation or other surface cover and is evaporated back to the atmosphere during and after the storm. The amount subtracted from the rainfall in this way is known as interception. Although it is not a major factor in most hydrologic calculations, the planner should be aware of interception because it is one of the ways in which altering the vegetation cover affects the water balance of an area. Replacement of one tree species with another, for example, might alter the amount of runoff available for water supply. The real importance of interception in such matters, however, is still debated, and we will return to this question in the last section of the chapter.

Little rain reaches the soil surface directly in densely vegetated areas. The falling rainwater is stored on the leaves, branches, and stems of vegetation. After these surfaces have been wetted, further rain will displace water on the lower edges of leaves, twigs, and other components of the cover, and this displaced water will fall to the next lower layer of vegetation or the forest floor litter and eventually to the mineral soil. The amount of water stored on the wetted surface of the cover is called interception storage, and its magnitude depends upon the form, density, and surface texture of the leaves, twigs, or other surfaces.

Rain moves through the vegetation canopy by two mechanisms: throughfall and stemflow. Throughfall penetrates the canopy directly through spaces between leaves, or by dripping from leaves, twigs, and branches. Stemflow reaches the ground by running down stems and trunks. These two processes deliver water to the forest floor where it must penetrate the litter layer before entering the mineral soil. A portion of the water reaching the litter is stored in that layer and never enters the soil.

During and after rainfall the water that has been stored in the canopy, on stems, and in the litter is returned to the atmosphere by evaporation. Even during rainstorms, when the atmosphere is very humid, the evaporative losses are appreciable because they occur from a large area of leaf surface.

The relations described above are shown in Figure 3-1 and can be summarized in the following equations:

$$\begin{aligned} R &= P - I \\ R &= P - (C + L) \\ R &= (T + S - L) \end{aligned} \tag{3-1}$$

where P = gross rainfall
R = net rainfall entering the soil
I = total interception from canopy and litter
C = canopy interception
L = litter interception
T = throughfall
S = stemflow.

The amount of interception depends on the characteristics of the precipitation and the nature of the cover. Although there is much discussion in the hydrologic literature about the effects of rainfall intensity, storm duration, storm frequency, evaporation conditions during the storm, and other variables, a survey of the literature indicates that the total volume of rainfall is used most commonly and most successfully to predict interception losses.

The nature of the cover is the second important control of interception losses. There is some difference in interception between urban and non-urban regions. In built-up areas the rough surfaces of roofs, walls, roads, and so on provide about 2 to 5 mm of storage opportunity. Higher temperatures in urban regions, exceeding those of rural regions, almost certainly increase the in-storm evaporation rate over that in vegetated areas. However, we know of no observations of interception in urbanized areas.

In vegetated areas the major controls of interception losses are vegetation type, density, form, and age. There are important differences between deciduous trees, which intercept different amounts in their growing and dormant seasons, and conifers, which have a constant foliage throughout the year. There are also differences between the broad types of cover: trees, shrubs, grasses, and crops. There is no interception, of course, over bare ground. Generally speaking, the denser the foliage the greater its intercep-

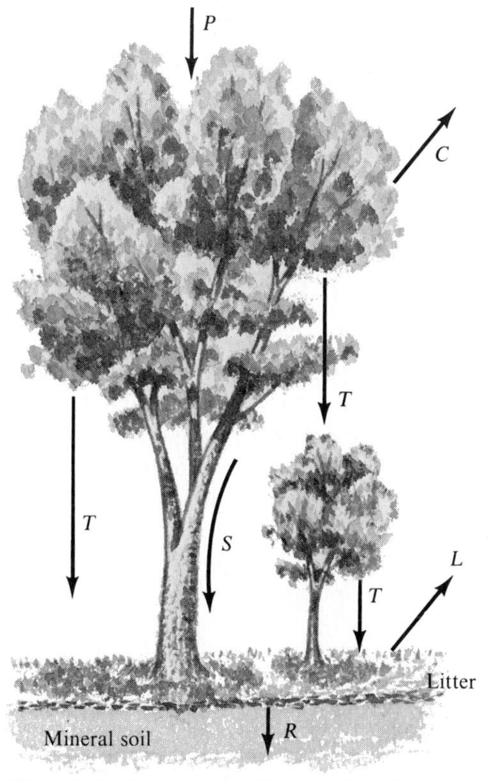

Figure 3-1 Components of interception. P is precipitation, C is canopy interception, L is litter interception, T is throughfall, S is stemflow, R is net rainfall entering soil.

tion storage. Density here refers not only to the topmost canopy but also to lower layers such as shrubs, forbs, and litter. Density changes with the age and management of the cover. The form of the leaves is important, particularly the difference between the broad leaves of deciduous trees and the needles of conifers. On broad-leaved trees the drops of rain run together on the leaves, forming larger drops that overcome surface tension, and drop off. On a coniferous tree, drops are held apart on or between individual needles and do not fall off so easily. One might expect large differences in interception between forest trees and grasses. This is the case on an annual basis, because grasses and crops provide a cover that changes seasonally. When in full leaf, however, some grasses have as much leaf area per unit area of ground as do most trees. Interception from agricultural crops varies with the type of crop, the density of planting, and the duration of the growing season relative to the occurrence of rainfall. A heavy crop such as alfalfa can intercept as much as a forest during the growing season.

Stemflow is affected by the character of tree bark. Species with rough bark yield much less stemflow than those with smooth bark. The character of tree branches is another variable. Trees whose branches slope upward and outward away from the main trunk are likely to concentrate more stemflow than trees with horizontal or pendant branches. The number of stems or trunks per unit area also controls stemflow, though for plants with the greatest stem densities, such as grasses and crops, stemflow is not measured separately.

Interception by the forest floor is largely an unknown quantity, though recent work by Helvey (1964, 1967) indicates that an understanding of litter interception is important in assessing the water budget of a forest. The ability of the litter to intercept and store rainfall depends largely upon its thickness and its water-holding characteristics (Helvey and Patric 1965). Frequency of rainfall in relation to the wetting and drying rates of litter also have an effect upon the amount of water intercepted by the litter. Helvey and Patric estimate that under mature hardwoods of the eastern United States, approximately 2.5 percent of gross rainfall is intercepted by the forest floor during the growing season, and 3.5 percent during the dormant season. Helvey (1967) computed litter interception under a pine forest as 2 to 4 percent of gross precipitation.

Measurement of Interception

The measurement of interception involves direct gauging of gross precipitation, throughfall, stemflow, and observations of changes in the moisture content of litter samples. Gross precipitation is measured in standard rain gauges above the canopy, or in a large clearing or adjacent field. Throughfall is measured by many standard gauges placed in a random fashion under several trees in a plot. These gauges are moved periodically. Stemflow is caught by collars which are fixed around a number of tree trunks to convey rainwater to a collecting can. Litter interception is more difficult to study. By repeatedly weighing samples of forest litter in the field, Helvey (1964) was able to define minimum and maximum water content of the litter and its rate of drying. By weighing during storms, the rate of wetting was determined. An accounting procedure could then be used to calculate the changing opportunity for storage within the litter between storms, and the rate at which rain was stored in the litter during rain.

Early studies of interception by grasses and crops were carried out by artificial sprinkling of herbaceous plants cut from plots of known area. More recently, combined throughfall and stemflow have been measured under artificial and natural rainfall by placing a collar around the plants, sealing the surface of the soil, and collecting all water reaching the ground surface in a pipe leading from the collar to a recording rain gauge (Crouse et al. 1966).

Interception by Forests

Probably because of the ease with which canopy interception can be measured, a large number of field studies have been made, particularly in forests. The usual result of such an investigation is a linear regression equation relating throughfall or stemflow to gross storm precipitation. In Figure 3-2 we have summarized a number of these results for deciduous and coniferous forests. The data were abstracted from the papers listed in the bibliography, and the types of trees sampled are apparent from the titles of the listed papers. More complete discussions of field studies are given by Helvey and Patric (1965) and by Zinke (1967).

Figure 3-2 shows that stemflow is a minor component of net precipitation, never rising to more than a few percent of gross rainfall. Throughfall is the dominant component. There is substantial agreement between throughfall measurements in deciduous forests throughout the world, indicating that there is little to be gained from further routine measurements. In the dormant season several percent more rainfall penetrates the canopy than during the growing season. Coniferous forests seem to intercept slightly more rainfall than deciduous trees, though the data from the conifers are more variable. Interception is particularly great under dense covers of fir and spruce in Alaska (Patric 1966) and northern Europe (Law 1958, Leyton et al. 1967). Conifers generally have greater masses of foliage and branches throughout the year, and their needles can hold more interception storage than broad leaves.

Examining interception on an annual basis, instead of during individual

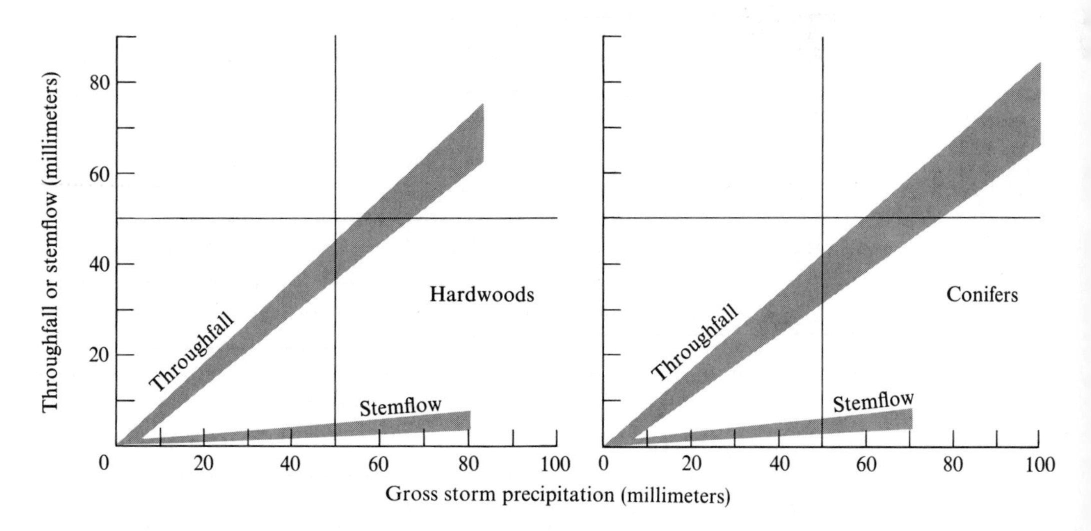

Figure 3-2 Relation of throughfall and stemflow to gross storm precipitation. Shaded zones include all the data provided by various papers listed in the bibliography.

Table 3-1 Median values of canopy interception as a percentage of annual or seasonal gross precipitation. (From various reports, the more accessible of which are listed in the bibliography.)

	NUMBER OF OBSERVATIONS	MEDIAN CANOPY INTERCEPTION (% OF GROSS PRECIPITATION)
Deciduous forest		
All data	10	13
Coniferous forest		
Rainfall only	11	22
Observations that include rain and snow	26	28
European data only	9	35
North American data only	27	27
Taiwan	1	8

storms, reveals the role of interception in the water balance of a region. Table 3-1 lists canopy interception $[P - (T + S)]$ as a percentage of gross rainfall. The summary is not exact because the results of interception studies are often not strictly comparable as a result of different measuring practices, different proportions of rain and snow, and variation in the definition of forest types. However, the frequency distribution of the values showed some consistency and a central tendency that is well described by the median. The data suggest that long-term canopy interception is greater under conifers than under broad-leaf hardwoods. This is presumably due to the greater density of the former, and possibly because of the more frequent occurrence of light rains and snow, which are more fully intercepted, in coniferous regions. The data suggest that under conifers more snow is intercepted than rain, although this result may be partly due to the difficulty of measuring throughfall from intercepted snow. There is also some suggestion that canopy interception is greater in European coniferous forests than in North American forests. The largest values given in the literature (more than 50 percent of gross precipitation in some cases) are all from Europe. It is possible that such differences result from the greater frequency of long rainstorms of low intensity there. The single observation from Taiwan, a country with torrential rainfalls, strengthens this suggestion.

Interception by Grasses

At their maximum growth, grasses may intercept as much precipitation as trees (10–20 percent of gross precipitation during individual storms). Because of their rapid seasonal changes of bulk, the total annual interception in grasslands is usually less than in forests, but is not usually reported.

Merriam (1961) sprinkled plots of rye grass with rains varying in magnitude from 25 to 150 mm, and found no significant relation between gross rainfall and total interception. He was, however, able to correlate the amount of interception with the surface area of the grass measured by a quadrat point-counting technique. The resulting relationship was

$$I\ (\text{mm}) = -0.028 + 0.122G \tag{3-2}$$

where G = a measure of vegetative surface area (twice the average number of grass blades touching each pin of the quadrat-sampler). Total interception was also roughly correlated with grass height, and varied between 0.51 and 2.8 mm for stems ranging in height from 10 cm to 48 cm.

Crouse et al. (1966) studied the effect of vegetative bulk on interception by brome grasses in the San Dimas watershed in California. Their data show that daily interception totals can be related to vegetation by the following approximate relationship:

$$I\ (\text{mm}) = 0.86HV \tag{3-3}$$

where H = average grass height (mm) and V = cover density (decimal fraction). For clover and meadow grass, Horton (1919) developed the following relationship:

$$I\ (\text{mm}) = 0.00042H + 0.00026HP \tag{3-4}$$

where H = average grass height (mm) and P = precipitation (mm).

Interception by Crops

Lull (1964) reported the results of U.S. Department of Agriculture experiments on total interception by various crops. These data are listed in Table 3-2. Stoltenberg and Wilson (1950) cut mature corn plants and sprinkled them, weighing before and after water application. They converted their

Table 3-2 Interception by various crops. (From U.S. Department of Agriculture, Soil Conservation Service.)

	DURING GROWING SEASON			DURING LOW-VEGETATION DEVELOPMENT
	RAINFALL (MM)	INTERCEPTION (MM)	INTERCEPTION (%)	INTERCEPTION (%)
Alfalfa	275	98	36	22
Corn	181	28	16	3
Soybean	158	23	15	9
Oats	171	12	7	3

Table 3-3 Summary of Russian data on interception by crops. (From Kontorshchikov and Eremina 1963.)

CROP	MEASUREMENT PERIOD	INTERCEPTION AS A PERCENTAGE OF GROSS PRECIPITATION
Spring wheat	Growing season	10–35
Rye (50–150 cm high)	–	4–6
Oats	July	16
Oats	August	23

measurements to interception from fields of known plant density, concluding that mature green plants intercepted 0.56–0.64 mm per storm, while field-dried corn intercepted 0.46 mm per storm. Kontorshchikov and Eremina (1963) made measurements and reviewed other Russian work on interception by crops. Their conclusions are summarized in Table 3-3.

The Importance of Interception

There has been much controversy about the importance of interception in the water budget. Foresters (e.g., Kittredge 1947) viewed interception as a loss from the land phase of the hydrologic cycle. Later, when energy-budget calculations were used to predict evapotranspiration losses (see Chapter 5), it was pointed out that the total energy available for evaporation of water from leaf surfaces was the same, whether evaporation was supplied from intercepted rainfall or from within the leaves of plants. It was suggested, therefore, that the evaporation of intercepted water was not a loss, but was balanced by a reduction of transpiration that would otherwise have occurred. An experimental verification of this claim was made by Burgy and Pomeroy (1958), who studied interception and evapotranspiration from grass under laboratory conditions. They found that watering reduced transpiration to such an extent that the savings balanced interception, and that the net interception as far as the soil was concerned was essentially zero. Their results were confirmed under field conditions by McMillan and Burgy (1960), who left open the possibility that net loss by interception may be appreciable if the wind is strong, soil moisture low, or vegetation dead, dormant, or woody.

This argument is supported by data in Table 3-4, which show losses of water from grassy plots in Zaire. The first plot, which received only rainwater, lost 115 mm of intercepted water by evaporation from leaves and 1003 mm of water drawn from the soil. The second plot received more frequent watering by irrigation, as well as by rainfall, and so lost 238 mm of intercepted water. The evaporation of this water, however, reduced the demand of evapotranspiration for soil water to 817 mm. The third plot was heavily fertilized and irrigated to produce a heavier grass cover, which together with the more frequent watering resulted in a canopy interception of 654 mm. The water loss from the soil was reduced to 521 mm. The total amount of water evaporated from each plot was approximately the same, whether the water came from interception or from the soil, because the evaporative demand imposed by the incoming energy was the same.

More recent evidence (Thorud 1967) indicated that although evaporation of intercepted water does reduce transpiration, the effect is small compared with the amount of water intercepted. Maximum savings of 14 percent of water intercepted by small pine trees were recorded, but the average was close to 9 percent, and was very small under cool conditions or in dry soil. Thorud concluded that under the conditions of his experiment, more than 90 percent of the intercepted water was evaporated from the foliage without reducing the transpiration rate, and was a true loss from the water budget of the soil. This agrees with the field data of Leyton et al. (1967). From both theory and field measurement, Rutter (1967) concluded that the rate of evaporation of intercepted water is, on the average, about four times as great as the transpiration rate from within the leaves under the same environmental conditions. Other writers have pointed out that in some parts of the year, the total amount of water intercepted exceeds the computed potential rate of evapotranspiration. The argument is by no means settled. Yet it has some significance for assessing the probable impact upon water supply when mixed hardwood forests are replaced by faster-growing coniferous trees in the eastern United States and many tropical countries. We shall review some field data after discussing evapotranspiration in Chapter 5.

The situation is further complicated by a kind of reverse interception process, known as fog precipitation, by which plants seem able to increase the water supply reaching the ground surface. If you have walked through grass or forest early in a morning, you will have been thoroughly soaked by water that has condensed on vegetation during the night. In a coastal coniferous forest this condensation process may be so intense that one can hear a steady, gentle fall of drops from the canopy to the forest floor. At present it is difficult to estimate the amount of water collected by vegetation in this way or to assess its importance in the hydrologic cycle. The measurement techniques used to study it are not very satisfactory. There are some

Table 3-4 Water loss under grass at Yangambi, Zaire. (From Bernard, quoted in Penman 1963.)

WATER LOSS (MM)	PLOT TREATMENT		
	RAIN ONLY	IRRIGATION AND RAIN	IRRIGATION, RAIN, AND FERTILIZER
Evaporation of intercepted water	115	238	654
Water drawn from soil	1003	817	521
Total water loss	1118	1055	1175

obvious ecological consequences, however. On the the arid littoral of northwest Africa, for example, the increased water supply from nighttime fog precipitation supports a 100-meter-wide strip of green vegetation in one of the world's driest areas. This form of precipitation has been discussed by Nagel (1956), Oberlander (1956), and Becking (1962).

The subtraction of intercepted water from gross precipitation becomes insignificant during very large rainstorms. Interception, therefore, has little effect upon the development of major floods. There is no doubt, however, of the value of interception to soil conservation during large, intense rainstorms. In considering erosion and sedimentation later in the text, we will stress the vital importance of interception processes in reducing the kinetic energy of raindrops before they reach the soil surface, and therefore, in reducing the amount of soil erosion. Before considering the geomorphic work of water, however, we must follow the precipitation further through the hydrologic cycle.

Bibliography

Becking, R. W. (1962) The fog belt rainforest of the Pacific Northwest (U.S.A.); *Proceedings of the Ninth Pacific Science Congress,* vol. 4, pp. 95–98.

Brown, J. H. and Barker, A. C. (1970) An analysis of throughfall and stemflow in mixed oak stands; *Water Resources Research,* vol. 6, pp. 316–323.

Burgy, R. H. and Pomeroy, C. R. (1958) Interception losses in grassy vegetation; *EOS, American Geophysical Union Transactions,* vol. 39, pp. 1095–1100.

Coffay, E. W. (1962) Throughfall in a forest; *International Association of Scientific Hydrology Bulletin,* vol. 7, no. 2, pp. 10–16.

Collings, M. R. (1966) Throughfall for summer thunderstorms in a juniper and pinyon woodland, Cibecue Ridge, Arizona; *U.S. Geological Survey Professional Paper 485-B.*

Crouse, R. P., Corbett, E. S. and Seegrist, D. W. (1966) Methods of measuring and analyzing rainfall interception by grass; *International Association of Scientific Hydrology Bulletin,* vol. 11, no. 2, pp. 110–120.

Eidmann, F. E. (1959) Die interception in buchen- und fichten-bestanden; *International Association of Scientific Hydrology, Symposium of Hannover, Publication no. 48,* pp. 5–25.

Helvey, J. D. (1964) Rainfall interception by hardwood forest litter in the southern Appalachians; *U.S. Forest Service Southeast Forest Experiment Station, Research Paper 8.*

Helvey, J. D. (1967) Interception by eastern white pine; *Water Resources Research,* vol. 3, pp. 723–730.

Helvey, J. D. and Patric, J. H. (1965) Canopy and litter interception of rainfall by hardwoods of eastern United States; *Water Resources Research,* vol. 1, pp. 193–206.

Horton, R. E. (1919) Rainfall interception; *Monthly Weather Review,* vol. 47, pp. 603–623.

Johnson, W. M. (1942) The interception of rain and snow by a forest of young Ponderosa pine; *EOS, American Geophysical Union Transactions,* vol. 23, pp. 566–570.

KITTREDGE, J. E. (1947) *Forest influences;* McGraw-Hill, New York, 394 pp.

KONTORSHCHIKOV, A. S. AND EREMINA, K. A. (1963) Interception of precipitation by spring wheat during the growing season; *Soviet Hydrology*, vol. 2, pp. 400–409.

LAW, F. (1958) Measurement of rainfall, interception and evaporation losses in a plantation of Sitka spruce trees; *International Association of Scientific Hydrology, General Assembly of Toronto, Publication No. 44,* pp. 397–411.

LAWSON, E. R. (1967) Throughfall and stemflow in a pine-hardwood stand in the Ouachita Mountains of Arkansas; *Water Resources Research,* vol. 3, pp. 731–736.

LEYTON, L., REYNOLDS, E. R. C. AND THOMPSON, F. B. (1967) Rainfall interception in forest and moorland; in *Forest hydrology* (eds. W. E. Sopper and H. W. Lull), Pergamon Press, Oxford, pp. 163–177.

LULL, H. W. (1964) Ecological and silvicultural aspects; Section 6 in *Handbook of applied hydrology* (ed. Ven te Chow), McGraw-Hill, New York.

MCMILLAN, W. D. AND BURGY, R. H. (1960) Interception loss from grass; *Journal of Geophysical Research,* vol. 65, pp. 2389–2394.

MERRIAM, R. A. (1961) Surface water storage on annual ryegrass; *Journal of Geophysical Research,* vol. 66, pp. 1833–1838.

NAGEL, J. F. (1956) Fog precipitation on Table Mountain; *Quarterly Journal of the Royal Meteorological Society,* vol. 82, pp. 452–560.

NEMEC, J., PASAK, V. AND ZELENY, V. (1967) Forest hydrology research in Czechoslovakia; in *Forest hydrology* (eds. W. E. Sopper and H. W. Lull), Pergamon Press, Oxford, pp. 31–33.

NYE, P. H. AND GREENLAND, D. J. (1960) The soil under shifting agriculture; *Technical Communication No. 51, Commonwealth Bureau of Soils,* Harpenden, Eng., 156 pp.

OBERLANDER, G. T. (1956) Summer fog precipitation on the San Francisco peninsula; *Ecology,* vol. 37, pp. 851–852.

PATRIC, J. H. (1966) Rainfall interception by mature coniferous forests of southeast Alaska; *Journal of Soil and Water Conservation*, vol. 21, pp. 229–231.

PENMAN, H. L. (1963) Vegetation and hydrology; *Technical Communication No. 53, Commonwealth Bureau of Soils,* Harpenden, Eng.

ROGERSON, T. L. (1967) Throughfall in pole-sized loblolly pine as affected by stand density; in *Forest hydrology* (eds. W. E. Sopper and H. W. Lull), Pergamon Press, Oxford, pp. 187–190.

ROGERSON, T. L. AND BYRNES, W. R. (1968) Net rainfall under hardwoods and red pine in Central Pennsylvania; *Water Resources Research*, vol. 4, pp. 55–58.

ROTHACHER, J. (1963) Net precipitation under a Douglas fir forest; *Forest Science,* vol. 9, pp. 423–429.

RUTTER, A. J. (1967) An analysis of evaporation from a stand of Scots Pine; in *Forest hydrology* (eds. W. E. Sopper and H. W. Lull), Pergamon Press, Oxford, pp. 403–418.

SHENG, T. C. AND KOH, C. C. (1967) Forest hydrology research in Taiwan; in *Forest hydrology* (eds. W. E. Sopper and H. W. Lull), Pergamon Press, Oxford, pp. 89–94.

SIM, L. K. (1972) Interception loss in the humid forested areas, with special reference to Sungai Lui catchment, W. Malaysia; *The Malayan Nature Journal,* vol. 25, pp. 104–111.

STOLTENBERG, N. L. AND WILSON, T. V. (1950) Interception storage of rainfall by corn plants; *EOS, American Geophysical Union Transactions,* vol. 31, pp. 443–448.

THORUD, D. B. (1967) The effect of applied interception on transpiration rates of potted ponderosa pine; *Water Resources Research,* vol. 3, pp. 443–450.

VOIGHT, G. K. (1960) Distribution of rainfall under forest stands; *Forest Science,* vol. 6, no. 1, pp. 2–10.

ZINKE, P. J. (1967) Forest interception studies in the United States; in *Forest hydrology* (eds. W. E. Sopper and H. W. Lull), Pergamon Press, Oxford, pp. 137–160.

Typical Problems

3-1 Interception by Forests

Using average values from Figure 3-2, compute the net rainfall under hardwoods and conifers during a sequence of 4 storms applying 50 mm each and a sequence of 20 storms supplying 10 mm each.

Solution

Total yields are obtained by multiplying the storm yields in Table 3-5 by the number of storms.

Hardwoods
- 10-mm storms $8 \times 20 = 160$ mm
- 50-mm storms $44 \times 4 = 176$ mm

Conifers
- 10-mm storms $8 \times 20 = 160$ mm
- 50-mm storms $37 \times 4 = 148$ mm

Table 3-5 Storm yields from Figure 3-2.

	STORM YIELDS UNDER "AVERAGE" HARDWOODS (MM)	
	THROUGHFALL	STEMFLOW
10-mm storms	7	1
50-mm storms	40	4
	STORM YIELDS UNDER "AVERAGE" CONIFERS (MM)	
	THROUGHFALL	**STEMFLOW**
10-mm storms	7	1
50-mm storms	37	4

3-2 Interception by Grass

Four 50-mm rainstorms are spread uniformly through a growing season, during which the height of a grass cover varies are shown in Figure 3-3. Calculate the net rainfall reaching the ground surface.

Solution

The heights of the grass cover during the storms were 200 mm, 400 mm, 500 mm, and 300 mm. Using these heights in Equation 3-4 gives interception values of 2.7 mm, 5.4 mm, 6.7 mm, and 4.0 mm for a total of 18.8 mm. Net rainfall reaching the ground surface is therefore 181 mm.

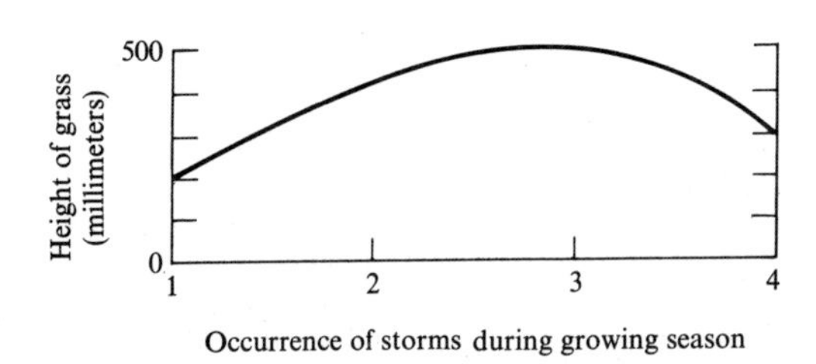

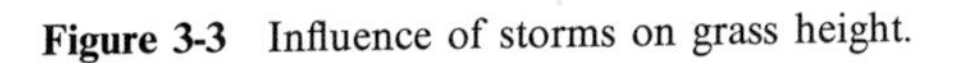

Figure 3-3 Influence of storms on grass height.

4

Water Loss from Lakes

The Importance of Studying Evaporation

Rainwater that evades interception follows various paths to rivers and lakes. During its residence in lakes, some of this water evaporates back to the atmosphere. During the past 50 years there has been a large increase in the number of artificial lakes for storage of irrigation water, municipal supply, stock watering, power generation, industrial cooling, or simply for ornamental purposes in urban residential areas. Evaporation from natural or artificial lakes, therefore, is an important and growing loss from the terrestrial water budget, particularly in arid regions (Figure 4-1). In the catchment of the lower Colorado River, for example, increased evaporation losses would largely offset any desired increase in water supply from new reservoirs. Before any new reservoir is impounded in such a region, engineers, planners, economists, and all concerned with water resources planning in the area should be aware of the net loss of water that will result from the increased areal extent of open water. They can compute the value of the lost water and incorporate this value in estimates of the value of the project. A knowledge of evaporation is also required for the efficient operation of reservoirs. This is particularly true of international reservoirs, where evaporative losses and other withdrawals must be apportioned by treaty.

Figure 4-1 Stock-watering pond at the Coal Draw Waterspreading Facility near Terry, Montana. The installation of large and small lakes in arid regions increases the amount of evaporation. (Photograph by N. J. King.)

In southern California more than 10,000 hectares of land were flooded in 1904–1907 by an uncontrolled diversion of the Colorado River; the Salton Sea was formed. The salinity of the water was approximately 500 mg/liter. Since then large quantities of salts have been washed into the lake by irrigation return flow and wastewater. Evaporation concentrated the water still further, and the salinity and water level have fluctuated greatly as evaporation and inflow have varied. Recreational development on the shores of the lake has increased interest in the water level and has raised the need for information on the hydrologic regimen, water chemistry, and the possibilities for regulation. The U.S. Geological Survey undertook a major study of evaporation rates as a part of this resource evaluation.

On the arid floor of the Great African Rift Valley in Kenya, Lake Nakuru is supplied by streams draining volcanic highlands flanking the rift. All water entering the lake is evaporated from its surface and the dissolved salts brought into the lake by streams are concentrated to several thousand milligrams per liter. The high salinity and sunny climate favor tremendous production of algae, which supports two million pink flamingoes. The flamingoes line the shore and provide one of the world's loveliest bird spectacles. Tourist revenues from the National Park on the lakeshore are important in the local and national economy. The inflow of stream water and the output of evaporation fluctuate, however, and both the water level and the salinity undergo major fluctuations. During wet years the salinity decreases, algal production declines, and the flamingoes leave the lake to search for a more saline environment. They often do not return for several years, during which time local revenues associated with the National Park are severely reduced. An ability to predict the variation of water level and salinity for Lake Nakuru and the other closed lakes of the Rift Valley in Kenya and Tanzania would be useful for understanding the ecology of these birds and for assessing the long-term fluctuations the park population is likely to undergo. Keeping one's eyes and mind open for new applications of hydrology like this can produce interesting insights on nature, as well as useful planning tools.

Evaporation from a water surface in the semi-arid tropics, for example, may be as high as 2000 mm per year. If the annual rainfall of such a region is 400 mm, there is a large net loss of water at the site. A 10,000-hectare lake surface would lose 160 million cubic meters of water each year, which is enough to support more than one million inhabitants of a modern industrial city, or to irrigate 10,000 to 15,000 hectares of cropland. On a smaller scale, the design of a farm pond or an urban lake require a quantitative appreciation of evaporation. In order to guard against washing out of the dam by large floods or its infilling by sediment, the designer might prefer to locate the impoundment below a small catchment. The drainage area, however, must be sufficiently large to generate enough runoff to offset losses to evaporation and seepage. The design calculations, therefore, should involve at least a rough estimate of evaporation rates, as indicated in standard design manuals on this subject (U.S. Soil Conservation Service 1969). In the United States annual evaporation rates from small ponds vary from about 400 mm in the cool, cloudy regions of the Pacific Northwest and northern New England to 2500 mm in the arid rangelands of the southwest.

Workers in other disciplines concerned with bodies of water often need a quantitative appreciation of evaporation rates. Students of the geochemistry of closed lakes need an estimate of evaporation rates for calculation of the water budget, and hence fluctuations of salinity (Langbein 1961, Phillips and VanDenburgh, 1971, Hely et al. 1967). Because evaporation is one of the ways by which the temperature of a body of water is regulated, the process becomes vitally important in situations where additions of waste heat from power plants cause an increase in the temperature of water body. The transfer of this heat to the atmosphere by evaporation is one of the most effective ways of cooling the water, and a quantitative understanding of evaporation is required for assessing the probable effects of such thermal discharges and for establishing controls. It is particularly important that conservationists and biologists understand how the calculations are made.

Factors Affecting Evaporation

Evaporation of water in the hydrologic cycle consists of the change of state from liquid to vapor and the net transfer of this vapor into the atmosphere. The process occurs when some molecules of the liquid attain sufficient kinetic energy to overcome the forces of surface tension and escape from the surface of the liquid. To do so, the molecules must obtain energy from an outside source. This energy comes from solar radiation, sensible heat transfer from the atmosphere, or from heat advected into the water body by inflowing warm water. Because the control of radiation is usually dominant, evaporation is a function of latitude, season, time of day, and cloudiness.

Water molecules moving out of the liquid produce a vapor pressure, which is a function of the rate at which water molecules leave the surface, and therefore of the temperature (energy) of the surface. Some molecules of water vapor move from the atmosphere into the liquid and condense. The rate at which they do so is directly proportional to the vapor pressure of water in the atmosphere, which in turn is controlled by its water vapor content. Evaporation is the net rate of movement out of the liquid and is therefore directly proportional to the difference in vapor pressure between the water surface and the atmosphere. This difference is called the vapor pressure deficit.

If the air above the lake surface were still, the addition of water vapor to the lower layer of air would soon raise the atmospheric vapor pressure to that of the water body and the air would be saturated with water vapor. If this condition is to be avoided and evaporation is to continue, the surface layer must be removed and continually replaced by unsaturated air. The major mechanism for replacing the surface layer of air is wind, which creates the turbulence responsible for moving air away from the water surface.

Three conditions are necessary, therefore, for continued evaporation from a free water surface: energy, a difference in vapor pressure between the water surface and the atmosphere, and a method of exchanging surface air for air capable of holding more water vapor. These factors will be discussed quantitatively later in the chapter.

Some characteristics of the water body are also important. Part of the energy received by a lake is used to raise the temperature of the water. In a deep lake, the increase of water temperature may extend to a considerable depth and involve large amounts of heat that would otherwise be used for evaporation. In winter, however, when the water is warmer than the atmosphere, heat stored in the lake is used to evaporate more water than one would expect from a consideration of radiation and sensible heat transfer. Over a period of a year, there is little net change of heat storage, so the total annual evaporation is not much affected by this mechanism. The monthly distribution, however, may be changed drastically. For the Elephant Butte Reservoir in New Mexico, Houk (1937) reported that seasonal temperature fluctuations extended to more than 30 m and the amount of heat stored between January and August would have been sufficient to evaporate a depth of more than 60 cm of water; the average annual lake evaporation in the region is only a little over 100 cm. Storage changes in the Great Lakes allow evaporation rates to reach a maximum in late autumn and winter (Morton 1967).

In small lakes, the area of the water body can also affect evaporation rates, particularly in arid and semi-arid regions. Hot, dry air moving from a land surface over a lake will supply sensible heat to the water and create a large vapor pressure deficit and high evaporation. If the lake is large, the vapor pressure deficit will decrease as moisture moves into the atmosphere. There should, therefore, be an inverse relationship between the evaporation rate and the size of the lake, at least up to some critical size at which equilibrium conditions are established.

The salinity of a lake affects evaporation rates, but only in very saline waters is this effect appreciable. Salinity reduces the vapor pressure of the water body. Turk (1970) has published the results of field observations on the depression of evaporation rate as salinity increases. Such considerations are important in studies of the hydrologic regime of closed-basin lakes (Phillips and VanDenburgh 1971).

Measurement of Evaporation: Water-Budget Method

Because there is no direct way of measuring evaporation from a natural water body, the most obvious way of assessing the process is to measure all inflows and outflows to and from the lake, and to calculate evaporation as a residual. This is the basis of the water-budget method of measuring evaporation, and for a specific time period the budget is expressed by the equation:

$$\text{Inflow} = \text{Outflow} + \text{Change in storage}$$

Expanding this statement, we can write

$$I_{\text{surf}} + I_{\text{sub}} + P = O_{\text{surf}} + O_{\text{sub}} + E_o + (V_2 - V_1)$$

where $I_{\text{surf}}, I_{\text{sub}}$ = volume of surface and subsurface inflows from streams and groundwater

P = volume of precipitation onto the lake surface

$O_{\text{surf}}, O_{\text{sub}}$ = volume of surface and subsurface outflows to streams and groundwater

E_o = volume of water evaporated

V_1, V_2 = volume of lake at the beginning and end of the measurement period.

Therefore,

$$E_o = I_{\text{surf}} + I_{\text{sub}} + P - O_{\text{surf}} - O_{\text{sub}} - V_2 + V_1 \qquad (4\text{-}1)$$

There are great difficulties in measuring each of the variables in Equation 4-1, and few studies of evaporation justify the costs of accurate determinations of the parameters referred to. Few lakes provide conditions that minimize errors in determination. The field methods for evaluating each of the components in Equation 4-1 are described later in this book and are fully treated in a report by the U.S. Geological Survey (1954).

The water-budget method is not feasible for routine measurement of evaporation, but has been used under favorable conditions as a control, against which methods of calculating evaporation can be checked. This was the reason for choosing Lake Hefner, Oklahoma, as the site for a detailed study of techniques for calculating evaporation (U.S. Geological Survey 1954). If a lake presents optimum conditions, errors in estimating monthly evaporation can be kept to ±10 percent. Successful application of the technique to small lakes can be cheaper if geological and hydrological conditions are favorable (McKay and Stichling 1961). Ideal conditions occur where subsurface flows are essentially zero and surface outflow is small relative to evaporation.

Measurement of Evaporation: Pans

In view of the difficulty of gauging evaporation from large water bodies, most direct measurements have been made by the use of small pans. There are many kinds of pans in use, varying in dimensions, materials, and conditions of exposure. Some are placed above ground, others are sunk below ground, and a third type floats. The various kinds of pans are reviewed elsewhere. In this text, only one type will be referred to because it has become the most widely used pan in many countries.

The U.S. Weather Bureau Class A pan (see Figure 4-2) is made of unpainted, galvanized iron, supported on a low wooden frame. It is 4 feet in

Figure 4-2 Class A evaporation pan of the U.S. Weather Bureau. A small cylinder standing in the center of the pan is used to support a hook gauge for precise measurements of water level. A three-cup anemometer for measuring windspeed is attached to the front of the pan.

diameter, 10 inches deep, and is filled with water to a depth of 8 inches. Evaporation is measured as fluctuations of the water level, corrected for additions from precipitation. The water-level variations are monitored by means of a point gauge in a stilling well or by adding water from a graduated cylinder to bring the water level in the pan up to a depth of 8 inches.

In contrast to a lake, the Class A pan receives large quantities of energy from radiation and conduction through its base and sides. Evaporation from the pan, therefore, will be larger than from a lake under the same meteorological conditions. Moreover, the difference between pan and lake will vary through the year because of seasonal differences in radiation, air temperature, wind, and heat storage within the larger body of water. A coefficient that varies through the year must, therefore, be applied to measurements of pan evaporation in order to estimate water loss from a lake. This *pan coefficient* varies geographically and with season. Table 4-1 is a compilation of data on pan coefficients from various sources. The coefficient shows greatest variation for large, deep lakes in areas with a large annual temperature range. In areas where pan coefficients have not previously been derived experimentally, an average annual value of 0.70 to 0.75 is generally assumed, and pan evaporation data are multiplied by this amount in calculating lake evaporation. If the lake is very small, such as a shallow stock pond, a coefficient of 0.90 or even higher is more appropriate. These pan coefficients are not usable in cases where large amounts of energy are brought into the lake by heated effluent from a power plant.

Some hydrologists have tried to relate pan evaporation directly to meteorological observations. Such relationships are valuable for extrapolating short pan records, or for estimating evaporation at meteorological stations where pan observations are not made. These techniques for predicting evaporation from small pans are referred to in Typical Problem 4-1.

Many countries have now produced maps of average annual Class A pan evaporation (see Figure 4-3), which are useful for design and operation of small reservoirs and for making preliminary studies for large reservoirs. They

Table 4-1 Pan coefficients for a Class A pan. (From Hughes 1967, Kohler 1954, Ficke 1972, U.S. Geological Survey 1958.)

LOCATION	TIME OF YEAR	MEAN COEFFICIENT	RANGE OF THE COEFFICIENT
Fort Collins, CO	Apr.–Nov.	0.70	0.60–0.82
L. Elsinore, CA	All year	0.77	0.63–0.97
Texas	All year	0.68	
Florida	All year	0.81	0.69–0.91
L. Hefner, OK	All year	0.69	0.35–1.32
L. Mead, AZ	All year	0.60	
Salton Sea, CA	All year	0.50	0.31–0.83
Pretty Lake, IN	All year	0.70	0.50–0.90

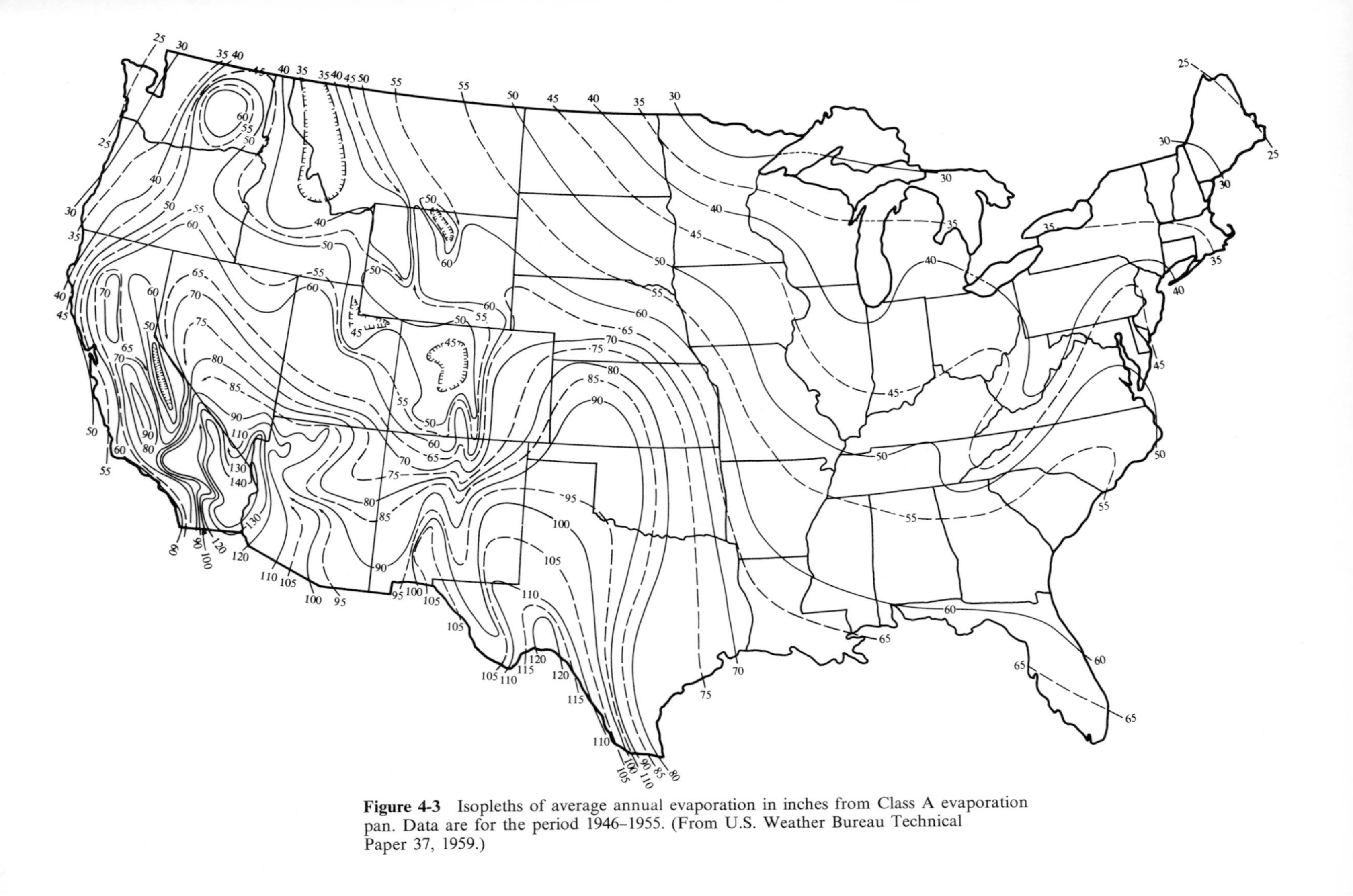

Figure 4-3 Isopleths of average annual evaporation in inches from Class A evaporation pan. Data are for the period 1946–1955. (From U.S. Weather Bureau Technical Paper 37, 1959.)

also provide a good means of estimating potential evapotranspiration (see Chapter 5). Dalinsky (1971) has shown that it is possible to predict monthly variations of pan evaporation anywhere in Israel by fitting a sinusoidal function to average monthly pan data at the 47 observing stations. Monthly values for any station can then be estimated from only a map of average annual evaporation.

In the United States evaporation from Class A pans is measured at several hundred U.S. Weather Bureau stations throughout the country, and the data are available from local offices. Monthly and annual summaries are published for many stations in the Weather Bureau publication *Climatological Data: National Summary*. Average monthly and annual data have been published in tabulated form in U.S. Weather Bureau, *Technical Paper 13;* and in map form in U.S. Weather Bureau, *Technical Paper 37;* and in the *National Atlas of the United States*. Monthly values of pan evaporation are also listed in many state climatic summaries.

Evaporation varies, of course, from year to year but is generally more conservative than precipitation or runoff. Annual values of pan evaporation in the United States are normally distributed with a standard deviation of 50 to 100 mm in the Pacific Northwest (where mean annual values are in the range of 600 to 900 mm). In the southern plains region, however, the standard deviation exceeds 300 mm around a mean of 1500 to 2300 mm.

Calculation of Evaporation: Energy-Budget Approach

The transformation of one gram of water into vapor at normal lake temperatures requires approximately 590 calories of heat energy.* Not all the heat received by the water body is used for evaporation, however. The energy budget of a water body for some interval of time is expressed in Equation 4-2:

$$Q_s - Q_{rs} - Q_{lw} - Q_h - Q_e + Q_v - Q_{ve} = Q_\theta \qquad (4\text{-}2)$$

where Q_s = incoming solar radiation
Q_{rs} = reflected solar radiation
Q_{lw} = net longwave radiation from the water body into the atmosphere
Q_h = sensible heat transferred by turbulent exchange from the water body to the atmosphere
Q_e = energy utilized for evaporation
Q_v = net energy advected into the lake by flows of water
Q_{ve} = energy advected out of the water body by the evaporated water
Q_θ = change of energy stored in the lake.

*This quantity, known as the latent heat of vaporization of water, varies from 596 calories per gram at 0°C to 580 calories per gram at 27°C, and to 540 calories per gram at 100°C.

All units of Equation 4-2 are calories per square centimeter of lake surface (langleys).

This equation and its use are described in detail in a report by the U.S. Geological Survey (1954) where the following manipulations of the energy balance are explained. The sensible heat transfer term is not measured directly but is incorporated into the dimensionless Bowen's Ratio (R), defined as

$$R = \frac{Q_h}{Q_e} = 0.00061p\frac{(T_s - T_a)}{(e_s - e_a)} \tag{4-3}$$

where p is the atmospheric pressure (mb), T_s and T_a are the temperature of the water surface and the atmosphere (°C), while e_s and e_a are the vapor pressure of the water surface and the atmosphere (mb). The vapor pressure of the water surface depends upon the temperature of the water, and the atmospheric vapor pressure can be measured directly with a sling psychrometer or a hygrothermograph (see Figure 4-4).

The energy transferred from the water by the evaporating water can be calculated from

$$Q_{ve} = Q_e c(T_s - T_b)/L \tag{4-4}$$

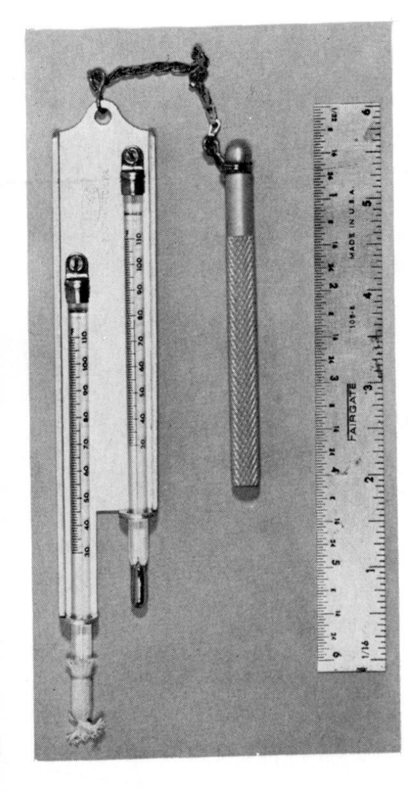

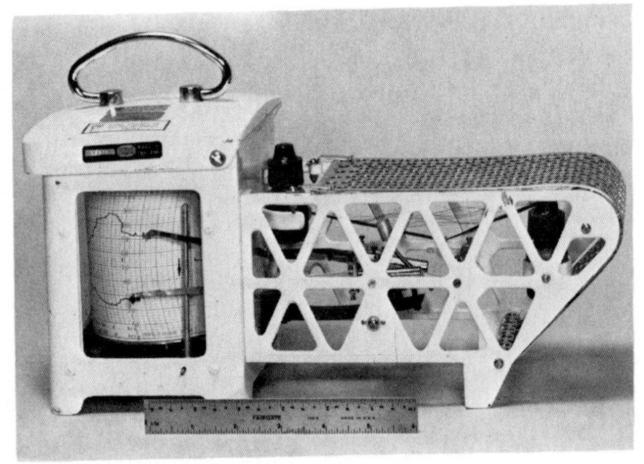

Figure 4-4 Standard instruments used to obtain vapor pressure. On the left is a psychrometer, which consists of two thermometers, one measuring the air temperature and one measuring the temperature of a piece of damp muslin wrapped around the left thermometer bulb. The difference in the two temperatures is a measure of the atmospheric vapor pressure. Above is a hygrothermograph, which records both air temperature and the relative humidity of the atmosphere.

where c is the specific heat of water (cal/gm/°C), T_b is an arbitrarily chosen base temperature for calculations (usually 0°C), and L is the latent heat of vaporization (590 cal/gm).

Using Equations 4-3 and 4-4, it is possible to rewrite Equation 4-2 as

$$Q_e = \frac{Q_s - Q_{rs} - Q_{lw} + Q_v - Q_\theta}{1 + R + c(T_s - T_b)/L} \qquad (4\text{-}5)$$

The amount of energy utilized for evaporation (Q_e) is related to the depth of evaporation (E_o) by the relationship

$$E_o = \frac{Q_e}{L\rho} \qquad (4\text{-}6)$$

where E_o is in cm, and ρ is the density of water (gm/cm^3). Combining Equations 4-5 and 4-6, we obtain:

$$E_o = \frac{Q_s - Q_{rs} - Q_{lw} + Q_v - Q_\theta}{\rho[L(1 + R) + c(T_s - T_b)]} \qquad (4\text{-}7)$$

An energy-budget calculation of evaporation, therefore, consists of evaluating each of the terms on the right-hand side of Equation 4-7 and solving for E_o.

The advection (Q_v) and storage (Q_θ) terms are evaluated by repeated measurements of the temperature and volume of inflows and outflows and of the water stored in the lake. The surface water temperature (T_s) is required for Equations 4-7 and 4-3, where it is also used to obtain e_s, which is a function of water temperature (see Figure 4-8). Air temperature (T_a), atmospheric pressure (p), and vapor pressure (e_a) in Equation 4-3 are obtained from direct measurements at the site or from published meteorological records for nearby stations. In the United States, daily and monthly averages are published each year in the U.S. Weather Bureau series *Climatological Data: National Summary,* while long-term averages for these intervals are published in the *Climatic Summary* series and in reports entitled *Climate of the States.*

The radiation terms in Equation 4-7 are more complicated, and we will discuss them in some detail because they will reappear in later discussions of the energy budget in evapotranspiration and snowmelt. Incoming solar radiation (Q_s) can be measured directly by means of a pyrheliometer (Figure 4-5), and daily and monthly summaries of this energy input are usually available in publications such as the U.S. Weather Bureau *Climatological Data* series. Long-term averages are included in U.S. Weather Bureau, Technical Publication 11, *Weekly mean values of total solar and sky radiation,* in the *Climatic Guide* series for large cities; in the publications of state agencies; and in various national atlases.

There are, however, relatively few stations throughout the world at which instrumental observations of solar radiation are made. In many regions where the water balance of natural and artificial lakes is becoming im-

(a)

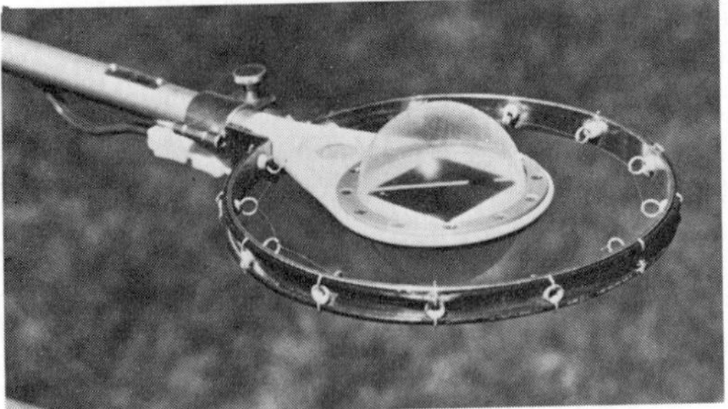

(b)

Figure 4-5 Instruments for measuring radiation. (a) Two pyrheliometers: the one facing the sky measures incoming solar radiation, and the second measures the shortwave radiation reflected from the ground surface. In the background stands a weather screen in which thermometers and a hygrothermograph are housed. (b) A net radiometer. (Photograph by L. J. Fritschen.)

portant in water-resource management, it is necessary to estimate solar radiation from latitude, date, cloud cover, or the duration of bright sunshine. These data are often tabulated and published for weather stations without recording instruments. Maps of cloud cover or hours of bright sunshine are commonly included in national atlases. In the United States, U.S. Weather Bureau Technical Publication No. 12, *Sunshine and cloudiness at selected stations in the United States and Puerto Rico,* is useful.

Using records at 150 stations throughout the world, Black (quoted in Chang 1968) developed a relationship for predicting mean monthly solar radiation:

$$Q_s = I_o(0.803 - 0.340C - 0.458C^2) \tag{4-8}$$

where Q_s = mean daily solar radiation for the month (cal/cm²/day)
I_o = solar radiation per day received on a horizontal surface at the exterior of the atmosphere (see Table 4-2)
C = mean monthly cloudiness (decimal fraction), estimated by weather observers and published routinely in weather records.

Another method of estimating Q_s is by means of the equation

$$Q_s = I_o\left(a + b\frac{n}{N}\right) \tag{4-9}$$

where a, b = empirical constants (see Table 4-3)
n = observed duration of sunshine (hours) published in atlases and in standard meteorological summaries
N = maximum possible duration of sunshine (hours) given in standard meteorological tables (see Table 4-4).

If direct measurements of solar radiation are available, they should be used of course. In some countries these data are available in monthly and annual summaries or in atlases.

Table 4-2 Average solar radiation received on a horizontal plane at the upper edge of the atmosphere (cal/cm²/day).

LATITUDE	JAN.	FEB.	MAR.	APR.	MAY	JUNE	JULY	AUG.	SEPT.	OCT.	NOV.	DEC.
90°N	—	—	—	465	880	1070	930	660	155	—	—	—
80°N	—	—	105	460	860	1050	970	625	235	10	—	—
70°N	—	65	255	540	800	1000	870	670	400	140	5	—
60°N	75	205	400	655	860	975	925	750	500	275	110	55
50°N	200	350	540	750	910	985	950	820	620	430	155	175
40°N	355	490	650	820	880	985	960	870	740	550	395	325
30°N	500	620	750	870	945	975	955	900	795	670	540	465
20°N	640	725	820	895	930	930	930	900	850	760	660	610
10°N	755	820	870	895	885	870	870	885	880	830	770	730
0	855	885	895	870	820	790	795	840	880	885	860	840
10°S	930	930	885	810	730	685	705	770	845	900	920	930
20°S	985	940	855	740	630	570	595	680	790	900	965	990
30°S	1015	930	800	640	505	445	465	575	725	870	985	1030
40°S	1020	895	715	525	375	305	335	450	630	810	960	1045
50°S	1000	835	620	400	240	175	200	315	505	735	950	1040

Table 4-3 Some published values of the empirical constants a and b in Equation 4-9.

LOCATION	a	b	SOURCE
World	0.23	0.48	Black et al. (1954)
World	0.29 cosine lat.	0.52	Glover and McCulloch (1958)
S.E. England	0.18	0.55	Penman (1948)
Virginia, U.S.A.	0.22	0.54	Quoted by Penman (1948)
Canberra, Australia	0.25	0.54	Quoted by Penman (1948)
Brisbane, Australia	0.23 to 0.35*	0.38 to 0.54	Cartledge (1973)
West Africa	−0.12 to 0.26*	0.99 to 0.50	Davies (1966)

*Values vary with the month.

Table 4-4 Maximum possible monthly duration of sunshine (hours).

LATITUDE	JAN.	FEB.	MAR.	APR.	MAY	JUNE	JULY	AUG.	SEPT.	OCT.	NOV.	DEC.
50°N	265	280	366	415	480	490	495	450	380	330	274	252
40°N	303	300	370	400	445	450	455	425	375	345	300	290
30°N	324	314	370	388	425	420	430	410	370	353	320	316
20°N	341	324	370	378	407	400	410	400	366	360	335	338
10°N	360	327	370	370	390	380	390	385	366	366	352	356
0	375	340	375	363	375	363	375	375	363	375	363	375
10°S	388	350	378	355	363	346	360	364	360	380	378	396
20°S	410	360	378	350	346	328	340	344	360	388	393	414
30°S	430	370	380	342	330	306	328	345	360	404	410	435
40°S	466	380	385	334	310	280	302	330	360	415	432	463
50°S	490	403	387	320	276	242	266	315	356	427	465	508

The reflectivity of a surface, henceforth called the *albedo*, can be measured in the field with an inverted pyrheliometer (see Figure 4-5(a)). Albedo varies as a power function of sun altitude, with the coefficient and exponent of the power equation depending on cloud type and the extent of cloud cover (Anderson 1954). In most studies, however, the albedo of water is usually assumed to be constant. Thus,

$$Q_{rs} = \alpha Q_s \tag{4-10}$$

where α is the albedo, or reflectivity of the water surface. Most studies assume that $\alpha \approx 0.06$ for water. Values ranging from 0.05 to 0.10 are commonly used.

Net longwave radiation (Q_{lw}) is more difficult to compute. The earth's surface emits longwave radiation into the atmosphere. The intensity of this terrestrial radiation depends mainly upon the temperature of the surface. Much of the radiation is absorbed by water vapor, clouds, and carbon dioxide in the atmosphere, and a portion is radiated back to the earth as atmospheric radiation. Atmospheric radiation is also generated by solar radiation that is absorbed by water vapor and carbon dioxide and then re-radiated at longer wavelengths. The intensity of atmospheric radiation depends upon the profile of air temperature, water vapor content, and cloud cover throughout the atmosphere. Because of the difficulty of obtaining measurements of these variables, there have been many attempts to develop relationships between net longwave radiation loss (Q_{lw}) and near-surface measurements of its three major controls.

Several empirical equations have been developed for estimating net longwave radiation; the most widely used is the Brunt Equation (Anderson 1954):

$$Q_{lw} = \sigma[T_s^4 - (c + d\sqrt{e_2})T_2^4](1 - aC) \tag{4-11}$$

where σ = the Stefan-Boltzmann constant (1.17×10^{-7} cal/cm^2/°K^4/day)
T_s = temperature of the surface (°K)
T_2 = air temperature at the 2-meter level (°K)
e_2 = vapor pressure of the air at the 2-meter level (mb)
c, d = empirical coefficients, which can vary geographically (see Table 4-5)
C = cloudiness (decimal fraction of the sky covered)
a = a constant depending upon cloud type: 0.25, 0.6, and 0.9 for high, medium, and low clouds, respectively.

If data on cloud type are not available $(1 - aC)$ may be replaced by $(0.10 + 0.9C)$ or by $(0.10 + 0.90\, n/N)$, where the ratio n/N is defined in Equation 4-9.

Estimates of net longwave radiation loss can also be made without using the surface temperature. The most common of the empirical equations for doing this is given by Chang (1968) as

$$Q_{lw} = \sigma T_2^4(0.56 - 0.08\sqrt{e_2})(1 - aC) \tag{4-12}$$

where T_2 is in °K, e_2 is in mb.

Estimates obtained from these empirical equations are not very precise, and errors often exceed ±25 percent, even when measurements are averaged over a day. Under uniform cloud conditions, the errors may be reduced to ± 15 to 20 percent for monthly values.

Net allwave radiation is that portion of incoming radiation that is not reflected or radiated back to the atmosphere; that is,

$$Q_n = Q_s(1 - \alpha) - Q_{lw} \tag{4-13}$$

This quantity can be measured directly with a net radiometer (Figure 4-5(b)), and such data are published for a few important meteorological stations. It

Table 4-5 Empirical values of constants for the Brunt Equation. (Compiled by Anderson 1954.)

PLACE	c	d
Sweden	0.43	0.082
Washington, DC	0.44	0.061
Austria	0.47	0.063
Algeria	0.48	0.058
California	0.50	0.032
England	0.53	0.065
France	0.60	0.042
India	0.62	0.029
Oklahoma	0.68	0.036

can also be calculated as the difference between net shortwave radiation and net longwave radiation evaluated by the procedure outlined above. A third method involves estimating net radiation as a linear function of Q_s for surfaces of a certain average albedo. Szeicz et al. (1969), for example, give the following relationship for water surfaces during the daylight hours:

$$Q_n = 0.890 \quad 56 \text{ cal/cm}^2\text{/day} \tag{4-14}$$

An equation such as 4-12 for a portion of a day must be used to estimate the net longwave loss when solar radiation is zero at night.

When the energy budget is applied for periods of one month with intensive direct measurement of all the terms, evaporation can be determined to an accuracy of 5 to 10 percent. This is expensive, however, and is only used as a means of calibrating less expensive methods (see next section) or for making final designs in large cooling ponds below power plants. Under less ideal conditions when the energy terms must be evaluated by empirical relationships or from observations at standard weather stations, the errors will range from 10 to 20 percent for monthly averages. This is sufficiently precise for preliminary calculations in reconnaissance evaluation of water resources or in the early stage of cooling pond design (Sefchovich 1970). Finally, the concept and evaluation of an energy budget will reappear several times throughout this text, and we find it useful to describe the approach at length in this early chapter.

Calculation of Evaporation: Mass Transfer Approach

The mass-transfer method of calculating evaporation is based on the assumption that evaporation is controlled by the windspeed and the vapor-pressure difference between the water surface and the atmosphere as follows:

$$E_o = N f(u)(e_s - e_a) \tag{4-15}$$

where N = a constant, known as the mass-transfer coefficient
$f(u)$ = a function of windspeed
e_s = vapor pressure of water surface
e_a = vapor pressure of the air.

This equation is an approximation but gives good results when applied in a semi-empirical fashion, as follows. Evaporation is first of all determined by either the water-budget or energy-budget methods. The measured rate of evaporation is then plotted against the product $u_2(e_s - e_2)$, windspeed u times vapor-pressure deficit, where the subscript refers to a 2-meter observation height (see Figure 4-6(a)). The slope of the resulting line is the value of N, the mass-transfer coefficient in Equation 4-15.

If the evaporation rate is to be determined by a water-budget method, it is likely that net inflow of surface and subsurface flow will cause fluctuations

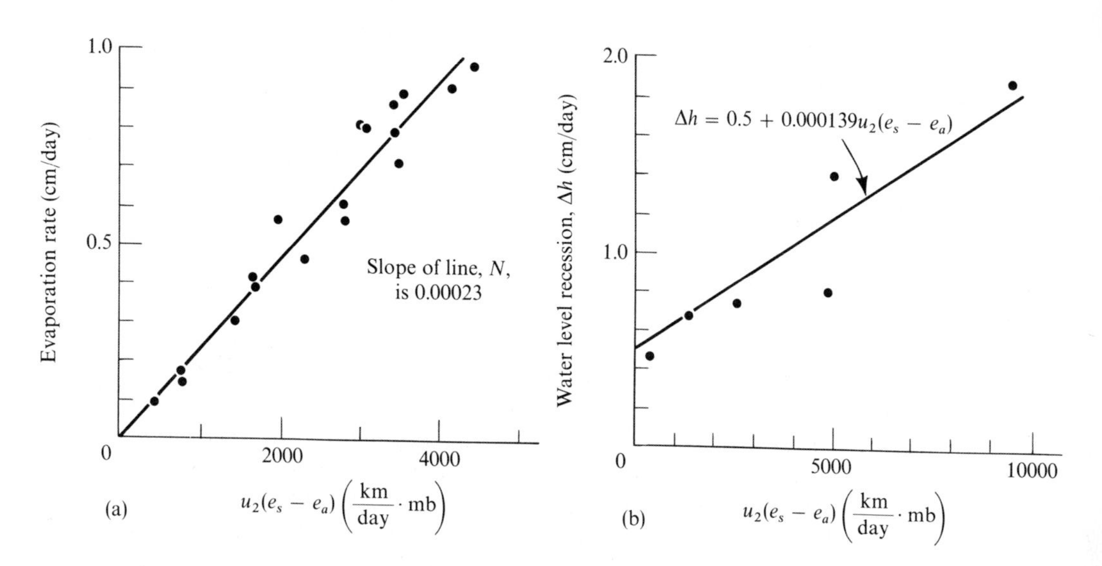

Figure 4-6 (a) Plot of measured or calculated evaporation rate against $u_2(e_s - e_a)$. (b) Plot of measured change of water level against $u_2(e_s - e_a)$ for a stock pond with seepage losses. (Data from Langbein et al. 1951, for Arizona.)

of the lake surface in addition to those due to evaporation. Under these circumstances, surface inflow and outflow and precipitation can be measured directly and used to correct the change of water level in the lake. The net subsurface seepage is still unknown but can be evaluated by considering Equation 4-16 and Figure 4-6(b).

$$\Delta h = E_o + S \tag{4-16}$$

where Δh = net change of water surface elevation adjusted for surface inflow and outflow and for precipitation onto the lake surface. An elevation-volume curve for the lake is needed for the adjustment. During the period of no surface inflow or outflow, this is simply the fall of the water surface, which can be evaluated through repeated observations of water level on a graduated staff set in the reservoir.

E_o = evaporation (depth units)

S = net groundwater seepage (depth units over the lake area).

If Δh is plotted against $u_2(e_s - e_a)$ as shown in Figure 4-6(b), the slope of the regression line is the value of N in Equation 4-15, in this case 0.000139. The water-level recession rate at the zero value of $u_2(e_s - e_a)$ is the seepage rate (S = 0.5 cm/day) in Equation 4-16.

This method of estimating both the mass-transfer coefficient and the net seepage rate was first applied by Langbein et al. (1951) to small stock-watering reservoirs, and is more reliable under conditions of zero surface flow and zero precipitation. Meyboom (1967) applied the technique successfully to small lakes maintained by groundwater seepage in western Canada, though he found that for ponds less than about one hectare, the value of N varied throughout the year. Turner (1966) found the method useful even for a large reservoir with appreciable surface flow in a humid region. He was also able to relate the net seepage derived in this way to the discharge of a gauged inflow stream.

The semi-empirical mass-transfer approach is increasing in popularity. When detailed measurements of evaporation are required for managing large reservoirs, the U.S. Geological Survey (Harbeck and Meyers 1970) carries out a program of energy-budget measurements for a period of approximately 15 months (including two summers). Using the values of evaporation calculated by this method as a control, the mass-transfer coefficient is derived as shown in Figure 4-6(a). The expensive energy-budget instrumentation can then be moved to another site, and evaporation from the reservoir can be calculated from routine observations of u_2, e_s, and e_2.

For small reservoirs, stock ponds, and urban lakes, rapid and cheap determinations of the mass-transfer coefficient and the seepage can be made using the technique illustrated in Figure 4-6(b). A graduated staff is placed in the lake to obtain Δh, and daily observations of the water level are made during periods of no surface inflow or outflow (or the water level change must be corrected for these). Because the rate of seepage varies with the amount and temperature of the water in storage (see Chapter 7), it may be necessary to repeat the measurements a few times throughout the year and to sketch an approximate annual curve of seepage. If estimates of seepage from one lake are to be transferred to others in the same region, the underlying geologic materials should be checked at each site. A seepage rate for clay cannot be extrapolated to a sandy site.

To apply the mass-transfer technique, a measure of the water surface temperature is required to derive e_s. If plans for a reservoir are being considered, it will not be possible to measure the water temperature because the lake does not yet exist. In such cases, air temperature is often used as a surrogate for the water surface temperature, or pan evaporation data must be used.

Values of N for reservoirs in the arid southwestern United States vary with lake area (A) according to the relation

$$N = 0.000169A^{-0.05} \tag{4-17}$$

where the mass-transfer coefficient is calculated for evaporation rates in cm/day, windspeeds in km/day, vapor pressures in mb, and lake area in sq km. Because of the variation of units in the literature, care should be taken when comparing the mass-transfer coefficients obtained by different authors. Figure 4-7 portrays the data available at this time. There is considerable scatter about the relationship described in Equation 4-17; errors of up to 25 percent are possible if N is estimated from this curve, even in the area for which it was developed. If the evaporation estimate needs to be fairly precise, therefore, a local field study is advisable.

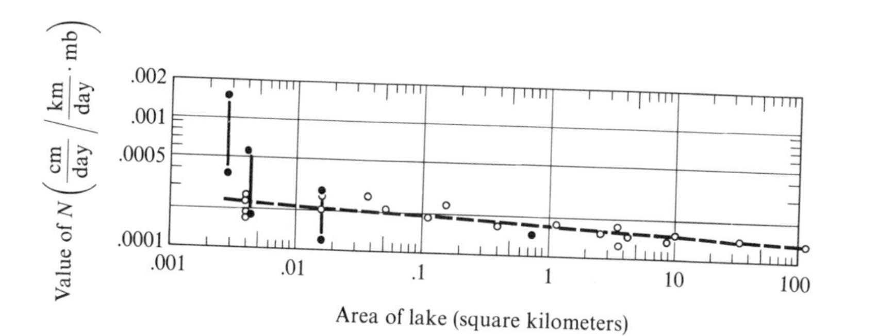

Figure 4-7 Mass-transfer coefficient, N, as a function of lake area. Open circles are data from lakes in the southwestern United States (Harbeck 1962); solid circles are for Indiana (Ficke 1972) and western Canada (Meyboom 1967); dashed line is Harbeck's mean relationship.

Calculation of Evaporation: Energy Balance of Small Pans and Shallow Lakes

The energy-budget and mass-transfer approaches can be combined to compute evaporation from small pans and shallow lakes, where energy-storage changes can be ignored. Such an analysis is useful not only for calculating lake evaporation but also as a basis for computing evapotranspiration in Chapter 5.

Penman (1948) studied the evaporation from a small sunken pan of water, ignoring heat-storage changes and the conduction of heat through the walls of the pan. The approximation allows the energy budget in Equation 4-1 to be written in the following simplified form:

$$Q_n = Q_h + Q_e \tag{4-18}$$

By dividing by (ρL), these energy components can be expressed in terms of equivalent depths (cm) of evaporation as

$$H = K + E_o \tag{4-19}$$

The equation states the obvious fact that in the absence of energy-storage changes or conduction through the walls of the pan, energy received from net radiation is partitioned between that used for evaporation and that transferred to the atmosphere as sensible heat.

Penman then derived the following expression for evaporation from the small sunken pan:

$$E_o = \frac{H\Delta + \gamma E_a}{\Delta + \gamma} \tag{4-20}$$

where E_o is evaporation rate in cm/day, H is net radiation in units of cm/day of evaporation, Δ (mb/°C) is the slope of the curve relating saturation

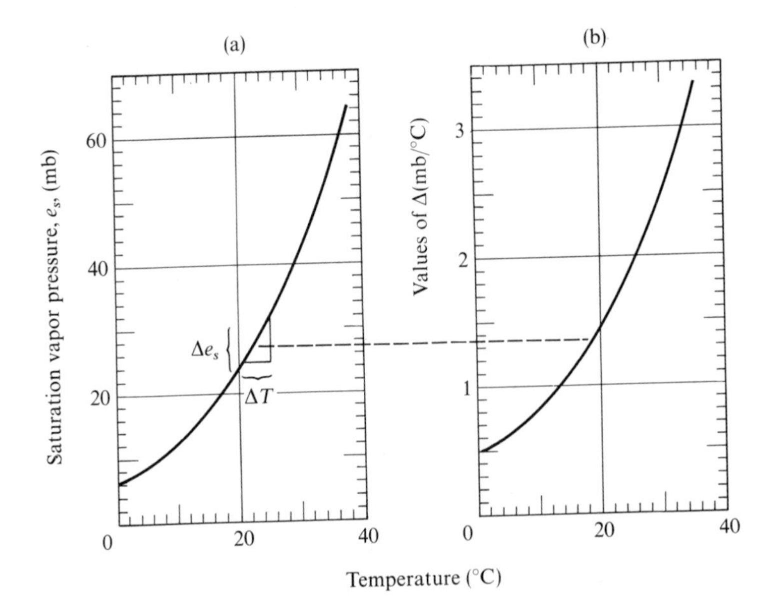

Figure 4-8 (a) Relation of saturation vapor pressure to temperature. (b) Variation of Δ (the slope of the curve in Figure 4-8(a)) with temperature.

vapor pressure to temperature (see Figure 4-8(b)), γ is known as the psychometric constant (0.66 mb/°C), and E_a is a term describing the contribution of mass-transfer to evaporation. This last term was determined empirically to be

$$E_a = (0.013 + 0.00016u_2)(e_{sa} - e_a) \tag{4-21}$$

where E_a is in cm/day, u_2 is the windspeed (km/day) measured at a height of two meters above the ground, e_{sa} (mb) is the saturation vapor pressure of a water surface at the air temperature, and e_a (mb) is the atmospheric vapor pressure.

As an aid to computation, the numerator and denominator of Equation 4-20 may be divided by γ, giving

$$E_o = \frac{\frac{\Delta}{\gamma}H + E_a}{\frac{\Delta}{\gamma} + 1} \tag{4-22}$$

The term Δ/γ is a function of temperature and is tabulated in Table 4-6.

Unlike Penman's sunken pan, there is significant transfer of heat by conduction and radiation through the walls and base of a Class A evaporation pan. Kohler et al. (1955), however, found that the combination of meteorological variables used by Penman could be related statistically to evaporation from a pan, and then, through the use of the pan coefficient, to

Table 4-6 Values of Penman's dimensionless parameter $\frac{\Delta}{\gamma}$ for various temperatures.

T(°C)	$\frac{\Delta}{\gamma}$	T(°C)	$\frac{\Delta}{\gamma}$
0	0.67	25	2.72
5	0.90	30	3.57
10	1.23	35	4.57
15	1.58	40	5.70
20	2.14		

evaporation from a lake. Their graph for computing lake evaporation is shown in Figure 4-9. Lamoreux (1962) developed from this graph an expression that can be used for the rapid processing of meteorological data by computer. His technique was used by Roberts and Stall (1966) for mapping lake evaporation throughout Illinois.

Maps of lake evaporation are useful in planning if only an approximate value is needed, or in the early stages of designing large reservoirs. Penman (1950) mapped evaporation over the British Isles with the use of his formula. Kohler et al. (1959) produced similar maps for the United States (see Figure 4-10). If the necessary meteorological stations are widely scattered, the computed evaporation may be correlated with some easily measured variable, such as elevation, which can then be used to interpolate evaporation values between meteorological stations (see Figure 4-11). In some water-management problems it is also useful to know the long-term temporal variation of evaporation, and Yu and Brutsaert (1969) generated a long record of evaporation from Lake Ontario with the aid of the meteorological record and the mass-transfer method. Evaporation is fortunately less variable than precipitation, runoff, and many other hydrologic processes. Its long-term average and variability, therefore, can be estimated closely from a relatively short record.

Evaporation from Bare Soils

Evaporation from fallow soils is important in dryland farming and in early-season irrigation. It is subject to the same controls as evaporation from an open water surface, and when the soil is wet, it is of approximately the same magnitude as water loss from a shallow buried pan. Penman (1948) found that evaporation from wet soil could be estimated closely by multiplying his open-water value in Equation 4-22 by 0.90. As the soil surface dries out, however, the remaining water is held tightly by capillary and osmotic forces in the soil, and the evaporation rate declines. Water must be evaporated

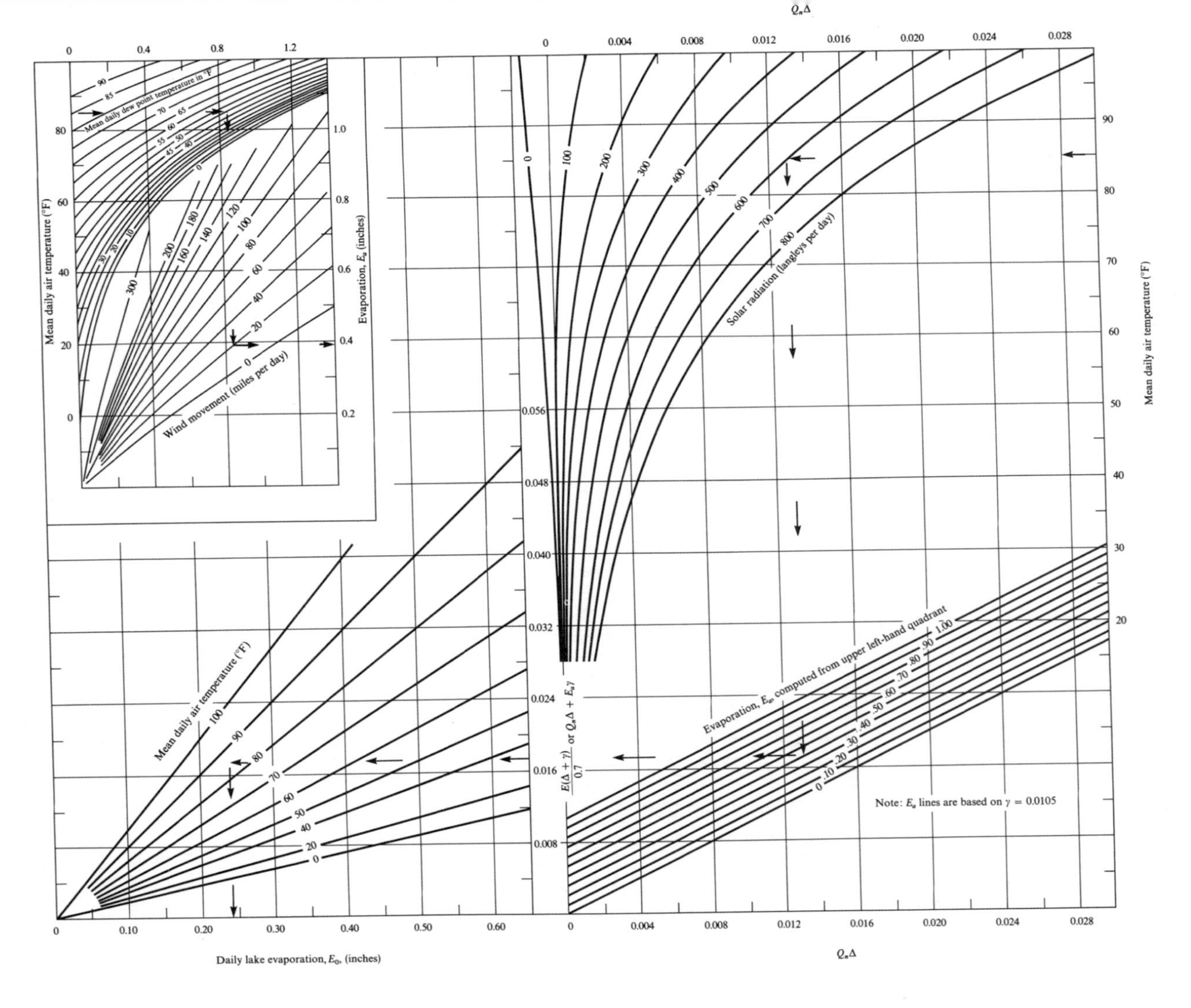

Mean daily air temperature (°F)
Evaporation, E_a (inches)
Wind movement (miles per day)
Mean daily dew point temperature in °F
$Q_n\Delta$
Solar radiation (langleys per day)
$\frac{E(\Delta + \gamma)}{0.7}$ or $Q_n\Delta + E_a\gamma$
Evaporation, E_a, computed from upper left-hand quadrant
Note: E_a lines are based on $\gamma = 0.0105$
Mean daily air temperature (°F)
Daily lake evaporation, E_0, (inches)

Figure 4-9 Computation of lake evaporation from meteorological data. To use the diagram: (1) enter upper left diagram with mean daily air temperature; (2) at mean daily dew-point temperature, read down to wind measurement; (3) read horizontally to right scale of E_a; (4) enter upper right diagram with mean daily air temperature, move left to value of solar radiation; (5) move downward to previously computed value of E_a in lower diagram; (6) thence left to lower left diagram to mean daily temperature; (7) thence downward to read answer, daily lake evaporation. The dew-point temperature is the temperature to which the atmosphere must be cooled before its water vapor will condense. It is therefore a measure of the vapor pressure and is routinely published with other weather records. (From Kohler et al. 1955.)

below the surface and the vapor must diffuse upward through the stagnant air in soil pores in order to reach the turbulent transfer regime of the atmosphere. This is a slow process, and usually the rate of evaporation from the soil declines to an insignificant value after 5 to 10 mm of water have been removed. If frequent rain or irrigation keeps the surface wet, the total evaporative loss is high. This is also true if the water table is close to the surface during dry periods. Otherwise a surface mulch of dry soil develops and greatly reduces the evaporation rate (Hide 1954).

Evaporation from Snow

Approximately 676 calories of energy are required to convert water from the solid state to the vapor state. The controls of this process are the same as those for evaporation from a free water surface, but there are some limitations that generally keep the evaporation from snow at a low rate.

The temperature of a snow surface has an upper limit of 0°C, and therefore its vapor pressure has an upper limit of 6.11 mb (see Figure 4-6(a)). Evaporation from a snowpack at 0°C can only take place, according to Equation 4-15, if the vapor pressure of the atmosphere is less than 6.11 mb, and this generally requires a very low relative humidity. If the term $(e_s - e_a)$ in Equation 4-15 is positive, and if there is sufficient energy being supplied by radiation or by the transfer of sensible heat from the atmosphere, evaporation will proceed. Because of the large amounts of energy required, however, these sources often do not supply enough energy to maintain evaporation and the deficit is made up by heat already stored in the snow. The snow surface is cooled by this process, and its vapor pressure falls to an even lower value, reducing the evaporation rate, usually to very low values. Williams (1961), for example, measured evaporative losses averaging 0.01 to 0.06 cm of water per day at Ottawa, Ontario. Dunne has measured 0.7 cm of evaporation in 15 days during the springtime melt season at Schefferville, Quebec. Only under conditions of bright sunshine, warm air temperatures, and strong winds does evaporation from snow become significant. When the snow temperature is below 0°C its vapor pressure is less

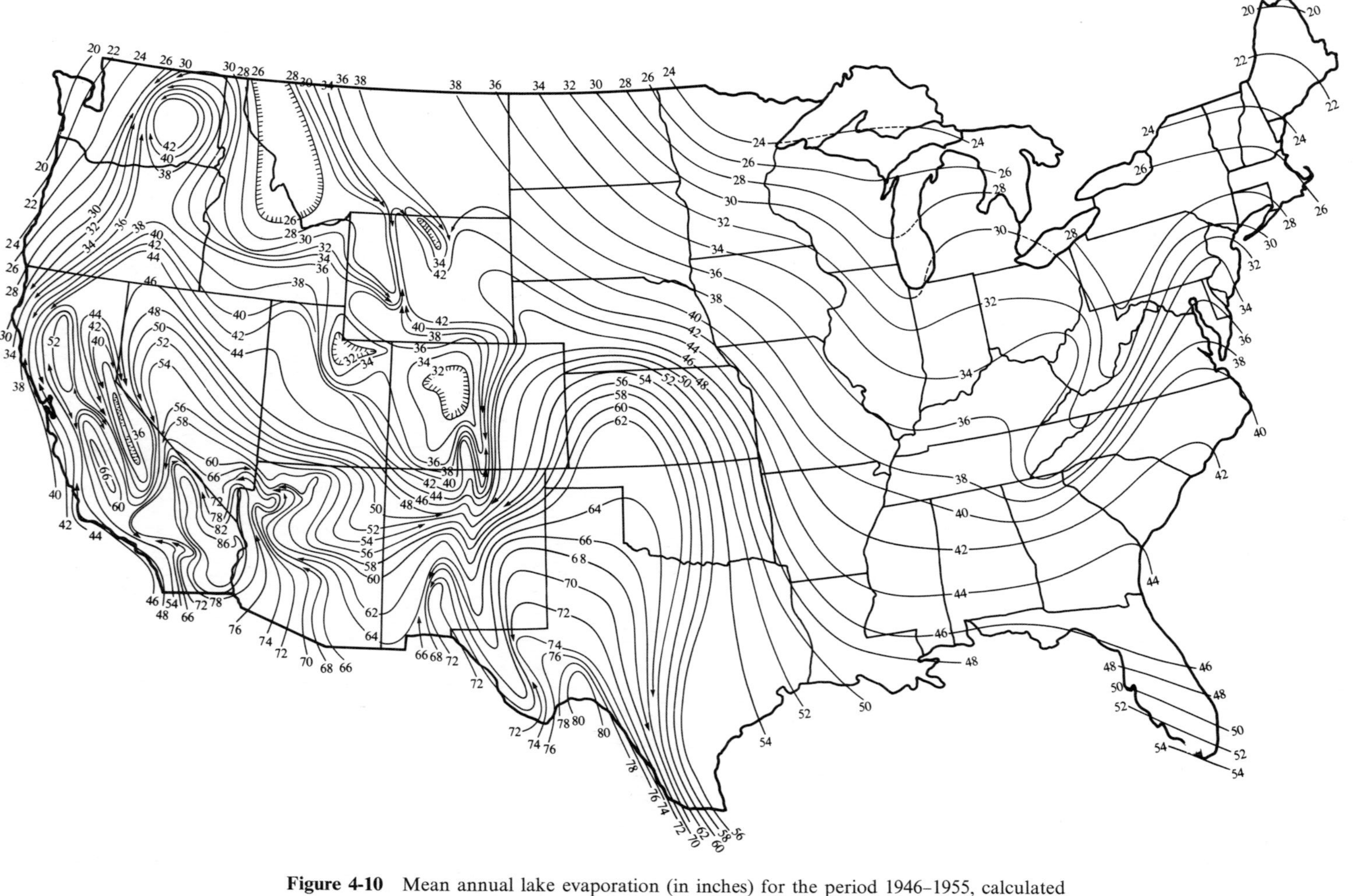

Figure 4-10 Mean annual lake evaporation (in inches) for the period 1946–1955, calculated with the aid of Figure 4-9. (From Kohler et al. 1959.)

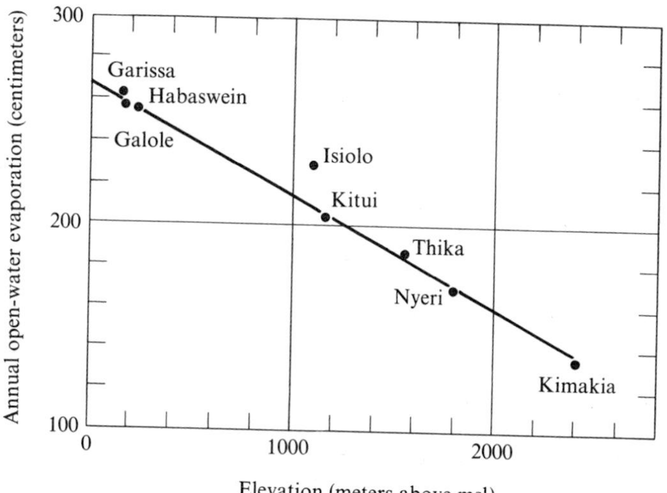

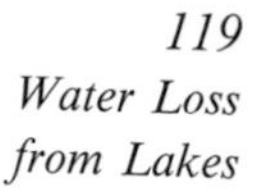

Figure 4-11 Average annual evaporation from open-water areas as a function of elevation in the Tana River Basin, Kenya. Computed from the Penman Equation 4-20 using data from the East African Meteorological Department. Names of meteorological stations are shown.

than 6.11 mb, and evaporation is even slower than under the conditions discussed above. The U.S. Army Corps of Engineers (1956) has suggested an equation of the form

$$E_s = 0.0231 z_a^{-0.33} u_a (e_s - e_a) \tag{4-23}$$

where E_s is evaporation in cm/day, z_a is the height (m) above the snow surface at which windspeed and vapor pressure are measured. Windspeed (u_a) is measured in km/day and vapor pressure in mb.

Kuzmin (1958) has proposed the formula

$$E_s = (0.018 + 0.00015 u_2)(e_s - e_a) \tag{4-24}$$

where the units are the same as in the preceding equation.

Bibliography

ANDERSON, E. R. (1954) Energy budget studies; *U.S. Geological Survey Professional Paper 269.*

BLACK, J. N., BONYTHON, C. W. AND PRESCOTT, J. A. (1954) Solar radiation and the duration of sunshine; *Quarterly Journal of the Royal Meteorological Society,* vol. 80, pp. 231–235.

CARTLEDGE, O. (1973) Solar radiation climate in a subtropical region; *Nature,* vol. 242, pp. 11–12.

CHANG, J. H. (1968) *Climate and agriculture: an ecological survey;* Aldine, Chicago, 296 pp.

DAVIES, J. A. (1966) The assessment of evapotranspiration for Nigeria; *Geografiska Annaler,* vol. 46A, pp. 139–156.

DALINSKY, J. S. (1971) The sinusoidal function of regional monthly average relative pan evaporation; *Water Resources Research,* vol. 7, pp. 677–687.

FERGUSON, H. L., O'NEILL, A. D. J. AND CORK, H. F. (1970) Mean evaporation over Canada; *Water Resources Research,* vol. 6, pp. 1618–1633.

FICKE, J. F. (1972) Comparison of evaporation computation methods, Pretty Lake, Lagrange County, northeastern Indiana, *U.S. Geological Survey Professional Paper 686-A.*

GLOVER, J. AND MCCULLOCH, J. S. G. (1958) The empirical relation between solar radiation and hours of sunshine; *Quarterly Journal of the Royal Meteorological Society,* vol. 84, pp. 172–175.

HAMON, R. W., WEISS, L. L. AND WILSON, W. T. (1954) Insolation as an empirical function of daily sunshine duration; *Monthly Weather Review,* vol. 82, pp. 141–146.

HARBECK, G. E. (1962) A practical field technique for measuring reservoir evaporation utilizing mass-transfer theory; *U.S. Geological Survey Professional Paper 272-E.*

HARBECK, G. E. AND MEYERS, J. S. (1970) Present day evaporation measurement techniques; *American Society of Civil Engineers, Journal of the Hydraulics Division,* vol. 96, HY7, pp. 1381–1390.

HARDING, S. T. (1942) Evaporation from free water surfaces; in *Hydrology* (ed. O. E. Meinzer), Dover, New York, pp. 56–82.

HELY, A. G., HUGHES, G. H. AND IRELAN, B. (1967) Hydrologic regimen of Salton Sea, California; *U.S. Geological Survey Professional Paper 486-C.*

HIDE, J. C. (1954) Observations on factors influencing the evaporation of soil moisture; *Soil Science Society of America Proceedings,* vol. 18, pp. 234–239.

HOUK, I. E. (1937) Water temperatures in reservoirs; *EOS, American Geophysical Union Transactions,* vol. 18, pp. 523–527.

HUGHES, G. E. (1967) Analysis of techniques used to measure evaporation from Salton Sea, California; *U.S. Geological Survey Professional Paper 272-H.*

KOHLER, M. A. (1954) Lake and pan evaporation; in *Water-loss investigations, Lake Hefner studies; U.S. Geological Survey Professional Paper 269,* pp. 127–148.

KOHLER, M. A., NORDENSON, T. J. AND FOX, W. E. (1955) Evaporation from pans and lakes; *U.S. Weather Bureau Research Paper 38.*

KOHLER, M. A., NORDENSON, T. J. AND BAKER, D. R. (1959) Evaporation maps for the United States; *U.S. Weather Bureau Technical Paper 37.*

KUZMIN, P. O. (1958) Hydrophysical investigations of land waters; *Bulletin of the International Association of Scientific Hydrology,* vol. 3, pp. 468–478.

LAMOREUX, W. W. (1962) Modern evaporation formulae adapted to computer use; *Monthly Weather Review,* vol. 90, pp. 26–28.

LANGBEIN, W. B. (1961) Salinity and hydrology of closed lakes; *U.S. Geological Survey Professional Paper 412.*

LANGBEIN, W. B., HAINS, C. H. AND CULLER, R. C. (1951) Hydrology of stock water reservoirs in Arizona; *U.S. Geological Survey Circular 10.*

MCKAY, G. A. AND STICHLING, W. (1961) Evaporation computations for prairie reservoirs; *National Research Council of Canada, Proceedings of Hydrology Symposium No. 2, Evaporation,* pp. 135–167.

MEYBOOM, P. (1967) Mass-transfer studies to determine the groundwater regime of permanent lakes in hummocky moraine of western Canada; *Journal of Hydrology,* vol. 5, pp. 117–142.

MEYERS, J. S. AND NORDENSON, T. J. (1962) Evaporation from seventeen western states; *U.S. Geological Survey Professional Paper 272-D.*

MORTON, F. I. (1967) Evaporation from large deep lakes; *Water Resources Research,* vol. 3, pp. 181–200.

PENMAN, H. L. (1948) Natural evaporation from open water, soil and grass; *Proceedings of the Royal Society of London,* Series A, vol. 193, pp. 120–145.

PENMAN, H. L. (1950) Evaporation over the British Isles; *Quarterly Journal of the Royal Meteorological Society,* vol. 76, pp. 372–383.

PENMAN, H. L. (1956) Estimating evaporation; *EOS, American Geophysical Union Transactions,* vol. 31, pp. 43–50.

PHILLIPS, K. N. AND VANDENBURGH, A. S. (1971) Hydrology and geochemistry of Abert, Summer and Goose Lake and other closed-basin lakes in south-central Oregon; *U.S. Geological Survey Professional Paper 502-B.*

ROBERTS, W. J. AND STALL, J. B. (1966) Computing lake evaporation in Illinois, *Water Resources Research,* vol. 2, pp. 205–208.

SEFCHOVICH, E. (1970) The preliminary thermal analysis of a body of water in power plant siting; in *Environmental effects of thermal discharges;* American Society of Mechanical Engineers, New York, pp. 19–25.

SZEICZ, G., ENDRODI, G. AND TAJCHMAN, S. (1969) Aerodynamic and surface factors in evaporation; *Water Resources Research,* vol. 5, pp. 380–394.

TURK, L. J. (1970) Evaporation of brine: a field study on the Bonneville Salt Flats, Utah; *Water Resources Research,* vol. 6, pp. 1209–1215.

TURNER, J. G. (1966) Evaporation study in a humid region, Lake Michie, North Carolina; *U.S. Geological Survey Professional Paper 272-G.*

U .S. ARMY CORPS OF ENGINEERS (1956) *Snow hydrology;* Portland, OR., 433 pp.

U.S. GEOLOGICAL SURVEY (1954) Water-loss investigations: Lake Hefner studies; *U.S. Geological Survey Professional Paper 270.*

U.S. GEOLOGICAL SURVEY (1958) Water-loss investigations: Lake Mead studies; *U.S. Geological Survey Professional Paper 298.*

U.S. SOIL CONSERVATION SERVICE (1969) *Engineering field manual for conservation practices;* Washington, DC.

U.S. WEATHER BUREAU (1949) Weekly mean values of daily total solar and sky radiation; *Technical Paper 11.*

U.S. WEATHER BUREAU (1950) Mean monthly and annual evaporation from free water surface for the United States, Alaska, Hawaii and the West Indies; *Technical Paper 13.*

U.S. WEATHER BUREAU (1951) Sunshine and cloudiness at selected stations in the United States, Alaska, Hawaii and Puerto Rico; *Technical Paper 11.*

VEIHMEYER, F. J. (1964) Evapotranspiration; Section 11 in *Handbook of applied hydrology* (ed. Ven te Chow), McGraw-Hill, New York.

WEBB, M. S. (1970) Monthly mean surface temperatures for Lake Ontario as determined by aerial survey; *Water Resources Research,* vol. 6, pp. 943–965.

WILLIAMS, G. P. (1961) Evaporation from water, snow and ice; *National Research Council of Canada, Proceedings of Hydrology Symposium No. 2, Evaporation,* pp. 31–45.

YU, S. L. AND BRUTSAERT, W. (1969) Generation of an evaporation time series for Lake Ontario; *Water Resources Research,* vol. 5, pp. 785–796.

Typical Problems

4-1 Pan and Lake Evaporation

Using Figure 4-3 and a pan coefficient, estimate the annual amount of evaporation from a proposed reservoir with a surface area of 100,000 acres in the vicinity of the Grand Canyon, northwestern Arizona. Subtract from this loss the precipitation falling onto the lake (12 inches/yr). Assuming a value for municipal water consumption of 150 gallons per person per day, what would be the size of the population that could be supplied by a water resource of the size that would be lost from the reservoir?

4-2 Calculation of the Radiation Balance over a Lake

Using the data in Table 4-7, calculate the mean monthly solar radiation, net longwave radiation, and net allwave radiation over a lake. The use of the resulting calculation will be illustrated in Typical Problem 4-4. The data are for Mokpo, Republic of Korea (latitude 35°N).

Both cloudiness and duration of sunshine are observed at this station, providing an opportunity for calculating solar radiation by two separate methods. Maximum possible duration of sunshine was obtained from Table 4-4. Air temperature in degrees Kelvin was obtained by adding 273° to the Celsius values. Vapor pressure was calculated from relative humidity (RH) measured with a hygrothermograph. The formula $RH = 100e_a/e_{sa}$ was used, where e_{sa} is the saturation vapor pressure at the temperature of the atmosphere, obtained from Figure 4-8(a).

Table 4-7 Climatological values for Mokpo, Republic of Korea. (Data from the Central Meteorological Office, Seoul, Republic of Korea.)

MONTH	CLOUDINESS (DECIMAL FRACTION)	MEASURED DURATION OF SUNSHINE (HR)	MAXIMUM POSSIBLE SUNSHINE DURATION (HR)	AIR TEMPERATURE (°K)	VAPOR PRESSURE (MB)
Jan.	0.6	125	313	273	4.4
Feb.	0.5	150	307	274	4.4
Mar.	0.5	180	370	279	6.7
Apr.	0.7	150	394	285	11.3
May	0.5	252	435	291	14.4
June	0.6	231	435	295	19.8
July	0.6	159	442	298	26.8
Aug.	0.3	293	418	301	31.5
Sept.	0.3	260	372	296	20.5
Oct.	0.3	259	349	290	12.9
Nov.	0.6	118	310	283	8.6
Dec.	0.7	97	303	274	4.9

Solution

For January the following calculations can be made.

1. From Equation 4-8, using values of I_o from Table 4-2,

$$Q_s = 428[0.803 - 0.340(0.6) - 0.458(0.36)]$$
$$= 186 \text{ cal/cm}^2\text{/day}$$

As a check, from Equation 4-9, using 0.23 and 0.48 for a and b,

$$Q_s = 428\left[0.23 + 0.48\left(\frac{125}{313}\right)\right]$$
$$= 180 \text{ cal/cm}^2\text{/day}$$

2. Net longwave radiation for January can be calculated from Equation 4-12:

$$Q_{lw} = (1.17 \times 10^{-7} \times 273^4)(0.56 - 0.08\sqrt{4.4})$$
$$\times [0.10 + 0.9(0.6)]$$
$$= 163 \text{ cal/cm}^2\text{/day}$$

3. Using Equation 4-13 with a value of $\alpha = 0.05$ for water, average daily net allwave radiation for the month is given by

$$Q_n = 186(1 - 0.05) - 163 = 14 \text{ cal/cm}^2\text{/day}$$

The reader should continue these calculations for the other months and plot an annual graph of each radiation flux. Alternatively local station values can be obtained from the records filed in most college libraries.

4-3 Mass-Transfer Calculation of Evaporation

You are planning a small urban lake for flood detention and for enhancement of a residential development and are concerned about whether enough water flows into the lake in summer to offset evaporation. A pilot study of evaporation rates on a nearby small lake on the same geologic material is advisable because the urban area is unlike the rural regions surrounding the lakes referred to in Figure 4-7. You install a staff gauge in the lake and wait until inflow and outflow cease in the summer. A set of observations of the daily fall in the water level are made (Table 4-8).

You also measure windspeed, water surface temperature (T_s), air temperature (T_a), and relative humidity (with a hygrothermograph) at a station near the lake. The vapor pressure of the water surface is obtained from T_s and Figure 4-8(a). Vapor pressure is obtained from relative humidity using the formula $RH(\%) = 100e_a/e_{sa}$, where e_{sa} is the saturation vapor pressure of the

Table 4-8 Field data on water loss from a small lake.

MEASUREMENT PERIOD	FALL IN WATER LEVEL (CM/DAY)	WINDSPEED (KM/DAY)	$(e_s - e_a)$ (MB)
1	0.30	46	5.0
2	0.43	135	5.7
3	0.48	112	8.8
4	0.48	77	14.8
5	0.56	147	10.5
6	0.58	85	17.2
7	0.76	216	11.6
8	0.86	185	16.3
9	0.94	348	11.0
10	1.22	429	12.3

atmosphere at the measured air temperature. If a weather station is located near the lake, all the data except the water surface temperature could be obtained from that facility. The field data are shown in Table 4-8.

Plot column 2 against the product of columns 3 and 4 for each measurement period. Estimate the mass-transfer coefficient, N, from the slope of the line, and the rate of seepage from the intercept on the ordinate for an abscissa value of zero.

If you were to repeat the exercise at another season, you might obtain a different seepage rate. Why?

How would you use the mass-transfer coefficient and seepage rate to answer the original question?

4-4 Use of Penman's Evaporation Formula

You need to calculate the evaporation from a shallow recreational lake proposed for a valley in a mountainous region where the geographic variation of climate is too intricate to be shown on generalized maps such as Figure 4-10. A good weather record is available at a town in the valley, providing you with the long-term average data in Table 4-9.

The net radiation values would be calculated as shown in Typical Problem 4-2. When these are divided by the latent heat of vaporization (590 cal/gm), H in Equation 4-22 is obtained. Values of Δ/γ are read from Table 4-6 using the air temperature each month. To find the E_a term in Penman's equation, e_{sa} must be read from Figure 4-8(a) for the appropriate air temperature. The values can then be tabulated as shown below, and E_o is computed from Equation 4-22. The reader can continue the calculation as tabulated and should obtain an annual total evaporation of approximately 215 cm.

MONTH	H (CM/DAY)	Δ/γ	E_a (CM/DAY)	E_o (CM/DAY)	E_o (CM/MO)
Jan.	0.19	1.16	0.43	0.30	9.33

Table 4-9 Climatological data for the Penman Formula.

MONTH	NET RADIATION OVER THE LAKE (CAL/CM^2/DAY)	MEAN 2-M WINDSPEED (KM/DAY)	MEAN AIR TEMPERATURE (°C)	MEAN 2-M VAPOR PRESSURE (MB)	SATURATION VAPOR PRESSURE AT AIR TEMPERATURE (MB)
Jan.	110	415	9	6.0	11.4
Feb.	135	420	10	6.3	12.0
Mar.	290	440	11	6.8	13.0
Apr.	460	425	14	8.4	15.8
May	610	393	18	10.4	20.3
June	630	382	24	11.5	29.5
July	690	359	22	13.0	26.0
Aug.	550	344	17	12.7	19.0
Sept.	390	351	15	12.6	16.7
Oct.	200	378	13	10.4	14.7
Nov.	70	394	11	8.6	13.0
Dec.	55	417	10	7.5	12.0

Using the following data, evaluate the daily lake evaporation from Figure 4-9.

Mean daily air temperature 25°C

Mean daily dew-point temperature 15°C

Mean daily windspeed 110 km/day

Mean daily solar radiation 600 langleys/day; i.e., cal/cm²/day.

Solution

Convert the data to the English units in Figure 4-9.

(25°C × 1.8) + 32 = 77°F

(15°C × 1.8) + 32 = 59°F

110 km/day = 69 mi/day

Entering the diagram with these values and following the arrows as described in the figure caption yields E_o = 0.215 in/day, or 0.55 cm/day.

5

Water Use by Vegetation

Definitions and Process

The annual amount of water leaving a drainage basin as runoff varies from less than 10 percent of the yearly precipitation in hot deserts to more than 90 percent in the Cascade Mountains of Washington. The difference between rainfall and runoff is largely explained by evapotranspiration. Because of its magnitude and its importance for plant growth, this link in the hydrologic cycle is so tremendously important to ecology, economic activity, and human welfare that it is difficult to understand why many texts on hydrology give the subject such cursory attention.

Transpiration is the loss of water from the cuticle or the stomatal openings in the leaves of plants. Water is vaporized within the leaf in the intercellular spaces and passes out of the stomata by molecular diffusion. The stomata are pores on the undersurface of a leaf. They open in sunshine, allowing the diffusion of carbon dioxide into the leaf during photosynthesis, and when they are open, water vapor can diffuse from wet cells into the atmosphere. This transpired water is replaced by water taken into the roots of the plant from the soil. When computing water loss from a vegetated surface, it is usually impossible to separate transpiration and evaporation from the soil surface, ponds, lakes, and rivers. The two processes are usually considered together under the title of evapotranspiration. This term is broadly synonymous with consumptive use in agronomy, a term that refers to the amount of water required to mature a crop.

When a vegetated surface is losing water to the atmosphere at a rate unlimited by deficiencies of water supply, the process is known as potential evapotranspiration. If water is in limited supply at some time during the year, actual evapotranspiration may be less than the potential rate. Penman (1948, 1961, 1963) has stressed that potential evapotranspiration is largely controlled by weather, with vegetation and soil factors playing only a minor role. Much of this emphasis stems from Penman's original definition of potential evapotranspiration as the evaporation from a short green crop completely shading the ground and with a nonlimiting supply of water. By specifying such stringent conditions, Penman eliminated soil and vegetation parameters from his analysis. The extension of his concepts to other crops and to forests has required the incorporation of variables describing cover characteristics. Vegetation and soil factors become increasingly important when water supply to the plant is limited. Actual evapotranspiration under dry conditions is a complicated process controlled by weather, vegetation, and soils. It is not fully understood at present, and computation procedures are the subject of lively debate.

Importance of Evapotranspiration

Through transpiration, plants control their temperature and perform other vital functions. A quantitative appreciation of the process, therefore, is fundamental to understanding plant growth and distribution. On a very practical level, the measurement or calculation of water use forms a basis for choosing crops that can be grown successfully in areas of unreliable rainfall (Dagg 1965). The calculation of irrigation water requirements is based on estimating evapotranspiration, both in the planning phase of a project and in the day-to-day control of water supply at the farm level (U.S. Soil Conservation Service 1970).

As a major subtraction of water from drainage basins, evapotranspiration dominates the water balance and controls such hydrologic phenomena as soil moisture content, groundwater recharge, and streamflow. More than two-thirds of the precipitation falling on the conterminous United States is returned to the atmosphere by evaporation from plants and water surfaces. In parts of Africa this proportion exceeds 90 percent. Although evapotranspiration is necessary for the growth of plants, it is usually viewed as a "loss" from the water budget in that it reduces the amount of streamflow, lake storage, and groundwater available for direct human use. Much attention has been given to methods of reducing this withdrawal in order to increase streamflow.

Some municipalities remove all trees from parts of their water-supply catchments on a rotational basis. The resultant reduction of evapotranspiration increases groundwater recharge and streamflow. Replacement of one type of vegetation with another can also change the transpirative loss. Along river floodplains grow phreatophytes, plants that withdraw their water supply from saturated soils. In arid regions this depletion of groundwater flow into streams significantly reduces water supplies. In the arid western United States phreatophytes cover 6.5 million hectares of valley floor and transpire over 30 billion cubic meters of water per annum. In a particular valley this discharge can be equivalent to a runoff yield of 0.02 $m^3/sec/km^2$ of drainage basin area, and may amount to 20 percent or more of the dry-weather yield of some streams.

For many years studies have been under way in the western United States to eradicate woody phreatophytes and replace them with a more productive vegetation cover such as alfalfa. Demonstration projects have produced significant gains in water yield, though at high cost. The gains are not without their drawbacks, since the phreatophyte woodlands provide good habitat for wildlife. The problem is discussed in detail by Dunford and Fletcher (1947) for humid regions, and by Robinson (1958), Van Hylckama (1963), and Culler (1970) for arid areas. The papers present valuable data on a topic we will discuss at the end of this chapter: the differences in evapotranspiration between cover types.

As the value of water increases, planners should be aware of the technical issues involved in manipulating vegetation for water yield. For this reason, and because of the tremendous importance of irrigation in many rural planning schemes, we analyze the methods for computing evapotranspiration in some detail in this chapter.

Measurement of Evapotranspiration: Evaporation Pans

Several methods of measuring evapotranspiration have been used. Some techniques indicate only potential evapotranspiration rates, while others measure actual evapotranspiration, which may or may not be at the potential rate.

Evaporation pans, such as the Class A pan, probably provide the best method of obtaining an index of potential evapotranspiration. They are more accurate than formulae, especially for indicating local and short-term fluctuations under conditions of strong heat advection. The pans are also cheap and easy to use. A coefficient relating pan evaporation to the rate of water use by a particular plant cover must be evaluated from some other direct method of measuring potential evapotranspiration. Figure 5-1 shows the variation of this coefficient through the growing season for a corn crop. Chang (1968) presents similar data for many other crops.

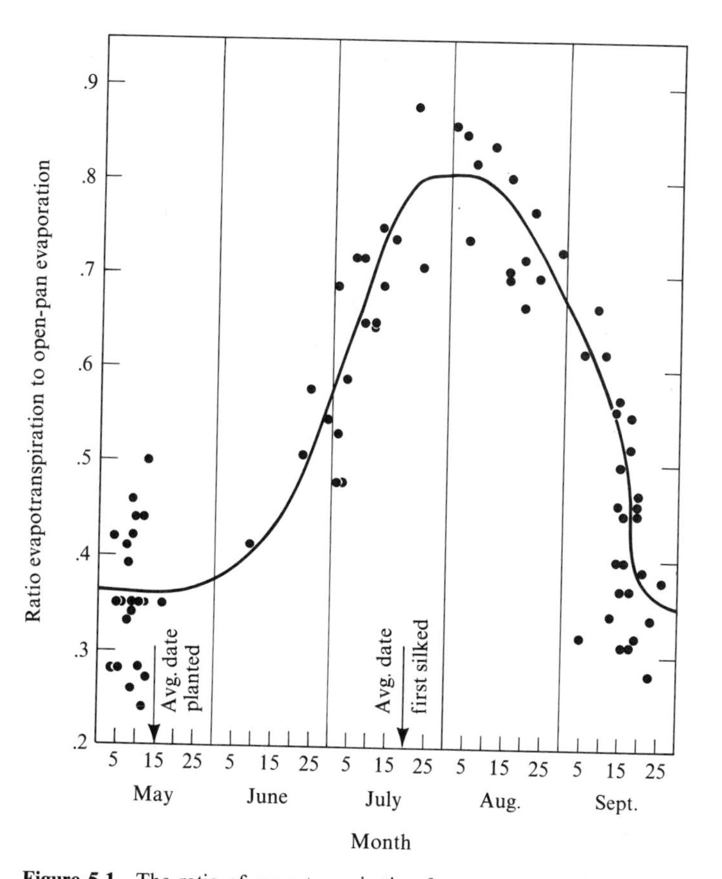

Figure 5-1 The ratio of evapotranspiration from corn to open-pan evaporation as a function of time during the growing season. (Reproduced from Denmead and Shaw, 1959, *Agronomy Journal,* vol. 51, pp. 725–726, by permission of the American Society of Agronomy.)

Measurement of Evapotranspiration: Lysimeters

A direct measurement of evapotranspiration can be obtained by evaluating the water balance for a block of soil. The soil is held in a porous-bottomed tank, called a lysimeter, buried in the ground. The tank itself should be as large and deep as possible to reduce boundary effects and to avoid restricting plant development. Lysimeters described in the literature vary in capacity from less than 1 cubic meter to more than 150 cubic meters. The soil profile, root development, and moisture conditions should be the same inside and outside the tank. Above the surface, the type, height, and density of vegetation should be the same as its surroundings. If the soil remains or is maintained in a wet condition, evapotranspiration will be measured at

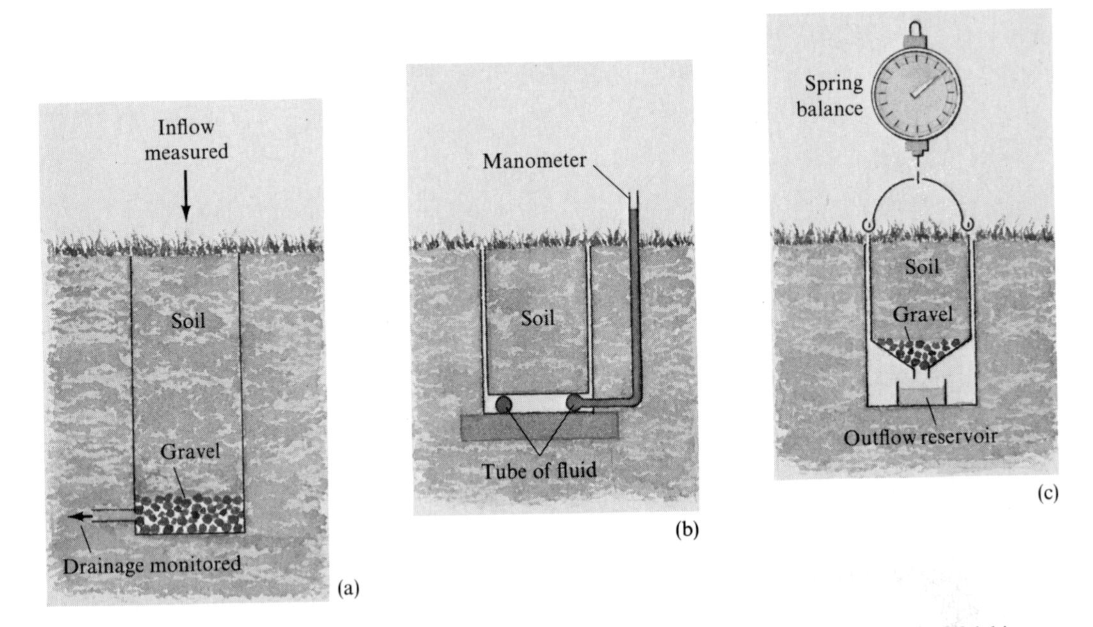

Figure 5-2 Schematic diagrams of types of lysimeters: (a) Drainage type. (b) Weighing float type. (c) Spring-balance weighing type.

potential rates. If the aim is to measure actual evapotranspiration, the soil moisture content of the lysimeter is allowed to fluctuate like that of the soil outside.

Lysimeters are generally classified as weighing types or drainage (non-weighing) types. In the drainage type (see Figure 5-2(a)) the water balance of the soil is assumed to be

$$\text{Evapotranspiration} = \text{Precipitation} + \text{Irrigation} - \text{Drainage}$$

The inflows and drainage are measured, but changes in storage within the soil are not measured.* The drainage-type lysimeter is, therefore, useful only during periods when the change in storage can be considered negligible. The length and timing of such periods will depend upon the frequency of wetting, the size of the lysimeter, and the rate of movement of water through the soil. This type of instrument is often irrigated to ensure that evapotranspiration can proceed at the potential rate. A buffer zone around the instrument should also be irrigated. Small drainage-type lysimeters are often referred to as evapotranspirometers.

---

*Changes of soil moisture storage could be measured by electrical or neutron-scattering techniques described in a later chapter, but usually such measurements are not made.

Weighing-type lysimeters allow an evaluation of each element in the water balance equation:

$$\text{Evapotranspiration} = \text{Precipitation} + \text{Irrigation} - \text{Drainage} \pm \text{Changes in storage}$$

Changes in storage are measured by weighing the soil block, which can be accomplished manometrically, as shown in Figure 5-2(b), or with an elaborate scale system (see Pelton 1961 for a review). Other, less elaborate weighing lysimeters can be made, as illustrated in Figure 5-2(c). The large weighing lysimeters are extremely expensive instruments used for research, especially for checking theoretical models of the evapotranspiration process. The other instruments, though less accurate, can be deployed in large numbers to obtain direct measurements in a wide range of environments. Like the drainage type, they can be managed to measure potential or actual evapotranspiration.

Measurement of Evapotranspiration: Plot Studies

Instead of enclosing the soil block in a lysimeter, it is possible to evaluate the water balance of undisturbed plots. The methods of measurement will vary with local details of soil and climate. For example, if the soil at some shallow depth is almost completely impervious, drainage can be collected by a subsurface drain. If, on the other hand, the water table is so deep that percolating water does not reach the groundwater zone but is stored temporarily in the root zone before being returned to the surface by plants, drainage is known to be zero. Changes in storage are evaluated by periodic measurements of vertical soil moisture profiles with instruments described in Chapter 6. The plots can be managed for measurement of either potential or actual evapotranspiration. Evaluation of the consumptive use of irrigated crops by measuring precipitation, irrigation, and return flow from the fields is an example of a plot study.

Measurement of Evapotranspiration: Catchment Studies

The water balance of even a small drainage basin is difficult to measure directly. There are uncertainties of precipitation input, referred to in a previous chapter, and difficulties of estimating areal averages of water stored in the soil and the groundwater body. Many drainage basins lose or gain water from neighboring catchments by groundwater seepage. Drainage basin estimates of actual evapotranspiration are usually made only for long periods, such as a year or several years when storage changes can be assumed to be negligible. For small catchments, seasonal or monthly estimates are possible if soil moisture and groundwater changes are monitored. The technique is described by Pegg and Ward (1971) for a small catchment in northeast England.

Calculation of Potential Evapotranspiration: Energy-Balance Approach

The energy budget for a vegetated surface can be written as

$$Q_s - Q_{rs} - Q_{lw} + Q_v = Q_{et} + Q_h + Q_\theta \tag{5-1}$$

where Q_s = incoming solar radiation
$Q_{rs} = \alpha Q_s$ = reflected solar radiation
α = albedo (reflectivity of the vegetative cover)
Q_{lw} = net longwave radiation from the vegetated surface to the atmosphere
Q_v = net energy advected to crop
Q_{et} = energy used for evapotranspiration
Q_h = energy transferred from crop to air as sensible heat
Q_θ = changes of energy stored in heating soil and crop.

All the units of Equation 5-1 are calories per square centimeter of ground surface, and the equation is the same as Equation 4-1, except that the surface is not open water.

The radiative terms in Equation 5-1 can be determined by the methods described in the previous chapter on evaporation. The situation here is a little more complex because albedo varies between vegetation types and between seasons for the same crop. Table 5-1 presents a sample of published values of albedo. Net longwave radiation is usually evaluated by means of Equation 4-12 and its variants.

The relation of incoming solar to net allwave radiation is also different from that for a water surface. Davies (1967) compiled data from 14 stations throughout the world from the arctic to the tropics. For the daytime hours he found the following relationship for vegetated surfaces (grass and crops with an albedo of 0.20–0.30):

$$Q_n = 0.62 Q_s - 24 \text{ cal/cm}^2\text{/day} \tag{5-2}$$

where Q_n is the net radiation.

Szeicz et al. (1969) found that this relationship varied somewhat with vegetation type, and presented the following equations:

Table 5-1 Some published values of mean daily albedo for various surfaces. The range given for crops represents the variation of reflectivity from bare ground or low growth to full cover.

TYPE OF SURFACE	ALBEDO	LOCALITY	SOURCE
Water	0.05–0.10	Various	Various
Bare soil (wet)	0.11	W. Europe	Monteith (1959)
Bare soil (dry)	0.18	W. Europe	Monteith (1959)
Spruce forest	0.05–0.08	W. Europe	Baumgartner (1967), Tajchman (1971)
Pine forest	0.10–0.12	W. Europe	Szeicz et al. (1969)
Bamboo forest	0.12	Kenya	EAAFRO
Evergreen forest	0.14	Kenya	EAAFRO
Tropical hardwood forest	0.18	Kenya	EAAFRO
Pineapple	0.05–0.08	Hawaii	Chang (1968)
Sugar cane	0.05–0.18	Hawaii	Chang (1968)
Tea	0.16	Kenya	EAAFRO
Potatoes	0.15–0.27	W. Europe	Various
Rye and wheat	0.10–0.25	W. Europe	Monteith (1959)
Corn (maize)	0.12–0.24	N. America	Chang (1968)
Sugar beet	0.14–0.25	W. Europe	Monteith (1959)
Short grass	0.14–0.25	Various	Various
Cotton	0.17–0.25	Various	Various
Alfalfa (lucerne)	0.19–0.25	Various	Various
Kale	0.19–0.28	W. Europe	Chang (1968)
Green vegetables	0.25	N. America	Monteith (1959)

For potatoes and lucerne ($\alpha = 0.25$):

$$Q_n = 0.66Q_s - 50 \tag{5-3}$$

which is close to the equation of Davies.

For pine forest ($\alpha = 0.11$):

$$Q_n = 0.83Q_s - 54 \tag{5-4}$$

For the nighttime hours, the net radiation must be calculated as the net longwave loss in Equation 4-12.

Advected energy (Q_v) is usually small when a large area of uniform natural vegetation is being considered. There are, however, situations in which this source of energy can dominate evapotranspiration. Any well-watered zone in the midst of an arid landscape will receive large fluxes of sensible heat from its surroundings. In considering irrigation need, water losses by natural valley-bottom vegetation, or by hydrophytes in prairie potholes, the energy advected by dry, hot winds must be evaluated separately. Nixon and Lawless (1968) showed that the opposite situation arises in the maritime region of coastal California, where cool oceanic air lowers evapotranspiration to 80–90 percent of that in humid interior valleys with the same net radiation on clear days. In general, however, Q_v will be small and thus will be neglected in the present discussion.

Changes of energy stored in the plant and soil (Q_θ) can also be ignored over periods of a day or longer. Penman (1961) calculated that over a four-month summer period in southeast England only 3 percent of the energy received as net radiation was used to heat the soil and make the plant grow.

With these simplifications, Equation 5-1 becomes

$$Q_n = Q_{et} + Q_h \tag{5-5}$$

which is the simplified energy balance given in Equation 4-18 from which Penman developed his expression for evaporation from an open water surface in a small sunken pan. Penman, in fact, conducted his research with the aim of predicting evapotranspiration from a vegetative cover. Water provided a convenient, reproducible surface of known properties. Referring back to Equation 4-22, Penman's expression can be written:

$$E_o = \frac{\frac{\Delta}{\gamma} H + E_a}{\frac{\Delta}{\gamma} + 1} \tag{5-6}$$

where E_o refers to evaporation from an open water surface.*

In order to use this expression as a predictor of water loss from vegetation, Penman then determined the potential evapotranspiration from a short grass cover by the water-budget method. He found that the ratio of potential evapotranspiration (E_t) to E_o varied somewhat with the length of daylight. In southeast England, summer values of the ratio were approximately 0.8, while in winter they averaged 0.6. He suggested 0.75 as the value most appropriate to the whole year. The ratio should converge to 0.7 in humid tropical regions, and to a value less than 0.7 in the semi-arid tropics. Pereira (1967) confirmed the value of 0.75 for a canopy of young pines in humid Kenya, though for mature evergreen rain forest in the same area, the ratio was 0.90. For crops whose height, density, and reflectivity change with the season, this ratio must be evaluated empirically. Pereira (1967), for instance, refers to ratios of E_t/E_o ranging from 0.5 to 0.8 for coffee bushes at different times of the year.

Another way of using the Penman Formula is to calculate potential evapotranspiration (E_t) directly by calculating the net radiation term ap-

---

*Because γ is not strictly a constant but is a function of atmospheric pressure, a correction must be applied for very high altitude catchments. McCulloch (1965) provides tables for making this correction, which makes a difference of 10 percent in the calculated value of E_o for East African catchments 3500 m above sea level.

propriate to the vegetation cover by use of Equation 4-13 with albedo values from Table 5-1, or from Equation 5-2 or its variants. Typical Problem 5-1 illustrates this method.

The mass-transfer term in Penman's equation has also undergone refinements which allow it to reflect differences between cover types. Van Bavel (1966), for example, circumvented the empiricism in Penman's value of E_a by deriving an expression based on the theory of turbulent transfer of water vapor between the vegetation cover and the atmosphere. His expression is

$$E_a = \frac{3.64}{T_a} \frac{u_a}{\left[\ln\left(\frac{z_a}{z_o}\right)\right]^2}(e_{sa} - e_a) \tag{5-7}$$

where E_a is the component of evapotranspiration due to mass transfer of water vapor, in cm/day, T_a is air temperature in °K, u_a is windspeed in km/day, the vapor pressure terms are as defined previously, and the subscript a refers to the height z_a (cm) above the vegetation cover at which all measurements are made. The parameter z_o characterizes the aerodynamic roughness of the surface to the wind. It is often called the *roughness length* of the surface, and as it increases so does the term E_a, because the turbulent transfer of water vapor into the atmosphere is more vigorous over a rough surface than over a smooth one. Our survey of literature of measured values of z_o indicates that the roughness parameter is approximately equal to one-tenth the height of the vegetation, and unless observations are available over the particular crop of interest, this approximation will be adequate for most purposes. Van Bavel (1966), Baumgartner (1967), Chang (1968), Szeicz et al. (1969), and Tajchman (1971) give values for particular crops and illustrate this refinement of the Penman Equation.

Calculation of Potential Evapotranspiration from Air Temperature

In view of the previous discussion, the use of empirical formulae based upon single climatic parameters is approximate, but sometimes necessary. Errors often result from using such formulae outside the climatic regime for which they were developed. On the other hand, their minimal data requirements make them useful for areas in which detailed meteorological records are lacking. Even in well-instrumented areas, an empirical equation may be useful for the rapid calculations necessary in operating irrigation schemes or other water management systems. When used for such purposes, the empirical methods should be checked against direct measurements or energy-balance methods. Many attempts have been made to compare the performance of various empirical formulae with one another or with direct measurements in pans, lysimeters, or drainage basins. It is difficult to state anything general about the findings, since results obtained in different areas do not usually agree (Dagg and Blackie 1970, McGuinness and Bordne

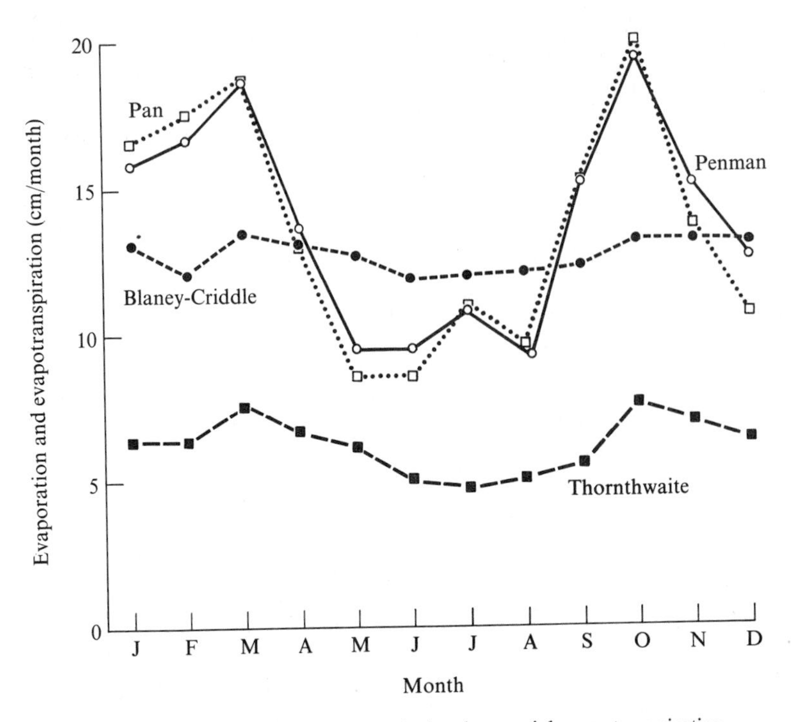

Figure 5-3 Measured pan evaporation and calculated potential evapotranspiration at Muguga, Kenya, for 1963. The curve marked "Pan" is measured evaporation. The other three curves are computed values of evapotranspiration using different methods. (After Dagg and Blackie 1970.)

1972). The methods based upon air temperature work best in the regions for which they were developed, namely, midlatitude continental climates, where air temperature is a fairly good index of net radiation. In the tropics, however, these methods often give erroneous results, and may seriously underestimate the amplitude of seasonal fluctuations of water demand (see Figure 5-3). In such areas it is preferable to use the energy-balance approach even if radiation must be estimated. In the tropical world, even data on wind and vapor pressure are relatively rare, but this problem is reduced by the fact that in the tropics the radiation term in the Penman Equation is usually dominant. The temperature methods are still in use, however, and planners concerned with rural areas should be familiar with them.

The Thornthwaite Method

The Thornthwaite method uses air temperature as an index of the energy available for evapotranspiration, assuming that air temperature is correlated with the integrated effects of net radiation and other controls of evapotranspiration, and that the available energy is shared in fixed proportion between heating the atmosphere and evapotranspiration. There is no correction for different vegetation types. The empirical formula Thornthwaite developed is

$$E_t = 1.6\left[\frac{10T_a}{I}\right]^a \tag{5-8}$$

where E_t = potential evapotranspiration in cm/mo
T_a = mean monthly air temperature (°C)

$$I = \text{annual heat index} = \sum_{i=1}^{12}\left[\frac{T_{ai}}{5}\right]^{1.5} \tag{5-9}$$

$$a = 0.49 + 0.0179I - 0.0000771I^2 + 0.000000675I^3. \tag{5-10}$$

Figure 5-4 can be used for the evaluation of Thornthwaite's E_t value, as described in the caption. The annual heat index, I, can be estimated directly from Figure 5-5, at least for stations in the United States. The relationship

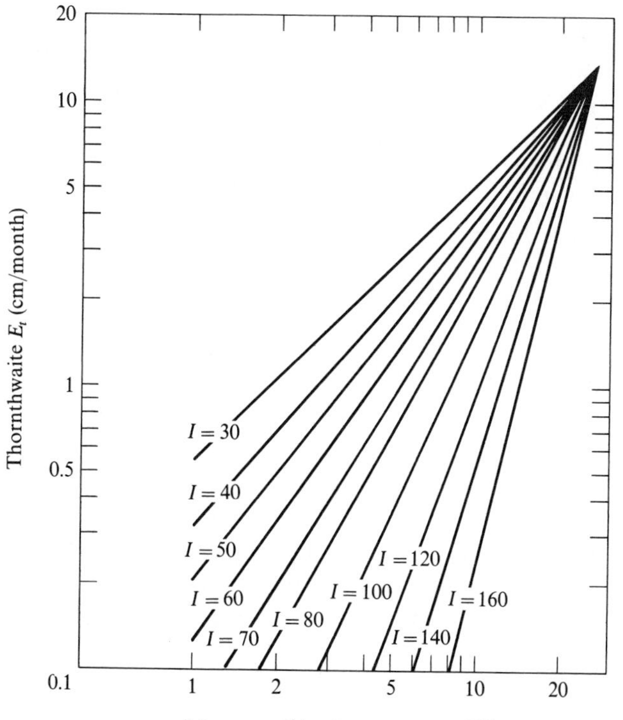

Figure 5-4 Graphical solution of the Thornthwaite formula for potential evapotranspiration, E_t, as a function of mean monthly air temperature for various values of annual heat index, I. The relation of the heat index I to mean annual temperature is shown in Figure 5-5.

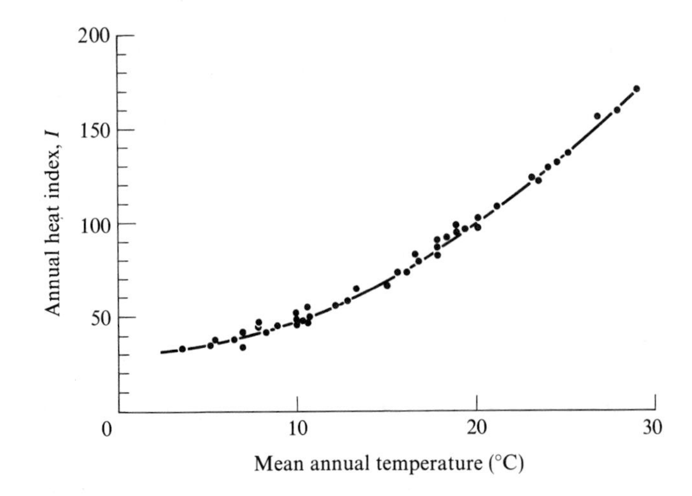

Figure 5-5 Annual heat index I of the Thornthwaite Equation as a function of mean annual temperature. (From Palmer and Havens 1958.)

should be checked before being used elsewhere. Daily or monthly potential evapotranspiration computed in Equation 5-8 or obtained from Figure 5-4 is for a standard month of 360 hours of daylight. It must be adjusted for the number of days per month and the length of day (a function of latitude). The standard potential evapotranspiration from Figure 5-4 should be multiplied by the appropriate factor given in Table 5-2 to make the adjustment for month and latitude.

Table 5-2 Correction factor for monthly sunshine duration for multiplication of the standard potential evapotranspiration from Figure 5-4.

LATITUDE	JAN.	FEB.	MAR.	APR.	MAY	JUNE	JULY	AUG.	SEPT.	OCT.	NOV.	DEC.
60°N	0.54	0.67	0.97	1.19	1.33	1.56	1.55	1.33	1.07	0.84	0.58	0.48
50°N	0.71	0.84	0.98	1.14	1.28	1.36	1.33	1.21	1.06	0.90	0.76	0.68
40°N	0.80	0.89	0.99	1.10	1.20	1.25	1.23	1.15	1.04	0.93	0.83	0.78
30°N	0.87	0.93	1.00	1.07	1.14	1.17	1.16	1.11	1.03	0.96	0.89	0.85
20°N	0.92	0.96	1.00	1.05	1.09	1.11	1.10	1.07	1.02	0.98	0.93	0.91
10°N	0.97	0.98	1.00	1.03	1.05	1.06	1.05	1.04	1.02	0.99	0.97	0.96
0	1.00	1.00	1.00	1.00	1.00	1.00	1.00	1.00	1.00	1.00	1.00	1.00
10°S	1.05	1.04	1.02	0.99	0.97	0.96	0.97	0.98	1.00	1.03	1.05	1.06
20°S	1.10	1.07	1.02	0.98	0.93	0.91	0.92	0.96	1.00	1.05	1.09	1.11
30°S	1.16	1.11	1.03	0.96	0.89	0.85	0.87	0.93	1.00	1.07	1.14	1.17
40°S	1.23	1.15	1.04	0.93	0.83	0.78	0.80	0.89	0.99	1.10	1.20	1.25
50°S	1.33	1.19	1.05	0.89	0.75	0.68	0.70	0.82	0.97	1.13	1.27	1.36

The Blaney-Criddle Formula

The Blaney-Criddle Formula was developed for estimating consumptive use of irrigated crops in the western United States. It is based on the same assumptions as Thornthwaite's method, and also uses temperature and day-length as the major independent variables. The most recent statement of the formula, as used by the U.S. Soil Conservation Service (1970), is

$$E_t = (0.142T_a + 1.095)(T_a + 17.8)kd \qquad (5\text{-}11)$$

where E_t = potential evapotranspiration (cm/mo)
T_a = average air temperature (°C). When T_a is less than 3°C, the first term in parentheses is set equal to 1.38.
k = empirical crop factor that varies with crop type and stage of growth (see Table 5-3). For perennial crops the coefficient is given for each month. For annual crops the coefficient is tabulated for various percentages of the growing season.
d = the monthly fraction of annual hours of daylight (see Table 5-4).

The crop factors were originally developed from plot studies in the western United States, and ideally should be checked against local conditions before application, though this is hardly ever done. The crop factors reflect differences in roughness, advection, and net radiation as affected by the structure of the crop during its various stages of growth and senescence. In general the crop factors increase with the height of the vegetation. They may also reflect differences in the methods of water application, or physiological differences among species. Coefficients for other crops are given in the U.S. Soil Conservation Service (1970) handbook.

For calculating water requirements for a whole growing season, the Blaney-Criddle Equation can also be used in the form

$$E_t(\text{cm}) = K \sum_{i=1}^{n} (1.8T_{ai} + 32)d_i \qquad (5\text{-}12)$$

where K = crop coefficient for whole growing season (Table 5-5)
n = number of months in growing season
i = individual month in the growing season, so T_{ai} and d_i are respectively the air temperature and the fraction of annual hours of daylight for each month.

The Blaney-Criddle Formula is now widely applied in agricultural development work, although it can produce large errors outside the western United States, where the information in Tables 5-3 and 5-5 was collected. The U.S. Soil Conservation Service (1970) has released a manual giving detailed instructions for the use of the Blaney-Criddle Formula in calculating irrigation requirements. The manual also describes how to estimate water losses from conveyance and storage works and many other important aspects of irrigation planning.

Table 5-3 Empirical crop factors (k) for the Blaney-Criddle Formula. In the Southern Hemisphere the monthly coefficients for perennial crops should be interchanged to correspond with the onset of growing seasons. (From U.S. Soil Conservation Service 1970.)

CROP	GROWING SEASON														
	BEGINNING MEAN TEMP. (°C)	END MEAN TEMP. (°C)	MAX. LENGTH (DAYS)	JAN.	FEB.	MAR.	APR.	MAY	JUNE	JULY	AUG.	SEPT.	OCT.	NOV.	DEC.
PERENNIALS															
Pasture grass	7	7	Variable	0.49	0.57	0.73	0.85	0.90	0.92	0.92	0.91	0.87	0.79	0.67	0.55
Alfalfa	10	−2 (frost)	Variable	0.63	0.73	0.86	0.99	1.08	1.13	1.11	1.06	0.99	0.91	0.78	0.64
Grapes	13	10	Variable	0.20	0.24	0.33	0.50	0.71	0.80	0.80	0.76	0.61	0.50	0.35	0.23
Deciduous orchard	10	7	Variable	0.17	0.25	0.40	0.63	0.88	0.96	0.95	0.82	0.54	0.30	0.19	0.15
				PERCENT OF GROWING SEASON											
ANNUALS				0–10	10–20	20–30	30–40	40–50	50–60	60–70	70–80	80–90	90–100		
Small vegetables	16	0	100	0.33	0.47	0.64	0.74	0.80	0.82	0.82	0.76	0.66	0.48		
Peas			100	0.54	0.65	0.80	0.97	1.08	1.12	1.11	1.08	1.04	0.98		
Potatoes	16	0 (frost)	130	0.36	0.45	0.59	0.85	1.09	1.26	1.35	1.37	1.34	1.27		
Sugar beet	−2 (frost)	−2 (frost)	180	0.46	0.54	0.69	0.87	1.03	1.16	1.24	1.24	1.18	1.10		
Grain corn (maize)	13	0 (frost)	140	0.46	0.54	0.64	0.82	1.00	1.08	1.08	1.03	0.97	0.89		
Silage corn (maize)	13	0 (frost)	140	0.45	0.50	0.59	0.71	0.90	1.03	1.07	1.07	1.04	1.00		
Sweet corn (maize)	13	0 (frost)	140	0.43	0.52	0.62	0.81	1.00	1.07	1.08	1.07	1.04	1.02		
Spring grain	7	0 (frost)	130	0.36	0.58	0.82	1.04	1.25	1.31	1.18	0.87	0.49	0.13		
Grain sorghum	16	0	130	0.32	0.47	0.72	0.93	1.07	1.04	0.94	0.82	0.70	0.60		
Soybeans			140	0.22	0.30	0.37	0.48	0.63	0.84	0.98	1.02	0.83	0.72		
Cotton	17	0 (frost)	240	0.22	0.28	0.40	0.64	0.90	1.01	1.00	0.88	0.73	0.57		

Table 5-4 Monthly fraction of annual hours of daylight (for use in the Blaney-Criddle Equation).

LATITUDE	JAN.	FEB.	MAR.	APR.	MAY	JUNE	JULY	AUG.	SEPT.	OCT.	NOV.	DEC.
60°N	.047	.057	.081	.096	.117	.124	.123	.107	.086	.070	.050	.042
50°N	.060	.063	.082	.092	.107	.109	.110	.100	.085	.075	.061	.056
40°N	.067	.066	.082	.089	.099	.100	.101	.094	.083	.077	.067	.075
20°N	.073	.070	.084	.087	.095	.095	.097	.092	.083	.080	.072	.072
10°N	.081	.075	.085	.084	.088	.086	.089	.087	.082	.083	.079	.081
0	.085	.077	.085	.082	.085	.082	.085	.085	.082	.085	.082	.085
10°S	.089	.079	.085	.081	.082	.079	.081	.083	.082	.086	.085	.088
20°S	.092	.081	.086	.079	.079	.074	.078	.080	.081	.088	.089	.093
30°S	097	.083	.086	.077	.074	.070	.073	.078	.081	.090	.092	.099
40°S	.102	.086	.087	.075	.070	.064	.068	.074	.080	.092	.097	.105

Table 5-5 Seasonal coefficients (K) for irrigated crops for use in the Blaney-Criddle Equation (5-12). (From U.S. Soil Conservation Service 1970.)

CROP	LENGTH OF NORMAL GROWING SEASON	K*
Alfalfa	Between frosts	2.0–2.3
Bananas	All year	2.0–2.5
Beans	3 months	1.5–1.8
Cocoa	All year	1.8–2.0
Coffee	All year	1.8–2.0
Corn (maize)	4 months	1.9–2.2
Cotton	7 months	1.5–1.8
Dates	All year	1.6–2.0
Grains, small	3 months	1.9–2.2
Grains, sorghum	4–5 months	1.8–2.0
Grass, pasture	Between frosts	1.9–2.2
Potatoes	3–5 months	1.6–1.9
Rice	3–5 months	2.5–2.8
Soybeans	5 months	1.6–1.8
Sugar beet	6 months	1.6–1.9
Sugar cane	All year	2.0–2.3
Truck crops	2–4 months	1.5–1.8

*The lower values of K are for more humid areas, and the higher values are for more arid climates.

Evapotranspiration When Soil Moisture Is Limiting

When moisture conditions are suitable, the actual rate of evapotranspiration is equal to the potential rate. Without frequent wetting by rainfall or irrigation, evapotranspiration will lower the soil moisture content until water loss will no longer occur at the potential rate. It is then necessary to consider other aspects of the soil and plants in addition to the factors controlling the energy balance. Rates of transpiration below the meteorologic potential must be estimated when calculating the water balance of drainage basins and soil profiles. The suitability of crops for various soils and climatic regions depends upon the actual evapotranspiration, which must be known in order to calculate whether the crop will survive without irrigation.

The most popular method of computing actual evapotranspiration is through the calculation of potential evapotranspiration. If there is abundant moisture in the soil, the two rates are equal. When moisture supply becomes limiting, the computed potential rate is modulated by a factor that depends upon the amount of water in the soil. The relationship can be expressed as

$$AET = PET \times f\left(\frac{AW}{AWC}\right) \tag{5-13}$$

where AET and PET = the actual and potential rates of evapotranspiration
$f(\ )$ = some function of the term inside the parentheses
AW = the available soil moisture = (soil moisture content − permanent wilting point) × rooting depth of vegetation (cm)
AWC = the available water capacity of the soil = (field capacity − permanent wilting point) × rooting depth of vegetation (cm).

The terms field capacity and permanent wilting point are explained in the following text.

As the soil dries during evapotranspiration, it becomes increasingly difficult for plants to extract water from soil pores. Dry soils hold water tightly, and the rate of movement of water toward roots decreases with moisture content. The reasons for this will be discussed in Chapter 6. For the time being, it is necessary only to introduce two aspects of soil moisture to continue the discussion of evapotranspiration.

If a soil is saturated by a rainstorm, water will drain quickly from the large pores. As the soil moisture content decreases, the rate of gravitational drainage will decrease and will eventually become zero for all practical purposes. In this state of balance between the gravitational force tending to produce drainage and the forces tending to hold water within the soil, the soil moisture content is called the field capacity. Further gravitational drainage is negligible, and any abstraction of soil moisture must take place by evapotranspiration. For water to be removed from the soil, the plants must exert a suction upon the soil moisture. As the soil dries, a greater suction is necessary to remove the remaining water from the soil pores. Plants can exert a maximum suction of approximately 15 atmospheres and, therefore, there is a limit to the amount of water that can be withdrawn from a soil by this process. When the plant cannot withdraw water at a rate sufficient to fulfill the demands of transpiration, the plant wilts and the soil moisture content at which this occurs is called the permanent wilting point.

The field capacity and permanent wilting point vary with soil characteristics, as will be discussed in Chapter 6. Any moisture in excess of the field capacity will drain rapidly out of the soil; moisture below the permanent wilting point cannot be withdrawn by the plant. Between these two limits is the water that is available to the plant for transpiration. If, for example, the field capacity of a certain soil lies at a moisture content of 30 percent by volume and the permanent wilting point at 10 percent, the available water capacity of the soil is 20 percent multiplied by the rooting depth of the vegetation. If the root system of the plants extend to 1.5 m, 30 cm of water are available for evapotranspiration if the soil is initially at its field capacity or wetter. The available water in the soil at any time is the amount of water held at a moisture content above the permanent wilting point.

There is little quantitative information about the way in which moisture moves from soil pores into plant roots and is withdrawn from the soil by evapotranspiration under conditions of limiting soil moisture. Many variables, such as soil texture, plant physiology, and rooting characteristics, and the rate of evapotranspiration itself, control the process, and it is difficult to make a general statement about the rate of movement from all soils at a certain intensity of potential evapotranspiration. In the absence of a generally applicable physical model, several compromises have been made between complex soil physics and the simplicity required in accounting procedures for evapotranspiration and soil moisture in agriculture, water management, and related fields. Several of these simplified models will be reviewed below and used to compute the actual rate of evapotranspiration under conditions of limiting moisture.

Veihmeyer (1964) and his co-workers conducted a series of experiments in California and concluded that evapotranspiration proceeds at the potential rate until soil moisture is depleted to the permanent wilting point. Plants then wilt and evapotranspiration rates fall sharply to near zero. Thornthwaite and Mather (1955), working on a loam in Nebraska, found that the ratio of actual to potential evapotranspiration decreases as a linear function of the amount of available water (see Figure 5-6). Other workers have proposed various compromises between these two extremes, most of them suggesting that evapotranspiration occurs at close to the potential rate until a considerable proportion of the available water is depleted, after which the actual evapotranspiration rate falls rapidly. One such example is incorporated in Figure 5-6(a).

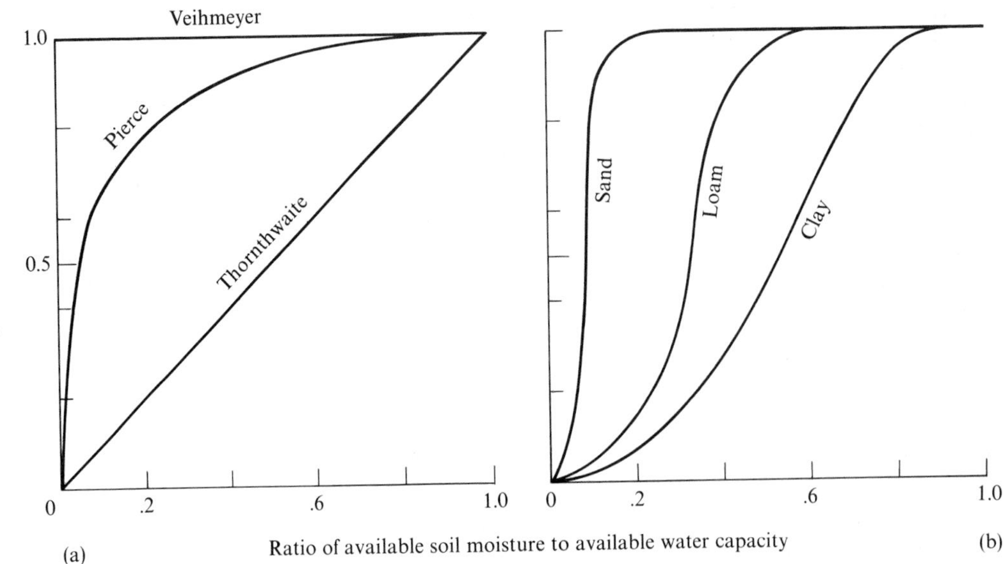

Figure 5-6 Relation of actual evapotranspiration to available soil moisture. (a) Actual evapotranspiration as a ratio to potential plotted against the ratio of available soil moisture to available water capacity according to models of three investigators. (After Holmes and Robertson 1959.) (b) Ratio of actual to potential evapotranspiration as a function of soil moisture for three soil textures. The ratio of available soil moisture to the available water capacity varies from zero at the permanent wilting point to 1.0 at field capacity. (After Holmes 1961.)

Holmes (1961) has pointed out that the accuracy of these approximations will vary with soil type and rooting characteristics. In a sandy soil, for example, plants with a dense root system can withdraw water rapidly from the pores, and evapotranspiration may proceed at close to the potential rate until soil moisture content is close to the wilting point. In a clay-rich, loamy soil, water is held more tightly at a given soil moisture content (see Chapter 6), and movement of water to the roots is slow. Under such circumstances moisture supply to the roots will not keep up with potential rates of loss, and actual evapotranspiration rates will decline throughout most of the range of available water. Holmes (see Figure 5-6(b)) presented a schematic relationship between actual and potential rates of evapotranspiration over the range of available water capacity for three soil textures. This relationship can be compared with the results of previously mentioned workers in Figure 5-6(a). The comparison shows that the Veihmeyer and Thornthwaite models should be adequate approximations for sandy and clay-rich soils, respectively. In the absence of detailed measurements, an estimation of the appropriate relationship for intermediate soils involves a good deal of guesswork. The situation is further complicated by the intensity of the evaporative

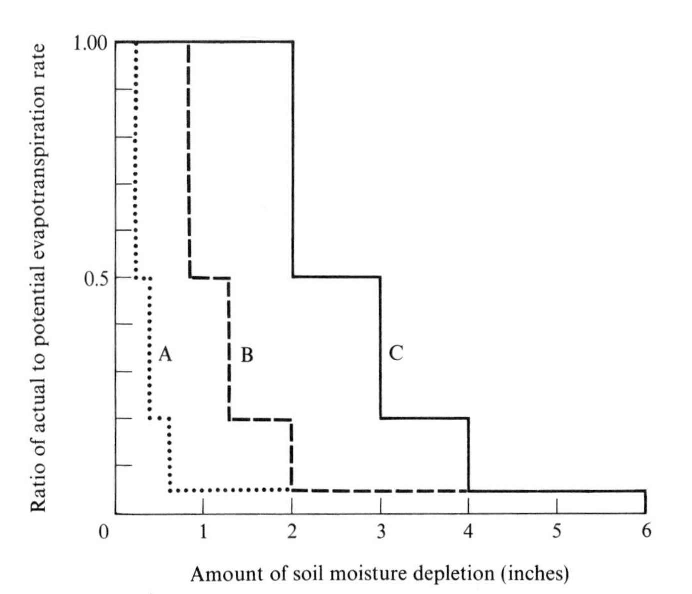

Figure 5-7 Ratio of actual to potential evapotranspiration rate as a function of soil moisture content for a wheat field in Alberta, Canada. Available water-holding capacity of the soil is 6 inches. Curve A applies to the period August 1–May 31. Curve B applies to the period June 1–June 30. Curve C applies to the period July 1–July 31, at which time the root system of the crop is fully developed. (From Holmes 1961.)

demand. Under low demand, soil moisture may be able to move to the root system rapidly enough to maintain transpiration at the potential rate. If the demand increases, the soil moisture movement may not be rapid enough to do this.

In view of the difficulties of applying the previously mentioned results to soils for the solution of particular problems, several simpler relationships have been used with some success. Ideally they should be checked under the field conditions to which they are to be applied by making measurements of actual soil moisture losses, but in practice this is not usually done. Often the hydrologist will use two extreme estimates and calculate the actual water loss under both conditions. Frequently, the results do not differ by a large amount.

Some examples of methods for modulating potential evapotranspiration are summarized in Figures 5-7 and 5-8. The method will be demonstrated in some of the typical problems in Chapter 8. In view of the uncertainty inherent in applying these methods to soil-vegetation associations in other regions, the best method of evaluating actual evapotranspiration is by direct measurement with lysimeters.

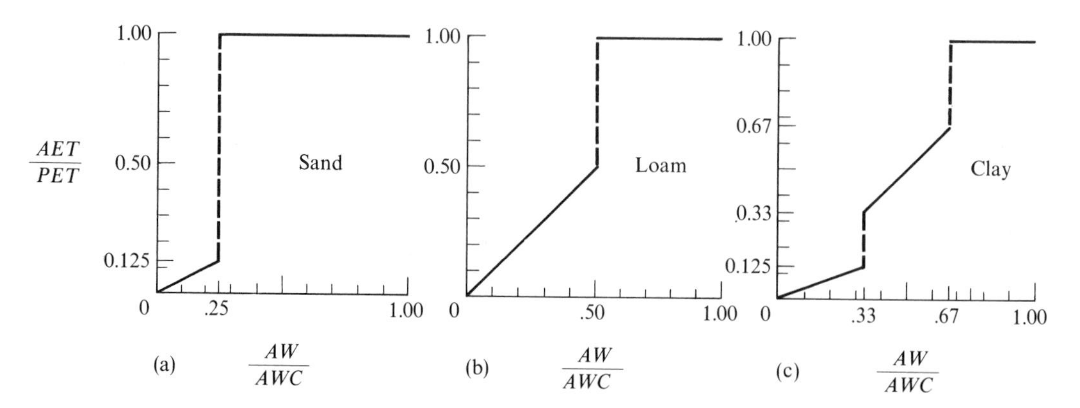

Figure 5-8 Ratio of actual to potential evapotranspiration as a function of soil moisture content for three textures of soil under forest. The soil moisture is expressed as the ratio of available moisture (AW) to water capacity of the soil (AWC). (From Zahner 1967.)

Weather, Vegetation, and Soil as Controls of Evapotranspiration

There is considerable disagreement in the literature about the relative importance of weather, vegetation, and soil as controls of evapotranspiration. The discussion is important in view of current interest in increasing the efficiency of irrigation practices, understanding the distribution of plants, and managing the vegetation of drainage basins for increased water yield.

Some workers have sought to simplify the discussion by defining potential evapotranspiration as water loss from a short green crop completely shading the ground and never short of water. This simplification led to the development of useful techniques for estimating potential evapotranspiration, but it is not directly useful to the watershed manager in deciding whether to replace a forest cover with grass, or to an agronomist planning supplemental irrigation of a crop that only covers the ground completely for a few weeks, is green only for a couple of months, and is growing at a rate of one meter per month. The situation becomes even more complex when periods of moisture deficiency must also be considered.

In order to make some generally useful statements about the role of weather, vegetation, and soil, it is necessary to review the various formulae proposed for the computation of evapotranspiration.

Meteorological Factors

All the formulas for computing evapotranspiration are dominated by meteorological terms. On this point there is no debate. Meteorologic factors dominate evapotranspiration. The most sophisticated formulas contain terms representing the basic energy input, net radiation, and also the "drying power" of the atmosphere. This ability of the atmosphere to transport water vapor from leaf surfaces depends upon air temperature (more strictly, leaf temperature), vapor pressure of the air, and wind velocity. Empirical equations that do not use these parameters use air temperature as a surrogate.

Several other meteorological factors affect evapotranspiration rates but can only be mentioned briefly here. The frequency and size of rainstorms control the amount of interception (see Chapter 3), and since this intercepted water evaporates more easily than water transpired from stomata, the rainfall regime affects total water loss. Length of growing season is another important variable of obvious significance.

Vegetative Factors

The discussion of vegetative controls is more complex. Some empirical formulas, such as Thornthwaite's, do not allow for vegetative influences. The Blaney-Criddle Formula includes an empirical constant to differentiate between the requirements of various crops. To see the physical reasons for such differences, however, it is necessary to examine the Penman and Van Bavel combination of energy-balance and mass-transfer terms in Equations 5-6 and 5-7.

The reflectivity of a vegetated surface determines the amount of solar radiation that is absorbed, and therefore, is an important control of the net amount of radiant energy available for evapotranspiration. Values of albedo given in Table 5-1 show important differences between species and between seasons for the same species. At full cover, most green crops have an albedo of approximately 0.25, but there is a marked seasonal trend from the albedo of bare soil at planting to a maximum at the time of full green cover, followed by a decline during senescence. There are also important differences between forests and crops. Baumgartner (1967) showed that if the net radiation over coniferous forests in Germany were taken as unity, the ratios of other cover types to conifers were deciduous forest 0.75–1.0, arable land 0.90, alfalfa 0.76, grassland 0.75, potato crops 0.64, and bare soil 0.60. He was able to show not only that forests transpired more water than agricultural lands, but also that there were differences in transpiration rates among tree species. If a spruce forest is taken as unity, the relative rates of evapotranspiration in other species were pine 0.59, beech 0.88, larch and birch 1.09, and Douglas fir 1.11. Tajchman (1971) demonstrated differences of evapotranspiration between forests and crops for the same reasons. Such variations may become important as the harvesting of water from catchments becomes a more critical affair.

To examine the influence of surface roughness on evapotranspiration, it is informative to use the Van Bavel Equation (5-7), substituted into Equation 5-6. In the Van Bavel modification of Penman's Formula E_a, the component of evapotranspiration due to mass transfer of water vapor, increases with the roughness parameter, z_0. As the roughness increases, greater turbulence causes an increase of water-vapor transport away from leaf surfaces. On these grounds alone, we might expect forests to transpire more than tall crops, and these to lose more water than grass. Rider (1957) found that roughness of vegetation did vary among green crops by amounts that significantly affected potential evapotranspiration. He reported that on a sunny day in England, typical water losses were 2 to 3 mm from grass, 4 mm from Brussels sprouts, and 2.5 to 7.5 mm from peas. Stanhill (1958) studied water losses from carrots and grass (vegetation types with similar albedos), and found that the greater roughness of the carrot plants caused more intense turbulent transport of heat and water vapor under the same weather conditions. Water losses were much greater from the carrot plants than from the grass. Tajchman (1971) also found that the turbulent exchange of water vapor was more intense over a spruce forest than over alfalfa and potatoes. The effects of interspecies differences of roughness can be examined under a range of meteorological conditions and reflectivities by substituting various roughness values (z_0 = one-tenth the height of the vegetation) into Equation 5-7.

Because of interspecies differences in stomatal resistance to the diffusion of water vapor from the leaves, there are important differences in potential evapotranspiration rates between species with the same albedo and height exposed to the same weather (Szeicz and Long 1969). It should be remembered, however, that stomatal resistance is not controlled solely by vegetative characteristics. Stomata close under high rates of potential evapotranspiration and under conditions of limited soil moisture and are, therefore, partly controlled by weather and soil. Stomatal control becomes increasingly important as evapotranspiration falls below the potential rate because of the limited availability of soil moisture.

When water is plentiful, the roots of plants probably do not control water loss. However, as soil moisture is depleted, the depth and density of roots attain major importance. When the water content of the topsoil is depleted, shallow rooted plants transpire at rates less than the potential rate. Deeper rooted plants can tap water in the subsoil and continue to transpire at the potential rate. Trees usually transpire more than grass for this reason. Rooting depth (and the texture of the soil) determine the available water capacity of the soil. Johnston (1970) showed that seasonal moisture depletion in the upper 2.75 m of a soil was 533 mm under deep rooted aspen, while herbaceous vegetation withdrew 390 mm, and 287 mm were lost from bare soils. The density of plant roots may become important if the rate of potential evapotranspiration is very high and is being limited by the rate at which water can migrate through the soil to the roots. Under such conditions the distance of travel within the soil will be minimized by a dense root system. Figure 5-7 shows how the development of the root system of a crop alters the rate at which actual evapotranspiration declines as the soil dries.

The rate at which evapotranspiration declines as soil moisture is depleted can vary with the rooting depths of vegetation, as shown for experimental

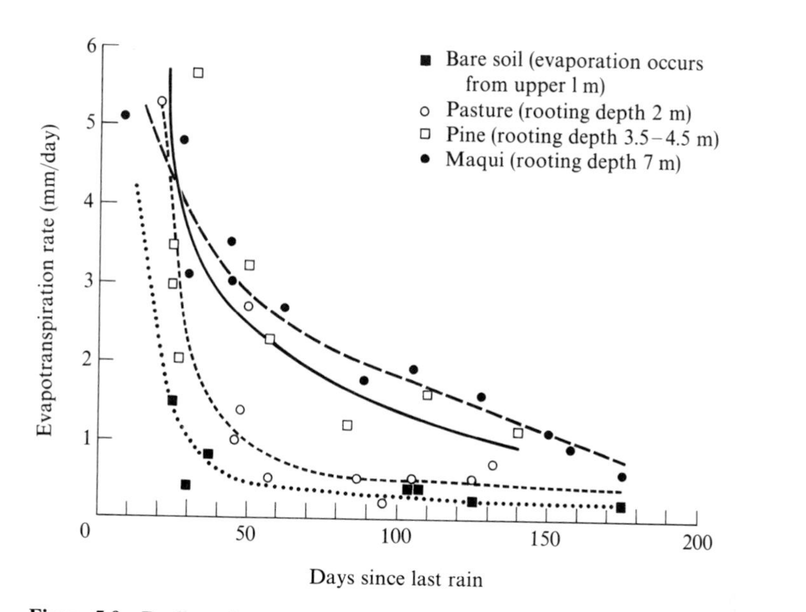

Figure 5-9 Decline of actual evapotranspiration in the Carmel Mountains, Israel, as the soil dries out under covers with different root depths. (From Schachori et al. 1967.)

plots in the Carmel Mountains of Israel (see Figure 5-9). Under these plots the water table lies at a depth of 9 m. Experiments were conducted on water use by various vegetative covers. Soil moisture measurements showed that on the bare soil, evaporation withdrew moisture from the upper 1 m of the soil, but that the rate of loss declined rapidly as the topsoil dried out (see Chapter 4). Pasture grass withdrew water from the upper 2 m of soil, and this loss also declined rapidly as the soil dried. Pines tapped water down to depths of 3.5 to 4.5 m, and maqui (a shrub vegetation similar to chaparral) exploited the subsoil to a depth of more than 7 m. The trees could maintain a high rate of water loss for a long time after the shallower rooted plants had become desiccated, although the evapotranspiration rate for the trees also declined drastically during the first 100 days of a dry period. The extra water loss from the deeper-rooted species is extremely important in a region with so little rainfall and groundwater recharge. In a dry year (630 mm of precipitation), no soil water percolated to recharge the groundwater under the trees and shrubs, whereas the pasture cover allowed a recharge of 73 mm and 84 mm of water drained from beneath the bare soil. In an average year (700 mm), 452 mm and 420 mm of water percolated to the groundwater beneath the bare soil and pasture cover, whereas the pine-lands supplied only 340 mm and the maqui 230 mm. In a country such as Israel with limited water resources such differences are critical and are carefully weighed in decisions that concern planning for watershed management, forage production, and water supply.

As with meteorological controls, those resulting from characteristics of vegetation show complex secondary relationships. There are variations in the amount of water intercepted by vegetation types. The length of the growing season is at least partly a characteristic of the plant. The layout of cropped lands is a characteristic that can be of significance locally. Particularly in irrigated plots amid an arid landscape, tall crops intercept more advected energy.

Soil Factors

The albedo of the soil is of some importance in determining the soil evaporation component of evapotranspiration. Soil texture affects the field capacity and permanent wilting point values of the soil, and with soil depth, these control the available water capacity. In a soil with limited available water capacity, moisture is soon depleted and evapotranspiration falls below the potential rate. The depth of the soil and the character of its profile also have an influence on the rooting characteristics of the plant cover. The effect of soil texture upon the migration of water to plant roots is mentioned in the discussion of Figure 5-6(b).

Summary

Meteorological factors dominate the rate of evapotranspiration, which therefore varies seasonally and spatially in response to regional climate, modified by local climatic influences and by irrigation. There is some variation in the rate of evapotranspiration among vegetation types exposed to the same weather. These variations are of a lower order of magnitude than those caused by meteorological factors, but are becoming increasingly important to those interested in the management of water resources. The variation of evapotranspiration among vegetation types assumes greater significance when moisture becomes limiting. Soil factors constitute relatively minor controls of evapotranspiration but are of some significance under conditions of limited moisture. Some of the typical problems in this chapter and in Chapter 8 on the water balance will illustrate the differences quantitatively.

Manipulation of Vegetation for Water Yield

As water shortages become more severe in some regions, attempts are being made to increase the water yield from drainage basins by manipulating vegetation. Because many important water-supply catchments are located in forest lands, most of the research has taken place in mountainous forests. Their remoteness from population centers also reduces some of the esthetic problems inherent in manipulating vegetation on a large scale, and some of the goals of water-yield augmentation can be linked to the goals of timber production. The methods used for increasing the water yields of forests include removal of the vegetation, replacement of one tree species by another, and structural alteration of existing stands (Figure 5-10). Research is also taking place on the use of antitranspirant chemicals. When sprayed on leaves, these chemicals cause a constriction of the stomatal openings and a reduction of water loss from the leaves. This technique has not yet shown sufficient promise for routine application; the others are more widely applied.

The volume of water transpired by a forest can be estimated fairly accurately by the methods described in this chapter. When the forest is cut down,

Figure 5-10 Drainage basin whose tree cover has been completely removed to study the effects of clearcut logging on volumes of runoff. (Photograph by Hans Riekerk.)

only logging slash and ground vegetation remain to lose water by interception and evapotranspiration. As these losses are reduced, soil moisture levels and streamflow increase. Some plants survive, however, and others begin to colonize the area. These plants use water, and more water is lost by evaporation from the soil which is exposed to sunshine and wind. Because of the uncertainty of our knowledge of soil shading, regrowth, rooting depths, and other variables in the cut area, it is not yet possible to calculate directly the amount of extra streamflow that will be yielded by cutting down a forest. There is, however, a good deal of empirical evidence about the increase of runoff that accompanies deforestation in various timber regions of the world. The typical experiment for evaluating these increases is the "paired watershed" experiment. Two forested catchments of similar physical geography are chosen, and their streamflow is gauged for a number of years until a correlation is established between their annual yields. Then one of the catchments receives some form of treatment, such as clearcutting or species conversion. The measured streamflow after treatment can be compared with the hypothetical streamflow in the undisturbed state, predicted from measurements on the untreated catchment. The difference between the predicted and observed runoff represents the effect of the treatment.

Figure 5-11 summarizes data on first-year increases in streamflow following clearcutting in various parts of the world. The increase is proportional to the reduction of the vegetation cover. There are few points for each region, but they cluster strongly about the lines shown in the figure. The best summary and detailed discussion of these experimental results are given by Hibbert (1967).

As the forest cover begins to regenerate, interception and evapotranspiration increase, and runoff declines. Few published studies give data for enough years of record after cutting to define this trend mathematically. An important exception is a 23-year record from a watershed in the southern Appalachians (Hibbert 1967). Here the increased water yield declined rapidly at first, and then more slowly, but still amounted to 60 mm after 23 years. Several other studies allow a very rough trend to be established by plotting the increase in water yield as a function of the logarithm of time on semi-logarithmic graph paper. The data indicate that increases of streamflow caused by deforestation decline exponentially with time, and have a half-life of about two to seven years.

The increased water yields from clearcutting have some disadvantages, and they should not be regarded as a panacea for solving water-supply problems on a large scale. The increased drainage of water through the soil causes a severalfold acceleration of the rate at which plant nutrients are leached from the soil. Because the increase is temporary, however, this leaching is unlikely to cause long-term damage to the soil or to the water quality of streams, but problems have arisen.

In many parts of western and southern Australia, where annual rainfall is 250 to 400 mm, clearing of dry forest and scrub has caused widespread disruption of soils and water supplies. In these areas the groundwater is

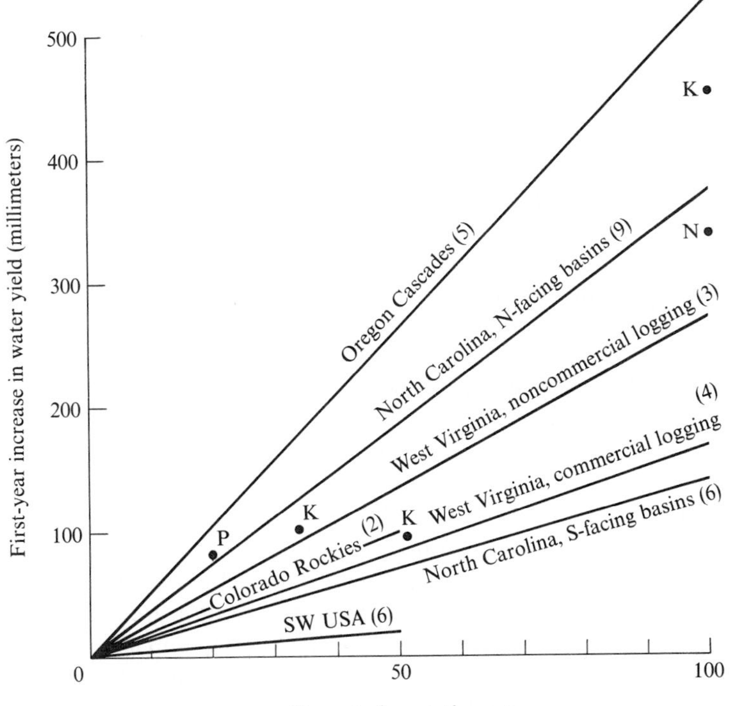

Figure 5-11 First-year increase in runoff as a function of percentage of forest vegetation removed by clearcutting. The points for the lines are omitted for clarity, but they define the lines well. Individual points are plotted, identified by letters, to show results of experiments on drainage basins in various regions: K = Kenya highlands; N = New Hampshire; P = Pennsylvania. (Data from Hibbert 1967 and many other reports.)

saline, and almost never recharged because the woody plants have long roots that can efficiently withdraw soil moisture during the dry season. Clearing of this vegetation to increase runoff and for agriculture has led to the replacement of deep-rooted plants by shallow-rooted ones. Reductions in evapotranspiration have allowed some water to percolate to the groundwater, which has risen into the surface layers of the soil. When the surficial soil is saturated in this way, evaporation from its surface is rapid, and high concentrations of salt are deposited in the soil, reducing its usefulness for cropping. The increased supply to the groundwater has also caused an increase of runoff, but because of the salinity of the groundwater, this runoff is of little use. Boughton (1970) estimates that more than 400,000 acres of land in Australia have been affected in this way.

Clearing of large areas of land usually accelerates mass wasting and erosion processes (Swanston 1971) and may cause sedimentation problems downstream. Serious increases in erosion, mass wasting, landslides, and channel aggradation from the clearcutting of redwoods in California have been documented by Janda (1975). If the forest cover is not rapidly re-established, sheet erosion and soil degradation will result (Dils 1953). In moist regions the rise of groundwater following the clearcutting of forests has severely limited regeneration, and has left regenerating trees prone to disease (Holstener-Jorgensen 1967).

The clearing of forest and scrub vegetation to increase water yields can, therefore, cause major and minor disruptions if not done carefully, or if carried out in areas that are totally unsuitable. This does not mean, however, that it is not a useful management tool in some situations. In a small municipal watershed, timber production and augmentation of water yields can be carried out successfully if the cutting is done on a rotational basis and is limited to gently sloping areas with stable soils, and if revegetation is encouraged. Potential problems should be considered carefully before a clearing program is undertaken.

Reforestation of abandoned farmlands accomplishes the opposite results; increased interception and evapotranspiration cause a decline of water yield. The decrease of water yield through time during reforestation is not easy to predict, but the final rates of interception and evapotranspiration from mature forest can be predicted with the techniques described in this text, and they can be taken into account when decisions are made to reforest an area. Few data are available from controlled experiments on this effect, so it is difficult to generalize quantitatively about trends, but over periods of 20 to 30 years of regrowth, reductions in annual water yield of 100 to 200 mm have been measured in several drainage basins. Over hundreds of square kilometers of the eastern United States, farm abandonment and recolonization of the land by pines, spruce, or cedar have been occurring throughout this century. There is reason to believe that this vegetation change has reduced streamflow by important amounts, and that it will continue to do so at a time when water supplies for some eastern cities are becoming critically short. Regrowth of conifers on the 1027-square-kilometer Sacandaga River catchment in the Adirondack Mountains, for example, caused increases in interception and evapotranspiration losses. The increase in the loss of water had risen to over 200 million cubic meters per year by 1950. This amount of water is large enough to supply more than one million people. There are water costs, then, in maintaining a forest cover for purposes of recreation, timber production, or soil stabilization, or in simply abandoning farms. Law (1956) pointed this out in relation to recent British policy of planting conifers on important water-supply catchments that serve industrial areas. His measurements showed that increased interception and evapotranspiration from a Sitka spruce plantation lowered the water yield by 280 mm per year compared with an adjacent grassland.

A related question is that of predicting effects upon the water yield of converting vegetation cover from one species to another. The results of controlled experiments on the effects of species conversion are variable, but can be rationalized qualitatively by reviewing the material on interception and evapotranspiration in earlier chapters. Conifers generally intercept more water than hardwoods, especially in the snow zone. They also transpire more because of their lower albedo (see Table 5-1) and longer duration in leaf. Urie (1967) found that an annual precipitation of 815 mm in Michigan produced 386 mm of groundwater recharge under oak forest, but only 320 mm under a pine plantation. After ten years of growth eastern white pine in North Carolina yielded 33–94 mm per year less than earlier deciduous covers on the same catchments (Swank and Miner 1968). Differences in rooting depth, as well as those of interception, albedo, and roughness, also produce differences in water yield. This is commonly true where soil moisture deficits limit evapotranspiration by shallow-rooted plants throughout much of the year. An Israeli demonstration of the effect is shown in Figure 5-9. The results of this work indicate that conversion of maqui to pasture would decrease evapotranspiration losses by an average of 125 mm/yr; conversion to pine would increase average water yields by 42 mm/yr. In the semi-arid (400–900 mm of rainfall) zone of Arizona, Hibbert (1971) showed that the removal of brush and its replacement by grass produced five- and six-fold increases in annual runoff on small catchments. Some of this increase, however, was overland flow, which through erosion is likely to do more harm than good, but most of the increase was ascribed to the reduction of evapotranspiration by the shorter, more shallow-rooted grass. In a slightly wetter region of the Central Californian Sierras, Lewis (1968) found that replacement of oak woodland by grass increased the average water yield by 114 mm/yr.

The results of many experiments, however, are quite ambiguous. In semi-arid areas where water reaching storage reservoirs has a high value, some people advocate the removal of vegetation in order to promote larger water yield in rivers downstream. In Arizona the principal vegetative type where clearing was proposed is the piñon–juniper woodland. The U.S. Geological Survey made an intensive and long-continued study to determine whether water yield is increased by woodland conversion carried out by methods actually employed on a large scale. This consists of knocking down trees by dragging a heavy chain between two tractors, followed by piling the downed trees in windrows and burning. Paired watersheds were instrumented, and measurements were made over a period of years. The researchers concluded that the variability of meteorologic and hydrologic factors was so large that decades of observation would be required before any results of statistical significance could be expected. In other words, after a detailed study, an increase in water yield could not be demonstrated.

"Juniper eradication," that is, destruction of piñon–juniper and the planting of grass, is carried out over large areas of the southwestern United States, but the justification is to increase forage production for grazing stock, not to increase water yield downstream. Furthermore, the trees tend to resprout, and there is doubt whether the replacement by grass can be permanent.

Also in Arizona, all the riparian cottonwood trees were cut in some mountain drainage basins with the expectation that streamflow downstream would thereby be increased. No such increase has been demonstrated, but the esthetic loss was immense. In a semi-arid area the riparian trees constitute an esthetic heritage, the social value of which is incalculable. Cottonwood trees, whose leaves turn yellow in the fall, are a special joy. Their destruction constitutes major environmental degradation.

As another example, a mature hardwood forest in a wet area of North Carolina was replaced by a grass cover, but there was no significant difference in water yield as long as the grass cover was maintained in a lush condition by heavy fertilization. When the grass was not fertilized, the water yield increased by as much as 300 mm/yr over the original yield of 830 mm/yr. In a similarly wet area of Kenya the replacement of evergreen tropical forest by lush, vigorous but lower tea bushes with shallower roots led to a 15-percent increase in water yield (Pereira 1962). The replacement of indigenous bamboo forest with pine trees in the same region did not alter the water yield once the pine canopy was fully established.

The manipulation of vegetation is a feasible way of altering the water budget of a catchment for useful purposes. For really large increases in runoff, however, drastic changes of the vegetation are necessary, and it is doubtful whether such changes will be justified for water supply alone. They can be usefully integrated into other strategies of land management, such as timber or forage production, however. In many areas, small municipal water catchments are being successfully managed for increased water yield and timber production by clearcutting on a rotational basis. Smaller and slower changes of vegetation may produce subtle effects on water yield, and these can be anticipated and taken account of in regional planning.

Large-scale, deliberate manipulation of vegetation, however, is costly. The vegetation that existed before human colonization resulted from long-continued operation of natural processes, and when changes are imposed, the vegetation succession appropriate to that locale may be expected to occur unless there is continuous effort to maintain the new type. Expectations of a permanent new vegetal association imposed by humans are in all probability not to be realized unless there is continual, costly maintenance by plowing, pruning, weeding, selective cutting, burning, or some combination.

Vegetation manipulation can also have important effects on esthetics, wildlife, recreation potential, and other social amenities. The problem is not merely one of hydrologic effects. The planner must take all of these into account.

Bibliography

BAUMGARTNER, A. (1967) Energetic bases for differential vapourization from forest and agricultural lands; in *Forest hydrology* (eds. W. E. Sopper and H. W. Lull), Pergamon Press, Oxford, pp. 381–389.

BLANEY, H. F. (1962) Utilization of water and irrigation in Israel; *Journal Irrigation and Drainage Division, American Society of Civil Engineers,* vol. 88, no. IR 2, pp. 55–65.

BOUGHTON, W. C. (1970) Effects of land management on quantity and quality of available water; *University of New South Wales Water Research Laboratory Report No. 120,* Australian Water Resources Council.

CARLSTON, C. W., ALESSI, J. AND MICKELSON, R. H. (1959) Evapotranspiration and yield of corn as influenced by moisture level, nitrogen fertilization, and plant density; *Soil Science Society of America Proceedings,* vol. 23, pp. 242–245.

CHANG, J-H. (1968) *Climate and agriculture: an ecological survey;* Aldine, Chicago, 304 pp.

CULLER, R. C. (1970) Water conservation by removal of phreatophytes; *EOS, American Geophysical Union Transactions,* vol. 51, pp. 684–689.

DAGG, M. (1965) A rational approach to the selection of crops for areas of marginal rainfall in East Africa; *East African Agriculture and Forestry Journal,* vol. 30, pp. 296–300.

DAGG, M. AND BLACKIE, J. R. (1970) Estimates of evaporation in East Africa in relation to climatological classification; *Geographical Journal,* vol. 136, pp. 227–234.

DAVIES, J. A. (1967) The assessment of evapotranspiration for Nigeria; *Geografiska Annaler,* vol. 48A, pp. 139–156.

DENMEAD, O. T. AND SHAW, R. H. (1959) Evapotranspiration in relation to the development of a corn crop; *Agronomy Journal,* vol. 51, pp. 725–726.

DILS, R. E. (1953) Influence of forest cutting and mountain farming on some vegetation, surface soil and surface runoff characteristics, *Station Paper 24, Southeastern Forest Experiment, U.S. Forest Service,* Asheville, NC, 55 pp.

DUNFORD, E. G. AND FLETCHER, P. W. (1947) Effect of removal of streambank vegetation upon water yield; *EOS, American Geophysical Union Transactions,* vol. 28, pp. 105–110.

EAAFRO (1962) Hydrologic effects of changes in land use in some East African catchment areas; *East African Agriculture and Forestry Journal,* vol. 27, pp. 1–131.

ESCHNER, A. R. AND SATTERLUND, D. R. (1966) Forest protection and streamflow from an Adirondack watershed; *Water Resources Research,* vol. 2, pp. 765–783.

FRITSCHEN, L. J. AND SHAW, R. H. (1961) Evapotranspiration for corn as related to pan evaporation; *Agronomy Journal,* vol. 53, pp. 149–150.

GOODELL, B. C. (1952) Watershed management aspects of thinned young lodgepole pine stands; *Journal of Forestry,* vol. 50, pp. 374–378.

HARR, R. D. AND PRICE, K. R. (1972) Evapotranspiration from a greasewood-cheatgrass community; *Water Resources Research,* vol. 8, pp. 1199–1203.

HIBBERT, A. R. (1967) Forest treatment effects on water yield; in *Forest hydrology* (eds. W. E. Sopper and H. W. Lull), Pergamon Press, Oxford, pp. 527–543.

HIBBERT, A. R. (1971) Increases in streamflow after converting chaparral to grass; *Water Resources Research,* vol. 7, pp. 71–80.

HOLMES, R. M. (1961) Estimation of soil moisture content using evaporation data; *Canadian National Research Council, Proceedings Hydrology Symposium No. 2,* pp. 184–196.

HOLMES, R. M. AND ROBERTSON, G. W. (1959) A modulated soil moisture budget; *Monthly Weather Review,* vol. 87, pp. 101–105.

HOLSTENER-JORGENSEN, H. (1967) Influences of forest management and drainage on groundwater fluctuations; in *Forest hydrology* (eds. W. E. Sopper and H. W. Lull), Pergamon Press, Oxford, pp. 325–334.

JANDA, R. J. (1975) Recent man-induced modifications of the physical resources of the Redwood Creek unit of Redwood National Park, California, and the processes responsible for those modifications; *U.S. Geological Survey Open File Report,* Menlo Park, CA, 10 pp.

JANDA, R. J., NOLAND, K. M., HARDEN, D. R. AND COLMAN, S. M. (1975) Watershed conditions in the drainage basin of Redwood Creek, Humboldt County, California, as of 1973; *U.S. Geological Survey Open File Report,* Menlo Park, CA, 257 pp.

JOHNSON, E. A. AND KOVNER, J. L. (1956) Effect on streamflow of cutting a forest understorey; *Forest Science,* vol. 2, pp. 82–91.

JOHNSON, R. S. (1970) Evapotranspiration from bare, herbaceous, and aspen plots: a check on a former study; *Water Resources Research,* vol. 6, pp. 324–327.

LAW, F. (1956) The effect of afforestation upon the yield of water catchment areas; *Journal of the British Waterworks Association,* vol. 35, pp. 489–494.

LAMBERT, J. L., GARDNER, W. R. AND BOYLE, J. R. (1971) Hydrologic response of a young pine plantation to weed removal; *Water Resources Research,* vol. 7, pp. 1013–1019.

LEE, R. (1967) The hydrologic importance of transpiration control by stomata; *Water Resources Research,* vol. 3, pp. 737–752.

LEWIS, D. C. (1968) Annual hydrologic response to watershed conversion from oak woodland to annual grassland; *Water Resources Research,* vol. 4, pp. 59–72.

MCCULLOCH, J. S. G. (1965) Tables for the rapid computation of the Penman estimate of evapotranspiration; *East African Agriculture and Forestry Journal,* vol. 30, pp. 286–295.

MCGUINNESS, J. L. AND BORDNE, E. F. (1972) A comparison of lysimeter-derived potential evapotranspiration with computed values; *U.S. Department of Agriculture Technical Bulletin 1452.*

MONTEITH, J. L. (1959) The reflection of shortwave radiation by vegetation; *Quarterly Journal of the Royal Meteorological Society,* vol. 85, pp. 386–392.

NIXON, P. R. AND LAWLESS, G. P. (1968) Advective influences on the reduction of evapotranspiration in a coastal environment, *Water Resources Research,* vol. 4, pp. 39–46.

O'CONNELL, P. F. AND BROWN, H. E. (1972) Use of production functions to evaluate multiple use treatments on forested watersheds; *Water Resources Research,* vol. 8, pp. 1188–1198.

PALMER, W. C. AND HAVENS, A. V. (1958) A graphical technique for determining evapotranspiration by the Thornthwaite method; *Monthly Weather Review,* vol. 86, pp. 123–128.

PEGG, R. K. AND WARD, R. C. (1971) What happens to the rain? *Weather,* vol. 26, pp. 88–97.

PELTON, W. L. (1961) The use of lysimetric methods to measure evapotranspiration; *Proceedings of Symposium No. 2, Evaporation, National Research Council of Canda,* pp. 106–134.

PENMAN, H. L. (1948) Natural evaporation from open water, bare soil and grass; *Proceedings of the Royal Society of London,* Series A, vol. 193, pp. 120–145.

PENMAN, H. L. (1961) Weather, plant and soil factors in hydrology; *Weather,* vol. 16, pp. 207–219.

PENMAN, H. L. (1963) Vegetation and hydrology; *Technical Communication No. 53, Commonwealth Agriculture Bureau,* Harpenden, Eng.

PENMAN, H. L. (1967) Evaporation from forests: a comparison of theory and observation; in *Forest hydrology* (eds. W. E. Sopper and H. W. Lull), Pergamon Press, Oxford, pp. 373–380.

PEREIRA, H. C., ed. (1962) Hydrological effects of changes in land use in some East African catchment areas; *East African Agricultural and Forestry Journal,* vol. 27, 131 p.

PEREIRA, H. C. (1967) Effects of land use on the water and energy budgets of tropical watersheds; in *Forest hydrology* (eds. W. E. Sopper and H. W. Lull), Pergamon Press, Oxford, pp. 435–450.

PIERCE, L. T. (1958) Estimating seasonal and short-term fluctuations in evapotranspiration from meadow crops; *Bulletin of the American Meteorological Society,* vol. 39, no. 2, pp. 73–78.

Rider, N. E. (1957) Water loss from various land surfaces; *Quarterly Journal of the Royal Meteorological Society*, vol. 83, pp. 181–193.

Robinson, T. W. (1958) Phreatophytes; *U.S. Geological Survey Water Supply Paper 1423.*

Robinson, T. W. (1970) Evapotranspiration by woody phreatophytes in the Humboldt River valley, near Winnemucca, Nevada; *U.S. Geological Survey Professional Paper 491-D.*

Rutter, A. J. (1967) An analysis of evaporation from a stand of Scots Pine; in *Forest hydrology* (eds. W. E. Sopper and H. W. Lull), Pergamon Press, Oxford, pp. 403–417.

Schachori, A., Rosenzweig, D. and Poljakoff-Mayber, A. (1967) Effect of Mediterranean vegetation on the moisture regime; in *Forest hydrology* (eds. W. E. Sopper and H. W. Lull), Pergamon Press, Oxford, pp. 291–311.

Schneider, W. V. and Ayer, G. R. (1961) Effect of reforestation on streamflow in central New York; *U.S. Geological Survey Water Supply Paper 1602.*

Stanhill, G. (1958) Evapotranspiration from different crops exposed to the same weather; *Nature*, vol. 182, p. 125.

Swank, W. T. and Miner, N. H. (1968) Conversion of hardwood-covered watersheds to white pine reduces water yield; *Water Resources Research*, vol. 4, pp. 947–954.

Swanston, D. N. (1971) Principal mass movement processes influenced by logging, road building and fire; *Proceedings of the Symposium on Forest Land Use and the Stream Environment*, Oregon State University, pp. 29–39.

Szeicz, G., Endrodi, G. and Tajchman, S. (1969) Aerodynamic and surface factors in evapotranspiration; *Water Resources Research*, vol. 5, pp. 380–394.

Szeicz, G. and Long, I. F. (1969) Surface resistance of crop canopies; *Water Resources Research*, vol. 5, pp. 622–633.

Tajchman, S. (1971) Evapotranspiration and energy balances of forest and field; *Water Resources Research*, vol. 7, pp. 511–523.

Thornthwaite, C. W. and Mather, J. R. (1955) The water balance; *Laboratory of Climatology, Publication No. 8*, Centerton, NJ.

Turner, S. F. and Skibitzke, H. E. (1952) Use of water by phreatophytes in a 2,000-foot channel between Granite Reef and Gillespie Dams, Mariopa County, Arizona; *EOS, American Geophysical Union Transactions*, vol. 33, pp. 66–72.

Urie, D. H. (1967) Influence of forest cover on groundwater recharge, timing and use; in *Forest hydrology* (eds. W. E. Sopper and H. W. Lull), Pergamon Press, Oxford, pp. 313–324.

U.S. Soil Conservation Service (1970) Irrigation water requirements; *U.S. Department of Agriculture, Technical Release No. 21.*

Van Bavel, C. H. M. (1966) Potential evapotranspiration: the combination concept and its experimental verification; *Water Resources Research*, vol. 2, pp. 455–467.

Van Hylckama, T. E. A. (1963) Growth, development and water use by salt cedar under different conditions of weather and access to water; *International Association of Scientific Hydrology, Symposium of Berkeley, Publication No. 62*, pp. 75–86.

Veihmeyer, F. J. (1964) Evapotranspiration; Section 11 in *Handbook of applied hydrology* (ed. Ven te Chow), McGraw-Hill, New York.

Zahner, R. (1967) Refinement in empirical functions for realistic soil-moisture regimes under forest cover; in *Forest hydrology* (eds. W. E. Sopper and H. W. Lull), Pergamon Press, Oxford, pp. 261–274.

Typical Problems

5-1 Penman Calculation of Potential Evapotranspiration

Planners must often review technical reports on the evaluation of water resources, and they should be able to check calculations that form the basis of proposals for development. If this can be done, some unwise schemes will be recognized as such at an early stage before there are large commitments of money or prestige.

Suppose that you are reviewing a report on the water supply of a large river basin in a remote region covered by evergreen tropical forest. The river has been gauged for one year during the appraisal period, and this record is offered as "at least a very rough estimate" of the annual water supply. (This is not as improbable an example as the reader may suppose.) You want to check the proposed runoff value by calculating evapotranspiration and subtracting it from basin rainfall in an average year and during a dry year with a recurrence interval of 10 years.

Solution

Maps of the mean annual precipitation and its standard deviation are available from a national atlas. The mean is 300 cm/yr. Using the graphical technique demonstrated in Typical Problem 2-3, you estimate the annual rainfall that is exceeded 90 percent of all years to be 225 cm/yr.

The atlas also indicates that the rainfall is well distributed throughout the basin. These conditions could be more complicated in a real situation, but the method used would be similar. Since the climate is moist throughout the year, evapotranspiration should continue at the potential rate, and can be calculated using Penman's Equation (5-6), modified for E_t.

As a first estimate, average values of the meteorological variables can be used for the whole year, though in a real situation, it is better to use individual monthly averages. We will also assume that potential evapotranspiration is constant from year to year. This is a reasonable approximation because evapotranspiration does not vary greatly, but it will probably overestimate the calculated runoff during a dry year, because during periods of low rainfall, conditions tend to be sunnier and less humid than average.

From the national atlas, from tabulations in the local meteorological office, or from a calculation like that shown in Typical Problem 4-2, the following data are obtained:

Incoming solar radiation	690 cal/cm^2/day
Duration of sunshine	78 percent of maximum possible
Mean air temperature	21°C (294°K)
Windspeed	120 km/day
Vapor pressure	15 mb

It is usually necessary to assume that the windspeed is representative of that at 2 m above the vegetation canopy, and this can lead to a significant error if the mass-transfer contribution to evapotranspiration is large.

From Table 5-1, a value of 0.15 seems reasonable for the reflectivity of tropical forest. Net outgoing longwave radiation is calculated from Equation 4-12 with the modification for using duration of sunshine instead of cloudiness:

$$\begin{aligned} Q_{lw} &= (1.17 \times 10^{-7} \times 294^4)(0.56 - 0.08\sqrt{15}) \\ &\quad \times [0.10 + 0.90(0.78)] \\ &= 175 \text{ cal/cm}^2\text{/day} \end{aligned}$$

Equation 4-13 gives net radiation as

$$Q_n = 690(1 - 0.15) - 175 = 412 \text{ cal/cm}^2\text{/day}$$

and this converts to an equivalent depth of evaporation as

$$H = \frac{Q_n}{\rho L} = \frac{412}{1 \times 590} = 0.70 \text{ cm/day}$$

Penman's mass-transfer component in Equation 4-21 is

$$\begin{aligned} E_a &= (0.013 + 0.00016 \times 120)(25 - 15) \\ &= 0.32 \text{ cm/day} \end{aligned}$$

where $e_{sa} = 25$ mb is read from Figure 4-8(a) for an air temperature of 21°C.

Inserting these values into Equation 5-6 and using a value of 2.26 for Δ/γ from Table 4-6,

$$E_t = \frac{(2.26 \times 0.70) + 0.32}{3.26} = 0.58 \text{ cm/day}$$

or 213 cm/yr.

During an average year, therefore, the runoff from the basin would be approximately 300 − 213 = 87 cm/yr (1 cm depth of runoff is equivalent to 10,000 m^3 from each sq km of basin area). During the driest 10 percent of the years, however, the runoff would decline to less than 12 cm/yr. Before there is any further commitment of funds, the planner can decide which water supply proposals are sound.

5-2 Van Bavel Estimate of Potential Evapotranspiration

Calculate the potential evapotranspiration from a grass pasture by the Van Bavel modification of Penman's equation using the following data from Charleston, South Carolina in June (data from U.S. Environmental Data Service):

Incoming solar radiation	590 cal/cm²/day
Net outgoing longwave radiation (from Equation 4-12)	100 cal/cm²/day
Average air temperature	23°C (296°K)
Windspeed (measured at 2 m)	95 km/day
Atmospheric vapor pressure	18 mb
Saturation vapor pressure at 23°C	26 mb

Solution

Grass has a reflectivity of approximately 0.25, according to Table 5-1. Therefore, the net radiation from Equation 4-13 is 342 cal/cm²/day, equivalent to 0.58 cm/day of evaporation.

The height of the vegetation would be approximately 15 cm, and therefore $z_o \approx 1.5$ cm. The mass-transfer term in Equation 5-7 therefore becomes

$$E_a = \frac{3.64}{296} \times \frac{95}{\left[\ln\left(\frac{200}{1.5}\right)\right]^2}(26 - 18)$$

$$= 0.39 \text{ cm/day}$$

Obtaining Δ/γ from Table 4-6, Equation 5-6 for E_t then becomes

$$E_t = \frac{(2.51 \times 0.58) + 0.39}{2.51 + 1.0} = 0.53 \text{ cm/day}$$

The publication by McGuinness and Bordne (1972) contains a large amount of meteorological data upon which practice calculations of E_t can be made.

5-3 Thornthwaite Calculation of Potential Evapotransportation

In a midlatitude region, a planner wants to make a rapid and rough check on an estimate of water need for an irrigated potato crop. The only data easily available are air temperatures, and so it is decided to use the Thornthwaite method.

The mean annual air temperature at the location is 15°C, and the average temperature during a particular month of interest (July) is 20°C. The latitude is 50°N.

Solution

From Figure 5-5 the annual heat index, I, is 68. Using this value with the monthly temperature in Figure 5-4 yields a value of $E_t = 8.5$ cm/mo for a standard month. Entering Table 5-2 for July at 50°N yields a correction factor of 1.33. The final Thornthwaite estimate, therefore, is 11.3 cm/mo.

This calculation gives only the potential evapotranspiration from a field. It does not take into account evaporation and percolation losses associated with conveyance of water to the field by canal or the extra water that must be applied to the field to leach away salts that would otherwise accumulate in the soil as water evaporated. These "nonproductive" water requirements can often be of the same general magnitude as the consumptive use, and so a rough estimate of the total water needed would involve a doubling of the calculated amount. The planner would have to consult an agricultural engineer familiar with local conditions to pursue this matter further. Or, the U.S. Soil Conservation Service (1970) handbook could be used.

5-4 Calculation of Irrigation Water Requirements

A rural planner is called upon to estimate the seasonal pattern of consumptive use for an irrigated crop of grain corn in a subtropical highland region. The duration of the growing season at the site is 110 days, mean air temperatures are available for the site, and the latitude is 28°S.

Solution

Using the air temperatures and data for grain corn from Table 5-3, Table 5-6 is set up for the calculation of the Blaney-Criddle Formula. Values of d are interpolated from Table 5-4. The tabulated values of E_t are for periods of 11 days (10 percent of the growing season).

In addition to knowing the average requirements, the planner may need to calculate, or to cooperate with an agricultural engineer on other matters of design. These may include the short-term peak consumptive use, the reduction of irrigation demand by rainfall, the efficiency of applied water in satisfying the irrigation demand, the amount of excess water that must be applied to leach away dissolved salts concentrated by evaporation, and seepage losses from conveyance channels. Methods for calculating all of these are outlined in the U.S. Soil Conservation Service (1970) technical release on irrigation water requirements, which contains several other typical problems.

Table 5-6 Calculation of the Blaney-Criddle Formula.

PERCENTAGE OF GROWING SEASON	APPROXIMATE DATE	T_a (°C)	k	d	E_t (CM)
0–10	Jan. 15–Jan. 25	15	.46	.093	1.7
10–20	Jan. 26–Feb. 5	19	.54	.089	2.5
20–30	Feb. 6–Feb. 16	21	.64	.083	3.1
30–40	Feb. 17–Feb. 27	23	.82	.084	4.5
40–50	Feb. 28–Mar. 10	25	1.00	.085	6.2
50–60	Mar. 11–Mar. 21	26	1.08	.085	7.1
60–70	Mar. 22–Apr. 1	28	1.08	.084	7.7
70–80	Apr. 2–Apr. 12	29	1.03	.080	7.4
80–90	Apr. 13–Apr. 23	30	.97	.076	6.9
90–100	Apr. 24–May 4	30	.89	.075	6.3
Total					53.4

6

Water in the Soil

Significance of Infiltration

Infiltration is the movement of water into the soil. There is a maximum rate at which the soil in a given condition can absorb water; this upper limit is called the infiltration capacity of the soil. If rainfall intensity is less than this capacity, the infiltration rate will be equal to the rainfall rate, whereas if rainfall intensity exceeds the ability of the soil to absorb moisture, infiltration occurs at the capacity rate. The excess of rainfall over infiltration collects on the soil surface and runs over the ground to streams (see Figure 6-1). Infiltration rates are expressed in units of depth per unit time, the same as rainfall intensities. They refer to the depth of a sheet of water that would soak into the soil in a chosen time interval.

The soil surface is a filter that determines the path by which rainwater reaches a stream channel. Water that does not infiltrate runs quickly over the ground surface, whereas water entering the soil moves much more slowly underground. The soil, therefore, plays a major part in determining the volume of storm runoff, its timing, and its peak rate of flow. These are all of importance to the hydrologist interested in the planning of culverts, bridges,

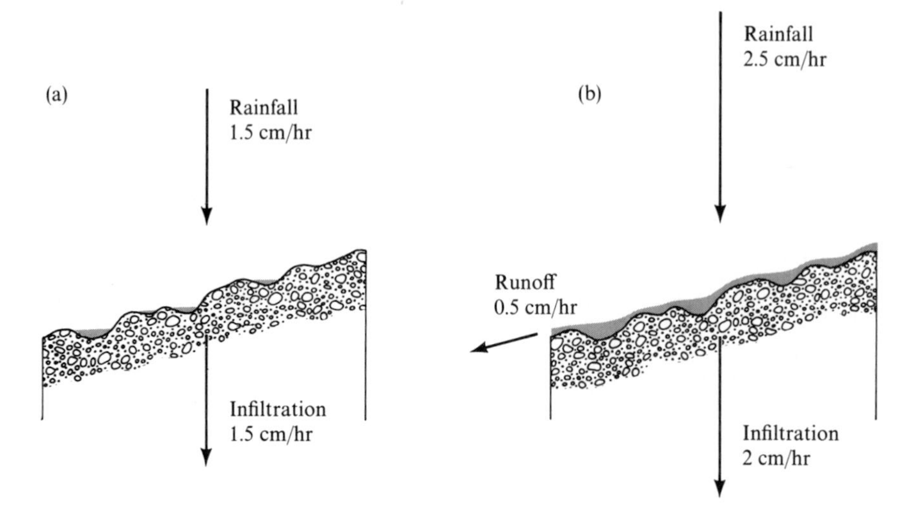

Figure 6-1 Surface runoff occurs when rainfall intensity exceeds infiltration capacity. (a) Infiltration rate = Rainfall rate, which is less than infiltration capacity. (b) Runoff rate = Rainfall intensity − Infiltration capacity.

and other small structures. In running over the ground surface, water is capable of eroding topsoil and important organic residues on the land surface (see Chapter 15). The soil conservationist is concerned with either inducing this overland flow to infiltrate or conducting it safely away from fields or farm structures. Geomorphologists are also concerned with the magnitude, frequency, and spatial characteristics of infiltration relative to rainfall intensity, because overland flow is an important agent of landscape development.

The water that infiltrates the soil controls to some extent the water available for evapotranspiration. This supply of soil moisture is the effective rainfall as far as plants are concerned. Ecologists and agriculturalists, therefore, need to refine their understanding of the relationships between plants and their water supply by considering infiltration and runoff. Infiltrated water that is not returned to the atmosphere by evapotranspiration reaches the groundwater system and supplies streamflow. Increasing the amount of infiltration may augment streamflow during dry weather, which is important for water supply, waste dilution, and other uses.

Process of Infiltration

Close examination of a lump of soil or the sides of a pit dug through a soil profile reveals that soils consist of millions of particles of sand, silt, and clay, separated by channels of different sizes (see Figure 6-2). These channels

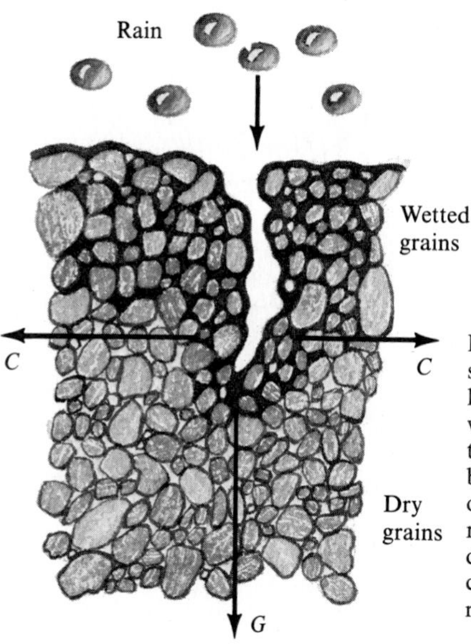

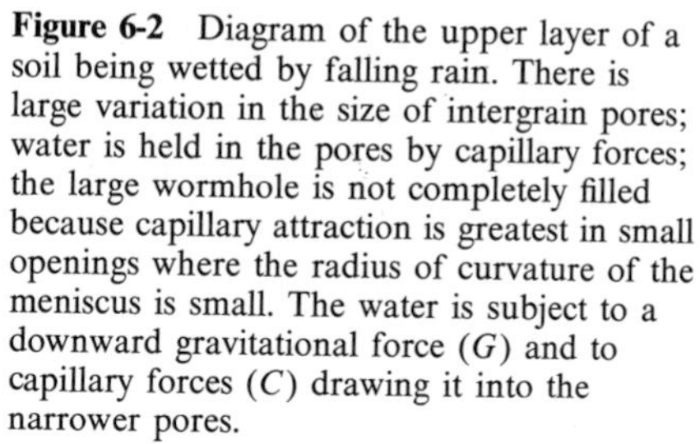
Figure 6-2 Diagram of the upper layer of a soil being wetted by falling rain. There is large variation in the size of intergrain pores; water is held in the pores by capillary forces; the large wormhole is not completely filled because capillary attraction is greatest in small openings where the radius of curvature of the meniscus is small. The water is subject to a downward gravitational force (G) and to capillary forces (C) drawing it into the narrower pores.

include shrinkage cracks, wormholes, root holes, spaces between lumps or "crumbs" of soil, and very fine sqaces between the individual particles themselves. Such cracks, holes, and fine spaces are called soil pores.

When rainfall reaches the ground surface, some or all of it enters soil pores. It is drawn into soil by the forces of gravity and capillary attraction. The rate of entry of water by free gravity flow is limited by the diameters of the pores. As water moves along such pores, it is subject to flow resistance, which increases as the diameter of the pore decreases. Under the influence of gravity, water moves vertically downward through the soil profile. On the other hand, capillary forces may move water vertically up or down or with a horizontal component. They act to draw water into the narrower pores, just as capillary forces drawing water up a narrow glass tube are greater than those in a wide tube. Although such forces are strongest in soils with very fine pores, the pores may be so small that there is considerable resistance to flow through them. In large pores, such as wormholes or root holes, capillary forces are negligible and water moves downward under free gravity drainage. While flowing down through these passages, water is subject to lateral capillary forces, which draw it into the finer intergranular spaces leading off from the larger channel (see Figure 6-2).

Infiltration, therefore, involves three interdependent processes: entry through the soil surface, storage within the soil, and transmission through the soil. Limitations on any of these processes can reduce infiltration rates. The last two processes will be treated more completely in the section on the storage and transmission of soil moisture. It is important to realize that

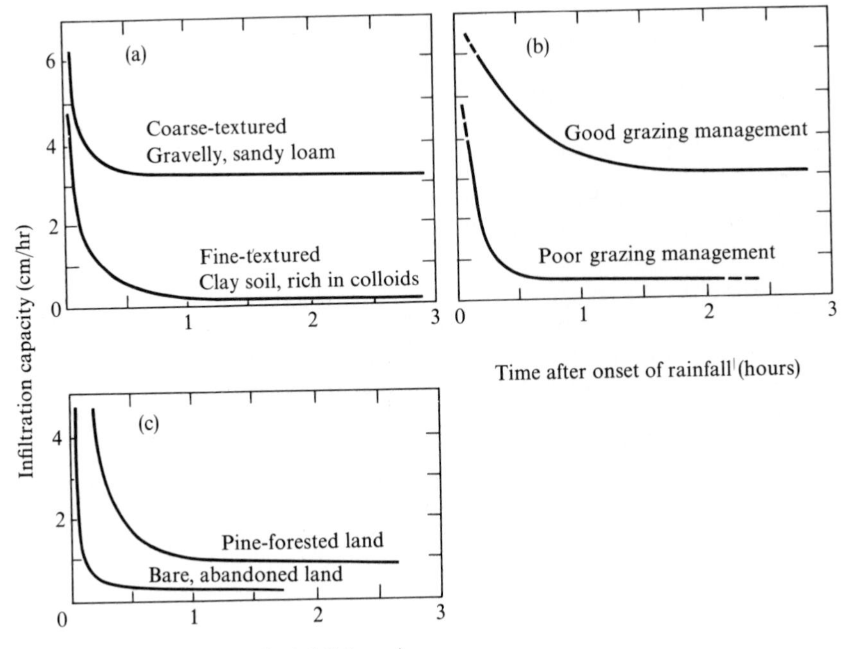

Figure 6-3 Infiltration capacity curves for soils of (a) different texture, (b) vegetative cover, and (c) land-use practice. (From Strahler 1975, p. 414.)

infiltration and the subsurface storage and movement of soil moisture are closely interrelated.

The rate of infiltration declines rapidly during the early part of a storm and reaches an approximately constant value after one or two hours of rain (see Figure 6-3). Several processes combine to reduce the infiltration capacity during a storm. The filling of fine pores with water reduces capillary forces drawing water into pores and fills the storage potential of the soil. Clay particles swell as they become wetter and reduce the size of pores. The impact of raindrops breaks up soil aggregates, splashing fine particles over the surface and washing them into pores where they impede the entry of water. In large rainstorms it is the final, low rate of infiltration that largely determines the amount of surface runoff that is generated.

Controls of Infiltration

Many factors influence the shape of the infiltration capacity curve, but the most important variables are rainfall characteristics, soil properties, vegetation, and land use.

Long, intense rainfall packs down the loose soil surface, disperses fine soil particles, and causes them to plug soil pores. Rainstorms of long duration fill up the storage potential of the soil and cause clays to swell or may even saturate the soil completely.

Coarse-textured soils such as sands have large pores down which water can easily drain, while the exceedingly fine pores in clays retard drainage. If the soil particles are held together in aggregates by organic matter or a small amount of clay, the soil will have a loose, friable structure that will allow rapid infiltration and drainage. The depth of the soil profile and its initial moisture content are important determinants of how much infiltrating water can be stored in the soil before saturation is reached. Deep, well-drained, coarse-textured soils with a large content of organic matter, therefore, will tend to have high infiltration rates, whereas shallow soil profiles developed in clays will accept only low rates and volumes of infiltration. If the soil is invaded by dense ice lenses known as "concrete frost," its infiltration capacity may be reduced almost to zero, although "porous frost," formed by the growth of disseminated ice crystals, does not reduce the rate of water intake (Post and Dreibelbis 1942).

Vegetative cover and therefore land use are very important controls of infiltration. Vegetation and litter protect soil from packing by raindrops and provide organic matter for binding soil particles together in open aggregates. Soil fauna that live on the organic matter assist this process by churning together the mineral particles and the organic material. The manipulation of vegetation during land use causes large differences in infiltration capacity under the same rainfall regime and soil type (Figure 6-4). In particular, the stripping of forests and their replacement by crops that do

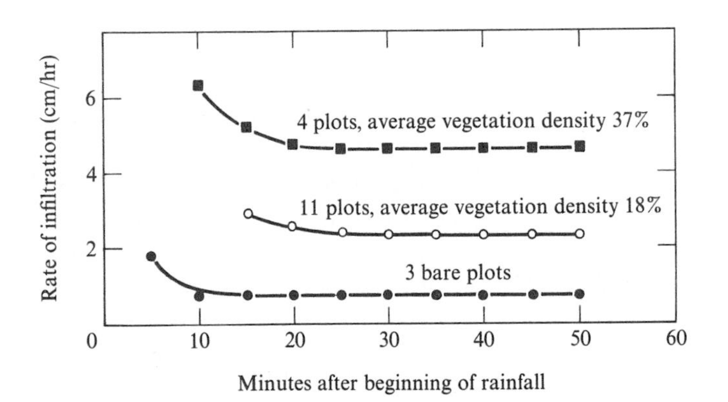

Figure 6-4 Infiltration capacity curves for grazed desert shrub having different densities of vegetation cover. Curves represent average values from 18 experiments with a sprinkled plot infiltrometer. Soil is wet clay loam, New Mexico. (From Smith and Leopold, © 1942 The Williams & Wilkins Co., Baltimore.)

not cover the ground efficiently and do not maintain a high organic content in the soil often lower the infiltration capacity drastically (Dils 1953). The problem is further aggravated by land management practices, such as plowing, which break up soil aggregates, or if the soil surface is compacted by vehicles or trampling by livestock. The most extreme reduction of infiltration capacity, of course, involves the replacement of vegetation by an asphalt or concrete cover in urban areas. Each of these reductions of infiltration capacity increases the amount of surface runoff, and can produce soil erosion, flooding, and other economic and social costs, including the cost of practices such as soil conservation, stream channel modification, and other attempts to minimize damages caused by excessive runoff.

Measurement and Estimation of Infiltration Capacity

The way infiltration capacity is measured or estimated depends upon the problem at hand. If detailed information is required for small areas in the design of some valuable installation or land management plan, direct field measurements may be made and used in the calculation of runoff rates. Even for high-cost projects, it is impossible to make direct measurements over large areas. In such a situation a few measurements of infiltration are made in the major soil types of the region and used only as indices. Most planning problems do not justify detailed field measurement, and the infiltration capacity must be estimated from soil properties and vegetative cover. Errors from these procedures are large and provide little more than an index of the runoff potential of a catchment. In the evaluation of infiltration rates, use is made of natural and artificial rainfall events or of water ponded on the surface of the soil.

A single-ring cylinder infiltrometer is simply a tube with a diameter of about 10 to 30 cm that is driven between 5 and 50 cm into the soil. Water is ponded 1–2 cm deep inside the tube and is maintained at a constant level by supplying water from a graduated reservoir. The rate of supply required to maintain a constant depth is measured in the graduated reservoir and indicates the rate at which water is entering the soil. Measurements with cylinder infiltrometers provide only indices of the variations between sites, generally yielding values that are 2–10 times as great as infiltration capacities measured during natural rainfall under the same conditions.

These measurements are of value, however, for some purposes. They simulate to a degree the conditions of infiltration in some irrigated fields and of seepage from ditches, septic systems, broken sewers, and broken water pipes. Cylinder infiltrometers provide relative rates of infiltration, which indicate the importance of certain controls of the process, such as land use, soil freezing, or soil texture. The main advantage of cylinder infiltrometers is that they are cheap, simple to install and use, and hand portable.

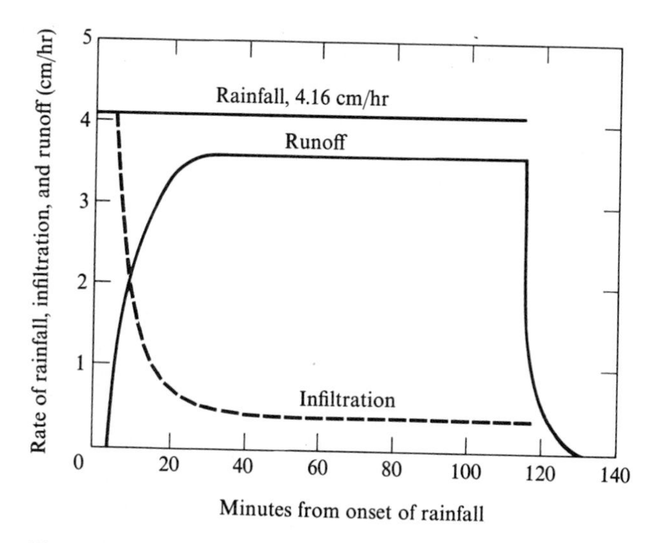

Figure 6-5 Rainfall intensity, runoff rate, and infiltration capacity for an artificial storm of constant intensity on an experimental plot.

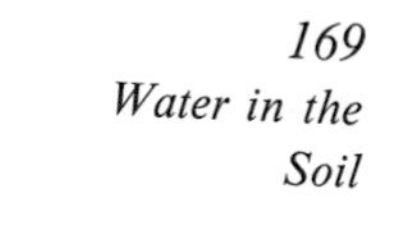

Several types of equipment have been designed for simulating rainfall on small plots; they are generally called sprinkling infiltrometers. The plots vary in size from less than one to more than 1000 square meters, and spray nozzles are designed to simulate closely the size and fall velocity of natural raindrops. Rainfall is generated at a constant intensity, and the rate of runoff from the plot is measured. The difference between these two rates is the infiltration capacity, as illustrated in Figure 6-5. Most sprinkling infiltrometers are expensive, and with their pumping equipment and water supplies are not easily portable. It is possible to make small, cheap systems that give realistic measurements.

Infiltration rates can be estimated from the response of runoff from plots or small drainage basins to rainfall. A very approximate technique, known as the "average infiltration" method, is most commonly used. The technique is demonstrated in Figure 6-6. Complex storms consisting of several bursts of rain are chosen. The total amount of rain in each burst is tabulated. The volume of stormflow resulting from each burst is separated from the hydrograph as shown by the shaded areas in Figure 6-6. The difference between the rainfall and runoff volumes indicates the volume of infiltration during each burst of rain. The time interval over which this infiltration occurred can be estimated as the duration of the burst of rain. Thus an average infiltration rate can be calculated for each part of the storm, and a composite infiltration curve can be drawn. The technique is obviously most reliable for small areas of uniform soil cover, vegetation, and land management.

The method is approximate and involves assumptions about the way storm runoff reaches a stream and moves down the channel to produce a

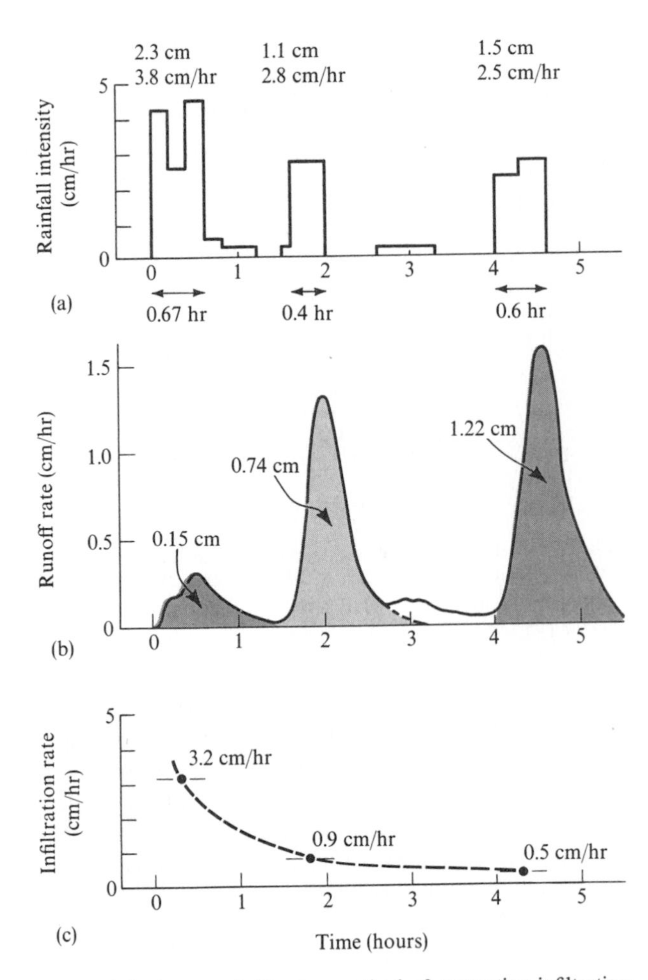

Figure 6-6 Average infiltration method of computing infiltration capacity curve for a small drainage basin. Bursts of rainfall plotted in the upper diagram (a) cause separate hydrograph rises (b). Each burst provides one point on the infiltration capacity curve (c).

storm hydrograph. This subject will be discussed more fully in Chapter 9. The procedure gives results that are useful for management purposes but is probably measuring something that may be vaguely defined as the "storm response" of the basin, instead of specifically measuring the infiltration rate.

Whichever technique is used for measuring infiltration capacity, difficult sampling problems arise in applying the measured rates to large drainage basins. Infiltration is controlled by many factors, and even within a uniform combination of, say, soil type and vegetation, there is great spatial variability in infiltration rates on any one day. Because of the large variations and the expense of defining them with an infiltrometer, it is not usually possible to obtain measured rates of infiltration for large areas. Yet such information might be desirable for calculating runoff for design purposes, for estimating the probable hydrologic effects of a land-use change, or for assessing the probable effective moisture supply to a new crop. For such purposes, it is necessary to rely on some empirical relationships, developed mainly by the U.S. Soil Conservation Service. These relationships are generalized correlations between infiltration, measured by one of the methods outlined in the previous section, and some property or properties of the soil–vegetation association. The methods (see Musgrave and Holtan 1964) need to be extended and tested for their geographical transposability.

For a number of soil–land use combinations, average infiltration capacity curves like those in Figure 6-3 are constructed. These standard infiltration curves are then used to represent the sampled soil–land use complex. Other estimates can be made simply from the soil texture. Infiltrometer experiments and hydrograph analysis have provided large amounts of data on the minimum rates of infiltration after a soil has been subjected to a period of prolonged wetting under a given land use. Musgrave and Holtan (1964) have grouped a large number of agricultural soils of the United States into the four classes shown in Table 6-1. The Soil Survey Reports, published by the U.S. Soil Conservation Service, indicate the appropriate group for each soil type in a county. The soils are also ranked against one another within each group. Other soils can be added on the basis of direct measurements or on the basis of indirect information on soil texture, profile depth, or organic content. In this way, the direct measurements on a few key soils can be rapidly extended.

Table 6-1 Classification of soils in order of minimum infiltration rates after a period of prolonged wetting when planted to row crops. (From Musgrave and Holtan, *Handbook of Applied Hydrology,* edited by Ven te Chow. Copyright © 1964 by McGraw-Hill, Inc. Used with permission of McGraw-Hill Book Company.)

GROUP	MINIMUM INFILTRATION RATE (MM/HR)	SOIL CHARACTERISTICS
A	8 to 12	Deep sands, deep loesses, aggregated soils
B	4 to 8	Shallow loess and sandy loams
C	1 to 4	Many clay loams, shallow sandy loams, soils low in organic matter, and soils high in clay
D	0 to 1	Soils of high swelling percentage, heavy plastic clays, and certain saline soils

Table 6-2 Relative influence of land use on the minimum rate of infiltration. (From U.S. Soil Conservation Service 1972.)

Highest infiltration	Woods, good
	Meadows
	Woods, fair
	Pasture, good
	Woods, poor
	Pasture, fair
	Small grains, good rotation
	Small grains, poor rotation
	Legumes after row crops
	Pasture, poor
	Row crops, good rotation (more than one quarter in hay or sod)
	Row crops, poor rotation (one quarter or less in hay or sod)
Lowest infiltration	Fallow

Table 6-2 is a list of various land-use practices in approximate order of their influence upon the infiltration capacities of soils. They are based on hydrograph analysis and infiltrometer measurements.

It is important, then, that the planner be able to recognize from field inspection, soil reports, and soil maps, the areas of a drainage basin that have high or low infiltration capacities. Assistance is available from the local office of the Soil Conservation Service. Because of the great temporal and spatial variability of infiltration capacity and because of the cost of measuring it with a sprinkling infiltrometer, calculations of runoff from drainage basins are usually made on a quasi-physical or a statistical basis, as we will describe in Chapter 10. Although better infiltration data are becoming available and are slowly being incorporated into physical models of the runoff process, there is reason to doubt the validity of infiltration-based runoff models in many regions (see Chapter 9). The results are of some use in engineering applications, however, in that they give useful design answers once the models have been calibrated for a particular drainage basin.

Soil Moisture

Water in the root zone of plants constitutes only 0.064 percent of the 39 million cubic kilometers of fresh water in the world. In the continental United States soil moisture makes up 0.43 percent of the 145,000 cubic kilometers of fresh water. Yet the significance of soil moisture is out of all proportion to its small size relative to deeper groundwater, ice caps, and lakes. For it is upon the meager supply of soil moisture that we depend for our food. Vegetation growth requires a moisture supply, as described in Chapter 5, and the physiological stresses placed upon plants during times of low soil moisture form one of the major determinants of plant and animal distributions and therefore of land use. In fact, drought is best defined in terms of how much moisture is in the soil, rather than how much rain has recently fallen.

High soil-moisture content can also be troublesome for society. If soils remain very wet for long periods of time, they are unsuitable for septic tank and drainfield operation. Plowing or harvesting by machinery is often thwarted by the poor trafficability of wet soils in New England, and many a corn crop has been lost as a result.

Soil moisture conditions can be easily upset. In some pastoral regions of northern Kenya, for example, heavy grazing and large rainstorms during the past 15 years have eroded the loose, shallow topsoil from extensive areas. The dense subsoil has a lower infiltration capacity so that the amount of water stored in the soil for plant growth is much lower than it formerly was. Trees that could withstand droughts when they were rooted in the loose topsoil now find the drier subsoil intolerable and are dying of desiccation, even though the rainfall has increased. In more humid areas soil moisture can be increased by some of the manipulations of vegetation referred to at the end of Chapter 5. Removal of a forest cover often increases soil moisture (Bethlahmy 1962), and new springs can sometimes be seen on newly cleared forestland.

The relationship between soil moisture and evapotranspiration has been reviewed in the previous chapter, and Chapter 8 describes how to make calculations of irrigation requirements to keep soil moisture at optimum levels for crop growth.

In the rest of this chapter we will describe the storage and movement of water in the zone above the main groundwater body. Here the soil and rock are usually unsaturated; i.e., not all the voids are filled with water. This unsaturated belt of subsurface water is often referred to as the vadose zone.

Storage of Water in Soils

When rain or meltwater enters the soil, a portion of it is stored in the pore spaces, while the remainder drains downward under the force of gravity. The maximum amount of water that can be stored in the soil is determined by the porosity (i.e., the percentage of the soil volume that is occupied by voids). In this condition of maximum storage, the soil is said to be saturated, and its saturation moisture content is equal to the porosity.

As water drains out of the profile, the soil pores begin to empty. Some of the water cannot drain out, however, because it is held in narrow pores by

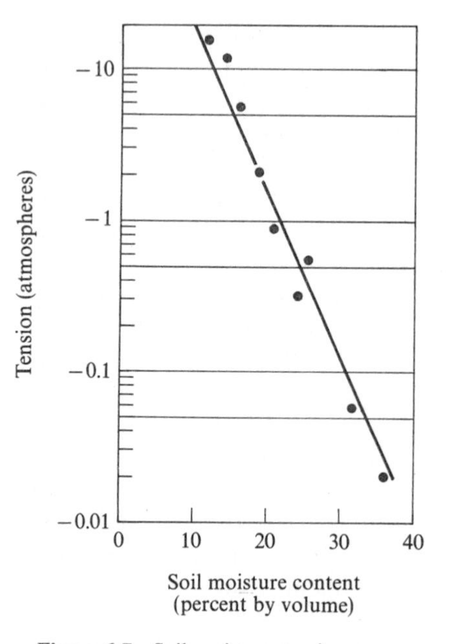

Figure 6-7 Soil moisture tension as a function of soil moisture content for a sandy loam. (After Hewlett 1961.)

capillary forces, like those that hold water up inside glass capillary tubes in a laboratory experiment (Satterly 1961). Because capillary forces are strongest in narrow tubes and pores, soil water is drawn most strongly into the finer pores. The largest pores drain more easily. In saline soils the water is also held by osmotic forces, like those that draw water through a semipermeable membrane between a concentrated and a dilute solution in the laboratory (Carter 1962, Yong and Warkentin 1966). Both capillary and osmotic forces pull water into the pores and cause it to be under a pressure less than atmospheric pressure; in other words, water in an unsaturated soil can be thought of as being held under tension. The tension, or negative pressure associated with various soil moisture contents, is shown for one soil in Figure 6-7.

The rate of soil drainage declines rapidly as the moisture content decreases and the remaining soil water is held more tightly in narrower pores. After one to several days, drainage ceases. When this happens, the water remaining in the soil is held under capillary and osmotic forces that are great enough to resist the force of gravity. Water is held in small amounts at the "necks" of soil pores and as thin films around each particle (see Figure 6-8). The moisture content in this condition is called the field capacity of the soil (again expressed as a percentage by volume), and it varies with soil texture, as shown in Figure 6-9. This is the maximum amount of water that a freely drained soil can store for long periods. The tension on soil water at field

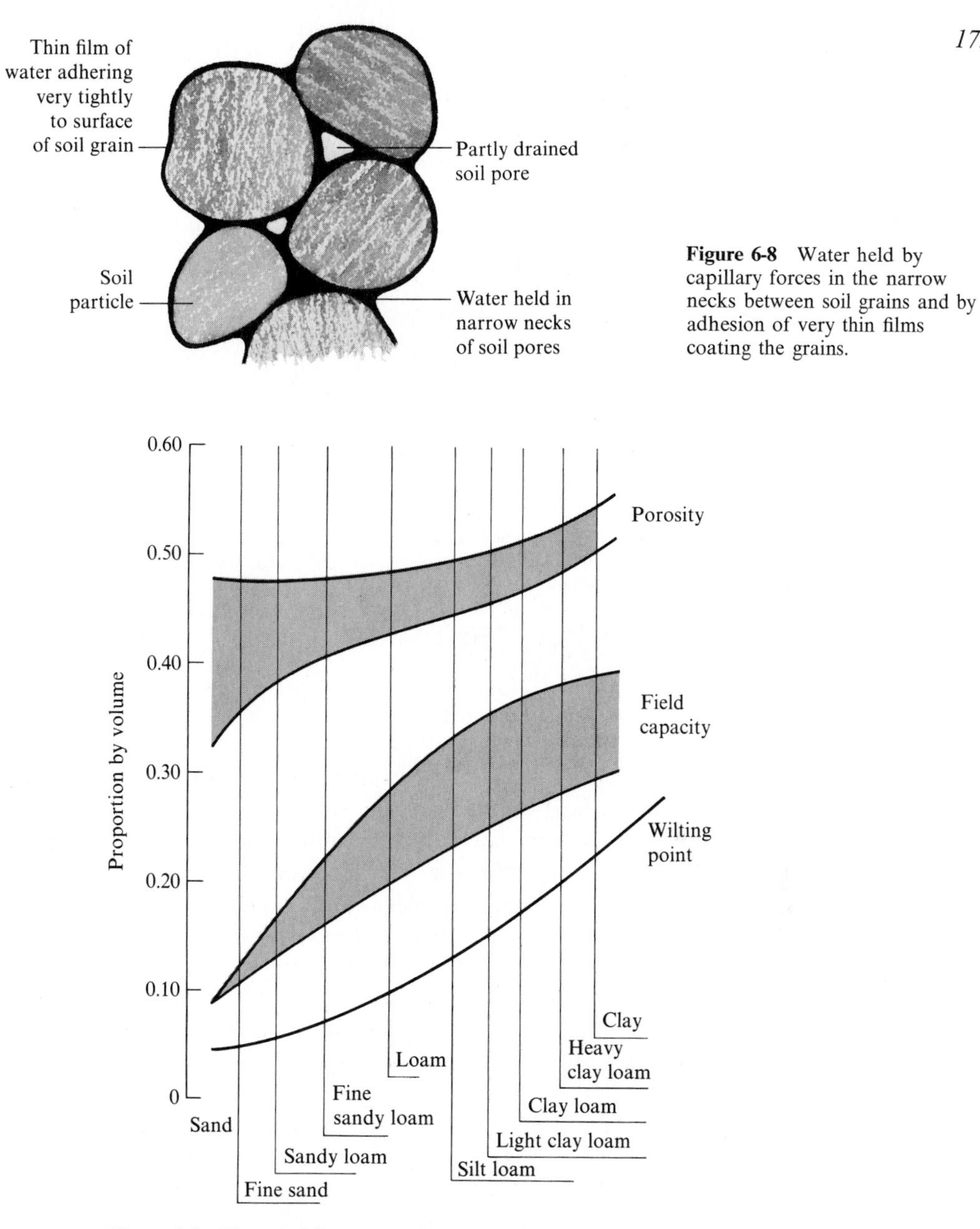

Figure 6-8 Water held by capillary forces in the narrow necks between soil grains and by adhesion of very thin films coating the grains.

Figure 6-9 Water-holding properties of various soils on the basis of their texture. For a fine sandy loam the approximate difference between porosity, 0.45, and field capacity, 0.20, is 0.25, meaning that the unfilled pore space is 0.25 times the soil volume. The difference between field capacity and wilting point is the available water capacity referred to in Chapter 5. Porosity and field capacity will be used later in this chapter. (After U.S. Department of Agriculture 1955, with additions.)

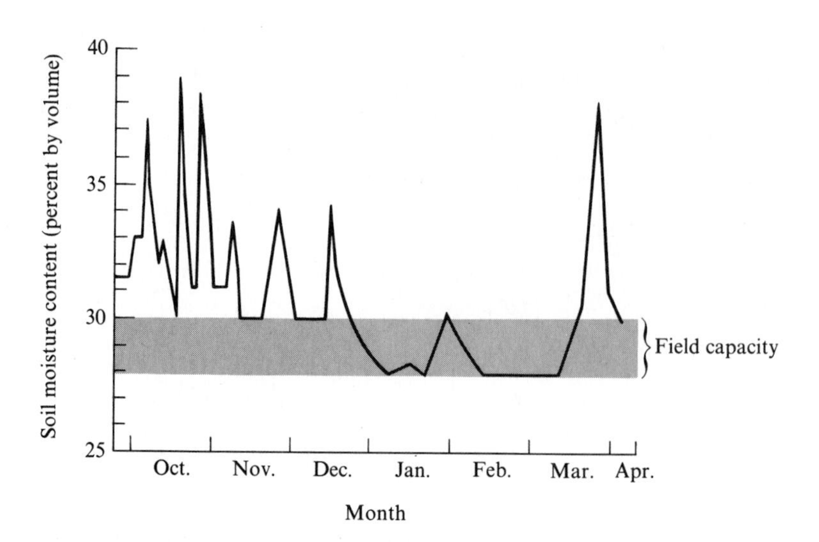

Figure 6-10 Estimation of field capacity of a soil by repeated measurements of soil moisture *in situ.* After each rainstorm or snowmelt period, the soil drained rapidly to a moisture content of 28–30 percent by volume, and field capacity is thus estimated as being within that range. The soil is a sandy loam at a depth of 0.3 m under a Vermont hillside.

capacity usually varies between 0.1 and 0.3 atmospheres. The moisture content of a soil at field capacity can be estimated then by measuring the volume of water which the soil can retain after it has been subjected to a suction of 0.1 to 0.3 atmospheres in the laboratory. Repeated measurement of the moisture content of a soil in place can also indicate the field capacity if the measurements are made during winter or below the root zone to avoid moisture withdrawals by evapotranspiration. Figure 6-10 shows how the soil moisture varied through autumn and winter at one site. After each rainstorm or snowmelt the soil drained rapidly to a moisture content of 28–30 percent by volume.

This concept of field capacity is an approximate one, because gravitational drainage can still occur at very low rates as the moisture content of the soil declines below field capacity. This drainage, though slow, is sufficient to supply the dry-weather flow of mountain streams in some regions (Hewlett 1961). The major abstraction of water below the field capacity, however, is by evapotranspiration. As the plants extract moisture from the soil, the water is drawn from progressively finer pores. The plant must therefore exert progressively greater suction to withdraw this water. Eventually, the moisture content of the soil becomes so low that plants cannot exert a great enough suction to remove any more water. At this point, the plants wilt, and the moisture content of the soil at which this occurs is known as its permanent wilting point. Most plants are able to exert a maximum suction of approximately 15 atmospheres, and so the wilting point of the soil is usually defined as the moisture content of the soil (percentage by volume) that can be maintained against a suction of 15 atmospheres.

There is some soil water, therefore, that is not available to plants because it is held too tightly. If the soil moisture content exceeds the field capacity, the excess drains rapidly out of the soil profile (if drainage is not impeded) and is also not available to plants. Vegetation can only use that portion of the soil moisture that is held between field capacity and the wilting point. Soils with large amounts of available water (i.e., large differences between field capacity and wilting point) are generally more favorable to plant growth. Because the amount of available water depends upon the size distribution of the soil pores, which in turn is related to the texture of the soil, the water availability of a soil generally varies with its texture in the manner shown in Figure 6-9. Chapter 5 discussed how the amount of water that can be stored by the soil is vital to plants because it controls the amount and rate of evapotranspiration. The water-holding characteristics of a soil, therefore, are a major control of the suitability of soils for plant growth.

Methods for calculating the amount of soil moisture in storage are discussed in Chapter 8.

Measurement of Soil Moisture

Soil moisture can be measured directly by collecting samples of soil in a cylinder driven into the soil. The sample is weighed, dried in an oven, and reweighed to evaluate the weight and therefore the volume of soil water it contained. Samples withdrawn from different depths allow the definition of a profile of soil moisture. Although it is simple and cheap, the usefulness of this method is limited by the variability of measurements from adjacent cores. Large numbers of samples must be taken to define the average soil moisture of even a small area, and the sampling technique destroys the measurement site. More precise, nondestructive measurements of soil moisture can be made with a nuclear probe, as illustrated in Figure 6-11. The nuclear probe is lowered into a metal tube installed in the soil, and it allows repeated measurements of a profile of moisture through the same undisturbed soil. The instrument, however, is expensive and will only be useful where repeated measurements must be made at a large number of sites.

Soil moisture can be monitored electrically with the aid of nylon soil moisture blocks, as illustrated in Figure 6-12. The moisture content of the porous block equilibrates with water in the surrounding soil, and is evaluated indirectly by measuring the electrical resistance of the nylon between two wires set into the block. These instruments are also nondestructive, and a set of blocks placed at various depths will define the soil moisture profile and its changes.

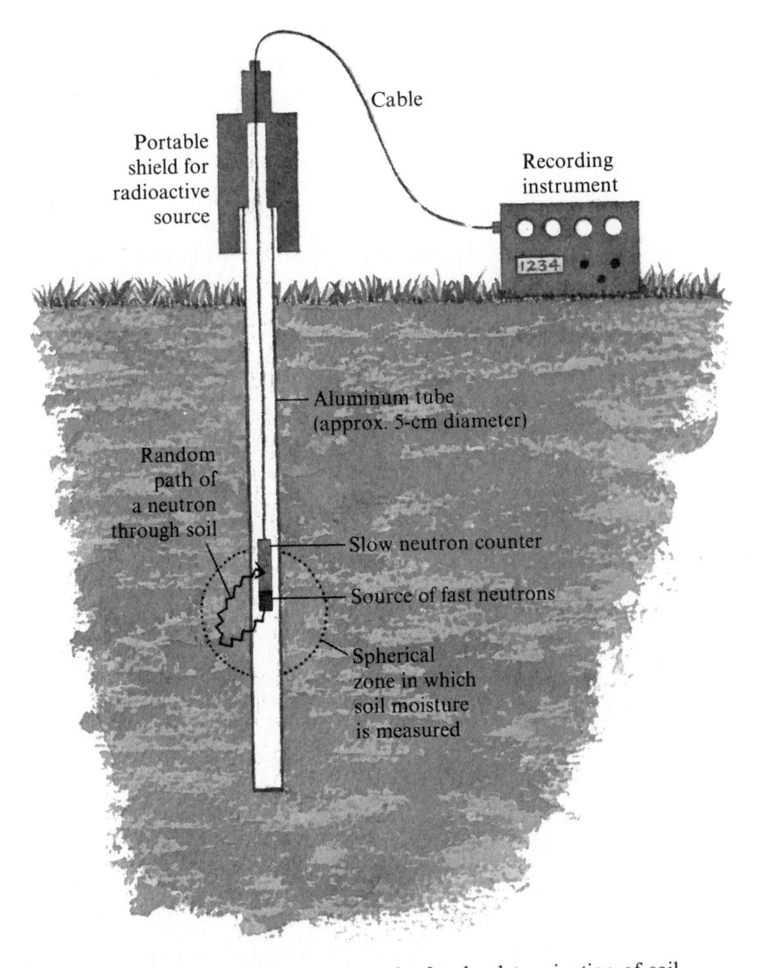

Figure 6-11 Diagram of a nuclear probe for the determination of soil moisture. A radioactive source lowered down the tube to the required depth emits fast neutrons that follow random paths through the soil and are slowed by collision with hydrogen atoms of soil water. The neutrons then migrate through the counter, which registers only slow neutrons. The number counted per unit of time is a measure of the amount of water in the soil within the spherical volume indicated.

Finally, soil moisture can be evaluated by measuring the tension under which the water is held in the soil. This is accomplished with the aid of a tensiometer, a small electrically powered pressure transducer installed in the soil. Moisture content can be evaluated from the negative pressure by using a diagram like that in Figure 6-7 for each soil type.

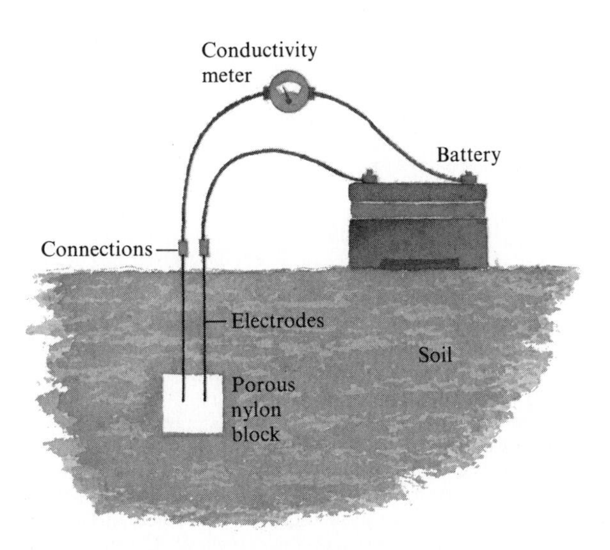

Figure 6-12 Nylon soil moisture block for measuring electrical resistance and, after calibration, the soil moisture content.

Movement of Water in Soil

When rainwater or meltwater enters the soil, a portion of it is stored in the soil pores, raising the moisture content. As the soil becomes wetter, its ability to conduct moisture increases until it is able to transmit water as fast as the rain is entering the ground surface. When this condition is reached, the soil moisture content of the surface becomes constant and a wave of moisture percolates downward, wetting successively deeper layers to a moisture content at which their hydraulic conductivity* is equal to the infiltration rate. Figure 6-13 shows the calculated movement of the soil moisture wave into an initially dry sand at two rainfall intensities and infiltration rates. When infiltration ceases, the soil begins to drain (see Figure 6-14) and the rate of downward percolation slows, eventually becoming very small as the soil moisture content approaches field capacity. This process is illustrated for several rainstorms in the report by Dunne (1970).

As water percolates down through the soil, it displaces water stored during earlier rainstorms. Laboratory experiments by Horton and Hawkins (1965) demonstrated this displacement, and Smith (1967) produced photographs

*A more formal discussion of the motion of subsurface water will be introduced in Chapter 7. For the present purpose, hydraulic conductivity is the ability of soil to transmit water, and it increases as soil moisture content increases.

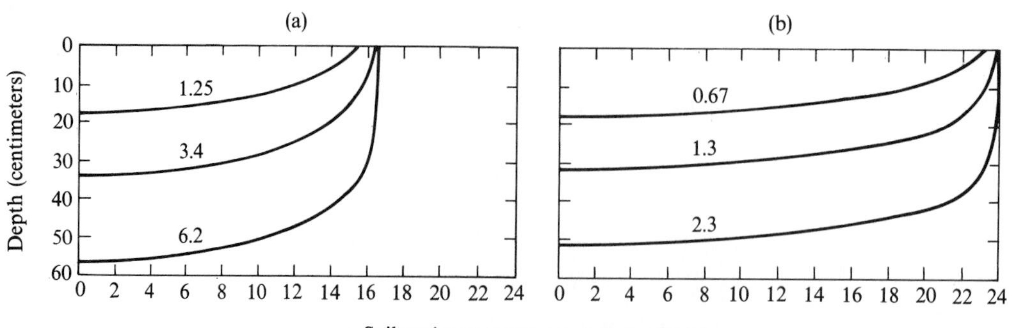

Figure 6-13 Changes of soil moisture with depth during infiltration into sand with an initial moisture content of 0.5 percent. Numbers on the curves indicate the duration of infiltration in hours. The rainfall intensity in (b) is 3.8 times greater than in (a) and as a result the soil moisture content had to be greater (about 24 percent in (b) versus 16 percent in (a)) in order to transmit the water at the applied rate. Note that these soil moisture contents are less than the saturated soil moisture content of 38 percent. (From J. Rubin 1966, *Water Resources Research,* vol. 2, pp. 739–750. Copyrighted by American Geophysical Union.)

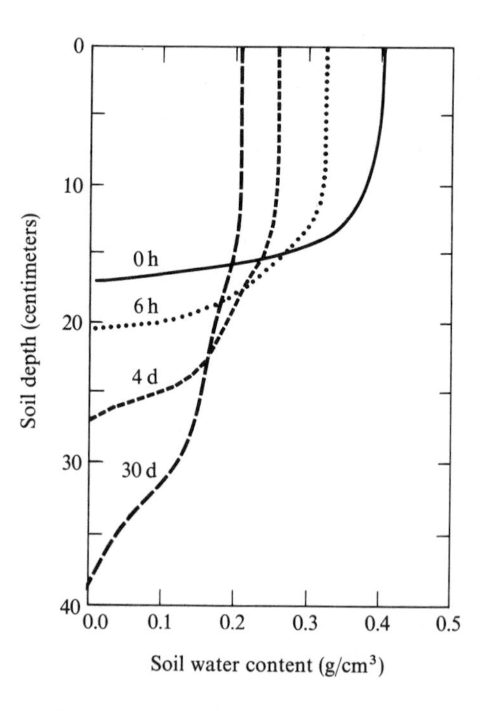

Figure 6-14 Soil water profiles during drainage of a silt loam. Each curve is numbered with the hours (h) or days (d) since the application of water at the surface ceased. (From T. D. Biswas et al. 1966, *Water Resources Research,* vol. 2, pp. 513–524. Copyrighted by American Geophysical Union.)

of the process in laboratory columns. Using a radioactive tracer, Zimmermann et al. (1966) showed that in a field soil, “A single rainstorm, labeled with isotope tracer, forms a tagged layer of water that, although blurred by diffusion effects, moves downward as a distinguishable water mass, between the older rainwater below and the young rainwater above.” The water eventually reaches the saturated zone, which is discussed in Chapter 7 on groundwater.

As the soil is dried out by evapotranspiration between rainstorms, moisture migrates slowly to plant roots or to the ground surface. This movement is slow and does not take place from great depths unless roots penetrate that far.

Soil Drainage

An aspect of soil conditions that is often important in planning is drainage. If the soil does not allow water to drain away freely, the soil will become saturated and unsuitable for construction, farming, or recreation without expensive artificial drains. Poor drainage places important constraints on locations of buildings with basements, septic tanks and filter fields, sanitary landfills, plowing, planting, harvesting, and recreation. Information on soil drainage is extremely useful, therefore, for zoning land as unsuitable for certain kinds of use (McHarg 1969).

Soil drainage conditions can vary markedly throughout the year; a broad swale that looks dry to a casual observer in summer can become a morass in winter. Houses and septic tanks located there can suffer damage as the ground saturates, even if the duration of waterlogging is not sufficient to leave pedologic or botanical evidence of its occurrence. At any early stage in the investigation of a site, therefore, the planner should be able to recognize those zones of the landscape that are saturated for long periods of the wet season and also those areas that are subject to brief but destructive waterlogging during single, large rainstorms. The indicators that we have found useful for recognizing zones subject to waterlogging on both of these time scales are described by Dunne et al. (1975).

Poor soil drainage may result from low permeability in dense, fine-textured soils or from a topographic position that concentrates water onto an area or does not let it drain away. The problem can be recognized in the field by direct observation or reports of seasonal flooding from local residents or soil scientists. Various kinds of vegetation such as sedges, willows, and in some areas cedars or spruce are associated with poor soil drainage. Regional variations in the water relations of plants are so great, however, that the planner should seek advice about the best vegetative indicators in his region of interest. Soil morphology is a good indicator of drainage, too. Poor drainage is indicated by thick surface organic layers of slightly decomposed vegetative material, or blue-gray coloring with or without rust-colored flecks (mottles) that can be seen in a shallow soil pit or auger hole. Soil Survey Reports of the U.S. Soil Conservation Service, and similar organizations in other countries, classify soils on the basis of drainage, and usually list the minimum depth below the surface that is subject to prolonged saturation. The planner should be familiar with the Soil Survey Reports for his or her region.

If there is also interest in the most extreme short-term conditions of soil saturation that might be encountered during a single rainstorm or snowmelt, the information cannot be obtained from the sources described above. A rough design calculation should be made for the site.

The problem is one of calculating the extent to which the storage capacity (i.e., unfilled pore space) will be exhausted by a rainstorm of a given magnitude. The storage capacity of the soil at the beginning of a rainstorm can be estimated from the depth of unsaturated soil and the porosity and field capacity of the soil. The distribution of soil moisture in the unsaturated zone varies with depth, being least at the surface and greatest near the zone of saturation. If information on this distribution is available, it can be used. But for routine computations on most soils it will not be, and it is an adequate approximation when critical wet seasons and large storms are considered to assume that all the unsaturated soil is at field capacity. If the porosity and field capacity have not been measured for the soil under consideration, they can be estimated from Figure 6-9. The depth to the saturated zone can be observed in the field at various seasons of the year, or it can be estimated for critical wet seasons from observations of the soil profile, or from the Soil Survey Reports, as previously described. The amount of unfilled pore space in the soil is then calculated by multiplying the depth of the unsaturated soil by the difference between porosity and field capacity. The planner can then calculate whether, say, a 5-cm rainstorm would exhaust this storage capacity and saturate the soil throughout its depth, causing surface flooding.

Alternatively, the designer may wish to know the height to which saturated conditions will rise during a storm of a given recurrence interval. If he wishes to know the maximum water-table elevation resulting from, say, the 100-year, 24-hour rainfall, which in the region of interest is found to be 15 cm, he can calculate the probable maximum height of the water table as follows. The amount of unfilled pore space per unit depth of soil is given by the difference between porosity and field capacity (see Figure 6-9). If the porosity is 0.50 and the field capacity is 0.25, the unfilled pore space as a proportion of the soil volume will be 0.25; i.e., 25 percent of each meter of soil will be occupied by unfilled pore space. Therefore, each meter of soil could store 25 more centimeters of water before becoming saturated, and a 15-cm rainstorm could fill up the unsaturated pore space in 60 cm of soil. The rise of the water table can therefore be calculated by dividing the depth of rainfall by the difference between porosity and field capacity.

These calculations assume that all the rain or meltwater infiltrates the soil; if this is not the case, an appropriate subtraction can be made for surface runoff. They also assume that there is no subsurface inflow or outflow occurring. This is not usually true, but for most situations the assumption does not introduce a large error. The simple procedure can therefore be used for designing the depths of tile fields and septic tanks, estimating the frequency of flooding of parks, building sites, or agricultural land, and similar design calculations. If local conditions are appropriate and if the need arises, the more general water-balance procedure described in Chapter 8 can be used to generate a continuous estimate of water-table elevations.

Soil Drainage and Septic Systems

As suburbs have expanded into rural land, their spread has often outstripped the extension of community sewage disposal facilities. On-site sewage disposal must therefore be incorporated into the planning of such suburban developments. Bad design or failure to anticipate the site limitations can lead to septic-tank failure, esthetic problems where sewage appears at the ground surface or in cut banks and ditches, and even public health dangers. Planners can avoid committing themselves to sites that have such limitations if they understand and use some widely available sources of information, and if they seek advice from their local Soil Conservation Service office or Health Department.

A conventional on-site sewage disposal system consisting of septic tank and filter field is shown in Figure 6-15. Sewage from the house enters the

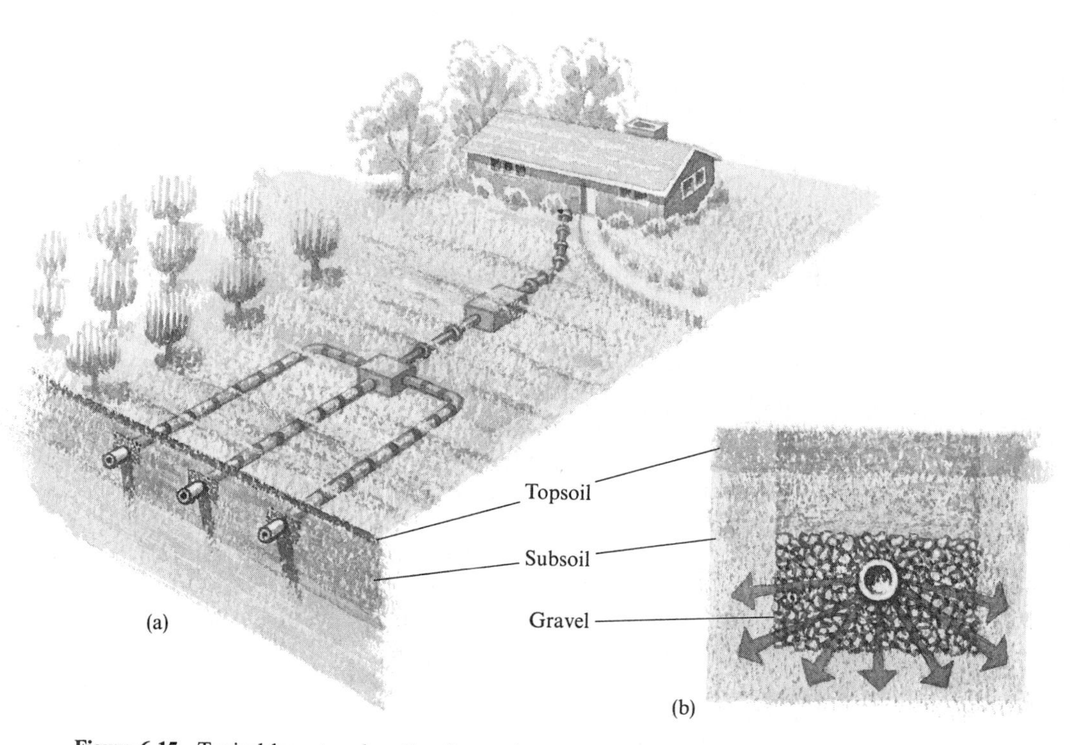

Figure 6-15 Typical layout and section for septic tank and tile field method of sewage disposal.

tank where its organic compounds are partially broken down by bacterial action. The effluent from the tank still contains many solid and soluble organic compounds, inorganic solutes, and bacteria. This effluent is conveyed slowly through the pipes in the filter field and allowed to drain into the soil over a large area. Bacterial action within the soil further breaks down the solid and soluble organic compounds, some of the inorganic solutes become bound to soil particles, and the remaining solids, including bacteria, are filtered out by the soil. The effluent then drains to groundwaters and streams with a greatly reduced potential for contamination. Some pollutants, particularly nitrates, remain in the effluent, however, even after it has undergone subsurface breakdown and filtering.

For efficient operation the filter field should be located in a large volume of soil into and through which the effluent can move easily. The drain pipe should be buried below the topsoil, usually at a depth of at least 60 cm. Below it there should be at least 1.5 m of soil above the bedrock so that the effluent will be adequately filtered before reaching the bedrock. Soil depths are given in the county Soil Survey Reports. The soil should pass water easily, and its capacity to do this varies with its texture. Sandy and gravelly soils are highly suitable (provided the gravel is not too coarse to filter the effluent), and silty-clayey soils are unsuitable. Textures are also listed in the Soil Survey Reports, and sometimes the report lists soil permeability values (see next chapter).

The soil should not become saturated at any time, for this backs up the system, causes it to flood the ground surface (see Figure 6-16) and upsets its functioning for a considerable period of time thereafter. During the wettest season the water table should be at least 60 cm below the base of the drain field. In the more recent Soil Survey Reports, depths to the seasonal high-water table are listed for each soil type; for extreme storms, the probable maximum short-term rise can be calculated as previously described. In older Soil Survey Reports the water-table depths are not given, but the soil is classed according to its drainage. Poorly drained soils are unsuitable for filter fields. For the same reasons, septic systems should not encroach upon areas prone to flooding by a river or lake, as this will soon destroy their effectiveness. In any case, filter fields should terminate at least 15 m from a lake or water course so that there is adequate distance for filtering of the effluent within the soil. The disposal system should also be located at least 30 m downslope from wells.

Filter fields ideally should not be located on hillslopes with gradients greater than about 0.10 (10 percent), and on such slopes should be laid along the contour and in a zig-zag pattern. In this way, flow velocities within the drain pipes will be slowed and the effluent will seep into the soil over the whole field. If the soil is shallower than is ideal, the effluent may return to the ground surface downslope of the drains on a steep hillslope. The hillslope gradient is also given in the Soil Survey Reports.

A great deal of information about the suitability of land for septic systems

Figure 6-16 The poorly drained soil on which these houses are being built will not be satisfactory for septic-tank filter fields. (USDA–Soil Conservation Service.)

can therefore be obtained from the local Soil Survey Reports. The more recent reports published by the U.S. Soil Conservation Service state explicitly whether each soil type is suitable for on-site sewage disposal. But even without this explicit statement, the descriptions of soils in older reports or in those of other agencies can be interpreted according to the specifications outlined above. It will be stressed at many points in this book, however, that maps of land suitability are approximate. At the scale of these maps, small hazard zones cannot be mapped, and those appealingly solid black lines on the map are actually diffuse zones in the field. Soil maps, therefore, are only suitable for outlining general areas which are suitable or unsuitable for on-site disposal. Somewhat more specifically, they may indicate qualitative degrees of limitation due to texture or permeability. In this way, one may judge roughly whether large lots or small lots will be necessary for sewage disposal. The Soil Survey Reports cannot be used for siting the disposal system within lots a few acres in extent. For this, one must rely upon an examination of the site and the soil profile, preferably in the company of a soil scientist.

The size of filter field that is needed for a septic system depends upon the amount of sewage to be filtered (i.e., the number of residents of the house) and the ability of the soil to transmit water. An index of the latter parameter can be obtained by means of a percolation test. Such a test should be carried out if there is any doubt about the permeability of the site, and many health departments require the test before a building permit is granted. The test involves digging a hole with a diameter of 10–30 cm to the depth of the proposed drain trenches. Six holes are usually employed to sample the variability of the site. The sides of the hole are scraped to remove any slickened clay surfaces, and the soil is thoroughly wetted for at least 4 hours. Water is then poured into the hole until it is 15-cm deep, and its rate of decline is measured over a 30-minute period. If the water drains away too rapidly, measurement over a 10-minute period at the end of a 1-hour soaking will suffice. An average is calculated for all the six holes at a site.

The average percolation rate can be used in conjunction with Figure 6-17 to calculate the area of the filter field required for adequate sewage disposal. Multiplication of the value on the ordinate of this graph by the number of bedrooms gives the required absorption area of the filter field. Since only the bottoms of the trenches provide effective absorption, this absorption area must be divided by the width of the proposed trench to obtain the total length of drain required. The trenches should be at least 2 to 3 m apart. From these figures it is possible to calculate the total area to be occupied by the filter field. It must be remembered, however, that the percolation test quantifies only the limitation due to soil permeability. It does not warn of seasonal high-water tables, steep slopes, flooding, or nearness to a water course. If the building lot is too small to accommodate the proposed tile field, and if the subsoil is quite permeable, a more compact arrangement of the tile drains in a seepage pit is possible. The required area of the pit is also calculated from a percolation test and Figure 6-17.

The reader should consult government agencies for methods required by law in the local area. For example, some state health agencies allow the

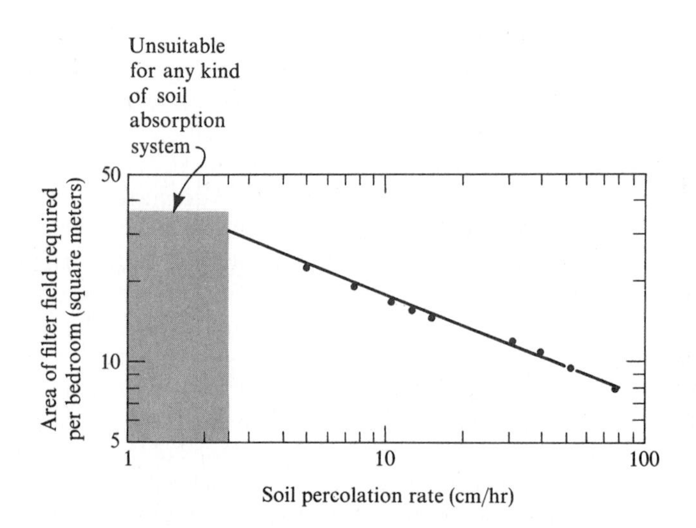

Figure 6-17 Area of filter field needed for private residences. The area is expressed in square meters of trench bottom required per bedroom.

Table 6-3 Minimum absorption areas for tile fields in various soils. (From U.S. Public Health Service, *Publication No. 526*.)

SOIL TEXTURE	ABSORPTION AREA (SQ FT PER BEDROOM)
Gravel or coarse sand	70
Fine sand	90
Sandy loam	115
Clay loam	150
Sandy clay	175
Clay with small amount of sand or gravel	250
Heavy clay	Unsuitable

absorption area per bedroom to be estimated from the texture of the soil, as indicated in Table 6-3. This technique also does not indicate hazard from seasonal waterlogging or steep slopes.

Bibliography

BERTONI, J., LARSON, W. E. AND SHRADER, W. D. (1958) Determination of infiltration rates on Marshall silt loam from runoff and rainfall records; *Soil Science Society of America Proceedings,* vol. 22, pp. 571–574.

BETHLAHMY, N. (1962) First-year effects of timber removal on soil moisture; *International Association of Scientific Hydrology Bulletin,* vol. 7, no. 2, pp. 34–38.

BISWAS, T. D., NIELSEN, D. R. AND BIGGAR, J. W. (1966) Redistribution of soil water after infiltration; *Water Resources Research,* vol. 2, pp. 513–524.

CARTER, D. G. (1962) Osmosis; *School Science Review,* vol. 43, p. 585.

DILS, R. E. (1953) Influence of forest cutting and mountain farming on some vegetation, surface soil, and surface runoff characteristics; *Station Paper 24, Southeastern Forest Experiment Station,* U.S. Forest Service, 55 pp.

DORTIGNAC, E. J. (1951) Design and operation of the Rocky Mountain infiltrometer; *Station Paper 5, Rocky Mountain Forest and Range Experiment Station,* U.S. Forest Service, 68 pp.

DUNNE, T. (1970) Runoff production in a humid area; *U.S. Department of Agriculture ARS-41-160,* 108 pp.

DUNNE, T. (1978) Field studies of hillslope flow processes; in *Hillslope hydrology* (ed. M. J. Kirkby), John Wiley & Sons, London.

DUNNE, T., MOORE, T. R. AND TAYLOR, C. H. (1975) Recognition and prediction of runoff-producing zones in humid regions; *Hydrological Sciences Bulletin,* vol. 20, no. 3, pp. 305–327.

FLETCHER, P. W. AND MCDERMOTT, R. E. (1957) Moisture depletion by forest cover on a seasonally saturated Ozark Ridge soil; *Soil Science Society of America Proceedings,* vol. 21, pp. 547–550.

FREE, G. R., BROWNING, G. M. AND MUSGRAVE, G. W. (1940) Relative infiltration and related physical characteristics of certain soils; *U.S. Department of Agriculture Technical Bulletin 729.*

HEWLETT, J. D. (1961) Soil moisture as a source of baseflow from steep mountain watersheds; *Station Paper 132, Southeastern Forest Experiment Station,* U.S. Forest Service.

HILLS, R. C. (1970) The determination of the infiltration capacity of field soils using the cylinder infiltrometer; *Technical Bulletin No. 3,* British Geomorphological Research Group, 24 pp.

HORTON, J. H. AND HAWKINS, R. H. (1965) Flow path of rain from the soil surface to the water table; *Soil Science,* vol. 100, pp. 377–383.

KINCAID, D. R., OSBORN, H. B. AND GARDNER, J. H. (1966) Use of unit-source watersheds for hydrologic investigations in the semi-arid Southwest; *Water Resources Research,* vol. 2, pp. 381–392.

LUSBY, G. C., TURNER, G. F., THOMPSON, J. R. AND REID, V. H. (1963) Hydrologic and biotic characteristics of grazed and ungrazed watersheds of the Badger Wash Basin in western Colorado, 1953–58; *U.S. Geological Survey Water Supply Paper 1533-B.*

MCHARG, I. (1969) *Design with nature;* Falcon Press, Philadelphia, 198 pp.

MUSGRAVE, G. W. (1955) How much of the rain enters the soil; *Water,* U.S. Department of Agriculture Yearbook, pp. 151–159.

MUSGRAVE, G. W. AND HOLTAN, H. N. (1964) Infiltration; Section 12 in *Handbook of applied hydrology* (ed. Ven te Chow), McGraw-Hill, New York.

NIELSEN, D. R., KIRKHAM, D. AND VAN WIJK, W. R. (1959) Measuring water stored temporarily above the field moisture capacity; *Soil Science Society of America Proceedings,* vol. 23, pp. 409–412.

PEARCE, A. J. (1973) Mass and energy flux in physical denudation, defoliated areas, Sudbury; Ph.D. Dissertation, McGill University, 235 pp.

POST, F. A. AND DREIBELBIS, F. R. (1942) Some influences of frost penetration and microclimate on the water relationships of woodland, pasture and cultivated land; *Soil Science Society of America Proceedings,* vol. 6, pp. 95–104.

REMSON, I. AND RANDOLPH, J. R. (1962) Elements of soil moisture theory; *U.S. Geological Survey Professional Paper 411-D.*

RUBIN, J. (1966) Theory of rainfall uptake by soils initially drier than their field capacity and its implications; *Water Resources Research,* vol. 2, pp. 739–750.

RUBIN, J. AND STEINHARDT, R. (1963) Soil water relations during rain infiltration: I. Theory; *Soil Science Society of America Proceedings,* vol. 27, pp. 246–251.

RUBIN, J. AND STEINHARDT, R. (1964) Soil water relations during infiltrations: III. Water uptake at incipient ponding; *Soil Science Society of America Proceedings,* vol. 28, pp. 614–619.

RUBIN, J., STEINHARDT, R. AND REINIGER, P. (1964) Soil water relations during rain infiltration: II. Moisture content profiles during rains of low intensity; *Soil Science Society of America Proceedings,* vol. 28, pp. 1–5.

SATTERLY, J. (1961) The rise of liquid in a capillary tube; *School Science Review,* vol. 43, pp. 111–120.

SMITH, H. L. AND LEOPOLD, L. B. (1942) Infiltration studies in the Pecos River watershed, New Mexico and Texas; *Soil Science,* vol. 53, pp. 195–204.

SMITH, W. O. (1967) Infiltration in sands and its relation to groundwater recharge; *Water Resources Research,* vol. 3, pp. 539–555.

STRAHLER, A. N. (1975) *Physical Geography,* 4th ed. Wiley, New York.

U.S. DEPARTMENT OF AGRICULTURE (1955) *Water;* U.S. Department of Agriculture Yearbook.

U.S. DEPARTMENT OF HEALTH, EDUCATION AND WELFARE (1958) Manual of septic tank practice, *Public Health Service Publication No. 526,* with 2 addenda: 1959, 1961.

U.S. HOUSING AND HOME FINANCE AGENCY (1954) Septic tank soil absorption systems for dwellings, *Division of Housing Research Construction Aid 5.*

U.S. SOIL CONSERVATION SERVICE (1967) Soils suitable for septic tank filter fields, *Agricultural Information Bulletin No. 243.*

U.S. Soil Conservation Service (1972) *Hydrology;* National Engineering Handbook, Section 4, Washington, DC.

Whipkey, R. Z. (1969) Storm runoff from forested catchments by subsurface routes; *International Association of Scientific Hydrology, Symposium of Leningrad,* Publication 85, pp. 773–779.

Yong, R. N. and Warkentin, B. P. (1966) *Introduction to soil behavior;* Macmillan, New York, 305 pp.

Zimmerman, U., Munnich, K. O., Roether, W., Krentz, W., Schuback, K. and Siegel, O. (1966) Tracers determine movement of soil moisture and evapotranspiration; *Science,* vol. 152, pp. 346–347.

Typical Problems

6-1 Effects of Land Use on Relative Infiltration Rates

Changes of infiltration capacity resulting from land use affect rates of runoff and soil erosion. A planner often needs to collect some field data on these effects to document a case for controlling land use or for recognizing where the effects are likely to be most pronounced. This typical problem is presented as a field exercise which the reader can use to investigate the effects of land management practices on the infiltration rates of soils.

Obtain a metal cylinder with a sharpened edge and a graduated water container that will allow you to supply water to the cylinder slowly. Set them up so that the infiltration rate can be measured as described in the text. Choose a uniform depth of penetration and a uniform depth of ponding for all measurements.

Design an experiment using the infiltrometer to show the effects of some land use within a uniform soil type. Consider, for example, making measurements in trampled and untrampled areas in a park, or on grazed versus ungrazed land, or on cultivated land versus undisturbed forest.

Make several measurements for each land use and soil type. Be sure, too, that the effect of the present use is not overshadowed by previous use of the land. If this exercise is used as a class project, several different uses and soil types can be evaluated and ranked according to impact on infiltration rates.

In conjunction with your infiltration measurements, dig a pit into the soil and attempt to describe the soil texture and structure. Look for such distinctive features as thinly laminated horizontal layering due to ground compaction, or large root holes and loose aggregates in undisturbed soil.

6-2 Estimation of the Water-Holding Properties of Soils

Several of the techniques discussed in other chapters of this book require the user to estimate the water-holding properties of a soil. The best method, of course, is direct measurement, and many soil survey reports list the porosity, field capacity, and permanent wilting point of soils. Often, however, the planner does not have such data, and the problem at hand may not warrant the expense of field sampling and laboratory measurement. Instead, Figure 6-9 or a similar graph can be used. Estimate the available water capacity of a silt loam from the figure.

Solution

From the diagram the typical range of field capacity values for a silt loam is from 0.22 to 0.32 (22 to 32 percent) of the volume of the soil. In one meter of silt loam, then, about 22 to 32 cm of water are stored when the soil is at field capacity. The wilting point for a silt loam is about 0.12 of the volume of the soil. The amount of water available to plants, therefore, is only about 10 to 20 cm of water in 1 m of soil. Although it is not shown in the diagram because of lack of data, the wilting point for a given soil texture also has a range of values. This technique, then, should be considered as supplying only an estimate of available water capacity.

Compute the available water supply to plants in a fine sand at field capacity. (Answer: 7 to 8 cm per meter of soil)

Also calculate the depth of rainfall that can be stored temporarily in 70 cm of a fine sandy loam above a layer of impervious bedrock if the soil is at field capacity when the rain begins. (Answer: 13 to 22 cm)

6-3 Recognition of Seasonally Waterlogged Soils

Soil that becomes seasonally waterlogged can only be farmed for a short period of the year. In addition the lack of recognition of such soils can result in construction of buildings where flooded basements and failure of septic-tank systems will be a common problem. As mentioned in this chapter, these soils have some distinct characteristics. The best way to learn to recognize them is to make a field study of soils that are subject to high-water tables.

At a local natural pond or swamp dig a sequence of three or four pits at least 2 feet deep along a line from the water's edge to a point more than 3 meters vertically above the water level. In these pits describe such soil features as size of particles, soil color, relative content of organic and mineral material, and other features related to soil drainage that are mentioned in the text or in the references on soil morphology at the end of this chapter. The sequence of pits will allow you to see the difference in the soil along the transition from a constantly saturated soil, to a seasonally saturated soil, and to a soil that is only rarely saturated. Alternatively, an auger could be used for examining the soil profile; this has advantages where a pit would be flooded, but elsewhere you will see more in a pit.

Be sure to use any available soil maps and descriptions to assist you in recognition of major soil features. Test your ability to recognize poorly drained soils by looking at other areas that are delineated as poorly drained organic or mineral soils on the map. If you do this as a class project, consider asking a soil scientist to join you for a day in the field. As long as you dig and backfill the pits he or she should be willing to help!

6-4 Soil Suitability for Septic Fields

You are asked to advise a large land-holding company about the purchase of land for subdivision and development of a recreation–home complex. After satisfying yourself that the site is free from geologic hazards such as landslides and floods, you turn your attention to more prosaic issues such as sewage disposal. How would you set about investigating the suitability of the site for septic-tank systems?

If there were published maps of soils and their suitability, how would you determine whether the data were detailed enough for your usage? If there were no soil maps, what would you do? What else would you do besides making percolation tests? Hints: Look for vegetative or topographic evidence for a potential high-water table. Are there any nearby springs or ephemeral ponds? Ask the local inhabitants if the land becomes waterlogged in the wet season. Do soil characteristics indicate seasonally saturated ground? Be sure to consider proximity of the land to rivers, lakes, and groundwater resources. Are there areas where the slope of land is too great?

6-5 Calculation of Drainfield Area

Calculate the absorption area of a drainfield for a three-bedroom house. Percolation tests give an average value of the percolation rate of 7.6 cm per hour.

Solution

From Figure 6-17, such a percolation index indicates that the absorption area should be 19.5 sq m per bedroom. For three bedrooms, 58.5 sq m of area is required. Suppose that the drain trenches are to be 0.75 wide, then the required length of drain tile would be 58.5/0.75 = 78 m. Such a system could be laid out as shown in Figure 6-18.

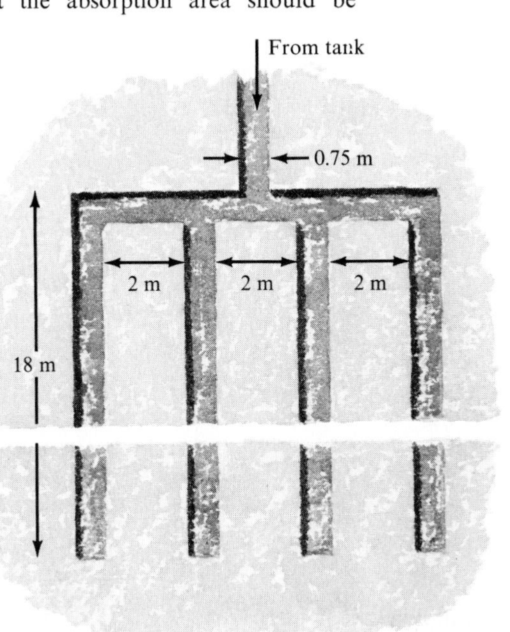

Figure 6-18 Suggested layout of tile field for a three-bedroom house.

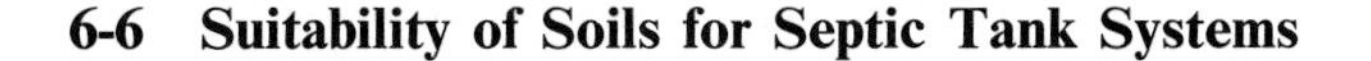

6-6 Suitability of Soils for Septic Tank Systems

As a field exercise, assess the septic field capability and design the sewage drainage field for the soil in a nearby housing development where septic tanks will be used. Use the hints mentioned above, and compare your methods and conclusions with guidelines established by your local government agency. If these guidelines allow you to determine septic field drainage area from the soil texture alone, as in Table 6-3, you can evaluate the reliability of such a method with the other field observations you have made. Also, if this exercise is used as a class project, choose several different soil types and compare notes on the suitability of the major soils of your area.

7

Groundwater

Importance of Groundwater

The saturated subsurface zone, or *phreatic zone,* contains the largest source of unfrozen fresh water in the world. It constitutes 21 percent of all the world's fresh water and 97 percent of all the unfrozen fresh water on earth. In the United States 80 to 90 percent of the total available water comes from this source. Groundwater can be found beneath some of the driest and coldest places on earth, although in deserts and in regions of frozen ground, problems of groundwater supply and chemical quality are frequently encountered. The phreatic zone, therefore, contains a major component of our water supply, and one which becomes of even greater significance as the more accessible surface supplies are used to capacity or polluted beyond usable limits. The contribution of groundwater to the total water supply is greatest in arid and semi-arid regions and in some places where geological conditions favor groundwater storage. Many large industrial plants depend entirely upon their own wells rather than upon a municipal water supply. Large regions of irrigated agriculture in arid areas are totally dependent on groundwater. The growing importance of subsurface water supplies demands an ability to develop and manage wisely this subsystem of the hydrologic cycle. Unfortunately, much development of groundwater has not been based on sound hydrologic principles, and the results have been costly.

An understanding of the storage and movement of groundwater is also necessary for appreciating other components of the hydrologic cycle. Chapter 1 stressed the interdependence of each of the hydrologic subsystems. Surface water and subsurface water are intimately associated, and they are in a continuous process of exchange. Drainage from the groundwater body maintains streamflow during dry periods. In some arid regions the opposite process occurs, as runoff from desert washes may be the major source of groundwater recharge. A knowledge of groundwater conditions, therefore, assists in understanding the fluctuations of streamflow, particularly in dry periods.

It is becoming increasingly difficult to dispose of household, industrial, and agricultural wastes. In some places the groundwater system has become a convenient sink for toxic industrial wastes, fertilizer leachates, wastes from stock-rearing areas and food-processing industries. This policy is obviously in conflict with the increased use of the phreatic zone for water supply, so minimizing pollution hazard is a growing concern of groundwater hydrologists and planners.

Because the limits of groundwater systems do not coincide with those of jurisdictional boundaries at the land surface, the use of the phreatic zone for water supply or waste disposal can lead to conflicts of interest between individuals or organizations with rights to the groundwater. The close relationship between surface and subsurface flow systems also gives rise to conflict if one of the systems is exploited at the expense of the other. The resolution of such conflicts depends upon a body of wise and realistic water law. Unfortunately, many water resources are governed by laws that were imported from different geographical regions or are the products of unwise political decisions. New laws governing groundwater development must be based upon sound hydrologic principles.

Groundwater is also of interest in other earth sciences. The geochemist and the geomorphologist have an interest in those conditions within the phreatic zone that control weathering reactions, water chemistry, or the development of subsurface landforms such as caves. As described in Chapter 15, groundwater conditions affect hillslope stability and the occurrence of landslides, a hazard that planners would like to avoid. Among ecologists there is an awareness of the importance of groundwater conditions in controlling the location of phreatophytic vegetation and its associated wildlife populations (Western and Van Praet 1971).

In this text we can give only a brief introduction to the simplest principles of groundwater hydrology. For further information, the reader should consult the excellent books by Davis and Dewiest (1966) and Walton (1970).

Introductory Definitions

Suppose that a well hole is drilled or dug into the ground, as shown in Figure 7-1. At some depth, water will enter the hole and will attain a static level. The surface of the water in this well defines the *water table.* The water below this level is held in intergranular pores in the soil or rock, or in joints or fractures within the rock. Immediately above the water table is a *capillary fringe,* or the zone in which the pores of the soil or rock are completely filled with water held up by capillary tension. If the pores are very narrow, as in a clay or silt, this capillary fringe can be several meters thick, while in the larger pores of coarse-textured materials, such as gravels or sandstones, capillary forces are weaker and the capillary fringe is negligibly small. Above the capillary fringe is the unsaturated soil-water zone already discussed. In this zone capillary forces at the air–water interfaces hold water in the soil at pressures less than atmospheric so that it cannot flow into the well. Below the water table the pores are filled, there are no air–water interfaces, capillary forces do not develop, and the water is under pressures greater than atmospheric pressure, so that water can enter the well. If several wells are drilled, the spatial variation in water-table elevation can be mapped. A cross section of the landscape might look like that shown in Figure 7-1. Within a uniform rock type, the water table has a configuration similar to that of the ground surface, but with a more subdued outline. Where the water table intersects the land surface, water drains out of the groundwater system to supply a stream, swamp, lake, or oasis.

The groundwater body receives rainwater or snowmelt that has percolated through the unsaturated zone. This *recharge* tends to increase the amount of groundwater in storage, and therefore to raise the water table, as shown in Figure 7-1. As the water table is raised, it is also steepened, and this causes water to drain out more quickly to streams. Such drainage lowers the elevation and gradient of the water table. The interaction of recharge and drainage results in seasonal and short-term fluctuations of the water table, as shown in Figures 7-1 and 7-2.

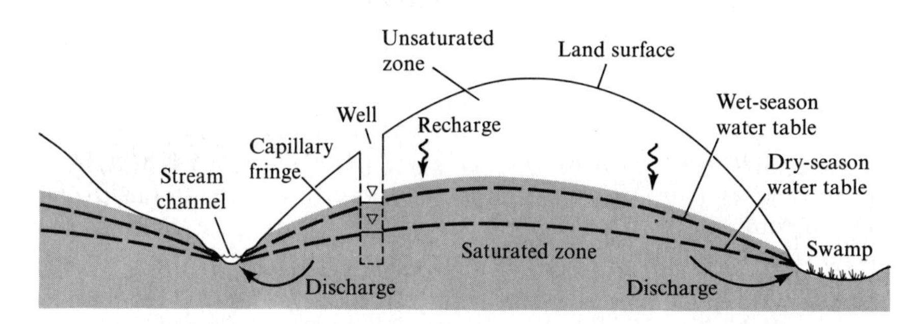

Figure 7-1 Diagram of an unconfined groundwater system.

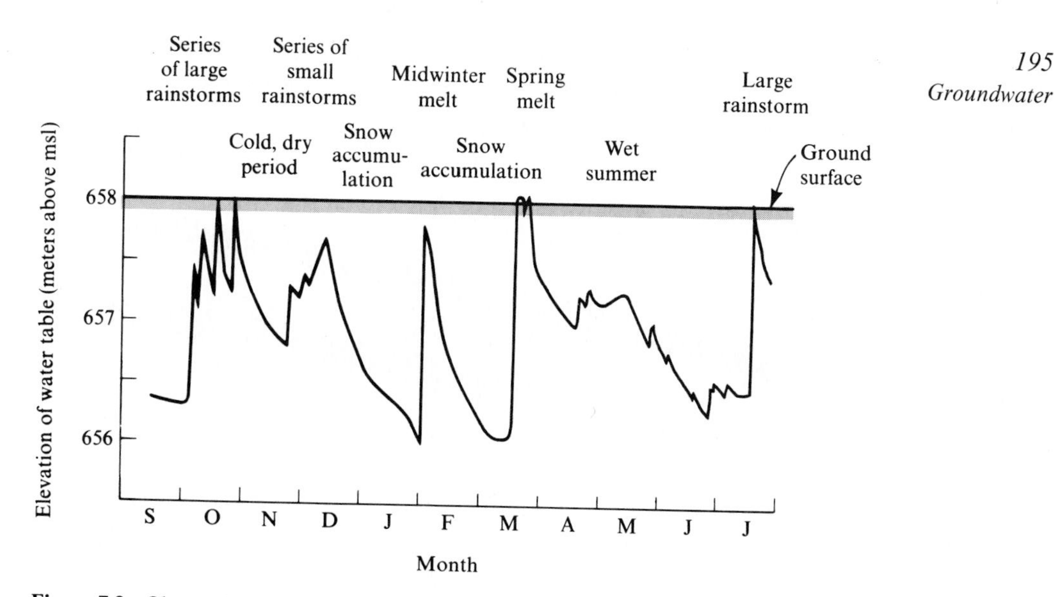

Figure 7-2 Observed fluctuations of the water table at a well in the middle of a steep, permeable hillside, Danville, Vermont.

If a groundwater body provides a good supply of water to wells, the soil or rock that contains the water is called an *aquifer.* Generally an aquifer has three characteristics. It must have a fairly large volume (in relation to the amounts of water that are being removed from it annually); its porosity, and particularly its drainable porosity, should be moderate to high; and it should allow easy movement of water toward a well. A geologic stratum through which water cannot move except at negligible rates is called an *aquiclude.*

If the groundwater is in direct vertical contact with the atmosphere through open pores of the aquifer (as in the upper member of the geologic sequence shown in Figure 7-3), the aquifer is said to be *unconfined,* and the top of the groundwater is the water table. If the aquifer is overlain by an aquiclude, as shown in the lower part of the geologic sequence in Figure 7-3, the aquifer is said to be *confined.* It is saturated throughout its thickness and does not have a free water surface. In practice it is difficult to classify many aquifers in this simple manner, but the principle is worth keeping in mind. The situation may be complicated locally by the existence of *perched* groundwater bodies, which develop above shallow aquicludes of limited extent (see Figure 7-3). These small systems are often ephemeral, developing during a single storm or wet season.

As described above, all groundwater is under positive pressures greater than atmospheric pressure (in other words, it is under positive pressure relative to the atmosphere). At any point in an aquifer, the sum of this pressure and of the elevation of the point above an arbitrary datum is called

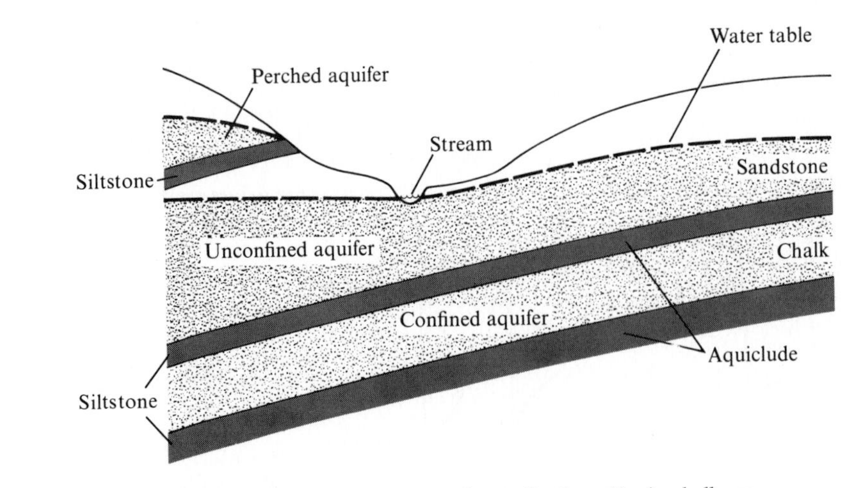

Figure 7-3 Geologic relations of a confined aquifer in chalk, an unconfined aquifer in sandstone, and a perched water table in the sandstone above a siltstone lens. The recharge area for the confined aquifer does not show on the diagram; it often may be quite limited in extent relative to the area underlain by the aquifer.

the *head,* or *piezometric potential.* This measure reflects the amount of potential energy per unit weight of water. The pressure in the water at a point is measured by inserting a solid-walled pipe (called a piezometer) to the point of measurement. Water can only enter the pipe at its end, and the height to which water rises in the pipe is called the piezometric pressure at that point (measured in terms of meters of water).

The components of head at several points in two groundwater systems are shown in Figure 7-4. The well at site A is perforated throughout its depth, and so measures only the elevation of water table in the unconfined aquifer, where, as described previously, the pressure is atmospheric (i.e., is zero relative to atmospheric pressure). The head at any place on the water table, therefore, is simply the elevation of the water table.

A network of wells and piezometers inserted to different depths allows the mapping of variations of head. Lines can then be drawn joining places with equal head on a diagram of the groundwater body. Such lines are called *equipotential lines,* or in three dimensions, *equipotential surfaces.* Figure 7-4 shows a set of such equipotential lines in a two dimensional geologic cross section. The piezometers at sites C and D, though indicating different pressures in the groundwater, measure the same total head and are therefore on the same equipotential line. The flow of groundwater in a uniform aquifer occurs at right angles to the equipotential lines and in a direction from high to low potential.

If the water in a piezometer tapping a confined aquifer stands above ground level, the aquifer is said to be artesian. Water will flow without pumping from a well in an artesian groundwater body, as shown in Figure 7-5. Several extensive artesian aquifers underlie the Sahara desert and are

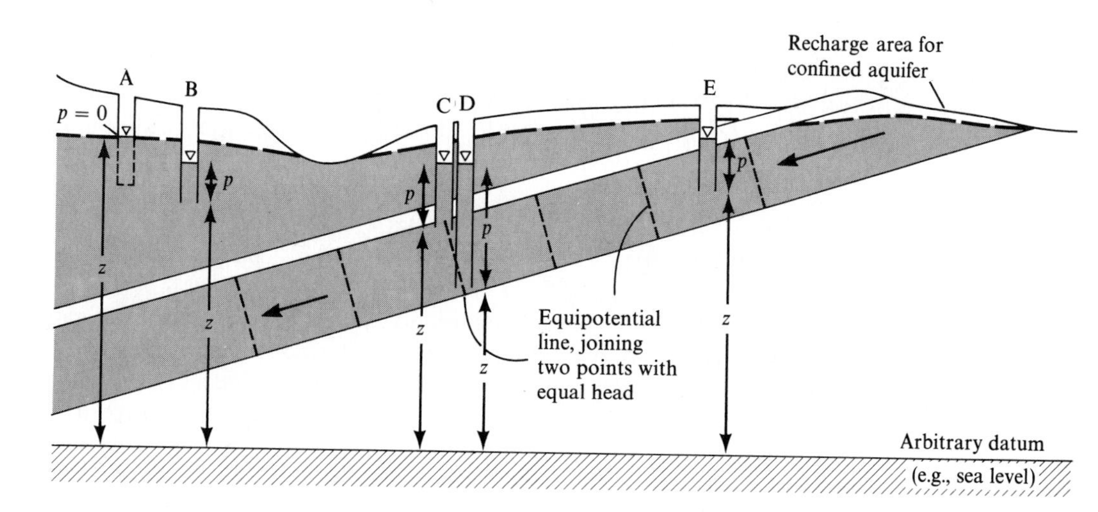

Figure 7-4 Measurement of head at various points in a confined and an unconfined aquifer. Head is the sum of pressure, p, and the elevation of the measuring point, z. Solid pipes B, C, D, and E open only at their ends are piezometers and only measure the water pressure at their lower ends. The perforated pipe at A is a well and measures only the elevation of the water table where $p = 0$.

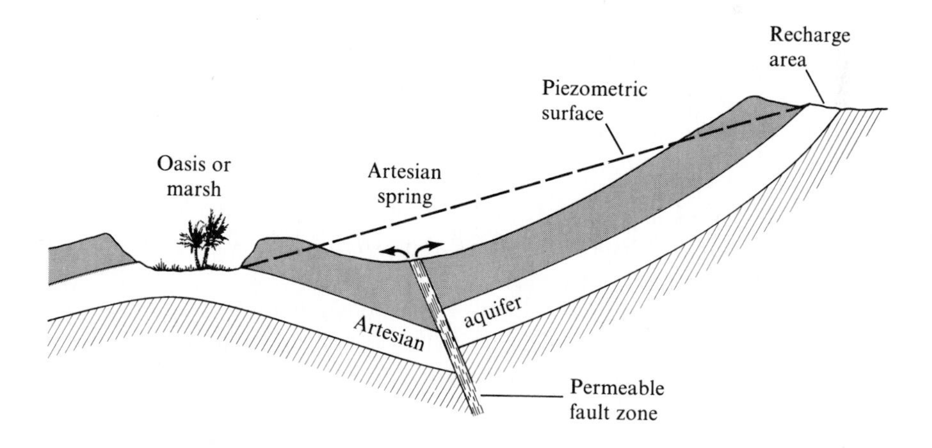

Figure 7-5 An artesian confined aquifer discharging groundwater at the land surface along a permeable fault zone. Where a fold brings the aquifer to the surface, groundwater emerges to supply a stream, a marsh, or, in a desert landscape, an oasis. The piezometric surface is the height to which water would rise in a pipe tapping the aquifer; it slopes downward from the recharge area as groundwater loses potential energy in flowing through the aquifer to the points of discharge. Water will flow freely to the land surface wherever it can breach the overlying aquiclude (including through a well) in the zone over which the piezometric surface lies above ground.

recharged by rain that falls on the moister hills on the desert margin. The water seeps out of the ground at oases, where folds or faults in the overlying aquiclude allow water to escape. Artesian aquifers are attractive because there are no pumping costs involved in developing the supply as long as the piezometric surface remains above ground.

Groundwater Storage

Groundwater is derived from precipitation that falls on the earth's surface. The only distinction between groundwater and surface water is the water's location at a particular time. Water on the surface in rivers, as overland flow, or in lakes may sink into the ground, at which time it would be called groundwater. Water that is underground but flows out to the surface as springs, as seeps, or as drainage into a stream channel would upon emergence become surface water. The water underground and on the surface are parts of a large plumbing system, all portions of which are interconnected.

For water to collect underground there must be an intake area or *recharge area,* that is, a unit of land surface into which the precipitation can infiltrate and thus charge or fill the storage space in the rock and soil.

The search for a favorable location to drill a well for water development consists, then, of the following considerations:

1. There must be an area that receives precipitation and provides a source of infiltrated water that fills some of the available pore space in the rock materials.

2. The materials into which precipitation can infiltrate must have sufficient porosity to contain an adequate volume of water.

3. The materials in which the water is stored must be sufficiently permeable to allow water to flow into the well hole to replace water pumped out and at a rate approximately equal to the rate of extraction.

Water is stored in the voids within rocks and soils. The total storage potential is governed by the thickness and extent of the aquifer and by its porosity. In unconsolidated rocks, the storage potential is provided by intergranular spaces whose volume is controlled mainly by texture, but also to some extent by the degree of sorting of the sediment and the shape of the grains. Figure 7-6 shows some examples of types of voids; they may vary in size and in the degree to which they are blocked by cements such as iron oxide, calcium carbonate, or silica. Figure 7-7(a) shows an inverse relationship between porosity and grain size for unconsolidated alluvial deposits in an area of the western United States. For the sake of comparison and to show the range of porosity to be expected in unconsolidated materials, more data are presented in Table 7-1. Generally speaking, well-sorted sediments (i.e., those with relatively little variation in their grain sizes) have the highest porosity.

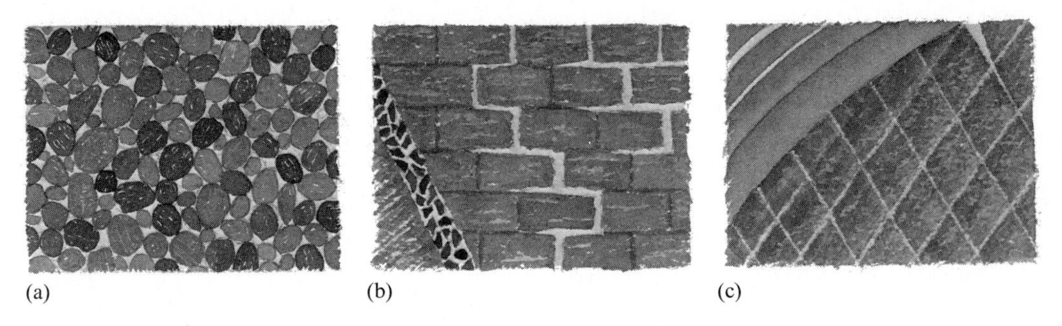

Figure 7-6 Types of rock interstices in which water can be stored. More than one type may be present in any aquifer. (a) Intergranular pores. Examples: consolidated and unconsolidated deposits such as sandstones, limestones, sands and gravels, weathered material originating from such rocks as granite. Some of the pores may be blocked by chemical cements such as calcium carbonate or iron oxide. (b) Joints between bedding planes and joints formed by flexing or cooling of the rock. If the joints are tightly closed, the porosity may be low, but if pressure-release or solution, or both, has occurred, the joints may have been widened from a fraction of a millimeter to many meters. Examples: basalts, some jointed limestones and sandstones. The rock shown is also cut by a fault zone, whose shattered breccia may also have a high porosity. (c) Pressure-release joints (curved) and tectonic joints. Porosity depends on the degree of fracturing and pressure-release and of exploitation by weathering processes.

Consolidated rocks have intergranular pores and larger openings produced by changes that occurred after the formation of the rock. Their intergranular porosity may be reduced by cementation or compaction, as in unconsolidated materials. The larger, secondary openings (see Figure 7-6) occur because of tectonic joints, cooling fractures, bedding planes, fault zones, and other structures. These depend upon the history of the rock and vary in density throughout the aquifer. They are generally narrow at depth, and are more or less absent below a depth of 300 to 400 m. In limestone and other particularly soluble rocks, structural openings may be enlarged into caverns by solution in the groundwater. The porosity of deeply weathered mantles such as occur on many granite rocks in the humid tropics may vary from 35 to 50 percent near the land surface to 1 to 10 percent at the bedrock surface. Some values of porosity for consolidated rocks are given in Table 7-2, but geologic history obviously plays a great part in controlling local conditions of porosity in such rocks.

Porosity governs the total amount of water a given thickness of aquifer can hold but not the amount that is available for supplying streamflow or well discharge. The amount of water available for these purposes is measured by the *storage coefficient* or *storativity* of the aquifer. The storativity is the volume of water released from or taken into storage per unit area of aquifer for a unit change of head. In unconfined aquifers the storativity is more usually called the *specific yield.* As the water table falls, water drains out of the larger pores, but the smaller pores retain water by capillary forces. The amount of water released by gravitational drainage is the specific yield. (Note the relationship of this parameter to the field capacity of soils described

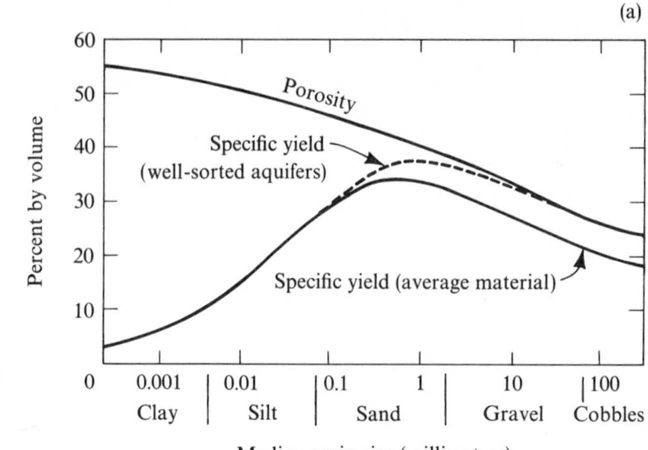

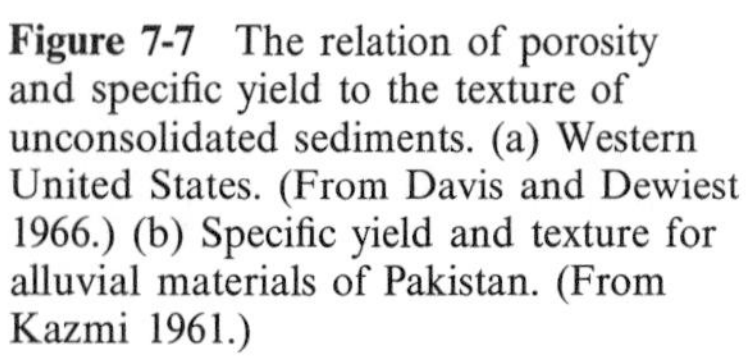
Figure 7-7 The relation of porosity and specific yield to the texture of unconsolidated sediments. (a) Western United States. (From Davis and Dewiest 1966.) (b) Specific yield and texture for alluvial materials of Pakistan. (From Kazmi 1961.)

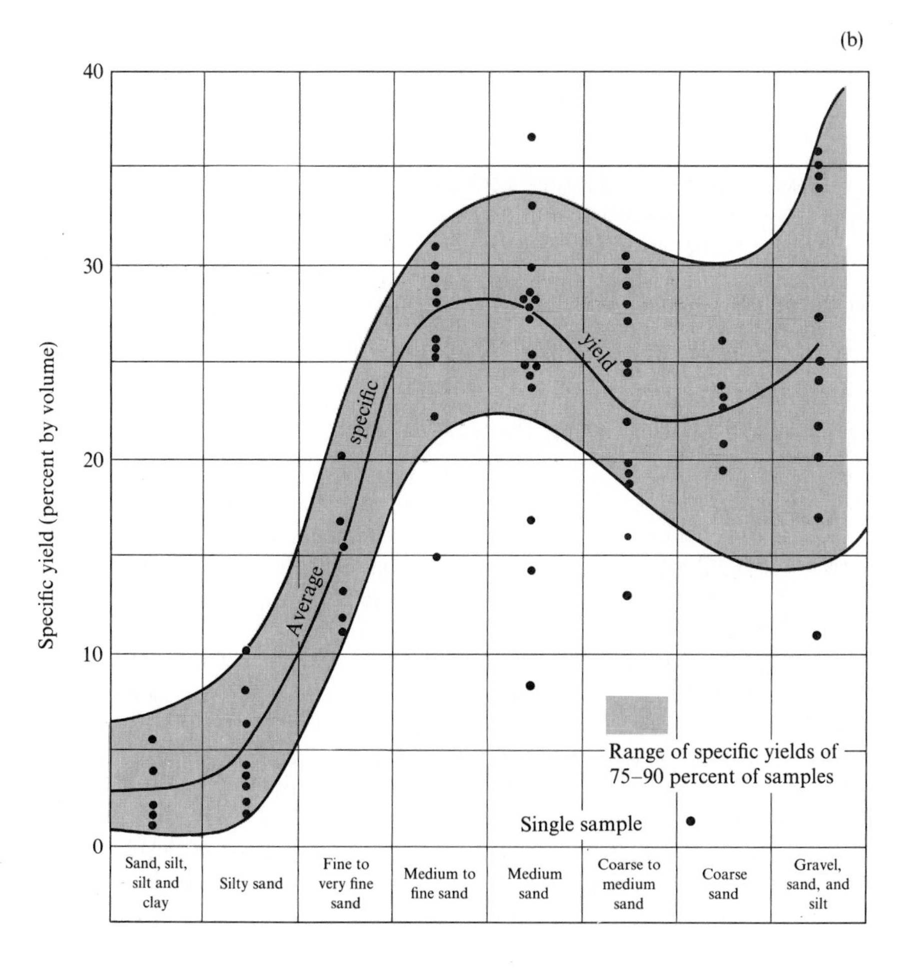

Table 7-1 Representative values of porosity for unconsolidated materials. Porosity can be expressed in terms of percent volume or as a decimal fraction, as given here. (From U.S. Geological Survey Water Supply Papers and many other sources.)

UNCONSOLIDATED MATERIAL	POROSITY
Soils	0.30–0.50
Weathered but undisturbed rock (saprolite)	0.01–0.50
Clays	0.45–0.55
Silt	0.40–0.50
Loess	0.40–0.55
Fine sand, old sediments	0.30–0.40
Fine sand, recent alluvium	0.45–0.52
Medium sand, old sediments	0.30–0.40
Medium sand, dunes	0.35
Coarse sand	0.30–0.35
Sand and gravel	0.20–0.30
Gravel	0.25–0.40
Glacial till	0.25–0.45
Dune sand	0.35–0.40

in Chapter 6.) The specific yield is usually calculated as the ratio of the volume of water draining out of the aquifer to the total volume of the aquifer that is drained. It may be expressed as a dimensionless fraction, a percentage, or in terms of the volume of water per unit area of aquifer per unit depth of lowering of the water table (e.g., cubic meters per square meter per meter). Some values of specific yield are given in Figure 7-7; they depend not only on the porosity but, more particularly, on the size of individual pores. Thus, highly porous clay has very fine pores in which strong forces can hold the water; the specific yield, therefore, is low. In well-jointed, consolidated rocks the voids are so large that no significant capillary attraction withholds water from a stream or well.

In confined aquifers, the coefficient of storage is not determined by gravitational drainage. When the head (or pressure) is lowered a unit distance by pumping, the pores remain full. Water is yielded by compaction of the aquifer as pressure within it is reduced, and by expansion of the water that is pumped out. The storativity in this case is proportional to the porosity, thickness, and compressibility of the aquifer. It is expressed in the same way as specific yield, but the values are generally much lower, indicating that large pressure reductions produce only small yields of water from confined aquifers. Values of storativity usually range from 10^{-6} to 10^{-3} cubic meters per square meter per meter, but for shallow confined aquifers they approach values of specific yield for the same rock in an unconfined

Table 7-2 Representative values of porosity for consolidated aquifers. (From U.S. Geological Survey Water Supply Papers and many other reports.)

CONSOLIDATED ROCK	POROSITY
SEDIMENTARY ROCKS	
Sandstone (depending largely upon degree of cementation and jointing)	0.05–0.30
Conglomerate (depending largely upon degree of cementation and jointing)	0.05–0.25
Siltstone (depending largely upon degree of compaction)	0.05–0.20
Shale (depending largely upon degree of compaction)	0.05–0.15
Old crystalline limestone (depending largely upon secondary porosity caused by jointing and solution)	0.01–0.10
Chalk and oolitic limestones (depending largely upon intergranular pore space, degree of cementation, and in some areas jointing and solution)	0.05–0.30
Marl	0.05–0.20
METAMORPHIC ROCKS	
Slate (depending largely on tectonic and pressure-release joints)	<0.01–0.05
Schist (depending largely on tectonic and pressure-release joints)	<0.01–0.05
Gneiss (depending largely on tectonic and pressure-release joints)	<0.01–0.05
IGNEOUS ROCKS	
Tuff (depending largely on degree of compaction, fusion, and number of gas bubbles)	0.10–0.80
Lavas (depending largely on cooling joints and tectonic joints)	0.01–0.30 Mostly <0.10, unless they contain abundant connected gas bubbles
Fresh granite (depending largely on tectonic and pressure-release joints)	<0.01–0.05
Weathered granite (depending on degree of weathering)	<0.01–0.10

situation. Some values are given in Table 7-3; they are derived from field pumping tests in which water is pumped from a well and measurements are made of the response of the piezometric head. Details of these tests can be obtained from a report by Ferris et al. (1962), or from the manual by Heath and Trainer (1968).

Table 7-3 Characteristics of aquifers. (From Ineson 1963, Maxey 1964, Frank 1973.)

MATERIAL	THICKNESS (M)	TRANSMISSIBILITY (CU M/M/DAY)	STORATIVITY (CU M/SQ M/M)
UNCONSOLIDATED ROCKS			
Glaciofluvial deposits, Hanford, WA	9–14	4,700–37,000	0.06–0.20
Alluvial sand and gravel Gallatin Valley, MT	8	1,240	0.006
Alluvial fan deposits Gallatin Valley. MT	19	450	0.06
Sand and gravel outwash, Mattoon, IL	5	320	0.0015
Sand and gravel outwash, Barry, IL	11	1,490	0.003
Valley train sand and gravel Fairborn, OH	24	3,470	0.0008
Glacial outwash, Bristol Co., RI	18	4,340	0.007
Glacial outwash, Providence, RI	17	1,860	0.20
Young, gravelly alluvium Willamette Valley, OR	6	11,000–25,000	0.05
Old silty, sandy, and gravelly alluvium Willamette Valley, OR	8–64	570–7,300	0.002–0.06
CONSOLIDATED AND SEMICONSOLIDATED CLASTIC ROCKS			
Carrizo sandstone, Lufkin, TX	37	400	0.000138
Aquia greensand, Coastal Plain, MD	6	125–250	0.00023
Spilsby sandstone, Lincolnshire, England	–	70	0.0002
Bunter sandstone, S. England	–	–	0.019
IGNEOUS AND METAMORPHIC ROCKS			
Weathered Wissahickon schist Baltimore Co., MD	20–30	40–125	0.002–.01
Snake River basalts, ID	–	1,240–223,000 mean = 50,000	0.02–0.06
CARBONATE ROCKS			
Fort Payne chert (limestone) Madison Co., AL	6–110	60–17,000	.00045–.0289 .005 (aver.)
Renault-St. Genevieve limestone Hopkinsville, KY	38–53	1560 (aver.)	.00029 (aver.)
Tymochtee dolomite, Ada, OH	70	100 (aver.)	002 (aver.)
Silurian dolomite, DePage Co., IL	75	750	.00035
Chalk, S.E. England	–	<110–4200	.015

Movement of Groundwater

Groundwater moves in response to differences in head and is retarded by its own viscosity (resistance to deformation) as it flows through the narrow, tortuous voids of the aquifer. The viscosity in turn is a function of the water temperature. These relationships are summarized formally in *Darcy's Law*, as follows:

$$Q = Au = wdK\left(\frac{\Delta h}{\Delta l}\right) \qquad (7\text{-}1)$$

where Q = the rate of flow (m^3/day)
A = cross-sectional area of flow (m^2)
u = mean flow velocity (m/day)
w = width of flow (m)
d = depth of flow (m)
K = coefficient of permeability (m/day), sometimes called the hydraulic conductivity
Δh = difference in head (m)
Δl = distance between measurement points in the direction of flow (m).

The ratio $\Delta h/\Delta l$ is usually referred to as the *hydraulic gradient*. Although we have omitted it, Equation 7-1 usually contains a negative sign on the right-hand side to indicate that flow takes place from regions of high to low head.

The dimensions of the groundwater body must be obtained from geologic maps showing the rocks within which most of the flow is occurring, or by drilling or digging pits to measure the thickness of the rock or soil. The difference in head can be measured, as previously explained, in piezometers if the groundwater is confined, or in water-table wells if unconfined. Quantitative surveys of groundwater conditions allow the construction of contour maps of head, which are published in various reports of the U.S. Geological Survey, state organizations, and similar agencies in other countries. When dealing with any question concerning groundwater, the planner should check the publications or open-file reports of these organizations.

The coefficient of permeability represents the ability of a rock to transmit water of a given temperature (viscosity). It is expressed in several different ways and can lead to confusion unless checked carefully. In metric usage the coefficient is commonly expressed in meters per day at a temperature of 15.6°C, or in English units, feet per day may be used. The U.S. Geological Survey, however, uses two other, somewhat confusing definitions of the coefficient, and since this organization is responsible for a large amount of groundwater exploration and analysis, the reader should keep them in mind. The definitions are as follows:

The laboratory, or standard, coefficient of permeability (K_s) is the flow of water with a temperature of 15.6°C in gallons per day, through a medium having a cross-sectional area of one square foot under a hydraulic gradient of one foot per foot. This unit is sometimes called the meinzer after Oscar E. Meinzer, a pioneer of modern groundwater hydrology.

The field coefficient of permeability (K_f) is the flow of water at field temperature in gallons per day, through a rock or soil one foot thick and one mile wide under a hydraulic gradient of one foot per mile.

The two coefficients are numerically equal, except for the effect of temperature (viscosity). This can be corrected for by use of the equation:

$$\frac{K_s}{K_f} = \frac{\mu_f}{\mu_s}$$

where μ_f and μ_s represent the dynamic viscosity of water at field temperature and at 15.6°C, respectively (see Table 7-4).

Table 7-4 Dynamic viscosity of water at various temperatures.

TEMPERATURE (°C)	DYNAMIC VISCOSITY (POISE)
0	0.018
10	0.013
15.6	0.011
20	0.010
30	0.008
40	0.007

Permeability depends mainly upon the size of openings in the soil or rock, but not necessarily upon the total volume of voids (the porosity). Clay, for example, has a greater porosity than other soils or rocks, but the individual spaces are so constricted that water can pass through them only very slowly.

Large voids, favorable to the easy passage of water are provided either by intergranular pore spaces, such as can be seen in gravel, or by joints, cracks, and bedding planes. Table 7-5 lists the range of permeability in several types of rock and some notes on the controls of permeability. A more comprehensive set of values is given by Davis (1969). Permeability is not necessarily constant throughout an aquifer, especially if the voids consist of joints, bedding planes, or similar structures that tend to be tightly closed at great depth. Figure 7-8 shows the variation of permeability with depth in three igneous rocks in southern British Columbia. Measurements made at shallow depths cannot therefore be used to estimate the permeability throughout a deep aquifer.

The ability of a rock to transmit water is often expressed by its *transmissibility,* which is defined as $T = Ky$ where y is the thickness of the aquifer in meters or feet and K is the coefficient of permeability. Both K and T are usually determined for large volumes of rock *in situ* by a *pumping test,* in which water is pumped from the groundwater and the reaction of the water

Table 7-5 Some values of permeability for geologic materials. (From U.S. Geological Survey Water Supply Papers and various other reports.)

ROCK TYPE	PERMEABILITY (M/DAY)	NOTES ON THE MOST COMMON CONTROL OF PERMEABILITY
Clay	<0.01	Very small pores
Silt	0.0001–1	Small pores
Loess	0.0001–0.5	Depending on texture and amount of cement
Fine sands	0.01–10	Depending on texture (pore size)
Medium to coarse sands	10–3,000	Depending on texture
Dune sand	2–20 (aver. 8–10)	Depending on texture
Gravels	1,000–10,000	Large pores
Sand and gravel	0.3–10	Poorly sorted; fine grains plug large pores in gravel
Glacial outwash deposits	Up to 1	Often poorly sorted. Up to 10m/day if very coarse and well sorted
Glacial till	0.001–10	Depending upon whether they are dense and silty ground tills, or sandy ablation tiles
Sandstones and conglomerates	0.3–3	Size of intergranular pores, degree of cementation and of jointing
Crystalline, unjointed limestone	0.00003–0.1	Very few pores; jointed limestones, however, can have very large and variable permeability
Gabbro	>0.0003	Few pores; permeability depends on degree of jointing
Granites and granodiorites	0.0003–0.003	Depends on degree of jointing. Deeply weathered granitic rocks, however, can have permeabilities in the range 0.003–3 m/day.
Volcanic tuffs	0.0003–3	Depends on depth of burial and compaction
Lavas	0.0003–3	Depends largely on degree of fracturing, but weathered surfaces may be highly permeable

table or piezometric head is monitored (Ferris et al. 1962, Heath and Trainer 1968, Walton 1970). Table 7-3 lists values of transmissibility for some important aquifers.

Maps of this parameter show large spatial variations, even in rocks of relatively uniform lithology. Ineson (1962), for example, mapped transmissi-

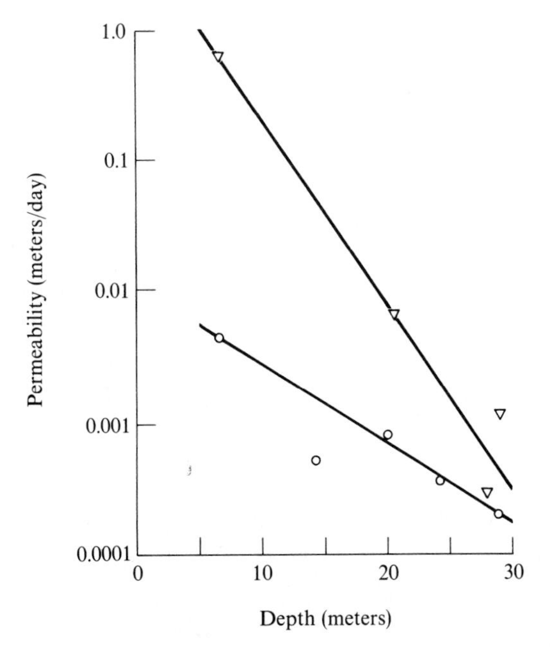

Figure 7-8 Variability of permeability with depth below the ground surface for two rock types in southern British Columbia. Open circles are measurements in granodiorite and gabbro, the closed circle is from a sheared granodiorite, and the triangles represent measurements in a volcanic tuff. (Modified from Lawson 1968. Reproduced by permission of the National Research Council of Canada from the *Canadian Journal of Earth Sciences,* vol. 5, pp. 813–824.)

bility in the Chalk, which is the most extensive and productive aquifer in England. The Chalk is a soft limestone broken by systems of relatively fine fractures. It does not usually support large, open solution joints or caverns of the kind found in harder, crystalline limestones. Yet Ineson's map (Figure 7-9) shows large differences of transmissibility. Most of the aquifer has a relatively low value of less than 115 $m^3/m/day$, but the highest value measured was 2,230 $m^3/m/day$. The higher values coincide with zones of fracturing along the axes of gentle anticlinal structure, and with small faults. Topographic valleys also coincide with these zones of higher transmissibility, and the junctions of valleys are particularly favorable sites for groundwater supply. Bedinger and Emmett (1963) mapped transmissibility in an alluvial aquifer along the Arkansas Valley (see Figure 7-10). Such maps can be used for choosing optimum locations of wells, for understanding differences in water movement that may cause problems after development, and for planning artificial recharge.

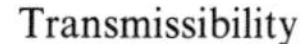

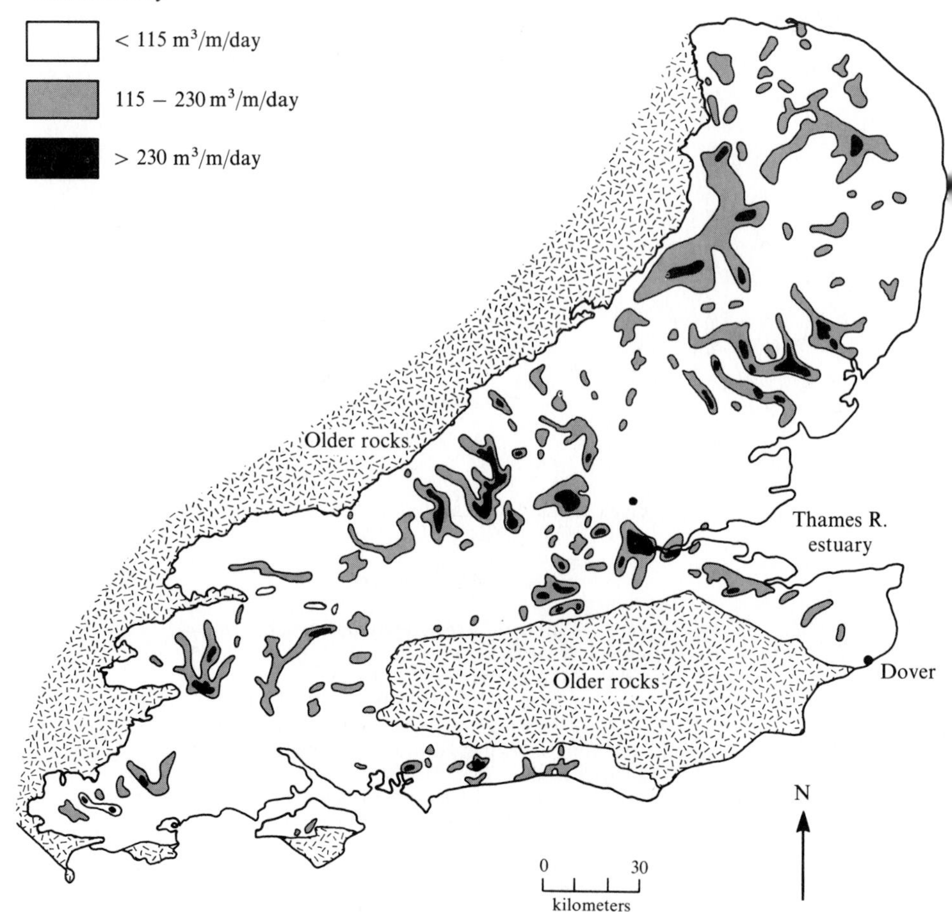

Figure 7-9 Variation of transmissibility in the chalk of southeastern England. (From Ineson 1962.)

Because of the difficulty and expense of measuring permeability, transmissibility, and storage coefficients in the field, the performance of an aquifer is often expressed in groundwater surveys by means of the *specific capacity* of wells drilled into it. This index is the rate of discharge of a well during pumping divided by the resulting drawdown of the water table due to pumping, and is expressed as m^3/day/m of drawdown, or in English units as gallons per minute per foot of drawdown. Efficient wells in good aquifers have high values of specific capacity. Figure 7-11 shows the frequency distribution of specific capacity values from aquifers in a region of Oregon. Such distributions are characteristically right-skewed with relatively few high values. This is especially true in consolidated, fractured rocks.

Statistical analysis and mapping of data on specific capacity can provide useful indicators of the probability of success in locating wells (Figure 7-12). The definition of aquifers by this method can also reveal which areas of the landscape should not be used for the disposal of toxic wastes or for housing and similar developments that may reduce recharge. Even data on sustainable well yields (obtained by pumping the groundwater until the head remains constant and then measuring the well discharge) can be used as indicators, though it is advisable to incorporate the depth of the wells into an analysis that uses these data. Because of the wide variations of specific capacity and well yield in some areas, a large number of observations should be made. The same is true, unfortunately, for values of permeability, though

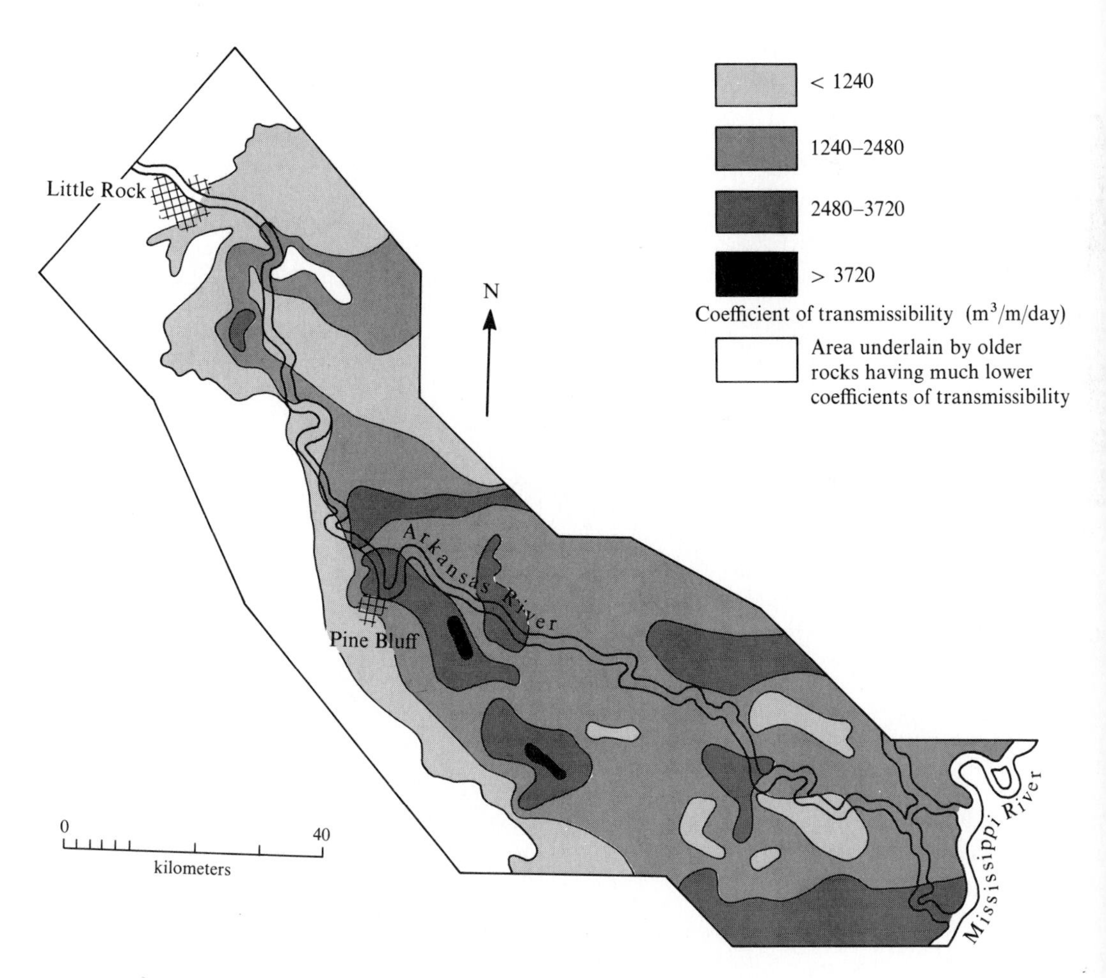

Figure 7-10 Variation of transmissibility in the alluvial aquifer of the Lower Arkansas River Valley, Arkansas. (From Bedinger and Emmett 1963.)

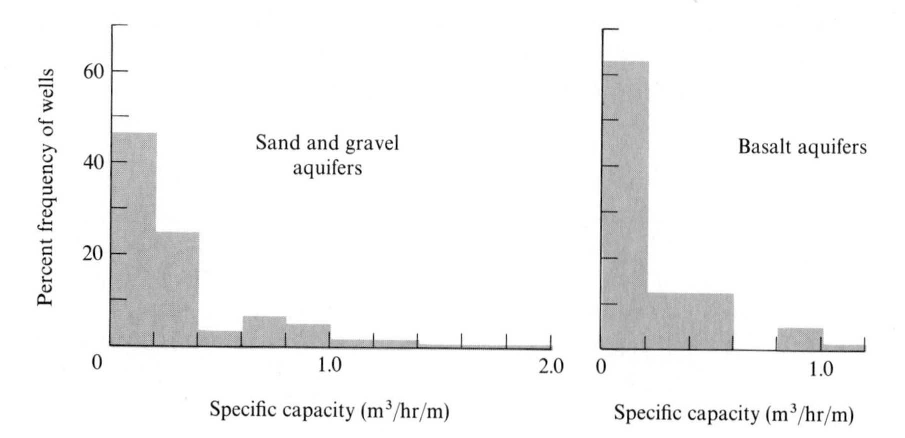

Figure 7-11 Frequency distribution of values of specific capacity for aquifers in two types of rock material, Eola-Amity Hills, Oregon. (Data from Oregon State Engineer, Portland.)

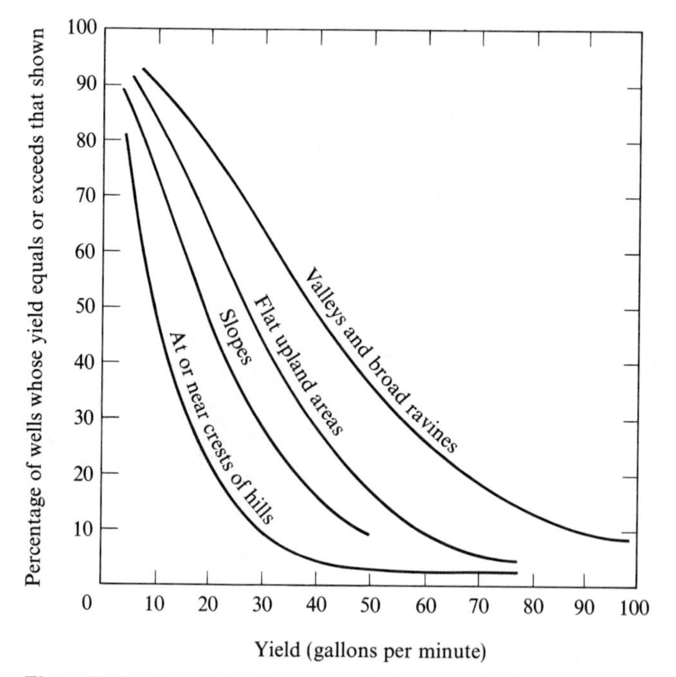

Figure 7-12 Cumulative frequency distribution of values of well yield for various topographic positions, Statesville area, North Carolina. (From Davis and Dewiest 1966, redrawn from LeGrand 1954.)

often the cost of pumping tests severely limits the number of measurements. The specific capacity data can be used to obtain a useful minimum value for the transmissibility of an aquifer (Theis et al. 1963).

Simple Models of Groundwater Movement

Here we consider quantitatively some of the simplest classes of groundwater flow in two dimensions and with no fluctuations in time. In spite of the compromises that must be made with the complexity of a field situation, the methods described below give surprisingly accurate and useful answers for many purposes. They also give the reader an insight into how more complex, three-dimensional models with temporal variations and complex structural boundary conditions can be constructed. The complex models are handled by computer and are being used increasingly to allocate or predict groundwater resources. Although the readers for whom this book is intended will probably never have to construct and operate such a computer model, it is important that they understand the conceptual basis of and the information required by the models. A brief introduction to the subject of groundwater movement will help them do this. More directly, there are some situations in which the simple models described can provide usable, approximate answers. In many field problems, it is useful to be able to calculate a maximum value for the groundwater discharge, or to sketch the pattern of pressures and flow directions within the phreatic zone.

Darcy's Law (Equation 7-1) indicates that groundwater flow occurs in response to a gradient of head and in a direction from high head to regions of low head. Head can be mapped either in a cross section of an aquifer (as shown in Figure 7-5) or over an area (as shown in Figure 7-13). The lines (or contours) joining places with equal values of head are called *equipotential lines.* Where the permeability is uniform, *flow lines* representing groundwater movement lie at right angles to equipotential lines, so that the two sets of lines form an orthogonal pattern of squares. More complicated situations are discussed in hydrogeology texts but can be ignored for the present purpose. The sides of the squares in Figure 7-13 are actually curved, but it is obvious that if all the other possible flow lines and equipotential lines were drawn between those shown in the diagram, the sides of the small squares would become straighter as the squares became infinitesimally small.

From a map of equipotential lines, therefore, a map of flow lines can be drawn, and if the hydraulic conductivity of the aquifer is known, a computation of the groundwater flow is possible.

In the simplest possible case of a confined aquifer, such as that shown in Figure 7-14, the flow lines are strongly constrained to parallel the boundaries of the aquifer. The equipotential lines are approximately normal to the boundaries of the aquifer. In this simple case, the hydraulic gradient

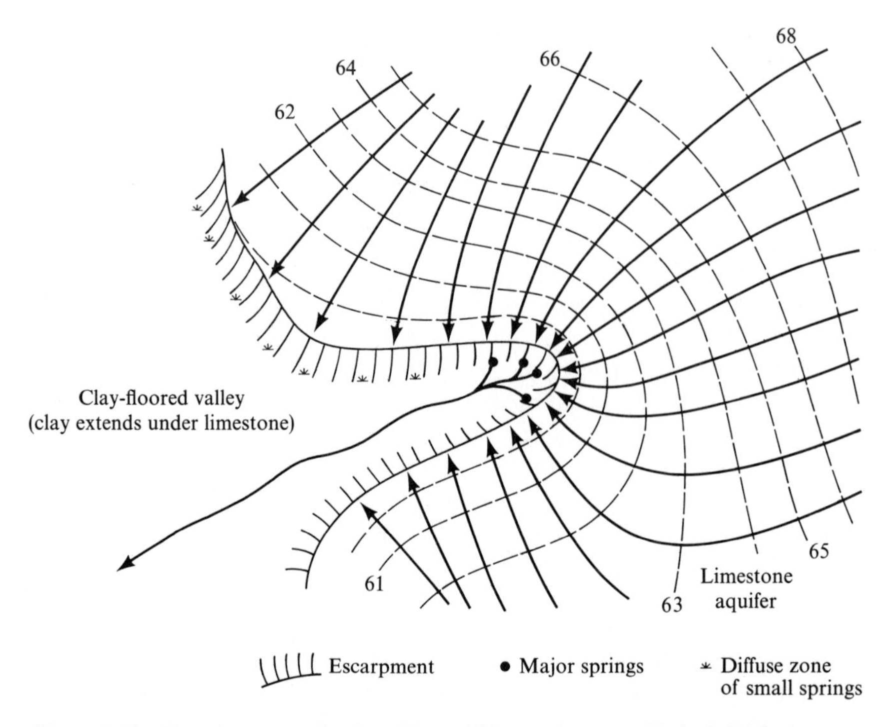

Figure 7-13 Plan of an unconfined aquifer overlying an impermeable bed. Dashed contours represent water-table elevation (piezometric head), and are equipotential lines. Solid arrows are direction of flow and are everywhere perpendicular to equipotential lines in a uniform aquifer. The foot of the scarp would be a zone of seeps or small springs.

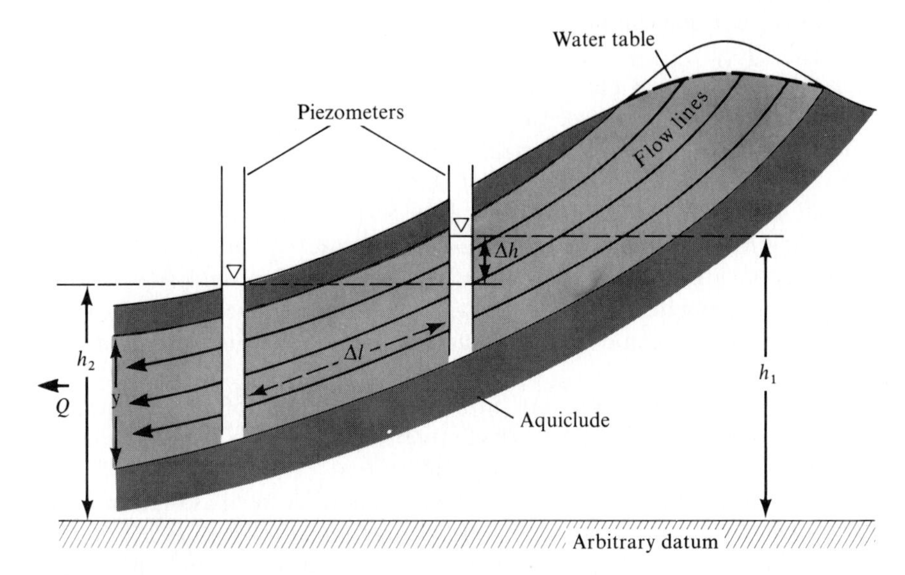

Figure 7-14 Flow in a confined aquifer. The piezometric gradient is $\Delta h/\Delta l$; aquifer thickness is y; flow lines are parallel to aquifer boundaries and equipotential lines normal to the boundaries.

can be measured at two piezometers, as shown in Figure 7-14. The flow would then be calculated directly from Darcy's Law as

$$Q = wyK\frac{\Delta h}{\Delta l} \qquad (7\text{-}2)$$

Alternatively, the flow per unit width of aquifer (q) can be obtained by using a value of unity for w, and Ky can be replaced by the transmissibility, T, so that

$$q = T\frac{\Delta h}{\Delta l} \qquad (7\text{-}2a)$$

The presence of the water table in an unconfined aquifer complicates the calculation of groundwater flow because the slope of the water table controls the magnitude of flow, and the flow in turn governs the shape of the water table. An approximate calculation of the rate of groundwater flow in an unconfined aquifer can be made by making some assumptions first introduced in 1863 by a French engineer named Jules Dupuit. If we assume that all the flow lines in the aquifer are horizontal and of uniform velocity in any vertical section, as shown in Figure 7-15, then the discharge per unit width (q) in any vertical section through the aquifer is given by the Dupuit–Forcheimer Equation:

$$q = \frac{K}{2x}(h_0^2 - h_1^2) \qquad (7\text{-}3)$$

where h_0, h_1, and x terms are defined in Figure 7-15.

The flow pattern assumed in the Dupuit method is not strictly possible in nature, but for thin aquifers with low slopes, the inaccuracies are not too important, and Equation 7-3 gives usable results.

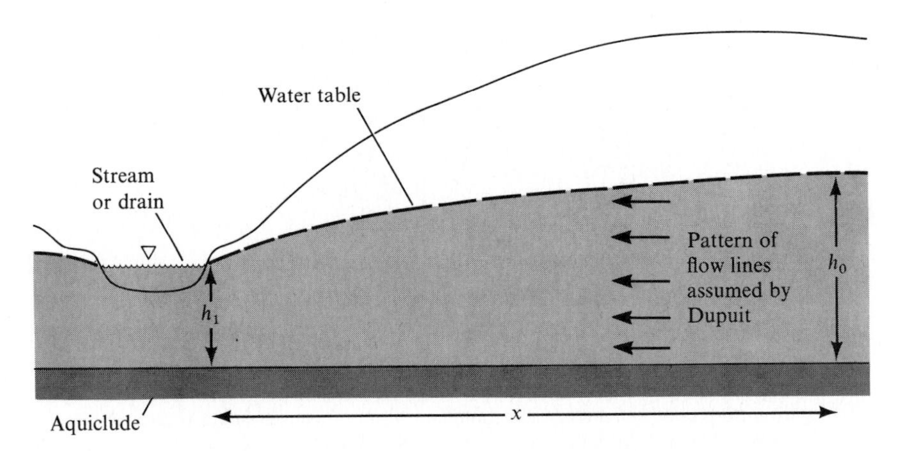

Figure 7-15 Constant flow from a shallow horizontal aquifer.

If the aquifer does not meet the rather stringent Dupuit conditions, a *flow net* like that sketched in Figure 7-13 must be constructed to estimate the pattern and magnitude of flows. This involves constructing the orthogonal network of equipotential lines and flow lines, subject to whatever boundary conditions are imposed upon the groundwater by aquicludes, water tables, and flow boundaries.

Suppose, for example, that geologic investigations and mapping of water-table elevations on a desert alluvial fan have shown that groundwater moves as a layer of relatively constant thickness (y) in a layer of sand and gravel over an impervious base. A map of water-table elevations has been constructed and contoured, as shown in Figure 7-16. We can envisage flow as occurring in tubes between flow lines. The flow lines must intersect the equipotential lines (i.e., the contours of water-table elevations in this case) at right angles. They can be spaced along each equipotential line such that Δw in Figure 7-16 is equal to Δl at that point (or in some small area). A set of curvilinear squares results. Let the flow rate through a square be Δq. For a groundwater body of constant thickness, y, Darcy's Law states that

$$\Delta q = \Delta w \, Ky \frac{\Delta h}{\Delta l}$$

Within each flow tube between a pair of stream lines, the values of Δq for all the squares will be equal because each square is simply passing water on to the next lower square at the same rate as it is receiving water. If the flow

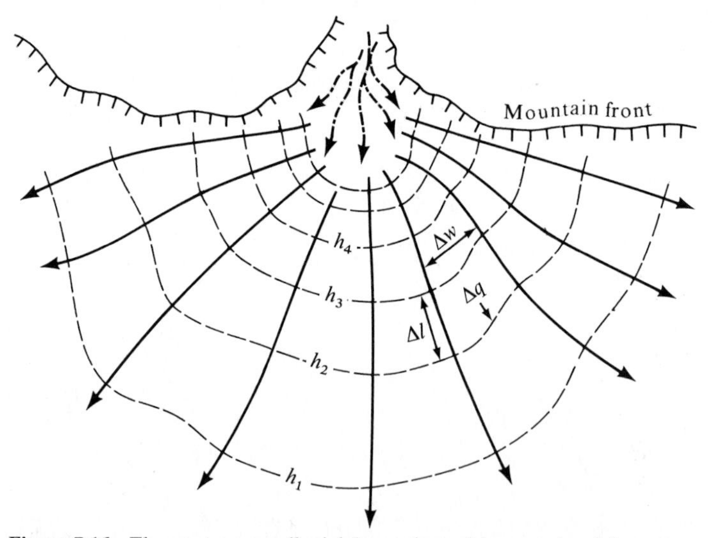

Figure 7-16 Flow net on an alluvial fan recharged by occasional flows from the mouth of a wash at a mountain front. Dashed lines are contours of water-table elevation; solid arrows are flow lines.

across each equipotential line is divided into n_w flow tubes (eight in the case shown in Figure 7-16), then the total flow through the aquifer is

$$Q = n_w \Delta q = n_w \Delta w \, Ky \frac{\Delta h}{\Delta l}$$

Since we made $\Delta w = \Delta l$,

$$Q = n_w Ky \, \Delta h \tag{7-4}$$

Δh is simply the contour interval of the water-table elevations on the map. Thus the flow rate is computed from the hydraulic conductivity and the geometry of the flow net, or more specifically from the number of flow lines that must be drawn on the water-table map in order to cross the equipotential lines in right angles and to maintain approximate equality of Δw and Δl (i.e., to form squares). Lewis and Burgy (1964) present another simple flow-net problem which they had to solve to understand the hydrology of a small watershed in California.

Geologic Relations of Groundwater

The occurrence of good groundwater supplies, given a source of natural or artificial recharge, depends upon geology. More particularly, the volume and texture of aquifers and their tectonic and solution history control the amount of water that can be stored and released to wells and streams in amounts and at rates that are beneficial for human purposes. Structural relationships between aquifers and aquicludes also affect the pattern of movement and entrapment of groundwater. In some rocks that generally form poor aquifers, the weathering history of the rock may control its water-supplying ability. A detailed treatment of the geologic relations of groundwater would be inappropriate here and is covered in the excellent text by Davis and Dewiest (1966).

Most igneous and metamorphic rocks have very low porosity and permeability and do not form good aquifers. If the rock has been fractured or weathered, however, such rocks may provide moderate supplies of water. Fracturing does not usually increase the porosity to more than a few percent, but the openings thus formed are often large and continuous enough to increase the permeability of the whole rock mass by 100 to 1000 times. At depths greater than about 30 m, however, such joints are usually tightly closed by the pressure of overburden, and the rock is not a good aquifer (see Figure 7-8). Fracture zones along faults may provide other valuable locations in rocks that otherwise provide relatively poor opportunities for groundwater development.

In semi-arid areas similar to the southwestern United States, where perennial streams are few, groundwater is of special importance particularly for small local supplies. The environmental planner is not likely to be called upon to find or develop supplies for a city, but he or she may have to assess the possibility of water for small stockponds, for an individual property, or for a recreational development. In such regions there are two prevalent conditions especially favorable for the development of small supplies of groundwater. These are in the valley alluvium and uphill of a fault.

Ephemeral channels, especially near the mouth of a canyon or valley emanating from a mountain front, are usually underlain by a valley fill of permeable sand and gravel overlying bedrock. A well drilled in the valley is likely to reach the water table at a reasonable depth. The closer the location to the mountain front, the greater is the likelihood that the valley alluvium is coarse and permeable. At greater distances from the mountain front, the valley alluvium tends to be silty. So given a choice, the best location for a well is near the mouth of a mountain valley and near the stream channel, even if the stream is dry most of the year. This does not mean that it is wise to locate a building where it is in danger of flooding, as we stress in Chapter 11.

The second favorable location in such regions is upstream of a fault, because the fault gouge (fine-grained shattered and weathered rock in the fault zone) is often a barrier to the downhill movement of groundwater. A large number of escarpments, cuestas, and footslopes of hills or mountains are caused by faulting that lifted the mountain relative to the valley floor. A fault can be mapped in the field more easily in the semi-arid climate where vegetation is sparse than in subhumid zones. On this situation Dr. Parry Reiche told a story. He had been asked to locate a well and, after mapping a fault as described above, had placed a stake indicating the proposed location at which he estimated water would be reached at 40 feet. Some weeks later he was informed that the client had drilled 400 feet and had only a dry hole. Reiche returned to the area and, after an inspection, asked the client why he had drilled in a different place than the one indicated by the geologist. The reply was, "I moved your stake a few hundred feet closer to the buildings because I did not see why the well had to be so far away. Surely that little distance could not hurt anything." Dr. Reiche told him to go back and drill where previously indicated. They struck water at 35 feet. The insignificant change in location was the difference between uphill and downhill of the fault.

In some regions of the humid tropics and subtropics, weathering of almost impervious igneous and metamorphic rocks has occurred to depths as great as 100 m, though the average depth is usually in the range of 5 to 15 m. The upper few meters of these weathered zones are lateritic soils with relatively high specific yields and permeabilities. Below the soil, however, is a thick *saprolite* or weathered zone that retains the structure of the rock and varies from highly weathered near the ground surface to almost fresh bedrock at depth. Its aquifer properties cover a similar range. The weathered zone tends to be deepest where the bedrock is most highly fractured or contains a zone of easily weathered minerals. Valley floors and the confluences of valleys at the base of an upland are also the locations of deeper weathered

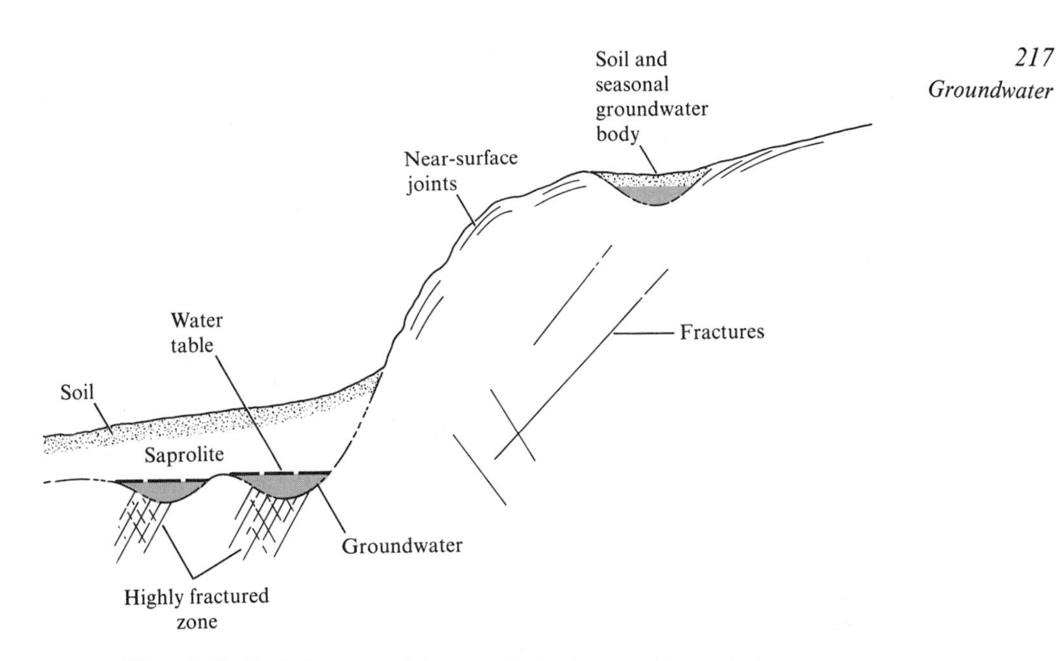

Figure 7-17 Typical pattern of deep weathering in a granitic terrain in a humid or seasonally wet climate in the tropics. Groundwater collects in hollows in highly fractured and weathered zones.

zones in some regions. In seasonally dry tropical regions, the location of adequate water supplies in these weathered zones is indicated by phreatophytic vegetation and spring lines. The depth of the transition from weathered to unweathered rock may be quite variable, and pockets of groundwater in depressions in the basal weathering surface can provide adequate yields to small wells (see Figure 7-17).

Most igneous and metamorphic regions then do not provide large supplies of groundwater. Careful geologic mapping, exploratory drilling, and an accumulation of experience in well development usually indicates those zones where fracturing, fault zones, and weathering provide usable aquifers, and supplies are often sufficient for small municipal and industrial needs. In many regions of igneous and metamorphic rocks, however, the only reliable groundwater supplies occur in recent unconsolidated sediments, such as glacial deposits or river gravels. In central Wisconsin, for example, most wells in crystalline rocks yield less than 10 liters per minute, while overlying sandy and gravelly outwash supplies 350–1500 liters per minute (Bell and Sherrill 1974). In northern New England, also, the uplands of crystalline metamorphic and igneous rocks supply very small yields to wells and springs, large enough for a single farm or a small village if a reservoir is impounded to supply water through the summer. Reliable aquifers with supplies large enough for small towns occur along major valleys where sand and gravel terraces were deposited by meltwater streams flowing from glaciers during the Pleistocene period.

Some volcanic igneous rocks are exceptions to the previous generalizations in that they are excellent aquifers. Volcanic rocks may consist of lavas, sills and dikes, ash deposits, and pumice; their properties are highly variable. Most lavas have very low intergranular porosities but have large specific yields and permeabilities because of dense joint systems developed by cooling and sometimes by tectonism. Many parts of the huge basaltic lava plateaus of the northwestern United States, for example, have transmissibilities ranging from 1,000 to 100,000 m^3/day/m. Other lavas contain vesicles, which, if they are connected, also contribute to high specific yields and permeability. Not all lavas are good aquifers, however, and many form aquicludes. In a sequence of such flows, developable groundwater supplies may only occur in thin buried soils that were formed on the lava between eruptions. These buried soil horizons are important aquifers between the generally dense and unfractured trachyte lavas on the backslope of the Rift valley escarpment north and east of Nairobi, Kenya. In areas of highly permeable lava, however, such buried soils, if clayey, may hold up perched groundwater bodies. Dike and sill rocks usually have low porosities and, unless highly fractured, can impede or even confine groundwater flow (see Figure 7-18). Volcanic ash sometimes provides good aquifer conditions, although many such deposits have a very low permeability if the particles were consolidated or fused together when the deposit was still hot after the eruption.

The groundwater conditions in sedimentary rocks also vary with lithology and structure, as described in the earlier discussion of specific yield, storativity, and permeability. Consolidated rocks such as sandstone and limestone can be excellent aquifers if the intergranular spaces are not clogged by cement. Their utility is generally enhanced if they are fractured, especially in limestones where solution of the rock by circulating groundwater may enlarge the openings. Fine-grained, consolidated rocks such as shale or siltstones usually function as aquicludes. Stratigraphic and structural features,

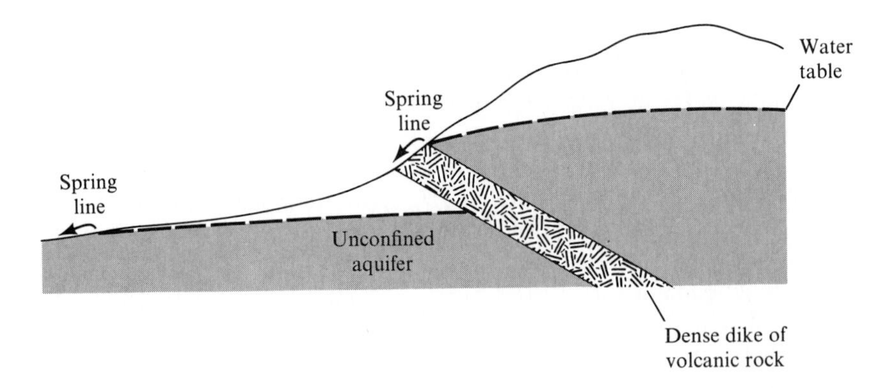

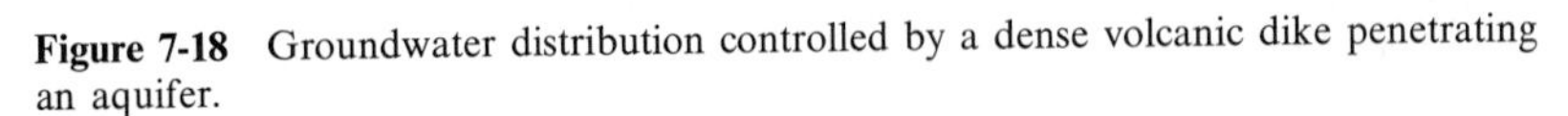

Figure 7-18 Groundwater distribution controlled by a dense volcanic dike penetrating an aquifer.

such as buried river valleys, anticlines, synclines, unconformities, and the sequence of strata are important in determining the occurrence of usable supplies of groundwater. The range of conditions in unconsolidated sediments such as recent glacial, fluvial, and eolian deposits is obvious from Figures 7-7 and 7-10, and Tables 7-1, 7-2, 7-3, and 7-5. In many terrestrial sediments of these kinds, rapid lateral changes in lithology can produce an intricate pattern of occurrence and movement that complicates the task of those trying to locate wells or simply assess the potential supply of groundwater in a region. The possibilities are too numerous to outline here, but the planner can soon gain some appreciation of them by examining the groundwater reports for the local area. The major aquifers of the United States are portrayed in a map of H. E. Thomas in his book *The Conservation of Ground Water* (1951), and reproduced widely in other publications, such as *The Water Encyclopedia* and the *National Atlas of the United States.*

The Groundwater Budget

When planning the management of a groundwater system one needs to know the limits to which water can be drawn without depleting the resource. In such plans the concept of the *safe yield* of the aquifer is often introduced. Safe yield is usually defined as the annual draft of water that can be withdrawn without producing some undesirable result. The problem in using this concept lies in the numerous undesirable results that can be produced by groundwater withdrawal. These include reducing the total amount of water available; lowering the water table, thereby increasing the cost of pumping or interfering with the withdrawal rights of others; and allowing the ingress of low-quality water. Because some of these involve legal and economic as well as hydrologic questions, there is often no general agreement on what constitutes safe yield.

In spite of these difficulties, however, it is possible to discuss the physical aspects of safe yield as expressed in the water budget for some definable groundwater system, as follows:

Input − Output = Change of storage (7-5)

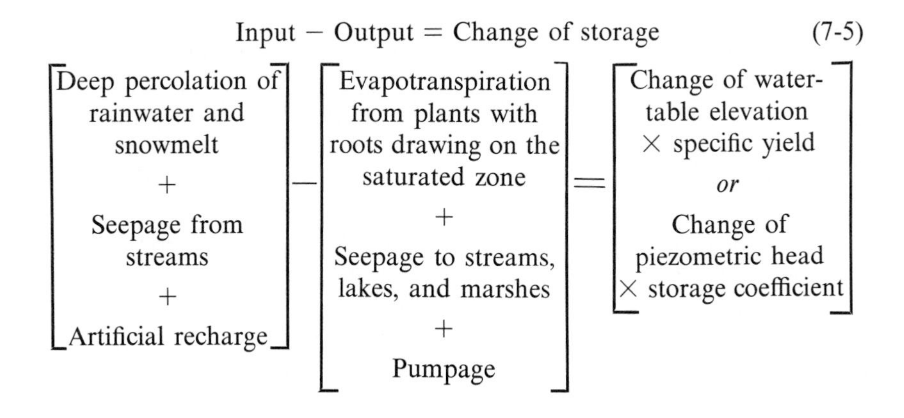

From a strictly physical definition of safe yield, the rates of recharge and pumping should be adjusted so that over a period of years (allowing for fluctuations of weather) the change in storage is zero and the resource is not depleted. Even this approach is complicated by the interdependence of the various input and output components. Pumping, for example, can lower the water table beyond the root range of phreatophytes, reducing the output of evapotranspiration. When the legal and economic issues are added, the safe yield of an aquifer becomes a debatable but useful concept.

The natural recharge from rainfall and snowmelt can be calculated after subtracting evapotranspiration from the amount of water which infiltrates the soil. The method for calculating seasonal variations of this input will be described in Chapter 8 on the water balance. Drastic alterations of the ground surface can reduce infiltration and thereby cause a reduction of groundwater recharge and of the dry-weather flow of streams. When the land is paved in urban areas, serious reduction of recharge and of summer streamflow results (Franke and McClymonds 1972). Land-use plans should take account of such deleterious effects of urbanization. The major zones of aquifer recharge in an area should be mapped and preserved in some use which will not restrict infiltration or contaminate the groundwater recharge. Over a 130-square-kilometer region of suburban Long Island, Franke (1968) has also documented a 3-meter decline of the water table since the installation of sanitary sewers in the mid-1950's. Before that time, the percolation of wastewater from septic tanks, though heavily polluted, recharged the groundwater and helped maintain streamflow.

In many arid regions the general water table lies below valley floors, and when the stream channels carry water during infrequent storms or during the snowmelt season in nearby mountains, water leaks through the bed and banks of the channels to recharge the groundwater, as shown in Figure 7-19.

Many groundwater bodies in arid regions are no longer recharged under the present climate. The water in storage originated as precipitation hundreds, or even thousands, of years ago. In the case of the artesian aquifers under the Sahara, dating of the groundwater by means of its radioactivity indicates that most of it was emplaced during a wetter period approximately 5000 years ago. Under the desert of western Egypt, 25,000-year-old water has been discovered. Groundwater stored under a wetter climate in the past has been found and exploited in the southwestern United States. The withdrawal of water from storage where it is being replaced at only very slow rates or not at all is called "groundwater mining." It can provide small supplies for a very long time, but if the resource is exploited for large-scale irrigation, its useful life is limited.

Where geologic and soil conditions are favorable, recharge may be artificially induced. We will describe in Chapter 20 the value of spraying biodegradable effluents from food-processing industries or sewage plants onto permeable soils where the pollution load is removed by biologic processes. The purified water percolates to the groundwater system. In some areas, such as Long Island, groundwater withdrawn for use in air conditioning is

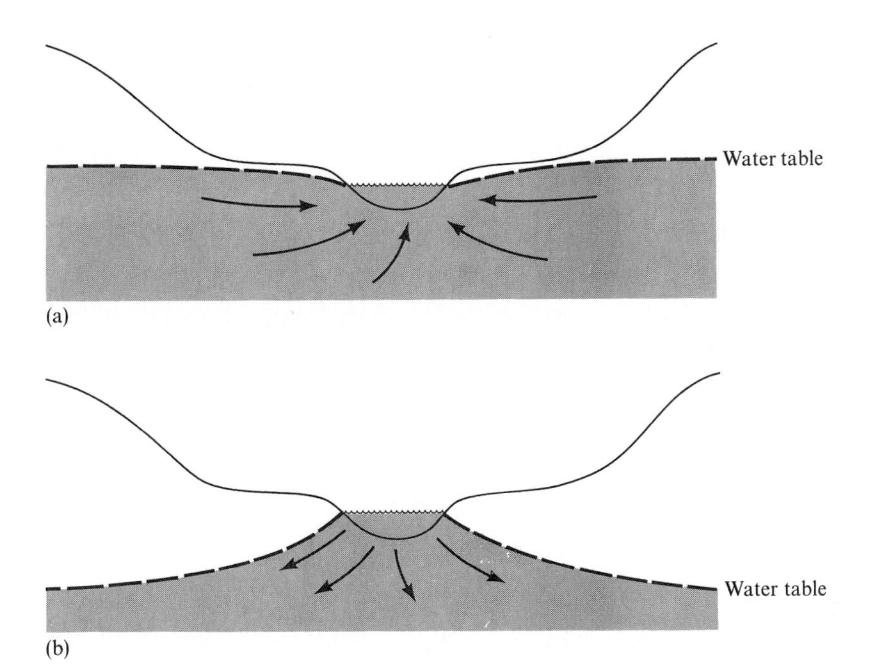

Figure 7-19 Effluent and influent streams. (a) In a humid region where there is considerable recharge of groundwater throughout the catchment, the water table slopes toward rivers that gain water, as shown by the arrows. Artificial drains act in the same way. (b) In arid areas, there is little or no direct recharge from the catchment; the water table lies below the riverbed. During storm runoff or snowmelt, water percolates through the bed and banks of the channel to supply groundwater at depth. Canals in arid areas lose water in the same way, as do artificial recharge systems such as seepage pits and water-spreading facilities.

pumped back into the phreatic zone through injection wells. In many urban areas and in deserts, storm runoff is impounded or diverted onto a permeable substrate and allowed to recharge the groundwater system. Long Island and Los Angeles County are two examples in the United States of areas where a major effort is made to use storm runoff in this way. The programs solve a part of the storm drainage problem and conserve groundwater (see Chapter 11).

The methods by which the water is applied vary with the volumes of water concerned, the local geologic conditions, and the capital available. In the case of waste disposal, the effluent is usually sprayed onto the land or allowed to run out of perforated pipes. Other methods of artificial recharge include impoundment or discharging the water in an unlined canal over a permeable substrate, or pumping the water down a well directly into the aquifer. A variation of the method of inducing recharge, illustrated in Figure 7-20, involves drawing down the water table by pumping water from wells in a valley floor until the water table slopes from the stream toward the well. The percolation of water from the river provides an assured recharge to the well, as long as the stream has a reliable flow. The city of

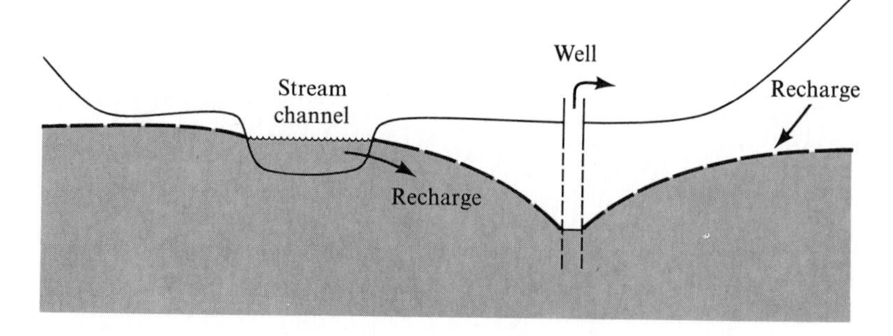

Figure 7-20 Pumping from a well near a stream can draw down the water table so that it slopes away from the river. Water then flows from the channel bed and banks into the well, ensuring a plentiful supply. Extremely high pumping rates can even lower the water table beneath the channel bed, causing the stream to dry up during the low-flow season.

Santa Fe, New Mexico, for example, is supplied in part by wells in the alluvium along the Rio Grande. Many wells along the Mississippi Valley are also supplied in this way.

Outputs from the groundwater system are equally numerous and subject to control. Discharge by evapotranspiration of phreatophytes whose roots tap the phreatic zone is a major loss to the groundwater. Although it occurs in humid regions as well, the loss is most obvious in arid areas with limited water supplies. Many of the phreatophytes growing on the valley floors of the western United States can develop extensive root systems that exploit water at great depth (Robinson 1958). Roots of mesquite, for example, have been observed penetrating 15–30 m below the surface, while greasewood and salt cedar can reach 3–5 m down. The magnitude of the losses through phreatophyte transpiration have stimulated attempts to replace these plants with more productive grasses. Although such land management is warranted in some places, there are situations in which phreatophytes are vitally important to the ecology of the region, and care should be taken to examine their value before the desire to harvest more water stimulates an eradication program. Many of the large mammal populations of East Africa, for example, depend upon phreatophytes for cover and forage during the long dry seasons. In the Rocky Mountains moose leave the high country to winter in the willows and cottonwoods of alluvial valleys, and in New England deer survive the coldest winter nights by taking cover under the heavy coniferous forest along rivers and marshes. The loss of groundwater, therefore, is often well compensated.

Discharge of groundwater into rivers and lakes varies seasonally with the amount of water that is stored in the phreatic zone and therefore with the slope of the piezometric surface, as indicated in the discussion of Figure 7-1. The rate of this drainage is indicated by the dry-weather flow of streams.

This water is not "lost" as far as human use is concerned because it maintains surface water resources.

Pumping is the most obvious withdrawal from the groundwater body, and in heavily developed regions is usually the largest. It is often surprisingly difficult to find out how much water is being pumped from an aquifer. Few records have been kept in the past, and even though permits are usually necessary to authorize maximum pumping rates, it is difficult to find out what actual withdrawals are being made.

Imbalance between inputs and outputs result in changes of storage in the aquifer, which are reflected in fluctuations of the piezometric surface, as indicated in Equation 7-5. In Figure 7-2, for example, changes of water-table elevation indicate the short-term differences between recharge by infiltration and outflow by evapotranspiration and groundwater flow to the nearby stream. Because of the uncertainty in estimating the recharge and discharge, changes in the piezometric surface usually give the first and most sensitive indication of an imbalance. In extreme cases the imbalance is great and the depletion of the groundwater resource is rapid. This is particularly common in confined aquifers that have low storage coefficients and that often have only limited recharge zones, as shown in Figures 7-4 and 7-5.

There are major uncertainties involved in estimating the long-term average annual recharge and discharge of an aquifer. The techniques are well described by Walton (1970). In general the groundwater manager aims to limit the total natural and artificial withdrawal from the aquifer to be in balance with the rate of recharge. The amount of usable groundwater under these conditions is the safe yield. The resource may be mined for a short period to lower the water table and reduce losses through transpiration and streamflow, but eventually the withdrawals must be brought into balance with the recharge, or a variety of untoward results will occur.

Environmental Effects of Groundwater Withdrawal

Because groundwater is not an obvious component of the hydrologic cycle, it is easy for planners and others to forget that some of their actions may have deleterious effects on water resources and ecology over a wide area. Careful consideration should be given to any schemes that might alter groundwater recharge or increase withdrawals. An appreciation of the limitations of the groundwater resource will save money and hardship in the long run.

When water is pumped from a new well, the initial withdrawal exceeds the rate at which groundwater flows into the vicinity of the well. The surrounding water table or piezometric surface is therefore lowered and slopes toward the well, forming a cone of depression, as shown in Figure 7-21. The increase in the slope of the piezometric surface increases the flow of water toward the well, according to Darcy's Law, until it balances the pumping rate. A new equilibrium piezometric surface develops, so long as the rate of recharge of the aquifer is sufficient to supply the pumpage. If it is not, the cone of depression will continue to steepen and enlarge, increasing the cost of pumping from greater depths. As the cost of power escalates, the depth from which water must be pumped in some mined aquifers, such as those in the Phoenix region of Arizona, is becoming a serious threat to the viability of irrigation agriculture. Beneath the City of London the piezometric surface in the confined Chalk aquifer has fallen more than 60 m over hundreds of square kilometers. In the nineteenth century some wells in this aquifer were artesian. Beneath Chicago, Illinois, pumping since the late nineteenth century has lowered the piezometric head by 200 m. The cost of lifting water becomes important.

The cones of depression from neighboring wells will eventually intersect if withdrawals continue to exceed recharge. The largest and deepest wells will draw water from below the shallower wells, taking away their water supply (see Figure 7-22). The shallow wells then typically are drilled deeper to compete with their neighbors, and the mining is accelerated amid legal battles over water rights and the ruination of landowners who lose their water supply. In an effort to prevent such competition and destruction of the resource, many government agencies are attempting to define the safe yield of major aquifers and to control pumping rates on the basis of a quantitative prediction of how each new withdrawal will affect the whole groundwater system. These predictions are usually made with the assistance of computerized models of the aquifer.

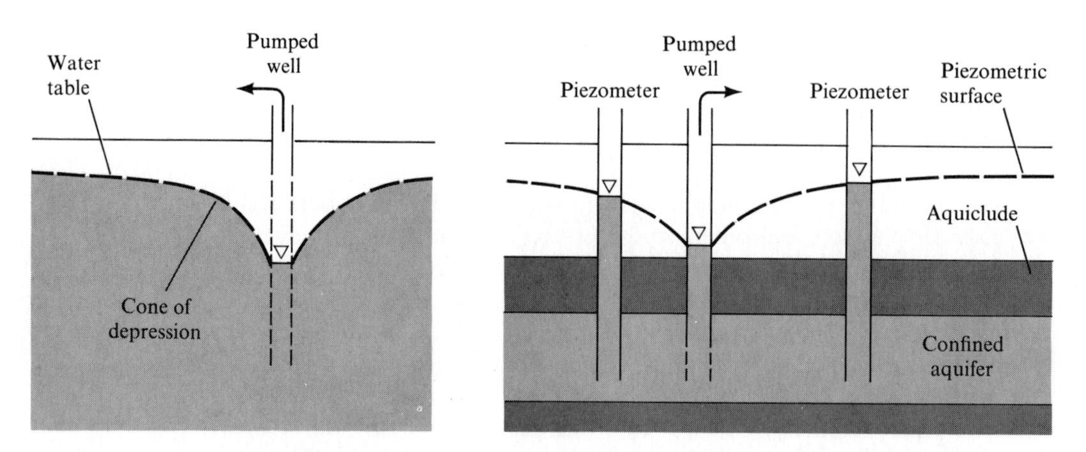

Figure 7-21 Cones of depression. Pumping from a well lowers the piezometric surface, which in the case of the unconfined aquifer on the left is the water table. The elevation changes are monitored in surrounding wells or piezometers. The cone spreads away from the well until its form is in equilibrium with the rates of recharge and pumping. Around large well fields the cone extends for several kilometers and takes years to equilibrate, even if there is an adequate recharge.

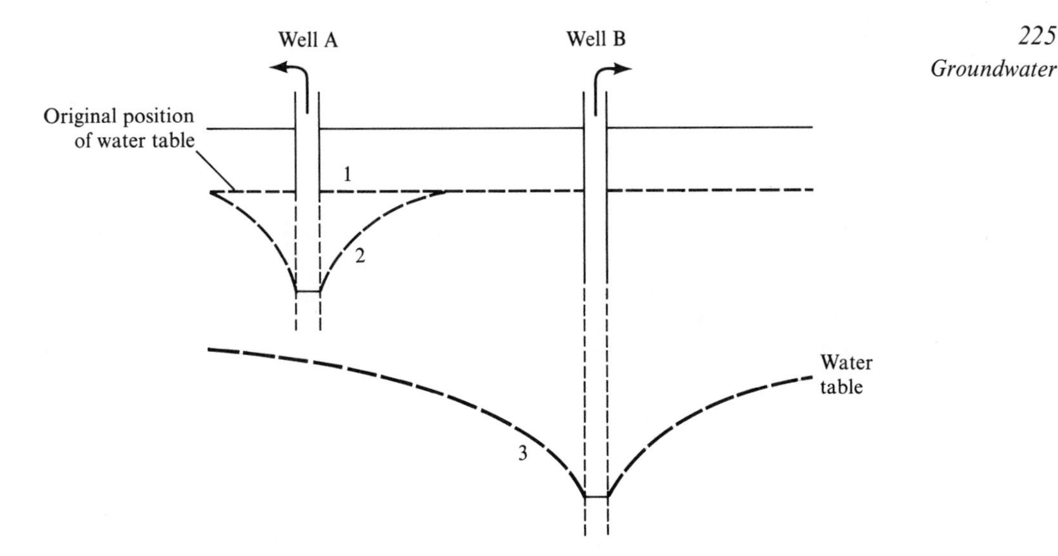

Figure 7-22 Competition between wells. Pumping from a shallow well, A, lowers the water table from its original position (1) into a cone of depression (2). A deeper well, B, is then drilled and is pumped at a higher rate than A. It generates a larger cone of depression (3) and draws the water table below well A, which dries up. To reinstate a water supply in the shallow well the owner has the option of suing to have the pumping rate in well B reduced so that the water table can recover. Alternatively, well A can be deepened, a larger pump installed, and the competition escalated one more step to the detriment of the groundwater balance.

It was mentioned early in this chapter that groundwater was intimately linked with surface water and that flow could take place between these two subsystems of the hydrologic cycle. The water table may be lowered by pumping until it falls below or slopes away from a surface water body. According to Darcy's Law, water will then flow from the river or lake, as shown in Figure 7-20. If the river has a large flow, the loss may not cause a significant reduction of discharge, and the results may be a clean, filtered water supply for a nearby town. If the aquifer is too permeable, however, bacteria in the river water can survive a journey to the well and pollute the water supply. If the river is heavily charged with dangerous chemicals, of course, this form of induced recharge is not useful. Furthermore, small rivers, lakes, and marshes may be depleted or even dried up in this way, causing the eradication of important wetland habitats. Before proposing the development of a water supply by pumping from a valley aquifer, the planner should seek advice from a hydrogeologist about the probable results of the scheme.

In coastal localities, drawdown of the water table or piezometric surface often allows sea water to enter wells. The process of *saltwater intrusion* is illustrated in Figure 7-23. In an unconfined aquifer near the sea, fresh groundwater occurs as a lens above the heavier sea water. The saline fluid may extend inland for about a kilometer. Because of the difference in

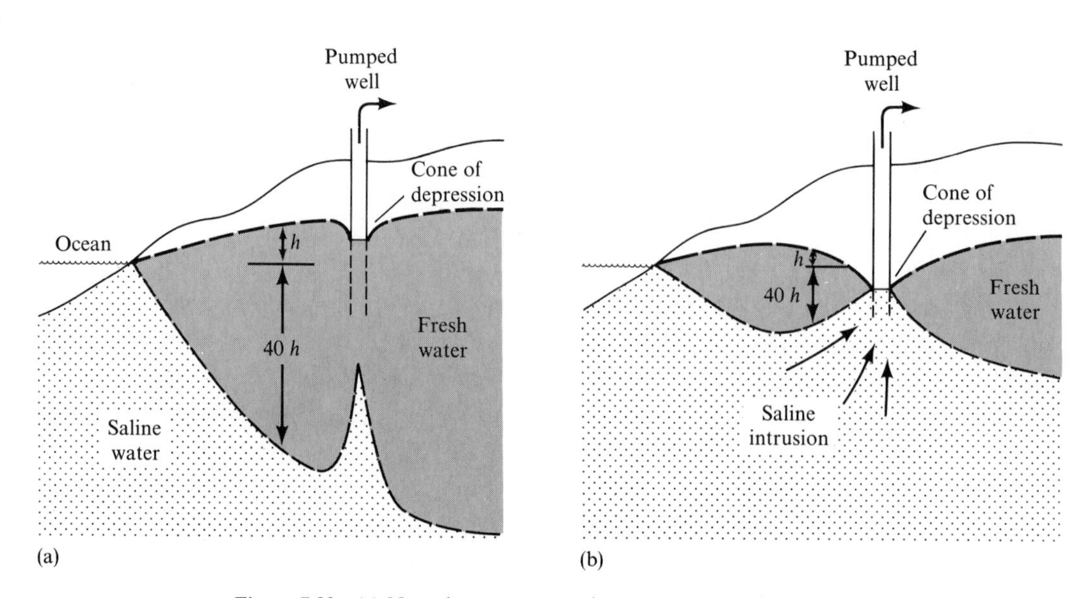

Figure 7-23 (a) Near the coast, groundwater occurs as a lens over saline water. The height of the lens above sea level is equal to one-fortieth of its depth below sea level. (b) Heavy pumping of the groundwater generates a large cone of depression in both the upper and lower boundaries of the freshwater lens, and eventually allows saline water to enter the well.

density, the depth of fresh water below sea level is approximately equal to 40 times the height of the water table above sea level, as shown in Figure 7-23(a). With these relative dimensions, the fresh and the saline waters are in a state of balance with the recharge rate and the flow of groundwater into the ocean.

If the water table is lowered by pumping, the cone of depression around the well is reflected in a rise of the boundary between fresh and salt water. Each meter decline of the water table, however, will cause a 40-meter rise of the lower boundary of the lens to maintain the balance referred to earlier. Heavy pumping, therefore, can produce such a large cone of depression that the saline water will eventually invade the well, as shown in Figure 7-23(b). The water supply becomes useless for most purposes, and the extent of the intrusion can be defined by mapping chloride-ion concentrations in well waters.

For the planner, saltwater intrusion is a serious problem on both a local and a large scale. Owners of summer homes on the New England coast are experiencing increasing degradation of the quality of water in their wells. Subdivision and sale of recreational property in this and many other coastal locations have occurred without any thought being given to the carrying capacity of the groundwater, which is ultimately controlled by the rates of recharge and of discharge to the ocean. Most wells along the Kenya coast near Mombasa suffered contamination by salt water during the last quarter century, and the city (population 250,000) is now supplied by a 230-kilometer-long pipeline stretching across arid country to a large spring in a volcanic aquifer. On many islands and peninsulas around the world, the supply of groundwater is severely limited by the need to guard against the intrusion of salt.

On an even larger scale the coast of California between Los Angeles and Long Beach provides just one example in the United States of saltwater intrusion resulting from massive groundwater withdrawals for industrial and municipal supply. Saline incursions now extend more than 6 km inland in some places, and have forced the abandonment of many wells. One response has been to reduce the withdrawal of groundwater, which necessitates importing water from large rivers to the east and north. Artificial recharge is also being tried, using stormwater impounded in seepage pits over permeable soils in urban areas. Treated sewage effluent is also being used. One large-scale experiment is being carried out to examine the technical and economic feasibility of forming a barrier to saline intrusions by injecting fresh water into the aquifer near the coast, as illustrated in Figure 7-24. Although this expensive tactic may alleviate the situation a little, there is no prospect of it making a large impact on the water supply of southern California or other coastal regions that have overtaxed their groundwater supply. Water-supply authorities in southern California are already looking to other regions of the country and are contemplating large-scale inter-basin transfers of water, which threaten to cause ecological and social disruption.

Water pressure in the pores of an aquifer supports some of the weight of the overlying rock. When water is withdrawn, the piezometric pressure declines and the particles of the rock settle together very slightly. This effect is eventually transmitted upward to the ground surface causing subsidence. For a given amount of lowering of the piezometric surface, the compaction is much greater in confined aquifers than in those with a water table. Fine-grained unconsolidated sediments are more compressible than sand, gravel, or bedrock aquifers. Most of the compaction occurs, therefore, as the piezometric head in the silty and clayey members of a geologic sequence declines. Because these members have low permeabilities, they release water slowly, so that the decrease of head and the resulting subsidence often take place over many years, even after pumping is curtailed (see Figure 1-6). The amount of subsidence depends mainly on the reduction of head and the compressibility and thickness of the silt and clay beds. Other factors are described by Poland and Davis (1969).

The damage resulting from subsidence and compaction throughout the world now amounts to hundreds of millions of dollars. Differential settling disrupts canal, drain, and river gradients, reducing their conveyance capacity or increasing their velocities and causing bank erosion. Drain pipes and well casings are fractured or closed. Buildings, bridges, and railroads are cracked or warped, and the flooding problem in Venice, described in Chapter 1, is only one example of a hazard that has occurred in other places.

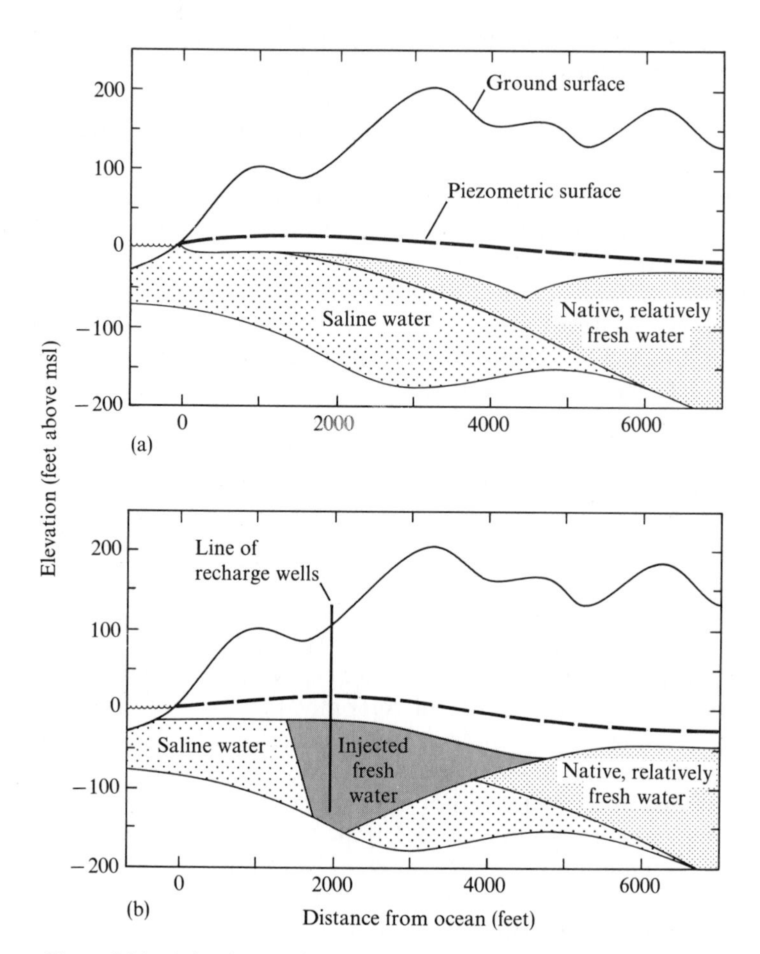

Figure 7-24 A barrier to saltwater intrusion can be constructed by pumping water (storm runoff, treated sewage, cooling water, and so on) down recharge wells seaward of the water-supply wells. (a) Saline intrusion into a confined aquifer on the coast of southern California, a region of significant groundwater pumping. (b) The result of pumping fresh water down a line of recharge wells parallel to the coast. (From Laverty and Van der Groot 1955. Reprinted from *Journal American Water Works Association,* Volume 47, by permission of the Association. Copyrighted 1955 by the American Water Works Association, Inc., 6666 W. Quincy Avenue, Denver, CO 80235.)

Another famous and catastrophic example of land subsidence due to groundwater withdrawal occurs in Mexico City (Loehnberg 1958). The city stands on an old lake bed into which volcanic sediments were washed during the last few million years. The volcanic ash has been altered to a highly compressible clay with a natural water content (porosity) of 80 to 90 percent. Fifty meters of sediments rich in these fine-grained materials overlie the major confined aquifer, which was originally artesian. Heavy groundwater pumping for municipal supply, accompanying the growth of the city, has lowered the piezometric surface by up to 2 m/yr. The pumping rate in wells as much as 300 m deep totaled about 800,000 m^3/day in the late 1950's, while the rate of natural recharge was approximately 200,000 m^3/day.

The lowering of piezometric pressure caused the land surface to subside at rates of up to 30 cm/yr by the late 1950's, and by that time had lowered the central city by 5 to 7 m. Sewers had broken, abandoned well cases protruded 5 m out of the ground, and a great deal of structural damage had occurred in buildings. The predicted maximum amount of settling if pumping continues is 20 m. Faced with this awesome prospect, the Mexican authorities have curtailed pumping, developed surface water supplies from the surrounding highlands, and attempted to recharge the groundwater artificially to halt further subsidence. Some of the strategies being tried are described by Loehnberg (1958).

Recharging the groundwater will halt the subsidence but can only reverse a small amount of the subsidence. Most of the settling is caused by permanent compaction of the fine-grained materials.

In the Central Valley of California, heavy pumping, mainly for irrigation, has lowered the ground surface by more than 30 cm over an area of 10,000 sq km, as the piezometric head has been drawn down by as much as 150 m. The situation is aggravated by two other subsidence problems related to water. Drainage of peat lands in the San Joaquin Delta has allowed compaction, oxidation, and wind erosion of the organic sediments, resulting in up to 5 m of lowering (Allen 1969). In other parts of the valley, irrigation of very dry sediments has caused settling, as described by Lofgren (1969).

Sources of Information on Groundwater

A variety of information is available to the planner concerned about groundwater. They range from summary technical reports to indirect evidence that can be easily compiled.

In the United States the U.S. Geological Survey has published reports on the groundwater conditions of many small regions of the country in its series of Water Supply Papers. A summary of the national situation was compiled by McGuinness (1963). The water agencies or geological divisions of state governments release similar reports (Giefer and Todd 1972). The situation is usually the same in other countries.

Many agencies also have on file groundwater information that can be examined upon request. Engineering consultants or their clients in industry or government frequently have reports of groundwater investigations around large structures.

These summary technical reports often include geologic maps outlining the major aquifers, as well as maps or tabulations of piezometric head;

direction and rates of groundwater motion; pumping-test data on transmissibility, storativity, and specific capacity; and the sustainable short-term yields of wells in each aquifer. Less frequently these publications deal with the more difficult problems of long-term safe yield and the environmental consequences of groundwater development. Most of the interpretations that a planner requires will usually be made in such reports, and he needs only to be able to recognize the important points in the report and the limitations of the available data.

If there are no interpretive hydrogeologic maps or reports available for the area of interest, a planner may have to rely upon interpretations of geologic maps by a geologist. In overseeing or reviewing such interpretations, the planner should remember all the requirements of an aquifer mentioned earlier in this chapter. A thin gravel deposit may be extensive on a geologic map but may not give sufficient yield for his purpose. A thick, permeable deposit may not have sufficient replenishment in an arid region. Even the presence of large groundwater bodies may not be sufficient evidence on an assured supply. In many areas of the world there are groundwater bodies that accumulated during wetter periods of the Pleistocene epoch (Ambroggi 1966).

A review of the geologic principles governing the location of aquifers can be found in E. E. Johnson's (1966) practical book *Groundwater and Wells,* and the hydrogeologic reviews by Meinzer (1923) and Maxey (1964). Meinzer (1927) has also described the use of vegetation as an indicator of groundwater conditions, while Meyboom (1962) and Toth (1966) have reviewed a variety of methods for recognizing zones of recharge and discharge of groundwater. The planner will find it useful to be able to recognize the location and temporal fluctuations of water tables in any field project with which he is associated. He should try to develop an awareness of groundwater in the field.

Water Witching (Dowsing)

Throughout human history it has been common among all cultures to ascribe inexplicable phenomena to supernatural agents. Groundwater is an example. It cannot be seen and therefore the principles of its occurrence are not obvious. So there has grown up around the subject an aura of mystery not dispelled even in the present technological age. An aspect of this is the still present widespread belief that underground water can be located by those few individuals who have supernatural powers.

All cults have insignia or badges of authority, whether it be the leather bag at the belt of the medicine man or its less obvious relatives the slide-rule in the pocket of an engineer and the stethoscope in the pocket of a physician. The tool as well as the badge of authority of the dowser is the forked stick, which is sincerely believed by practitioners to jerk uncontrollably downward when over a hidden water body. A touchstone to the mythology of dowsing is the frequent reference to an "underground river." Knowledge of geology and hydrology makes it obvious that anything that could possibly be called an "underground river" is scarce indeed and occurs only in solution cavities in a few rocks such as limestone. On the other hand, the principles of physics are general. The groundwater developed for water supply in all countries derives from precipitation on the earth's surface and moves according to the principles described earlier.

Some dowsers, of course, can find groundwater. Many persons appreciate the natural conditions under which water enters the ground, migrates, and collects. They also recognize the association between the places in which groundwater collects and various topographic, geologic, and botanical indicators. Those who have lived in the same region for a long time have observed many successes and failures in well drilling. For these reasons many farmers and well drillers can locate groundwater as well as the most experienced dowser, and in the same manner, namely, from observing natural characteristics of the landscape.

For locating a groundwater supply, therefore, we recommend, first of all, consulting the geologic and hydrologic reports for the region of interest. In some areas these will often indicate that a successful well can be located almost anywhere in the region and that there is no need for a dowsing service. If the situation is more complicated, a groundwater geologist should be consulted. Various governmental agencies and private firms offer this kind of service. Only if the planner wants a little color in his life do we recommend consulting a dowser.

Leopold had an aunt who was the strong-willed matriarch of a Spanish family. He heard that she had paid a dowser $25 to locate a well on her ranch. The little ranch house is high on a hill underlain by a thick sequence of sand and gravel, with no drainage area upstream of the house location and, to a hydrologist, no obvious recharge or intake area to supply water to ground storage. Yet only 200 m away was an ephemeral wash draining more than two square kilometers and underlain by alluvium that probably would yield some water to a shallow well. The dowser located a well site near the ranch house. It was drilled to 200 m and was a dry hole. Leopold exclaimed to his aunt, "Tía, if you asked me, my colleagues who are among the most competent groundwater scientists in the world would have been glad to locate a drilling spot for you." She replied, "But you and your friends wouldn't have put the well where I wanted it."

Bibliography

ALLEN, A. S. (1969) Geologic settings of subsidence; *Reviews of Engineering Geology*, vol. 2, pp. 306–342.

AMBROGGI, R. P. (1966) Water under the Sahara; *Scientific American*, vol. 214, no. 5, pp. 21–29.

AMERICAN SOCIETY OF CIVIL ENGINEERS (1972) Groundwater management; *Manuals and Reports on Engineering Practice*, no. 40, 216 pp.

BEDINGER, M. S. AND EMMETT, L. F. (1963) Mapping transmissibility of alluvium in the Lower Arkansas River Valley, Arkansas; *U.S. Geological Survey Professional Paper 475-C*, pp. C188–C190.

BELL, E. A. AND SHERRILL, M. G. (1974) Water availability in central Wisconsin, an area of near-surface crystalline rocks; *U.S. Geological Survey Water Supply Paper 2022.*

CEDERSTROM, D. J. (1972) Evaluation of yields of wells in consolidated rocks, Virginia to Maryland; *U.S. Geological Survey Water Supply Paper 2021.*

Davis, S. N. (1969) Porosity and permeability of natural material; Chapter 2 in *Flow through porous media* (ed. R. J. M. Dewiest), Academic Press, New York, pp. 54–90.

DAVIS, S. N. AND DEWIEST. R. J. M. (1966) *Hydrogeology;* John Wiley & Sons, New York, 463 pp.

DAVIS, S. N. AND TURK, L. J. (1964) Optimum depth of wells in crystalline rocks; *Ground Water,* vol. 2, no. 2, pp. 6–11.

FERRIS, J. G., KNOWLES, D. B., BROWNE, R. H. AND STALLMAN, R. W. (1962) Theory of aquifer tests; *U. S. Geological Survey Water Supply Paper 1536-E.*

FRANK, F. J. (1973) Groundwater in the Eugene–Springfield area, Southern Willamette Valley, Oregon; *U.S. Geological Survey Water Supply Paper 2018.*

FRANKE, O. L. (1968) Double-mass-curve analysis of the effects of sewering on groundwater levels on Long Island, New York; *U.S. Geological Survey Professional Paper 600-B*, pp. B205–B209.

FRANKE, O. L. AND MCCLYMONDS, N. E. (1972) Summary of the hydrologic situation on Long Island, New York, as a guide to water management alternatives; *U.S. Geological Survey Professional Paper 627-F.*

GIEFER, G. J. AND TODD, D. K., EDS. (1972) *Water publications of state agencies;* Water Information Center, Port Washington, NY, 319 pp.

HEATH, R. C. AND TRAINER, F. W. (1968) *Introduction to groundwater hydrology;* John Wiley & Sons, New York, 284 pp.

INESON, J. (1962) A hydrogeologic study of the permeability of the Chalk; *Journal of the Institution of Water Engineers,* vol. 16, pp. 449–463.

INESON, J. (1963) Applications and limitations of pumping tests: hydrogeological significance; *Journal of the Institution of Water Engineers*, vol. 17, pp. 200–215.

JOHNSON, A. I. (1967) Specific yield: compilations of specific yields for various materials; *U.S. Geological Survey Water Supply Paper 1662-D.*

JOHNSON, E. E. (1966) *Groundwater and wells, a reference book for the water-well industry,* E. E. Johnson, Inc., St. Paul, MN 440 pp.

KAZMI, A. H. (1961) Laboratory tests on test drilling samples from Rechna Doab, West Pakistan, and their application to water resources evaluation studies; *International Association of Scientific Hydrology, Symposium of Athens, Publication No. 57,* pp. 496–500.

LAVERTY, F. B. AND VAN DER GOOT, H. A. (1955) Development of a freshwater barrier in Southern California for the prevention of sea-water intrusion; *American Water Works Association Journal,* vol. 47, pp. 886–908.

LAWSON, D. W. (1968) Groundwater flow systems in the crystalline rocks of the Okanagan Highland, British Columbia; *Canadian Journal of Earth Sciences*, vol. 5, pp. 813–824.

LEGRAND, H. E. (1954) Geology and groundwater in the Statesville area, North Carolina; *North Carolina Division of Mineral Resources Bulletin 68*, 68 pp.

LEWIS, D. C. AND BURGY, R. H. (1964) Hydrologic balance from an experimental watershed; *Journal of Hydrology*, vol. 2, pp. 197–212.

LOEHNBERG, A. (1958) Aspects of the sinking of Mexico City and proposed countermeasures; *American Water Works Association Journal*, vol. 50, pp. 432–440.

LOFGREN, B. E. (1969) Land subsidence due to the application of water; *Reviews of Engineering Geology*, vol. 2, pp. 271–303.

LUSCZYNSKI, N. J. AND SWARZENSKI, W. V. (1960) Saltwater encroachment in southern Nassau and southeastern Queens Counties, Long Island, New York; *U.S. Geological Survey Water Supply Paper 1613-F.*

MAXEY, G. B. (1964) Hydrogeology; Section 4 in *Handbook of applied hydrology*, (ed. Ven Te Chow), McGraw-Hill, New York.

MCGUINNESS, C. L. (1963) The role of groundwater in the national water situation; *U.S. Geological Survey Water Supply Paper 1800.*

MEINZER, O. E. (1923) Outline of groundwater hydrology with definitions; *U.S. Geological Survey Water Supply Paper 494.*

MEINZER, O. E. (1927) Plants as indicators of groundwater; *U.S. Geological Survey Water Supply Paper 577.*

MEYBOOM, P. (1962) Patterns of groundwater flow in the prairie profile; *Proceedings of Hydrology Symposium No. 3, Groundwater*, National Research Council of Canada, pp. 5–20.

PETTYJOHN, W. A. AND RANDICH, P. C. (1966) Geohydrologic use of lithofacies maps in glaciated areas; *Water Resources Research*, vol. 2, pp. 679–689.

POLAND, J. F. AND DAVIS, G. H. (1969) Land subsidence due to withdrawal of fluids; *Reviews in Engineering Geology*, vol. 2, Geological Society of America, pp. 187–269.

ROBINSON, T. W. (1958) Phreatophytes; *U.S. Geological Survey Water Supply Paper 1423.*

SEABURN, G. E. (1970) Preliminary results of hydrologic studies at two recharge basins on Long Island, New York; *U.S. Geological Survey Professional Paper 627-C.*

SISSON, W. H. (1955) Recharge operations at Kalamazoo; *American Water Works Association Journal*, Vol. 47, pp. 914–922.

THEIS, C. V., BROWN, R. H. AND MEYER, R. R. (1963) Estimating the transmissivity of aquifers from the specific capacity of wells; *U.S. Geological Survey Water Supply Paper 1536-I*, pp. I331–I341.

THOMAS, H. E. (1951) *The conservation of ground water;* McGraw-Hill, New York, 327 pp.

TOTH, J. (1966) Mapping and interpretation of field phenomena for groundwater reconnaissance in a prairie environment, Alberta, Canada; *International Association of Scientific Hydrology Bulletin*, vol. 11, no. 2, pp. 20–68.

WALTON, W. C. (1970) *Groundwater resource evaluation;* McGraw-Hill, New York, 664 pp.

WESTERN, D. AND VAN PRAET, C. (1971) Cyclical changes in the habitat and climate of an East African ecosystem; *Nature*, vol. 241, pp. 104–106.

Typical Problems

7-1 Calculation of Groundwater Flow

Calculate the flow of groundwater to a stream through a confined aquifer 2 km wide and 7 m thick. Two piezometers 900 m apart show water-elevation differences of 10 m, and the coefficient of permeability measured in the field by a pumping test is 1.3 m/day.

Solution

In Equation 7-1, $w = 2000$ m, $d = 7$ m, $K = 1.3$ m/day, $\Delta h = 10$ m, $\Delta l = 900$ m. Therefore

$$Q = 2000 \times 7 \times 1.3 \times 10/900 = 202 \text{ m}^3/\text{day}$$

7-2 Direction of Groundwater Flow, I

A low-density residential development on a flat site will be served by septic-tank disposal systems. Before agreeing to the installation of the systems a planner wants to know the direction in which the groundwater flow at the site will carry pollutants. The aquifer is unconfined, so three shallow wells are ordered drilled in a rough equilateral triangle on the site, as shown in Figure 7-25. Find the flow direction under the site.

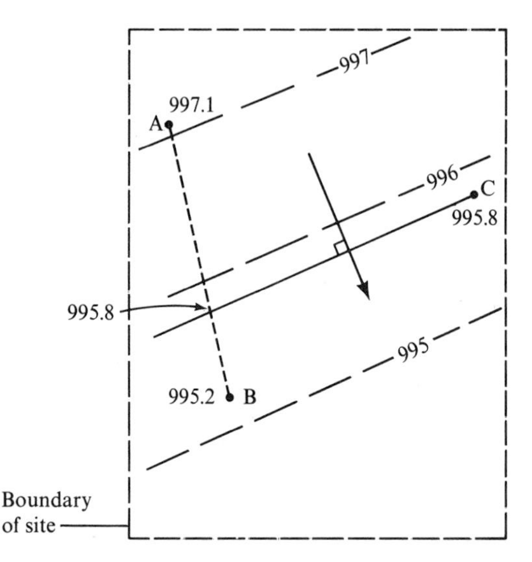

Figure 7-25 Use of well data to compute direction of groundwater flow.

Solution

The three well locations are plotted on a map as shown in Figure 7-25. When the water table has reached an equilibrium after drilling, its elevation is measured relative to some bench mark. This requires that the elevations of the top of the wells be surveyed in to the bench mark with an engineering level. A weighted tape or a chalked stick can then be lowered to the water level. The elevations are plotted on the map.

Along a line joining wells A and B a value of 995.8 is interpolated and joined to well C. The line is the 995.8 m contour of water-table elevation, and if the aquifer has no major oriented joint system or other constraint upon flow, the water will be moved at right angles to this equipotential line, as shown. Other water-table contours can be sketched parallel to the first one so that the designers of the septic systems will know the depth to the water table at any place within the development. In this case we have added the 995, 996, and 997 m contours, but still others could be interpolated.

A planner is considering a proposal to dispose of the treated effluent from a small canning factory by spraying it on a gently rolling, forested plateau. Around the borders of the low plateau is a line of springs that supply various small streams and the water requirements of a village. The planner wants to avoid contaminating the springs and thus facing a lawsuit. He would prefer to dispose of the effluent on a part of the plateau from which it would migrate slowly by a long path to an uninhabited area.

Solution

A network of wells is drilled in the unconfined aquifer of the plateau and water-table elevations are measured, as shown in Figure 7-26. Contours are interpolated between the measured water-table elevations, and the flow lines are drawn at right angles to these equipotential lines. The elevations of spring heads and river channels taken from topographic maps also provide useful clues to the elevation of the water table because they represent locations where the saturated zone intersects the ground surface.

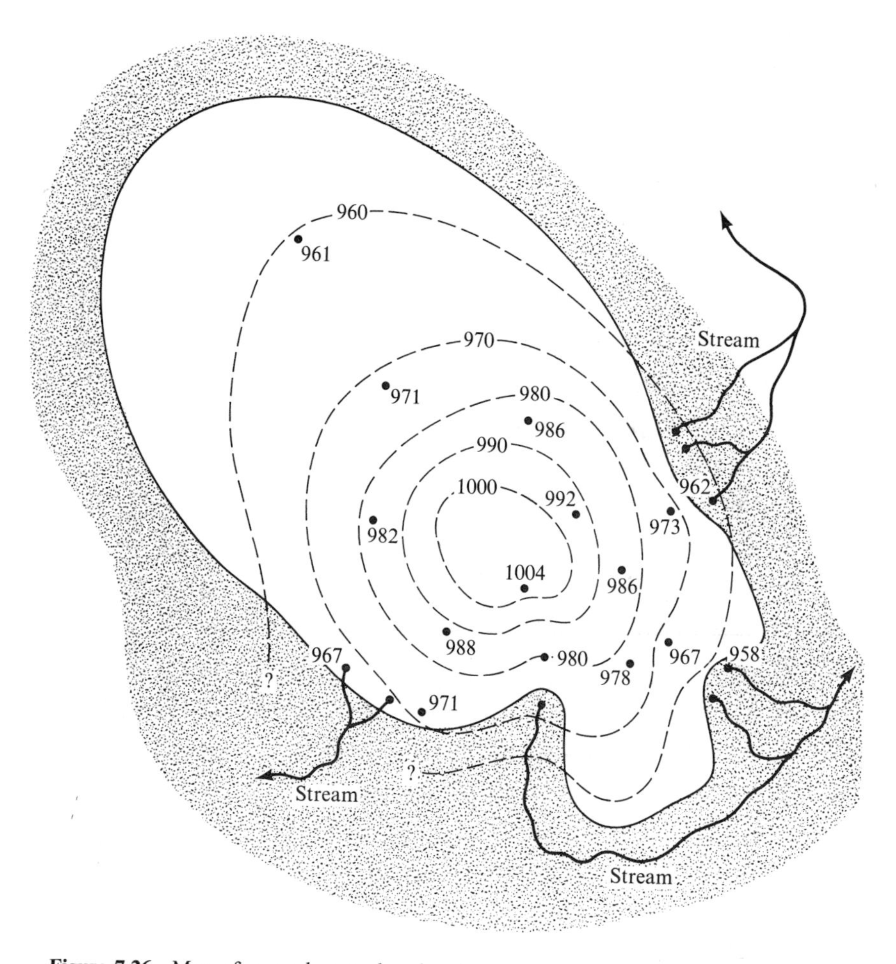

Figure 7-26 Map of groundwater elevation from well data and occurrence of springs.

8

The Water Balance

Usefulness

Previous chapters have treated hydrologic processes more or less separately. We are now in a position to draw together hydrometeorology, soil physics, and groundwater hydrology by means of a very useful tool known as the *water balance.* The term water balance was used in 1944 by the meteorologist C. Warren Thornthwaite to refer to the balance between the income of water from precipitation and snowmelt and the outflow of water by evapotranspiration, groundwater recharge, and streamflow. The budget can be computed for a soil profile or for a whole drainage basin. The method allows the planner to compute a continuous record of soil moisture, actual evapotranspiration, groundwater recharge, and streamflow from a meteorological record and a few observations on the soil and vegetation. Its use will be illustrated by means of examples and by reference to published accounts from various areas of the world.

The power of such a technique in planning is obvious. The water balance has been used for computing seasonal and geographic patterns of irrigation demand, the soil moisture stresses under which crops and natural vegetation can survive, the prediction of streamflow and water-table elevations, the flux of water to lakes, and therefore the variations of water level and salinity. Because commonly available meteorologic records are the usual basis of the method, long periods of data on soil moisture, groundwater, and streamflow can be generated and subjected to probability analysis for studying the economic and ecologic feasibility of various schemes for using or manipulating land and water resources.

The water balance is also useful for predicting some of the human impacts on the hydrologic cycle. The hydrologic effects of weather modification or changes of vegetation cover can be quickly estimated at a very early stage in the planning. Although the predictions may be approximate, they are sufficiently accurate to indicate whether a scheme is hydrologically sound or foolhardy. Also the water balance can be refined to meet the most sophisticated design needs if sufficient time and money for instrumentation are available. Finally, the method is valuable for helping to phrase precise questions about the chances of success, mode of operation, and environmental impact of proposed changes. It is, therefore, a valuable tool in the analysis of water problems in a region.

Direct Measurement of the Water Balance

The water balance of a small drainage basin underlain by impervious rock at depth can be represented by Figure 8-1 and expressed in the following equation:

$$P = I + AET + OF + \Delta SM + \Delta GWS + GWR \tag{8-1}$$

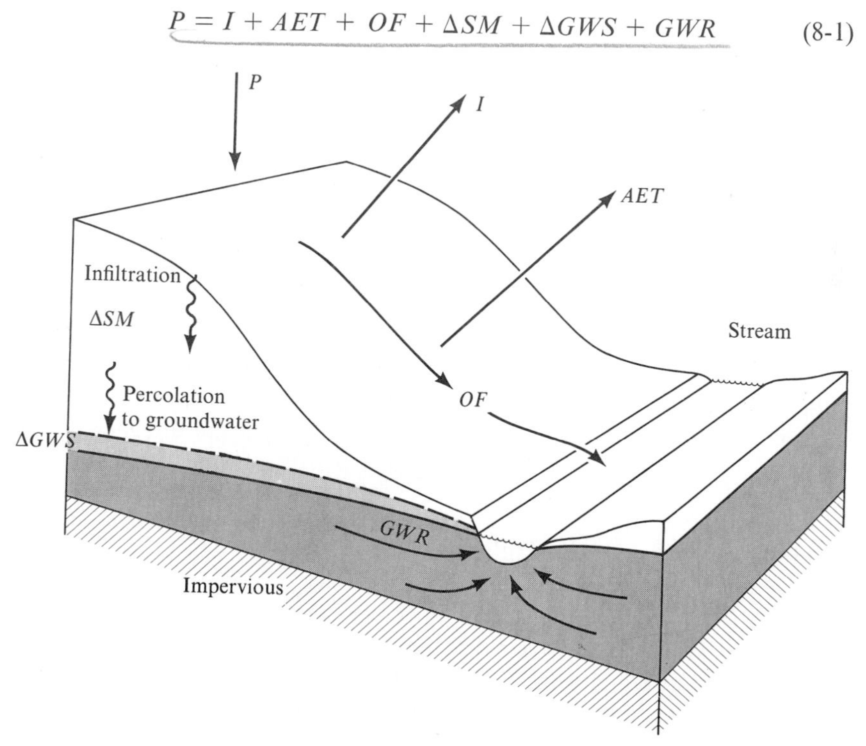

Figure 8-1 Components of the water balance on a hillside or a small catchment. P = precipitation; I = interception; AET = actual evapotranspiration; OF = overland flow; ΔSM = change in soil moisture; ΔGWS = change in groundwater storage; GWR = groundwater runoff.

where the symbols, expressed as equivalent depths of water for some time interval, represent precipitation, interception, evapotranspiration, overland flow, change of soil moisture storage, change of groundwater storage, and groundwater runoff.

If time and funds are available, each of these terms can be evaluated directly in the field by techniques that have already been discussed. Precipitation gauging and the measurement of interception are straightforward. Actual evapotranspiration can be evaluated from Equation 8-1 if all the other elements are measured. Small troughs below hillside plots would catch overland flow, or this component could be estimated from the streamflow hydrograph by methods described in Chapter 10. Soil moisture changes can be monitored with soil moisture blocks, a neutron meter, or a weighing-type lysimeter; and groundwater storage changes are reflected in fluctuations of the water table. Groundwater runoff can be calculated from Darcy's Law or from the streamflow hydrograph (see Chapter 10 on hydrograph separation). If calculations are made on an annual basis, it is often assumed that there is no net change of soil moisture or groundwater storage over the year and that the right side of Equation 8-1 reduces to the sum of interception, evapotranspiration, and streamflow.

Field measurements of this kind are valuable if the rural planner needs a detailed understanding of moisture regimes and their controls. They also allow the direct measurement of transpiration and the definition of the functional relationship between potential and actual water loss as expressed in Equation 5-13 and Figures 5-6, 5-7, and 5-8. A few weeks or years of direct measurement of the water balance can yield a large amount of hydrologic information about the interrelationships of climate, land use, groundwater, and runoff. Lewis and Burgy (1964) have given an excellent description of their field measurements of the water balance. During the International Hydrological Decade (1965–1974), the water agencies of many countries initiated water-balance studies in their main climatic regions, and a great deal of useful information was generated by these studies. In urbanized regions, the method can be extended to incorporate processes such as pumping and artificial recharge of groundwater and many other factors. Franke and McClymonds (1972), for example, present an analysis of the effects of urbanization on the water budget of Long Island.

Calculation of the Water Balance

A method for calculating the water balance was introduced by Thornthwaite in the early 1940's. It has been used for a great variety of purposes and has been modified many times. We will describe Thornthwaite's original method, as outlined in a very useful handbook by Thornthwaite and Mather (1957). We suggest that the reader who wishes to use the water balance extensively obtain this valuable publication. In the following discussion we suggest modifications of the basic technique that may be necessary or prudent under some circumstances.

The method is best explained by means of a worked example. The area under discussion is Kericho in the highlands of western Kenya. The problem may be phrased as follows. The area is relatively dry from December through February but especially wet in April and May. What is the magnitude of the evapotranspiration demand that is not satisfied by rainfall during the dry part of the year? When would one expect runoff derived from the excess of moisture over the demand of evapotranspiration, and what is the magnitude of this excess, if any exists? The results are summarized in Figure 8-2 and Table 8-1.

Monthly *precipitation* values for Kericho are listed in row 1 of the table. The next row contains values of *potential evapotranspiration,* calculated by the Penman method, though any other method, including pan evaporation data, could be used where appropriate. In row 3 the differences between precipitation and potential evapotranspiration define two seasons: a 9-month wet season when rainfall exceeds evapotranspiration and a 3-month dry season when the meteorologic demand is not satisfied by precipitation that has fallen in the same month. The severity of the dry season increases during the sequence of months with excessive potential evapotranspiration, and this is expressed in row 4 as the *accumulated potential water loss,* which is the cumulation of negative values of $(P - PET)$ for the dry season only. The summation begins at the end of the wet season.

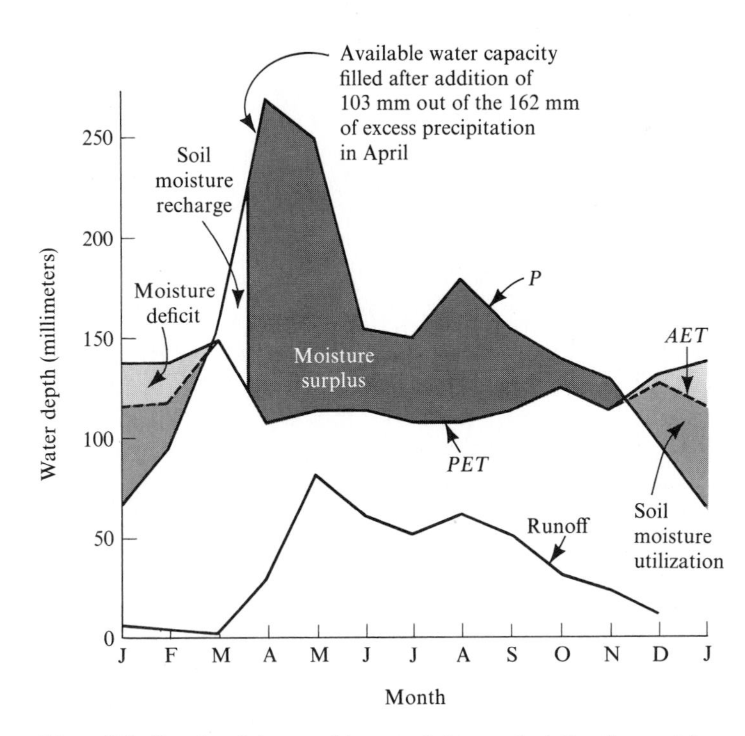

Figure 8-2 Results of the monthly water balance calculation for a catchment with a clay loam soil under woodland at Kericho, Kenya. Available water capacity of the root zone is 200 mm.

Table 8-1 Long-term average monthly water balance at Kericho, Kenya, for a soil with an available water capacity of 200 mm. The soil is a clay-loam under brushy regenerating woodland with a rooting depth of 0.80 m. All values in the table are in millimeters.

(MM)	JAN.	FEB.	MAR.	APR.	MAY	JUNE	JULY	AUG.	SEPT.	OCT.	NOV.	DEC.	YEAR
1. P*	65	95	155	270	250	155	150	180	155	140	130	95	1840
2. PET	138	138	150	108	114	114	108	108	114	126	114	132	1464
3. $P - PET$	−73	−43	5	162	136	41	42	72	41	14	16	−37	376
4. $Acc\ Pot\ WL$	−110	−153										−37	
5. SM	115	92	97	200	200	200	200	200	200	200	200	166	
6. ΔSM	−51	−23	+5	+103	0	0	0	0	0	0	0	−34	
7. AET	116	118	150	108	114	114	108	108	114	126	114	129	1419
8. D	22	20	0	0	0	0	0	0	0	0	0	3	45
9. S	0	0	0	59	136	41	42	72	41	14	16	0	421
10. Total avail. for runoff	13	7	3	59	165	124	104	124	103	65	49	25	
11. RO	6	4	2	30	82	62	52	62	52	32	24	12	420
12. Detention	7	3	1	29	83	62	52	62	51	33	25	13	

*P = precipitation; PET = potential evapotranspiration; $P - PET$ is difference by subtraction; $Acc\ Pot\ WL$ = accumulated potential water loss derived by accumulating negative values in row 3; SM = soil moisture; ΔSM = change in soil moisture during the month; AET = actual evapotranspiration; D = soil moisture deficit; S = soil moisture surplus; RO = runoff.

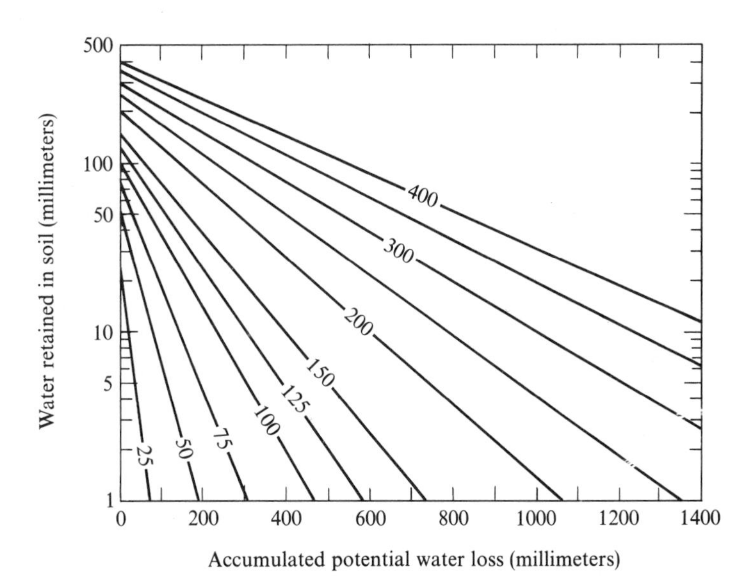

Figure 8-3 Water retained in the soil against an accumulated potential water loss. The number on each curve is the available water capacity for the soil in millimeters. (Data from Thornthwaite and Mather 1957.)

The difference between P and PET will be at least partly made up by the withdrawal of soil water. In Chapter 5, however, we described how evapotranspiration falls below the potential rate as the soil dries, and here it is necessary to choose which functional relationship to use in Equation 5-13. The earlier discussion of this subject will help the user with his choice. We have decided to use the method given in the original instruction manual by Thornthwaite and Mather. Their tabulated information is summarized in Figure 8-3, which shows how much water will be retained in the soil after various amounts of accumulated potential water loss. The lines represent soils with different available water capacities, which in turn depend upon the texture of the soil and the rooting depth of the vegetation. The available water capacity can be measured, or it can be estimated from Figure 6-9 or Table 8-2. In humid regions, however, the rooting range of crops can be much less than those indicated in the table, and therefore field observations are preferable. The table tells us that the available water capacity of a clay loam is approximately 25 percent (250 mm of water per m depth of soil). The depth of root penetration under the brushy regenerating woodland that we observed in the field was 0.80 m, indicating that the root zone could hold (0.80 m × 250 mm/m =) 200 mm of water at field capacity.

Using Figure 8-3, we find that if a soil with an available water capacity of 200 mm is subjected to a potential water loss of 37 mm, the amount of water that will be retained by the soil is 166 mm. This value of *soil moisture* is entered for December in the fifth row of Table 8-1, and the process is repeated for the other dry-season months. Soil moisture values for the wet

Table 8-2 Suggested available water capacities for combinations of soil texture and vegetation. (From Thornthwaite and Mather 1957.)

VEGETATION	SOIL TEXTURE	AVAILABLE WATER CAPACITY (% VOLUME)	ROOTING DEPTH (M)*	AVAIL. WATER CAP. OF ROOT ZONE (MM)
Shallow rooted crops	Fine sand	10	0.50	50
(spinach, peas, beans beets, carrots, etc.)	Fine sandy loam	15	0.50	75
	Silt loam	20	0.62	125
	Clay loam	25	0.40	100
	Clay	30	0.25	75
Moderately deep rooted crops	Fine sand	10	0.75	75
(corn, cereals, cotton, tobacco)	Fine sandy loam	15	1.00	150
	Silt loam	20	1.00	200
	Clay loam	25	0.80	200
	Clay	30	0.50	150
Deep rooted crops	Fine sand	10	1.00	100
(alfalfa, pasture grass, shrubs)	Fine sandy loam	15	1.00	150
	Silt loam	20	1.25	250
	Clay loam	25	1.00	250
	Clay	30	0.67	200
Orchards	Fine sand	10	1.50	150
	Fine sandy loam	15	1.67	250
	Silt loam	20	1.50	300
	Clay loam	25	1.00	250
	Clay	30	0.67	200
Mature forest	Fine sand	10	2.50	250
	Fine sandy loam	15	2.00	300
	Silt loam	20	2.00	400
	Clay loam	25	1.60	400
	Clay	30	1.17	350

*In some places where we have worked, the rooting depths given by Thornthwaite and Mather are too high. Field observations in the area for which the water balance is being computed are preferable to the use of this table.

season are obtained by adding the excess precipitation from row 3 to the soil moisture level at the end of the dry season. Thus, the 5-mm excess occurring in March will raise the soil moisture storage from 92 mm at the end of February to 97 mm at the end of March. In April 162 mm will be added to the 97 mm. Because the soil can only hold 200 mm of water, however, soil moisture values cannot exceed this value, and so for April and succeeding wet months the moisture content remains at field capacity. Surplus water which cannot be held in the soil by capillary and osmotic forces drains out of the root zone. This subject will be dealt with later.

From row 5 the *change in soil moisture* can be entered in row 6. When precipitation exceeds potential evapotranspiration, the *actual evapotranspiration* in row 7 equals the potential rate because the rainwater is considered to be easily available to the plant, even if the soil moisture of the whole root zone is not raised to the available water capacity. When the meteorologic demand must be partially satisfied from the stored soil water, actual evapotranspiration is the sum of precipitation and the amount of soil moisture withdrawn from storage (95 + 34 =) 129 mm in the case of December.

The amount by which the actual and potential evapotranspiration differ in any month is called the *soil moisture deficit,* and is listed in row 8. In this case, because the available water capacity is high and the dry season is short, evapotranspiration continues almost at the potential rate throughout the year.

Once a moisture deficit has developed, it can be reduced when excess precipitation is stored in the soil at the beginning of the wet season. The soil moisture will eventually attain field capacity, however, and further precipitation excess must leave the soil by gravitational drainage. The amount of water that cannot be stored (the excess from line 5) is termed the *moisture surplus* and is listed in row 9.

The moisture surplus drains to the groundwater body and eventually to streams. It cannot all drain out in the same month that it is generated, however, and a certain proportion remains in the soil and is carried over into later months. The fraction which will leave a river basin as *runoff* varies with the depth and texture of the soil and the physiography of the basin. Ideally this proportion should be evaluated for the region of application. It can be obtained by direct field measurement of the water balance for a few months of the wet season. A less exact estimate can be made by examining the relative amounts of streamflow during successive wet months. In some cases the use of widely differing proportions does not lead to very different answers, so there is not much reason to refine the estimate. For some applications of the water balance, however, such refinements are necessary. Thornthwaite and Mather give only the following advice: For large catchments approximately 50 percent of the surplus water that is available for runoff in any month actually runs off. The rest of the surplus is detained in the subsoil, groundwater, small lakes, and channels of the basin and is available for runoff during the next month. If the water balance is calculated for periods shorter than one month, or for catchments of only a few square kilometers, the proportion detained will often be less than 50 percent. In Table 8-1, however, we have used the 50 percent figure.

Beginning in April, 59 mm of moisture surplus were available and therefore 30 mm should run off. Twenty-nine millimeters will be detained in the catchment to be added to the moisture surplus of May for a total of 165 mm. Therefore, 82 mm of water will run off during May, and 83 mm will be detained until June. Repetition of this calculation completes rows 10, 11, and 12.

The results of this set of calculations are portrayed in Figure 8-2, which summarizes a great deal of information on the hydrology of Kericho. Besides showing the seasonal pattern of precipitation, evapotranspiration, and runoff, it indicates the times of moisture deficit, soil moisture recharge, and moisture surplus. The moisture deficit indicates that the plants are under some stress, and if the cover were a valuable crop, the calculations would indicate the timing and magnitude of irrigation necessary to remove the stress. The water balance is therefore valuable for understanding the ecology, agronomy, and economics of growing crops.

Figure 8-1 indicates a component of the water balance that is ignored in the Thornthwaite method, namely, storm runoff. In the region for which we have made the calculation, this problem is not an important one; storm runoff amounts to 3 to 5 percent of rainfall, and this abstraction is not sufficient to lower the evapotranspiration much below the potential rate. In other regions or other vegetation covers, however, the stormflow component is important. To take this into account, an additional step in the calculation should be added between rows 1 and 2 in the tables, in which storm runoff is subtracted from precipitation. Estimation of the volume of stormflow can be based on the infiltration capacity of the soil, or on the Soil Conservation Service method described in Chapter 10. This requires rainfall data from individual storms, but unfortunately, such data are often not available. If a few continuous streamflow records are available for small catchments in the region, monthly storm runoff can be separated from the hydrographs, as described in Chapter 10, and expressed as a fraction of the rainfall for each month. This method provides a rough idea of how much rainfall should bypass the calculation of the soil-water balance and be treated immediately as runoff.

Another modification is necessary if the climate of a place is so dry that the soil never reaches field capacity, that is, if the available water capacity of the root zone is never filled. Under these conditions it is not obvious when the accumulation of potential water loss in row 4 should begin. Instead, the user must find a value of the potential water loss for the beginning of the first month when the precipitation falls below the rate of potential evapotranspiration (June in Table 8-3). This is done by successive approximations as follows.

The accumulated potential water loss is estimated for the first month with a negative value of $(P - PET)$ that is, for the beginning of the dry season. We first sum the negative values of $(P - PET)$ for the dry season; in Table

Table 8-3 Long-term average monthly water balance calculation for Nairobi, Kenya. The soil is a fine sandy loam, and the vegetation is woodland with a rooting depth of 1.33 m. $AWC = 200$ mm.

(MM)	JAN.	FEB.	MAR.	APR.	MAY	JUNE	JULY	AUG.	SEPT.	OCT.	NOV.	DEC.	YEAR
1. P	59	51	126	219	160	51	21	23	29	57	143	91	1030
2. PET	140	140	116	120	100	99	96	90	118	88	145	126	1378
3. $P - PET$	−81	−89	10	99	60	−48	−75	−67	−89	−31	−2	−35	−348
4. *Acc Pot WL*	−446	−535			(−18)	−66	−141	−208	−297	−328	−330	−365	
5. SM	21	13	23	122	182	143	98	70	44	38	38	32	
6. ΔSM	−11	−8	+10	+99	+60	−39	−45	−28	−26	−6	0	−6	
7. AET	70	59	116	120	100	90	66	51	55	63	143	97	1030
8. D	70	81	0	0	0	9	30	39	61	25	2	29	348
9. S	0	0	0	0	0	0	0	0	0	0	0	0	0
10. RO	0	0	0	0	0	0	0	0	0	0	0	0	0

8-3 this total is −517 mm. Using Figure 8-3 with an available water capacity of 200 mm, we find the value of soil moisture that would be retained against this accumulated potential water loss by the end of the dry season. The value read from the ordinate of the figure is 15 mm, which is the moisture that would remain in storage at the end of the dry season if the soil had contained 200 mm of available water at the beginning of the dry period. The true value was less than 200 mm.

If we then add the sum of the positive ($P - PET$) values to the 15 mm, we obtain a soil moisture value of (15 + 169 =) 184 mm at the end of the wet season. Figure 8-3 indicates that the accumulated potential water loss up to the end of the wet season that would be necessary to result in a soil moisture storage of 184 mm is 16 mm. If the total negative ($P - PET$) value for the dry season is then added to this value, we would have an accumulated potential water loss of (16 + 517 =) 533 mm at the end of the dry period. Entering Figure 8-3 again gives us a value of 14 mm for the soil moisture value at the end of the dry season, and if the total positive ($P - PET$) values are added, we estimate that the soil moisture would be 183 mm at the end of the wet season, which would be in equilibrium with an accumulated potential water loss (from Figure 8-3) of 17 mm. The negative ($P - PET$) values are again added to this value, and the soil moisture retention at the end of the dry season is found to be 13 mm. Addition of the positive values gives a value of 182 mm at the end of the wet season, which would require an accumulated potential water loss of 18 mm up to that time. Therefore by the end of the dry season the accumulated potential loss would be 533 mm and the soil moisture retention would be 13 mm, the same as it was at the last estimate. Repetition will not change the figures, and we have determined that the correct value of the accumulated potential water loss at the end of the last month of the wet season is 18 mm. This value is entered in row 4 for May, the last month with a positive value of ($P - PET$). From this point on the water balance calculation proceeds normally. The results are shown in Figure 8-4.

The Nairobi water-balance calculation indicates large moisture deficits and no runoff. When rainfall exceeds evapotranspiration, the excess simply recharges the soil moisture but does not exceed the field capacity so that no moisture surplus or runoff is generated. This does not mean that there is no streamflow around Nairobi. First of all, stormflow occurs. We have not included it in the computations because techniques for assessing the volume of storm runoff will not be introduced until Chapter 10. Under woodland in the Nairobi region, however, this component accounts for 10 to 15 percent of the rainfall, generating some streamflow but increasing the moisture deficit. A second source of runoff in the region is groundwater that migrates through lava beds from higher, wetter regions to the west. This should remind the planner that the hydrogeological conditions of an area should be examined before the local water balance is blindly used to predict runoff.

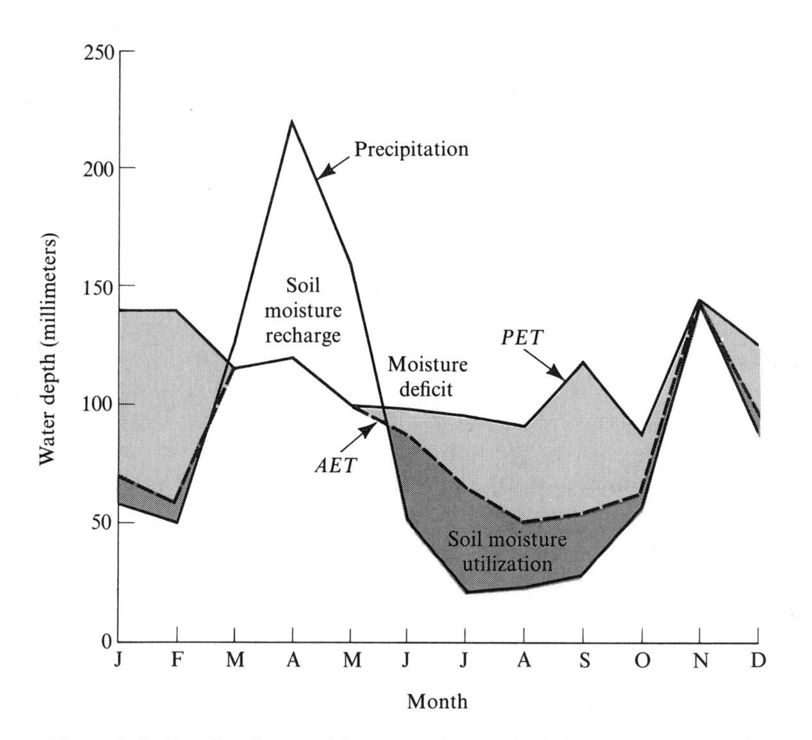

Figure 8-4 Results of a monthly water balance calculation for woodland on a fine sandy loam soil at Nairobi, Kenya.

Finally, water balance calculations are usually made on long-term average climatic data in order to obtain a rough picture of the seasonal march of rainfall, evapotranspiration, soil moisture, irrigation need, and runoff. Long-term averages tend to underestimate extreme deficits and surpluses. In a wetter-than-average year, a moisture surplus is often developed around Nairobi and the groundwater receives some recharge. Even during an average year an increase in the wet-season rainfall can generate a moisture surplus at the expense of soil moisture storage during the dry season. In order to examine the probability of runoff generation and also the probability of various degrees of moisture deficit, a meteorological record of many years can be processed in sequence. Moisture deficits or surpluses developed in one year are simply carried forward into the next. The accumulation of potential water loss is initiated after some period when runoff is known to have occurred. The resulting long-term water balance can then be examined statistically, by counting the frequency of moisture deficits, moisture surpluses, or runoff greater than various levels. An example is shown in Figure 8-5. Such statistical statements are valuable for calculating the chances of irrigation demand or of crop failure. The methodology is discussed by Van Bavel (1953).

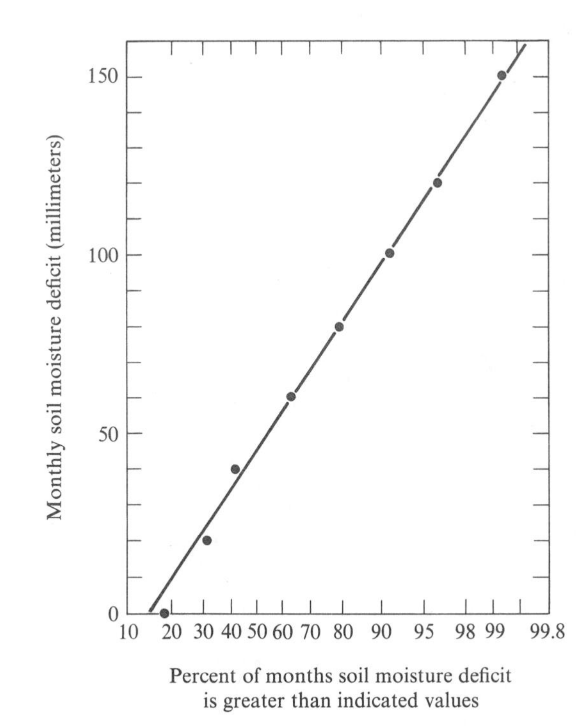

Figure 8-5 Results of statistical analysis of the water balance at Nairobi. The graph shows the frequency of soil moisture deficits for the period 1931–1940. Other results of the water balance calculations can be analyzed in the same way and are useful in agricultural planning.

The water balance can be refined in several other ways. The computations can be made on a daily basis to obtain better definition of short-term moisture deficit or surplus. Daily computations are particularly useful if deficit and surplus occur in the same month. These and other modifications to be made at stations with two dry seasons or a snowmelt input are described in the instruction manual of Thornthwaite and Mather, although their method of calculating snowmelt can be improved upon with the material in Chapter 13.

Applications

The most obvious use of the water balance is in a basic description of the hydrology of a place or region. Maps can be drawn of the annual extreme water deficit or surplus or of the total annual irrigation need or groundwater runoff. Spatial patterns within a region or a large river basin can be employed in planning the distribution of resources. Al-Khashab (1958) was

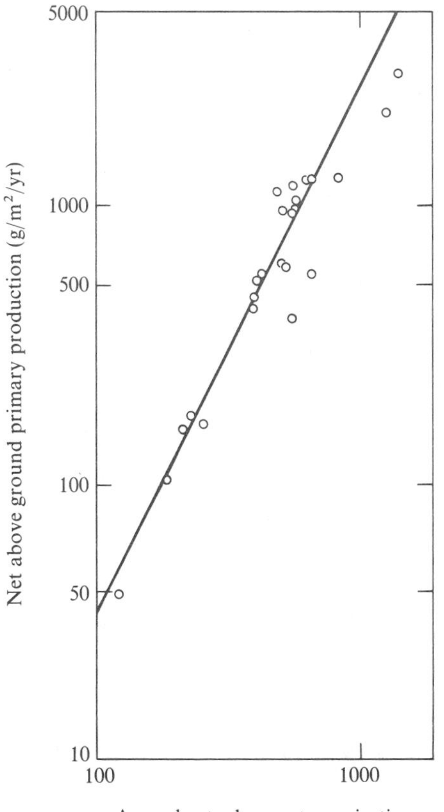

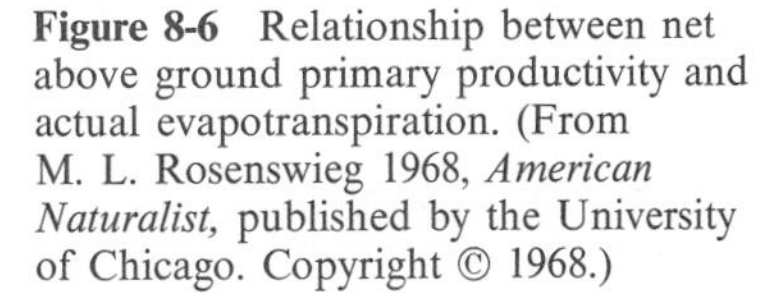
Figure 8-6 Relationship between net above ground primary productivity and actual evapotranspiration. (From M. L. Rosenswieg 1968, *American Naturalist,* published by the University of Chicago. Copyright © 1968.)

able to use the water balance to define the seasonal and geographical pattern of water supply and irrigation demand at an early stage in the planning for the development of the water resources of the 785,000-square-kilometer Tigris–Euphrates Basin in Iraq.

Actual evapotranspiration calculated from the water balance is correlated with primary productivity of vegetation and so with the value of rangeland or cropland (see Figure 8-6). Knowledge of the probable moisture deficit is useful for planning irrigation schemes and for predicting crop yields on lands that are not irrigated (Palmer 1964). Statistical maps of moisture deficit can be developed to indicate drought probabilities based on how much water is held in the soil rather than on the rainfall total. If, for example, a critical soil moisture deficit of 100 mm for two consecutive months in early summer is known to cause a catastrophic crop failure in a particular soil, then the long-term water balance can be examined and the frequency of years in which such a deficit occurred in that soil can be counted. These figures allow the construction of contour maps of the frequency of drought hazard. Such maps assist the rural planner in determining the economic viability of introducing new cash crops in a region, and the value of crops to be expected.

The calculated moisture surplus is also a useful planning tool for indicating regions and frequencies of useful runoff. A large moisture surplus may indicate conditions too wet for some crops that are susceptible to disease. Moisture surplus is also an index of the intensity with which plant nutrients are leached from the soil by excess precipitation. Nutrient-deficient, acid soils are often found in regions of high surplus, and fertilizers and liming are necessary but are leached relatively quickly in such areas. For other purposes the moisture surplus can be incorporated into a simple accounting technique for predicting the level of the water table (Boersma 1967).

It is clear then that the water balance is a valuable planning tool, one that deserves further application and refinement.

Bibliography

Ahlgren, L., Basso, E. and Jovel, R. (1969) Preliminary evaluation of the water balance in the Central American isthmus; *Proceedings of Symposium on the Water Balance in North America,* American Water Resources Association, pp. 288–298.

Al-Khashab, W. H. (1958) The water budget of the Tigris and Euphrates basin; *University of Chicago, Department of Geography Research Paper 54.*

Boersma, L. (1967) Water-table fluctuation in the soil series of the Willamette catena; *American Society of Agricultural Engineers, Transactions,* vol. 10, pp. 405–406, 510.

Dunin, F. X. and Costin, A. B. (1970) Analytical procedures for evaluating the infiltration and evapotranspiration terms of the water balance equation; *International Association of Scientific Hydrology, Symposium of Wellington, Publication No. 96,* pp. 39–55.

EAAFRO (1962) Hydrological effects of changes in land use in some East African catchment areas; *East African Agricultural and Forestry Journal,* vol. 27, pp. 1–131.

Franke, O. L. and McClymonds, N. E. (1972) Summary of the hydrologic situation on Long Island, New York, as a guide to water management alternatives; *U.S. Geological Survey Professional Paper 627-F.*

Hursh, C. R. and Pereira, H. C. (1953) Field moisture balance in the Shimba hills, Kenya; *East African Agriculture and Forestry Journal,* vol. 18, pp. 139–145.

Lewis, D. C. and Burgy, R. H. (1964) Hydrologic balance from an experimental watershed; *Journal of Hydrology,* vol. 2, pp. 197–212.

Ligon, J. T., Benoit, G. R. and Elam, A. B. (1965) Procedure for estimating occurrence of soil moisture deficiency and excess; *American Society of Agricultural Engineers, Transactions,* vol. 8, pp. 219–222.

Palmer, W. C. (1964) Climatic variability and crop production; in *Weather and our food supply,* Centre for Agricultural and Economic Development, Iowa State University, Ames, IA.

Palmer, W. C. (1964) Meteorological drought; *U.S. Weather Bureau Research Paper 45.*

Rosensweig, M. L. (1968) Net primary production of terrestrial communities: predictions from climatological data; *American Naturalist,* vol. 112, pp. 67–74.

THORNTHWAITE, C. W. AND MATHER, J. R. (1955) The water balance; *Laboratory of Climatology, Publication No. 8,* Centerton, NJ.

THORNTHWAITE, C. W. AND MATHER, J. R. (1957) Instructions and tables for computing potential evapotranspiration and the water balance; *Laboratory of Climatology, Publication No. 10,* Centerton, NJ.

VAN BAVEL, C. H. M. (1953) A drought criterion and its application in evaluating drought incidence and hazard; *Agronomy Journal,* vol. 45, pp. 167–172.

WARD, R. C. (1972) Checks on the water balance of a small catchment; *Nordic Hydrology,* vol. 3, pp. 44–63.

ZAHNER, R. (1967) Refinement in empirical functions for realistic soil moisture regimes under forest cover; in *Forest hydrology* (eds. W. E. Sopper and H. W. Lull), Pergamon Press, Oxford, pp. 261–274.

Typical Problems

8-1 Calculation of Monthly Runoff

Repeat the water-balance calculation of monthly runoff for Kericho, illustrated in the text, but assume that 80 percent of the surplus water available for runoff in any month runs off. This requires altering only the last three lines of Table 8-1. Plot your monthly runoff values on a tracing of the runoff curve from Figure 8-2, and describe the kinds of drainage basins to which the two calculations might apply. Suggest an application of the water balance for which the difference would be important and another application for which it would be unimportant.

8-2 Effect of Available Water Capacity on the Water Balance

From your local library, water resources agency, or atlas, obtain monthly values of precipitation and potential evapotranspiration.

1. Go out and look at a woodland near your home and measure (in a road cut) the average depth of the root zone, and note the soil texture. Estimate the available water capacity of the root zone from the local soil survey report or from Figure 6-9. Assume that the storm runoff is negligible.

2. Calculate the water balance of a small catchment underlain entirely by the single soil–vegetation complex and by impervious bedrock, so that all the soil moisture surplus leaves the basin as runoff.

Repeat the calculation for a grass cover whose potential evapotranspiration is equal to that of the forest, but whose root zone is only one-half as deep. Assume that storm runoff from the grass amounts to a constant ten percent of each month's runoff. Note that each of these approximations could easily be refined.

3. Plot an annual graph of moisture deficit, moisture surplus, and runoff for each cover type, and write a short essay interpreting your calculations.

8-3 Modification of Weather and Vegetation Cover

Advocates of particular technologies frequently make proposals for manipulating land and water resources. These proposals must be carefully reviewed before committing scarce resources or before raising hopes about quick solutions to land development problems.

Imagine an almost uninhabited 50-square-kilometer drainage basin covered with savanna woodland on a sandy loam soil having 200 mm of soil-moisture storage capacity in the root zone. A proposal is made to increase soil moisture levels (i.e., decrease soil moisture deficits) and increase runoff by a combination of weather modification and vegetation change. Bush clearance would allow the growth of grass for stock rearing or the cultivation of subsistence cash crops. The increased supply of precipitation should assist these developments by reducing soil moisture deficits. The increased runoff could be used for irrigation of crops along the floor of the main valley or for water spreading to improve pasture.

The proposal is one of those "very preliminary" ones that agencies frequently become committed to before they really know what is involved. This is particularly common if the funds for the weather modification are to be supplied for several years by another organization.

As a planner, you are asked to evaluate the feasibility and wisdom of the proposal. You decide to begin by making some hydrologic calculations of the degree to which soil moisture deficits would be changed and runoff increased. A survey of the literature in a variety of fields indicates that the following data are relevant for the region under study:

1. Weather modification can increase rainfall by perhaps 10 percent when atmospheric conditions are favorable during the 6-month wet season.

2. Changing the woodland to grassland would decrease the moisture storage capacity of the topsoil from 200 mm to 75 mm, because the root zone would be shallower. Under maize, the most likely crop, the soil moisture storage capacity would be 150 mm.

3. Under savanna woodland roughly 20 percent of the rainfall runs off as stormflow. This would be increased to about 30 percent with a grass cover and to 25 percent under maize.

4. Fifty percent of the land is cultivable, and 65 percent could be used for grazing.

5. Monthly rainfall values are published, and you are able to calculate potential evapotranspiration from evaporation pan data. The data are tabulated in Table 8-4.

Compute water balances for: (a) savanna woodland at present; (b) savanna woodland with weather modification; (c) grassland with weather modification; and (d) maize with weather modification. Plot annual graphs of your results. Summarize the effect of each scheme on the maximum soil moisture deficit.

Predict the mean monthly discharge of the main stream of the basin in m^3/sec if: (a) 65 percent of the basin is converted to grassland, the remainder being left in woodland; and (b) 50 percent of the basin is planted to maize, with woodland on the other half.

After making the hydrologic calculations,

Table 8-4 Precipitation and potential evapotranspiration values for a proposed region.

	JAN.	FEB.	MAR.	APR.	MAY	JUNE
P (mm)	52	34	17	19	38	37
PET (mm)	115	117	131	145	86	97

	JULY	AUG.	SEPT.	OCT.	NOV.	DEC.
P (mm)	165	175	165	155	85	135
PET (mm)	87	62	52	43	41	78

review all the other aspects of such a project that you can think of. List the kinds of information you would need to continue the review. This part of the problem may best be treated in a discussion by an interdisciplinary class with each person being assigned a particular aspect of the proposal to review. The problem could also be repeated with a different set of meteorologic data, and the results compared and interpreted in a class discussion.

Solution

For the savanna woodland under present weather conditions, storm runoff is first subtracted from the rainfall because it does not enter the soil moisture budget. The stormflow is added into the runoff total later in the computation. The net amount of infiltration should be entered as a second line in the water budget table. The third line includes the calculated potential evapotranspiration. On line 4 the difference between the water entering the soil and the potential evapotranspiration is entered. The total potential demand of evapotranspiration exceeds the amount of water entering the soil, and the soil moisture content does not reach field capacity. The successive approximation method must be used to find an accumulated potential water deficiency from which to start cumulating negative values of $(P - SRO - PET)$. The method, described in detail in the text, should lead the reader to find an accumulated potential water loss of about 77 mm for November, the last month of the wet season (i.e., the last month with a positive value of $(P - SRO - PET)$.

The calculation then proceeds as described in the text.

8-4 Water Balance for Various Soil Types

Using the data on precipitation and potential evapotranspiration from Table 8-1, calculate the water balance for a loam with an available water capacity of 150 mm. Instead of using the Thornthwaite and Mather curves from Figure 8-3, obtain the actual evapotranspiration rate by the method shown in Figure 5-8. Assume that storm runoff is negligible.

Solution

During months of excess rainfall, evapotranspiration occurs at the potential rate. When potential evapotranspiration exceeds precipitation, the rainwater is evaporated at the potential rate and the deficit is exerted as a demand upon soil moisture. As explained in Chapter 5, this demand is not exerted at the potential rate but at a rate that depends on the amount of soil moisture in storage. At the beginning of each month therefore the ratio of available water content (AW) to available water capacity (AWC) must be calculated and used to modulate the potential evapotranspiration that is unsatisfied by rainfall. Note that in Table 8-5 values of soil moisture content (SM) refer to the amount of water held at a level above the permanent wilting point, and thus SM is also the numerator AW on line 5 of the table.

Line 3 in Table 8-5 lists the unsatisfied transpirative demand as a negative value. At the end of the wet season (beginning of December) the soil moisture is at field capacity, the ratio AW/AWC is one, and evapotranspiration, occurring at the potential rate, withdraws 37 mm of water from the soil during December. This value is listed as the change of soil moisture for December on line 6. Actual evapotranspiration on line 7 is the sum of precipitation and the change of soil moisture.

The change of soil moisture for December yields the soil moisture content at the end of that month that will control the ratio AET/PET for January. In January the unsatisfied transpirative demand is 73 mm, and since the ratio AW/AWC is 0.75, Figure 5-8 indicates that evapotranspiration will continue at the potential rate. The change of soil moisture, therefore, is 73 mm.

By February 1, the soil moisture content will have declined to 40 mm, and the ratio AW/AWC to 0.27. After the month's precipitation has been evaporated, therefore, the ratio AET/PET will decline to 0.27 according to Figure 5-8, and the

Table 8-5 Calculating water balance for loam.

	JAN.	FEB.	MAR.	APR.	MAY	JUNE
P	65	95	155	270	250	155
PET	138	138	150	108	114	114
$P - PET$	−73	−43	5	162	136	41
SM	113	40	28	33	150	150
AW/AWC	0.75	0.27				
ΔSM	−73	−12	+5	+145	0	0
AET	138	107	150	108	114	114
D	37	110	122	127	0	0
S	0	0	0	17	136	41
	JULY	AUG.	SEPT.	OCT.	NOV.	DEC.
P	150	180	155	140	130	95
PET	108	108	114	126	114	132
$P - PET$	42	72	41	14	16	−37
SM	150	150	150	150	150	150
AW/AWC						1.0
ΔSM	0	0	0	0	0	−37
AET	108	108	114	126	114	132
D	0	0	0	0	0	0
S	42	72	41	14	16	0

decline of soil moisture will be ($0.27 \times 43 =$) 12 mm. Actual evapotranspiration ($P + \Delta SM$) will be 107 mm. From March onward, precipitation exceeds the PET demand, evapotranspiration occurs at the potential rate and the calculation proceeds as indicated in Table 8-1.

The reader can obtain further understanding and practice by repeating this water balance calculation for the other soil types shown in Figure 5-8, and by using a set of data from some other area of interest.

9

Runoff Processes

Relation to Planning Problems

When rain and meltwater reach the surface of the ground, they encounter a filter that is of great importance in determining the path by which hillslope runoff will reach a stream channel. The paths taken by water (see Figure 9-1) determine many of the characteristics of a landscape, the generation of storm runoff, the uses to which land may be put, and the strategies required for wise land management.

A knowledge of runoff generation is useful material for the planning process for several reasons. First, it provides background understanding for the computations of runoff that will be introduced in the next chapter. Second, it allows one to recognize those parts of a landscape that are likely to be major contributors of either storm runoff or groundwater recharge. Precautions may have to be taken to avoid these zones or to detain water generated upon them. Zones that allow groundwater recharge and therefore supply streamflow during dry weather should be conserved so that they might continue this function, instead of being paved or polluted (McHarg 1969). In many situations the controls of runoff are very sensitive to disturbance. The removal of vegetation from a forested area during construction, for example, can lower the infiltration capacity enough to generate large amounts of storm runoff where the previous runoff process was a slow subsurface percolation. Developments such as deforestation, road construction, or the spreading of wastes or sewage effluent, which lead to poor soil drainage, can promote growth of the saturated zones that yield storm runoff in many humid regions. Such increases can be very damaging and costly on a local scale but can be avoided with foresight.

Zones that produce storm runoff also yield sediment, plant nutrients, bacteria, and other pollutants. An understanding of storm runoff production, then, indicates the processes by which pollutants reach streams and also indicates the management techniques that might be used to minimize the discharge of these materials into surface waters.

An appreciation of runoff processes, therefore, allows the planner to recognize present constraints, to predict the consequence of some form of development, and to avoid possible problems. Information about runoff-producing areas is useful input to zoning regulations. Planners, therefore, would do well to familiarize themselves with and map these runoff-producing zones using a combination of fieldwork and published maps of topography, soils, and land use. An example of such a map for a small drainage basin in Pennsylvania is given in Figure 9-2. It was compiled from field inspection and information provided by members of the Tri-County Conservancy for the Brandywine. Such an exercise should form part of the training of physical planners, landscape architects, and others interested in land and water management. Methods for field recognition of zones that produce storm runoff are described by Musgrave and Holtan (1964) and Dunne et al. (1975).

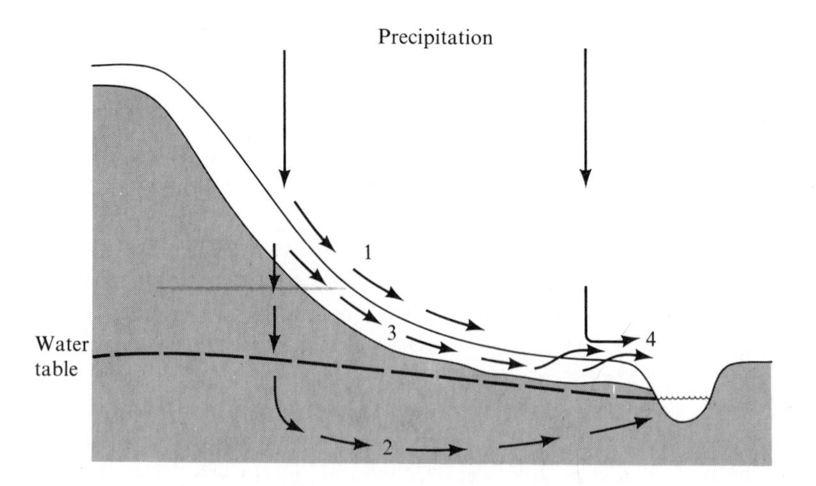

Figure 9-1 Possible paths of water moving downhill: path 1 is Horton overland flow; path 2 is groundwater flow; path 3 is shallow subsurface stormflow; path 4 is saturation overland flow, composed of direct precipitation on the saturated area plus infiltrated water that returns to the ground surface. The unshaded zone indicates highly permeable topsoil, and the shaded zone represents less permeable subsoil or rock.

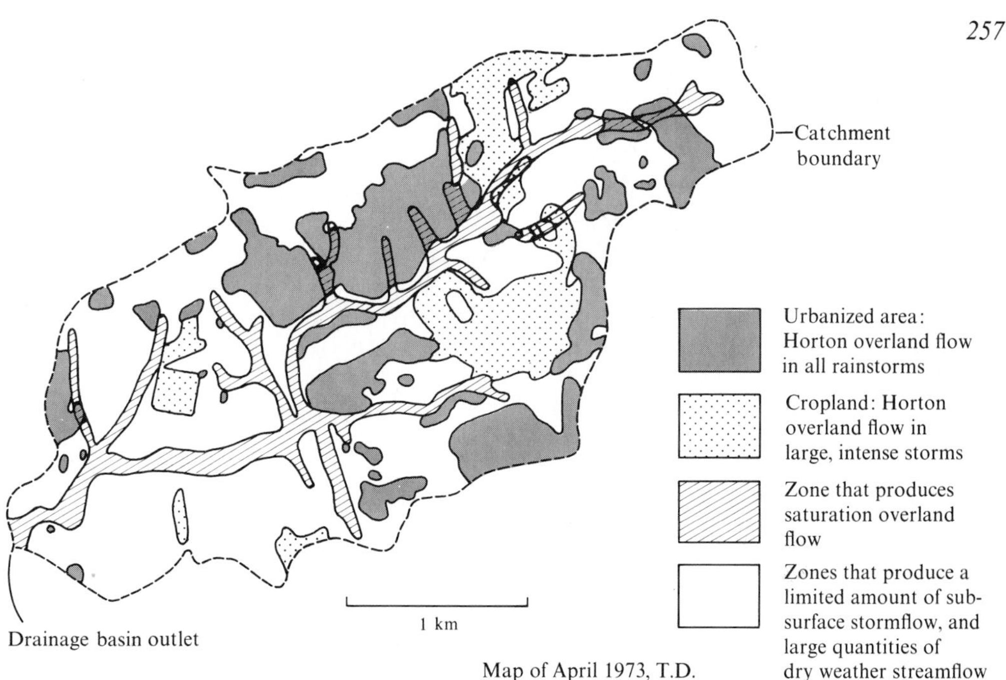

Figure 9-2 Runoff-producing areas in the catchment of Broad Run, Pennsylvania (April 1973). Terms are explained in the text. Note that land-use changes will cause changes in the spatial pattern of runoff processes.

Stormflow and Baseflow

If hillslope runoff reaches a stream channel during or within a day or so of rainfall, it causes high rates of discharge in the channel and is usually classified as *storm runoff*, or *direct runoff*. Water that percolates to the groundwater moves at much lower velocities by longer paths and reaches the stream slowly over long periods of time. It sustains streamflow during rainless periods and is usually called *baseflow*, or *dry-weather flow*. Hewlett and Hibbert (1967) proposed the terms *quickflow* and *delayed flow* for the two components. A distinction is often necessary when one is using the computational methods to be described in Chapter 10. Accepted methods for separating a hydrograph into stormflow and baseflow will be introduced in Figure 10-4. It should be kept in mind, though, that the separation is arbitrary and refers only to the place for which it is made. Today's stormflow in a headwater basin may become baseflow one week from now at a downstream gauging station.

A graph of the rate of runoff (i.e., discharge) plotted against time for a point on a channel or hillside is called a *hydrograph* (see Figure 9-3). Discharge is usually expressed in terms of volume per unit time; e.g., cubic

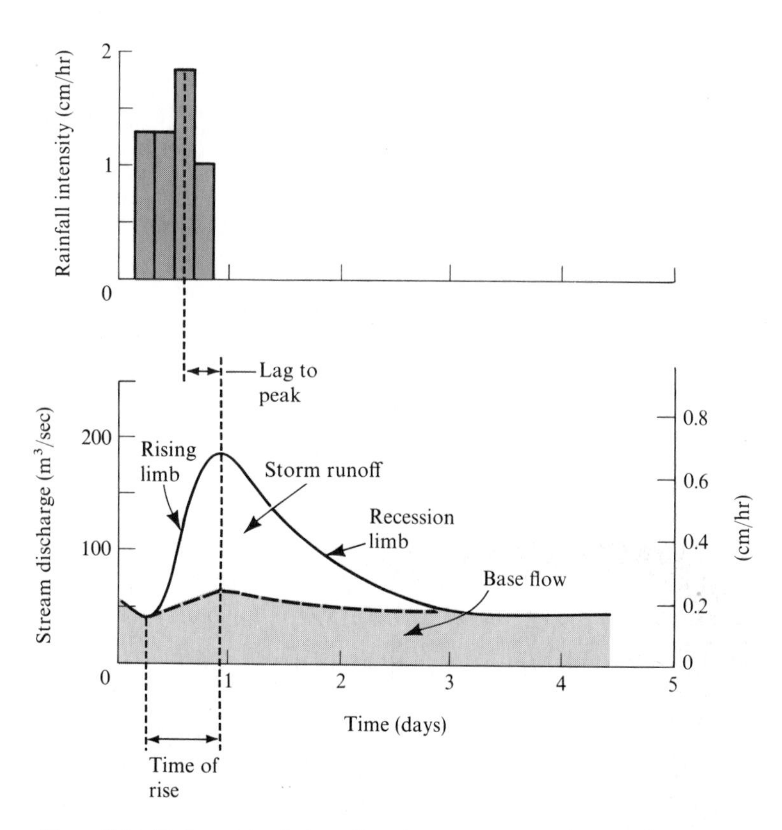

Figure 9-3 Hydrograph of streamflow in response to a rainstorm from a 100-square-kilometer basin. Methods of separating storm runoff and baseflow are described in Chapter 10.

meters per second, cubic feet per second (cfs), or acre-feet per day. If this volume per unit time is divided by the area of the catchment in appropriate units, the runoff can be expressed as depth per unit time (cm/hr, in/day), which is convenient for comparing with rates or rainfall, infiltration, and evaporation. In Figure 9-3, various characteristics of the storm hydrograph are defined. The *lag to peak* is the time difference between the center of mass of rainfall and the peak runoff rate.

Runoff Processes: General

Several processes produce storm runoff (see Figure 9-1). They are described here under the titles *Horton overland flow, subsurface stormflow, return flow,* and *direct precipitation onto saturated areas.* The last two processes are also known together as *saturation overland flow.* Each one has a different response to rainfall or snowmelt in the volume of runoff produced, its peak rate, and the timing of contributions to the channel. The relative importance of each process in a region (or more particularly on each hillslope) is affected by climate, geology, topography, soil characteristics, vegetation, and land use. The dominant process may vary between large and small storms, and this should be considered in overall planning and in individual design problems. Here we will divide the discussion into two parts: runoff in regions (or storms) in which the rainfall intensity exceeds the infiltration capacity of the soil, and runoff when the infiltration capacity is not a limiting factor.

Horton Overland Flow

As described in Chapter 6, there is a maximum limiting rate at which a soil in a given condition can absorb rainfall. Robert E. Horton (1933), whose keen observations and analytical ability were the foundations of modern quantitative hydrology, called this limit the *infiltration capacity* of the soil, and it declines with time after the onset of rainfall, reaching a fairly constant rate one-half to two hours into the storm. If rainfall at any time during the storm exceeds the infiltration capacity (see Figure 9-4), water will accumulate on the soil surface and fill small depressions. The water stored in these depressions (see Figure 9-5) is called *depression storage,* and its maximum volume may vary from 0.1 cm on steep, smooth hillslopes to 5 cm on agricultural lands of low gradient that have been furrowed or terraced to catch this water. Depression storage does not contribute to storm runoff; it either evaporates or infiltrates later. The depression storage capacity is eventually exhausted, and water spills over to run downslope as in irregular sheet of *overland flow*. The sheet increases in depth and velocity as more and more excess precipitation is added to it while it flows downslope. Velocities range from 10 to 500 m/hr as the depth of flowing water ranges up to one centimeter. The amount of water stored on the hillside in the process of flowing downslope is called *surface detention.*

Once rainfall intensity exceeds the infiltration capacity, runoff rises rapidly to a sharp peak at the end of rainfall, followed by a rapid decline as soon as rainfall intensity decreases. Water stored as surface detention during the storm drains away to provide the steep recession limb of the hydrograph. During the succeeding burst of rainfall, the process is repeated. In areas where Horton overland flow is the dominant producer of storm runoff, it can be seen as a thin film or sheet or as a series of tiny rivulets over large areas of hillslopes (see Figure 9-6). It does not necessarily occur over a whole drainage basin during a storm, however, since the infiltration capacity of soils may vary in even a small catchment. Betson (1964) pointed out that the area contributing Horton overland flow may be only a small portion of the catchment. This idea has become known as the *partial-area concept* of storm runoff and is a modification of the original Horton model.

Horton overland flow occurs anywhere rainfall intensity exceeds the infiltration capacity of the soil. As described in Chapter 6, this occurs most frequently on areas devoid of vegetation or possessing only a thin cover.

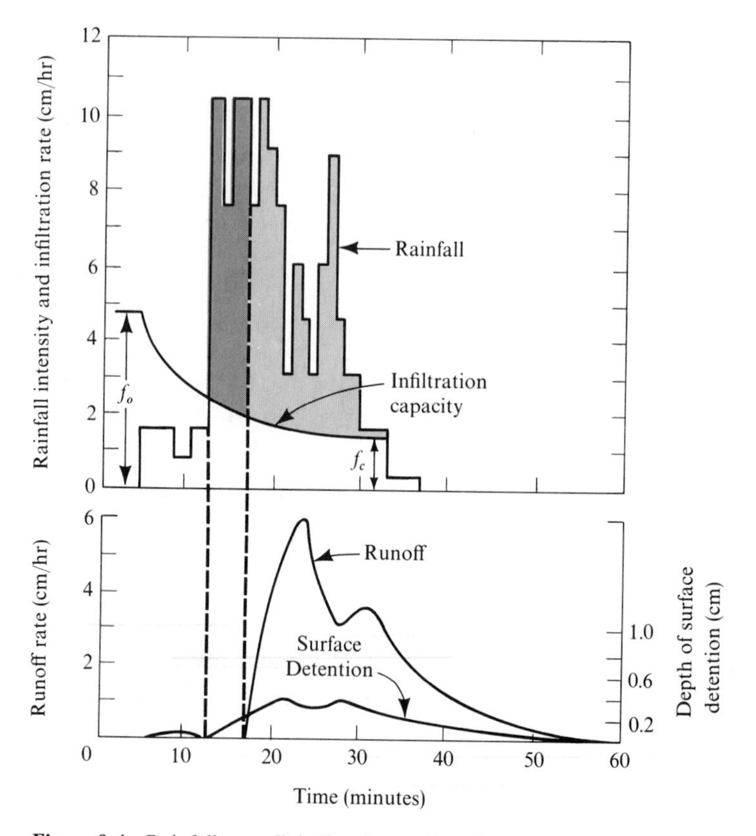

Figure 9-4 Rainfall, runoff, infiltration, and surface storage during a natural rainstorm on a hillside plot. The shaded areas under the rainfall graph represent precipitation falling at a rate exceeding the infiltration rate. The dark gray area represents depression storage, which is filled before runoff occurs. Light gray shading represents overland flow. The initial infiltration rate is f_o, and f_c is the final constant rate of infiltration approached in large storms. (Modified from Horton 1940, *Soil Science of America Proceedings*, vol. 5, pp. 399–417, by permission of the Soil Science Society of America.)

Semi-arid rangelands and cultivated fields in regions with high rainfall intensity are places where this process can be observed. It can also be seen where the soil has been compacted or the topsoil removed on lawns, tracks, skid trails, and haul roads. The flow is particularly obvious on paved urban areas. Barren spoil heaps and construction sites also experience Horton overland flow, as do areas around some smelters and other sources of serious air pollution, where the vegetation has been destroyed by acidic fumes.

Volumes and peak rates of Horton overland flow vary with storm size and intensity, and with the factors that affect infiltration. Thus, great differences occur between regions and between storms. On catchments of less

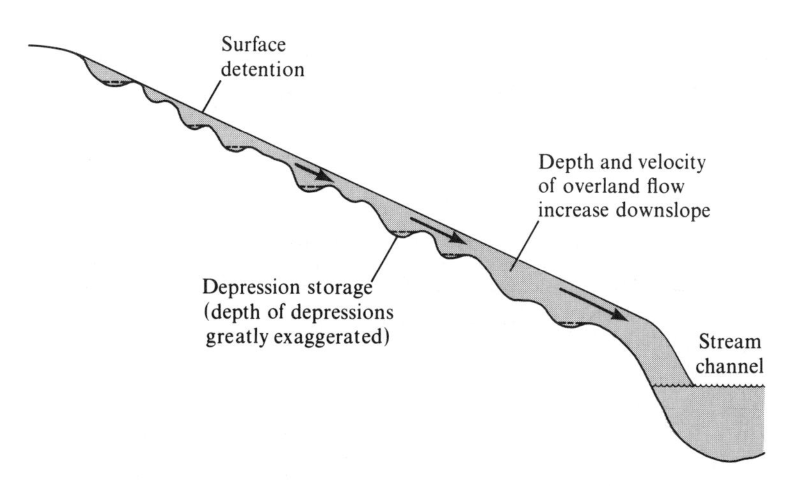

Figure 9-5 Overland flow moves downslope as an irregular sheet.

Figure 9-6 The feet of W. B. Langbein exploring Horton overland flow, during a rainstorm on a semi-arid landscape in a piñon–juniper woodland, where most of the ground has only a sparse cover of bunch grass. Rio Pescado, near Zuñi, New Mexico.

than one square kilometer, however, intense storms sometimes yield over 50 percent of the rainfall as Horton overland flow, and in the largest storms, the yield approaches 100 percent from hillsides underlain by fine-textured, bare soils or pavement. Beyond one square kilometer, the yield frequently declines with increasing drainage area, reflecting the lower average rainfall intensities and volumes over large areas (Chapter 2) and seepage losses in sandy floodplains and channel alluvium in arid regions. Peak rates of runoff by this process can also be high (see Chapter 10), ranging up to 20 cm/hr for individual hillsides, and even one-square-kilometer basins can generate Horton overland flow at rates up to 10 cm/hr. For larger catchments, peak runoff rate declines approximately with the square root of drainage area. The lag between a burst of intense rainfall and the resulting hydrograph peak for this runoff varies from a few minutes for individual hillsides to one-half hour for one-square-kilometer catchments and several hours at 100 square kilometers (Dunne 1978).

Storm Runoff in Regions of High Infiltration Rate

In most humid regions, infiltration capacities are high because vegetation protects the soil from rain-packing and dispersal, and because the supply of humus and the activity of microfauna create an open soil structure. Under such conditions rainfall intensities generally do not exceed infiltration capacities, and Horton overland flow does not occur on large areas of the landscape. Water infiltrates the surface and percolates downward as described in Chapter 6.

Figure 9-7 shows an idealized cross section of a valley with straight hillslopes and no floodplain. In a simplified situation with uniform soils, the water table before a rainstorm has an approximately parabolic form, and soil moisture content decreases with increasing height above the water table. The exact form of the soil moisture profile would vary with the texture of the soil, specifically with the amount of water that can be held by capillary forces at any elevation above the water table. Additions of water by recent storms and removal of water from the root zone by evapotranspiration would cause other complications, but the simple picture is adequate for the

Figure 9-7 Runoff processes in an area of high infiltration rate. Diagram shows a hillslope with a stream at its base, the water table, and graphs of soil moisture with depth at three locations. The moisture content of the soil at saturation is θ_s and 0 is zero soil moisture content. (a) Condition before rainfall begins; *BF* = baseflow. (b) Initial response to rainfall. Additional moisture (dark shading) accumulates in the soil near the surface. Subsurface stormflow, *SSSF*, adds to baseflow, *BF*. (c) After continuing rain, the water table has risen to the surface over the lower part of the hillside, and the saturated area is expanding uphill. Water emerges from the ground and runs downslope as return flow, *RF*; precipitation falling onto the saturated zone also runs over the ground surface, *DPS*.

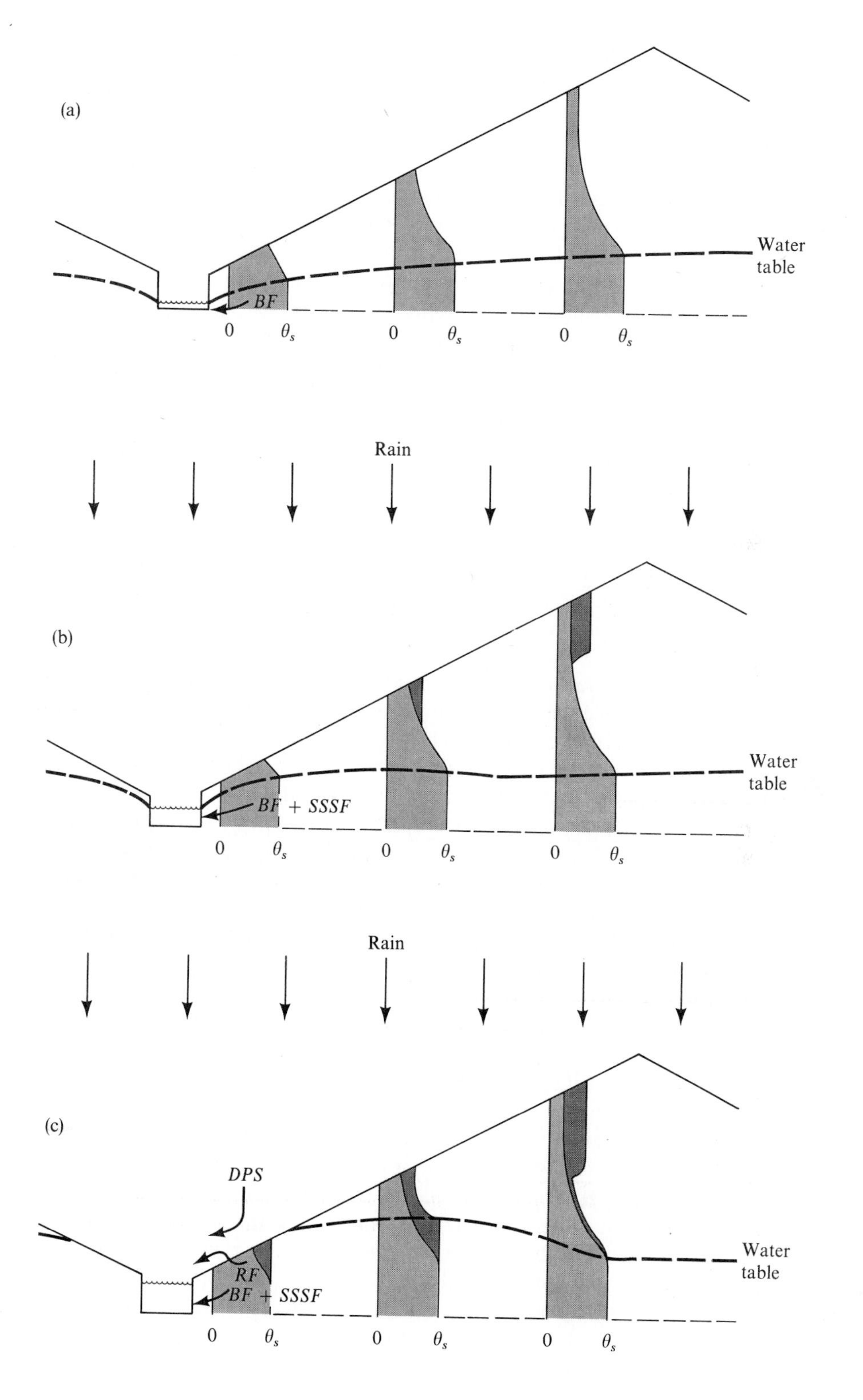
(a)
Water table
BF
0
θ_s
0
θ_s
0
θ_s
Rain
(b)
Water table
BF + SSSF
0
θ_s
0
θ_s
0
θ_s
Rain
(c)
DPS
RF
BF + SSSF
Water table
0
θ_s
0
θ_s
0
θ_s

present purpose. Because the water table in this example slopes gently toward the channel, there will be a slow flow of groundwater to maintain the baseflow (*BF*) of the stream, as shown in Figure 9-7(a).

At any depth below the soil surface, the moisture content of the soil profile will increase with distance from the hilltop. Near the base of the slope, soil moisture content is high, and when rainfall infiltrates the surface, water soon will be displaced downward into the saturated zone. This and the smaller depth to the water table near the stream combine to ensure that vertical percolation will cause the water table near the base of the slope to rise early in a storm. Further upslope, the surface soil is drier, and some rainfall is stored in the profile before the displacement of soil moisture occurs. Since distance to the water table is also greater, the transmission of water into the saturated zone occurs more slowly than at the bottom of the slope. If the water table is deep enough, all the infiltrating water may go into storage in the unsaturated zone and not reach the water table for many days after the storm. In a rainstorm, therefore, one might expect a situation like that portrayed schematically in Figure 9-7(b). The water table has remained stationary under the higher part of the hillside but has risen near the stream. This rise will also cause a steepening of the water table, as shown in the figure, and according to Darcy's Law an increase of subsurface flow should result. The runoff produced in this manner is termed *subsurface stormflow* (*SSSF*).

A slightly more complicated but common situation is that in which the topsoil is underlain by a horizon of lower permeability. Percolation is impeded by this horizon, and water accumulates above it and flows downhill through the soil, as shown by path 3 in Figure 9-1. Moving along a relatively short route through the permeable topsoil, this subsurface stormflow reaches the stream channel quickly enough to dominate the storm hydrograph in some regions, while in others it moves slowly and contributes only small amounts of runoff to the receding limb of the hydrograph. Subsurface stormflow generated by both of the processes described above has been measured in the field by several hydrologists referred to in the bibliography.

Where subsurface stormflow is the dominant contributor of storm runoff, the volumes of stormflow are much lower than those from Horton overland flow, being generally less than 20 percent of rainfall, but with an occasional study indicating approximately 50 percent for very large storms on optimum conditions for subsurface stormflow. Most of the rain is stored in the soil and in the groundwater zone and is released slowly to supply the copious baseflow of these humid regions. By comparison with velocities of Horton overland flow, subsurface stormflow is very slow, the highest rates so far measured being approximately 11 m/day in a highly permeable sandy loam on a steep hillside. It reaches channels only slowly and does not attain the same peak discharge rates as those generated by Horton overland flow. The highest rates of runoff so far measured in field studies of this process are 0.1 to 1.0 cm/hr for small hillside plots, but most observers have measured

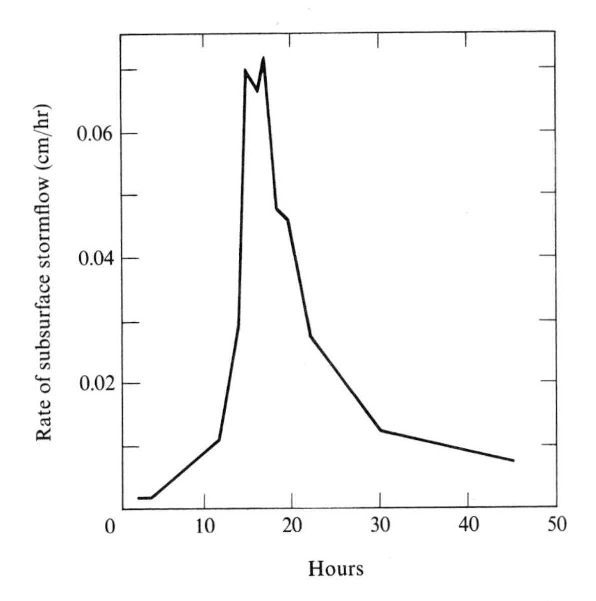

Figure 9-8 Subsurface stormflow from a 74-square-meter plot covered with a sandy-loam forest soil, 2.1 m deep on 35-percent slope at Coweeta Forest Experiment Station, North Carolina. The hydrograph was produced by a rainstorm of 101 mm. (From Hewlett and Nutter 1970.)

peak rates of less than 0.1 cm/hr even in large storms. Hydrograph peaks lag rainfall by 1 to 20 or 30 hours, even for small catchments, and the recession limbs of hydrographs are much flatter than those for Horton overland flow. Figure 9-8 presents measured subsurface stormflow from a sandy loam on a forested hillside in North Carolina. The importance of subsurface stormflow in the generation of runoff has been documented in the Southern Appalachians by Hewlett and his co-workers at the Coweeta Forest Experiment Station. The process also dominates hydrographs in other areas with deep, very permeable soils, such as the Nebraska Sand Hills and the forested Coast Range of Oregon.

If the rainstorm is large enough or the water table or impeding horizon shallow enough, infiltration and percolation will cause the water table to rise to the ground surface, as shown in Figure 9-7(c). When this happens, subsurface water can escape from the soil and flow to the channel over the land surface. This kind of runoff process was termed *return flow* by Musgrave and Holtan (1964) and is indicated by the symbol *RF* in Figure 9-7(c). That portion of the hillside over which return flow emerges from the surface is impervious to rain falling onto it. The rain must flow off the area as *direct precipitation onto saturated areas* (*DPS* in Figure 9-7(c)). This runoff, together with return flow, is called *saturation overland flow.* It often moves at

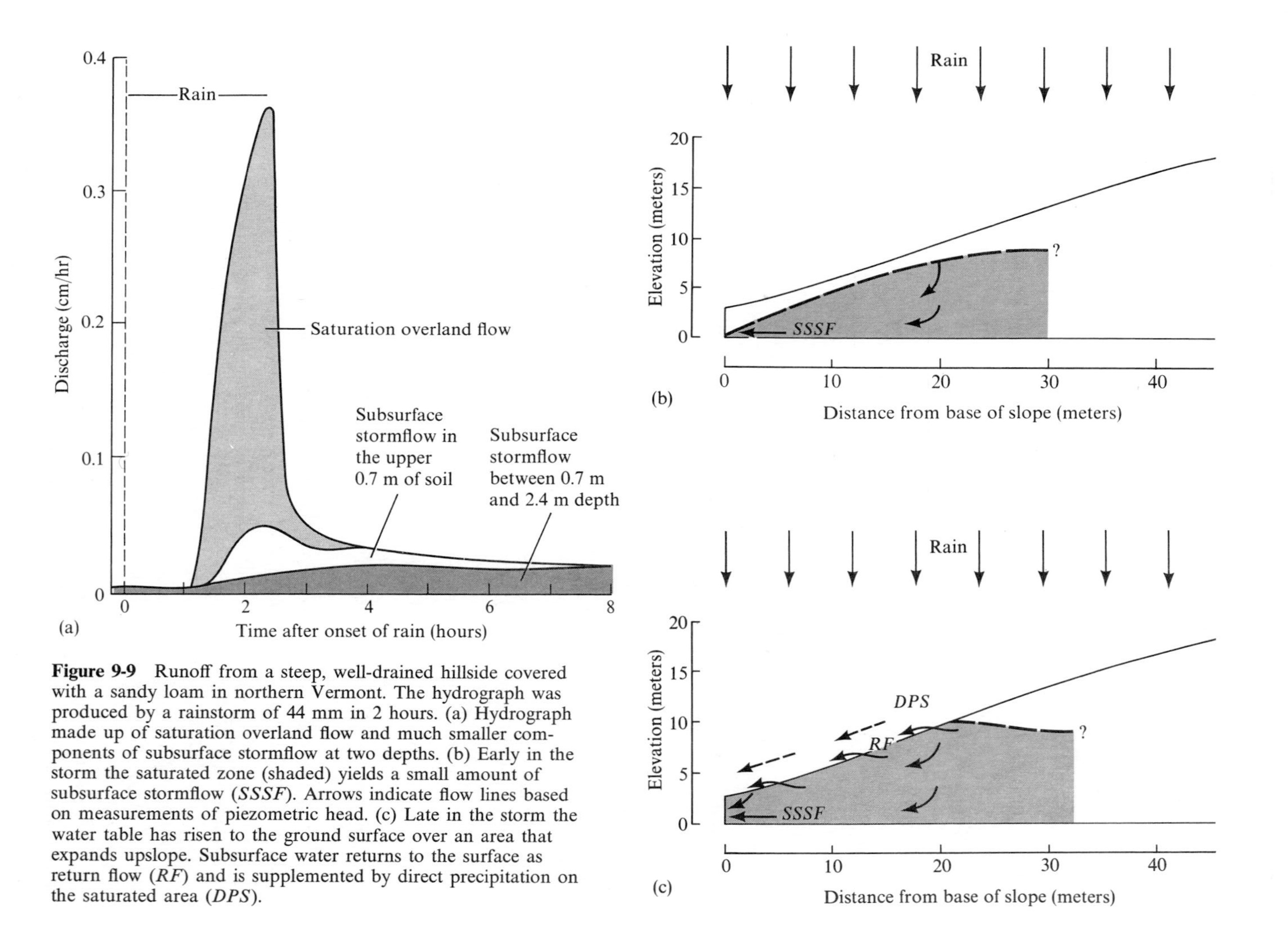

Figure 9-9 Runoff from a steep, well-drained hillside covered with a sandy loam in northern Vermont. The hydrograph was produced by a rainstorm of 44 mm in 2 hours. (a) Hydrograph made up of saturation overland flow and much smaller components of subsurface stormflow at two depths. (b) Early in the storm the saturated zone (shaded) yields a small amount of subsurface stormflow (*SSSF*). Arrows indicate flow lines based on measurements of piezometric head. (c) Late in the storm the water table has risen to the ground surface over an area that expands upslope. Subsurface water returns to the surface as return flow (*RF*) and is supplemented by direct precipitation on the saturated area (*DPS*).

velocities one hundred times as fast as those of subsurface stormflow and can supply runoff to the channel more quickly than water following the subsurface route. Hydrographs of this flow have higher peaks and much shorter lag times than those of subsurface stormflow. Velocities of saturation overland flow cover the lower range of values given earlier for Horton runoff because it usually flows on gentle footslopes which are thickly vegetated and rough.

The peak rates of generation of saturation overland flow in Vermont vary with catchment size from 5 cm/hr for individual hillslopes to 1 cm/hr for basins of one square kilometer. In regions with higher rainfall intensities these rates will be greater, but they are less than those for Horton overland flow, because only a portion of the drainage basin is contributing saturation overland flow, namely, the lower parts of swales and hillsides that become saturated during a rainstorm or snowmelt season. Lag times between a burst of rainfall and the resulting hydrograph peak range from a few minutes for single hillsides to about one hour at one square kilometer (Dunne 1978).

As the storm progresses, the saturated area expands upslope, causing more of the catchment to contribute saturation overland flow. Figure 9-9 shows the measured runoff components for a storm together with the rise of the water table, the expansion of the saturated area, and the direction of flow lines on a deep, well-drained sandy-loam soil on a steep hillside during a 44-mm, 2-hour rainstorm. Figure 9-10 shows the runoff produced by various processes during a 54-mm, 3.5-hour rainstorm on a shallow, moist silt loam on a neighboring hillslope of low gradient. Because of the wetter antecedent condition, the gentler gradient, and the shallowness of the silt

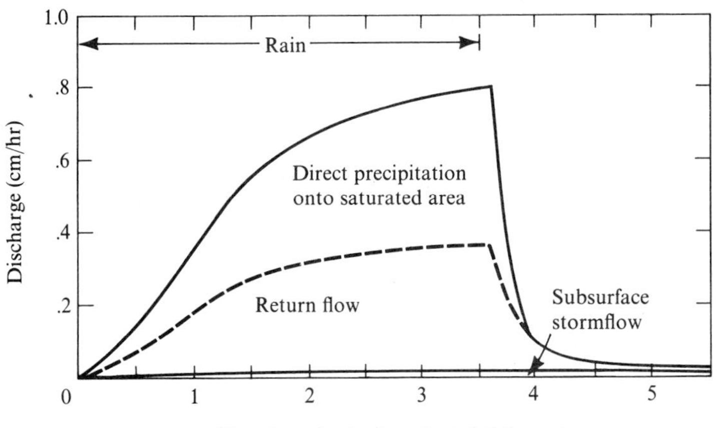

Figure 9-10 Runoff from a shallow, poorly drained silt loam on a gentle hillslope during an artificial rainstorm of 54 mm in 3.5 hours in northern Vermont.

loam, it became saturated more quickly, and the saturation spread over a larger area than on the steep, well-drained hillslope shown in Figure 9-9. Consequently, return flow and direct precipitation onto saturated areas were more important in this situation than on the permeable, well-drained soil.

The expansion of the saturated area during a single 46-mm rainstorm in a basin with steep, well-drained hillsides and a narrow valley bottom is shown in Figure 9-11. The expansion took place mainly into swales where pre-storm drainage of groundwater keeps the water table high, and the soils are shallow and poorly drained. In the same region (northeastern Vermont), many other basins have shallower soils on gentler hillslopes with long, concave footslopes. Here a much greater lateral expansion of the saturated area was measured during storms.

The saturated area also expands and contracts seasonally, as shown by

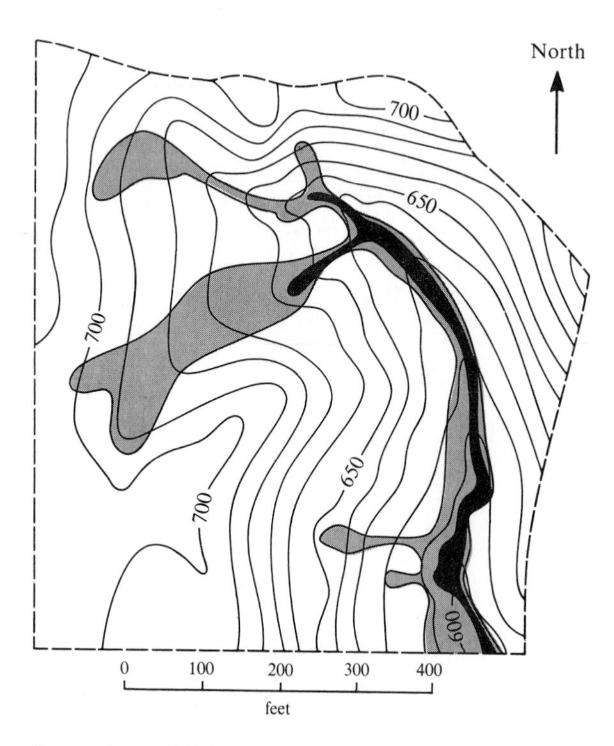

Figure 9-11 Map of saturated areas showing expansion during a single rainstorm of 46 mm. The basin has steep, well-drained slopes and a narrow valley floor. The solid black shows the saturated area at the beginning of rain; the lightly shaded area is saturated by the end of the storm and is the area over which the water table had risen to the ground surface.

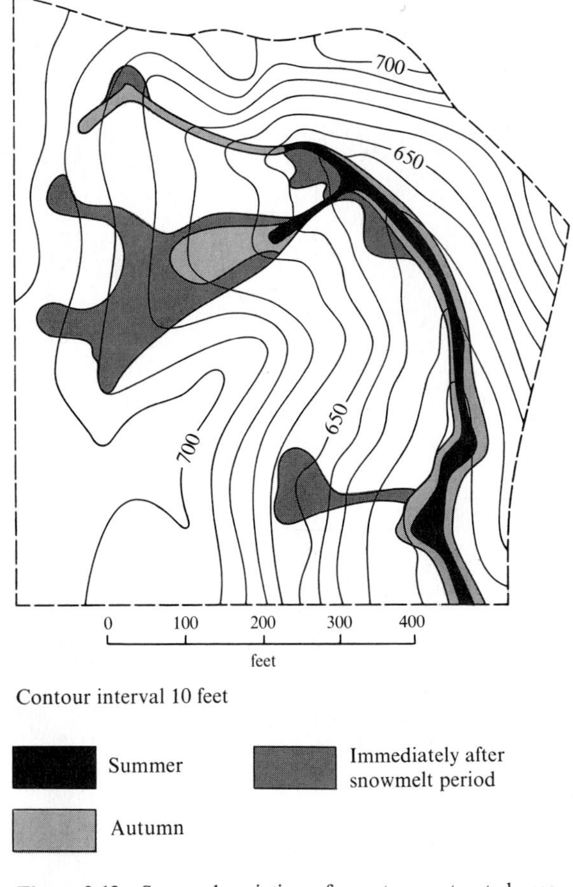

Figure 9-12 Seasonal variation of pre-storm saturated area in a catchment with steep, well-drained hillsides and a narrow valley floor, Danville, Vermont. Compare this seasonal change with the changes during a single rainstorm on the same area in Figure 9-11.

Figure 9-12, which portrays the pre-storm saturated area at three times of the year for the basin shown in Figure 9-11. Because this runoff-producing zone occupies only a small proportion of the watershed, even small changes can cause important differences in the volume and rate of runoff when rainfall occurs. An even larger seasonal fluctuation of the pre-storm saturated zone can be seen in Figure 9-13 for a catchment with gentle concave hillslopes and poorly drained soils.

The saturated areas may not be easy to see, especially in dense forests. They are best seen during warm weather in early spring (see Figure 9-14). At that time the dead grass on the well-drained areas dries quickly to a light

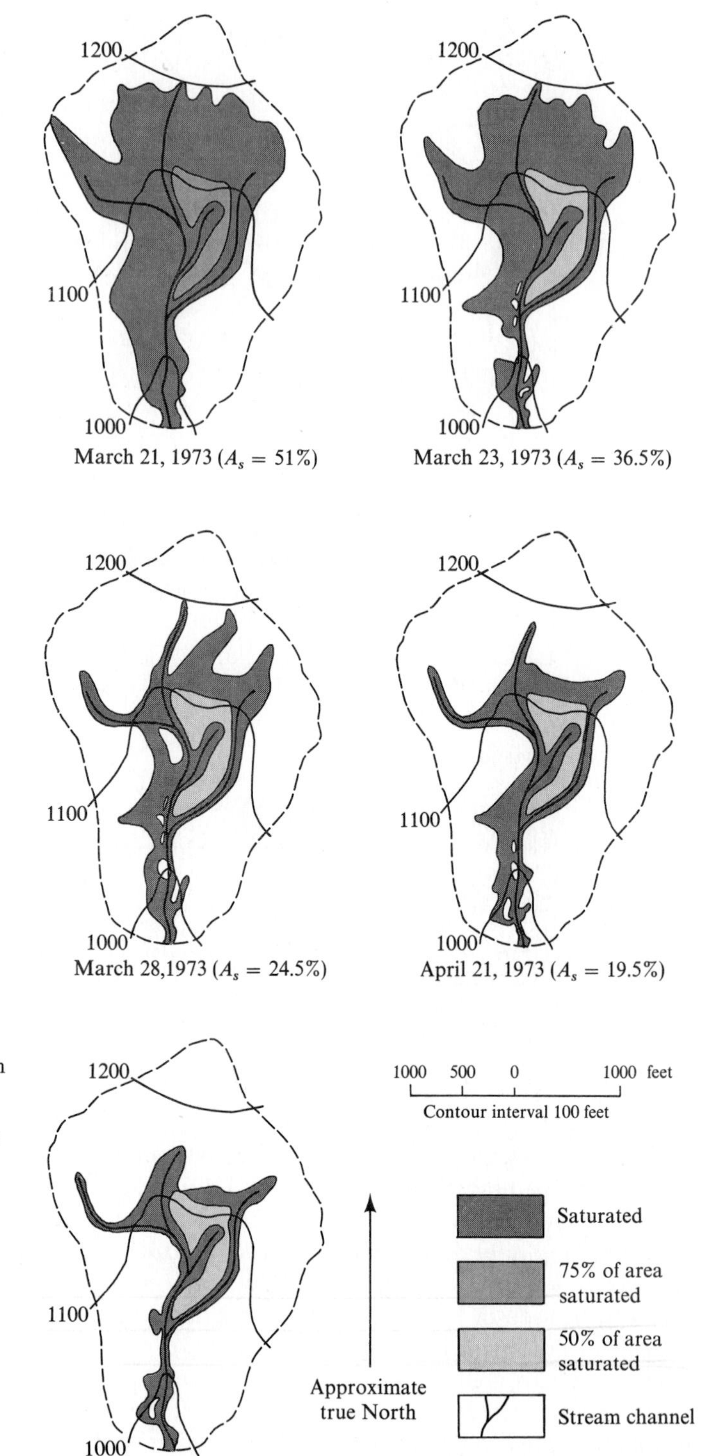

Figure 9-13 Seasonal variation of pre-storm saturated area in a basin with gentle footslopes and moderately to poorly drained soils, Danville, Vermont. For each date, A_s is the percentage of the catchment area that is saturated.

color. The saturated zones remain dark and are the sites of the earliest regeneration of grass. Careful examination of these areas reveals a thin film of saturation overland flow emerging from the soil and moving downslope through the thick grass mat at velocities of about 1 cm/sec. In the summer, the pre-storm saturated area contracts and is usually covered with thick, rough vegetation such as sedges, cattails, and willows (see Figure 9-15), because the land is not trafficable by farm machinery. During a large storm, however, the saturated area can expand upslope into the slightly better drained soils that are saturated in Figure 9-12. The recognition of these saturated zones, and some suggestions for mapping and predicting their temporal and spatial variation are given in a paper by Dunne, Moore, and Taylor (1975).

The original conceptual model of the generation of storm runoff in humid areas was developed by hydrologists of the U.S. Forest Service (1961) and the Tennessee Valley Authority (1964). The model was termed the *variable source concept* of storm runoff (Hewlett and Hibbert 1967). The processes by which water actually reached the stream according to this model, however, were not defined very precisely until field measurements were made.

Figure 9-14 Saturated variable source (dark areas) immediately after the snowmelt season in northern Vermont. The photograph was taken in the southeastern corner of the basin mapped in Figure 9-13 and shows the two small extensions of the saturated zone above the 1000-foot contour on the map of March 21.

Figure 9-15 Saturated zone after dry weather in northern Vermont. The rough vegetation of sedges, cattails, and coarse grasses in the center of the picture covers an area that remains saturated throughout the summer. During large storms and wet seasons the saturated area expands into the foreground and into the hayland on the opposite side of the valley.

Ragan (1968) made the first field study of the variable source concept and defined several of the processes and their relative importance and timing in a small catchment near Burlington, Vermont. The field evidence gathered by several other hydrologists is summarized in the paper by Dunne (1978).

Runoff Processes in Rural Areas: Summary

Water can reach a stream by several routes. The processes that deliver stormflow and the volumes and timing of their contributions vary with climate, vegetation, land use, soil properties, topography, and rainfall characteristics. The relation of the various runoff processes to their major controls is summarized in Figure 9-16. In arid and semi-arid regions and those disturbed by humans (through agriculture, urbanization, and mining), infiltration capacity is a limiting factor and Horton overland flow is the dominant storm runoff process. In most humid regions where infiltration is not a limiting factor, the variable source model of storm runoff is appropriate. But there are important differences within and between humid regions in the relative importance of the two major runoff processes at work: subsurface stormflow and saturation overland flow.

Where soils are well-drained, deep, and very permeable, and cover steep hillsides bordering a narrow valley floor, subsurface stormflow dominates the volume of storm runoff. The saturated zone is more or less confined to the valley floor, and saturation overland flow is limited, though even in such situations, it frequently generates the peak rates of runoff from small catchments. Subsurface stormflow achieves its greatest importance in areas such as the forested highlands of the Southern Appalachians; in the deep, permeable forested soils on volcanic tuffs and sandstones of the Oregon Coast Range; and in the deep, permeable volcanic ash deposits of central Kenya. In most other humid regions, where the saturated and near-saturated valley bottoms are more extensive, and where footslopes are gentler and soils thinner, the saturated area is more extensive before and throughout a storm or snowmelt period. Although subsurface stormflow occurs in such regions,

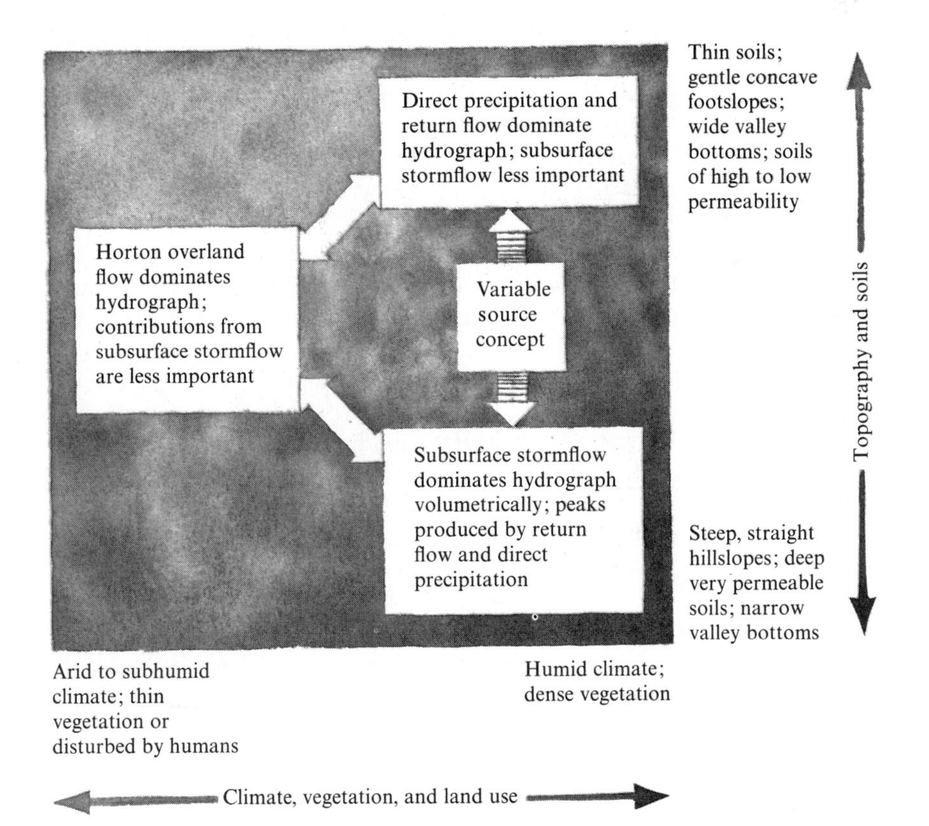

Figure 9-16 Runoff processes in relation to their major controls.

it is less important to the storm hydrograph than are return flow and direct precipitation onto saturated areas, which produce saturation overland flow from limited areas of the catchment. A range of topographic and pedalogic conditions exists between those that tend to produce a preponderance of subsurface stormflow and those that favor the occurrence of saturation overland flow.

The saturated zones do not often cover more than half of a catchment, and usually cover much less. They are extremely important, however, not only for the generation of storm runoff but for other matters of concern to planners. Saturation overland flow is capable of washing into a stream pollutants that would be filtered out of runoff if the water remained below the soil surface. One day before the artificial rainstorm that generated the runoff shown in Figure 9-10, a 0–20–20 commercial fertilizer was spread on the hillside plot with an application rate of 400 kg/ha. Concentrations of total phosphorus and potassium were measured in the runoff collected during the artificial rainstorm. The concentrations remained low until saturation overland flow developed and then rose rapidly to peaks of 118 mg/liter for potassium and 32 mg/liter for phosphorus. Fifteen percent of all the potassium applied and 5 percent of the phosphorus were lost to the stream during this storm.

Just before the rainfall ended, the zone subject to saturation overland flow was outlined by inserting colored pegs into the ground, and this area was later mapped. Eighty-eight percent of all the potassium applied to this zone had been lost in the runoff along with 30 percent of all the phosphorus. Several weeks after the experiment, the area subject to saturation overland flow retained its usual cover of poor grass, whereas the surrounding area had a luxurious growth. The sharp boundary between the two areas followed precisely the line of colored pegs emplaced during the storm. Ideally, this experiment should have been repeated using different time lags between the fertilizer application and the rainstorm, but work in another place intervened. The tentative conclusion to be drawn from the single experiment, however, was that on soils used for growing crops, saturation overland flow can develop frequently and carry fertilizers into the stream.

Kunkle (1970) repeated this experiment in his study of bacterial water pollution in the Vermont catchment. His conclusions were of the same kind. Bacteria from cattle feces reach the stream from zones that generate saturation overland flow.

Because the restricted zones of saturation are important sources of chemical and bacterial water pollutants, their recognition is important to the planner concerned with the management of land and water quality. In parts of Washington, saturated zones contribute sediment and adsorbed fertilizers after the spring snowmelt. The saturation also affects the trafficability of croplands, limiting the plowing or harvesting of many fields in northern New England and southern Canada. As described in Chapter 6, septic-tank performance can also be limited in the zones of occasional saturation into which many new housing subdivisions are spreading.

Runoff Processes in Urban Areas

Modifications of the land surface during urbanization produce changes in the type or magnitude of runoff processes, and cause the planner many complex problems. The increased storm runoff leads to difficulties of storm drainage control, stream channel maintenance, groundwater recharge, and stream-water quality. Solutions to these problems are costly, although many of the difficulties and costs can be avoided if planners understand runoff processes in an urban region and take them into account early in the planning process.

The major change in runoff processes results from covering parts of the catchment with impervious roofs, sidewalks, roadways, and parking lots (Figure 9-17). An indication of the proportion of a catchment that is rendered impervious can be obtained from population density, as shown in Figure 9-18. The infiltration capacity of these areas is lowered to zero, and many areas that remain soil-covered are trampled to an almost impervious state, so that the volume and rate of Horton overland flow is increased. In other regions this process is introduced into areas that formerly contributed only low volumes of slowly responding subsurface flow.

Gutters, drains, ans storm sewers are laid in the urbanized area to convey runoff rapidly to stream channels. Natural channels are often straightened, deepened, or lined with concrete to make them hydraulically smoother. Each of these changes increases the efficiency of the channel, so that it transmits the flood wave downstream more quickly and with less storage in the

Figure 9-17 Aerial view of the commercial district of Seattle showing an almost completely impervious cover.

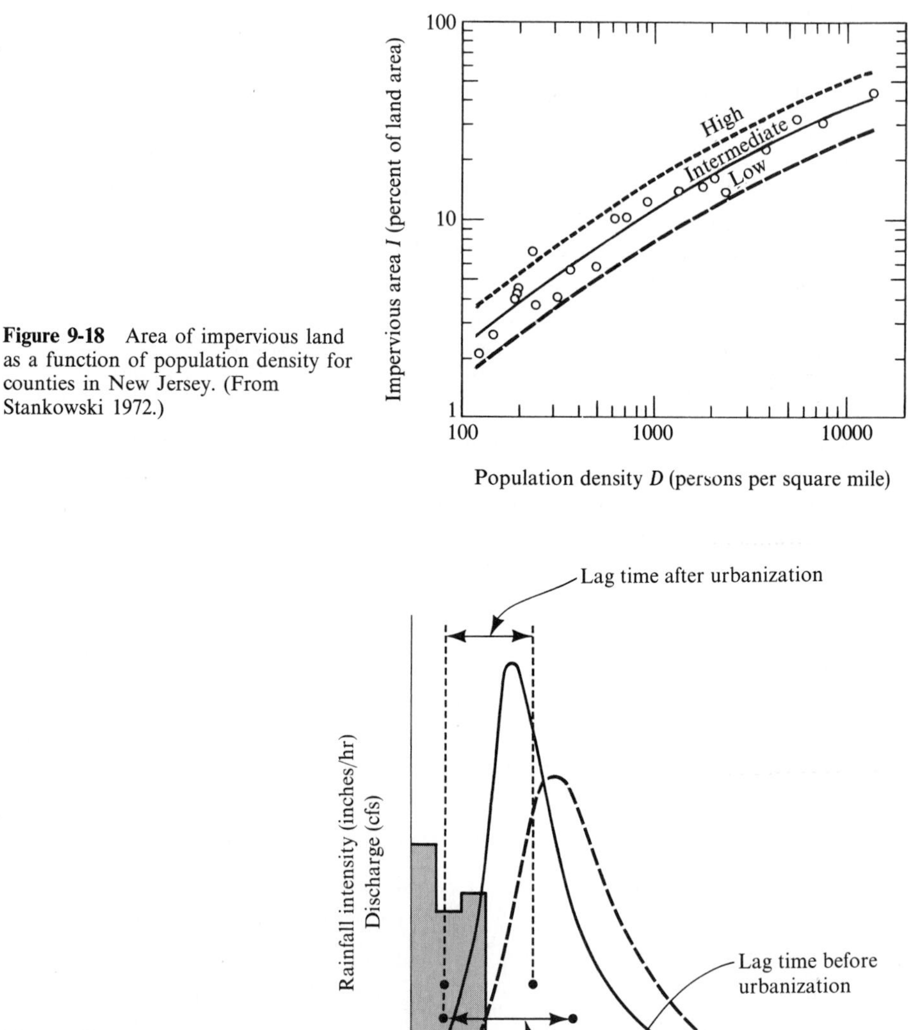

Figure 9-18 Area of impervious land as a function of population density for counties in New Jersey. (From Stankowski 1972.)

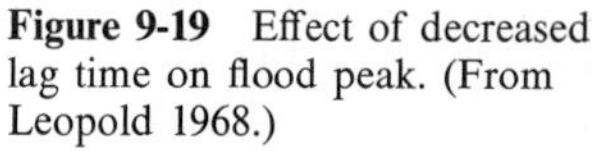

Figure 9-19 Effect of decreased lag time on flood peak. (From Leopold 1968.)

channel. Stormwaters can therefore accumulate downstream more quickly than in natural river systems and produce high flood peaks. Even if the total volume and peak rate of runoff from the land surface were not increased by urbanization (which of course they are), the increased velocities in the channels would still decrease the lag between rainfall and runoff. Figure 9-19 shows that if the lag time of a hydrograph of fixed volume is decreased, the peak rate must increase to keep the volume constant.

The increase of storm runoff has many costly consequences in urban areas. Frequent overbank flooding damages houses and gardens or disrupts traffic.

The capacities of culverts and bridges are overtaxed. Channels become enlarged in response to the larger floods (see Chapter 18), and residential lots suffer erosion and reduction of their value.

The increased amounts of water that generate storm runoff are not available for recharging the groundwater to supply baseflow during dry weather. Low flows are therefore reduced, lowering water quality during the summer months. For a review of the effects of urbanization on storm runoff, the reader is referred to the reports of Leopold (1968), Seaburn (1969), and Anderson (1970).

Because of the importance to planners of being able to predict the hydrologic impact of urbanization and to make simple design calculations, we will deal with the subject of flood prediction in urban environments at several places in Chapter 10. We will also describe how to predict the effect of other land-use changes, but will pay particular attention to the impact of urbanization. Then in Chapter 11, we will describe various strategies for minimizing flood damage, again paying particular attention to the urban environment.

Bibliography

ANDERSON, D. G. (1970) Effects of urban development on floods in northern Virginia; *U.S. Geological Survey Water Supply Paper 2001-C.*

BETSON, R. P. (1964) What is watershed runoff?; *Journal of Geophysical Research,* vol. 68, pp. 1541–1552.

DUNNE, T. (1970) Runoff production in a humid area; *U.S. Department of Agriculture Publication ARS-41-160,* 108 pp.

DUNNE, T. (1978) Field studies of hillslope flow processes; Chapter 7 in *Hillslope hydrology* (ed. M. J. Kirkby), John Wiley & Sons, London.

DUNNE, T. AND BLACK, R. D. (1970a) An experimental investigation of runoff production in permeable soils; *Water Resources Research,* vol. 6, pp. 478–490.

DUNNE, T. AND BLACK, R. D. (1970b) Partial-area contributions to storm runoff in a small New England watershed; *Water Resources Research,* vol. 6, pp. 1296–1311.

DUNNE, T., MOORE, T. R. AND TAYLOR, C. H. (1975) Recognition and prediction of runoff-producing zones in humid regions; *Hydrological Sciences Bulletin,* vol. 20, no. 3, pp. 305–327.

HEWLETT, J. D. AND HIBBERT, A. R. (1967) Factors affecting the response of small watersheds to precipitation in humid regions; in *Forest hydrology* (eds. W. E. Sopper and H. W. Lull), Pergamon Press, Oxford, pp. 275–290.

HEWLETT, J. D. AND NUTTER, W. L. (1970) The varying source area of streamflow from upland basins; *Proceedings of the Symposium on Interdisciplinary Aspects of Watershed Management.* American Society of Civil Engineers, pp. 65–83.

HORTON, R. E. (1933) The role of infiltration in the hydrologic cycle; *EOS, American Geophysical Union Transactions,* vol. 14, pp. 446–460.

HORTON, R. E. (1940) An approach toward a physical interpretation of infiltration capacity; *Soil Science Society of America Proceedings,* vol. 4, pp. 399–417.

KUNKLE, S. H. (1970) Sources and transport of bacterial indicators in rural streams; *Proceedings of the Symposium on Interdisciplinary Aspects of Watershed Management,* American Society of Civil Engineers, pp. 105–132.

LEOPOLD, L. B. (1968) Hydrology for urban land planning: a guidebook; *U.S. Geological Survey Circular 554.*

MCHARG, I. (1969) *Design with nature;* Falcon Press, Philadelphia, 198 pp.

MUSGRAVE, G. W. AND HOLTAN, H. N. (1964) Infiltration; Section 12 in *Handbook of applied hydrology* (ed. Ven te Chow), McGraw-Hill, New York.

RAGAN, R. M. (1968) An experimental investigation of partial-area contributions; *International Association of Scientific Hydrology, Symposium of Bern, Publication No. 76,* pp. 241–251.

SEABURN, G. E. (1969) Effects of urban development on direct runoff to East Meadow Brook, Nassau County, Long Island, New York; *U.S. Geological Survey Professional Paper 627-B.*

STANKOWSKI, S. J. (1972) Population density as an indirect indicator of urban and suburban land-surface modifications; *U.S. Geological Survey Professional Paper 800-B,* pp. B219–B224.

TEMPLE, P. H. (1972) Measurements of runoff and soil erosion at an erosion plot scale, with particular reference to Tanzania; *Geografiska Annaler,* vol. 54A, pp. 203–220.

TENNESSEE VALLEY AUTHORITY (1964) Bradshaw Creek-Elk River, a pilot study in area-stream factor correlation; *Research Paper No. 4, Office of Tributary Area Development,* Knoxville, 64 pp.

U.S. FOREST SERVICE (1961) Some ideas about storm runoff and baseflow; *Annual Report, Southeastern Forest Experiment Station,* pp. 61–66.

WHIPKEY, R. Z. (1965) Subsurface stormflow from forested slopes; *International Association of Scientific Hydrology Bulletin,* vol. 10, no. 2, pp. 74–85.

WHIPKEY, R. Z. (1969) Storm runoff from forested catchments by subsurface routes; *International Association of Scientific Hydrology, Symposium of Leningrad, Publication No. 85,* pp. 773–779.

Typical Problem

9-1 Mapping of Runoff-Producing Zones

Choose a small drainage basin with a mixture of land use and an area between 5 and 10 sq km. Obtain an aerial photograph of the area at a scale of at least 1:10,000. Study the soils map of the basin, which can be obtained from the County Soils Survey Report, or a similar publication.

Walk over the basin, and map the areas that produce various kinds of runoff. The exercise is best done during a wet season when you can observe the runoff processes or recent evidence of them. You will need to make judgments and rough calculations throughout the exercise, and some of the suggestions in the paper by Dunne, Moore, and Taylor (1975) may be useful.

Write a short report explaining the evidence and the reasoning you used. Describe any problems that the runoff processes you mapped are likely to cause for those concerned with land and water management.

10

Calculation of Flood Hazard

Few people need to be convinced that floods should be a major concern of planners. Annual losses from floods are staggering. Even in Washington, a sparsely populated state with relatively low flood hazard, the annual monetary loss is $25 million. In the state of California, the flood loss between 1970 and 2000 is expected to total $6.5 billion if there are no improvements of existing flood-control practices, and $3.1 billion if all feasible control measures, costing $2.7 billion, are employed (California Division of Mines and Geology 1973). Even these losses pale into insignificance, however, by comparison with the huge death tolls that have resulted from large floods in some of the major alluvial lowlands of the world, such as the Ganges and Yangtze valleys. On a smaller scale, new housing developments and factories continue to be located on land subject to flooding, indicating that many planners have a limited awareness of the hazard. The situation is further complicated when changes of land use increase the flood hazard, causing problems to downstream areas originally outside the hazard zone. The costs of damage, inconvenience, and control then remain with the planner, developer, landowner, or society for a very long time.

We will reserve further discussion of the impact of flooding until the next chapter, however, and will there describe methods of minimizing flood damage. In this chapter we will introduce some hydrologic principles about flood magnitude, which should help the planner make sense of the many conflicting claims about the factors that influence floods and about the effects of land-use changes, dam construction, and other alterations of the hydrologic regime. We will then describe various sources of information on floods and will review a variety of methods for predicting flood discharges and heights. Planners increasingly need to make simple calculations of probable flood magnitudes for small catchments when the project with which they are concerned is too small to warrant a full-scale study by a hydrologic consultant. In other situations, the planner may need to make some rough flood calculations at a very early stage in the planning process to find out whether the flooding problem is likely to be a constraint important enough to demand a more sophisticated analysis.

It has been the attitude of planners that, because they are not hydrologists, they must accept the engineering reports of large or specialized organizations as fact. Far better would be the adoption of the attitude that a report on engineering hydrology is based on certain assumptions that even the nonspecialist can understand readily. Planners can, with a little experience, isolate and examine these assumptions and can also make rough calculations to obtain results based on somewhat different assumptions.

The growing concern about the hydrologic impact of various developments is causing planners to make quantitative predictions of the effects of their installations. We will show how it is possible for the planner to calculate the hydrologic changes that will ensue from various kinds of land-use modification. Considering these changes at an early stage in the plan can save money and avoid a great deal of social, economic, and ecologic disruption.

Finally, we hope to provide planners with a simple description of what is involved in even the more complicated types of flood prediction used on large-scale design problems. It is important that planners know what these techniques are based upon and what their limitations are. They will not then be overwhelmed by the tables and graphs in the usual hydrologic report, for they are competent to digest and review such information.

Storage and Transmission of Floodwater

In the preceding chapter we discussed the processes by which runoff is generated on hillsides during storms. A 2-inch rainstorm, for example, occurring on a one-square-mile catchment the soils of which could absorb 0.5 inch would supply to the channels of the area almost 3.5 million cubic feet (100,000 cubic meters) of water in perhaps 2 hours. In Chapter 13, we will discuss the generation of snowmelt runoff, which occurs at lower rates than rainfall but which lasts for a longer time and is often generated from larger portions of a catchment. These vast amounts of water are accommodated within the stream channel system, and if the channel capacity is overtaxed, water moves out over the valley floor and a flood is said to have occurred. Other processes such as dam failures and outbursts of meltwater from beneath glaciers can also supply water to the stream at rates that exceed the storage capacity of the channel system.

Runoff supplied to the channel moves downstream as a wave of increasing and then decreasing stream discharge. As this flood wave moves down the channel it is subject to two processes that alter its character. The first of these processes is *uniform, progressive flow,* or *translation,* whereby the wave moves downstream without changing its shape (see Figure 10-1(a)). This tendency is dominant in steep, straight mountain streams and desert washes during intense rainstorms, where flow velocities are high and remain relatively

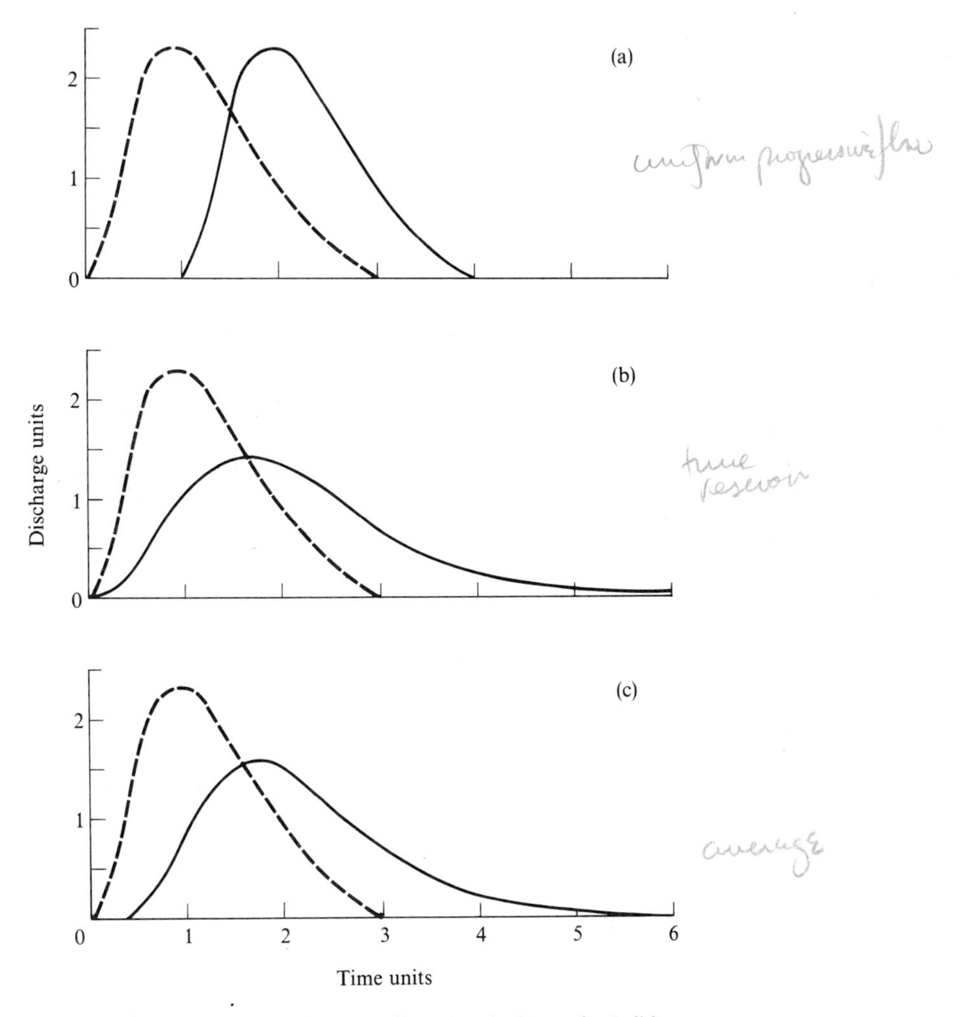

Figure 10-1 Comparison of computed outflow hydrographs (solid line) from a reach of channel for a given inflow (dashed line). The diagrams represent the effects of (a) simple translation or uniform progressive flow, (b) true reservoir action, and (c) "average" river channel storage, a combination of translation and reservoir action. (From W. B. Langbein 1940, *EOS, American Geophysical Union Transactions,* vol. 21, pp. 620–627. Copyrighted by American Geophysical Union.)

constant throughout the range of flood discharge. The second process operating on the flood wave is *reservoir action,* or *pondage,* whereby the wave is attenuated by storage within the channel and valley bottom. When a rapid pulse of water enters a true reservoir (such as a lake or a bathtub), it does not flow out or displace a similar amount of water immediately. Rather, most of the input is stored within the reservoir. In such water bodies there is a relationship between the amount of water in storage (i.e., the height to which water stands above the outlet, whether it is a stream channel or a plug hole) and the rate of outflow from the system. Consequently, as water flows in, it is stored, raising the water level progressively, and causing the outflow to increase. But because some of the input must be stored to increase the output rate, the peak outflow rate cannot be as high as that of the inflow. When the inflow ceases or declines significantly, the water in storage then drains out slowly, maintaining an outflow in excess of the input rate (see Figure 10-1(b)). These relations will be treated more formally and quantitatively in a later section, but for now this qualitative description will suffice.

In most river channels the flood wave operates in a manner that is intermediate between the two extreme situations described above. As the flood wave is being translated downstream, a part of the water is stored in the channel and the wave is attenuated by this reservoir action (see Figure 10-1(c)). From the diagram it can be seen that an "average" river has a very important reservoir function in attenuating flood peaks. The valley floor acts upon the flood wave in the same general manner but with an even greater reservoir component (see Figure 10-2). During extreme floods the channel and the valley floor store a considerable part of the total volume of flood runoff generated on hillslopes by large storms. Table 10-1 shows the magnitude of this storage for some large storms in the eastern United States. The volume of storage in the Ohio River flood of 1937 was approximately 2.3 times the capacity of Lake Mead, the largest artificial reservoir in the United States. We will have more to say about this important function of channels and valley floors in the next chapter.

On a smaller scale, one can see the effects of the translation and storage mechanisms in a channel system by comparing storm hydrographs measured at various places down a channel. Figure 10-3 shows a set of hydrographs from catchments of various size on the Sleepers River Experimental Watershed in northeastern Vermont. It can be seen that on the smallest catchment (0.2 sq mi), runoff responds to each of the three bursts of rainfall; at the outlet of the 3.2-sq-mi basin, the peaks have become attenuated, the smaller, early peaks are becoming damped out and absorbed into the main rising limb, and the time base of the storm hydrograph is increasing. At 16 sq mi, the peaks have become one, and at 43 sq mi this single peak has become attenuated. The damping effect decreases the peak rate of runoff expressed on an area–inch basis as the drainage area increases, although the absolute discharge increases as floodwaters drain from larger and larger areas. Each of these hydrographs is a composite of discharge from the upstream gauge and runoff generated on hillslopes in the drainage area between gauges.

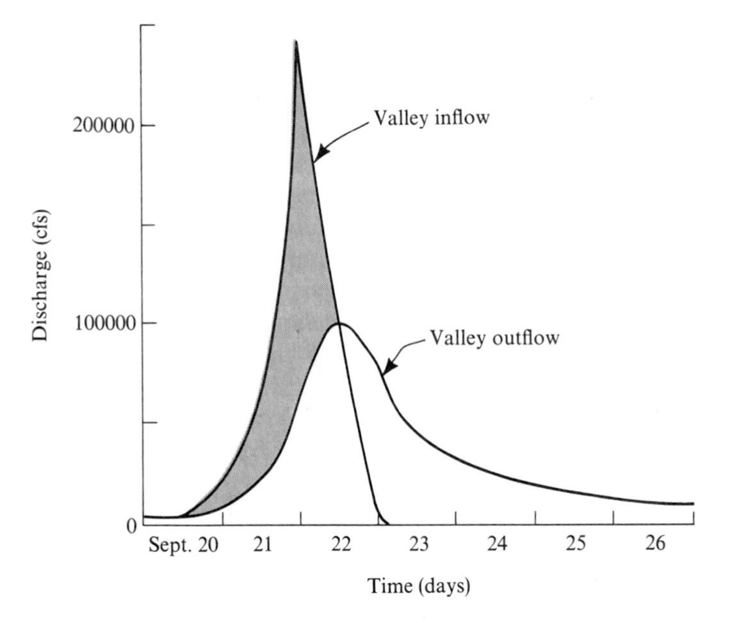

Figure 10-2 Inflow and outflow hydrographs for the valley bottomlands and the channel of the Delaware River above Port Jervis, New York, during the hurricane flood of September 1938. The drainage area is 3076 sq mi. The inflow hydrograph represents the computed time distribution of water provided from the basin to the valley bottomlands. The outflow is the measured flow in the channel and shows the attenuation of peak due to storage in the valley and the channel. The shaded area represents the volume of water stored in the channels and valley floors of the catchment before the peak outflow occurred. (From H. K. Barrows 1942, *EOS, American Geophysical Union Transactions,* vol. 23, pp. 483–488. Copyrighted by American Geophysical Union.)

Table 10-1 Volumes of channel and valley-floor storage in relation to rainfall and runoff during major floods. (From U.S. Geological Survey.)

					MAXIMUM VOLUME OF STORAGE IN CHANNEL AND VALLEY FLOOR		
BASIN	DATE	DRAINAGE AREA (MI^2)	MEAN AREAL RAINFALL (IN)	DIRECT RUNOFF (IN)	(IN)	(FT^3)	(% OF DIRECT RUNOFF)
Muskingum River above McConnellsville, OH	8/8/35	7,411	4.15	2.3	1.83	3.14×10^{10}	80
Ohio River above Metropolis, IL	1/26/37	203,000	12.95	8.9	5.1	2.44×10^{12}	57
Susquehanna River above Marietta, PA	8/26/33	25,990	4.13	1.39	1.1	7.84×10^{11}	79
Connecticut River above Hartford, CT	9/22/38	10,480	7.55	4.05	2.9	4.79×10^{10}	72

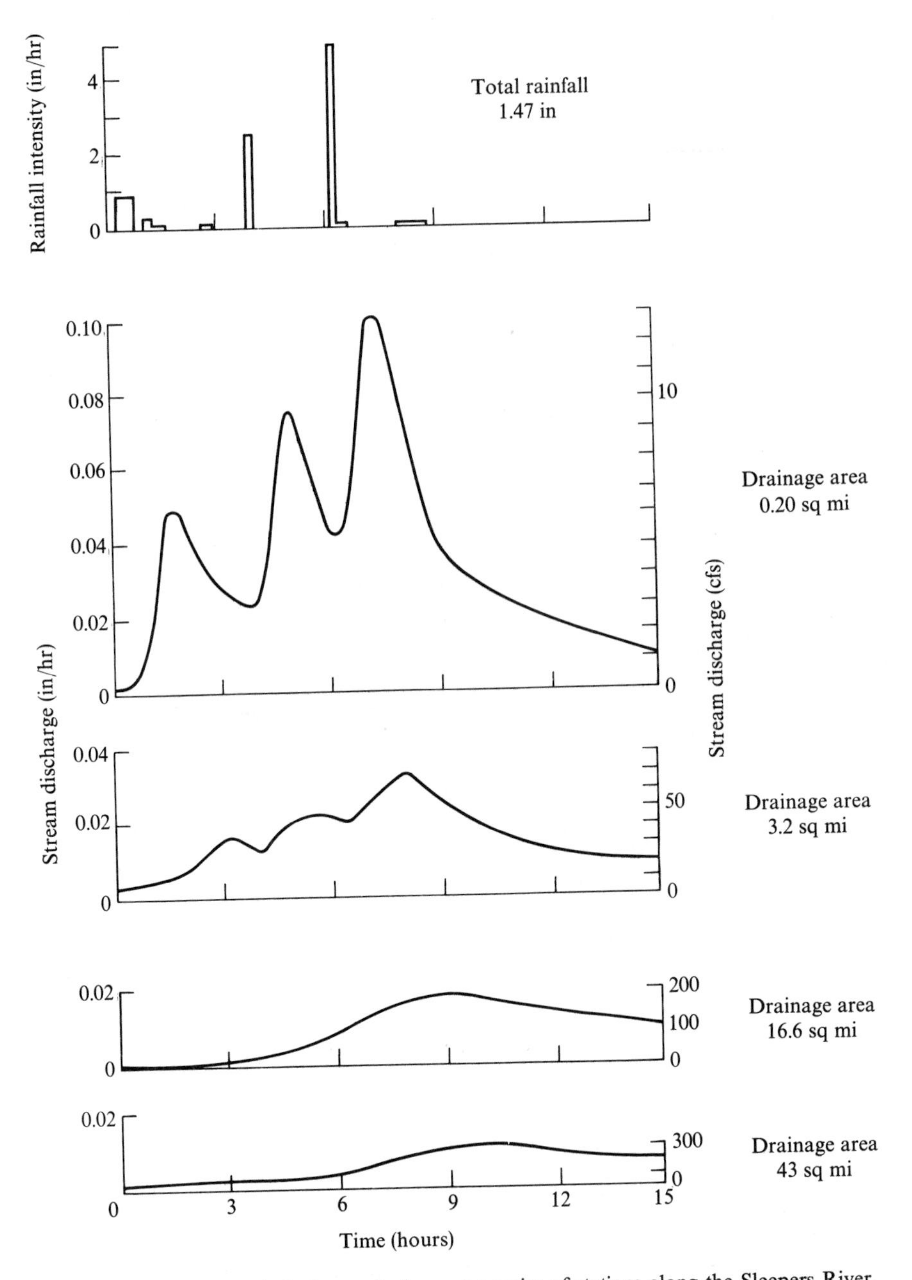

Figure 10-3 Changes in hydrograph shape at a series of stations along the Sleepers River, near Danville, Vermont. (Data from the Agricultural Research Service, U.S. Department of Agriculture.)

Flood Prediction

In predicting the occurrence of floods and the damage they are likely to produce, the hydrologist, engineer, or planner may need to assess one or more of the following features of a flood:

1. Volume of storm runoff, which is necessary for the design of storage works for flood control, water supply, and irrigation.

2. Peak flood discharge, which is necessary for designing bridges, culverts, spillways on dams, and other storage devices.

3. Flood height, to determine whether and where stream banks or artificial embankments will be overtopped, as well as the area of the valley bottom that will be flooded and the depth of such flooding. The depth of flooding is a critical determinant of the chances of survival of structures and crops and of the cost and feasibility of rendering structures resistant to flooding.

4. Time distribution of the whole storm hydrograph, which is necessary for determining the duration of inundation, for reservoir design, and for adding hydrographs together to assess the effects of various tributary inputs, and channel characteristics upon flood discharge at a downstream point. The rate of rise of a flood, in particular, affects the success of warning and evacuation procedures.

5. Area inundated; methods of determining this will be described in the following chapter.

6. Velocities of flow across the valley bottom, which affect the amount of damage done to structures and channel improvements or the amount of scouring of soil from agricultural land. We will discuss flow velocities in Chapter 16.

Flood Records

The best estimates of probable future floods require local information on past flooding, suitably adjusted for changes that may be occurring or that are projected for the basin in question. Information on flood discharges obtained at river-gauging stations like the one shown in Figure 16-2 is published by several agencies of the United States Government and is available in the files of state agencies, the engineering departments of local governments, conservancy districts, and other planning agencies. Records are kept by corresponding agencies in other countries. Local newspapers can be a useful source of historical information on the extent of major floods before the installation of stream gauges. Local officials and inhabitants of valley floors may provide historical information, though memories of floods are remarkably short; this information should be treated with caution and checked wherever possible.

The main sources of information on river discharges in the United States are the publications of the U.S. Geological Survey. Until 1960 the Survey published an annual series of Water Supply Papers entitled *Surface Water Supply of the United States,* which contain streamflow records from each of 14 regions of the United States. Since 1961 these publications have been replaced by one annual compilation for each state, published under the title *Water Resources Data for* (*Arizona*): *Part I, Surface Water Records,* and available from the local District Office of the Geological Survey. Both the Water Supply Papers and the state compilations contain mean daily discharges (in cubic feet per second), the maximum instantaneous flow for the year, and the corresponding water surface elevation (gauge height). Notes are also included on the location of the gauging station, the drainage area of the river basin, the extreme flows of record, and any artificial modifications of streamflow. The data on peak flows provide the means of constructing a rating curve (the relation of gauge height to discharge) and a flood-frequency curve.

Most of the gauging-station data referred to above are for large or intermediate streams. There is very little information on the flow of streams draining 10 square miles or less. This deficiency is being rectified, however, and so more data are becoming available for rivers of the size of interest in most planning problems.

Two federal agencies routinely collect good stream records from small catchments, and their data can be very useful to hydrologists, engineers, and planners. The Soil and Water Conservation Research Division of the Agricultural Research Service, U.S. Department of Agriculture, has established a network of approximately 70 experiment stations across the United States. At each of these stations, 5 to 30 catchments are gauged continuously, their drainage areas ranging from a fraction of an acre to more than 50 square miles. At each station office, detailed streamflow records are available for periods as short as a few minutes. In 1958, the Agricultural Research Service released a mimeographed compilation entitled *Annual Maximum Flows from Small Agricultural Watersheds in the United States.* It included annual maximum discharges and annual maximum volumes of runoff for time intervals of 1, 2, 6, and 12 hours, and 1, 2, and 8 days for 322 small catchments at 50 stations in 27 states from 1923 to 1957. Descriptions of the catchments and their land-use changes were released in a 1957 publication, *Monthly Precipitation and Runoff for Small Agricultural Watersheds in the United States.* More recent data on monthly precipitation and runoff, annual maximum discharges, annual maximum volumes, and typical hydrographs and rainfall intensities for selected storms are published for almost 200 catchments at 20 to 30 locations. These publications appear in the U.S. Department of Agriculture Miscellaneous Publication Series and are entitled *Hydrologic Data for Experimental Agricultural Watersheds in the United States.* Publication intervals are erratic, and the one for each year is released several years later, but this little-known series is a valuable one for the planner and others interested in floods. Not all the streamflow records from each experiment station are released, however, and the published information may be supplemented by applying to the director of each experiment station.

The U.S. Forest Service also maintains experiment stations, where small catchments are gauged on forest- and rangelands across the country. These results are not published.

A third source of flood information on small catchments are the engineering offices of the Bureau of Public Roads, of state highway departments, and of some large cities. Most of these drainage basins are urbanized, and though the data are not published routinely, they are of great value.

Hydrograph Separation

The stream hydrograph is a plot of discharge rate against time at some gauging station. It has a characteristic shape, the rising limb (period before the peak) being steeper than the falling limb (period after the peak). Because a storm hydrograph comprises both stormflow and baseflow, and we wish to examine only storm runoff here, it is necessary to have a technique for separating these two components. By examining hydrograph records it is possible to separate and discard the baseflow and analyze the volume, peak flow rate, and timing of storm runoff. To predict future flood hydrographs, the stormflow is first computed and then added to the baseflow, which is calculated separately.

The techniques of hydrograph separation are all arbitrary and have little or nothing to do with the processes by which stormflow is generated, but if one method is employed consistently, then usable results are obtained. Some of the accepted methods of hydrograph separation are illustrated in Figure 10-4. Whichever method is chosen, it should be checked against observed hydrographs from the catchment or the region, or at least against some qualitative field observations of the approximate duration of storm runoff.

Estimation of Storm Runoff Volume

The estimation of storm runoff volumes is required in planning certain engineering structures, such as reservoirs for water supply, flood detention, or power generation. It is also necessary for applying the unit hydrograph (see later) to the prediction of hydrographs from particular storms.

The simplest method of predicting the volume of storm runoff is by direct correlation with the volume of rainfall. An example of such a rainfall–runoff relation is shown in Figure 10-5. The scatter of points on such graphs is generally great because of differences in storm intensity and duration and in

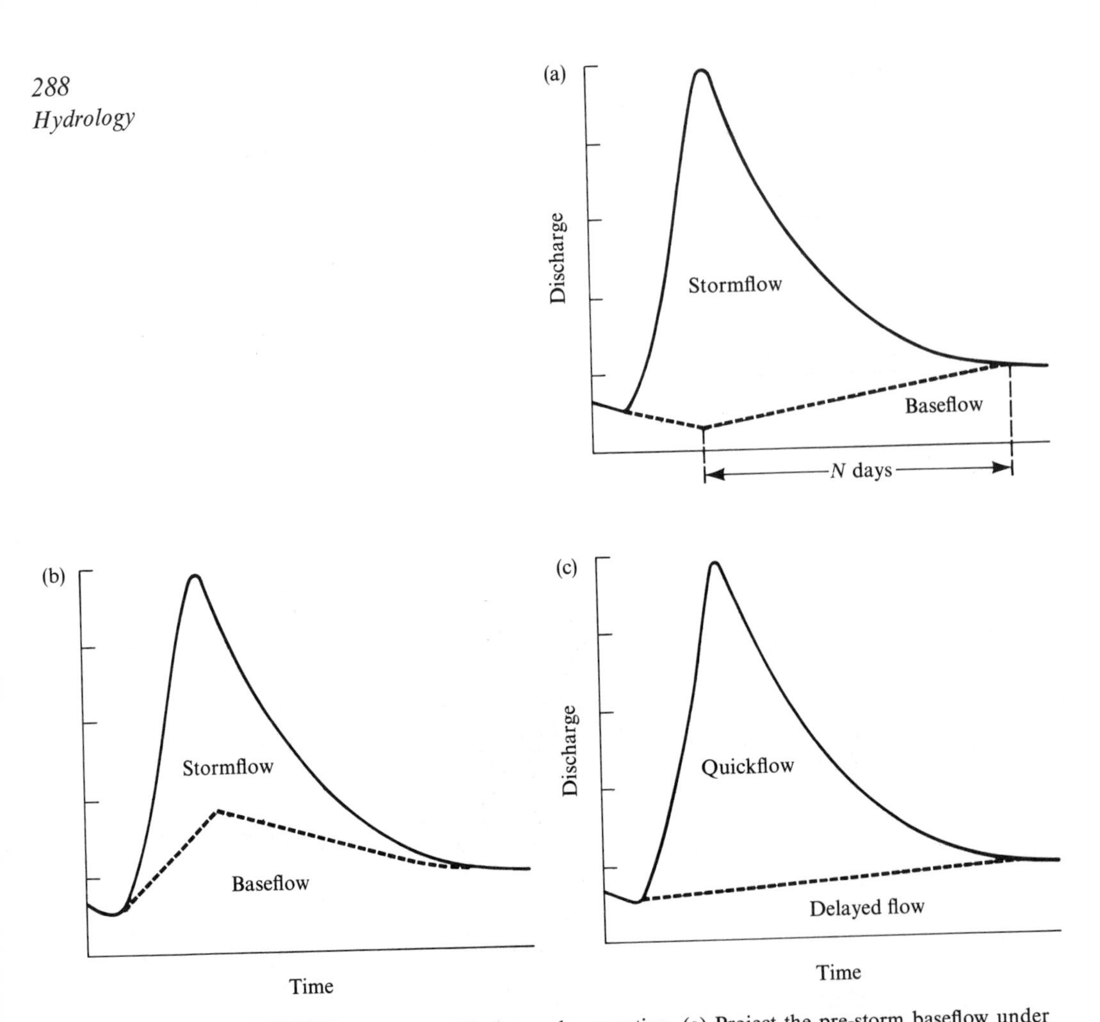

Figure 10-4 Methods of hydrograph separation. (a) Project the pre-storm baseflow under the peak. Draw the separation line rising from beneath the peak to a point on the recession limb that is N days after the peak, where N (days) $= A^{0.2}$ (sq mi). (b) Plot the hydrograph on semi-logarithmic paper with discharge on the logarithmic scale. Fit a straight line to the lower part of the recession limb on this paper and project it backward under the peak. Transfer the values on this line to arithmetic graph paper. Sketch a rising limb for the baseflow to meet the projected curve. (After Barnes 1939.) (c) From the point of initial rise, draw a line rising at a rate of 0.05 cfs per square mile of drainage basin per hour. For catchments smaller than 20 sq. mi. (After Hewlett and Hibbert 1967).

antecedent moisture conditions in the basin. Their predictive power can often be improved by incorporating an index of the antecedent wetness of the catchment. Such an index is often developed from the pre-storm baseflow (see Figure 10-6). In other applications the curves in Figure 10-6 could be labeled with values of an *antecedent precipitation index,* which indicates the effect of previous rainfall in wetting the soil and of natural drainage and

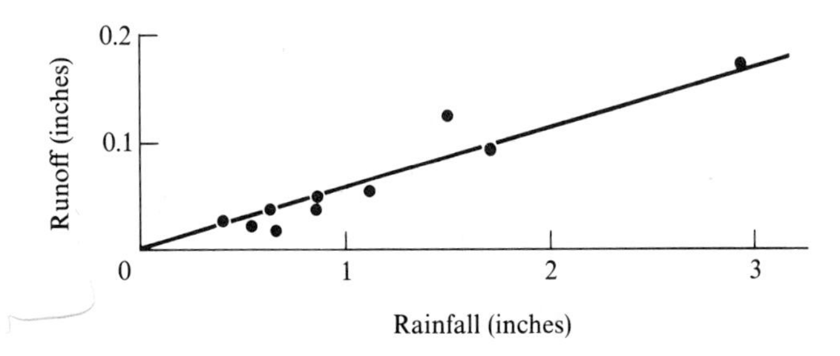

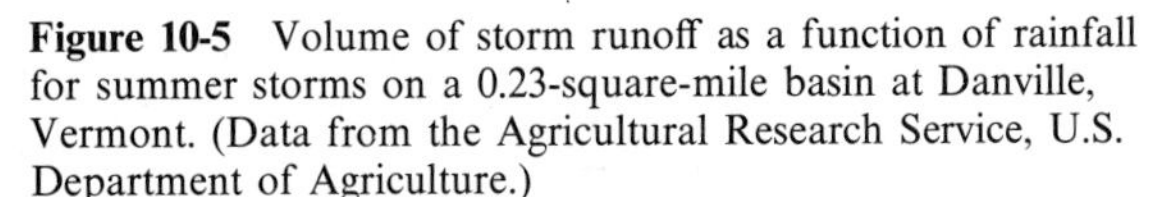

Figure 10-5 Volume of storm runoff as a function of rainfall for summer storms on a 0.23-square-mile basin at Danville, Vermont. (Data from the Agricultural Research Service, U.S. Department of Agriculture.)

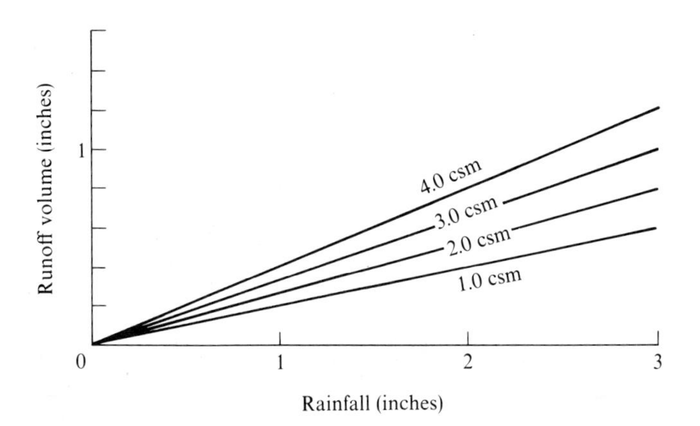

Figure 10-6 Relation between volume of rainfall and storm runoff for a range of antecedent moisture conditions, represented by baseflow. The family of lines represent antecedent baseflow in units of cubic feet per second per square mile (csm).

evapotranspiration in reducing the soil moisture at a logarithmically decreasing rate over time. The index is calculated from the equation:

$$I_t = I_0 k^t \tag{10-1}$$

where I_t and I_0 are values of the antecedent precipitation index on day t and at the beginning of the calculation period (inches); k is a constant usually varying between 0.85 and 0.95, indicating the rate of reduction of soil wetness; and t is the time (in days) since the last rainfall. On successive days the index will take on values of I_0, I_0k, I_0k^2, I_0k^3, and so on. Therefore the index for any day is obtained by keeping a running calculation in which the previous day's value is multiplied by k. If rain occurs on any day, the amount of rain (in inches) is added to the index, t is set equal to zero again, and the daily multiplication procedure is continued. The initial value of I_0 must be estimated and is often taken to be the amount of available moisture

(Chapter 6) in the soil profile on the first day of calculation. The procedure is not sensitive to variations in this estimate. The use of antecedent precipitation indices for predicting storm rainfall is covered in detail by Linsley and Kohler (1951).

Storm runoff volumes can also be estimated by subtracting from rainfall the volumes of infiltration, interception, and depression storage (see Chapters 3, 6, and 9). Under the usual conditions of fluctuating rainfall intensity, heterogeneous soil cover, and variable antecedent moisture, however, it is very difficult to apply the Horton infiltration model to this subtraction for areas larger than a few acres. This technique does not take into account subsurface stormflow or saturation overland flow. There have been some attempts to generate hydrographs by modeling the whole runoff process (Crawford and Linsley 1966), but the techniques are time-consuming and beyond the scope of most small-scale planning problems. Instead, it is common to subtract a value from the storm rainfall that represents the sum of all the processes that retain water during a storm. Such a value is the Φ index, which is defined as the amount of rainfall that is retained by the basin divided by the duration of the rainstorm.

Values of Φ can be obtained from the rainfall and runoff records of past storms in the basin of interest or in a nearby similar basin. Results from a number of basins can be regionalized, as shown by Rantz (1971), for rural catchments around the San Francisco Bay, but his values seem very low when compared with our own field experience. Figure 10-7 shows Rantz's results, indicating that Φ is a function of both the relative magnitude of the storm and the mean annual precipitation (which represents the effects of vegetation upon processes such as interception, infiltration, and surface retention). We present Figure 10-7 only as an indication of one method of

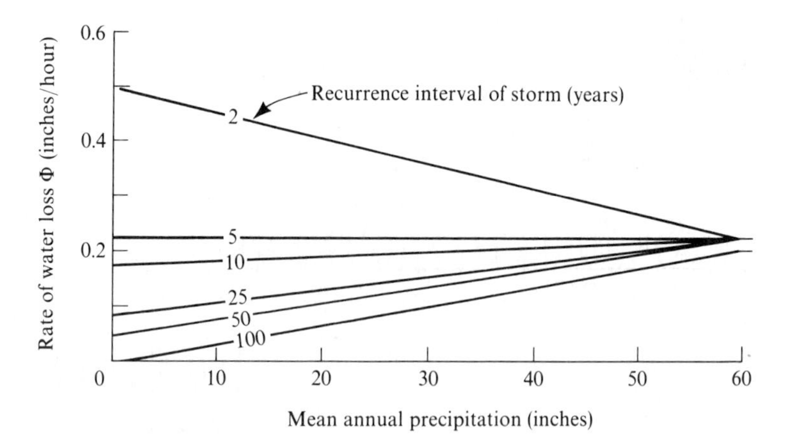

Figure 10-7 Relation of Φ index to mean annual rainfall for storms of various recurrence intervals in the San Francisco Bay region. (From Rantz 1971.)

regionalizing Φ values. For other regions, Φ values should be estimated from a few measured hydrographs, as described in Typical Problem 10-2.

In urban regions, storm runoff volumes are higher than in rural catchments, and to quantify this effect Rantz suggested coefficients by which the rural Φ values should be multiplied for various degrees of urbanization (see Table 10-2). This simplification should be approximately applicable to other urbanized areas.

Table 10-2 Coefficients to convert Φ values for rural catchments to those for urbanized catchments. (From Rantz 1971.)

PERCENTAGE OF CATCHMENT THAT IS URBANIZED*	COEFFICIENT
0	1.00
10	0.95
20	0.90
30	0.85
40	0.80
50	0.75
60	0.70
70	0.65
80	0.60
90	0.55
100	0.50

*Urbanization and impervious cover are not synonymous as used here. "100% urbanized" is roughly equivalent to 50% of the area having an impervious cover.

The U.S. Soil Conservation Service has developed a method for estimating storm runoff volumes from small agricultural catchments with various kinds of soil and land use. The technique is based on a simplified infiltration model of runoff and a good deal of empirical approximation. The necessary graphs for calculating runoff are presented in Figure 10-8. For each catchment and storm, a *curve number* is chosen for use in the diagrams. The curve number is an empirical rating of the hydrologic performance of a large number of soils and vegetative covers throughout the United States.

Runoff curve numbers for various combinations of soil, cover, and land-use practice can be read from Table 10-3. The hydrologic soil groups and cover types are defined in Tables 10-4 and 10-5. Major soils of the United States have been classified into the hydrologic groups described in Table

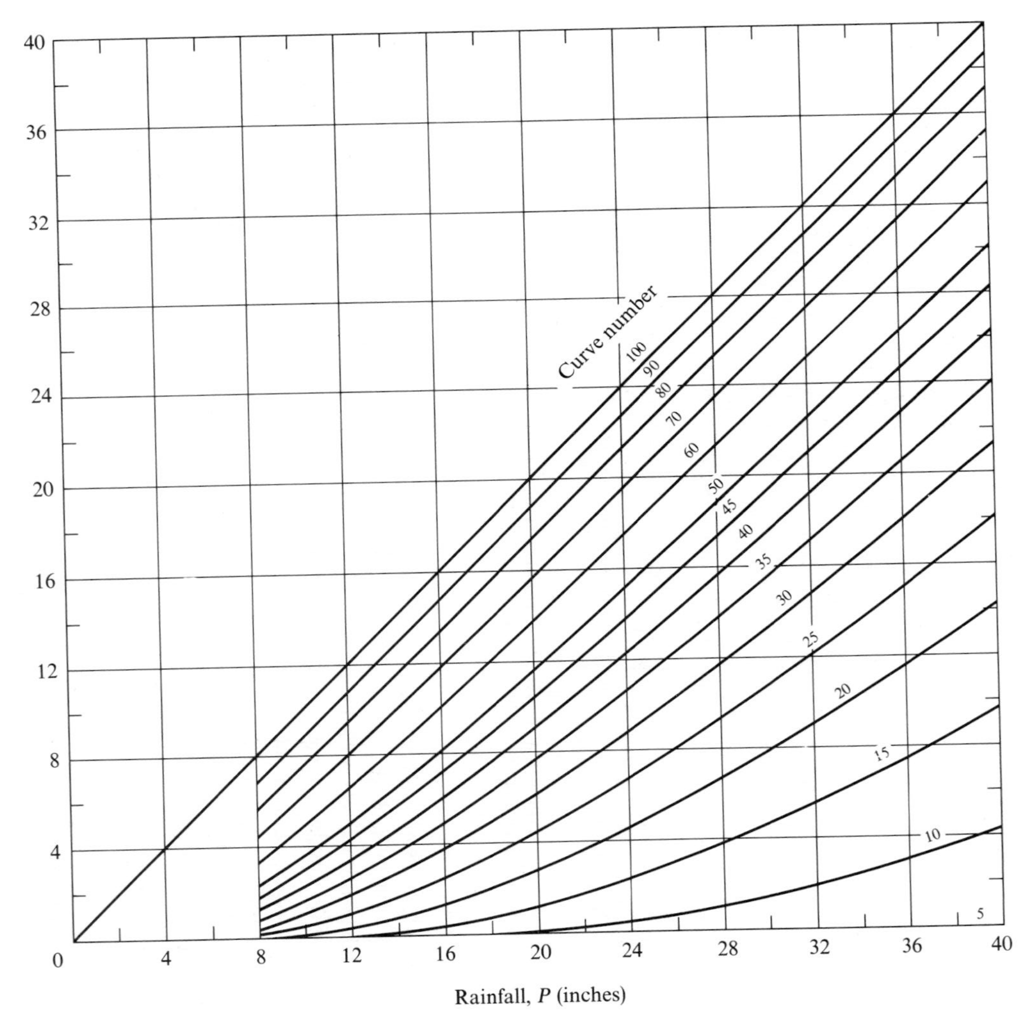

Figure 10-8 Charts for estimating the volume of storm runoff from rainfall for various hydrologic soil-cover complexes indicated by the curve numbers. Use the curves on the opposite page for rainfall less than 8 inches. (From U.S. Soil Conservation Service 1972.)

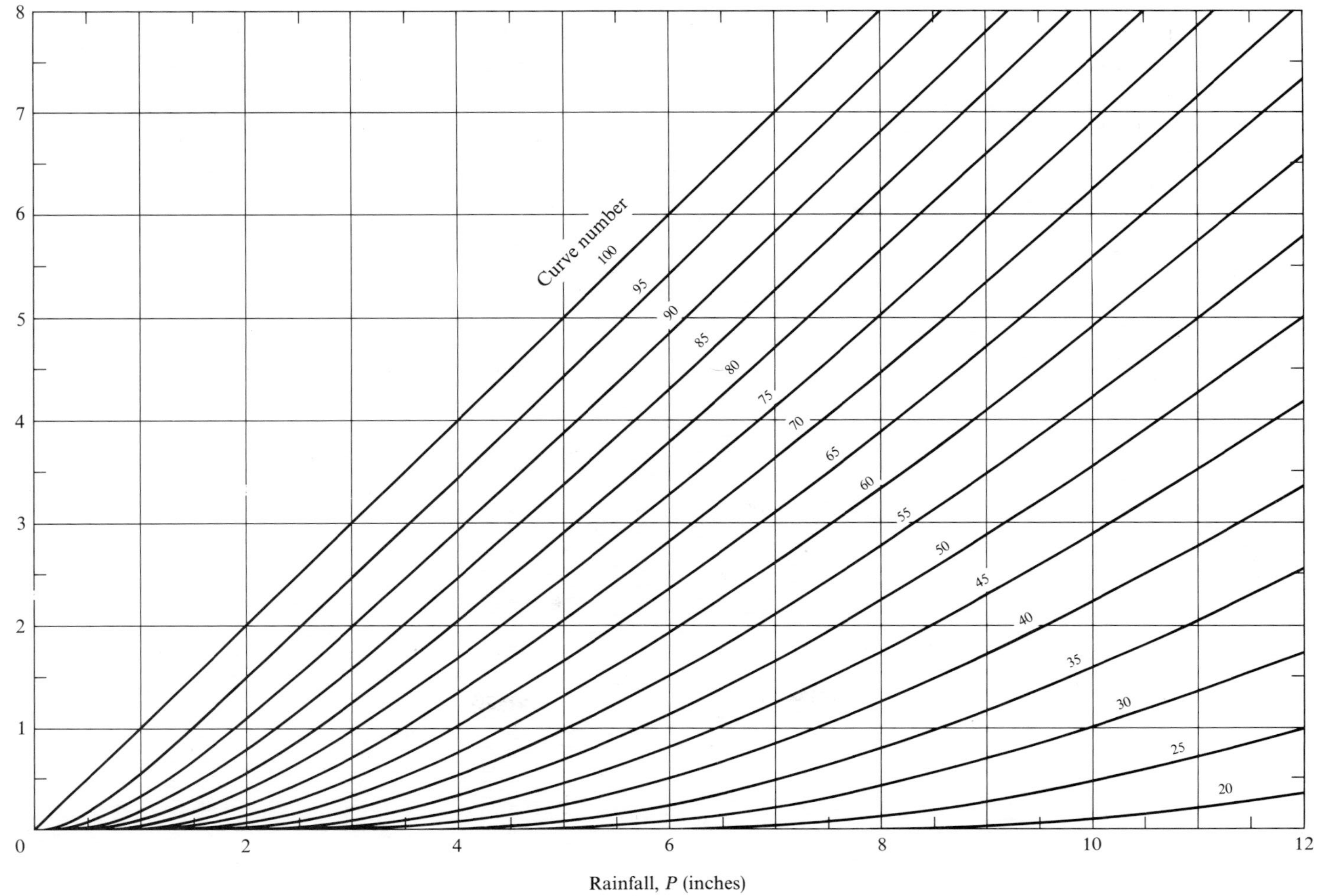

Storm runoff, Q (inches)
Rainfall, P (inches)
Curve number
100
95
90
85
80
75
70
65
60
55
50
45
40
35
30
25
20
0
2
4
6
8
10
12
1
3
5
7

10-4, and are listed in a U.S. Soil Conservation Service (1972) handbook and in county Soil Survey Reports. Soils in other countries can be similarly classified.

The curve numbers in Table 10-3 apply to average soil moisture conditions. To make runoff estimates for drier or wetter conditions requires the use of Table 10-6, as follows: The antecedent moisture levels are classified into three groups in Table 10-6 on the basis of total precipitation occurring

Table 10-3 Runoff curve numbers for hydrologic soil–cover complexes under average conditions of antecedent moisture. (From U.S. Soil Conservation Service 1972.)

LAND USE OR COVER	TREATMENT OR PRACTICE	HYDROLOGIC CONDITION	HYDROLOGIC SOIL GROUP			
			A	B	C	D
Fallow	Straight row	Poor	77	86	91	94
Row crops	Straight row	Poor	72	81	88	91
	Straight row	Good	67	78	85	89
	Contoured	Poor	70	79	84	88
	Contoured	Good	65	75	82	86
	Contoured and terraced	Poor	66	74	80	82
	Contoured and terraced	Good	62	71	78	81
Small grain	Straight row	Poor	65	76	84	88
	Straight row	Good	63	75	83	87
	Contoured	Poor	63	74	82	85
	Contoured	Good	61	73	81	84
	Contoured and terraced	Poor	61	72	79	82
	Contoured and terraced	Good	59	70	78	81
Close-seeded legumes or rotation meadow	Straight row	Poor	66	77	85	89
	Straight row	Good	58	72	81	85
	Contoured	Poor	64	75	83	85
	Contoured	Good	55	69	78	83
	Contoured and terraced	Poor	63	73	80	83
	Contoured and terraced	Good	51	67	76	80
Pasture or range		Poor	68	79	86	89
		Fair	49	69	79	84
		Good	39	61	74	80
	Contoured	Poor	47	67	81	88
	Contoured	Fair	25	59	75	83
	Contoured	Good	6	35	70	79
Meadow (permanent)		Good	30	58	71	78
Woodlands (farm woodlots)		Poor	45	66	77	83
		Fair	36	60	73	79
		Good	25	55	70	77
Farmsteads			59	74	82	86
Roads, dirt			72	82	87	89
Roads, hard-surface			74	84	90	92

Table 10-4 Classification of soils by their hydrologic properties. (From U.S. Soil Conservation Service 1972.)

CLASSIFICATION	TYPE OF SOIL
A (low runoff potential)	Soils with high infiltration capacities, even when thoroughly wetted. Chiefly sands and gravels, deep and well drained.
B	Soils with moderate infiltration rates when thoroughly wetted. Moderately deep to deep, moderately well to well drained, with moderately fine to moderately coarse textures.
C	Soils with slow infiltration rates when thoroughly wetted. Usually have a layer that impedes vertical drainage, or have a moderately fine to fine texture.
D (high runoff potential)	Soils with very slow infiltration rates when thoroughly wetted. Chiefly clays with a high swelling potential; soils with a high permanent water table; soils with a clay layer at or near the surface; shallow soils over nearly impervious materials.

Table 10-5 Classification of vegetative covers by their hydrologic properties. (From U.S. Soil Conservation Service 1972.)

VEGETATIVE COVER	HYDROLOGIC CONDITION
Crop rotation	Poor: Contain a high proportion of row crops, small grains, and fallow.
	Good: Contain a high proportion of alfalfa and grasses.
Native pasture or range	Poor: Heavily grazed or having plant cover on less than 50% of the area.
	Fair: Moderately grazed; 50–75% plant cover.
	Good: Lightly grazed; more than 75% plant cover.
	Permanent Meadow: 100% grass cover.
Woodlands	Poor: Heavily grazed or regularly burned so that litter, small trees, and brush are destroyed.
	Fair: Grazed but not burned; there may be some litter.
	Good: Protected from grazing so that litter and shrubs cover the soil.

within the preceding 5 days. The curve numbers of Table 10-3 refer to *antecedent moisture condition II* in Table 10-6. Find the curve number for this condition, and if the antecedent wetness is less or greater than average, convert the curve number to the value for condition I or III, as shown in columns 2 and 3 of Table 10-7. If a storm continues for several days, the

Table 10-6 Rainfall limits for estimating antecedent moisture conditions. (From U.S. Soil Conservation Service 1972.)

ANTECEDENT MOISTURE CONDITION CLASS	5-DAY TOTAL ANTECEDENT RAINFALL (INCHES)	
	DORMANT SEASON	GROWING SEASON
I	Less than 0.5	Less than 1.4
II	0.5–1.1	1.4–2.1
III	Over 1.1	Over 2.1

Table 10-7 Conversion of runoff curve numbers (*CN*) for antecedent moisture condition II to those for conditions I (dry) and III (wet). (From U.S. Soil Conservation Service 1972.)

CN FOR ANTECEDENT MOISTURE CONDITION	*CN* FOR ANTECEDENT MOISTURE CONDITION	
II	I	III
100	100	100
95	87	98
90	78	96
85	70	94
80	63	91
75	56	88
70	51	85
65	45	82
60	40	78
55	35	74
50	31	70
45	26	65
40	22	60
35	18	55
30	15	50
25	12	43
20	9	37
15	6	30
10	4	22
5	2	13

rainfall should be broken down into daily totals, the antecedent moisture class changed daily and Figure 10-8 used for each day's precipitation.

It is worth pointing out in closing this section on storm volumes, that most of the computational techniques described above assume that the major storm runoff process is Horton overland flow. We have described other processes in Chapter 9. The techniques still seem to work under other runoff conditions, presumably because the major variables (rainfall, antecedent moisture, soil conditions, and topography) function in the same direction to control the magnitude of stormflow, whatever the runoff process.

The Soil Conservation Service method has recently been extended to apply to small urbanized catchments, for which curve numbers can be estimated from Table 10-8. A weighted average curve number can be computed using the proportions of each land-use type, as illustrated in Typical Problem 10-4. Figure 10-9 facilitates the computation of composite curve numbers for differing amounts of impervious area in a catchment whose curve number in the undeveloped state is first estimated from soils and land use.

Table 10-8 Runoff curve numbers for urban and suburban land use for antecedent moisture condition II. (From U.S. Soil Conservation Service 1975.)

		HYDROLOGIC SOIL GROUP			
LAND USE		A	B	C	D
Open spaces, lawns, parks, golf courses, cemeteries, etc.					
good condition: grass cover on 75% or more of the area		39	61	74	80
fair condition: grass cover on 50% to 75% of the area		49	69	79	84
Commercial and business area (85% impervious)		89	92	94	95
Industrial districts (72% impervious)		81	88	91	93
Residential*					
Average lot size	Average % Impervious†				
1/8 acre or less	65	77	85	90	92
1/4 acre	38	61	75	83	87
1/3 acre	30	57	72	81	86
1/2 acre	25	54	70	80	85
1 acre	20	51	68	79	84
Paved parking lots, roofs, driveways, etc.‡		98	98	98	98
Streets and roads					
Paved with curbs and storm sewers‡		98	98	98	98
Gravel		76	85	89	91
Dirt		72	82	87	89

*Curve numbers are computed assuming the runoff from the house and driveway is directed toward the street with a minimum of roof water directed to lawns where additional infiltration could occur.

†The remaining pervious areas (lawn) are considered to be in good pasture condition for these curve numbers.

‡In some warmer climates of the country a curve number of 95 may be used.

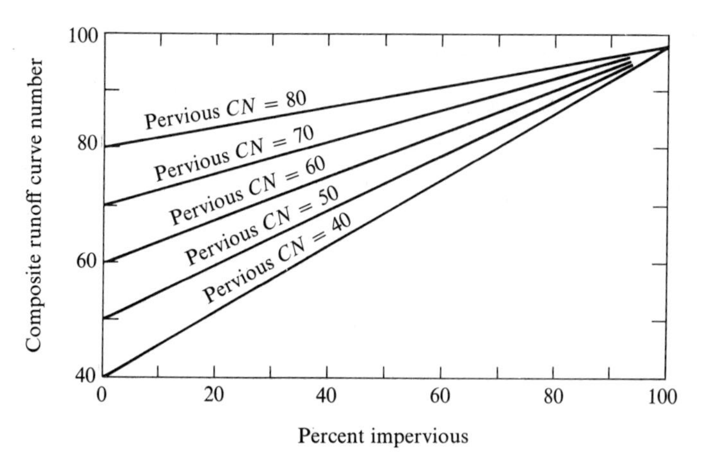

Figure 10-9 Composite runoff curve numbers for various combinations of impervious area and curve number for the remaining unpaved area. (From U.S. Soil Conservation Service 1975.)

Calculating Flood Peak Discharges

In spite of what may seem a large number of sources of flood information, when one is trying to obtain local data for a design problem there is usually no appropriate streamflow record. Peak discharges must be estimated from the size of rainstorms to be expected and from the characteristics of the catchment. Whenever this is done, the results should be checked against field information such as the size of the stream channel, historical flood data, or the flood peaks from a gauged catchment, which, though different from the one in question, is known to produce larger, similar, or smaller flood peaks.

In some situations, the planner must estimate the size of a future flood at a station for which there is a streamflow record; in others a prediction must be made for an ungauged site. Because the methods of flood prediction from gauged and ungauged catchments are intertwined, they will be treated here in order of increasing sophistication under the following headings: The rational method, Probability analysis, Unit hydrograph methods, and Flood routing (coupled with the other methods).

The Rational Method

The *rational runoff method* predicts peak runoff rates from data on rainfall intensity and drainage-basin characteristics. Ideally, it should be used only for catchments of less than 200 acres, but it is frequently used for basins of up to one square mile and is a widely accepted method for the design of storm sewers. The method assumes that a rainstorm of uniform intensity covers the whole basin. Runoff will increase as water from more and more distant parts of the catchment reaches the outlet. When the whole drainage area is contributing, a steady state is reached, and discharge becomes a constant maximum. The time required to reach this steady state is called the *time of concentration* of the basin, and after this time, stormflow discharge is a fixed proportion of the rainfall intensity, and is equal to

$$Q_{pk} = CIA \tag{10-2a}$$

where in English units, Q_{pk} is the peak rate of runoff (cfs), C is the rational runoff coefficient, I is the rainfall intensity (in/hr), and A is the drainage area (acres). If these units are used, the peak discharge is calculated in cfs because 1 in/hr of runoff from 1 acre is approximately equal to 1 cfs.

If metric units are to be used, the formula becomes

$$Q_{pk} = 0.278CIA \tag{10-2b}$$

where Q_{pk} is in m^3/sec, I is in mm/hr, and A is in km^2.

Some accepted values of C are listed in Table 10-9. These values reflect soil type, topography, surface roughness, vegetation, and land use, and they are usually assumed to remain approximately constant during and between large storms for a given basin. The method is most commonly applied to small urban catchments, and so the majority of published C values relate to urban conditions. If there are important variations of topography, soil, or vegetation within the basin, a weighted average value of C is obtained by weighting the coefficients from each area according to the proportion of the total area they occupy. If a lake or reservoir lies within the catchment, the C values do not apply and some other method of flood prediction must be used.

The values for urban conditions given by the American Society of Civil Engineers in Table 10-9 are for storms with recurrence intervals of 5 to 10 years, and should be adjusted upward for larger storms. A brief survey of data from small forested mountainous catchments, for example, shows that for sandy-loam soils, the value of C can be as high as 0.40 to 0.50 for long storms with a recurrence interval of 100 years. Rantz (1971) takes this factor into account in urban regions by providing a relationship (Figure 10-10) between the percentage of impervious surface in a catchment and the appropriate value of C for a range of recurrence intervals. If the planner has information only on the average lot size in a proposed development, he can estimate the amount of impervious area from Figure 10-11 and use this value in Figure 10-10. If the amount of impervious area is to be directly evaluated for the study, it can be obtained by counting houses and measuring road widths and lengths on topographic maps or air photographs or by a field survey on sample areas.

The appropriate rainfall intensity (I) is chosen with reference to the recurrence interval of the storm to be designed for. The method assumes that

Table 10-9 Values of the rational runoff coefficient, C. (From American Society of Civil Engineers 1969, Rantz 1971, and elsewhere.)

	C
URBAN AREAS	
Streets: asphalt	0.70–0.95
concrete	0.80–0.95
brick	0.70–0.85
Drives and walks	0.75–0.85
Roofs	0.75–0.95
Lawns: sandy soil, gradient $\leq$ 2%	0.05–0.10
sandy soil, gradient $\geq$ 7%	0.15–0.20
heavy soil, gradient $\leq$ 2%	0.13–0.17
heavy soil, gradient $\geq$ 7%	0.25–0.35
The values listed above can be used, together with areas of each type of surface measured from a map or aerial photograph, to compute weighted average values of C. Alternatively, the following overall values apply to most North American urban areas.	
Business areas: high-value districts	0.75–0.95
neighborhood districts	0.50–0.70
Residential areas: single-family dwellings	0.30–0.50
multiple-family dwellings, detached	0.40–0.60
multiple-family dwellings, attached	0.60–0.75
suburban	0.25–0.40
apartment buildings	0.50–0.70
Industrial areas: light	0.50–0.80
heavy	0.60–0.90
Parks and cemeteries	0.10–0.25
Playgrounds	0.20–0.35
Unimproved land	0.10–0.30
RURAL AREAS	
Sandy and gravelly soils: cultivated	0.20
pasture	0.15
woodland	0.10
Loams and similar soils without impeding horizons: cultivated	0.40
pasture	0.35
woodland	0.30
Heavy clay soils or those with a shallow impeding horizon; shallow soils over bedrock: cultivated	0.50
pasture	0.45
woodland	0.40

the recurrence interval of a flood peak is the same as that of the rainfall that caused it. This is not strictly true, especially for rural areas where the recurrence interval of a flood is slightly greater than that of the rainfall. The duration of the design storm is taken as the time of concentration of the basin, i.e., the time required for overland and channel flow to reach the basin

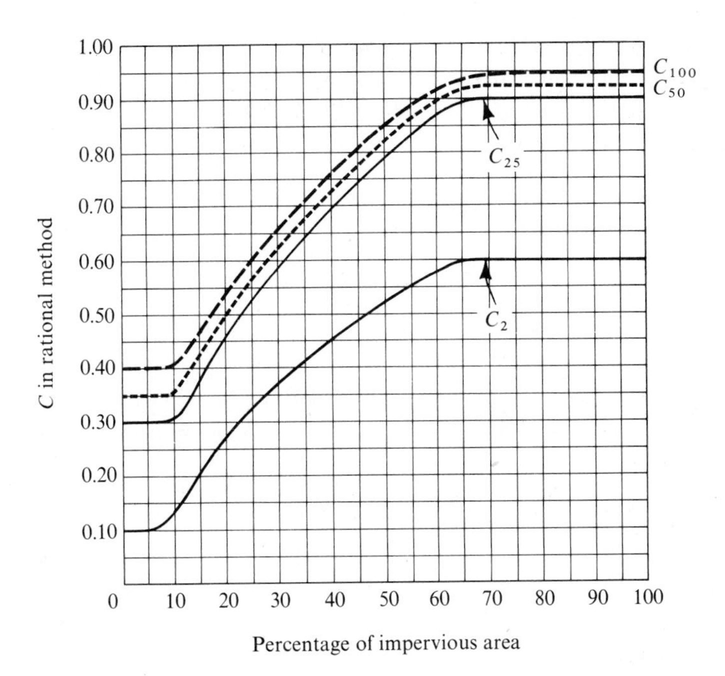

Figure 10-10 Relation of C in the Rational Formula to percentage of impervious area. Curves labeled C_2 to C_{100} refer to recurrence intervals, and values for other recurrence intervals can be interpolated from them. (From Rantz 1971.)

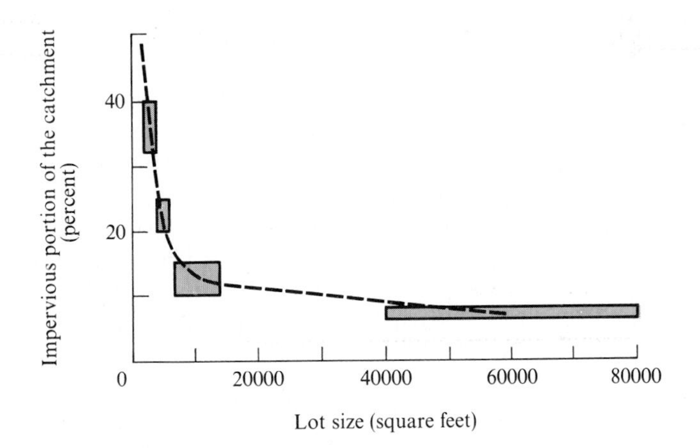

Figure 10-11 Variation of the impervious area with lot size in urban areas. (Data from American Society of Civil Engineers 1969; Rantz 1971.)

outlet from the hydraulically most distant part of the catchment. If the duration of a rainstorm does not equal or exceed the concentration time, the rational method will overestimate the flood peak.

The time of concentration of the catchment can be estimated in various ways. Statistical studies of the time of concentration on small agricultural basins has led to the development of the following formula (U.S. Soil Conservation Service 1972):

$$t_c = \frac{L^{1.15}}{7700H^{0.38}} \tag{10-3}$$

where t_c is the time of concentration (hr), L is the length of the catchment along the mainstream from the basin outlet to the most distant ridge (ft), and H is the difference in elevation between the basin outlet and the most distant ridge (ft).

As an independent check, it is wise to assess the time of concentration from estimates of the velocities of overland flow and channel flow. If the overland flow traverses more than one kind of surface, the travel times across them should be added up. For paved areas, Jens and McPherson (1964) recommend velocities of 0.33 ft/sec for hillslopes of the order of 100 feet in length, ranging up to 0.82 ft/sec for 500-foot-long areas. For turf they recommend velocities of less than 0.2 ft/sec for the shorter slopes, and 0.25 ft/sec for the longer. Bare areas should be intermediate between these values, depending on surface roughness. Emmett (1970) measured velocities of overland flow ranging from 0.02 to 0.05 ft/sec for rangeland hillslopes. Channel velocities can also be computed from the Manning Equation (see Chapter 16), but regardless of the computed value, one should not accept values greater than 8 ft/sec for small artificial channels and 6 ft/sec for small natural channels of the kind likely to be encountered in planning problems.

In urban areas, constant values of concentration time for overland flow on paved areas are often used, as indicated in Table 10-10. Rantz (1971) presented Figure 10-12 for deriving overland travel time as a function of hillslope length and gradient and the C value for the catchment (as an index of the amount and, therefore, depth of overland flow).

The assumptions of the rational method are not well met in practice, but the method has gained popularity because it gives usable results and because it allows one to assess the probable impact of future land-use changes. The method works best for urban and suburban areas and other

Table 10-10 Constant times of concentration for overland flow used in urban hydrology.

TYPE OF CATCHMENT	t_c (MIN)
Individual parking lots, yards, and streets with closely spaced drains	≤5
Areas of commercial development on gentle slopes and greater drain spacing	10–15
Flat residential districts with few drains	20–30

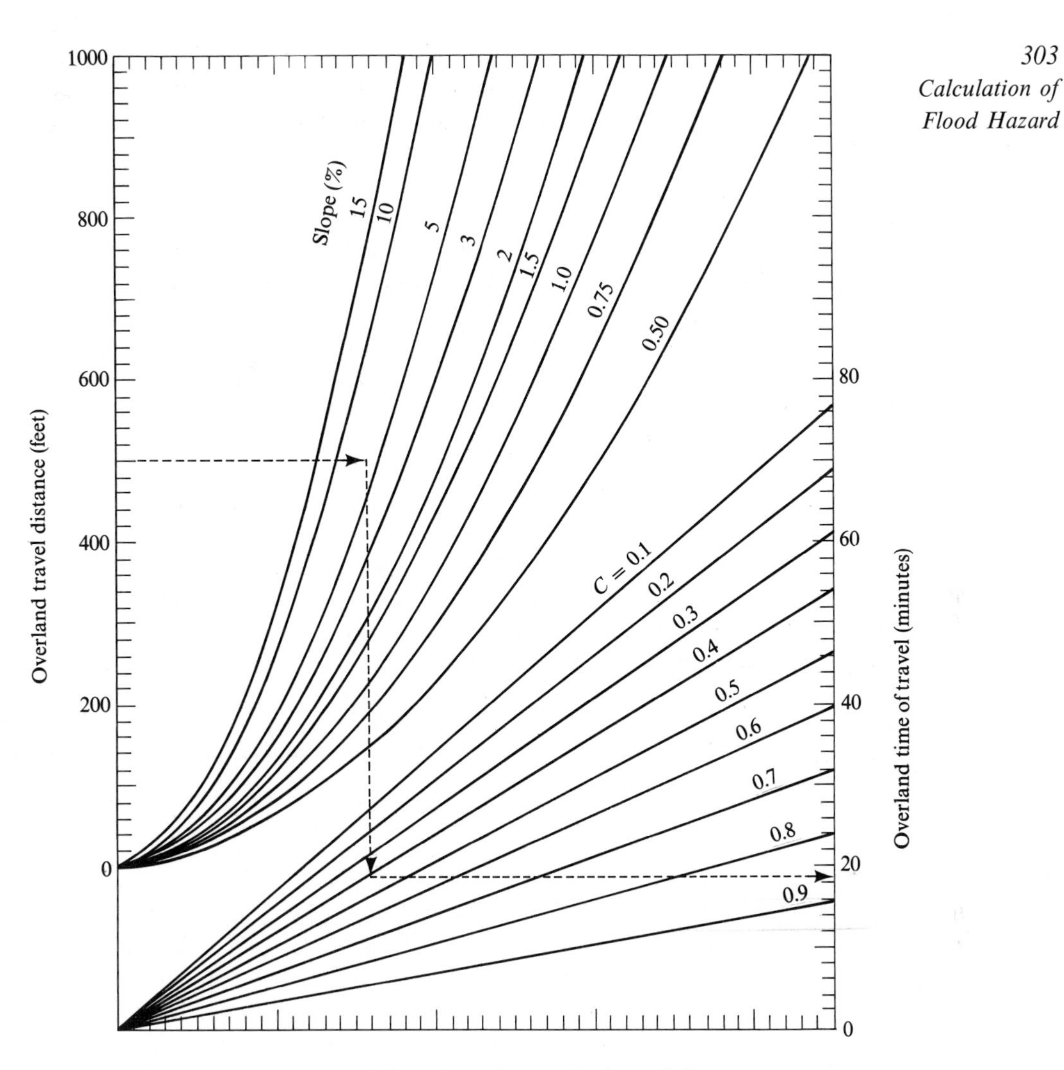

Figure 10-12 Relation of overland time of travel to overland travel distance, average overland slope, and the rational runoff coefficient, *C*. (From Rantz 1971.)

areas with high runoff rates, moderately steep channels, limited channel storage, and no lakes. It assumes that the generation process is Horton overland flow with the whole catchment contributing. In practice, values of *C* have been obtained by measuring peak runoff rates and dividing by the product of rainfall intensity and drainage area. This method is appropriate for uniform Horton overland flow.

The *C* values given in Table 10-9 for forested and other well-vegetated areas, however, have a different meaning. Here most of the catchment does not yield any stormflow while a small proportion of the area is contributing a high percentage of its rainfall. Exactly what this percentage is, we cannot generalize at present, but 100 percent seems to be the best estimate, particularly during high intensity storms when there is insufficient time for large subsurface or return-flow contributions. In this case C in Equation 10-2 becomes equal to 1.0 and

$$Q_{pk} = IA_s \tag{10-4}$$

where A_s, the saturated area (acres), is a function of the size and duration of the storm, topography, soils, and antecedent moisture conditions. The problem then becomes one of choosing the appropriate contributing area. Unfortunately there are few data on this subject at present. The values of C for forests in Table 10-9 presumably represent average values of the proportion of the area contributing runoff in large storms. Figures 9-11 and 9-13 suggest that in Vermont the maximum size of the contributing area varies from about 10 percent of the whole catchment in basins with limited valley floors and steep sideslopes mantled with deep, well-drained soils to about 50 percent in catchments with gentle sideslopes, thin, moderate-to-poorly-drained soils, and extensive valley bottoms. After a period of dry weather, the contributing area in well-drained watersheds may cover only 2 to 5 percent of the catchment. Dunne et al. (1975) treat the recognition and prediction of contributing areas in greater detail.

When the rational method is applied to variable-source contributions, the meaning of the time of concentration must also change, and it presumably means the time required for water to travel from the hydraulically most distant contributing area to the basin outlet. Since statistical studies of this delay have not yet been made, we can only mention a few of our own field observations. First, the velocities of saturation overland flow are generally much lower than those of Horton overland flow, because the contributing areas tend to have low gradients; dense, grassy, shrubby, or marshy vegetation; and rough surfaces. Only on a slope of 40 percent have we measured velocities of 0.1 to 0.5 ft/sec. Elsewhere, 0.1 ft/sec seems to be a maximum achieved only in large storms. Flow velocities in the small, winding channels that drain these marshy areas vary from 0.5 to 1.0 ft/sec as channel size increases. Because of the generally longer times of concentration and the greater detention storage of overland flow on rough contributing areas, the rational method will probably overestimate peak rates of variable-source runoff, but this will be partly offset by the fact that return flow and subsurface stormflow have been overlooked. Until more field measurements become available, this is the best approximation that can be made. An approximate method for calculating peak rates of variable-source runoff is described by Dunne et al. (1975).

As a cautionary note to potential users of the rational formula, Jens and McPherson (1964) present some data on the precision of peak runoff estimates by this method (see Table 10-11). These values indicate the level of variability in answers obtained for urban areas. Peak flows calculated by the rational method were consistently low.

Table 10-11 Comparison of measured peak discharges and those calculated by the rational formula. (From Jens and McPherson, *Handbook of Applied Hydrology,* edited by Ven te Chow. Copyright © 1964 by McGraw-Hill, Inc. Used with permission of McGraw-Hill Book Company.)

DRAINAGE AREA	TOTAL NUMBER OF ESTIMATES	NUMBER OF ESTIMATES FALLING OUTSIDE 20% LIMITS	MEAN ABSOLUTE DEVIATION (%)
Baltimore 1	25	17	27
Baltimore 2	19	6	17
Baltimore 3	4	3	55
Baltimore 4	7	5	34
Baltimore 5	4	2	19
St. Louis	3	3	79
Los Angeles	1	1	23
Hertfordshire, England	3	1	22
	66	38	Mean = 34

Probability Analysis of Flood Records

A statement of the probability of floods greater than certain limits (or their average frequency of occurrence) is the basis of much planning that concerns river channels and valley floors. Such information is required for engineering design, planning flood-insurance schemes, and land-use zoning of flood-prone areas.

The concepts and methods of probability analysis have already been introduced in the chapter on rainfall intensity. There are two sets of problems to be reviewed. The first arises when a record of floods exists for a station in the reach of the river in question. The second arises when there is no flood record for the particular site, but when other stations in the same area provide records that can be regionalized and applied to the prediction of floods at the ungauged site.

The general method of analysis is the same as for rainfall intensity: either the annual-maximum series or the partial-duration series is used in the analysis, the difference between results obtained from either series being essentially the same for recurrence intervals beyond 10 years (see Table 2-2). Momentary peak discharges should be used rather than average daily discharges, except for large rivers where the two are nearly identical. A probability distribution is fitted to a sample of floods observed at a gauging site, and the estimated parameters of the distribution are then used to predict the average recurrence intervals of floods of chosen magnitudes or the magnitudes of events of chosen frequencies at the site.

Several theoretical probability distributions are commonly used for fitting the observed sample distributions of annual maximum floods. They are:

1. The lognormal distribution
2. The Gumbel Type I extreme-value distribution, as used in Chapter 2 for rainfall
3. The Gumbel Type III extreme-value distribution (a logarithmic transformation of Gumbel Type I)
4. The Pearson Type III distribution.

The only criteria for choosing one of the four are convenience and goodness-of-fit. The Pearson Type III distribution is slightly more complicated to use and, although increasingly in use by U.S. federal agencies, will not be treated here. The method is clearly described by Benson (1971). We have not found its results to be any better, or even very different, from those of other methods.

The use of each of the other distributions is simplified by the availability of various graph papers, the scales of which are designed so that the cumulative frequency curve of the field values plots as a straight line if the observed data fit the appropriate theoretical frequency curve. Fitting a straight line to a set of plotted points aids in the comparison of curves between stations and in extrapolation.

Figures 10-13 to 10-15 are examples of the probability papers in most common use in flood studies. A cumulative frequency curve of observed floods on the Tana River at Garissa, Kenya is plotted on each of the papers for comparison. On each graph, the plotting formula

$$T = \frac{n + 1}{m} \tag{2-4}$$

is used. The line drawn through the plotted points is the *flood-frequency curve* for the station. We feel that graphical curve-fitting is preferable to the analytical computation of a best-fit line by computer alone, because it allows one to see whether an individual flood lies well off the general trend defined by all other floods. Such an outlying point can cause the computed best-fit line to be shifted into an unrepresentative position. It is not unusual for a record to include an extreme flood whose recurrence interval is much greater than the length of the record. Because of the uncertainties inherent in a short record, however, it is not wise to move the extreme point to the right until it falls on the curve unless there is some independent confirmation of the true recurrence interval.

Comparison of Figures 10-13 to 10-15 shows that there is little difference between flood magnitudes predicted for various frequencies within the observed range of the data. Extrapolation of the lines drawn on the different papers, however, can lead to widely different estimates of the magnitudes of rare floods.

Such variability brings up the whole question of the validity of extrapolating from a short hydrologic record to estimate rare events. It is obviously risky, but frequently there is no alternative, and the best one can do is to

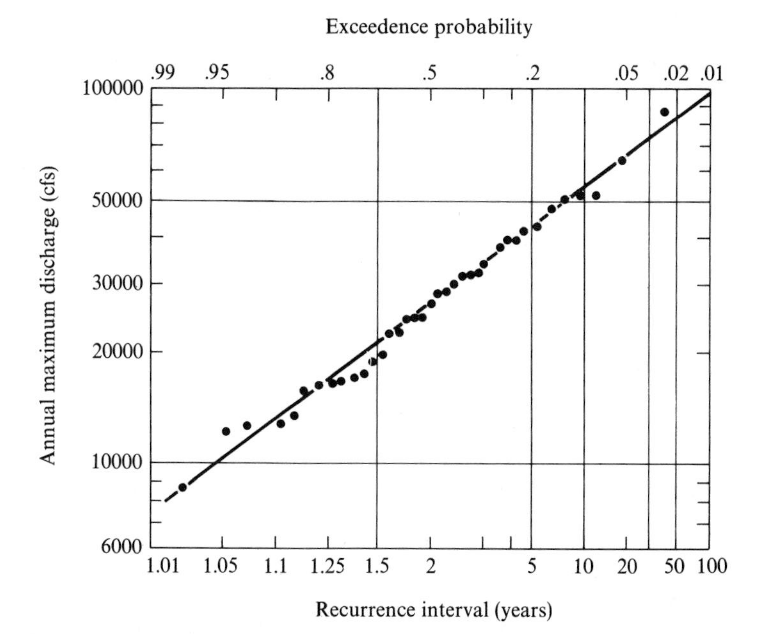

Figure 10-13 Flood-frequency curve plotted on logarithmic probability paper, Tana River at Garissa, Kenya, 1934–1970. The scale at the top is the probability that the discharge is equaled or exceeded in any given year. The bottom scale, recurrence interval, is the average number of years in which the annual peak equals or exceeds the discharge given on the ordinate. (Data from the Ministry of Water Development, Nairobi.)

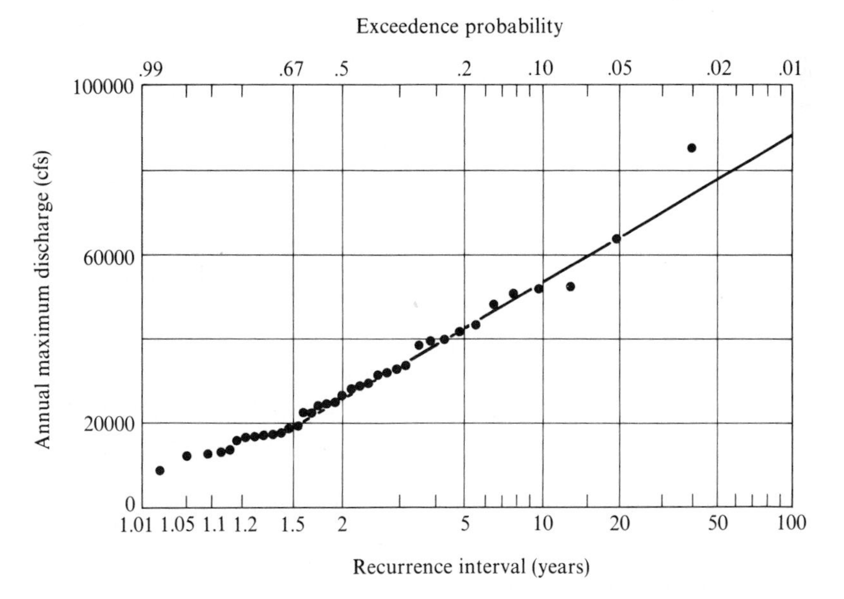

Figure 10-14 Flood-frequency curve plotted on arithmetic Gumbel Type I graph paper, Tana River at Garissa, Kenya, 1934–1970. (Data from the Ministry of Water Development, Nairobi.)

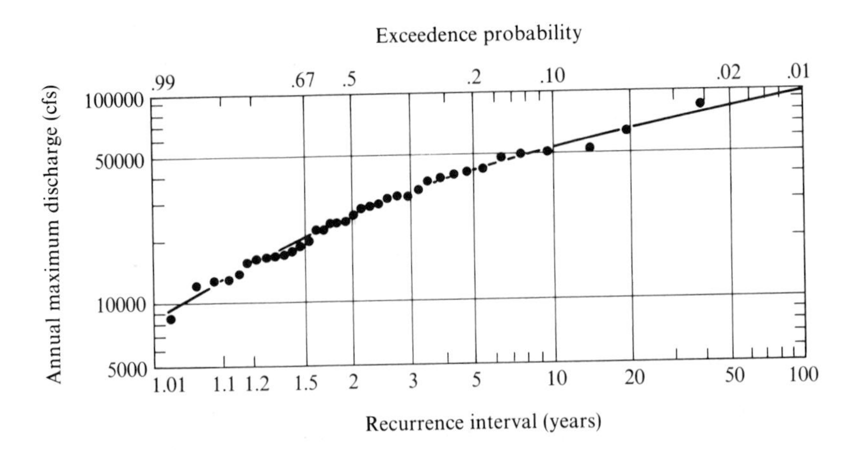

Figure 10-15 Flood-frequency curve plotted on logarithmic Gumbel Type III graph paper, Tana River at Garissa, Kenya, 1934–1970. (Data from the Ministry of Water Development, Nairobi.)

estimate rare events by such extrapolation (possibly using more than one of the theoretical probability distributions). It is wise then to interpret the reasonableness of the result in terms of one's knowledge of the regional hydrology, historical records, or other information such as the dimensionless rating curve illustrated in Chapter 16. The degree of uncertainty that one accepts in such estimates depends upon ethical and economic issues with respect to the risk of loss of life and property or the inconvenience in areas affected by the flooding. Ogrosky (1964) has outlined the policy of the U.S. Soil Conservation Service with regard to the design of small floodwater-retarding structures. In the design of some major dams, the risk of failure has been reduced by designing spillways to accommodate the 1000-year or 10,000-year flood. It is doubtful whether these floods have any physical meaning in view of the rapidity with which climate, and therefore hydrology, changes, but their estimation minimizes the problem of uncertainty by providing a large safety factor.

Some hydrologists have proposed the drawing of *confidence bands* around flood-frequency curves. Suppose we are concerned about the reliability of point x_m (representing the annual maximum discharge for a recurrence interval of T_m years) in Figure 10-16. Its plotted position, obtained from the line through the set of points in a sample, is the best estimate we can make of its true position. But there is a possibility of sampling error, and the true position might lie above or below (i.e., there is a chance that an error has been made in estimating the discharge of a flood with a recurrence interval of T_m).

If we assume that such sampling errors are normally distributed about the mean flood frequency curve, we should be able to draw an interval $\pm\Delta x_m$ within which there is a probability of, say, 90 percent that the true

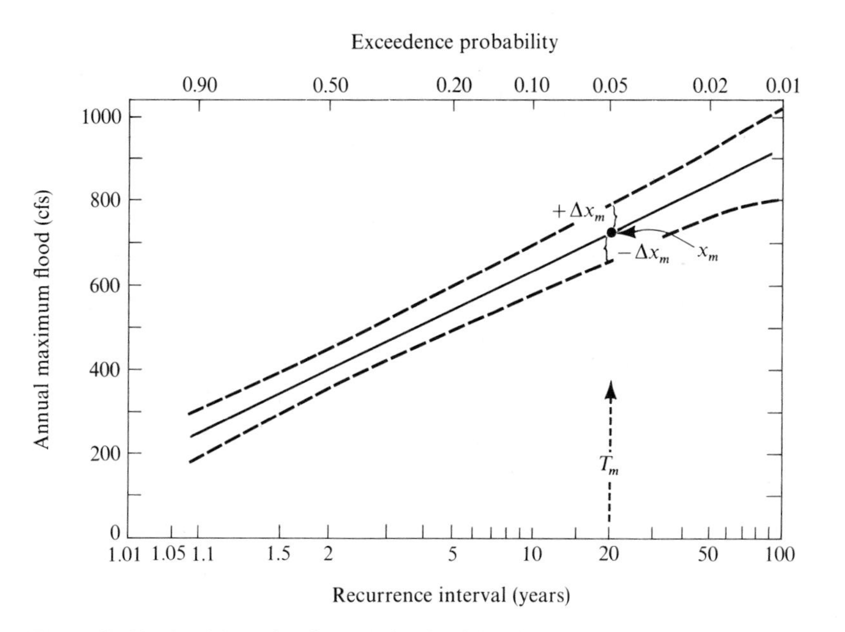

Figure 10-16 Confidence bands around a flood-frequency curve.

position of x_m lies. If this confidence interval is computed for a selection of plotted points in Figure 10-16 and curves are drawn as shown, bands are defined within which there is a 90-percent certainty that the true discharge for a chosen recurrence interval should lie.

Beard (1962) has presented a method for drawing confidence bands within which 90 percent of all floods with a chosen recurrence interval should lie. He published coefficients (see Table 10-12) for various levels of confidence, which should be multiplied by the standard deviation of the sample and then added to and subtracted from the discharges of the flood-frequency curve for various recurrence intervals. If the ordinate scale of the graph is logarithmic, the standard deviation of the logarithms is multiplied by each coefficient. The antilogarithm of this product is then added to or subtracted from the flood-frequency curve for various recurrence intervals.

In Figure 10-17 these intervals have been calculated for a flood-frequency curve and the confidence bands drawn for a 90-percent level of probability. They provide a warning to those who extrapolate the flood-frequency curve, for at the higher end of the curve the bands are widening sharply, indicating considerable uncertainty in estimating rare floods.

At this point it is also useful to recall (from Chapter 2) that even a well-defined recurrence interval is only the average frequency with which storms occur. The 50-year flood is *not* something that will occur 50 years from now, nor is it something that will occur 50 years after the last event of similar magnitude and at 50-year intervals thereafter. There is a 2-percent chance that the 50-year flood will occur in any given year, and if it occurred this

Table 10-12 Coefficients for the calculation of the 90% and 75% confidence intervals around the flood-frequency curve. The standard deviation of the sample of annual maximum floods is multiplied by these coefficients and added to or subtracted from the discharge values of the flood-frequency curve at the appropriate recurrence interval. (From Beard 1962.)

		RECURRENCE INTERVAL (YR)					
CONFIDENCE BAND	YEARS OF RECORD	1000	100	10	2	1.1	1.01
90%, upper	5	4.41	3.41	2.12	.95	.76	1.00
	10	2.11	1.65	1.07	.58	.57	.76
	15	1.52	1.19	.79	.46	.48	.65
	20	1.23	.97	.64	.39	.42	.58
	30	.93	.74	.50	.31	.35	.49
	40	.77	.61	.42	.27	.31	.43
	50	.67	.54	.36	.24	.28	.39
	70	.55	.44	.30	.20	.24	.34
	100	.45	.36	.25	.17	.21	.29
90%, lower	5	−1.22	−1.00	−.76	−.95	−2.12	−3.41
	10	−.94	−.76	−.57	−.58	−1.07	−1.65
	15	−.80	−.65	−.48	−.46	−.79	−1.19
	20	−.71	−.58	−.42	−.39	−.64	−.97
	30	−.60	−.49	−.35	−.31	−.50	−.74
	40	−.53	−.43	−.31	−.27	−.42	−.61
	50	−.49	−.39	−.28	−.24	−.36	−.54
	70	−.42	−.34	−.24	−.20	−.30	−.44
	100	−.37	−.29	−.21	−.17	−.25	−.36
75%, upper	5	1.41	1.09	.68	.33	.31	.41
	10	.77	.60	.39	.22	.24	.32
	15	.57	.45	.29	.18	.20	.27
	20	.47	.37	.25	.15	.18	.24
	30	.36	.29	.19	.12	.15	.20
	40	.30	.24	.16	.11	.13	.18
	50	.27	.21	.14	.10	.12	.16
	70	.22	.17	.12	.08	.10	.14
	100	.18	.14	.10	.07	.09	.12
75%, lower	5	−.49	−.41	−.31	−.33	−.68	−1.09
	10	−.39	−.32	−.24	−.22	−.39	−.60
	15	−.34	−.27	−.20	−.18	−.29	−.45
	20	−.30	−.24	−.18	−.15	−.25	−.37
	30	−.25	−.20	−.15	−.12	−.19	−.29
	40	−.22	−.18	−.13	−.11	−.16	−.24
	50	−.20	−.16	−.12	−.10	−.14	−.21
	70	−.18	−.14	−.10	−.08	−.12	−.17
	100	−.15	−.12	−.09	−.07	−.10	−.14

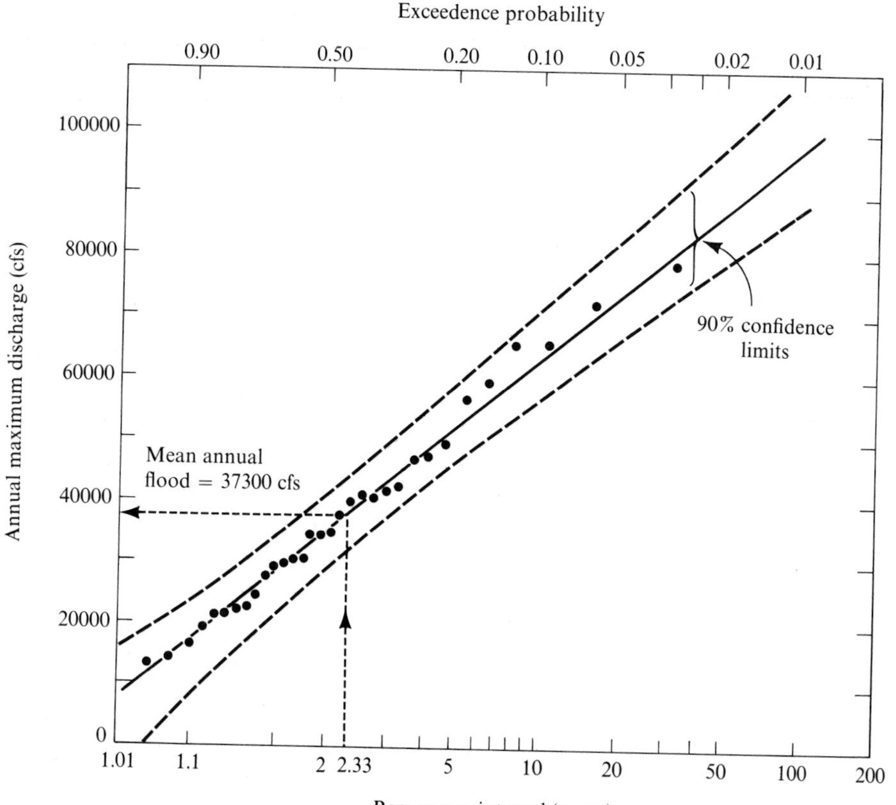

Figure 10-17 Flood-frequency curve for annual flood values on Skykomish River at Gold Bar, Washington. Dashed lines show zone within which there is a 90-percent chance that the true value for that recurrence interval will lie. (Data from U.S. Geological Survey.)

year, there is still a 2-percent chance that it will occur next year. Generalizing this, we can say (from Equation 2-5) that the probability of a flood with a recurrence interval of T years occurring or being exceeded within the next n years is

$$q = 1 - \left(1 - \frac{1}{T}\right)^n \tag{2-5}$$

There is a 21-percent chance that the 200-year flood will occur in the next 50 years. Figure 2-15 is equally applicable to floods and to rainfall in this context.

Climatic fluctuation is still another factor that can cause uncertainty and lead to misjudgment and misuse of flood-frequency curves. The data fitted by the flood-frequency curve must be homogeneous in time, in the sense that the sample record used should not include floods from two different rainfall–

runoff regimes. Many parts of East Africa, for example, have undergone a striking change of rainfall since 1960, and it is doubtful whether floods from before and after this date should be mixed in constructing a flood-frequency curve. As an illustration, the data compiled in Figure 10-13 have been segregated into two sets of years and replotted in Figure 10-18 as two flood-frequency curves that are strikingly different. Unfortunately there are no hard and fast rules to guide the hydrologist in such a situation. So he must make a judgment about the significance of the separation of these two curves. Does it represent only a short run of wet years? Or has the hydrologic regime of the basin undergone a radical change? If the latter hypothesis is correct, use of the longer but mixed record in Figure 10-13 could lead to a serious underestimation of floods in the new regime. In this case the shorter record, although subject to grave sampling errors, would be the one to use for planning. A question would also arise about the probable duration of the new regime, and again there are no precise statistical answers. The hydrologist would have to consult climatologists. We raise the problem here, not because we can give answers, but so that the hydrologist and planner can see that flood-frequency curves and the statistics they yield are subject to large uncertainties, and that they should be treated conservatively.

For such reasons, flood-frequency curves should be checked and updated from time to time. If the record remains homogeneous, its increasing length will reduce the standard deviation of the sample and thus narrow the confidence bands in Figure 10-17. Land-use changes, dam construction, and channel changes could cause a lack of homogeneity in flood records, and in

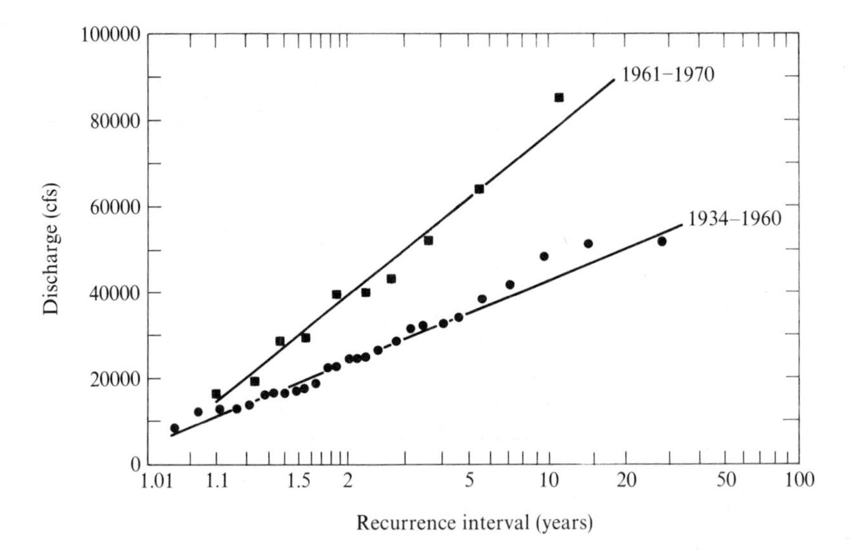

Figure 10-18 Flood-frequency curves derived from two periods of time within the same long record of Tana River at Garissa, Kenya. (Data from the Ministry of Water Development, Nairobi.)

many regions these changes are rendering flood records of little value. Another factor that may cause a lack of homogeneity in a flood record is the variation of the causative meteorological event. In New England, for example, some annual maximum floods are generated by summer rainstorms, others by autumn hurricanes, others by snowmelt, and still others by rain on melting snow, sometimes coupled with surges following the breakup of ice jams. Usually all such floods are included in the flood-frequency analysis. Whether they should be or not is a subject for debate.

On Figure 10-17, we have labeled one discharge on the ordinate as the *mean annual flood.* This is simply the arithmetic mean of all the annual maximum discharges. If the sample flood-peak distribution fits a Gumbel Type I theoretical frequency distribution, the recurrence interval of a discharge equal to the mean annual flood is 2.33 years. For convenience, the mean annual flood is usually read from the flood-frequency curve as the discharge with a recurrence interval of 2.33 years. The same can be done for the Gumbel Type III flood-frequency graph, where the 2.33-year recurrence interval should correspond to a discharge equal to the logarithmic mean of the annual flood peaks. Sometimes, small discrepancies occur between the computed and the graphically determined values of the mean annual flood, when the sample record is not closely fitted by the Gumbel distribution. Dalrymple (1960) states that for U.S. Geological Survey practice at least, "The mean annual flood for a gauging station is by definition the 2.33-year flood from the graphic-frequency curve . . . ." Because of the widespread usage of this definition, it is probably best to employ the graphical determination of the mean annual flood.

Graphical estimation of the mean annual flood, the uncertainty of extrapolating a flood-frequency curve, and the width of the confidence bands make flood prediction a risky enterprise. Often the observed flood distribution is not fitted well by a straight line on any of the graph papers, and the hydrologist must sketch a curve to fit the points. He should be fully aware of the possible errors when using the information so gained. It is also advisable not to rely on one method of flood prediction, but to use several methods in an attempt to obtain a consensus.

Use of Historical Information on Floods

Suppose the highest flood observed in a 50-year record is also known to be the highest in a much longer period of time (perhaps since colonization, or since the beginning of newspaper publication). If, for example, the historical record of floods is 220 years long, then the recurrence interval of the highest recorded flood should be set as 221 years rather than as 51 years. The second highest recorded flood should then be computed as the second highest of a 50-year record (25.5 years), as usual. Benson (1950) has treated the use of historical data more extensively.

Stage-Frequency Curves for a Station

Often, one is concerned not so much with the magnitude of floods at a station but with their height or stage. The estimation of recurrence intervals for flood heights can be made directly from records of stage or indirectly through the stage–discharge relation (the *discharge rating curve* for the gauging station, as described in Chapter 16). A problem arises because the stage–discharge relation may vary, being subject to both short-term random fluctuations and to long-term trends. If the rating curve has remained virtually stable or has changed only randomly and by small amounts, frequencies of flood stages can be computed directly from the record of annual maximum stages, or they can be obtained by using the flood-frequency curve to estimate the recurrence interval for discharges and then converting the discharges to heights by means of the stage–discharge relationship.

If the discharge rating curve has changed by some large amount, one can only assume that the most recent relation will hold in the future, but some estimate should also be made of the possible errors inherent in this assumption. If the stage–discharge curve is showing a definite trend because of aggradation, degradation, or widening of the channel, the discharge frequencies should first be determined and converted to heights by means of an assumed stage–discharge relationship based on the previous trend. Again, an estimate of possible errors should be made, and where possible the stage–discharge relationship should be checked periodically.

The Partial-Duration Flood Series

In using the partial-duration series (see Chapter 2), all flood peaks above a certain base magnitude are used. The base is usually chosen equal to the lowest annual maximum flood of record, or of a magnitude such that the partial-duration series contains only as many floods as there are years of record. If peaks occur so close together that they cannot be considered independent events, only the larger is included in such a listing. Using the plotting-position formula given in Equation 2-4, the partial-duration series is usually plotted on semi-logarithmic paper with recurrence interval on the logarithmic scale, or on double-logarithmic paper.

There is a relationship between recurrence intervals obtained from the annual-maximum series and the partial-duration series, as shown in Table 10-13. The differences are negligible for return periods greater than 10 years. But there is a distinction between the meaning of recurrence interval of floods obtained from the two series. For the annual-maximum series the recurrence interval is the average interval within which a flood of a given size will occur *as an annual maximum.* The recurrence interval obtained from the partial-duration series is the average frequency of occurrence between floods of a given size irrespective of their relation to the year. It is

Table 10-13 Relation between recurrence intervals of the annual-maximum series and the partial-duration series. (From Langbein 1960.)

RECURRENCE INTERVALS (YR)	
ANNUAL-MAXIMUM SERIES	PARTIAL-DURATION SERIES
1.16	0.5
1.50	0.9
1.58	1.0
2.00	1.45
2.54	2.0
5.52	5.0
10.50	10.0
20.50	20.0
50.50	50.0
100.50	100.0

the average time between flows equal to or greater than a given discharge. The usual method of obtaining return periods for the partial-duration series is to obtain them for the annual-maximum series and then to convert the frequencies by use of Table 10-13.

In Chapter 16, it is shown that the bankfull discharge for most rivers has a recurrence interval on the annual flood series of 1.5 years. This means that 1 year out of 1.5 or 2 years out of 3, the highest discharge for the year will be equal to or will exceed the bankfull capacity of the channel. Table 10-13 shows that a flow having a recurrence interval of 1.5 years in the annual flood series will have a recurrence interval of 0.9 years in the partial-duration series. Therefore a discharge equal to or greater than bankfull may be expected to occur on the average once every 0.9 years, or 100 times every 90 years. This is slightly more often than once a year. On the average, then, the bankfull stage will be equaled or exceeded about once a year.

Maximum Probable Flood

In the design of large flood-control dams and other structures, it is often necessary to consider the possibility of a flood that would result from the most critical combination of flood-producing conditions. The assessment of this maximum probable flood is usually based on the consideration of the probable maximum precipitation (see Chapter 2) and measurements of the largest historical floods. In some areas such as the Pacific Northwest or New England, estimates of the maximum probable flood must also take into account snowmelt occurring during large rainstorms and the possibility of ice jams on the river. The derivation of a maximum probable flood for a large river in the Indus Valley of Pakistan is described by Binnie and Mansell-Moulin (1966), who used the probable maximum precipitation for the basin, the worst conceivable antecedent moisture and runoff conditions, and the unit hydrographs from earlier large storms.

A different design frequency is usually chosen for the spillway and for the reservoir storage in large structures. If a large dam storing water for irrigation overflowed once in 50 years, the overflow water might cause damage in the valley downstream but would not be catastrophic. If, however, the dam itself were to wash out, the flood wave caused by the sudden release of the stored water might wipe out cities downstream and cause great loss of life. Therefore, a reasonable risk of overflow could be accepted and the recurrence interval of such an event might be chosen as once in 50 years. But the spillway design would be chosen to accommodate the largest possible flood or maximum probable event because the risk of dam failure should be as close to zero as scientific calculation allows. The engineering hydrologist and the design engineer may have to deal with computing the maximum probable flood, but this is not within the purview of the planner to whom the present book is directed.

Regional Flood-Frequency Curves

As we have seen, flood-frequency analysis for single stations is subject to large errors because of the brevity of most records, the inherent variability of floods, and the difficulty of fitting theoretical frequency distributions to the sample record. The sampling variability has been analyzed by Benson (1960) for a hypothetical 1000-year record that is fit perfectly by an extreme-value distribution. Possible errors are shown to be uncomfortably large. One method of reducing the variability due to sampling is to combine the records from many gauging stations in a region. In effect, this involves substituting space for time to increase the size of the sample. Flood-frequency characteristics of various catchments can also be correlated with meteorologic or physiographic parameters. Floods in ungauged basins can then be estimated from the physical geography of the catchment. The relations developed with this method are called *regional flood-frequency curves.*

The development of regional flood-frequency curves is based on the empiricism that for large regions of homogeneous meteorologic and physiographic conditions, individual basins covering a wide range of drainage areas have flood-frequency curves (on, say, Gumbel Type III paper) of approximately the same slope. This is illustrated in Figure 10-19 for some short records from six basins of the Sleepers River Experimental Watershed in northeastern Vermont. The mean annual flood for each basin can be read at a recurrence interval of 2.33 years, as shown. If each flood is divided by

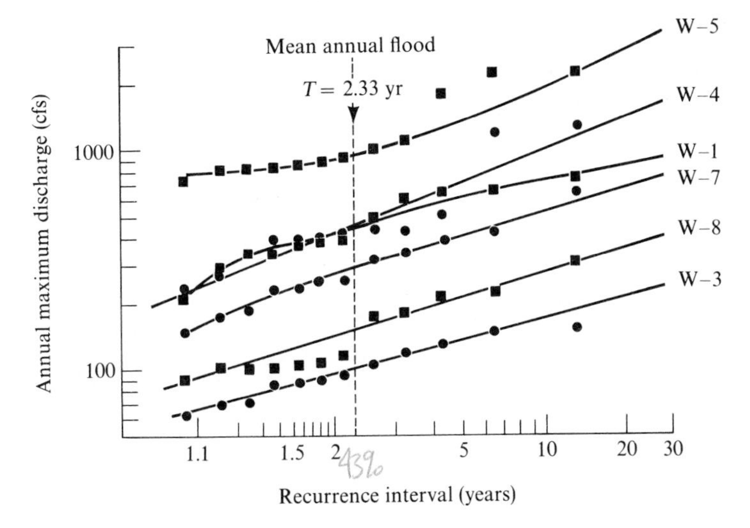

Figure 10-19 Flood-frequency curves derived from measurements on six small catchments in a homogeneous region of northeastern Vermont (Sleepers River Experimental Watershed). (Data from the Agricultural Research Service, U.S. Department of Agriculture.)

the mean annual flood, as shown in Table 10-14, the flood-frequency curves can be replotted, as shown in Figure 10-20(a), with the ordinate now being a ratio of each flood to the mean annual flood. The curves for all the stations will pass through a ratio of 1.0 at a recurrence interval of 2.33 years.

A question then arises whether the differences in slope in Figure 10-20(a) are due to chance sampling variations among stations within a homogeneous region, or whether they indicate significant differences among basins with respect to the factors controlling flood characteristics. To test whether the variation is significant, we use a *homogeneity test,* the rationale of which is discussed by Dalrymple (1960). The test, which is illustrated in Typical Problem 10-10, shows that the six catchments in our example are sufficiently homogeneous to use for the definition of a single regional flood-frequency curve. This regional curve is defined by computing the median ratio for values of each recurrence interval and drawing an average curve, as shown in Table 10-14 and Figure 10-20(b).

We now have a graph that, if we know the mean annual flood for a basin, can be used to estimate floods of any other recurrence interval. The final part of the problem, then, is the development of a technique for relating the mean annual flood to some characteristic of the basins, so that the regional flood-frequency curve can be used to estimate floods from ungauged basins. The most obvious and widely used basin characteristic for such a correlation is the drainage area above the stream gauge. Figure 10-21 shows the relationship for the Vermont stations, where there is a strong correla-

Table 10-14 (a) Tabulation of the *T*-year flood for various basins taken from the accompanying flood-frequency curves. The floods are expressed in cubic feet per second. The data are for 6 basins in the Sleepers River Experimental Watershed, Vermont.
(b) Ratio of *T*-year flood to the mean annual flood for the basins.
(a)

DRAINAGE BASIN	W-1	W-3	W-4	W-5	W-7	W-8
Area (sq mi)	16.58	3.23	16.80	42.91	8.35	6.04
Mean annual flood	455	105	465	1000	290	157
1.1-year flood	240	67	234	770	155	93
1.5-year flood	370	86	340	880	230	122
2-year flood	430	99	420	940	270	144
3-year flood	500	117	540	1100	330	175
5-year flood	600	140	710	1400	400	215
10-year flood	720	173	1000	2000	520	280
15-year flood	780	198	1210	2500	600	330
25-year flood	860	230	1550	3050	720	400

(b)

DRAINAGE BASIN	W-1	W-3	W-4	W-5	W-7	W-8	MEDIAN
1.1-year flood	0.53	0.64	0.50	0.77	0.54	0.60	0.57
1.5-year flood	0.82	0.82	0.74	0.88	0.79	0.78	0.81
2-year flood	0.95	0.94	0.91	0.94	0.93	0.92	0.94
3-year flood	1.10	1.11	1.17	1.10	1.14	1.12	1.12
5-year flood	1.32	1.34	1.53	1.40	1.39	1.38	1.38
10-year flood	1.58	1.65	2.15	2.00	1.80	1.80	1.80
15-year flood	1.71	1.89	2.62	2.50	2.07	2.10	2.08
25-year flood	1.90	2.20	3.35	3.05	2.48	2.56	2.52

tion between drainage area and mean annual flood. The development of a regional flood-frequency curve is now complete, and it can be used as illustrated in Typical Problem 10-11.

For other regions, where the basins are larger, less steep, and contain larger channels and more extensive floodplains, the channel storage mechanisms referred to earlier tend to produce a less rapid increase in mean annual flood downstream, and the slope of the line in Figure 10-22, for example, is lower than that of the preceding diagram. This slope represents the exponent in the equation relating mean annual flood to drainage area, and for many regions it lies close to 0.75. The tendency for intense rainstorms to be localized (see Chapter 2) also contributes to a less rapid increase in mean annual flood with catchment size, but in northeastern Vermont most floods are produced by large general rainstorms or snowmelt.

The Vermont basins have a single area-to-flood relationship, but this may not be the case elsewhere. Even if the homogeneity test shows that all the gauges in a region are similar with respect to the slope of the flood-frequency curve, they may need to be divided into groups to obtain correlations of drainage area and mean annual flood. Particularly in mountainous regions, it may be necessary to add mean basin altitude as a second variable to predict the mean annual flood (see Figure 10-22). Where strong gradients of precipitation occur in a region, it is usually necessary to use mean annual precipitation or mean annual runoff as a second variable along with drainage area. These factors can be related to the mean annual flood by multiple regression techniques (Rantz 1971).

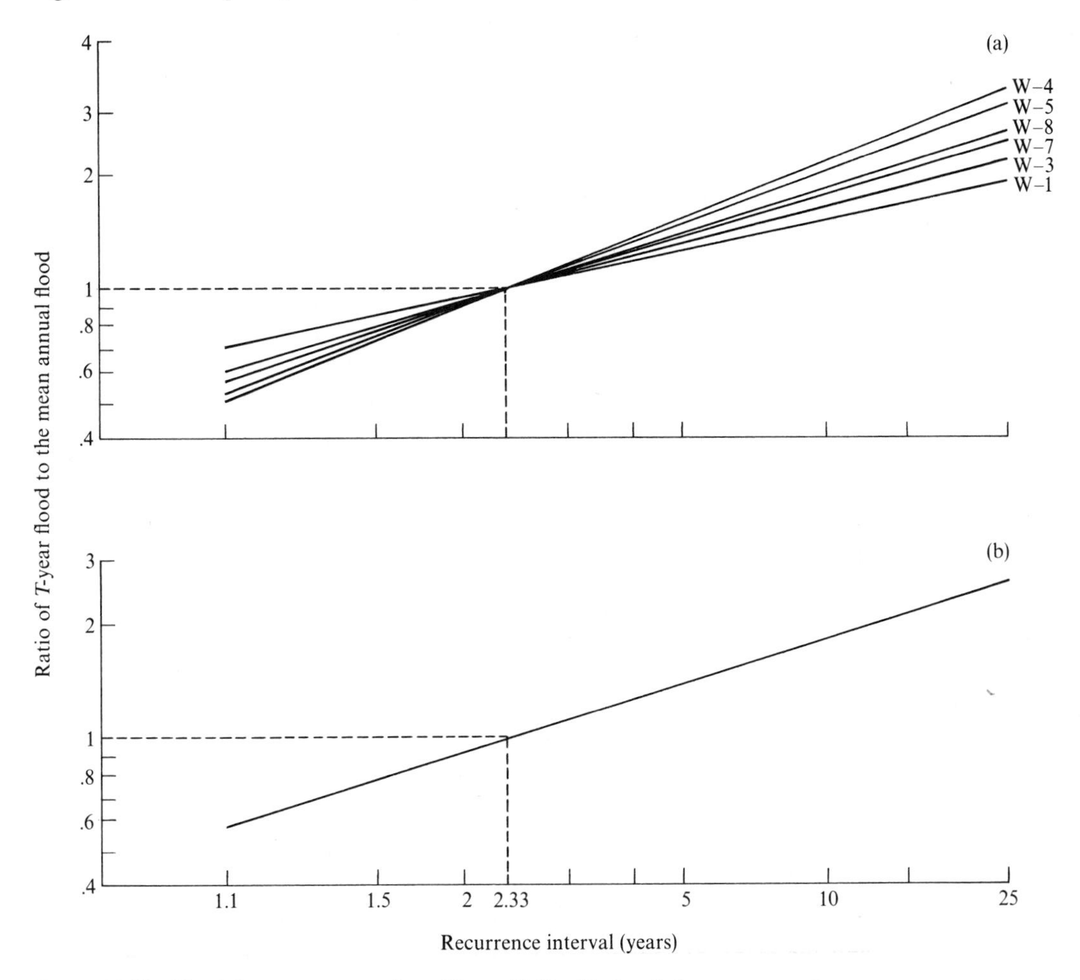

Figure 10-20 Flood-frequency curves from Figure 10-19 with the discharge expressed as a ratio to the mean annual flood. (a) Individual or station curves for the six catchments. (b) Regional curve for northeastern Vermont, constructed from median values of curves in (a) above.

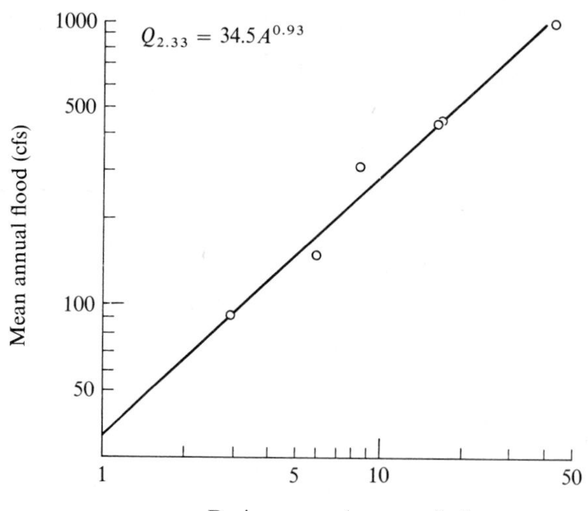

Figure 10-21 Relationship of mean annual flood to drainage basin area for catchments of the Sleepers River Experimental Watershed near Danville, Vermont.

In a large proportion of the flood problems faced by the planner or environmentalist, there will not be a gauging station at the location under study, and estimates must be made from surrounding areas and usually from data applying to larger basins. To make these estimates, the regional flood-frequency curve becomes, then, one of the planner's most important tools. Fortunately, most of the continental United States has been included in regional studies the data for which are readily available in published form. The publications include tabulated lists of annual floods for most stations in the United States, and curves from which the value of the mean annual flood and frequency of various discharges can be computed. The location and the size of the basin for which estimates are needed are the parameters required to use the published curves. The whole set of data and the curves are published in the Water Supply Papers of the U.S. Geological Survey under the title *Magnitude and Frequency of Floods in the United States.* Each volume of the series covers a different region. The volumes are Water Supply Papers No. 1671 to No. 1688, and are available in nearly all large public and university libraries.

Unfortunately, most of the curves presented in this massive library of data do not include drainage basins of less than 10 square miles, so the curves must sometimes be extended or extrapolated to smaller areas. When such extrapolation is used, the results should be considered estimates only and should be checked against other data. For example, results from an extrapolated regional curve may be checked against those from computing frequency curves for individual stations close to the area of interest. In making a final choice of a discharge value, extra weight can be given to

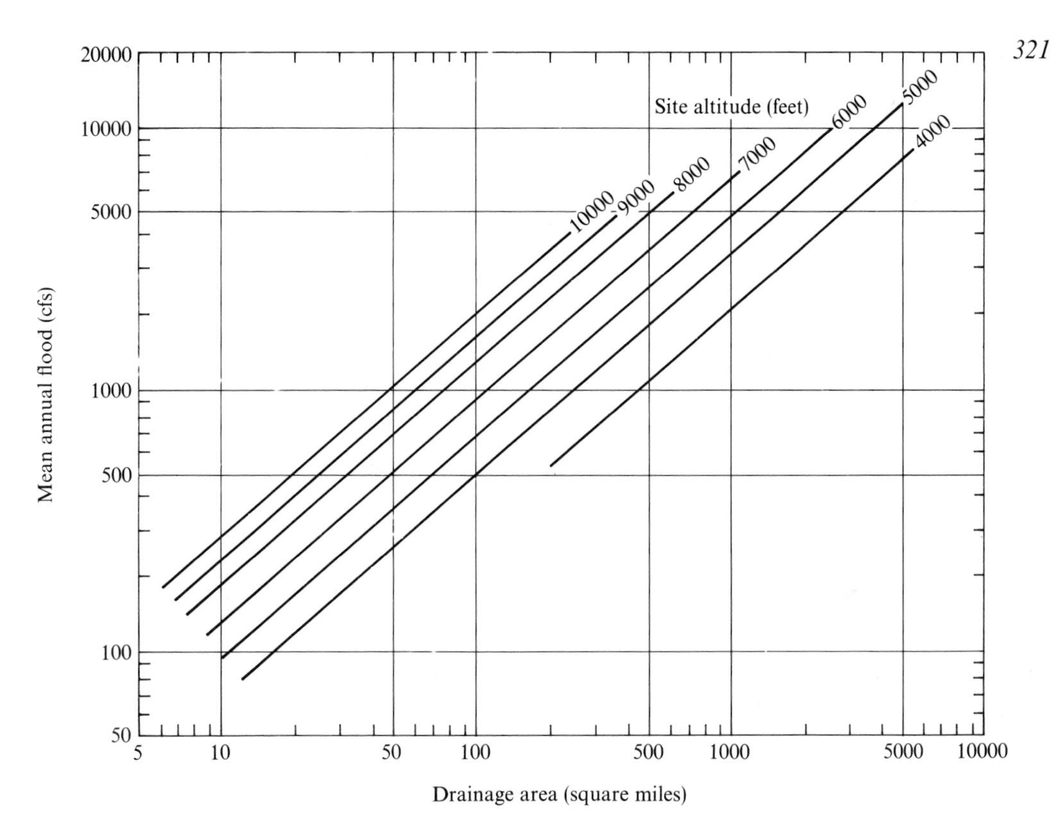

Figure 10-22 Variation of mean annual flood with drainage area and altitude of the gauging site for a region in northern Wyoming. Note that for a fixed altitude, $Q_{2.33}$ is proportional to $A^{0.85}$. (From Carter and Green 1963.)

values derived from nearby stations having the same topographic aspect, size, and vegetation characteristics as the area under study. Local data for the individual basin may be considered. Local residents can be asked how often the stream flows out of its banks and whether the frequency of overbank flow has changed in past decades.

Dalrymple's manual gives detailed instructions for the compilation of regional flood-frequency curves, including such matters as how to fill in gaps in the record of annual floods, how to adjust records of unequal length to a common base period, and how to adjust a short record by means of a longer one.

The regionalization of flood data for an area of uniform physiography can be extended to the analysis of flood heights. An example of the results of such an exercise is presented in Figure 10-23. Planners will immediately recognize the utility of such a graph. At any streamside site on which development is proposed, the drainage area can be measured from a topographic map. Reading upward from the drainage area on the abscissa of the

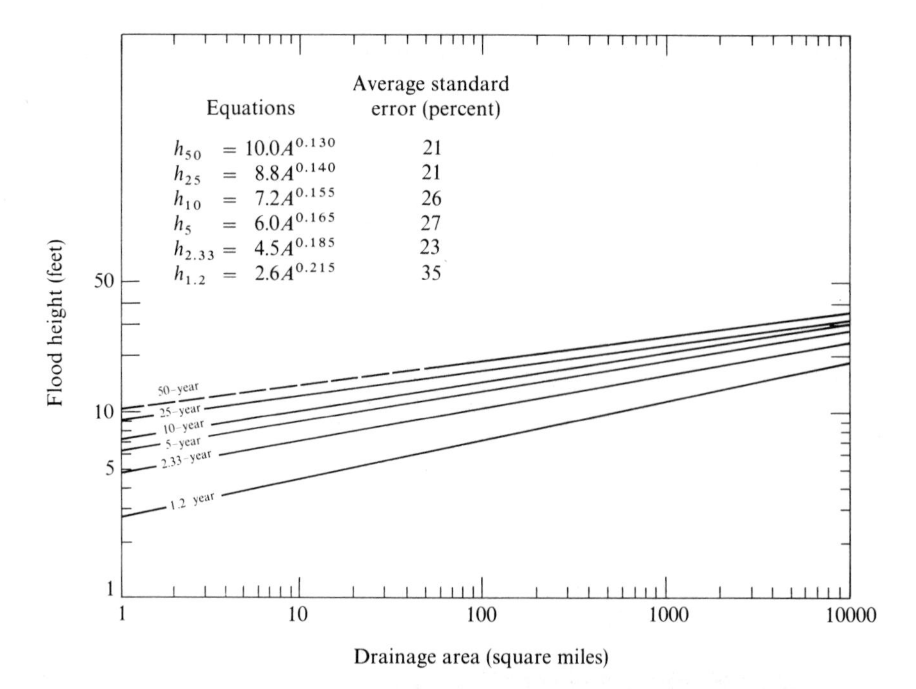

Figure 10-23 Regional flood height–drainage area–frequency relations for the plains area in Missouri. Height refers to the elevation of the water surface above the bed of the channel at the average (median) discharge. (From Gann 1968.)

diagram, it is possible to estimate the heights of floods above the bed of the channel without an extensive hydrologic analysis.

We refer to this technique of regionalizing hydrologic and geomorphic information in the text because it has great value for the planner. The analysis of records collected at a few stations within a hydrologically homogeneous region produces relationships like those shown in Figures 10-20 through 10-23, and others in later chapters. From such a summary it is possible to estimate the hydrologic and geomorphic characteristics of ungauged sites where the planning must be done. Errors of estimation, of course, can be large and should be checked, but usually the results are sufficiently precise for most planning problems. We recommend the technique strongly.

Flood-Frequency Curves for Large Rivers

Large rivers receive runoff from several flood-frequency regions and so do not fit into the patterns required for regional curves to be useful. In these cases it is usual to plot frequency curves for several stations along the main stem of the river in the manner shown in Figure 10-24. Each of the vertical dashed lines in the figure represents the flood-frequency curve at one station. Curves are drawn to join points representing the mean 1-year, 10-year, 25-year, and 50-year floods. The data for most rivers when plotted on such a diagram would show steadily increasing discharge with increasing distance downstream, because flood values for a given frequency usually increase as the 0.75 power of the drainage area. The Rio Grande from Colorado to Texas, illustrated in Figure 10-24, drains a semi-arid to arid area downstream of the Jemez River. Note that the discharge of a given frequency does increase from Lobatos to Jemez River, a generally mountainous area. Through the next 325 miles the discharge remains about constant, even decreasing slightly below the Rio Salado. The decrease is a logical result of the fact that in that lower reach any flood would spread out over a wide, brushy, and in places, marshy valley and lose water to infiltration.

Multiple Regression Analysis of Floods

Although many flood records pass the homogeneity test, indicating that the slopes of their flood-frequency curves are approximately equal, more detailed examination often indicates that the slopes decline slightly, but

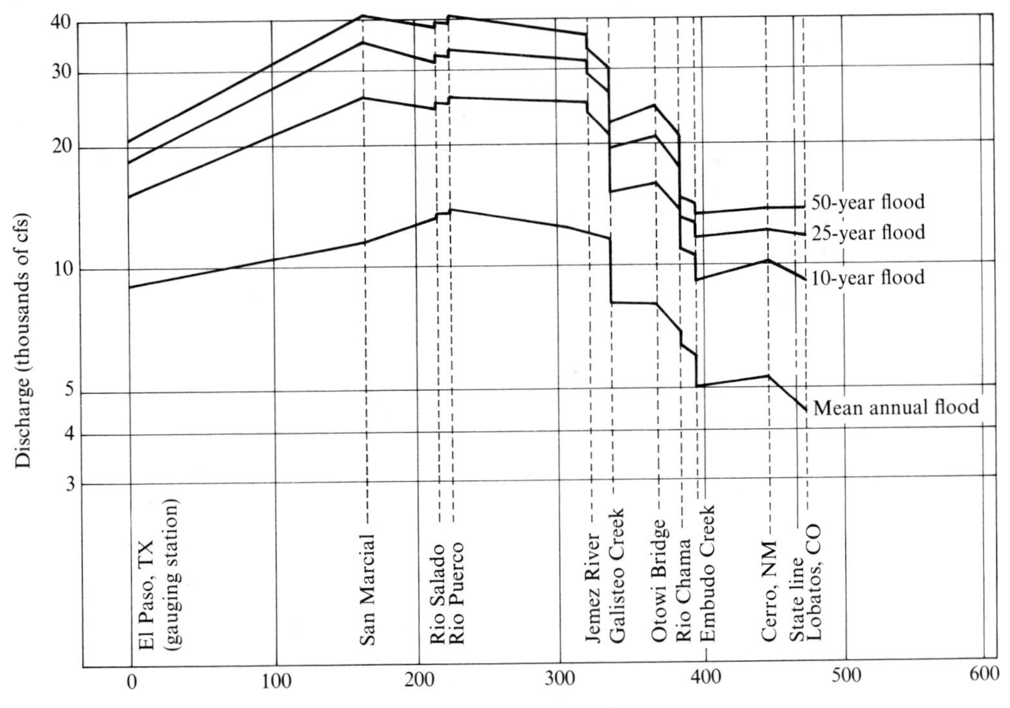

Figure 10-24 Variation of discharge for selected flood frequencies with distance above the El Paso gauging station, Rio Grande main stem below Lobatos, Colorado. (From Wiard 1962.)

significantly, as drainage area increases. This variation does not render the regional flood-frequency method useless, but it introduces a bias. To circumvent this and other minor difficulties, several hydrologists have proposed the use of multiple regression techniques to predict floods of various recurrence intervals from measures of controlling factors such as precipitation, drainage area, and topography. Rantz (1971), for example, was able to predict the magnitude of floods of 2-year, 5-year, 10-year, 25-year, and 50-year recurrence intervals from the drainage area and mean annual rainfall. Basin size is an obvious determinant of flood size, as mentioned for the Vermont regional flood-frequency analysis. Mean annual precipitation is a good index of the relative magnitude of storms of any frequency in a region (see Figure 2-18).

The multiple regression equations, listed in Table 10-15, were obtained by statistical analysis of data from 40 gauging stations in rural catchments of the San Francisco Bay region. This analysis is simply another kind of regionalization of flood-frequency information and can be used for estimating annual floods from ungauged basins.

Benson (1962, 1964) has made extensive multiple regression studies of floods in New England and in the southwestern United States. His equations incorporate such factors as drainage area, channel slope, amount of lake storage, rainfall intensity, air temperatures, elevation, frequency of thunderstorms, and mean annual runoff. Where such data are available, the resulting multiple regression equations usually give good predictions. But in most planning studies the necessary data would be too costly and time-consuming to collect. An increasing number of these multivariate studies are appearing, however, and those requiring few input variables are very useful.

Use of Flood-Frequency Analysis in Urban Catchments

As explained in Chapter 9, urbanization causes an increase in the size of floods in the small catchments with which planners are usually concerned. In the planning process it is important, therefore, to be able to assess the probable impact of urbanization upon the magnitude of flood peaks. In an earlier section we described how such an assessment could be made by the rational method. Here flood-frequency relations are applied to the same problem.

Carter (1961) and Anderson (1970) presented analyses of the effects of urbanization on hydrographs from drainage basins in the Piedmont and Coastal Plain physiographic provinces around Washington, DC. They first computed the change of peak discharge to be expected after urbanization due to the change of impervious area alone. An average peak runoff rate for rural parts of basins in the region was about 30 percent of the rainfall intensity, while on the impervious areas it was approximately 75 percent.

Table 10-15 Regression equations used to estimate floods of various recurrence intervals on rural drainage basins of the San Francisco Bay region. (From Rantz 1971.)

RECURRENCE INTERVAL (YR)	MULTIPLE REGRESSION EQUATION*	STANDARD ERROR OF ESTIMATE			
		LOGARITHMIC UNITS	PERCENT		
			PLUS	MINUS	MEAN
2	$Q_2 = 0.069A^{0.913}P^{1.965}$	0.226	68.3	40.5	54.4
5	$Q_5 = 2.00A^{0.925}P^{1.206}$	0.175	49.6	33.2	41.4
10	$Q_{10} = 7.38A^{0.922}P^{0.928}$	0.168	47.2	32.1	39.6
25	$Q_{25} = 16.5A^{0.912}P^{0.797}$	0.178	50.7	33.6	42.2
50	$Q_{50} = 69.6A^{0.847}P^{0.511}$	0.192	55.6	35.7	45.6

*Q_T = the T-year flood in cfs.
A = drainage area in square miles (range 0.2 to 196)
P = mean annual basinwide precipitation in inches (range 13 to 60)

(These values are really coefficients in the rational formula, as listed in Table 10-9). If the percentage of impervious area in an urbanized basin is Z, and the rainfall intensity during a storm is I, then the peak discharge, Q_{pk}, is

$$Q_{pk} \text{ (after urbanization)} = I\left[\frac{0.30(100 - Z)}{100} + \frac{0.75(Z)}{100}\right] \qquad (10\text{-}5)$$

which simplifies to $I\ (0.30 + 0.0045Z)$. The change of flood peaks due to impervious area alone can then be expressed as

$$\frac{Q_{pk} \text{ (after urbanization)}}{Q_{pk} \text{ (before urbanization)}} = K = \frac{0.30 + 0.0045Z}{0.30} \qquad (10\text{-}6)$$

The effect of urbanization depends on the nature of the soils in the undisturbed region, as well as the extent of impervious area. Areas that experience high runoff rates before urbanization will be least affected. Substitution of values of Z from Figures 9-18 and 10-11 and values of C for soils from Table 10-9 will give the reader a feel for the effects of increasing the impervious area alone. Surprisingly, these effects are rather small in most basins.

Carter and Anderson went on to show that the dominant influence of urbanization was on the lag time of the catchment, as described in Chapter 9. They used as their index of timing, l_c, the lag between the centroids of rainfall and of runoff (as defined in Figure 10-25). For rural basins around Washington, DC, they showed that this lag could be correlated with the ratio $L/\sqrt{S}$, where L is the length of the mainstream channel in miles and S is the channel slope (in feet per mile) between points 10 percent and 85 percent of the mainstream length above the basin outlet. Their relationship for rural basins is shown as the upper line in Figure 10-26.

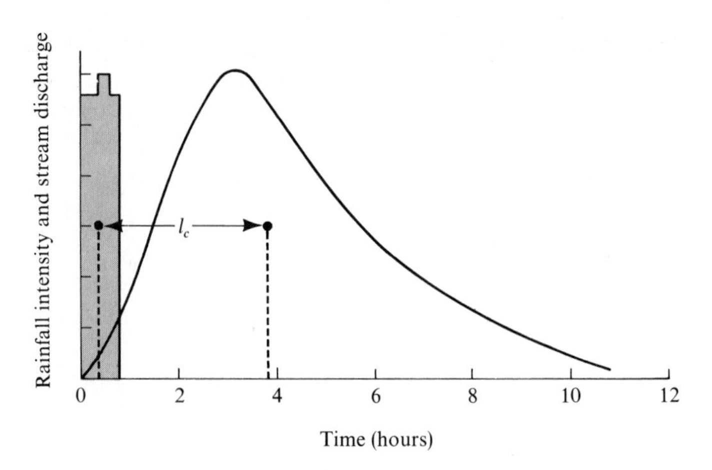

Figure 10-25 The centroid lag, l_c, is the time difference between the center of mass of rainfall and the center of mass of runoff. It can be obtained by plotting accumulation curves for rainfall and for runoff. The times at which 50 percent of the total accumulation of both variables are located on the graph and the time difference between them is the centroid lag.

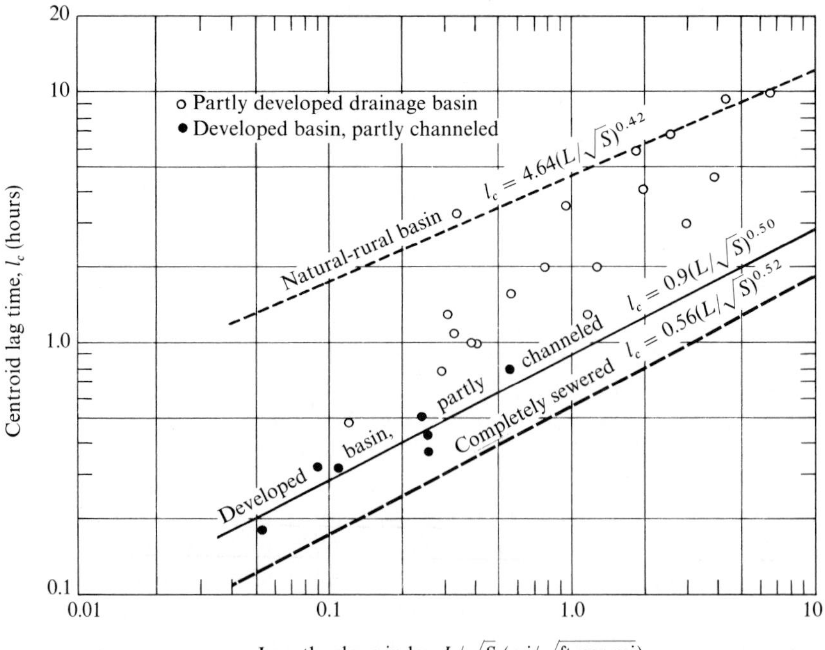

Figure 10-26 Relation of centroid lag, l_c, to a length–slope index for basins with different degrees of storm sewering near Washington, DC. L is the length of the main channel in miles; S is the slope of the channel between points 10 and 85 percent of the channel length above the outlet. (From Anderson 1970.)

For basins whose stream channels had been lined, straightened, or diverted into pipes, the centroid lags were reduced to 12–15 percent of their rural values. In many suburban areas, however, the ultimate degree of development involved the alteration (sewering) of all small channels, with only moderate or no alteration of the larger channels. The lag times in these partly sewered basins are shown by the central line in Figure 10-26 and average about 20–25 percent of the rural values. Basins currently undergoing development lie between the upper two lines on the graph.

Multiple regression was then used to define a relationship between the previously discussed basin parameters and the mean annual flood ($Q_{2.33}$):

$$Q_{2.33} = 230KA^{0.82}l_c^{-0.48} \tag{10-7}$$

where flow is expressed in cfs, the area (A) is in square miles, K is given by Equation 10-6, and l_c (in hours) is obtained from Figure 10-26 for any degree of urbanization. Using this equation, the mean annual flood can be predicted for any basin from its degree of imperviousness, its drainage area, and the length and slope of the main channel.

Anderson then derived regional flood-frequency curves for rural basins and for completely impervious basins around Washington, DC. For the latter he used the regional rainfall intensity-frequency curves and the rational method. By interpolation he estimated regional flood-frequency curves for basins with intermediate degrees of imperviousness and obtained Figure 10-27.

If the 50-year flood is to be predicted for a basin with 40-percent impervious cover, the mean annual flood of the basin is first computed from Equation 10-7. Figure 10-27 is then entered at an abscissa value of 50 years,

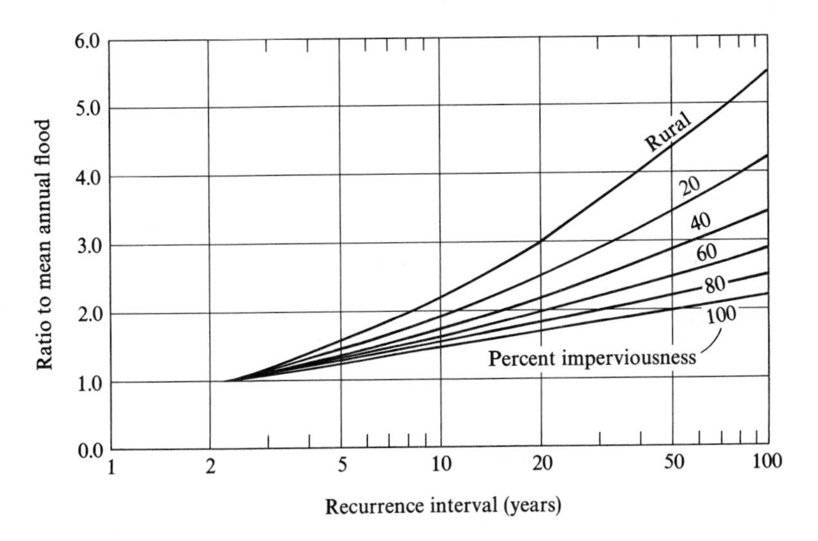

Figure 10-27 Regional flood-frequency curves for catchments with various percentages of impervious area near Washington, DC. (From Anderson 1970.)

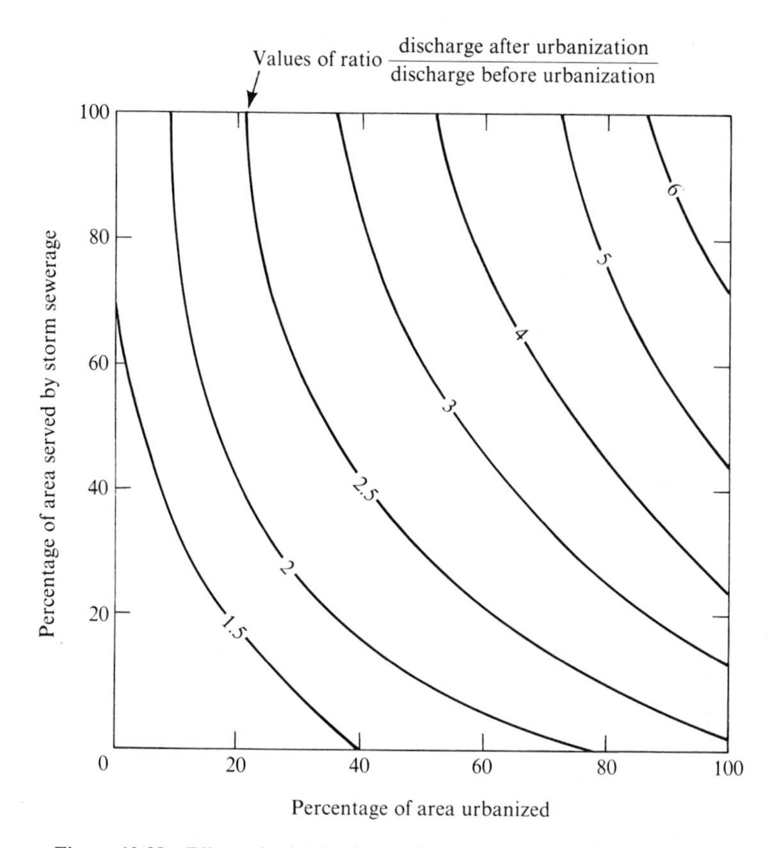

Figure 10-28 Effect of urbanization and storm sewerage on mean annual flood for a one-square-mile basin. Family of lines define the ratio of peak discharge after urbanization to peak discharge before urbanization. Complete (100%) urbanization is approximately equivalent to 50 percent of the area being impervious. (From Leopold 1968.)

and from the 40-percent curve the ratio of the 50-year flood to the mean annual flood is read.

In regions similar to the Washington area, this technique could be used in the manner described above. Elsewhere, the regional flood-frequency curves for rural and completely impervious basins would be derived and a new interpolation made.

Leopold (1968) drew together the results of various studies on the hydrologic impact of urbanization. Figure 10-28 shows the effect of urbanization on the size of the mean annual flood from one-square-mile basins. Two parameters are used to describe the extent of urbanization, the percentage of area served by storm sewers and the percentage of the area urbanized. Complete, or 100 percent, urbanization is generally equivalent to 50 percent of the area rendered impervious. The graph would have to be changed for different drainage areas and for flows of different frequencies, but it is a useful illustration of the effects of urbanization on catchments of a size of interest to planners. Rantz (1971) developed a similar set of graphs shown in Figure 10-29 from work by James (1965). They are to be used in conjunction with the regression equations listed in Table 10-15 for floods of various recurrence intervals in natural catchments. Floods of various recurrence intervals for the rural condition are estimated from the table and are then multiplied by the number on the appropriate curve in Figure 10-29. The appropriate values for "percentage of channels sewered" and "percentage of the basin developed" can be taken from a map of proposed land use to predict the hydrologic impact of urbanization.

The Unit Hydrograph

The flood-prediction techniques described previously only allow estimates of storm runoff volumes and peak flows. For many planning purposes one needs to predict the form of the storm hydrograph, that is, the time distribution of runoff throughout the storm. Hydrographs are used for engineering planning such as reservoir design or for assessing the influence of flood-detention structures in reducing flood peaks. The whole hydrograph is also necessary when hydrographs from dissimilar tributary areas are added and routed downstream to a channel reach of interest. The duration of flooding, a critical factor in many planning problems, can also be studied from a prediction of the hydrograph.

The most common technique of hydrograph prediction involves construction of the *unit hydrograph.* The method is approximate and is subject to many theoretical difficulties, but for more than 30 years has given answers that are sufficiently accurate for most planning purposes, usually predicting flood peaks within ±25 percent of their true value. Another advantage of the method is that it allows predictions to be made from only a short record of rainfall and runoff. Results can be regionalized to allow predictions for ungauged basins. After describing the construction of unit hydrographs for rural catchments, we will illustrate their application to urban areas so that the planner can use the technique for predicting the hydrologic impact of various degrees of urbanization.

The unit hydrograph of a basin is the hydrograph of one inch of storm runoff generated by a rainstorm of fairly uniform intensity occurring within a specific period of time. The specified duration of the rainstorm varies and will be discussed later. There is some difference in practice and in the hydrologic literature about the characteristics of the rainstorm referred to in the definition. Some hydrologists require that the storm be distributed uniformly over the catchment, while others refer only to a rainstorm whose spatial distribution reflects fixed basin characteristics such as topography or usual storm tracks. The latter usage is more realistic.

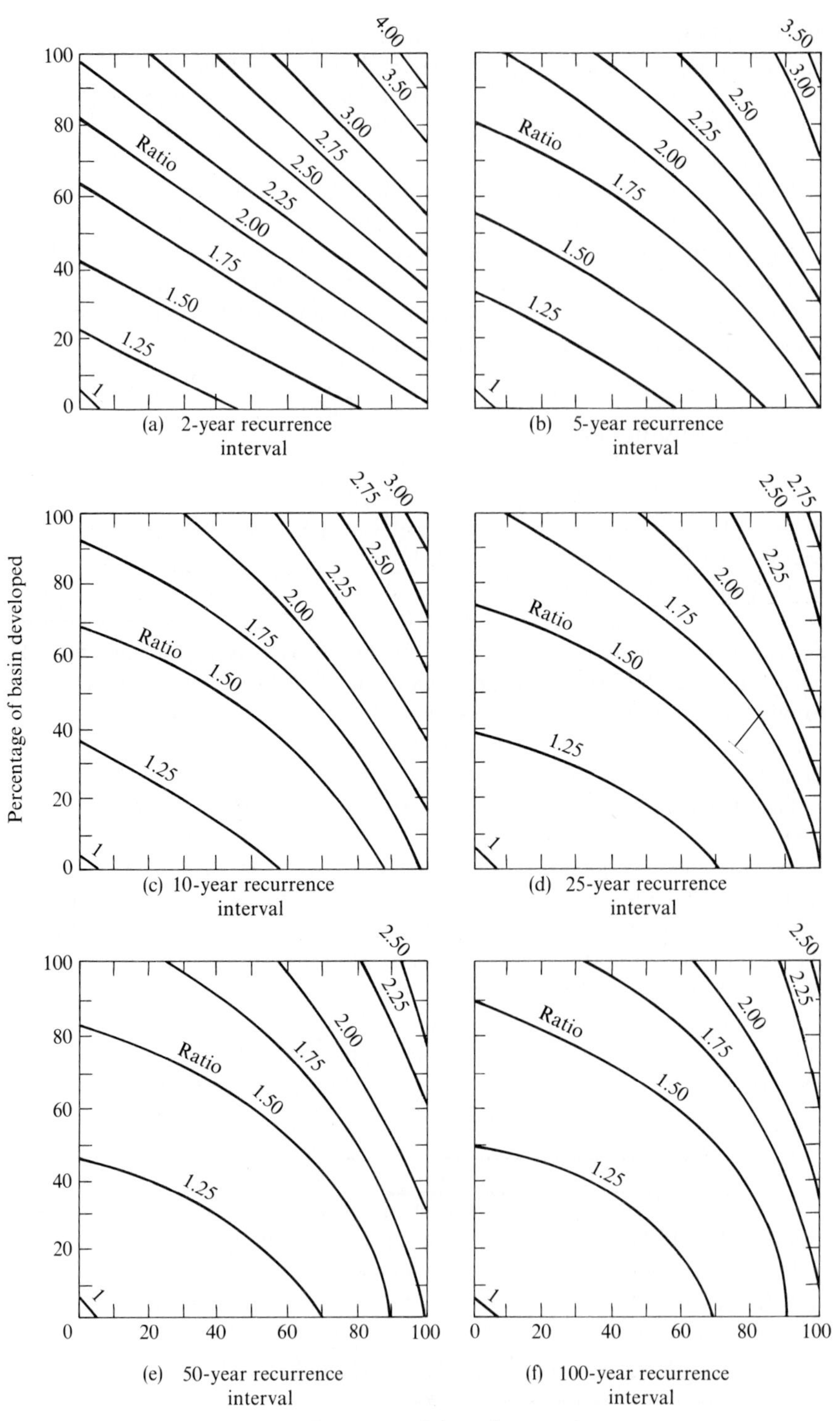

(a) 2-year recurrence interval

(b) 5-year recurrence interval

(c) 10-year recurrence interval

(d) 25-year recurrence interval

(e) 50-year recurrence interval

(f) 100-year recurrence interval

Figure 10-29 Ratios of flood-peak magnitude for urbanized basins to that for unurbanized basins for floods of various recurrence intervals. One hundred percent of the basin is roughly equivalent to 50 percent of the area being impervious. (From Rantz 1971.)

Many of the catchment characteristics that affect the timing of runoff (i.e., the shape of the flood hydrograph) are fixed from storm to storm. These include the size and shape of the drainage basin; the pattern, density, gradient, and size of stream channels; and the pattern of soils and land use. Because of the constancy of these controls, one might expect rainstorms of a specific duration to produce hydrographs of the same shape and duration. If the time base, X, of the hydrograph in Figure 10-30 is fixed for rainstorms of a specific duration, D_r, and the temporal pattern of runoff is determined by the constant drainage-basin characteristics referred to above, then the ordinates of the hydrograph (i.e., the discharge at various times) should be proportional to the total volume of storm runoff generated. If the shaded hydrograph in Figure 10-30 encompasses one inch of runoff, it is defined to be the unit hydrograph for the rainstorm with a duration D_r. All hydrographs generated by rainstorms of duration D_r will have the same time base. If the total volume of storm runoff is 2 inches, the ordinates of the hydrograph will be twice those of the unit hydrograph. If we can predict the volume of direct runoff that will be produced by a storm of similar duration, we can compute the hydrograph by altering the ordinates of the unit graph

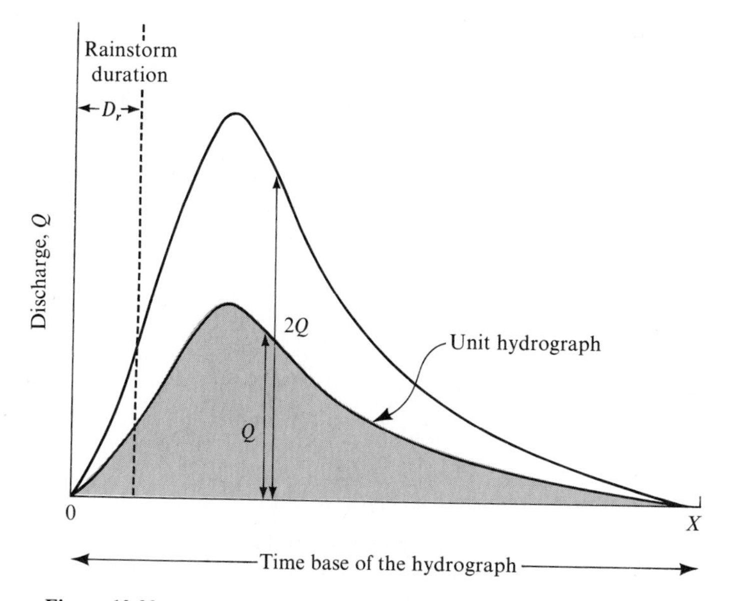

Figure 10-30 A unit hydrograph (shaded) and a hydrograph consisting of 2 inches of runoff, obtained by doubling the ordinates of the unit hydrograph on a fixed time base.

by the ratio of the storm runoff volume to one inch of runoff, as indicated in Figure 10-30.

Theoretically, one should compute a unit hydrograph for each storm duration. In practice, this is not necessary, though sometimes several unit hydrographs of different durations are derived for a basin by grouping storms into categories according to their length. We will show later how unit hydrographs of any duration can be calculated once a graph is derived for one duration. Unit hydrographs are often derived for durations of the most common or most critical storms in a region. Figure 10-31 shows the durations of storms for which unit hydrographs have been successfully constructed by various authors. For basins of less than 10 square miles, the storm duration is commonly taken as one-third to one-quarter of the time of concentration, or as 20–25 percent of the lag to peak. These are only rough guidelines, however, and should be checked in the region of application. The choice of an appropriate storm duration comes down to a matter of the one that gives the most consistent unit hydrographs over a number of storms. It should be remembered that the duration to which we are referring is that of the *rainstorm* generating the unit hydrograph, *not* to the duration of the hydrograph. A 6-hour unit graph is one that results from a 6-hour rainstorm; the duration of the resulting storm runoff may exceed 24 hours.

Unit hydrographs are constructed using the following steps (see Typical Problem 10-15):

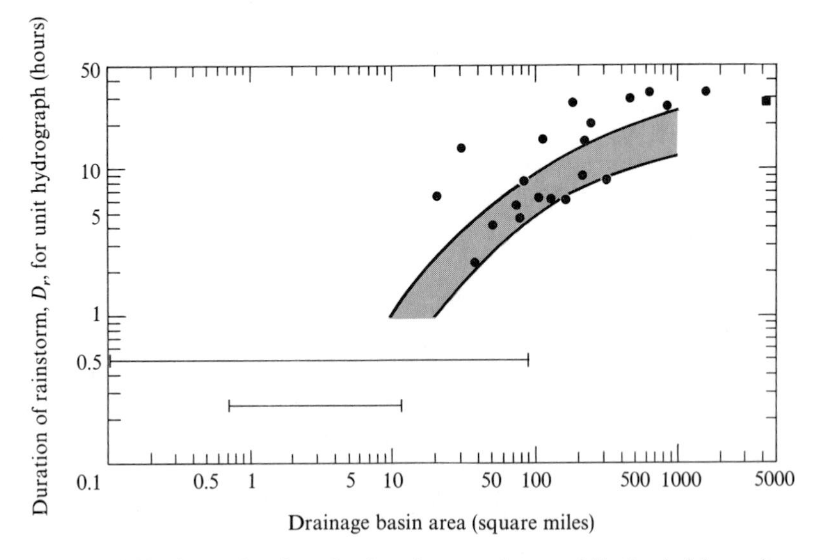

Figure 10-31 Storm durations that have been used successfully for deriving unit hydrographs from catchments of various sizes. The shaded area indicates a range of commonly used durations; the other data are from particular basins in the South Atlantic states (solid square, ■), the North Atlantic states (solid circles, ●). Texas and California (bars). (Data from Snyder 1938, Taylor and Schwarz 1952, Espey et al. 1966, Rantz 1971.)

1. Choose four or five hydrographs from storms of intense, moderately uniform rain.

2. Plot each hydrograph on a sheet of graph paper, and separate the stormflow and baseflow by one of the techniques illustrated in Figure 10-4.

3. By use of a planimeter or by counting squares on the graph paper, or from the tabulated stream record if one is available, compute the total amount of stormflow in inches.

4. Reduce the ordinates of the stormflow graph to their equivalent values for 1 inch of runoff by dividing each ordinate by the ratio of the total amount of stormflow to 1 inch.

5. Plot the reduced hydrographs and superimpose them, each hydrograph beginning at the same time.

6. Fix the peak of the unit hydrograph by computing the average discharge of all the peaks, and their average time of occurrence.

7. Sketch the unit hydrograph to conform to the average shape of the reduced hydrographs, passing through the computed peak and having a volume of 1.0 inch.

If hydrographs with more than one peak are used, the method of derivation becomes rather complicated (see Collins 1939).

Prediction of the hydrograph of a future storm from the unit hydrograph requires an estimate of the probable volume of stormflow by the techniques discussed earlier in this chapter. If a prediction is made that a certain rainstorm will generate 2.5 inches of stormflow, the ordinates of the unit hydrograph are multiplied by 2.5, and the storm runoff thus computed is added to the estimated baseflow to produce the desired hydrograph.

Unit Hydrographs for Storms of Various Durations

If a 2.5-hour storm generating 1 inch of storm runoff is followed immediately by another of equal length, producing the same amount of runoff, the result will be as shown in Figure 10-32. A second unit hydrograph is produced, which lags the first one by 2.5 hours, and the discharge of the second is added to that of the first. The result is a hydrograph of 2.0 inches of runoff generated by a 5-hour rainstorm. If the ordinates of this graph are halved, one obtains the unit hydrograph (1.0 inch of stormflow) for a 5-hour storm.

This method can be extended by adding many successive 2.5-hour unit hydrographs to obtain Figure 10-33. The resulting hydrograph is known as the *summation-curve,* or *S-curve* for a 2.5-hour unit hydrograph. It represents the hydrograph from an infinitely long rainstorm generating direct runoff at a rate of 0.4 in/hr (i.e., the same rate of runoff as in a 2.5-hour unit hydrograph). From the *S*-curve can be obtained the unit hydrograph for any other

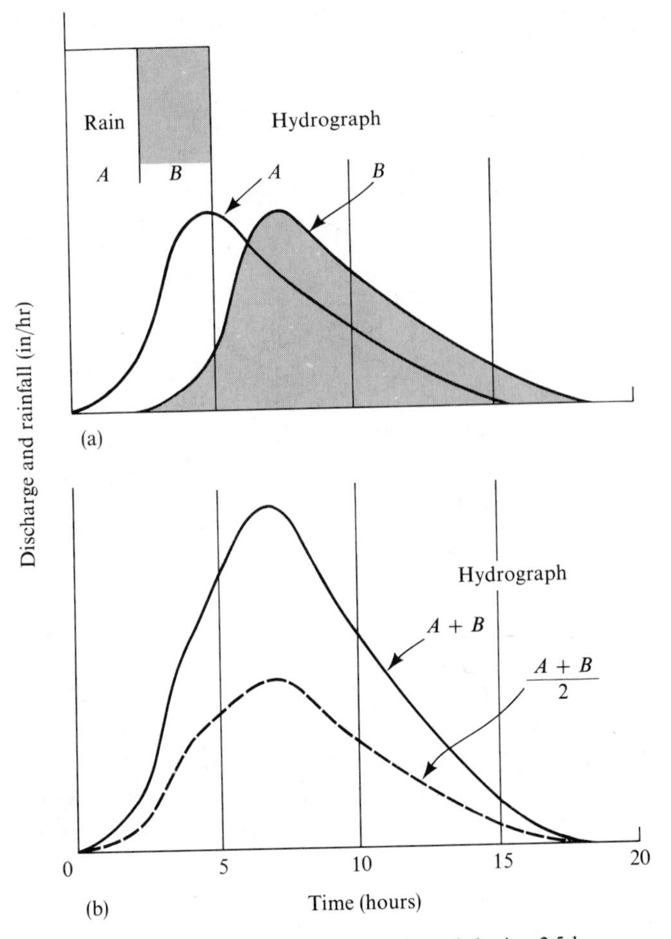

Figure 10-32 (a) Two consecutive storms, each lasting 2.5 hours, produce unit hydrographs A and B. (b) In the channel these are added to produce hydrograph $A + B$, the area under which is 2 inches. Halving the ordinates of the 2-inch hydrograph gives the unit hydrograph for a 5-hour storm $(A + B)/2$.

duration. For periods greater than 2.5 hours the procedure of adding unit hydrographs to obtain one with a longer duration is outlined above. Hydrographs with shorter durations (say, 1 hour) can be obtained by a process of subtracting S-curves. Two such S-curves, representing runoff rates of 0.4 in/hr, can be offset by 1 hour, as shown in Figure 10-34(a). The differences between the ordinates of the two curves gives the ordinates of a hydrograph consisting of 0.4 inches of runoff (Figure 10-34(b)). The multiplication of these ordinates by 2.5 gives the 1-hour unit hydrograph for the basin (Figure 10-34(c)).

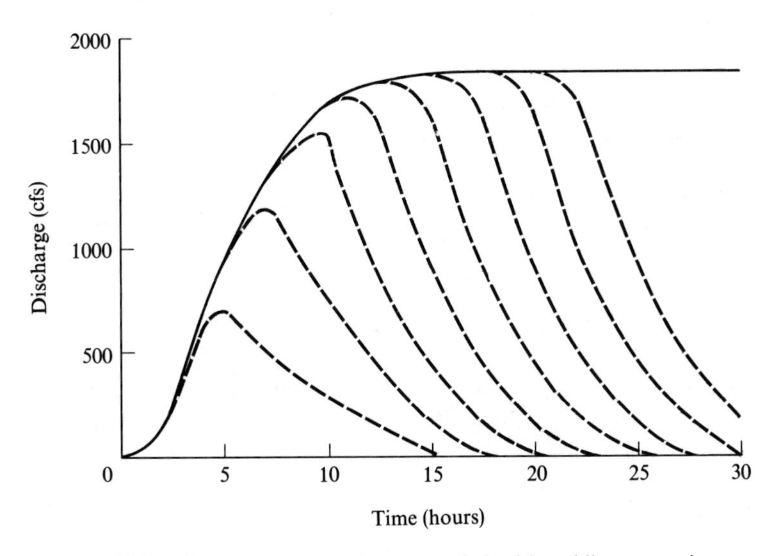

Figure 10-33 *S*-curve, or summation-curve, derived by adding successive 2.5-hour unit hydrographs from Figure 10-32.

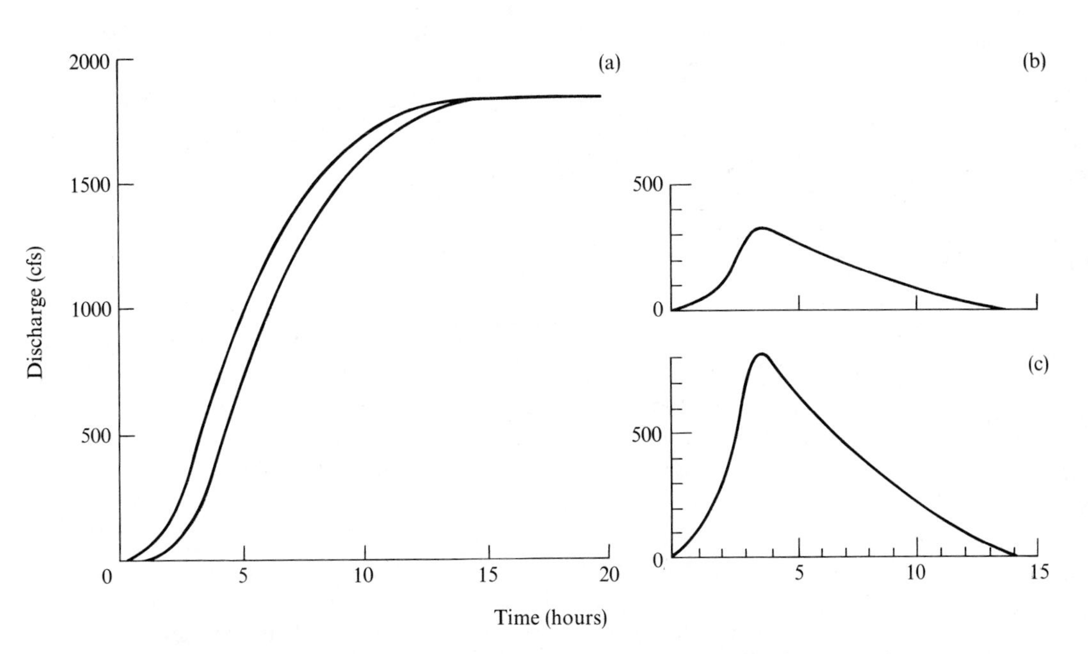

Figure 10-34 Construction of unit hydrographs for a chosen storm duration of one hour. (a) Two *S*-curves for a 2.5-hour storm offset by 1 hour. (b) Subtraction of *S*-curves in (a) gives a hydrograph of 0.4 inches of runoff produced by a 1-hour storm. (c) Multiplication of ordinate values of (b) by 2.5 gives a 1-hour unit hydrograph (volume of runoff is 1 inch).

Synthetic Unit Hydrographs

Despite the large number of streams that are gauged in the world, most planning problems involve ungauged basins. Estimates of the unit hydrograph for an ungauged stream can be made if the information derived from the analysis of gauging records on other streams is regionalized. This is done by constructing *synthetic unit hydrographs.* Unit hydrographs are computed for the gauged streams in an area, and the lag, peak, and duration of these hydrographs are related to geomorphic parameters of the catchment, such as drainage area, channel gradient, and drainage density. These geometric characteristics of the catchment represent the constant factors that affect the storage and transmission of a volume of runoff generated by a rainstorm, and thereby control the temporal distribution of storm runoff. Once correlations have been established between geomorphic variables and the characteristics of the unit hydrograph, an estimate of these latter characteristics can be made for ungauged basins from a few measurements of their physical geography. If the correlations include some variables related to human activity, then the synthetic unit hydrograph technique can also be used to predict the hydrologic consequences of development. This will be illustrated later by examples of synthetic unit hydrographs for urban catchments.

The procedure for developing synthetic unit hydrographs was introduced by Snyder (1938), who correlated the timing and peak rates of hydrographs derived for basins in the Appalachian Mountains with measures of the physiography of the catchments. The most important parameter that must be estimated for a synthetic unit hydrograph is the time lag (t_p) between the center of mass of the rainstorm and the peak of the hydrograph. For catchments varying in size from 10 to 10,000 square miles, Snyder was able to correlate this lag to peak with the length of the drainage basin in a form expressed in the following equation:

$$t_p = C_t(LL_c)^{0.3} \tag{10-8}$$

where t_p is the lag to peak (hours), L is the length of the mainstream from outlet to divide (miles), and L_c is the distance from the outlet to a point on the stream nearest the centroid of the basin (miles). The coefficient, C_t, varied from 1.8 to 2.2 in Snyder's study of Appalachian basins, with the lower values generally being associated with steeper basins, as expected.

In other regions the same general form of equation has been used and a new coefficient obtained to represent local physiographic conditions. Values of C_t mentioned in the literature vary from 0.3 for very steep mountain basins to 8 or 10 for lowlands. In many regions, however, we have been able to obtain good correlations between lag to peak and the area of the catchment, as shown for two regions in Figure 10-35. Such a graph can be rapidly constructed after measuring the lags on a few hydrographs generated by moderately uniform storms of the appropriate duration. In that figure it can be seen that the runoff process affects lag time for basins of the same drainage

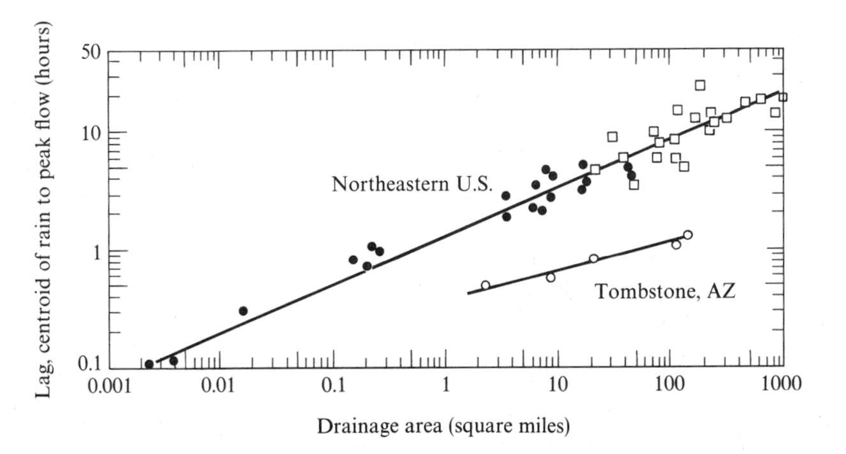

Figure 10-35 Lag to peak as a function of drainage area. The upper curve is for some basins in northeastern United States where variable-source contributions with large amounts of saturation overland flow occur. The lower curve is for basins in Arizona, where Horton overland flow occurs. Data apply to rural conditions. The open squares represent data from Taylor and Schwartz (1952). Solid circles are from the Sleepers River Experimental Watershed, northeastern Vermont, and the open circles are from Experimental Watersheds of the U.S. Department of Agriculture at Tombstone, Arizona.

area. In the northeastern United States, moist soils near stream channels become saturated and contribute the storm runoff. In Arizona, Horton overland flow dominates, and runoff follows rainfall more quickly. The lag of storm runoff generated by different processes is discussed further by Dunne (1978).

The duration of the rainstorm generating each unit hydrograph in Snyder's Appalachian study was related to the lag to peak in the following manner:

$$D_r = 0.18 t_p \tag{10-9}$$

where D_r is duration in hours. Again, this relationship is not necessarily appropriate for other regions. The data of Taylor and Schwarz (1952), for example, suggest that for basins larger than 20 square miles, D_r and t_p should be approximately equal in New England. The peak discharge of the unit hydrograph for storms of duration D_r was found by Snyder to be correlated with the drainage area as

$$Q_{\text{pk}} = \frac{C_p A}{t_p} \tag{10-10}$$

where Q_{pk} is the peak discharge of the unit hydrograph (cfs), A is the drainage area (mi^2), and C_p is a coefficient ranging from 370 when C_t is high to 440 when C_t is low, and averaging 405.

The time base, or duration, of the unit hydrograph was correlated with the lag to peak as

$$t_b = 72 + 3t_p \qquad (10\text{-}11)$$

where t_b is the time base of the unit hydrograph (hours). This equation is obviously not valid for small catchments that have hydrographs of about one day's duration, but in large basins its use will not usually cause large errors, especially in humid regions with significant drainage of subsurface stormflow. Taylor and Schwarz (1952), who worked in the central and northern Appalachians, suggest the expression

$$t_b = 5(t_p + 0.5D_r) \qquad (10\text{-}12)$$

In small catchments the Soil Conservation Service method described on page 342 should give better results since it was developed from data on small catchments. Again, however, the estimated values should be checked against some measured hydrographs from basins in the region of interest.

In order to define a unit hydrograph from a storm of different duration, say D_R, the lag to peak t_{pR} was found by Snyder to equal

$$t_{pR} = t_p + 0.25(D_R - D_r) \qquad (10\text{-}13)$$

This lag time is then substituted into Equations 10-10 and 10-11 to obtain the peak and duration of the new unit hydrograph. Snyder's equations have been tested elsewhere, and their general form is applicable to unit hydrographs from other regions. To obtain good estimates of the lag and peak, however, unit hydrographs from gauged basins in the region should be used to evaluate C_t, C_p, and D_r.

Some treatments of the synthetic unit hydrograph evaluate the lag from the centroid of rainfall to the centroid of runoff. This lag is often correlated with the ratio $L/\sqrt{S}$, where L is the mainstream length and S (ft/mi) is the slope of the main channel between points located 10 and 85 percent of the length (L) upstream from the basin outlet. This quotient can then be used to obtain centroid lags for synthetic unit hydrographs (see, for example, Carter 1961, and Figure 10-26).

The centroid lag can also be estimated from the drainage-basin area as indicated in Figure 10-36, in which the lag to centroid is plotted. Again the differences between regional curves reflect differences of runoff processes and various physiographic characteristics. Plotting of centroid lags for a few hydrographs from a range of catchment sizes will define the curve for a region.

There is a wide enough variation among the empirical formulas for lag time that it is not obvious which should be used in a particular circumstance. For example, in Figure 10-36 there is a threefold difference in lag time for an area of one square mile, depending on whether the runoff is generated by saturation overland flow from variable sources or by Horton overland flow. Even larger lags occur where subsurface stormflow is the dominant source

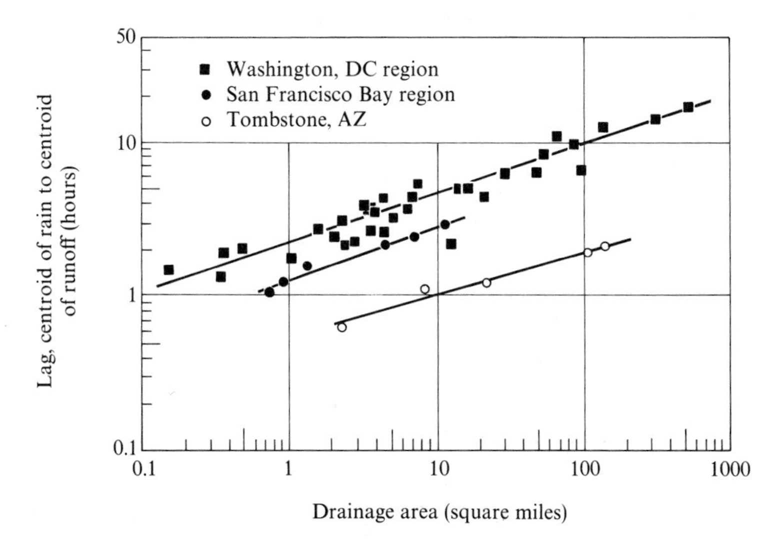

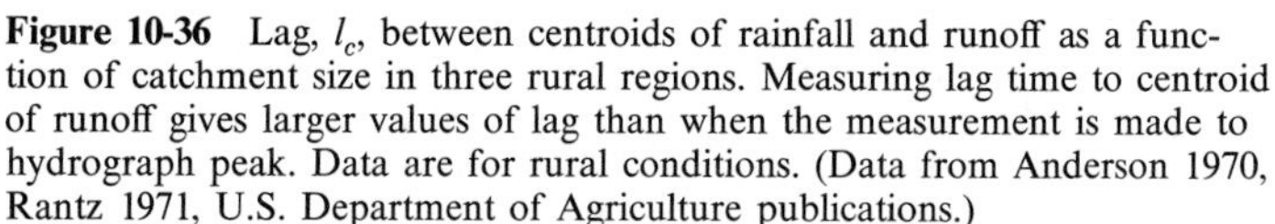

Figure 10-36 Lag, l_c, between centroids of rainfall and runoff as a function of catchment size in three rural regions. Measuring lag time to centroid of runoff gives larger values of lag than when the measurement is made to hydrograph peak. Data are for rural conditions. (Data from Anderson 1970, Rantz 1971, U.S. Department of Agriculture publications.)

of storm runoff. A planner may not know which of the processes dominates in the area of concern. Streamflow records may be numerous, but if climatic conditions vary greatly over short geographic distances as in parts of California, the choice among gauging stations may be difficult. Also, the gauging station records probably apply to drainage areas considerably larger than those with which many planners deal. Confidence can be increased greatly by a direct observation of lag time on the area being studied. The technique is easy on very small basins. A single rainstorm is sufficient to obtain usable results.

A staff gauge is placed in the channel of the basin of interest. The gauge plate need not be of enameled metal as at gauging stations. A yardstick, meter stick, or stick of wood marked with a scale having graduations to 0.2 foot can be used. It is driven into the channel bed in a straight reach of channel and supported by diagonal braces above the expected water surface.

Lag to peak can be determined by plotting the gauge height and rainfall against time through a storm. A plastic rain gauge is installed in an open area, not shaded by trees, in the vicinity of the gauge plate. If a cheap plastic rain gauge is not available, any tin can may be used.

During a rainstorm, rain gauge readings should be made at 10-minute or 15-minute intervals, and if the basin under study is 0.1 of a square mile or less, the staff gauge should be read about every 3 minutes. The time of reading the staff gauge and the rain gauge should be recorded.

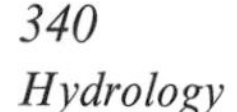

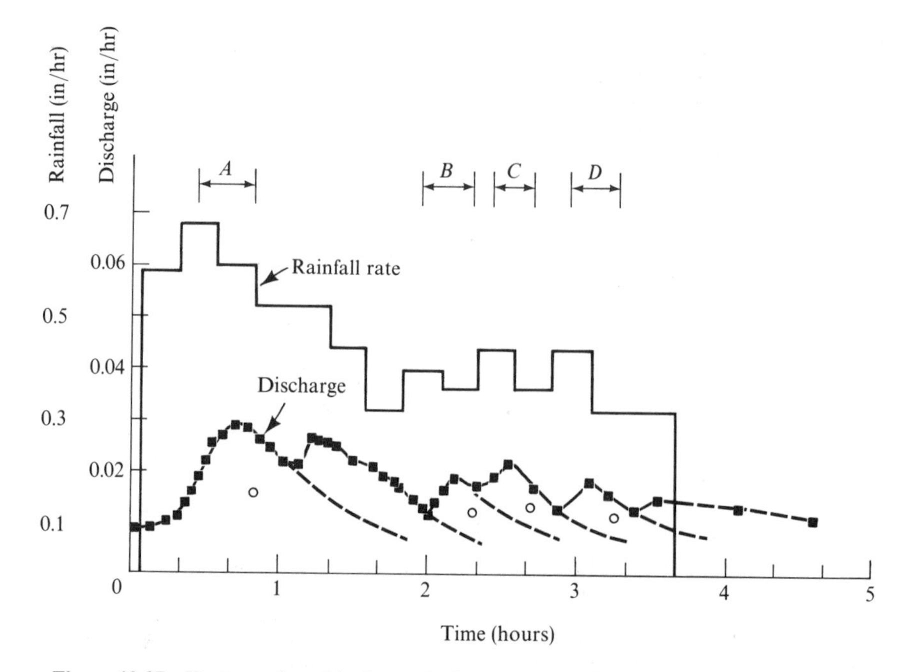

Figure 10-37 Hyetograph and hydrograph for a storm on Cerrito Creek at Leopold's house, Berkeley, California, an urbanized area of 0.068 sq mi. The data from this storm furnished four estimates of centroid lag, *A–D*, the average of which is 20 minutes. The lag to peak is 11 minutes.

Figure 10-37 presents data obtained in this way for a storm on February 29, 1976. The basin observed is Cerrito Creek at Leopold's home, Berkeley, California, where the area of the basin, mostly urbanized, is 0.068 square miles. The rain gauge was read at intervals of 15 minutes and the staff gauge every 3 to 4 minutes during a storm that lasted 3 hours. Each solid square on the diagram represents a staff gauge reading. Four bursts of rainfall caused identifiable rises in the stream discharge.

The dashed lines are an estimate of how runoff would have receded following each rise if the rainfall had ceased at that time. The open circles within the hydrograph area indicate the positions of the centroid of runoff for each individual hydrograph rise.

The time lags between the center of each rainfall burst and the centroid of resulting runoff are labeled *A* to *D*. The average of these four estimates of lag is 20 minutes. Plotting this lag in Figure 10-36 at a drainage area of 0.068 square miles places the point at a slightly smaller lag time than would be expected from the extrapolation of the San Francisco Bay region curve. This is logical because Cerrito Creek Basin is urbanized, whereas the Bay region curve applies to rural conditions.

The procedure for developing synthetic unit hydrographs by the methods of Snyder and others, then, is as follows:

1. The required geomorphic parameters of the basin are measured from maps.

2. The lag to peak of the unit hydrograph is calculated from an equation with the same form as Equation 10-8, from a regional relationship such as those shown in Figure 10-35, or from a field measurement.

3. The duration of the unit storm is calculated from Equation 10-9. If this is the duration of interest in the planning problem, proceed to the next step. If not, the lag to peak is calculated for the desired duration using Equation 10-13.

4. The peak discharge of the unit hydrograph is calculated from Equation 10-10, with or withour the modification for storm duration. The value of C_p should be checked for the region if at all possible.

5. The time base of the unit hydrograph is calculated from one of the methods described above and is checked against field observation if possible.

6. The centroid of rainfall ($0.5D_r$ after the onset of rain), the lag to peak (t_p), peak discharge (Q_{pk}) and time base (t_b) are plotted on a sheet of graph paper, and the unit hydrograph is sketched through the three points to encompass an area representing a storm runoff volume of one inch, as shown in Figure 10-38. As an aid to sketching the hydrograph, the U.S. Army Corps of Engineers has developed two empirical formulas for estimating the width

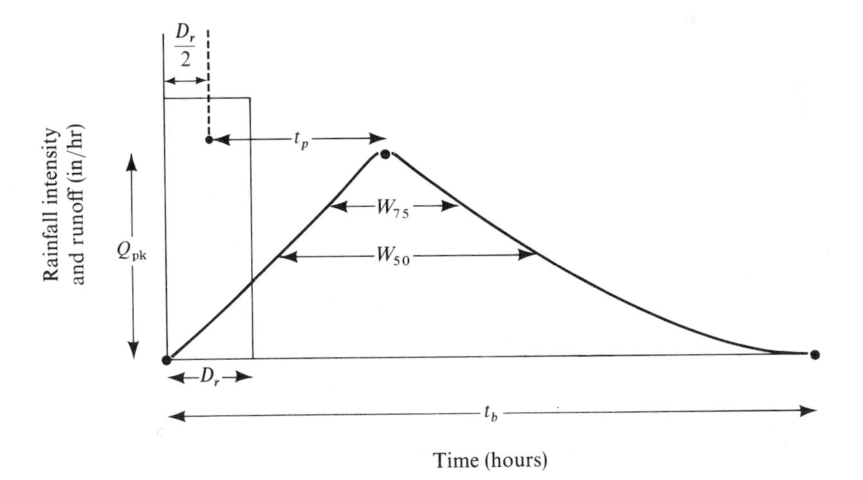

Figure 10-38 Construction of the synthetic unit hydrograph. It is obtained by plotting the three points shown by solid circles, which are evaluated by the methods described in the text. The hydrograph is then sketched through these points and is adjusted to encompass an area representing one inch of runoff. All symbols are defined in the text.

of the unit hydrograph at discharges equal to 50 percent and 75 percent of the peak. The relationships are

$$W_{75} = \frac{440A}{Q_{pk}^{1.08}} \tag{10-14}$$

$$W_{50} = \frac{770A}{Q_{pk}^{1.08}} \tag{10-15}$$

where W_{75}, W_{50} = width of the unit hydrograph at discharges equal to 75 and 50 percent of the peak (hours), A is the drainage area (square miles), and Q_{pk} is the peak discharge.

7. The desired baseflow is added to the storm hydrograph.

The U.S. Soil Conservation Service (1972) by the study of unit hydrographs from a large number of small drainage basins has developed modifications of the synthetic unit hydrograph technique. These methods are most suited to drainage areas of less than 100 square miles. The simplest method is an approximation of the unit hydrograph by a *triangular unit hydrograph*, illustrated in Figure 10-39 and derived as follows:

From the area of the triangle in Figure 10-39, the volume of runoff, R (inches), is equal to

$$R = \frac{Q_{pk}T_p}{2} + \frac{Q_{pk}T_r}{2}$$

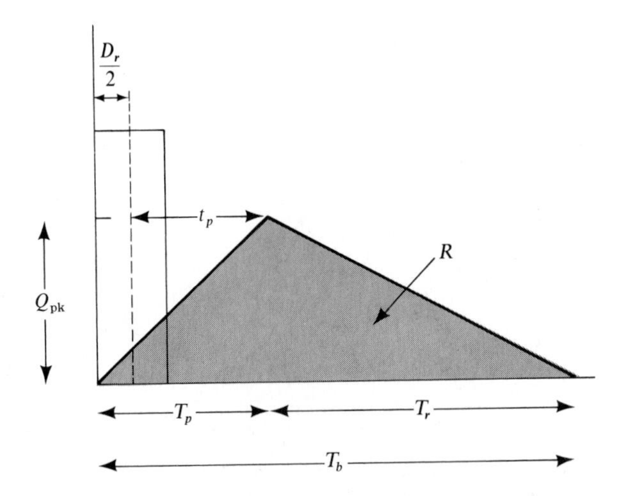

Figure 10-39 The triangular unit hydrograph as defined by the U.S. Soil Conservation Service, where R is the volume of storm runoff (in), Q_{pk} is the peak rate of runoff (in/hr), T_p the time of rise to the peak (hr), T_r the duration of the recession limb (hr), T_b the time base or duration of rainfall excess. Note the slight differences in the symbols used in this diagram and in Figure 10-38.

where Q_{pk} is peak discharge (in/hr), T_p is time duration of the rising limb, and T_r is duration of the recession limb. Therefore,

$$Q_{\mathrm{pk}}(\mathrm{in/hr}) = \frac{2R}{T_p + T_r} \tag{10-16}$$

Examination of a large number of hydrographs from small agricultural basins throughout the United States led to the empirical generalization that

$$T_r \approx 1.67 T_p \tag{10-17}$$

This generalization should be checked before application to a specific region, but since the coefficient 1.67 is low by comparison with coefficients from most regions in which we have worked, its use will give results on the safe side for design purposes.

Substituting Equation 10-16 into Equation 10-17 yields

$$Q_{\mathrm{pk}}(\mathrm{in/hr}) = \frac{0.75R}{T_p} \tag{10-18}$$

The coefficient varies from about 0.5 for flat, swampy catchments to 0.9 for mountains, as the ratio of T_p to T_r changes somewhat from that indicated in Equation 10-17. To convert from inches per hour to cfs, Equation 10-18 is multiplied by 645.3 and the basin area (1 in/hr = 645 cfs/sq mi), giving

$$Q_{\mathrm{pk}}(\mathrm{cfs}) = \frac{484AR}{T_p} \tag{10-19}$$

The time of rise, T_p, consists of one-half the duration of rainfall ($0.5D_r$) plus t_p, the lag between the centroid of rainfall and the peak of the hydrograph. Another empirical generalization developed by the Soil Conservation Service for small drainage basins is that the lag to peak, t_p, is approximately equal to $0.6t_c$, where the time of concentration can be estimated from Equation 10-3. Therefore,

$$T_p = 0.5D_r + 0.6t_c \tag{10-20}$$

and

$$Q_{\mathrm{pk}} = \frac{484AR}{0.5D_r + 0.6t_c} \tag{10-21}$$

Again, it is usually better to measure t_p or to estimate it from the drainage area, as indicated in Figure 10-35. A good indication of the form of such a relationship for small agricultural basins can be obtained from the hydrographs published by the U.S. Department of Agriculture in their annual publication *Hydrologic Data for Small Agricultural Watersheds in the United States.* Some of the lag values derived in this way are strikingly different from those obtained by using the hypothetical time of concentration. If the duration of the appropriate unit storm is not known, it can be estimated as one-third to one-quarter of the lag to peak.

The triangular unit hydrograph is a useful tool for small basins, especially on agricultural lands and where Horton runoff conditions dominate the storm hydrograph. For most design purposes the triangular approximation to the true hydrograph form is acceptable. The technique is deceptively simple, however, and the results should be checked against field observations of flows wherever possible. When more detail is needed in the form of the unit hydrograph, the Soil Conservation Service uses a *dimensionless unit hydrograph,* illustrated in Figure 10-40, and derived from a variety of small catchments, differing in size and physiography. It is used by first calculating Q_{pk} and T_p from one of the methods described previously and obtaining other values of time and discharge for the hydrograph as ratios to Q_{pk} and T_p from the figure.

The dimensionless unit hydrograph discussed above is comparable to the dimensionless hydrograph of Langbein (1940) (see Figure 10-41) in which the time scale is in percent of centroid lag (l_c) rather than percent of the rising limb duration (T_p). In the Langbein graph the peak of the hydrograph occurs at a time after the beginning of runoff equal to 60 percent of the lag between centroid of rain and centroid of runoff. The hydrograph base, or total time for runoff, is about 4.0 times the centroid lag. To use this average unit graph, one estimates lag time from a formula or from the graph in Figure 10-36, and having chosen the volume of runoff, the distribution of this volume into hydrograph form is accomplished by the use of the average unit graph (see Typical Problem 10-20).

The Soil Conservation Service triangular and dimensionless unit hydrographs employ a standard hydrograph form to all catchments. If the hydrologist does not want to rely on such a generalized procedure, he may adopt the approach described above but derive triangular hydrographs or a dimensionless unit hydrograph for his own region.

Rantz (1971) developed a slightly more complicated technique for rural and urban basins in the San Francisco Bay region, using a tool known as the *instantaneous unit hydrograph.* The next few pages draw heavily from Rantz's paper. In an earlier section referring to Figure 10-34, we introduced the procedure for changing the duration of the unit hydrograph by subtracting two *S*-curves. If the *S*-curves are brought closer and closer together, they will eventually be separated by only an infinitesimally small time increment. The difference between their ordinates (when multiplied by an appropriate factor) would then represent a hydrograph of one inch of runoff generated in an infinitesimal time period. The purpose of an instantaneous unit hydrograph is to provide some general relationships among principal hydrograph factors that can be applied to a storm of any short duration to obtain the unit hydrograph for that storm.

The development of such hydrographs is beyond the scope of this book, but Rantz obtained approximate instantaneous unit hydrographs for small basins from 15-minute unit graphs by subtracting a time interval equal to D_r (15 minutes) from T_b and subtracting $0.5D_r$ from T_p (see Figure 10-42). This does not change the centroid lag.

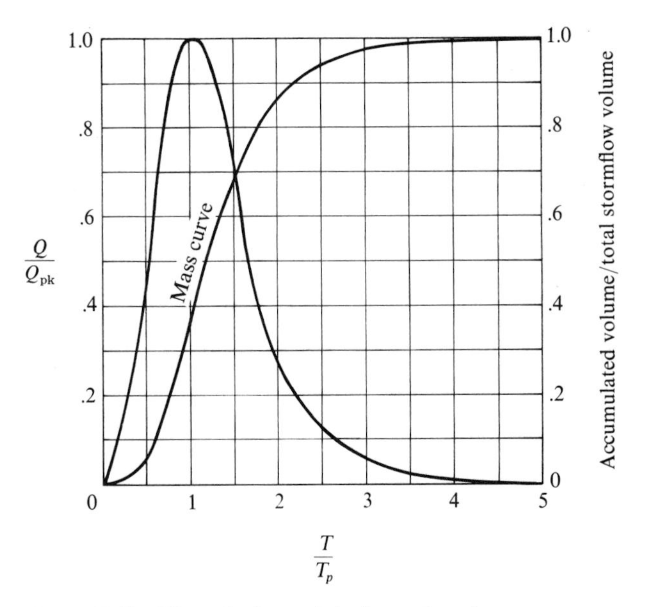

Figure 10-40 Dimensionless unit hydrograph and mass curve; the latter represents the accumulated area under the hydrograph. (From U.S. Soil Conservation Service 1972.)

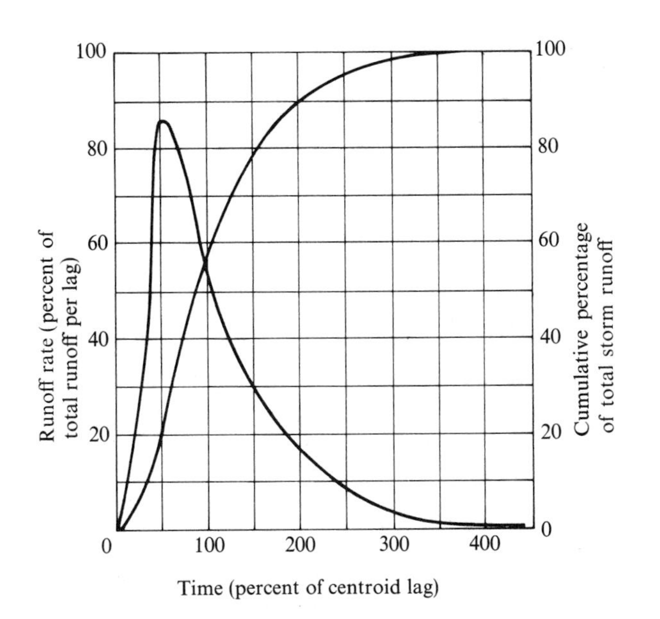

Figure 10-41 Langbein's summation graph and the derived unit hydrograph. (From W. B. Langbein 1940, *EOS, American Geophysical Union Transactions,* vol. 21, pp. 620–627. Copyrighted by American Geophysical Union.)

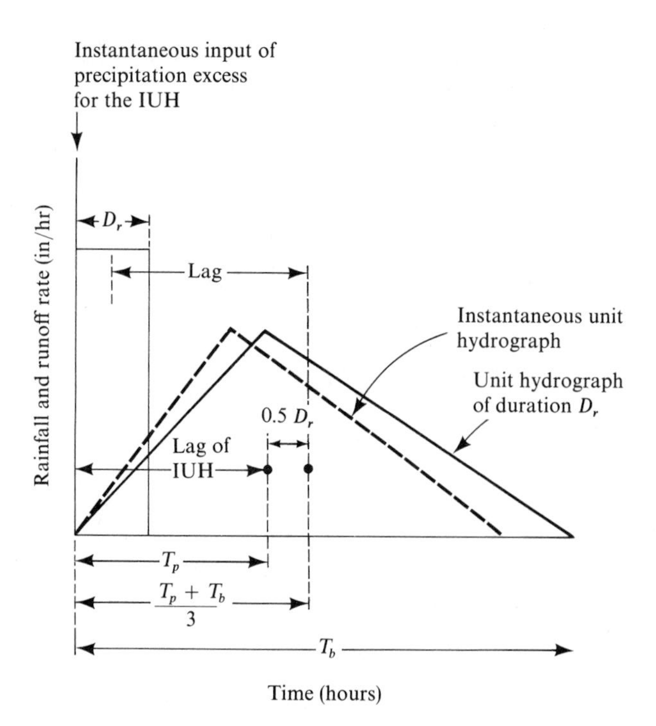

Figure 10-42 Basic triangular unit hydrograph and instantaneous unit hydrograph (IUH) used by Rantz (1971). The symbols are defined in the text. The solid circles represent the centroids of the two unit hydrographs, and it can be seen that the two hydrographs have the same lag between the centroids of rainfall and runoff.

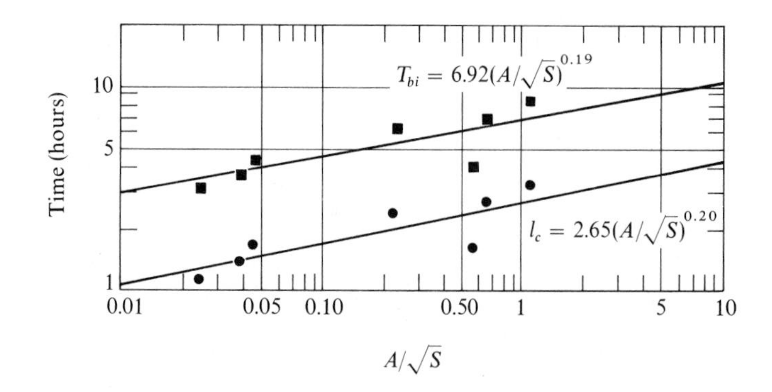

Figure 10-43 The centroid lag (l_c) and the duration of the unit hydrograph (T_{bi}) of the instantaneous unit hydrograph are related to the area and channel slope in rural catchments of the San Francisco Bay region. Drainage basin area (A) is expressed in square miles, and the channel slope (S) is measured in ft/mi between points located 10 and 85 percent of the distance along the channel from the outlet. (From Rantz 1971.)

Rantz then correlated the centroid lag of the instantaneous unit hydrographs from gauging stations with the characteristics of the gauged basins, as shown in Figure 10-43. With an estimate of the lag, Rantz was then able to construct the instantaneous triangular unit hydrograph. His usage, however, is slightly different from that of the Soil Conservation Service described previously. Recall that he could estimate the lag from basin characteristics. Because the hydrograph is a triangle, the time from the start of runoff in Figure 10-42 to the centroid of the hydrograph is equal to $(T_p + T_b)/3$. Therefore,

$$l_c = \frac{T_p + T_b}{3} - \frac{D_r}{2}$$

But since D_r for the instantaneous unit hydrograph is infinitesimally small, the last term is zero and

$$T_{pi} = 3l_c - T_{bi}$$

where the *i*-subscripts refer to the instantaneous unit hydrograph.

When the instantaneous unit hydrograph is used to obtain a unit hydrograph for a storm of finite duration, the subtractions of D_r and $0.5D_r$ described earlier are simply reversed and

$$T_b = T_{bi} + D_r \qquad (10\text{-}22)$$

$$T_p = T_{pi} + 0.5D_r \qquad (10\text{-}23)$$

The duration of precipitation excess, D_r, that should be used is one that lies between $0.2T_p$ and $0.33T_p$, according to Rantz.

Because the area under the triangle in Figure 10-42 represents one inch of runoff, T_b can now be used to calculate Q_{pk} from equations like 10-16 and 10-19:

$$Q_{\text{pk}}(\text{cfs}) = \frac{2 \times 1(\text{in}) \times 645.3(\text{cfs/mi}^2/\text{in/hr}) \times A(\text{mi}^2)}{T_b(\text{hr})}$$

$$Q_{\text{pk}} = \frac{1290.67A}{T_b} \qquad (10\text{-}24)$$

In applying the triangular synthetic unit graph for predictions the hydrologist must examine some local hydrologic records and decide what storm durations produce the critical floods in the drainage basins of interest. Most often, in the small drainage basins of interest to planners, the critical design storms are those of relatively short duration and high intensity. It seems appropriate to use a storm with a duration approximately equal to the centroid lag because by this time the area contributing runoff to the outlet will be a maximum. Note that we are now discussing the duration of some future design storm, *not* the duration, D_r, of the storm for which the triangular unit hydrograph was constructed. Rantz suggests using a storm duration of the next hour larger than the centroid lag. After choosing the appropriate duration, one reads the total storm depth for the chosen recurrence interval from the rainfall intensity-duration-frequency curves described in Chapter 2.

The chosen rainstorm duration will then be divided into time increments equal to D_r (between $0.2T_p$ and $0.33T_p$), and the unit hydrograph will be constructed as described above and applied to successive increments of precipitation as previously described. It is necessary, therefore, to decide how the precipitation will be distributed during the design storm. In Chapter 2 we described mass curves of storm rainfall and presented a table used by Rantz, and based upon data from the U.S. Soil Conservation Service, for estimating the time distribution of precipitation during storms with durations ranging from one to six hours. The total storm duration can be divided into increments of D_r, and from the average mass curves, the rain falling in each increment can be obtained. It is wise to find characteristic mass curves of rainfall for one's own region from the National Weather Service, the Soil Conservation Service, or a similar agency.

The next step is to compute the volume of runoff (precipitation excess) that will be generated during each time increment by the chosen rainfall. This was described in an earlier section. The lag and time base of the instantaneous unit hydrograph for the basin are determined from Figure 10-43, and are altered to the corresponding parameters of the synthetic hydrograph for a storm of duration D_r by means of Equations 10-22 and 10-23. The peak of this synthetic unit hydrograph is then obtained from Equation 10-24, and the triangle is defined. Its ordinates are multiplied by the volume of stormflow generated in the first time increment D_r of the storm, as previously calculated.

We now have a hydrograph that would be generated by rainfall occurring in the first increment of the storm. The process is repeated for successive increments, and the hydrographs are added together to define the runoff pattern for the entire design storm.

The baseflow should then be added. Because critical large storms usually occur on wet antecedent conditions, the baseflow may be quite large and should be estimated with care, using the statistical or water-balance techniques discussed elsewhere in this text or preferably some field observations. For the San Francisco Bay region, Rantz considered baseflow as a fixed percentage of peak flow, as shown in Table 10-16.

By using design rainfalls of various recurrence intervals, the investigator can construct hydrographs for storms of various frequencies, including the probable maximum storm. The method is illustrated in Typical Problem 10-21.

Synthetic Unit Hydrographs for Urban Areas

Espey et al. (1966) derived 30-minute unit hydrographs for rural and urban basins in Texas. They then selected several parameters of these unit hydrographs (such as T_p, T_b, and Q_{pk}) and used multiple regression techniques to relate each of these parameters to the length and gradient of the main stream, the extent of impervious cover, and the degree of alteration of the channels. The resulting equations can be used to predict changes in the unit hydrograph as the impervious area and channel improvements spread through the basin.

Rantz (1971) stated that almost all the studies of the effects of urbanization upon the unit hydrograph had shown centroid lags for completely urbanized basins to range from 10 to 50 percent of those for rural catchments. He therefore chose an average of 25 percent of the rural value for both the time to peak and the time base of the instantaneous unit hydrograph (i.e., for T_{pi} and T_{bi}). For partly urbanized basins he interpolated coefficients between 100 percent for a rural watershed and 25 percent for complete urbanization, as shown in Table 10-17. Note that "100 percent urbaniza-

Table 10-16 Baseflow for design floods in the San Francisco Bay region. (From Rantz 1971.)

RECURRENCE INTERVAL (YR)	BASEFLOW EQUALS PEAK DISCHARGE OF SURFACE RUNOFF TIMES THE PERCENTAGE LISTED BELOW
2	5
5	5
10	10
25	15
50	20
100	25

Table 10-17 Coefficients to convert T_{pi} and T_{bi} of the instantaneous unit hydrograph for rural catchments to those for urbanized catchments. (From Rantz 1971.)

PERCENTAGE OF CATCHMENT THAT IS URBANIZED	COEFFICIENT
0	1.00
10	0.92
20	0.85
30	0.78
40	0.70
50	0.62
60	0.55
70	0.48
80	0.40
90	0.32
100	0.25

tion" does not mean "100 percent impervious." Rantz suggests that in a fully urbanized region with a mixture of residential and business areas, about 50 percent of the basin is impervious. Once T_{pi} and T_{bi} are found for the urbanized basin, the time dimensions of the unit hydrograph for a finite time interval, D_r, can be obtained from Equations 10-22 and 10-23. The value of D_r selected should lie between $0.2T_p$ and $0.33T_p$, as in the rural case. The calculation of the unit hydrograph and the design hydrograph for a storm of any recurrence interval then proceeds as for rural basins. The volume of storm runoff for the urbanized catchment is calculated by the methods described in an earlier section of this chapter.

The synthetic unit hydrograph is a valuable tool for assessing the hydrologic effects of urbanization. For ungauged basins, the method allows the calculation of a spectrum of flood peaks if rainstorms of various frequencies and durations are used. From a planning map showing the future spread of urbanization through a catchment, it is possible to predict future changes in the storm hydrograph. Such predictions can be used to anticipate and avoid many problems, such as channel alteration, bridge and culvert failure, and the expansion of the flood-prone area into residential property that was previously safe. Synthetic unit hydrographs computed for various combinations of urban land use and flood-control methods can provide the planner with a quantitative indication of how best to limit the hydrologic impact of urbanization. We will consider some of the other tools he might need for this in the next section, and in Chapter 11 we will review some methods of flood control.

Flood Routing

Once a storm hydrograph has been calculated, other questions may be asked about it. What will happen to the peak rate and timing of runoff if the flood wave passes through a lake or reservoir or along a reach of channel? What will happen if flood waves move down two dissimilar tributary valleys and coalesce downstream? Such questions must often be asked when the probable effects of dams upon downstream floods are being considered, or when flood forecasts are made. Many engineering design problems, such as the planning of reservoir sites and of spillways on dams, the construction of levees, or the assessment of the probable effects of a flood-control dam on discharges past some community far downstream, also require a knowledge of the modifications of hydrographs as the flood wave moves downstream. Most planners, of course, will not be involved in the detailed design work for large engineering structures. They should, however, know the basic principles underlying such work so that they can communicate with the design team.

Some planners concerned with small areas are also finding that they need to make calculations of how flood waves move through small catchments and how they are affected by stormwater detention basins, farm ponds, and similar structures. To do this, the planner must be familiar with the simplest techniques used by engineers for larger reservoirs and channels. For these reasons, the subject of *flood routing* is now introduced.

The beginning of this chapter described qualitatively how river channels, lakes, and artificial reservoirs modify the form of a flood wave. The modifications involve both storage and translation of the wave, and the group of methods used for calculating these effects are known as *flood-routing techniques.* Some of the methods available require detailed knowledge of the hydraulics of river channels, information that is costly and time-consuming to collect. Such information is not used in most planning problems, except where costly ventures are being considered. The most frequently used routing methods are based on the simple statement that the mass of storm runoff remains constant. During any time interval, the water that flows into a reach of channel or a lake must contribute both to the outflow and to a change in the volume of water stored in the channel or lake. The simplest case of routing a flood wave through a reservoir is considered first.

Reservoir Routing

For a chosen time interval, Δt, the equation of continuity for a reservoir can be written as

$$I\,\Delta t = O\,\Delta t + \Delta S \tag{10-25}$$

where I and O are the average rates of inflow and outflow for the time interval (cfs), ΔS is the change in storage during the time interval (ft^3, or in cfs-hours), and Δt is the time increment being considered (hours).

The inflow and outflow rates are the averages of rates at the beginning and end of each time increment. The change in storage is the difference between the volumes in storage at the beginning and end of the time interval. If we use subscripts 1 and 2 to denote the beginning and end of each interval, we can rewrite Equation 10-25 as

$$\left(\frac{I_1 + I_2}{2}\right) = \left(\frac{O_1 + O_2}{2}\right) + \left(\frac{S_2 - S_1}{\Delta t}\right) \tag{10-26}$$

Equation 10-26 can be rewritten as Equation 10-27, which despite its odd-looking form is useful for our purpose:

$$\left(\frac{S_2}{\Delta t} + \frac{O_2}{2}\right) = \left(\frac{S_1}{\Delta t} - \frac{O_1}{2}\right) + \left(\frac{I_1 + I_2}{2}\right) \tag{10-27}$$

For each time increment, all but one of the values in Equation 10-27 are known and can be used to compute the single unknown value, which is O_2, the rate of outflow from the reservoir at the end of the time interval. I_1 and I_2 are the known rates of input to the reservoir (calculated by the unit hydrograph or a similar method). The storage at any time is related to the

Table 10-18 Evaluation of the storage–outflow relationship for a reservoir.

WATER ELEVATION ABOVE BOTTOM OF SPILLWAY (FT)	STORAGE ABOVE BOTTOM OF SPILLWAY (FT^3)	RATE OF OUTFLOW (CFS)	$\frac{S}{\Delta t}$ (CFS)	$\left(\frac{S}{\Delta t} - \frac{O}{2}\right)$ (CFS)	$\left(\frac{S}{\Delta t} + \frac{O}{2}\right)$ (CFS)
0	0	0	0	0	0
0.5	2,500,000	38	347	328	366
1.0	5,000,000	103	694	642	745
1.5	7,500,000	187	1041	947	1134
2.0	10,000,000	294	1388	1241	1535
2.5	12,500,000	390	1735	1540	1930
3.0	15,000,000	515	2082	1824	2339
3.5	17,500,000	645	2429	2107	2752
4.0	20,000,000	780	2776	2386	3166
4.5	22,500,000	930	3123	2658	3588
5.0	25,000,000	1080	3470	2930	4010

height of the water surface in the lake. The outflow rate is also related to the height of the water surface above the bottom of the outlet channel. Since both storage and outflow are related to the water level, they are related to one another. The definition of this relationship will be explained in the following text.

In summary, reservoir routing involves (1) the evaluation of the storage–outflow relationship for the reservoir; and (2) the application of the continuity equation in the form of Equation 10-27 for small intervals of time to relate inflow, storage, and outflow.

Suppose that the discharge rating curve* for the outlet of a reservoir has been evaluated by a set of simultaneous measurements of discharge and of the height of the water surface above the bottom of the outlet. From the discharge rating curve, we choose a range of water levels and tabulate them in the first column of Table 10-18. When the water level stands at each of these elevations, a calculable volume of water is in storage above the outlet. If the reservoir has vertical sides, this volume is simply the area of the water surface multiplied by the height of the water surface above the outlet. For gently sloping sides the volume–height relationship must be obtained from a topographic or hydrographic survey of the reservoir. To illustrate, we consider a vertical-sided reservoir with an area of five million square feet. The volumes in storage for each water surface elevation can then be calculated, as shown in column 2 of Table 10-18. The third column of the table can be read from the discharge rating curve for each selected value of the water

*A discharge rating curve for a channel cross section defines the relationship between discharge and water surface elevation; it will be discussed in more detail in Chapter 16.

level. The last three columns of Table 10-18 are derived from the first three for a fixed chosen time increment, Δt; they are required for use in Equation 10-27. For the time increment in this problem we have chosen a value of 2 hours. Therefore, Δt in column 4 is 7200 seconds. For various water elevations we now have both the outflow rates and the two terms involving storage. We can plot these quantities against one another to define the storage–outflow relations as shown in Figure 10-44. Column 3 in Table 10-18 is the abscissa and columns 5 and 6 are ordinate values in the figure.

We are now ready to route an inflow through the lake. Suppose the one to be routed is that shown in Figure 10-45. From this inflow hydrograph we tabulate values of I_1 and I_2, and of the average inflow rate $(I_1 + I_2)/2$ for each 2-hour increment, as shown in Table 10-19. If inflow and outflow rates are approximately equal before the flood wave enters the lake, the outflow rate is 90 cfs. These values are entered in the first line of Table 10-19. At the beginning of the second time interval, therefore, the discharge, O_1, is 90 cfs and, according to Figure 10-44, this implies a value for $[S_1/\Delta t - O_2/2]$ of 580 cfs, which is entered in column 4 of the second line of Table 10-19. According to Equation 10-27, we obtain $[S_2/\Delta t + O_2/2]$ in the fifth column of the table by summing $(I_1 + I_2)/2$ and $[S_1/\Delta t - O_1/2]$ from columns 3 and 4. Returning to Figure 10-44 with the newly calculated value of $[S_2/\Delta t + O_2/2]$, we can read off the outflow rate at the end of the time interval and enter it in column 6 of Table 10-19. This outflow rate is the second point on our routed hydrograph and the new value of O_1 for the second time interval. It allows us to obtain a new value of $[S_1/\Delta t - O_1/2]$ for the next time interval. The process is repeated for each period as illustrated in Table 10-19, and the routed outflow hydrograph is plotted as the dashed line in Figure 10-45.

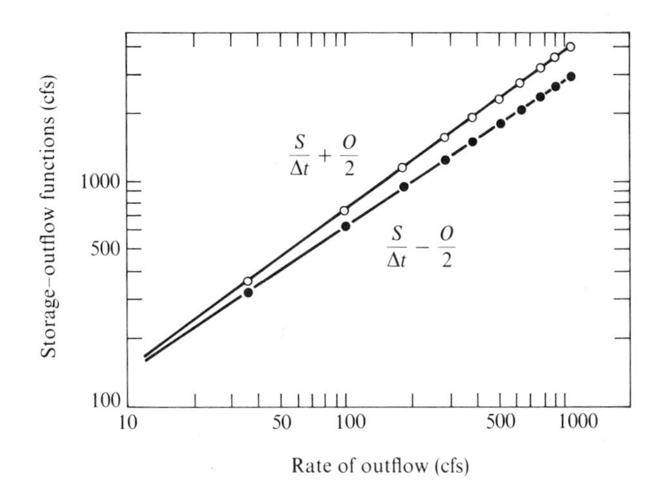

Figure 10-44 Storage–outflow relationships for a reservoir.

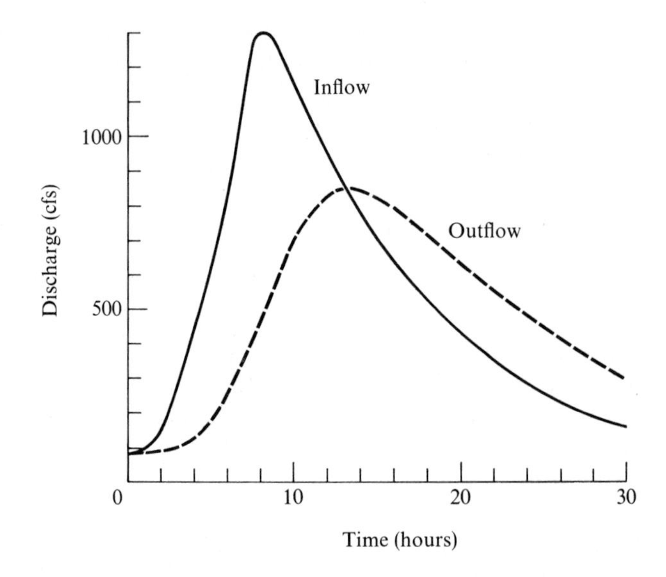

Figure 10-45 Hydrograph of inflow to a reservoir, solid line; and of outflow from reservoir, dashed line.

Table 10-19 Hydrograph routing of inflow through the reservoir.

TIME (HR)	INFLOW (CFS)	AVERAGE INFLOW $\left(\frac{I_1 + I_2}{2}\right)$ (CFS)	$\left(\frac{S_1}{\Delta t} - \frac{O_1}{2}\right)$ AT BEGINNING OF TIME INTERVAL (CFS)	$\left(\frac{S_2}{\Delta t} + \frac{O_2}{2}\right)$ AT END OF TIME INTERVAL (CFS)	OUTFLOW (CFS)
0	90	—	—	—	90
2	145	118	580	698	94
4	440	292	595	887	132
6	835	638	741	1379	240
8	1300	1068	1100	2168	455
10	1175	1238	1690	2928	695
12	970	1072	2210	3282	815
14	790	880	2450	3330	840
16	650	720	2510	3230	800
18	535	592	2420	3012	725
20	445	490	2300	2790	640
22	365	405	2100	2505	560
24	300	332	1930	2262	485
26	245	272	1750	2022	420
28	205	225	1600	1825	360
30	170	188	1430	1618	305

If the lake level is below the outlet at the beginning of inflow, the water is simply accumulated in storage until it raises the level of the lake above the outlet.

Two main factors control the effect of the reservoir on the routed storm hydrograph. The first is the area of the reservoir, which controls the volume of storage for a given change of height. The second is the nature of the relationship between storage (represented by the elevation of the water surface above the bottom of the outlet) and the rate of outflow. If the outlet is a natural stream channel, the form of this discharge rating curve varies widely. On artificially controlled outlets, however, the storage outflow relationship can be designed to suit the purposes of the planner. If he wishes to pass the flood wave through the reservoir with relatively little attenuation, he might design a broad, low weir on the outlet. If he must strongly attenuate stormflow in a reservoir with a small surface area, he might decide to construct a high, narrow outlet, which would force the water to accumulate to a high level in the reservoir in order to pass flood discharges. Some stage–discharge relationships for typical outlets are listed in Table 10-20. On natural stream outlets the stage–discharge relationship is given by the rating curve of the channel (see Chapter 16).

Table 10-20 Stage–discharge relationships for typical reservoir outlets.

TYPE OF OUTLET	STAGE–DISCHARGE RELATIONSHIP*
Rectangular, sharp-crested weir without contractions	$Q = 3.33LH^{1.5}$
Rectangular, sharp-crested weir with contractions	$Q = 3.34LH^{1.47}$
Trapezoidal, sharp-crested (Cipoletti) weir with sides sloping at 4:1	$Q = 3.37LH^{1.5}$
Triangular, sharp-crested weir	
120° V-notch	$Q = 4.43H^{2.45}$
90° V-notch	$Q = 2.48H^{2.48}$
30° V-notch	$Q = 0.67H^{2.5}$
Rectangular, broad-crested weir	$Q = 2.7LH^{1.5}$, but with many variations depending upon the form of the crest
Natural stream, small channel below lake in Labrador	$Q = 14.3H^{1.34}$, but with many variations depending upon the form of the channel

*Q = outflow (cfs)
H = water elevation above bottom of outlet (ft)
L = width of rectangular outlet (ft)

Channel Routing

The most commonly used method of routing a flood wave from one point on a channel to some site downstream is the *Muskingum channel-routing method*, developed by G. T. McCarthy of the U.S. Army Corps of Engineers. As with the case of a reservoir, routing along a channel requires the use of the continuity equation and of a relation between storage (water surface elevation) and outflow. In a reach of channel, however, storage is a function of both inflow and outflow. Figure 10-46 shows a flood wave moving along a channel. Consider the volume of water in storage in the reach between stations A and B when the outflow from the reach (at A in Figure 10-46) is 10,000 cfs. On the rising limb of the hydrograph, the volume of water in storage is greater than at the same discharge on the falling limb. The difference between the two storage volumes on the rising and falling limbs will depend upon the form of the flood wave, and therefore upon both the rate of inflow to the reach and the rate of outflow.

It is possible to measure directly the volume of water in a long reach of channel (and valley floor at overbank flows) by making detailed topographic surveys and computing the volume of water in storage when the water surface stands at various elevations. More often, however, the amount of storage is obtained directly, through an anlysis of measured hydrographs from the same storm at the upper and lower ends of the channel reach in question.

In natural channels the exact relationship between inflow, outflow, and storage is usually quite complicated, but the Muskingum channel-routing

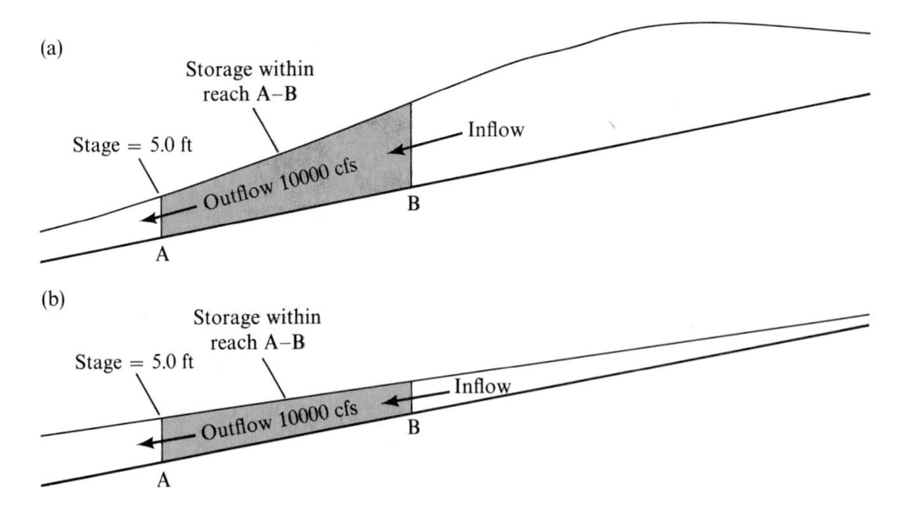

Figure 10-46 Profiles of water flowing in a channel reach during the rising limb of a flood wave (a) and the recession limb (b). Storage of water in the reach between A and B is greater on the rising limb than at the same outflow rate on the falling limb.

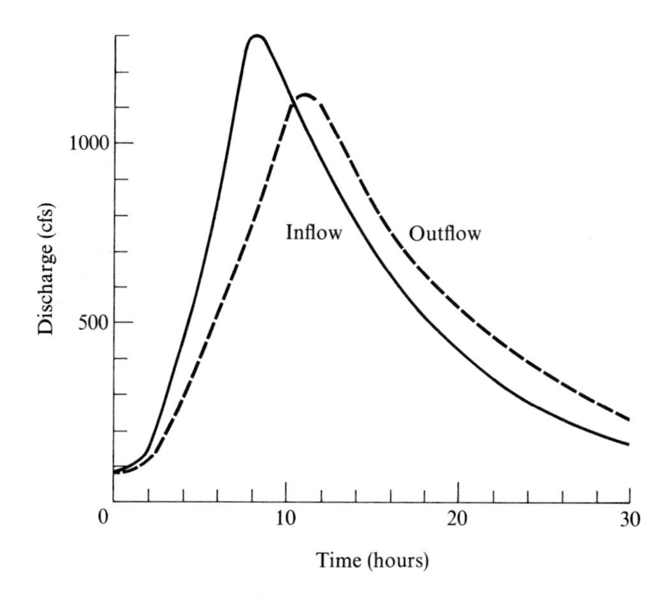

Figure 10-47 Inflow and outflow hydrographs for a reach of channel used to evaluate K and x in Equation 10-28.

procedure makes use of the simplifying assumption that the relationship can be approximated as

$$S = K[xI + (1 - x)O] \tag{10-28}$$

where K is a constant, whose units are those of time, and x is a factor that weighs the relative influences of inflow and outflow upon the storage.

The application of this equation to a large number of rivers has shown that the value of K approximates the travel time of the flood wave through the reach. This is to be expected because Equation 10-28 shows that K is a quotient between storage within the channel and some function of the rate of flow through the channel; i.e., K is an approximate "residence time." The weighting factor, x, expresses the amount of attenuation of the flood wave within the reach. In the extreme case of a true reservoir, the value of x is zero; i.e., storage is not related to inflow, and only the storage–outflow rating curve for the outlet need be defined. At the opposite extreme, if $x = 0.5$, inflow and outflow affect the storage equally, and the flood wave is simply translated through the reach without attenuation. These two extreme cases were discussed qualitatively at the beginning of this chapter. They can now be handled quantitatively. For most river channels, x lies between 0.1 and 0.3, indicating both attenuation and translation. The modal value is about 0.2, but the factor must be evaluated empirically for the reach in question. The constants, K and x, can be evaluated from simultaneous measurements of inflow and outflow for the reach, as illustrated below.

At any time the storage within the reach can be computed as the difference between the cumulative amounts of inflow and outflow. Given the observed inflow and outflow at the ends of a channel reach as plotted in Figure 10-47 and tabulated in Table 10-21, the storage is computed in columns 2 through 10 of the table. The average inflow and outflow rates for each time interval are listed in columns 3 and 7, the volumes of inflow and outflow are listed in columns 4 and 8, and the cumulative volumes of flow up to each time are entered in columns 5 and 9. Subtracting cumulative outflow from cumulative inflow gives the volume of storage in the reach at any time. The remainder of the table is used to calculate values of the weighted discharge $[xI + (1 - x)O]$ in Equation 10-28, using arbitrarily chosen values of x.

The numbers in column 10 can be taken as values for the left-hand side of Equation 10-28. Values of the right-hand side of that equation can be obtained by using various values of x, as shown in columns 11 through 16 of Table 10-21. Then, for each chosen value of x, S can be plotted against $[xI + (1 - x)O]$. According to Equation 10-28, there should be a linear relationship between the two variables, and the slope of the straight line should be K. This will only be true for the correct value of x; other values of x will produce hysteresis loops, as shown in Figure 10-48(a).

Table 10-21 Evaluation of storage and weighted discharge for a reach of river.

(1)	(2)	(3)	(4)	(5)	(6)	(7)	(8)
HOUR	I (CFS)	$\left(\frac{I_1 + I_2}{2}\right)$	$\left(\frac{I_1 + I_2}{2}\right)\Delta t$ (FT³)	CUMUL. INFLOW (FT³)	O (CFS)	$\left(\frac{O_1 + O_2}{2}\right)$ (CFS)	$\left(\frac{O_1 + O_2}{2}\right)\Delta t$ (FT³)
0	90	–	–	0	90	–	–
2	145	118	849,600	849,600	120	105	756,000
4	440	292	2,102,400	2,952,000	175	148	1,065,600
6	835	638	4,593,600	7,545,600	510	342	2,462,400
8	1300	1068	7,689,600	15,235,200	760	635	4,572,000
10	1175	1238	8,913,600	24,148,800	1070	915	6,588,000
11	1060	1118	4,024,800	28,173,600	1140	1105	3,978,000
12	970	1015	3,654,000	31,827,600	1100	1120	4,032,000
14	790	880	6,336,000	38,163,600	930	1015	7,308,000
16	650	720	5,184,000	43,347,600	750	840	6,048,000
18	535	592	4,262,400	47,610,000	640	695	5,004,000
20	445	490	3,528,000	51,138,000	550	595	4,284,000
22	365	405	2,916,000	54,054,000	470	510	3,672,000
24	300	332	2,390,400	56,444,000	400	435	3,132,000
26	245	272	1,958,400	58,402,800	340	370	2,664,000
28	205	225	1,620,000	60,022,800	280	310	2,232,000
30	170	188	1,353,600	61,376,400	260	270	1,944,000

The required value of x is indicated, then, by the narrowest loop, and the calculation and plotting must be repeated until a narrow loop is derived, as shown in Figure 10-48(b). A straight line through the points of this curve has a slope of K time units. In the example, two trials were sufficient to produce a tight loop indicating that 0.15 is the appropriate value of x. The slope of the line indicates that the value of K is 800,000 cubic feet per 100 cfs, or 8000 seconds. We have now defined the relationship between inflow, outflow, and storage and are prepared for the development of the routing technique itself.

Again, the method is based on the continuity equation, expressed for some time interval as

$$\frac{I_1 + I_2}{2} - \frac{O_1 + O_2}{2} = \frac{S_2 - S_1}{\Delta t} \qquad (10\text{-}29)$$

From Equation 10-28, we know that

$$S_2 - S_1 = K[x(I_2 - I_1) + (1 - x)(O_2 - O_1)] \qquad (10\text{-}30)$$

(9)	(10)	(11)	(12)	(13)	(14)	(15)	(16)
		$[xI + (1 - x)O]$					
CUMUL.		$x = 0.10$			$x = 0.15$		
O (FT³)	S (FT³)	xI	$(1 - x)O$	$[xI + (1 - x)O]$	xI	$(1 - x)O$	$[xI + (1 - x)O]$
0	0						
756,000	93,600	12	94	106	18	89	107
1,821,600	1,130,400	29	133	162	44	126	170
4,284,000	3,261,600	64	308	372	96	291	386
8,856,000	6,379,200	106	572	678	160	540	700
15,444,000	8,704,800	124	823	947	186	778	964
19,422,000	8,751,600	112	994	1106	168	939	1107
23,454,000	8,373,600	102	1008	1110	152	952	1104
30,762,000	7,401,600	88	914	1002	132	863	995
36,810,000	6,537,600	72	756	828	108	714	822
41,814,000	5,796,000	59	626	685	89	591	680
46,098,000	5,040,000	49	535	584	74	506	580
49,770,000	4,284,000	41	459	500	61	434	495
52,902,000	3,542,000	33	390	423	50	370	420
55,566,000	2,836,800	27	333	360	41	314	355
57,798,000	2,224,800	23	279	302	34	264	298
59,742,000	1,634,400	19	243	262	28	230	258

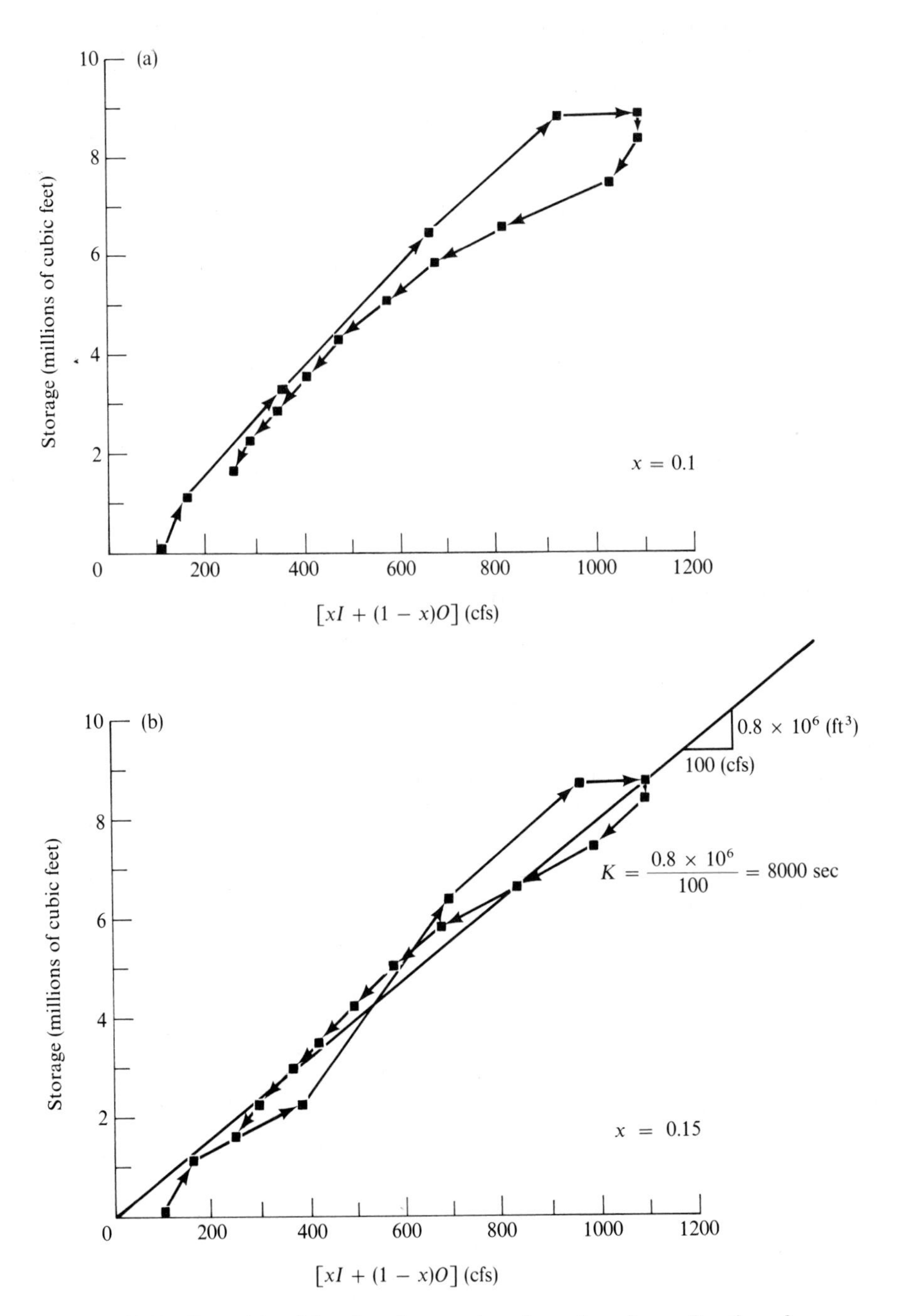

Figure 10-48 Two trials of the plot of storage in a channel reach as a function of weighted discharge, $[xI + (1 - x)O]$, for two assumed values of x. (a) $x = 0.1$. (b) $x = 0.15$. The data for these graphs are listed in Table 10-21.

and combining this with Equation 10-29, we obtain

$$O_2 = C_0I_2 + C_1I_1 + C_2O_1 \tag{10-31}$$

where

$$C_0 = -\frac{Kx - 0.5\,\Delta t}{K - Kx + 0.5\,\Delta t}$$

$$C_1 = \frac{Kx + 0.5\,\Delta t}{K - Kx + 0.5\,\Delta t}$$

$$C_2 = \frac{K - Kx - 0.5\,\Delta t}{K - Kx + 0.5\,\Delta t}$$

and where, as a check, $C_0 + C_1 + C_2 = 1.0$.

The value of O_2 for one time period becomes the value of O_1 for the next period. In the foregoing it is assumed that K and x are constant throughout the range of flows to be expected. If this is not so, they may have to be changed as the river rises, and the line shown in Figure 10-48 may become a curve. Once the values of K and x have been evaluated for a reach of river, they can be used to route hypothetical design hydrographs downstream.

The situation considered above is for a major channel with no important tributaries. If the unmeasured tributary area is small, its contribution to the inflow in Table 10-21 can be estimated without significant error. If large tributaries enter the channel, the inflow hydrograph on the main channel must be routed to the junction and added to the hydrograph from the tributary, and then the combined hydrograph is routed downstream.

A review of approximate methods of channel routing in more complicated field situations on large rivers is given by Lawler (1964).

In Figure 10-49 the solid curve represents the hydrograph of a flood entering the upper end of a channel reach, whose K and x values have been evaluated from an earlier runoff event, as shown in Table 10-21 and Figure 10-48. The new flood can then be routed along the reach of channel by using Equation 10-31 with the following values for the three constants:

$$C_0 = -\frac{8000(0.15) - 0.5(7200)}{8000 - 8000(0.15) + 0.5(7200)} = 0.23$$

$$C_1 = \frac{8000(0.15) + 0.5(7200)}{8000 - 8000(0.15) + 0.5(7200)} = 0.46$$

$$C_2 = \frac{8000 - 8000(0.15) - 0.5(7200)}{8000 - 8000(0.15) + 0.5(7200)} = 0.31$$

$$C_0 + C_1 + C_2 = 0.998$$

The routing procedure is illustrated in Table 10-22, and the result is plotted as the outflow graph in Figure 10-49.

Channel routing is widely used in flood forecasting and in the design of major structures. For the planner the principles involved should be under-

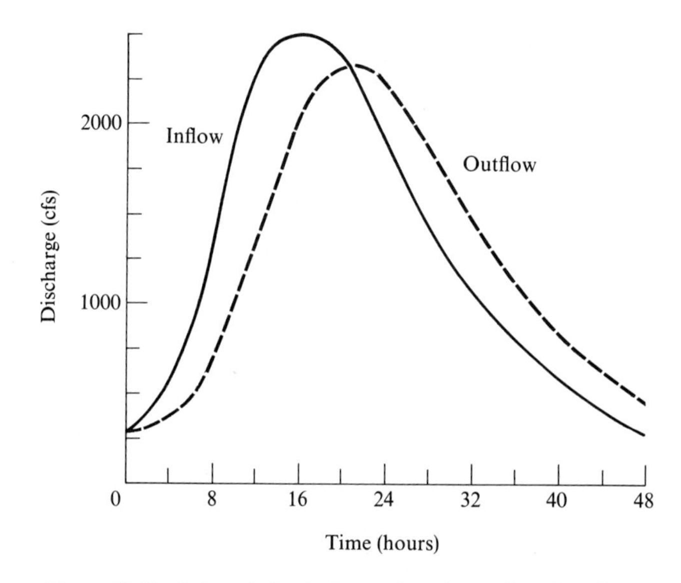

Figure 10-49 Inflow design hydrograph and routed outflow from a reach of channel.

Table 10-22 Calculation of the outflow hydrograph from the reach of channel with $K = 8000$ seconds and $x = 0.15$. The inflow is a previously calculated design hydrograph for the drainage basin above the inflow point at the upper end of the channel reach of interest.

HOUR	I (CFS)	C_0I_2	C_1I_1	C_2O_1	O (CFS)
0	300	—	—	—	300
4	540	124	138	93	355
8	1340	308	248	110	666
12	2260	520	616	206	1342
16	2480	570	1040	416	2026
20	2370	545	1141	628	2314
24	1850	426	1090	717	2233
28	1410	324	851	692	1867
32	1050	242	649	579	1470
36	790	182	483	456	1121
40	590	136	363	291	790
44	420	97	271	245	613
48	290	67	193	190	450

stood because they provide a guide to the effect of changes made on small basins by urbanization, channelization, and other land alterations. But the technique is difficult to use in small basins because there seldom are two gauging stations along a small stream from which inflow and outflow hydrographs can be obtained for analysis. Even if by good fortune there exists a gauging station on the stream in the area of study, it is unlikely that a second station also exists on the same stream at another point sufficiently nearby to be used in deriving the storage–discharge relations.

The authors are strongly of the opinion that the environmental planner should and can make field observations appropriate to the solution of a hydrologic problem even when the standard network of gauging stations, climatic stations, or other desirable data collection points do not exist in the study area. If the problem at hand includes the necessity of determining how a change of land use upstream will be felt at some downstream point on a channel, it is quite practical to make the observations necessary to obtain at least one inflow and outflow hydrograph for the channel length involved. To be sure, it requires waiting for a storm to occur and observations during the storm, but the results obtained are worth the effort if a good solution to the routing problem is important. There is also great satisfaction in the accomplishment. The procedure involves the following considerations:

Channel routing is easiest and most satisfactory if the reach of channel under consideration does not have any major tributaries entering it. A staff gauge should be installed at each end of the reach chosen. For a small basin it is often possible to find a channel length of one-quarter to one-half mile in which no important tributary enters. The staff gauges should be placed as far apart as tributary entrances allow. During low or moderate flow, at least one discharge estimate should be made at each staff gauge. A cross section at each staff is surveyed. For each an estimated rating curve can be constructed (see Typical Problems 16-1 and 16-3).

The investigator then waits until a storm of moderate size occurs, preparation having been made in advance for an observer to reach each staff gauge as early in the storm as possible. With synchronized watches, the observers read the water level at the staff gauges, plotting the observed readings as they are made until at least one simple hydrograph has been observed.

With one good hydrograph rise observed at the two locations and a rating curve available for each, the data necessary for flood routing are at hand. The procedure shown in Table 10-21 and Figure 10-48 is carried out, leading to a determination of a storage–discharge relation for the channel reach. If tributaries enter the reach of interest, they can also be gauged and their discharge added to the inflow to the reach.

In preparing for such an observational period, it is well to know in advance the probable time period elapsing from beginning of the hydrograph to the peak in order to plan the logistics of being at the gauge early enough in the storm. The relation of drainage area to lag time, Figure 10-35, provides a time estimate that is helpful.

Bibliography

American Society of Civil Engineers (1969) Design and construction of sanitary and storm sewers; *Manuals and Reports on Engineering Practices, No. 37,* 332 pp.

Anderson, D. G. (1970) Effects of urban development on floods in northern Virginia; *U.S. Geological Survey Water Supply Paper 2001-C.*

Barrows, H. K. (1942) A study of valley storage and its effect upon the flood hydrograph; *EOS, American Geophysical Union Transactions,* vol. 23, pp. 483–488.

Beard, L. R. (1962) *Statistical methods in hydrology*; U.S. Army Corps of Engineers, Sacramento, CA.

Benson, M. A. (1950) Use of historical data in flood-frequency analysis; *EOS, American Geophysical Union Transactions,* vol. 31, pp. 419–424.

Benson, M. A. (1960) Characteristics of frequency curves based on a theoretical 1000-year record; *U.S. Geological Survey Water Supply Paper 1543-A,* pp. A51–A77.

Benson, M. A. (1962) Evolution of methods for evaluating the occurrence of floods; *U.S. Geological Survey Water Supply Paper 1580-A.*

Benson, M. A. (1962) Factors influencing the occurrence of floods in a humid region of diverse terrain; *U.S. Geological Survey Water Supply Paper 1580-B.*

Benson, M. A. (1964) Factors affecting the occurrence of floods in the Southwest; *U.S. Geological Survey Water Supply Paper 1580-D.*

Benson, M. A. (1971) Uniform flood-frequency estimating methods for federal agencies; *Water Resources Research,* vol. 4, pp. 891–908.

Binnie, G. M. and Mansell-Moulin, M. (1966) The estimated probable maximum storm and flood on the Jhelum River: a tributary of the Indus; *River flood hydrology,* Institution of Civil Engineers, London, pp. 184–210.

California Division of Mines and Geology (1973) Urban geology: master plan for California, *Bulletin 198,* Sacramento, 112 pp.

Carter, J. R. and Green, A. R. (1963) Floods in Wyoming: magnitude and frequency; *U.S. Geological Circular 478.*

Carter, R. W. (1961) Magnitude and frequency of floods in suburban areas; *U.S. Geological Survey Professional Paper 424-B,* pp. B9–B11.

Collins, W. T. (1939) Runoff distribution graphs from precipitation occurring in more than one time unit; *Civil Engineering,* vol. 9, pp. 559–561.

Crawford, N. H. and Linsley, R. K. (1966) Digital simulation in hydrology: Stanford Watershed Model IV; Department of Civil Engineering, Stanford University, *Technical Report 39,* 210 pp.

Dalrymple, T. (1960) Flood-frequency analyses; *U.S. Geological Survey Water Supply Paper 1543-A.*

Dunne, T. (1978) Field studies of hillslope flow processes; in *Hillslope hydrology* (ed. M. J. Kirkby), John Wiley & Sons, London.

Dunne, T., Moore, T. R. and Taylor, C. H. (1975) Recognition and prediction of runoff-producing zones in humid regions; *Hydrological Sciences Bulletin,* vol. 20, pp. 305–327.

Emmett, W. W. (1970) The hydraulics of overland flow on hillslopes; *U.S. Geological Survey Professional Paper 622-A.*

Espey, W. H., Morgan, C. W. and Masch, F. D. (1966) Study of some effects of urbanization on storm runoff from a small watershed; *Texas Water Development Board Report 23,* 109 pp.

FELTON, P. N. AND LULL, H. W. (1963) Suburban hydrology can improve watershed conditions; *Public Works,* vol. 94, pp. 93–94.

GANN, E. E. (1968) Flood height-frequency relations for the plains area in Missouri; *U.S. Geological Survey Professional Paper 600-D,* pp. D52–D53.

GUMBEL, E. J. (1945) Floods estimated by the probability method; *Engineering News-Record,* vol. 134, pp. 833–837.

HARRIS, E. E. AND RANTZ, S. E. (1964) Effect of urban growth on streamflow regimen of Permanente Creek, Santa Clara County, California; *U.S. Geological Survey Water Supply Paper 1591-B.*

HELLEY, E. J. AND LAMARCHE, V. C. (1973) Historic flood information for northern California streams from geological and botanical evidence; *U.S. Geological Survey Professional Paper 485-E.*

HOYT, W. G. AND LANGBEIN, W. B. (1955) *Floods,* Princeton University Press, Princeton, NJ, 468 pp.

JAMES, L. D. (1965) Using a computer to estimate the effects of urban development on flood peaks; *Water Resources Research,* vol. 1, pp. 223–234.

JENS, S. W. AND MCPHERSON, M. B. (1964) Hydrology of urban areas; Section 20 in *Handbook of applied hydrology* (ed. Ven te Chow), McGraw-Hill, New York.

LANGBEIN, W. B. (1940) Channel storage and unit hydrograph studies; *EOS, American Geophysical Union Transactions,* vol. 21, pp. 620–627.

LANGBEIN, W. B. (1949) Annual floods and the partial-duration flood series; *EOS, American Geophysical Union Transactions,* vol. 30, pp. 879–881.

LANGBEIN, W. B. (1960) Plotting positions in frequency analysis; *U.S. Geological Survey Water Supply Paper 1543-A,* pp. A48–A51.

LAWLER, E. A. (1964) Flood routing; Section 25-II in *Handbook of applied Hydrology* (ed. Ven te Chow), McGraw-Hill, New York.

LEOPOLD, L. B. (1968) Hydrology for urban land planning: a guidebook on the hydrologic effects of urban land use; *U.S. Geological Survey Circular 554.*

LEOPOLD, L. B. AND MADDOCK, T., JR., (1954) *The flood-control controversy;* Ronald Press, New York, 275 pp.

LINSLEY, R. K. AND KOHLER, M. R. (1951) Predicting the runoff from storm rainfall; *U.S. Weather Bureau Research Paper 34.*

OGROSKY, H. O. (1964) Hydrology of spillway design: small structures: limited data; *American Society of Civil Engineers, Journal of the Hydraulics Division,* vol. 91, HY3, pp. 295–310.

RAPP, A., BERRY, L., AND TEMPLE, P. H. (1972) Soil erosion and sedimentation in Tanzania; *Geografiska Annaler,* vol. 54A, pp. 105–379.

RANTZ, S. E. (1971) Suggested criteria for hydrologic design of storm-drainage facilities in the San Francisco Bay Region, California; *U.S. Geological Survey Open File Report,* Menlo Park, CA.

RICHARDSON, D. (1968) Glacial outburst floods in the Pacific northwest; *U.S. Geological Survey Professional Paper 600-D,* pp. D79–D86.

RODDA, J. C. (1967) A country-wide study of intense rainfall in the United Kingdom, *Journal of Hydrology,* vol. 5, pp. 58–69.

SCOTT, K. M. AND GRAVLEE, G. C. (1968) Flood surge on the Rubicon River, California: hydrology, hydraulics, and boulder transport; *U.S. Geological Survey Professional Paper 422-M.*

SNYDER, F. F. (1938) Synthetic unit hydrographs; *EOS, American Geophysical Union Transactions,* vol. 19, pp. 447–454.

STANKOWSKI, S. J. (1972) Population density as an indirect indicator of urban and suburban land-surface modifications; *U.S. Geological Survey Professional Paper 800-B,* pp. B219–B224.

TAYLOR, A. B. AND SCHWARZ, H. E. (1952) Unit-hydrograph lag and peak flow related to basin characteristics; *EOS, American Geophysical Union Transactions,* vol. 33, pp. 235–246.

THOMAS, D. M. (1964) Height-frequency relations for New Jersey floods; *U.S. Geological Survey Professional Paper 475-D,* pp. D202–D203.

U.S. SOIL CONSERVATION SERVICE (1972) *Hydrology,* Section 4, *National engineering handbook,* Washington, DC.

U.S. SOIL CONSERVATION SERVICE (1975) Urban hydrology for small watersheds; *Technical Release No. 55,* Washington, DC.

WIARD, L. (1962) Floods in New Mexico, magnitude and frequency; *U.S. Geological Survey Circular 464.*

Typical Problems

10-1 Antecedent Precipitation Index

You need to develop an index of the wetness of an area and show how it varies over a season or a year. The index could be used as part of a prediction of storm runoff volumes as mentioned in this chapter, or it could be correlated with infiltration capacity, extent of saturated area, or other hydrologic quantity, and used as a predictor of these variables.

Compute the antecedent precipitation index for a fine sandy-loam soil under the following rainfall regime for a 15-day period. Rain occurs as shown in column 2 of Table 10-23.

Solution

Equation 10-1 defines one kind of antecedent precipitation index as

$$I_t = I_0 k^t$$

You decide to use the available water capacity (from Figure 6-9) multiplied by the depth of the soil profile, a first estimate of I_0. You also decide to use a value of 0.9 for k.

Set up a table like Table 10-23. At the beginning of the calculation period, $t = 0$ and $I_t = I_0$. On succeeding days t changes by increments of one, and the values of I_t listed in column 3 are obtained by multiplying the previous day's value

Table 10-23 Calculating the antecedent precipitation index.

DAY (t)	DAILY RAINFALL (IN)	$I_0 k^t$ (IN)	ADD RAINFALL (IN)
0		30	
1		27	
2		24.3	
3	1.6	21.9	23.5
4		21.2	
5		19.1	
6		17.1	
7	2.2	15.4	17.6
8		15.8	
9		14.3	
10		12.8	
11	1.9	11.6	13.5
12	0.5	12.1	12.6
13		11.4	
14		10.2	
15		9.2	

by k. When rain occurs, the value of I_t is first calculated from that of the previous day and entered in column 3. The rainfall is added for the day, and the new value of I_t is entered in column 4. The value in the last column is then used for the continuation of the calculation. The next day's value in column 3 is obtained by multiplying column 4 by k. The results are plotted in Figure 10-50.

Values of I_t can then be taken from the curve and correlated with the ratio of storm runoff volume to rainfall, with infiltration capacity, or with any other hydrologic variable of interest. The correlation might be improved by altering the values of I_0 or k on a trial-and-error basis. The reader should repeat the calculation for the same rainfall regime with a value of $k = 0.8$, and plot the resulting data on the curve relating I_t to t to observe the effect of such a change.

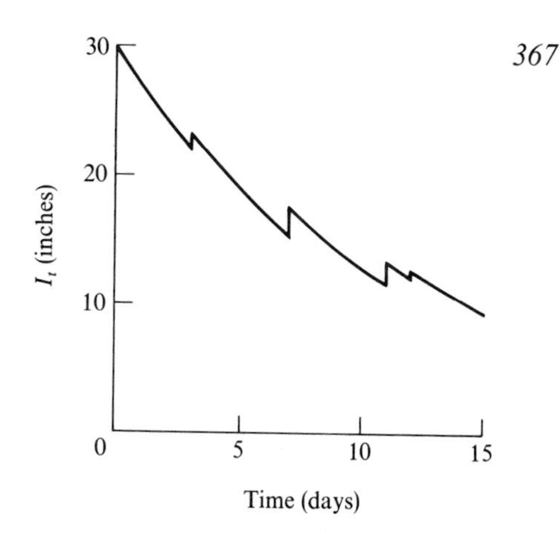

Figure 10-50 Plotting the antecedent precipitation index.

10-2 Calculation of Storm Runoff Volume by the Φ Index

As a member of the Mulkilt County planning department you are reviewing the environmental impact statement for a proposed industrial park in a small catchment. It is claimed that there will be minimal impact of the proposed park on storm runoff into a nearby creek and that only minor detention storage need be designed for. No numbers are supplied, so you decide to use a design storm appropriate for your area and size of catchment (say, a 50-year, 2-hour rainstorm) and to compute the volume of storm runoff into the creek. From your examination of 2-hour storms from the weather record for the region, the mass curve of Figure 10-51 was developed for the 50-year storm.

The drainage area of the creek is two square miles. It is entirely rural, and by 1990 forty percent of the catchment will be covered by the industrial park.

Solution

An estimate of the storm runoff volume before and after construction of the industrial park can be made by computing an appropriate Φ value and subtracting it from the rainfall intensity of the chosen design storm.

You decide that your area is sufficiently similar to the San Francisco Bay region for the use of Figure 10-7 as a means of estimating Φ. Mean annual rainfall in your catchment is 30 inches. As a first attempt, obtain a value of Φ from Figure 10-7. Then divide your design storm into four 30-minute periods, calculate the rainfall intensity for each, subtract the Φ value from each rainfall intensity, and compute the volume of precipita-

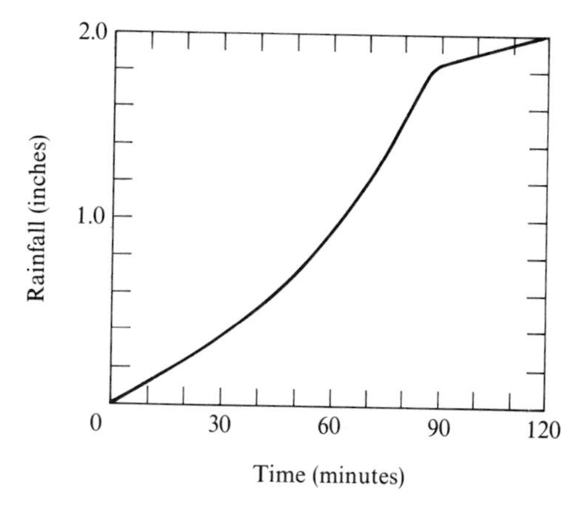

Figure 10-51 Mass curve of rainfall as a function of time for the 50-year storm.

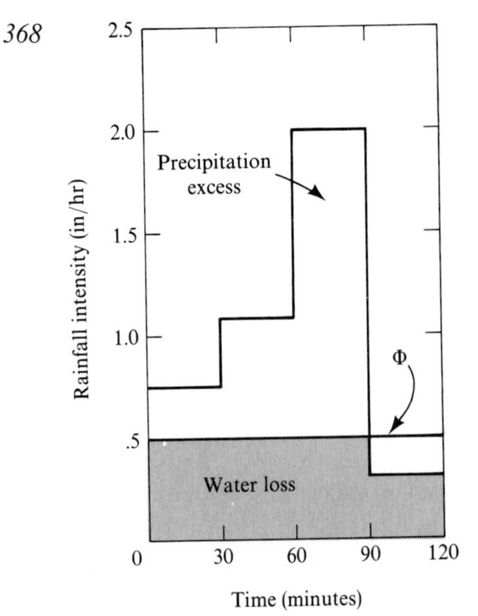

Figure 10-52 Calculated rainfall intensity in four 30-minute periods to represent a design storm.

tion excess as in Figure 10-52. You should get approximately 1.87 inches of runoff from the 2.0-inch storm. As you will learn from working other problems, this is a surprisingly large runoff volume for the size of rainstorm, and so you decide to check it.

You obtain several storm hydrograph and rainfall intensity records for a small gauged basin near the proposed park. One storm event, for example, consisted of a 2.5-inch rainstorm in 3.5 hours and produced 0.75 inches of runoff. The water-loss rate (precipitation − runoff) for the event was 1.75 inches over the 3.5-hour storm, or 0.5 inches per hour. Take this rate as a value of the Φ index for the basin that is to be industrialized, and compute the runoff volume for the 2-inch design storm as shown in Table 10-24. Thus, a 2-inch rainstorm that occurs in 2 hours produces 1.17 inches of runoff from the rural basin.

To estimate the impact on runoff by urbanization, use Table 10-2, whose coefficients seem reasonable for urban areas, as Rantz derived them. For this problem the Φ value is multiplied by 0.8 and results in a runoff of 1.36 inches. What will be the volume of storm runoff in cubic feet?

Table 10-24 Calculating storm runoff volume by the Φ index.

PERIOD (MIN)	ACCUMULATED RAINFALL (IN)	RAINFALL FOR PERIOD (IN)	RAINFALL INTENSITY (IN/HR)	RATE OF EXCESS (IN/HR)	VOLUME OF RUNOFF (IN)
0–30	0.38	0.38	0.76	0.26	0.13
30–60	0.92	0.54	1.08	0.58	0.29
60–90	1.92	1.00	2.00	1.50	0.75
90–120	2.00	0.08	0.16	0	0
				Total	1.17

10-3 Estimation of Storm Runoff Volumes

Estimate the volume of direct runoff from a 2.0-inch storm during the growing season on a small Vermont catchment that has 40 percent of its area in contoured corn and 60 percent in permanent meadow. The soil type is Cabot silt loam. The 5-day antecedent rainfall totaled 1.7 inches.

Solution

Cabot silt loam, which has a dense horizon that impedes drainage at a depth of one foot, is in hydrologic soil group C, according to the local soil survey report. This agrees with Table 10-4.

Corn is a "poor" soil cover from the hydrologic point of view, as indicated in Table 10-5. In Vermont, corn is not usually rotated with a better cover.

The 5-day antecedent rainfall indicates that the catchment should be in an average moisture condition (II). Set up a working table as in Table 10-25.

The value of runoff R is obtained by entering Figure 10-8 with an abscissa value of 2.0 inches and reading up to interpolated curve numbers of 84 and 71. In the last column the runoff depths are weighted according to the areal extent of the two conditions and are summed to give the answer 0.47 inches of storm runoff. This computed value compares favorably with the volumes of runoff from large summer storms on average conditions of wetness at the Sleepers River Experimental Watershed in northern Vermont, where Cabot soils are common.

The reader may repeat this calculation for a 5-day antecedent rainfall of 3.0 inches and, by using Table 10-7 to alter the curve numbers to 93 and 86, should obtain a storm runoff volume of approximately 1.05 inches.

Table 10-25 Working table for estimating storm runoff volume for small rural catchment.

LAND USE	TREATMENT	HYDROLOGIC CONDITION	CURVE NUMBER (FROM TABLE 10-3)	AREA (% OF BASIN)	RUNOFF (R) (IN)	$R \times \frac{A}{100\%}$
Corn	Contoured	Poor	84	40	0.75	0.30
Meadow	Permanent	Good	71	60	0.28	0.17
					Total	0.47

10-4 Storm Runoff Volumes from Urban Catchments

Suppose the small catchment in Typical Problem 10-3 were to be developed for residential use in a way that would render 40 percent of the basin impervious, with the remainder of the area being covered by heavily used grassy open space. Compute the runoff from a 2.0-inch rainstorm for average conditions of antecedent moisture.

Solution

From Table 10-8 the hydrologic soil group C has a value of 98 if it is covered by pavement and roofs, and 79 if it has a thin grass cover. These curve numbers are entered into a working table (Table 10-26) and used as before.

Runoff in column 4 is obtained from Figure 10-8 as before, and the weighted average runoff depth is calculated in the last column. The runoff volume would increase by a factor of more than two. If the drainage basin covered an area of half a square mile, the increased volume of storm runoff would be about 650,000 feet^3, or 15 acre-feet.

Table 10-26 Working table for estimating storm runoff volume for urban catchment.

COVER	CURVE NUMBER (FROM TABLE 10-8)	AREA (% OF BASIN)	RUNOFF (IN)	$R \times \frac{A}{100\%}$
Impervious	98	40	1.8	0.72
Open space	79	60	0.52	0.31
			Total	1.03

10-5 Rational Runoff Formula

Calculate the peak runoff rate from a 25-year rainstorm on a 202-acre drainage basin in central Maryland. The information is to be used in designing a culvert under a highway.

Solution

1. From a recent soil map and land-use map, aerial photograph, or field survey, determine the proportion of the catchment under various land uses, as shown in Table 10-9. Also, confirm that there are no important storage impoundments or natural lakes in the basin.

2. From Table 10-9, choose appropriate values of the runoff coefficient for each of these categories, and compute a weighted average value of C for the whole catchment, as shown in Table 10-27.

3. From a topographic map, determine the area and length of the catchment with a planimeter and a map measurer, and the vertical relief from the contours.

Area = 202 acres
Length = 3400 feet
Relief = 15 feet

4. Estimate the time of concentration, from Equation 10-3.

$$t_c = \frac{(3400)^{1.15}}{7700(15)^{0.38}} = 0.53 \text{ hours}$$

5. Estimate the intensity of the 25-year rainfall with a duration of 0.53 hours. For this problem, the 30-minute duration can be used. From Figure 2-18, plot the 30-minute rainfalls in central Maryland for recurrence intervals of 2 years, 10 years, and 100 years on a sheet of logarithmic paper, and interpolate the 30-minute rainfall for the 25-year event, as shown in Figure 10-53.

30-minute rainfall
for the 25-year event = 2.4 in
30-minute intensity
for the 25-year event = 4.8 in/hr

6. Compute the peak flow rate:

$$Q_{pk} = CIA$$
$$= 0.38 \times 4.8 \times 202 = 368 \text{ cfs}$$

7. Before making a recommendation about the culvert, check whether any plans exist for changing the land use of the catchment within the foreseeable future.

Table 10-27 Computing a weighted average value for C.

COVER AND SOIL	AREA (% OF BASIN)	C VALUE	$C \times \frac{\text{AREA}}{100\%}$
Woodland on thin stony-clay loam	27	0.40	0.11
Pasture on loam	52	0.35	0.18
Cultivated land on loam	20	0.40	0.08
Buildings and roads	1	0.75	0.01
	100		0.38 = Weighted average value of C

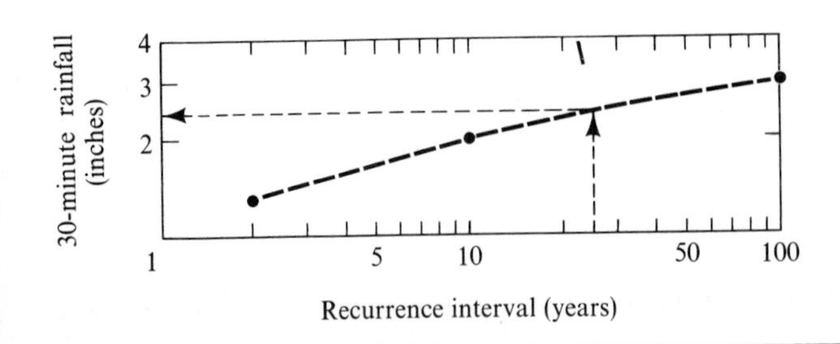

Figure 10-53 Recurrence interval graph for central Maryland for 30-minute rainfall amounts.

10-6 Rational Runoff Formula for Predicting Effects of Urbanization

You are planning the development of a residential region outside Nairobi, Kenya. The developed area will cover the upper portion of a small drainage basin. You are concerned about its impact upon the hydrology of the basin because last year you located another community at the outlet of the basin and you had an engineer outline the flood-prone area and design culverts and bridges to accommodate floods to be expected from this agricultural basin. Calculate the probable impact of partial urbanization on the peak runoff rate generated by a 10-year rainstorm in the catchment. *Note:* In a real urbanized area, you would usually design for a recurrence interval larger than 10 years, but this value will suffice to demonstrate the method.

Solution

1. Estimate the rational runoff coefficient from observations of the soil type and land use.

Soil is loam; land use is smallholder agriculture: $C = 0.40$.

(Interpreted from data by Rapp et al. 1972)

2. From a topographic map, determine the drainage area (with a planimeter), the length and average gradient of the hillslopes (measuring the length and vertical fall of a small number of the longer hillslopes on the map), and the maximum length and average channel gradient (using a map measurer to obtain the channel length, and reading the fall from the contours).

Area = 960 acres
Hillslope length = 500 feet
Average hillslope gradient = 0.06
Maximum length of channel = 12,000 feet
Average channel gradient = 0.015

3. By rapid field survey, determine the mean bankfull channel depth at a number of stations, and estimate Manning's n for the channel (see Table 16-1).

Mean bankfull depth = 1.5 feet
Manning's $n = 0.06$
(channel small and winding with abundant weeds)

4. Compute the time of concentration as the sum of overland and channel travel times. Using Figure 10-12, the overland travel time is obtained from the hillslope length, hillslope gradient, and C value, as indicated by the arrows. The value derived in this way is 19 minutes. The Manning Equation for the average velocity of channel flow is

$$u = \frac{1.5}{0.06}(1.5)^{0.67}(0.015)^{0.5} = 4.0 \text{ ft/sec}$$

and the channel travel time is

$$\frac{12{,}000 \text{ ft}}{4 \text{ ft/sec}} \times \frac{1 \text{ min}}{60 \text{ sec}} = 50 \text{ min}$$

Therefore, $t_c = 19 + 50 \text{ min} = 69 \text{ min}$.

5. Knowing the time of concentration of the basin, choose the 10-year intensity for that duration. Figure 2-19 can be used for a rough estimate of the 10-year, 1-hour storm, which is approximately the duration required. The map shows that Nairobi (which is close to the equator in East Africa) has a 2-year, 1-hour rainstorm of approximately 2 in/hr. The graph (Figure 2-19(a)) indicates that for a region with moderate annual rainfall, the ratio of the 2-year storm to the 1-year storm is 1.2. Furthermore, the ratio of the 10-year and 1-year storms is 2.1, so the 10-year event should be (2.1/1.2 =) 1.75 times as large as the 2-year storm. The necessary 10-year, 1-hour intensity, therefore, is 3.5 in/hr.

6. Compute the peak flow rate under cultivation as

$$Q_{pk} = CIA$$
$$= 0.40 \times 3.5 \times 960 = 1344 \text{ cfs}$$

7. From the planning document proposing the urbanization, obtain (a) the extent and location of urbanization within the catchment, and (b) the character of urbanization.

(a) It is proposed that within 10 years, 30 percent of the basin will be urbanized, with the built-up area confined to the ridgetops and upper parts of the hillslopes. No storm drains are to be installed.

(b) The built-up area will be residential; lot sizes will average 3000 square feet.

8. Estimate C after urbanization.

C for non-urbanized area = 0.40 for 70 percent of area

C for urban area: From Figure 10-11, lot sizes of 3000 square feet indicate an average value of about 40 percent for the impervious area. Figure

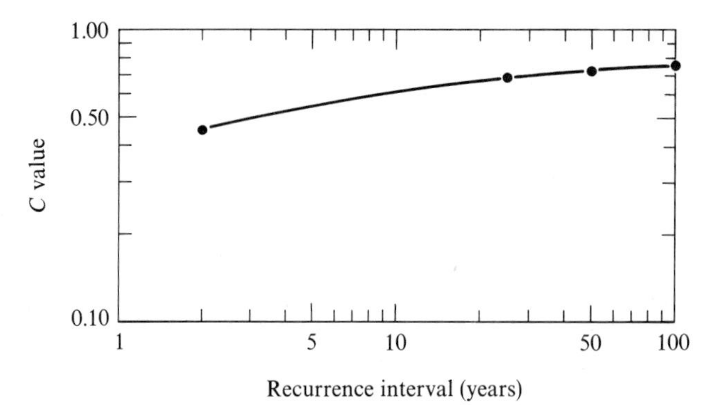

Figure 10-54 Estimating values of C as a function of recurrence interval using an appropriate value for impervious area for urban lots of a given size.

10-10 can then be used to obtain a C value for the 10-year storm. For a 40-percent impervious area, four values of C are read from this graph and replotted on logarithmic graph paper, as shown in Figure 10-54. The 10-year value can be interpolated as 0.62 for the urban area. The weighted average value of C for the whole basin is

$$C = (0.70 \times 0.40) + (0.30 \times 0.62) = 0.47$$

9. Because only the upper one-third of the basin on relatively gentle gradients (ridgetops) is to be urbanized, the time of concentration probably would not change much, though it could be calculated by the methods previously described. The hillside would be divided into urban and rural sections, the velocity of overland flow calculated for each, and the values added. In this problem, however, we judge that the time of concentration would be affected hardly at all, and we will use the same design duration and intensity as in the rural calculation.

10. $Q = 0.47 \times 3.5 \times 960 = 1579$ cfs

11. As an exercise, calculate the 25-year peak runoff rate for a small rural basin with which you are familiar. Imagine that it is to be developed for multiple-family dwellings, and that the main channel is to be modified into a straightened, concrete-lined storm drain with a Manning's n value in the range 0.015 to 0.020.

10-7 Construction of a Flood-Frequency Curve

Compute the flood-frequency curve for watershed 994 at the Coshocton Experimental Watershed (U.S. Department of Agriculture) from the annual flood peaks in Table 10-28. Estimate the 25-year flood and express it in cfs. The drainage area is 27.2 square miles. Compute the probability of the 25-year flood occurring within the next five years. Evaluate the mean annual flood.

Solution

The peaks are ranked in order of their magnitude, beginning with the largest. The recurrence interval is calculated from Equation 2-4, as shown in the fourth column of Table 10-28. Discharge from column 2 and the recurrence interval are then plotted against one another on logarithmic Gumbel extreme-value paper, as shown in Figure 10-55. The reader can repeat the plotting on arithmetic Gumbel paper traced from Figure 10-14, for comparison.

Reading up to the curve from an abscissa value of 25 years yields 0.35 in/hr for the 25-year flood. From an area of 27.2 square miles this is equivalent to a discharge of 6140 cfs (1 in/hr = 645 cfs/sq mi).

The probability of the 25-year flood occurring within the next five years is obtained from Figure 2-15 with $n = 5$ and a recurrence interval of 25 years.

The mean annual flood (0.084 in/hr) is read from the Gumbel graph for a recurrence interval

Table 10-28 Constructing a flood-frequency curve.

YEAR	ANNUAL PEAK DISCHARGE (IN/HR)	RANK (m)	RECURRENCE INTERVAL (YR) $T = \frac{n+1}{m}$	YEAR	ANNUAL PEAK DISCHARGE (IN/HR)	RANK (m)	RECURRENCE INTERVAL (YR) $T = \frac{n+1}{m}$
1937	0.15	8	4.0	1953	0.04	26	1.23
1938	0.05	20	1.6	1954	0.03	31	1.03
1939	0.04	23	1.39	1955	0.06	18	1.78
1940	0.07	15	2.1	1956	0.19	6	5.3
1941	0.08	14	2.3	1957	0.44	1	32.0
1942	0.03	30	1.07	1958	0.05	22	1.45
1943	0.07	16	2.0	1959	0.25	3	10.7
1944	0.04	24	1.33	1960	0.14	9	3.6
1945	0.21	5	6.4	1961	0.22	4	8.0
1946	0.13	10	3.2	1962	0.04	27	1.18
1947	0.07	17	1.88	1963	0.27	2	16.0
1948	0.09	12	2.7	1964	0.18	7	4.6
1949	0.04	25	1.28	1965	0.04	28	1.14
1950	0.09	13	2.5	1966	0.06	19	1.69
1951	0.05	21	1.52	1967	0.04	29	1.10
1952	0.12	11	2.9	Total $n = 31$			

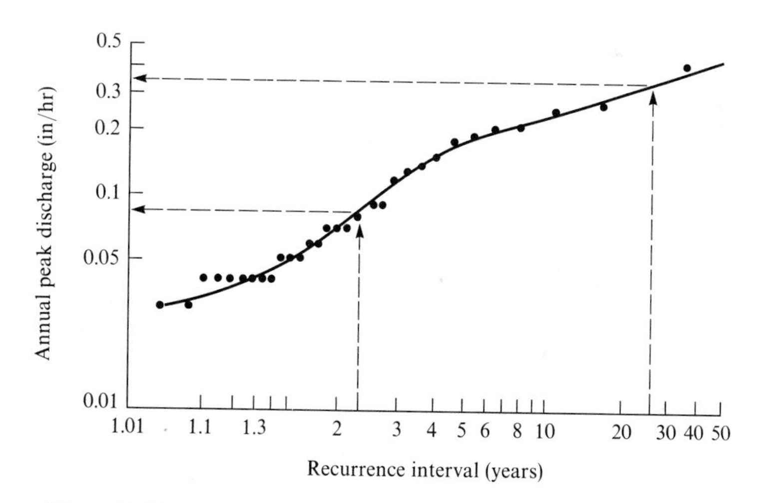

Figure 10-55 Plot of recurrence interval graph of annual peak discharge.

of 2.33 years. Note that the value obtained by averaging the values in column 2 differs from the graphical estimate. The latter is the most commonly used method, however, as mentioned in the text.

As an exercise in locating published data sources, the reader should now obtain a set of annual maximum flood discharges from a local library or water resource agency. Use them to compute the 1.5-year, 10-year, and 50-year floods for the station. In the United States lists of annual floods and gauge heights are available in the *U.S. Geological Survey Water Supply Papers.* Annual values after 1960 can be obtained from the annual publications. Water Supply Papers 1671 through 1688 cover rivers in various parts of the country. Unfortunately they do not contain records for many drainage basins smaller than 30 square miles.

10-8 Confidence Bands around a Flood-Frequency Curve

Figure 10-17 shows a flood-frequency curve based on 32 years of record for the Skykomish River at Gold Bar, Washington. Calculate values for the 90-percent confidence bands around this curve.

Solution

The standard deviation of annual flood peaks is first calculated for the 32 values. In the Skykomish case the standard deviation is 18,314 cfs.

From Table 10-12, obtain the coefficients for the upper and lower 90-percent confidence bands for 32 years of record. By interpolating between the fifth and sixth rows of the table for the upper band and between the fourteenth and fifteenth rows for the lower band, the coefficients in Table 10-29 are obtained. Multiplying these coefficients by 18,314 cfs yields the standard deviations in Table 10-29. These deviations from the best-fit flood-frequency curve are then plotted in Figure 10-17, and the confidence bands are sketched as shown. The interpretation is that there is a 90-percent chance that the 100-year flood lies within the range 88,000 and 109,000.

Table 10-29 Coefficients and standard deviations for the Skykomish River.

	RECURRENCE INTERVAL (YR)				
	100	10	2	1.1	1.01
Coefficient for upper band	0.71	0.48	0.30	0.34	0.46
Coefficient for lower band	−0.46	−0.34	−0.30	−0.48	−0.71

	DEVIATION (CFS)				
	100 YR	10 YR	2 YR	1.1 YR	1.01 YR
Deviation for upper hand	13,000	8800	5500	6200	8400
Deviation for lower band	−8400	−6200	−5500	−8800	−13,000

10-9 Flood Frequency and the Partial-Duration Series

Using data for, say, the New Fork River at Boulder, Wyoming, published in the *U.S. Geological Survey Water Supply Paper 1683,* tabulate all the discharges above a chosen base. There will be several in some years. An appropriate base discharge is usually given in the Water Supply Paper; the lowest annual peak of record is often used. This tabulation is the partial-duration series.

Rank the discharge values and compute the recurrence interval of each from Equation 2-4. Plot the points on your flood-frequency graph paper, and sketch a curve through them. Convert the recurrence intervals to those that would be obtained with an annual maximum flood series using factors shown in Table 10-13.

The discharge having a recurrence interval of 1.5 years in the annual flood series is an approximation to the bankfull discharge of the channel. Enter this discharge in the curve of the partial-duration series, and obtain a recurrence interval. Interpret the results you obtain from the two curves.

10-10 Illustration of the Homogeneity Test for Flood-Frequency Curves

In order to judge whether the six flood-frequency curves used in Figure 10-20 are from a homogeneous hydrologic region, it is necessary to carry out Dalrymple's (1960) homogeneity test as follows.

Solution

Compile Table 10-30 by the following computations.

1. List mean annual floods ($Q_{2.33}$) and 10-year floods (Q_{10}) for each station (columns 2 and 3).

2. Compute the 10-year ratio as the ratio of column 3 to column 2 (column 4).

3. Compute the mean 10-year ratio (= 1.85).

4. Multiply each $Q_{2.33}$ by the average 10-year ratio (column 5).

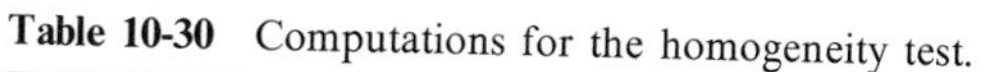

Table 10-30 Computations for the homogeneity test.

CATCHMENT (1)	MEAN ANNUAL FLOOD $Q_{2.33}$ (CFS) (2)	10-YR FLOOD Q_{10} (CFS) (3)	RATIO $\frac{Q_{10}}{Q_{2.33}}$ (4)	$Q_{2.33} \times 1.85$ (5)	RECURRENCE INTERVAL FOR Q IN COL. 5 (6)	PERIOD OF RECORD (YR) (7)
W-1	455	720	1.58	843	24.0	12
W-3	105	173	1.65	194	16.0	12
W-4	450	1000	2.22	834	7.4	12
W-5	1000	2000	2.00	1850	9.0	12
W-7	285	520	1.82	528	12.0	12
W-8	155	282	1.82	287	11.5	12
Average 10-year ratio			1.85			

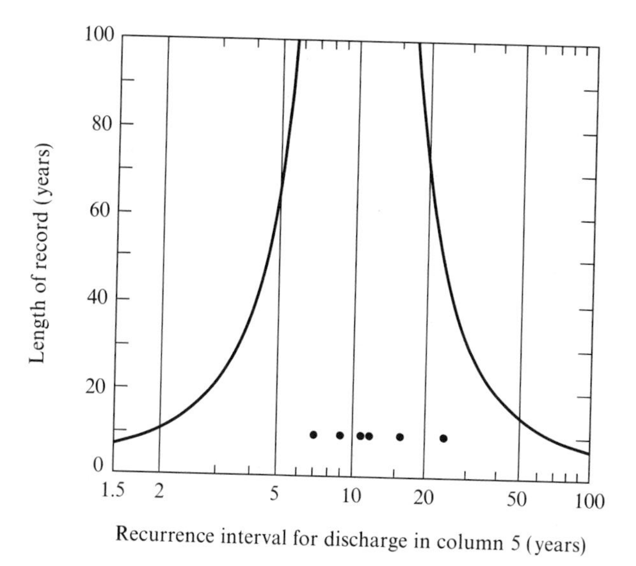

Figure 10-56 Homogeneity test of data for Vermont basins.

5. From the flood-frequency curves in Figure 10-19, read the recurrence interval for a flood equal to that in column 5 (column 6).

6. Tabulate the length of the record (column 7).

7. Plot column 6 against column 7 on the homogeneity test chart in Figure 10-56.

8. Points that fall between the two curves are accepted as representing basins from a region that is homogeneous with respect to flood-producing characteristics, and the basins may be grouped together to produce a regional flood-frequency curve.

9. If some points fall outside the curves, the physical geography of the basins should be re-examined and the stations divided into two or more groups. Then a separate regional flood-frequency curve can be defined for each group.

10-11 Regional Flood-Frequency Curve

A bridge is to be built across an ungauged river in the upland of northern Vermont. At that location the river drains a 10-square-mile catchment. The bridge will be designed to pass a flood having a recurrence interval of 25 years. Estimate the discharge of this flood.

Solution

From Figure 10-21, the mean annual flood of a 10-square-mile basin in the region is 290 cfs.

From Figure 10-20(b), the ratio of the 25-year flood to the mean annual flood is 2.6.

The 25-year flood for a 10-square-mile basin is 2.6×290 cfs $=$ 754 cfs.

As another exercise, the reader should find the published regional flood-frequency curve for a local region and use it to estimate several floods. In the Water Supply Papers referred to in Typical Problem 10-7, a large number of regional flood-frequency curves are published. Use *Water Supply Paper 1683,* for example, to estimate the 50-year flood for the New Fork River at Boulder, Wyoming. Compare this value with the 50-year flood obtained as suggested in Typical Problem 10-9. Results from the regional curve will not be identical to those from the station data, but they should agree approximately.

10-12 Use of Multiple Regression Equations for Flood Prediction

A small water-supply reservoir is to be impounded in a rural portion of the San Francisco Bay region. It will hold runoff from a 72-square-mile drainage basin. The dam must be provided with a spillway that will conduct away floodwaters and protect the structure from being overtopped and breached. Because the dam is a small one in a rural region and its failure would not cause much damage, the planning committee decides that the spillway should accommodate the 25-year flood peak if the reservoir is already filled to capacity by winter rains. You are asked to estimate the magnitude of this 25-year flood.

Solution

Obtain the mean annual basinwide precipitation, either from an isohyetal map of the region, or from data at several rain gauges, using one of the methods described in Chapter 2 for obtaining estimates of average precipitation over an area.

$$P = 33 \text{ inches}$$

Using the multiple regression equation given in Table 10-15, compute the 25-year flood from this basin.

$$Q_{25} = 16.5(72)^{0.912}P^{0.797} = 13{,}230 \text{ cfs}$$

The standard error of estimate for 25-year floods is -33.6 percent and $+50.7$ percent of the mean.

Answer: $8800 < 13{,}230 < 20{,}000$ cfs

If it is necessary to calculate a flood for a recurrence interval not represented in Table 10-15, all the equations in the table should be used to calculate values that are then plotted on Gumbel graph paper and used to define the flood-frequency curve. The required value can then be obtained by interpolation.

The reader should repeat this problem for a local basin, using the multiple regression equations describing regional flood-frequency curves in Water Supply Papers 1671 through 1688 or a similar source.

10-13 Statistical Estimation of Floods for Urban Areas

Estimate the 50-year flood for a basin in the Mid-Atlantic region of the United States that is expected to be developed so that 50 percent of its area becomes impervious, and the small tributaries will be placed in storm sewers while the main channel is left in its natural state.

Solution

From topographic maps, measure the drainage area, the length of the main channel, and the slope of the channel between two points 10 percent and 85 percent of the distance above the outlet.

$A = 4.5$ sq mi $\qquad S = 40.0$ ft/mi $\qquad L = 4.05$ mi

Use of these values in Figure 10-26 gives a centroid lag of 0.7 hr for partly sewered basins. In Equation 10-6 with $Z = 50$ percent,

$$K = \frac{0.30 + 0.0045 \times 50}{0.30} = 1.75$$

Substituting these values into Equation 10-7, the mean annual flood after urbanization will be

$$Q_{2.33}(\text{urban}) = 230 \times 1.75 \times (4.05)^{0.82}(0.7)^{-0.48}$$
$$= 1504 \text{ cfs}$$

Now, entering the regional flood-frequency curve of Figure 10-27 with a value of 50 percent for the impervious area, we find that the 50-year flood should be 2.7 times as large as the mean annual flood. Therefore,

$$Q_{50} = 4061 \text{ cfs}$$

Repetition of this last calculation for other frequencies will give values that can be plotted to define the flood-frequency curve for this urban catchment.

10-14 Effect of Urbanization on the Flood-Frequency Curve

Fifteen years after a water-supply reservoir was installed in a 20-square-mile rural drainage basin in the San Francisco Bay region, the catchment undergoes rapid development. The regional plan calls for eventual urbanization of 60 percent of the catchment, and for 50 percent of the channels to be sewered or lined with concrete. You become concerned that the spillway on the dam, which was originally designed to accommodate the 50-year flood, will be endangered just at the time when its failure would be catastrophic because of the increasing density of population. To illustrate your concern to local authorities and to display the magnitudes of floods they may wish to guard against, you decide to plot the rural and post-urbanization flood-frequency curves for the basin.

Solution

From the records of several rain gauges in the area, the mean annual rainfall is found to be 15 inches. The drainage area is 20 square miles. Using these values in the equations of Table 10-15, obtain discharges for floods with recurrence intervals ranging from 2 to 50 years. Plot these values on Gumbel paper as shown below, and sketch a rural flood-frequency curve.

Extrapolate the curve to estimate the 100-year flood, because the local authorities may be interested in enlarging the spillway to pass this larger discharge in view of the increasing population density.

To estimate the effect of the predicted urbanization, enter each graph in Figure 10-29 with a value of 60 percent for the ordinate and 50 percent for the abscissa. Interpolate the ratio of urban to rural flood-peak magnitude. The values should range from 1.47 at a recurrence interval of 100 years to 2.25 for the 2-year flood.

Multiply the rural flood magnitudes by these values, plot the resulting numbers on the graph, and sketch an urban flood-frequency curve, as shown in Figure 10-57.

The graph shows that the original 50-year flood (3500 cfs) is now to be expected with a frequency of about 21 years. The 50-year peak under the proposed urban conditions is 5340 cfs.

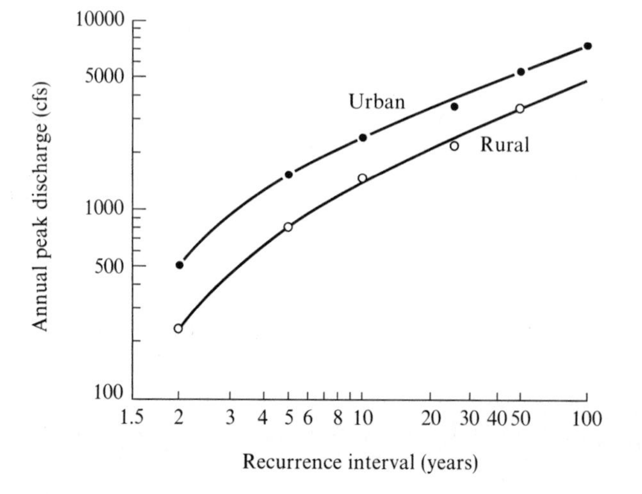

Figure 10-57 Change in annual peak discharge for various frequencies when area changes from rural to urban.

10-15 Derivation of a Unit Hydrograph

Suppose we wish to derive a unit hydrograph for W-994, a 27.2-square-mile drainage basin of the Agricultural Research Service Experimental Watershed at Coshocton, Ohio.

Solution

Figure 10-31 indicates that a unit hydrograph plotted from a rainstorm with a duration of about 2.5 hours is likely to be most useful for a 27-square-mile basin.

Searching through the published hydrographs in *Hydrologic Data for Experimental Watersheds in the United States,* we find only four hydrographs for this basin based on rainstorms with durations between 1.5 and 3.5 hours. Small amounts of low-intensity rainfall at the beginning or end of the storm were ignored in judging the correct storm duration. The unit hydrograph is not very sensitive to variations in storm duration, and so we decided to use the four storms. The temporal pattern of rainfall intensity during the storms was quite variable when expressed as 5-minute averages, or as averages for variable but short periods. When 30-minute or 60-minute averages were calculated, however, the hyetographs were as uniform as could be hoped for. The short-term fluctuations are damped out by storage within the basin, and each hydrograph used has a single peak. These four hydrographs are shown in Figure 10-58.

An average unit hydrograph was produced by first plotting the peak and then sketching the rising and falling limbs from the average shape of the four reduced hydrographs (Figure 10-59). The volume of rainfall was obtained by counting squares on graph paper and was found to be 0.99 inches after the first attempt at sketching.

As an exercise, draw the hydrograph generated by a 4-inch rainstorm with a duration of 2.5 hours for watershed 994. The Φ index for this storm of uniform intensity is 0.88 in/hr.

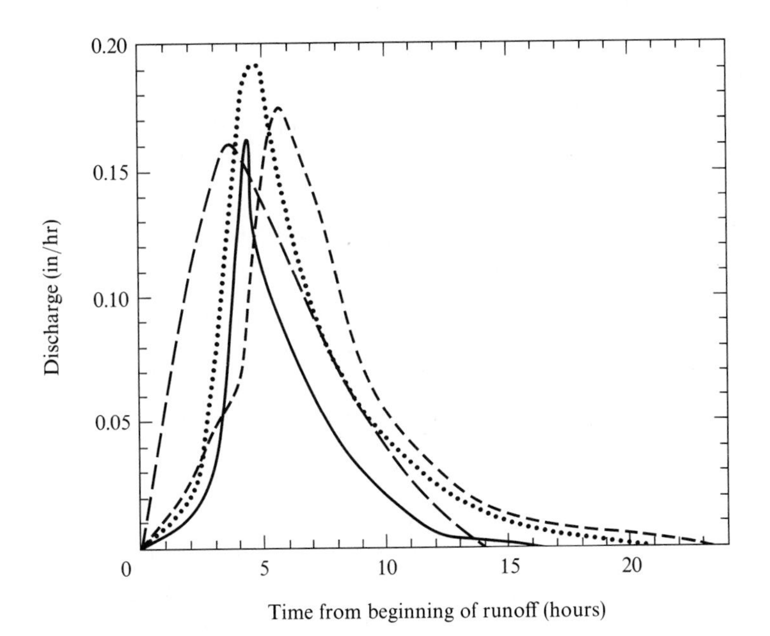

Figure 10-58 Hydrographs selected to use in the construction of the unit hydrograph for a basin.

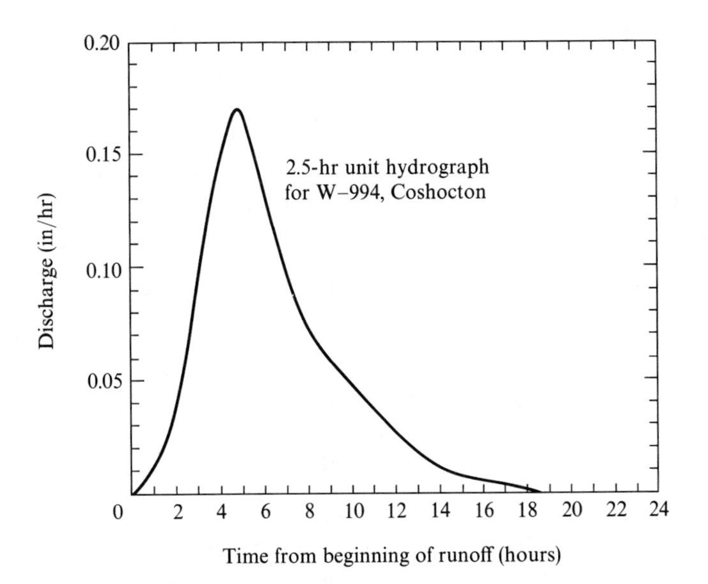

Figure 10-59 Unit hydrograph constructed from the selected observed hydrographs.

10-16 Addition and Subtraction of Unit Hydrographs

Suppose you need to predict the runoff from a long, uniform rainstorm, with the aid of a 6-hour unit hydrograph, for the basin in Typical Problem 10-15. Construct a 6-hour unit hydrograph from the 2.5-hour unit hydrograph derived in the last problem.

Solution

Add together two 2.5-hour unit hydrographs, as illustrated in Figure 10-33. Construct a summation curve for watershed W-994 by adding the 2.5-hour unit hydrographs as shown in Figure 10-33. Then obtain a 1-hour unit hydrograph by lagging *S*-curves in the manner shown in Figure 10-34. Add the 1-hour graph to the earlier sum, beginning 5 hours after the start of the first storm. The 6-hour composite hydrograph will encompass a runoff volume of three inches and is reduced to a unit hydrograph when its ordinate values are divided by three.

10-17 Triangular Synthetic Unit Hydrograph

Construct a three-quarter-hour triangular unit hydrograph for a 5-square-mile agricultural catchment with a time of concentration of one hour.

Solution

1. Using Equation 10-21,

$$Q_{\text{pk}} = \frac{484 \times 5 \times 1.0}{0.5 \times 0.75 + 0.6 \times 1.0} = 2482 \text{ cfs}$$

2. Using Equation 10-20,

$$T_p = 0.5 \times 0.75 + 0.6 \times 1.0 = 0.975 \text{ hr}$$

3. Using Equation 10-17,

$$T_b = T_p + 1.67T_p = 2.6 \text{ hr}$$

Plotting these three values on a graph allows the construction of the triangular unit hydrograph, as shown in Figure 10-60.

Repeat the derivation for a 10-square-mile basin in Vermont, using Figure 10-35.

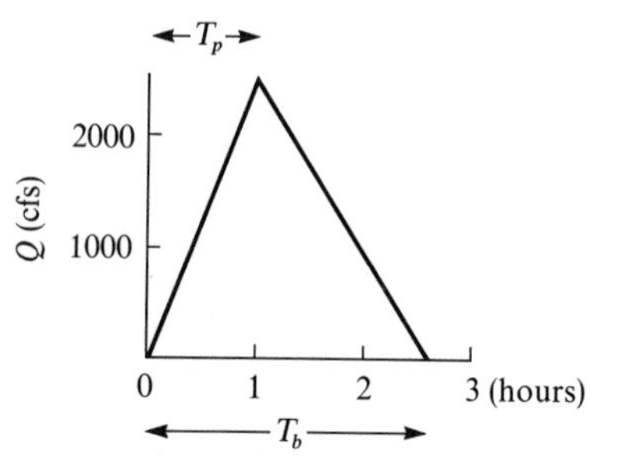

Figure 10-60 Triangular unit hydrograph for a 5-square-mile basin having a time of concentration of one hour.

Finally, in a small drainage basin, make a field measurement of the lag to peak and the centroid lag as described in the text. Plot your answers on Figures 10-35 and 10-36, and discuss the similarities or differences from the data summarized therein.

10-18 Composite Triangular Unit Hydrograph

Compute the 3-hour unit hydrograph for the catchment in Typical Problem 10-17.

Solution

From Equation 10-20,

$$T_p = 0.5(3.0) + 0.6(1.0) = 2.1 \text{ hr}$$

The resulting hydrograph would be of the form shown in Figure 10-61. The duration of the rainstorm exceeds the rise time of the drainage basin, and under these conditions the unit hydrograph method does not provide accurate results if applied directly. Instead, a composite hydrograph must be constructed by adding unit hydrographs of shorter duration.

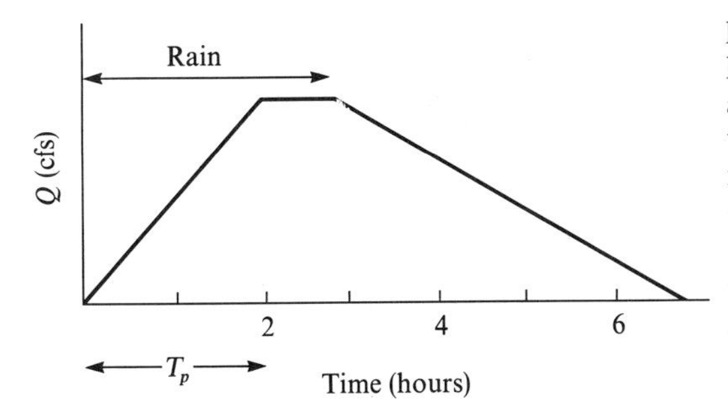

Figure 10-61 Hydrograph constructed from equations for rise time and resulting peak discharge.

The duration of effective rainfall should be divided into time increments that are less than the time of concentration. In this case, we might use the previously constructed three-quarter-hour unit graph. Four of these are plotted (or tabulated) to lag each other by three-quarters of an hour. The ordinates of the hydrographs are then added at various times to obtain the ordinates of the composite, 3-hour hydrograph that would result from four inches of runoff (the summation of four unit hydrographs). The ordinates of the composite hydrograph, Figure 10-62(a), are then divided by four to obtain the 3-hour unit hydrograph shown in Figure 10-62(b).

The unusual shape of the recession limb probably results from Equation 10-17 underestimating the length of the recession limb in many small catchments. The error, however, is on the side of safety because it should cause a slight overestimate of the hydrograph peak. If local observations indicate the recession limb to be much longer than the value used here, the calculations can be modified accordingly. The addition of the baseflow to the recession limb would also give the hydrograph a more "usual" shape.

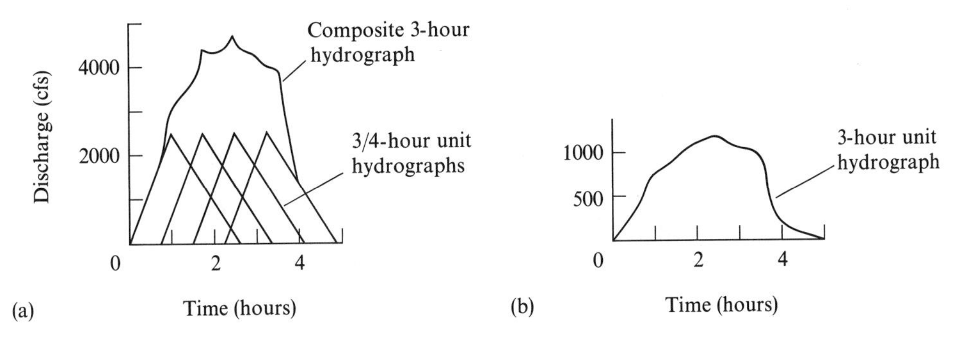

Figure 10-62 (a) Composite 3-hour hydrograph constructed from triangular hydrographs. (b) 3-hour unit hydrograph derived from composite graph.

10-19 Dimensionless Unit Hydrograph

Compute the one-hour unit hydrograph from an 8-square-mile agricultural drainage basin whose length from outlet to divide (L) is 5.2 miles and whose total relief (H) is 500 feet.

Solution

1. From Equation 10-3,

$$t_c = \frac{L^{1.15}}{7700H^{0.38}} = \frac{(5.2 \times 5280)^{1.15}}{7700(500)^{0.38}} = 1.6 \text{ hr}$$

Therefore, from Equation 10-20,

$$T_p = 0.5(1.0) + 0.6(1.6) = 1.5 \text{ hr}$$

As a check on this value, plot t_p ($= 0.6 \times 1.6$ hr) on Figure 10-35 at a drainage area of 8 square miles. The point falls between the lines for the southwestern rangelands and the northeastern basins, which have a mixed cover of forest, pastureland, and cultivated fields. Since the basin in this problem is entirely agricultural and is steep (96 ft/mi), the intermediate lag to peak seems reasonable. It is left for the reader to examine whether the lag to peak is reasonable by comparison with the centroid lag from Figure 10-36.

2. Compute Q_{pk} from Equation 10-19:

$$Q_{pk} = \frac{484 \times 8 \times 1.0}{1.5} = 2580 \text{ cfs}$$

3. Prepare a table, listing selected time ratios T/T_p and corresponding discharge ratios Q/Q_{pk} from Figure 10-40 (Table 10-31).

4. Compute the values of time and discharge from the ratio columns and the previously calcalculated values of T_p and Q_{pk}.

5. Plot the values of time and discharge against one another, and sketch the unit hydrograph.

6. Plot the dimensionless unit hydrograph on a sheet of graph paper, and superimpose a triangular unit hydrograph calculated for the same storm.

Table 10-31 Calculations for the dimensionless unit hydrograph.

TIME RATIO T/T_p	DISCHARGE RATIO Q/Q_{pk}	TIME (HR)	DISCHARGE (CFS)
0	0	0	0
0.25	0.12	0.38	310
0.50	0.43	0.75	1109
0.75	0.80	1.12	2064
1.00	1.00	1.50	2580
1.25	0.87	1.88	2245
1.50	0.66	2.25	1703
1.75	0.45	2.62	1161
2.00	0.32	3.00	826
2.50	0.15	3.75	387
3.00	0.075	4.50	194
3.50	0.030	5.25	77
4.00	0.018	6.00	46
5.00	0.004	7.50	10

10-20 The Summation Graph in Terms of Lag Time

Use the average summation graph to provide a hydrograph to compare with the observed storm hydrograph.

Table 10-32 represents the average summation graph (also called a distribution graph) derived by Langbein (1940) from river data. This study showed that hydrographs for different basins tend to be alike if expressed in terms of the centroid-to-centroid lag time for each basin. A storm will run off in a time period equal to about 4 times the lag of the basin. About 50 percent of the runoff will have passed the mouth in a time period equal to 85 percent of the lag time after runoff begins.

Because the summation graph is expressed in percent of total runoff as a function of time in percent of lag, when the total runoff is known, one can compute a synthetic hydrograph. When this is plotted at the same scale as the observed hydrograph, the two can be compared. The exact dimensions of the synthetic graph will depend, then, only on the choice of lag time.

If the peak of the synthetic hydrograph is lower than the observed hydrograph, a recomputation may be made using a slightly shorter lag time and the comparison made again. It is clear that the shorter the lag time for a given runoff volume, the higher will be the peak flow.

Table 10-32 Summation graph in terms of lag interval. Columns 4 and 5 are from the slope of the cumulative curve.

TIME (% OF LAG)	CUMULATIVE VALUES OF Q (% OF TOTAL)	TIME AVERAGE OF INTERVAL BETWEEN TIMES IN COL. 1 (% OF LAG)	RUNOFF INCREMENT (%)	RUNOFF RATE (% OF TOTAL RUNOFF PER LAG)
0	0			
		10	3.8	19.0
20	3.8			
		30	9.0	45.0
40	12.8			
		50	16.2	81.0
60	29.0			
		70	17.5	87.5
80	46.5			
		90	12.8	64.0
100	59.3			
		110	8.9	44.5
120	68.2			
		130	7.2	36.0
140	75.4			
		150	6.0	30.0
160	81.4			
		170	4.9	24.5
180	86.3			
		190	4.0	20.0
200	90.3			
		210	3.1	15.5
220	93.4			
		230	2.3	11.5
240	95.7			
		250	1.7	8.5
260	97.4			
		270	1.0	5.0
280	98.4			
		290	0.7	3.5
300	99.1			
		310	0.03	1.5
320	99.4			

Discussion

If the total volume of the storm runoff is known, it can be distributed through time by use of the summation graph. For example, the summation graph data show that the increment of runoff will be 9 percent of the total runoff (column 4) during the time period between 20 percent and 40 percent of lag (column 1). The mean point in time for this interval is 30 percent (column 3).

If the total runoff of the storm is one inch, and the lag is 60 minutes, the point in time representing 30 percent of lag is $0.3 \times 60 = 18$ minutes after runoff begins. The runoff rate at that time will be 9.0 percent of the total, occurring in 20 percent of the lag time, that is,

$$\frac{0.09 \times 1 \text{ inch}}{0.20 \times 60 \text{ min}}$$

expressed in inches per minute. The inches per hour will then be 60 times as great or 0.45 inches per hour.

The rate of runoff can be more conveniently computed by expressing the slope of the hydrograph as percent of total runoff per lag interval. Because 9.0 percent of the runoff went by the observation station in 20 percent of the lag time, then at the same rate 5 × 9.0 or 45.0 percent would pass the station during a period equal to one lag. To compute the runoff rate, q, multiply the slope of the hydrograph, s, times the total volume of runoff divided by the lag time in hours, for

$$\frac{\text{inches}}{\text{hour}} = \%\text{ of runoff} \times \frac{\text{inches of runoff}}{\text{lag time in hours}}$$

so the units are correct. Thus,

$$\text{runoff rate } q = \frac{s}{100}\,\frac{(\text{total runoff})}{\text{lag time}}$$

To compute the runoff rate in inches per hour at the time shown in column 3, compute

$$\frac{\text{column 5} \times \text{total inches of runoff}}{100 \times \text{lag in hours}}.$$

Procedure

Using the data in Table 10-33 for the gauge on North Fork Strawberry Creek, January 2, 1977, plot the hyetograph and the hydrograph in inches per hour. Measure the total runoff in the single hydrograph between 4:00 and 5:00 PM.

Estimate the lag time. Using columns 3 and 5 in the summation graph, compute a synthetic hydrograph. Plot on the same graph.

Data on the relation of centroid lag time as a function of drainage area applicable to the San Francisco Bay region is given in Figure 10-36, where lag is in hours and drainage area in square miles. Though this may not be representative of other areas, it will be used for present purposes. Read the lag time from this figure for the drainage area of 0.25 square miles, the area of North Strawberry Creek.

Table 10-33 Readings from North Fork Strawberry Creek near Haviland Hall, University of California Campus. Jan. 2, 1977. Drainage area = 0.25 sq mi. Rating curve gh (ft) = $4.0Q^{0.35}$ (in/hr). The rain gauge was emptied after each reading.

TIME (HR)	GAUGE HEIGHT (FT)	RAIN (IN)	TIME (HR)	GAUGE HEIGHT (FT)	RAIN (IN)
2:45 PM	0.54		3:33 PM	0.71	0.04
:47	0.57		:37	0.76	
:48	0.70	0	:40	0.87	
:49	0.78		:42	0.93	0.04
:50	0.84		:48		Very heavy
:52	0.89		:54		0.06
:53		0.04	4:04		0.16
:55	0.88		:06	1.95	
:57	0.86		:07	2.05	
:59	0.87		:08	2.25	
3:01	1.00		:09	2.40	
:03	1.02	0.02	:10	2.45	
:05	1.00		:11	2.55	
:07	0.97		:12	2.60	
:09	0.95		:13	2.65	
:12	0.90		:14		0.14
:13		0.015	:15	2.70	
:15	0.86		:16	2.75	
:20	0.80		:17	2.80	
:23		0.017	:18	2.65	
:24	0.75		:24		0.07
:29	0.72		:35	1.70	

Using this new lag for natural (rural) conditions, and the same volume of runoff measured earlier, compute a new synthetic hydrograph that will approximate that which would have been observed during this storm if the basin were natural rather than urbanized. Plot this synthetic hydrograph on the same sheet as the observed hydrograph.

10-21 Use of an Instantaneous Unit Hydrograph

Compute a unit hydrograph that would be generated by a 50-year rainstorm on a rural drainage basin in the San Francisco Bay region with the following characteristics.

Drainage area (A), 10 sq mi
Mean annual precipitation, 40 in
Channel slope index (S), 20 ft/mi

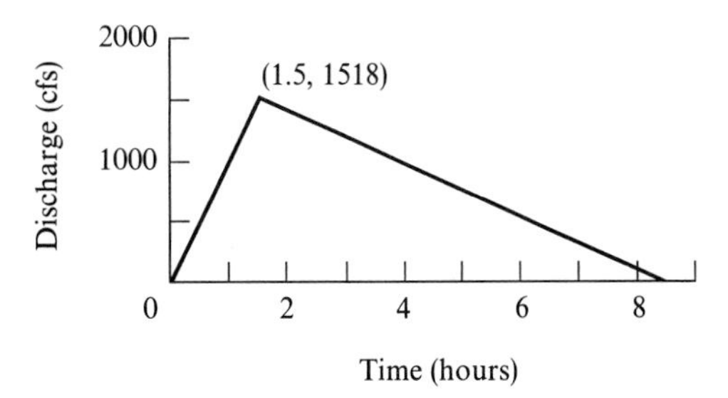

Figure 10-63 30-minute triangular hydrograph.

Solution

1. Construct the instantaneous unit hydrograph.

It is decided to use the instantaneous unit hydrograph (IUH) method, the parameters for which are obtained from Figure 10-43:

$$l_c = 3.11 \text{ hr}$$

$$T_{bi} = 8.04 \text{ hr}$$

and therefore,

$$T_{pi} = 3l_c - T_{bi} = 1.29 \text{ hr}$$

2. Transform the IUH into the unit hydrograph from a storm of some finite duration, D_r.

From Equation 10-23,

$$T_p = 1.29 + 0.5D_r$$

where $0.2T_p < D_r < 0.33T_p$. Choose $D_r = 0.5$ hr to obtain $T_p = 1.54$ hr. From Equation 10-22,

$$T_b = 8.04 + 0.5 = 8.5 \text{ hr}$$

From Equation 10-24, $Q_{\text{pk}} = 1518$ cfs. The 30-minute triangular unit hydrograph for the basin can now be constructed, as shown in Figure 10-63.

3. Compute the design storm.

The rainstorm most likely to produce a critical high discharge is one with a duration slightly longer than the centroid lag of the basin. A duration of 4 hours should therefore be used, but to save space we will use a design storm duration of 3 hours, which is almost equal to the lag. We leave it as an exercise for the reader to check what difference that makes to the final answer. The runoff produced by a 3-hour storm will be calculated by lagging six 0.5-hour hydrographs by increments of 0.5 hours and adding the runoff from all of them.

From the rainfall intensity-duration-frequency curves for the San Francisco Bay region, the 50-year, 3-hour storm in an area having a mean annual rainfall of 40 inches is 2.16 inches. Using Table 2-3, we then compute the mass curve of this 3-hour storm in increments of $D_r = 0.5$ hr, as shown in Table 10-34.

Compute the runoff generated by the precipitation in each 0.5-hr increment. From Figure 10-7, during a 50-year storm the rate of abstraction from rainfall (Φ) in a region with a mean annual rainfall of 40 inches is 0.17 inches per hour, or 0.085 inches per 0.5-hour time increment. Therefore in Table 10-34, 0.085 inches can be subtracted from the rainfall in each time increment, as shown in the last column of the table.

Compute the design hydrograph using the incremental values of the runoff generation in each 30-minute increment (last column of the table) and the previously calculated 30-minute unit hydrograph. This involves setting up Table 10-35. The increments of runoff from column 6 of Table 10-34 are listed across the top line.

The first increment of precipitation excess listed at the top of the table is multiplied by the ordinates of the unit hydrograph to yield in the third column the hydrograph of storm runoff that would be generated by such a pre-

Table 10-34 Computing the mass curve of storm rainfall and of precipitation excess.

CUMULATIVE HOURS	CUMULATIVE % OF STORM DURATION	CUMULATIVE % PRECIPITATION	INCREMENTAL PRECIPITATION (%)	INCREMENTAL PRECIPITATION (IN)	INCREMENTAL PRECIPITATION EXCESS (IN)
		0	0	0	0
0.5	16.6	10	10	0.216	0.131
1.0	33.3	31	21	0.454	0.369
1.5	50.0	69	38	0.821	0.736
2.0	66.7	80	11	0.238	0.153
2.5	83.7	90	10	0.216	0.131
3.0	100.0	100	10	0.216	0.131
				2.16	1.651

Table 10-35 Computing the design hydrograph.

TIME (HR)	ORDINATES OF 30-MIN UNIT HYDROGRAPH (CFS)	PRECIPITATION EXCESS (IN) 0.131	0.369	0.736	0.153	0.131	0.131	SUM OF STORMFLOW (CFS)	BASEFLOW (CFS)	TOTAL DISCHARGE (CFS)
0	0	0						0	304	304
0.5	500	66	0					66	304	370
1.0	1000	131	185	0				316	304	620
1.5	1520	199	369	368	0			936	304	1240
2.0	1410	185	561	736	77	0		1559	304	1863
2.5	1300	170	520	1119	153	66	0	2028	304	2332
3.0	1190	156	480	1038	233	131	66	2104	304	2408
3.5	1080	141	439	957	216	199	131	2083	304	2387
4.0	970	127	399	876	199	185	199	1985	304	2289
4.5	860	113	358	795	182	170	185	1803	304	2107
5.0	750	98	317	714	165	156	170	1620	304	1924
5.5	640	84	277	633	148	141	156	1439	304	1743
6.0	530	69	236	552	132	127	141	1257	304	1561
6.5	420	55	196	471	115	113	127	1077	304	1381
7.0	310	41	155	390	98	98	113	895	304	1199
7.5	200	26	114	309	81	84	98	712	304	1016
8.0	100	13	74	228	64	69	84	532	304	836
8.5	0	0	37	147	47	55	69	355	304	659
9.0			0	74	31	41	55	201	304	505
9.5				0	15	26	41	82	304	386
10.0					0	13	26	39	304	343
10.5						0	13	13	304	317
11.0							0	0	304	304

cipitation excess. The second increment of precipitation excess is also multiplied by the ordinates of the unit hydrograph, but the products are lagged behind the first hydrograph by one time-increment and are placed one row lower in the table. This process is repeated for each increment of precipitation excess.

For each time in column 1 the discharges generated by each precipitation increment are summed horizontally to yield the storm runoff discharge in column 9.

Baseflow should then be added to the storm runoff value. Different investigators may wish to assign other values here, depending on their knowledge of the regional hydrology. We will use the suggestion of Rantz given in Table 10-16, and estimate the antecedent baseflow for the 50-year storm to be 20 percent of the peak discharge of storm runoff, or in this case, 304 cfs. The addition of stormflow and baseflow is accomplished in the last column of the table, which lists the ordinates of the hydrograph of the 50-year storm. The computed hydrograph is shown in Figure 10-64.

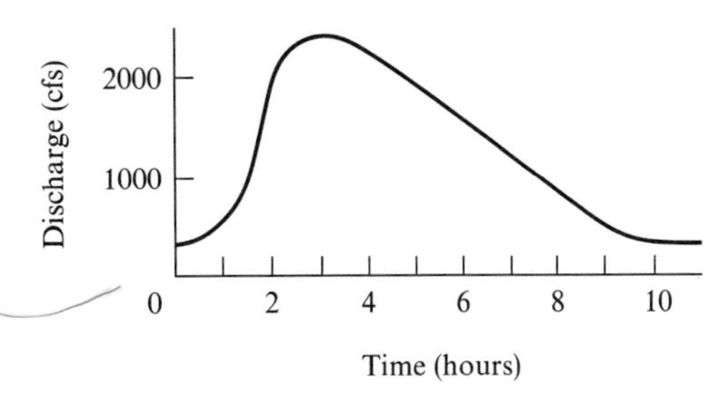

Figure 10-64 3-hour, 50-year design hydrograph.

For comparison, compute the Soil Conservation Service triangular unit hydrograph for a 30-minute storm if the basin were to have a mainstream length of 6.25 miles and a relief of 150 feet.

10-22 Synthetic Unit Hydrographs for Urban Areas

The basin used in Typical Problem 10-21 is to be urbanized over 60 percent of its area. Compute the resulting hydrograph during the 50-year storm and compare it with the response of the rural basin.

Solution

Estimate the effect of urbanization on the time of rise and the time base of the instantaneous unit hydrograph with the aid of Table 10-17. The original values were

$$T_{pi} = 1.3 \text{ hr} \qquad T_{bi} = 8.0 \text{ hr}$$

From Table 10-17, the urban values will be

$$T_{pi} = 0.7 \text{ hr} \qquad T_{bi} = 4.4 \text{ hr}$$

It is possible to check whether these values are reasonable for a partly urbanized basin by use of Figure 10-26. The centroid lag can be calculated from T_{pi} and T_{bi}, and in this case it is 1.7 hr. As indicated in the last problem, the mainstream length of the basin is 6.25 mi. A point representing the basin therefore plots on Figure 10-26 among the points representing partly developed drainage basins, but close to the completely urbanized line. Our numbers are therefore reasonable by comparison with those from another urban region.

The duration of the unit storm for the urban catchment must change as T_p changes. It can be obtained from

$$T_p = T_{pi} + 0.5D_r$$

Since D_r should lie between $0.2T_p$ and $0.33T_p$, we have

$$0.9T_p = 0.7 \qquad \text{and} \qquad 0.84T_p = 0.7$$

therefore T_p should be about 0.8 hr and D_r must be about 0.25 hr.

A 15-minute synthetic unit hydrograph is therefore constructed, as illustrated in the last problem. A new 50-year design storm must be chosen (a 2-hour duration is slightly longer than the centroid lag), and the calculation proceeds as outlined in Typical Problem 10-21. The estimation of the storm runoff volume after urbanization requires choosing a new value of Φ from Table 10-2.

We leave it to the reader as an exercise to complete this problem and also to estimate the 50-year hydrograph peak for the urbanized basin by flood-frequency methods for comparison.

Another approach to the prediction of urban runoff effects with synthetic unit hydrographs is described in detail by the U.S. Soil Conservation Service (1975).

10-23 Flood Routing through a Reservoir

1. Route the inflow in Figure 10-45 through the lake if the size and shape of the outlet remain the same but the area of the lake is doubled.

2. Route the inflow in Figure 10-45 through the lake if the area remains at five million square feet, but the natural stream outlet is controlled by a 120° sharp-crested V-notch weir.

10-24 Lag Time from Direct Measurement of Rainfall and Gauge Height

Extrapolation of regional data to an individual site is usually necessary, but it is highly desirable to obtain direct measurements at the site. Even data for a single storm add greatly to the confidence in an extrapolated figure for hydrologic conditions.

Plot the observed data on rainfall and gauge height in the form of a hydrograph and a hyetograph. Determine the average lag time and infiltration rate. Data were observed on El Cerrito Creek at 400 Vermont Avenue, Berkeley, California. The drainage basin is 80 percent urbanized and is 0.068 square miles (43 acres) in area.

Procedure

1. On log-log paper, plot the rating curve given at the top of Table 10-36. Plot a new rating curve for which discharge is in inches per hour rather than cfs. Memorize this constant: 1 inch per hour from 1 acre = 1 cfs.

2. Complete Table 10-36 so that for each time of observation or time unit, there is a value of discharge in inches per hour and of rainfall in inches per hour. The rainfall was read at about 10-minute intervals, and the gauge emptied. Thus, the inches per hour are inches measured, divided by hours or tenths of hours between observations.

3. Plot the hyetograph and the hydrograph, each in inches per hour versus time in hours. Use different scales for rainfall and runoff. A reasonable scale for runoff might be 1 inch vertically equal to 0.02 inches per hour. For rainfall 1 inch vertically might be 0.2 inches per hour.

The volume of runoff can be computed by estimating the mean runoff rate from the plotted hydrograph for each hour of the storm, multiplying by the time period and adding the products.

4. For each distinct peak in the runoff hydrograph, draw a dashed line to indicate an estimated recession limb if no more rainfall occurred.

5. At least some of these separate hydrographs appear to be related to particular bursts of rainfall. Estimate the position of centers of mass of rainfall and corresponding runoff, and measure the lag or time between each pair. Average these to obtain a mean estimate of lag time for the basin.

6. Sum the runoff quantities and the rainfall. Subtract the two; divide this remainder by the total storm hours, and estimate a mean rate of infiltration (Φ) for the storm. Also note the ratio of runoff to rainfall. This is a mean runoff coefficient. Plot the infiltration rate on the hyetograph.

7. On log-log paper, plot values of lag time in hours (abscissa) against drainage area in square miles (ordinate), using data from Rantz (1971:29, Table 5).

A general relation to estimate centroid lag time in hours from drainage area in square miles is

$$l_c = 1.3A^{0.43}$$

On the same graph, plot the estimated lag time for El Cerrito Creek. This is an urbanized area and should have a somewhat smaller value of lag for its basin area than do the basins tabulated by Rantz.

Table 10-36 Readings from El Cerrito Creek at Leopold's, 400 Vermont, Berkeley. Drainage area = 0.068 sq mi. Preliminary rating curve, $gh = 0.56Q^{0.42}$, gauge height in feet, discharge in cfs.

TIME (STORM OF 1/11/76)	*gh* (FT)	Δ RAIN (IN)	*Q* (IN/HR)	RAINFALL TIME (HR)	RAINFALL RATE (IN/HR)	Δ RUNOFF (IN)
9:00 AM		0				
:09	0.40					
:15	0.40					
:20	0.42	0.35				
:24	0.46					
:25	0.48					
:28	0.52					
:31	0.55					
:34	0.57					
:36		0.18				
:38	0.58					
:43	0.60					
:47	0.60					
:50		0.15				
:52	0.57					
:56	0.57					
10:01	0.55					
:05		0.13				
:07	0.54					
:13	0.58					
:16	0.57					
:18	0.57					
:20		0.13				
:23	0.56					
:30	0.55					
:35		0.11				
:38	0.54					
:42	0.52					
:47	0.50					
:49		0.08				
:54	0.47	Stopped				
:58	0.44					
11:00	0.43	Began				
:03	0.45					
:05		0.10				
:07	0.49					
:10	0.51					
:20	0.50	0.09				
:27	0.52					
:33	0.54					
:35		0.11				
:43	0.49					
:50		0.09				
:52	0.44					

continued

Table 10-36, *continued*

TIME (STORM OF 1/11/76)	gh (FT)	Δ RAIN (IN)	Q (IN/HR)	RAINFALL TIME (HR)	RAINFALL RATE (IN/HR)	Δ RUNOFF (IN)
12:05	0.50	0.11				
:08	0.50					
:13	0.48					
:20		0.08				
:23	0.44					
:32	0.47					
:35		0.08				
(STORM OF 1/2/77)						
10:30 AM	0.30	0.03				
11:05	0.44	0.10				
:20	0.45					
:40	0.62	0.12				
:45	0.77					
:50	0.76	0.105				
:55	0.76					
12:00	0.67	0.04				
:16	0.47	0.02				
:28	0.40	0.01				
:45	0.34					

10-25 Hydrograph and Lag Time

Procedure

The measured hydrograph and rainfall data for the storm at Tombstone, Arizona, W-3, August 17, 1973 are given in Table 10-37. Its drainage area is 2220 acres.

1. Plot the hydrograph and hyetograph (rainfall rate versus time). For the latter, use an ordinate scale of one-tenth the scale of the hydrograph because the number of inches per hour of runoff is much smaller than the inches per hour of rainfall. For rainfall, use 1 inch = 1 inch per hour, for runoff 1 inch = 0.1 inches per hour. For time, 1 inch = 40 minutes.

2. Compute the total storm runoff in inches by multiplying the runoff rate during a short period by the time. For example, if 0.025 inches per hour of runoff lasted five minutes, runoff volume is $5/60 \times 0.025 = 0.0104$ inches. Add the incremental volumes of runoff to get the total in inches. Keep a record of the incremental values. Divide the total runoff in inches by the total rain in inches to obtain the total runoff coefficient for the storm.

Across the top of the tabulation sheets, write the column headings for the whole problem as shown below. Data in columns 1, 2, 5, and 6 are given in Table 10-37. All the other columns are computed.

3. Estimate the lag time in hours—the center of mass of rainfall to the center of mass of runoff (centroid to centroid).

4. Having measured the lag time from the plotted data, compare the result with the data in Figure 10-36.

Table 10-37 Data for Tombstone, Arizona, Watershed W-3, storm of August 17, 1963. Basin area = 3.47 sq mi, rain gauge R-38. Assume that each rainfall rate was constant for the period of time represented by the time shown and the preceding time (e.g., 0.77 in/hr for the time period 9:20 to 9:27).

RAIN RECORD		RUNOFF RECORD	
TIME	RATE (IN/HR)	TIME	RATE (IN/HR)
9:20 PM	0	9:42	0
:27	0.77	:43	0.008
:33	2.20	:44	0.015
:35	3.30	:45	0.024
:39	2.25	:50	0.049
:42	2.60	:51	0.103
:47	2.04	:53	0.121
:50	3.20	:54	0.163
:53	2.00	:55	0.241
:58	1.20		
		10.00	0.282
10:22	0.05	:08	0.311
		:15	0.268
		:25	0.206
		:35	0.183
		:45	0.116
		11:00	0.067
		:15	0.048
		12:00	0.025
		1:00	0.013
		2:00	0.006
		3:00	0.001

RAIN RECORD				RUNOFF RECORD			
1	2	3	4	5	6	7	8
TIME	RATE (IN/HR)	MINUTES IN PERIOD (MIN)	Δ RAIN (IN)	TIME	RATE (IN/HR)	MINUTES IN PERIOD (MIN)	Δ RUNOFF (IN)

11

Human Occupance of Flood-Prone Lands

Flood Losses

On the evening of June 9, 1972, a group of thunderstorms over the Black Hills of South Dakota poured more than 10 inches of water within 6 hours onto an area of 60 sq mi. Rivers draining the steep valleys of the region rose quickly, and around midnight devastating floods swept through campgrounds in the valley bottom and through towns and villages situated where the rivers drain from the hills onto the Great Plains. The largest of these towns is Rapid City, situated on Rapid Creek.

Floodwaters from the upper 320 sq mi of the catchment of Rapid Creek were stored in a reservoir and did not contribute to the flood at the townsite. The damaging storm runoff was generated in an area of 51 sq mi draining to 9 mi of the main channel. The rates of runoff in this lower area were extreme. Measurements of peak discharge give values of 1.29 in/hr from 4.28 sq mi; 1.59 in/hr from 6.71 sq mi; and 1.57 in/hr (50,000 cfs) from the whole 51 sq mi.

Immediately above Rapid City was a small multipurpose storage reservoir, impounded by a 500-foot-long, earthfill dam. When the inflow to the reservoir began to rise, the gates on the spillway were opened to lower the lake level. The gates were soon blocked by floating debris, including boats and a floating dock. The impounded water eventually overtopped the dam, eroding its face. When the dam was breached, a flood wave swept through Rapid City with a peak discharge of 50,000 cfs, and traveled downstream at 6.3 mi/hr.

The valley floor in Rapid City was mostly occupied by homes and businesses. Despite a warning broadcast by the mayor, most people stayed in their homes and were caught there when the flood hit the town. The river eventually rose 14 feet in about 4 hours, and 3.5 feet during one period of 15 minutes. Two hundred and thirty-seven people died in the flood, 3057 were injured, and the total cost of the damage was estimated by the Corps of Engineers to exceed $160,000,000 of which less than $300,000 was insured. Five thousand automobiles and 1335 homes were destroyed, while 2820 homes suffered major damage (Figures 11-1 and 11-2). This catastrophe occurred during a period in which various parts of the United States suffered heavily from flooding. Hurricane Camille generated floods that killed 152 people in Virginia during August 1969. Floods spawned by Hurricane Agnes ravaged the east coast from Virginia to New York and left

Figure 11-1 Flood damage by Rapid Creek, in western Rapid City, South Dakota, June 10, 1972. (Photograph by Perry H. Rahn.)

Figure 11-2 House moved during flood of Rapid Creek, Rapid City, South Dakota, June 10, 1972. (Photograph by Perry H. Rahn.)

117 people dead. Failure of a dam impounding coal waste had sent a flood wave down a narrow valley in West Virginia, killing 118 people. Minor tragedies are played out each flood season in all regions of this and other countries as people are killed, or homes are washed away or furniture ruined by even minor flooding. But even these pale by comparison with the major floods of history, which have killed hundreds or thousands of people in a single cataclysm. The 1887 floods in Hunan Province, China, killed an estimated one million people, and the Yangtze River flood of 1911 killed more than 100,000.

In this section, we review some of the economic and social aspects of flooding, along with some of the options open to the planner for avoiding or minimizing the large, and often unnecessary, costs of this hazard. For a much fuller discussion of these matters, the reader should consult the works of Gilbert F. White and his co-workers, listed in the bibliography, as well as the books by Hoyt and Langbein (1955) and Leopold and Maddock (1954).

Despite the danger of flooding, valley floors are attractive areas for human settlement. They have often provided avenues for colonization through and into mountainous regions and even today may provide the most convenient, or even the only, transportation corridors. The soils of valley floors are usually fertile and irrigable and are attractive to agriculturalists in spite of drainage problems in some areas. The rivers themselves provide water supply, waste disposal, and transportation facilities, and the flat land of the valley floor allows residential and commercial development close to transportation. In the United States there are more than 50,000 square miles of flood-prone land, much of it highly productive. In some other countries, the valley floors are the only areas fit for agricultural or commercial development. For these and other reasons, human societies have usually borne the dangers and costs of periodic flood losses.

Kates (1962) has analyzed the perception of hazard among various groups of people and has summarized many of the reasons for which they inhabit flood-prone areas in the face of such danger. The perception of flood hazard waxes and wanes. Immediately after a flood, the danger is appreciated and even exaggerated. As the flood lapses into history, and the evidence of its power and extent is eradicated, the danger fades in people's minds. Information is suppressed in some cases by unscrupulous land developers or others with interests in economic growth and local pride. Yet the danger remains, for as we stressed in Chapter 10, the occurrence of a large flood last year in no way diminishes the chances of another large flood this year.

It is difficult to gauge flood losses, and the various agencies responsible for assessing flood losses have frequently obtained widely differing figures for the same flood. The problems of making such surveys, and comparisons of the methodologies of various U.S. federal agencies, are given by White (1945), Kates (1965), and Grigg and Helmeg (1975).

Whichever methodology is accepted, however, some generalizations can be made. There are large differences in average annual flood damages in various regions of a country, as indicated for the United States in Figure 11-3. These differences depend upon the nature of the economy, the topography, and the flood hydrology of the regions. The costs of floods are also rising, even when allowance is made for the effects of inflation (see Figure 11-4). Even though more than $7 billion has been spent on flood control in the United States since 1936, the annual cost of flood damage has increased since its reporting by the National Weather Service began in 1903. A part of this increase is due to the improvement of report coverage; another part is due to an overall increase in the flooding of rivers because of weather fluctuations. A third factor is the increase of property values, and a fourth is the change in occupance of flood-prone lands. In 1955, Hoyt and Langbein suggested that 45 percent of the increase in flood damage reported between 1903 and 1951 in the United States was due to an increase of property values; 25 percent was due to an increased amount of flooding between 1928 and 1951 as compared with the period 1903–1927. Thirty percent of the increase was due to building in flood-prone lands.

Some of the change of occupance has occurred because the increasing value of valley floors has made the risks worth taking, but in other cases the changes have occurred because of ignorance of the flood hazard. In other situations, people have moved in with a sense of security generated by upstream flood-control works. Few people realize that flood-control works never can give absolute protection.

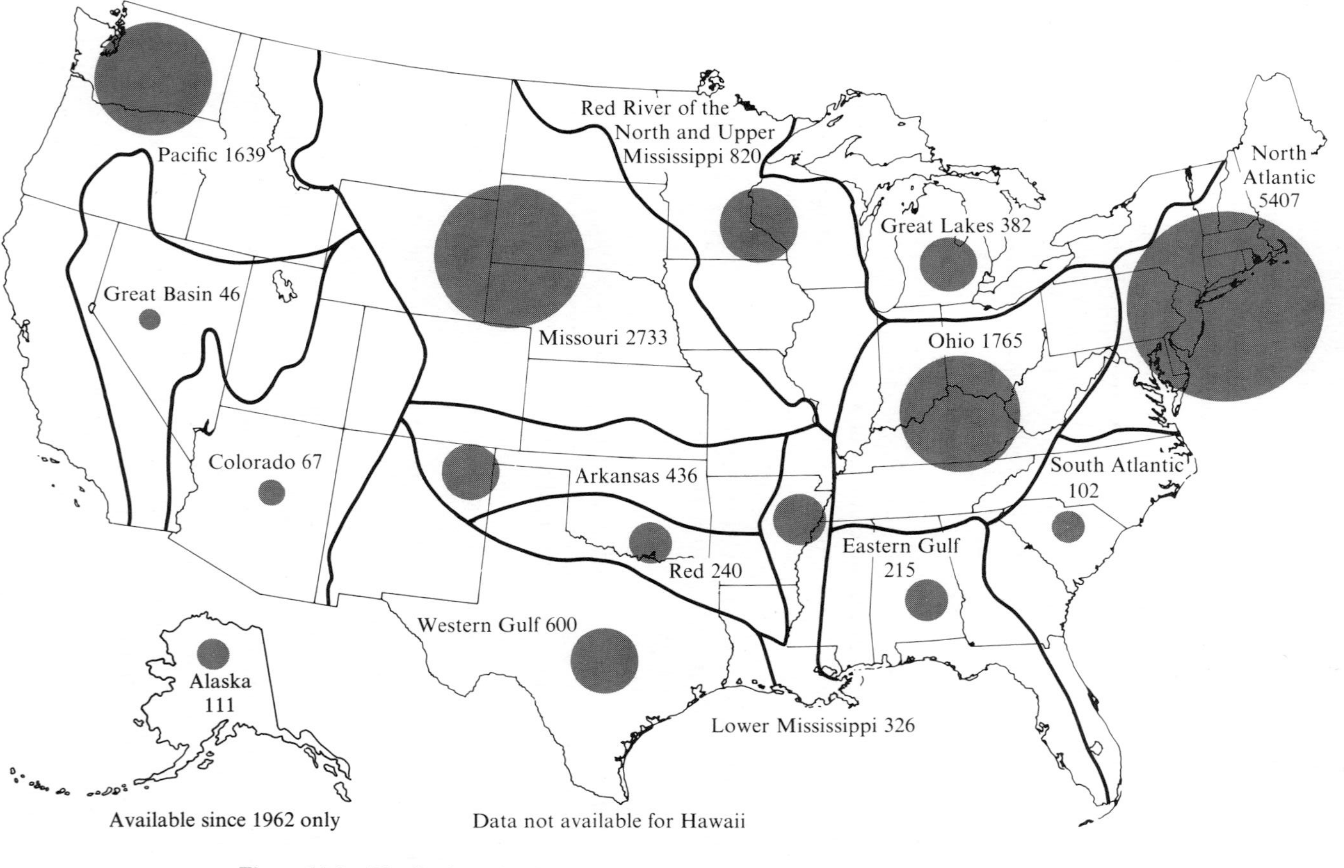

Figure 11-3 Distribution of estimated flood damage costs in the United States by major river basins, 1925–1972. The numbers represent costs in millions of dollars. (From National Weather Service.)

The type of flood loss differs with the activities occupying the valley floor. Agricultural settlement suffers most frequently from the loss of both stored and growing crops which are inundated. In this case, the type of crop and its stage of development (season) are important, as are the depth and duration of flooding, the velocity of water and its sediment concentration. Less

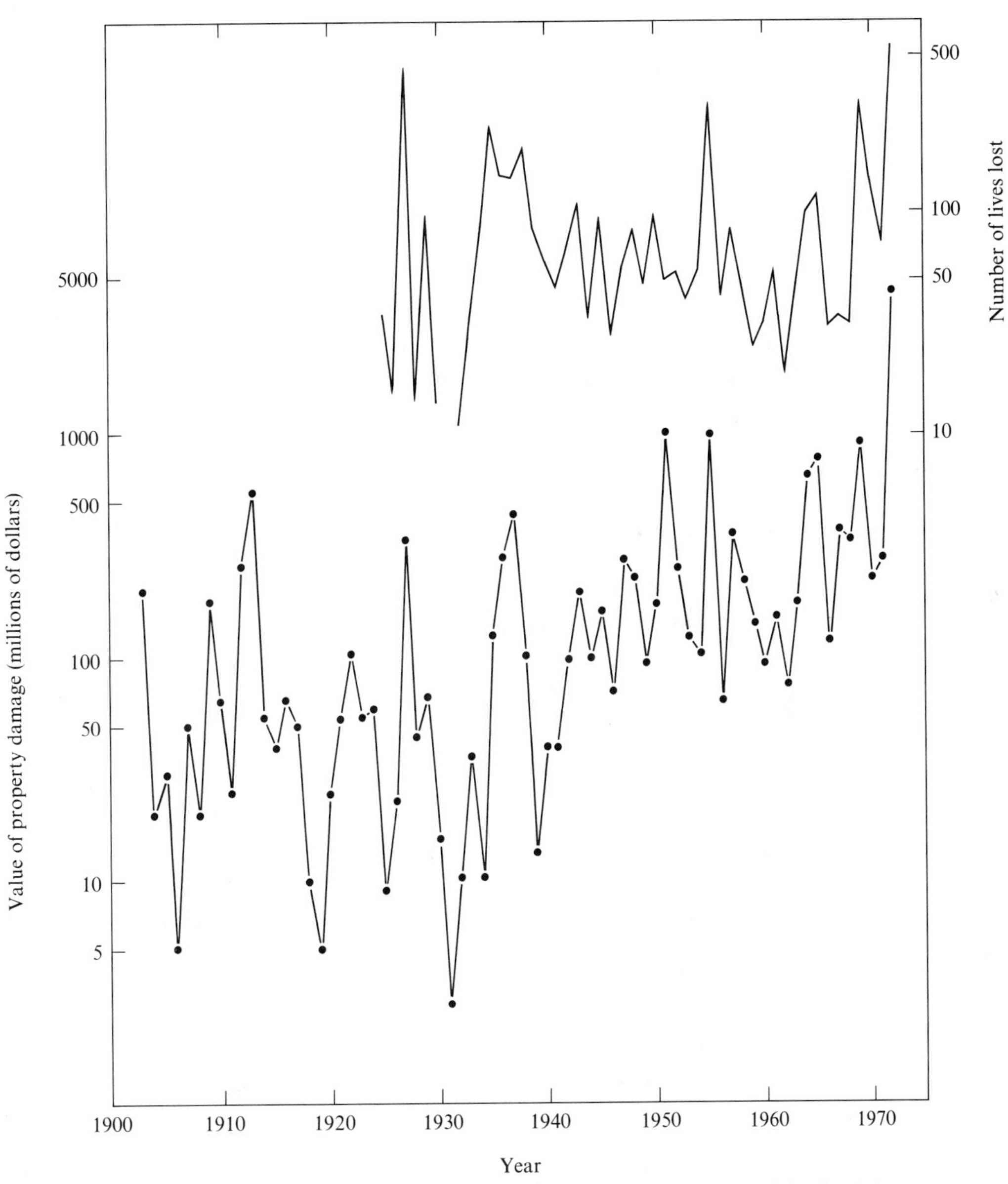

Figure 11-4 Annual property damage costs and number of deaths caused by floods in the United States, 1903–1972. (From National Weather Service.)

frequently, livestock losses are considerable, or fertile topsoil can be ruined by the scouring of its surface or the deposition of gravel. Drainage and irrigation works frequently suffer from scouring or siltation in some regions, while water damage to farm machinery and dwellings parallels those in other sectors of the economy.

In manufacturing and commercial areas, inundation of machinery and stocks of products or raw materials is generally responsible for the largest flood losses. Business interruptions and loss of production are also costly, though they can be partly made up after the flood. The major determinants of costs here are the height and duration of flooding, the degree of sediment damage, and the degree of warning that can be given so that evacuation and other preventative measures can be taken.

In residential areas the major damage is associated with soaking and sediment damage to dwellings, furnishings, and automobiles. Some buildings may be scoured off their foundations, and sewage, water supply, and electrical utilities are often damaged. Water pollution frequently results from the flooding of sewage and treatment plants and solid waste disposal sites located on valley floors. The costs depend upon the height and duration of flooding, the rate of rise, degree of warning, velocity of water, its sediment concentration, and the amount of floating debris that can damage buildings.

Human Adjustments to Floods

An important series of papers by Gilbert F. White and his co-workers at the University of Chicago and the University of Colorado have reviewed the strategies pursued by various societies to reduce the net costs of flood damage. Some of the principal papers are listed in the bibliography of this chapter as Research Papers of the Department of Geography at the University of Chicago. The major strategies for minimizing flood damage, as described in much greater detail by these authors, are reviewed briefly below.

Perhaps the most obvious technique for minimizing flood losses is to *raise the flood-prone unit* above the probable high water level (see Figure 11-5). Methods of doing this include placing artificial landfills under railroads, highways, and towns, and building on stilts as is done in some villages in southeast Asia. This strategy works well in some areas, though it is costly and has a major disadvantage in that the filled land occupies space that would otherwise be used to store floodwater; for this reason the method is now generally discouraged.

Floodproofing of buildings on valley floors is another means of reducing flood losses. It requires investment to render buildings and their contents less vulnerable to flood damage. Figure 11-6 shows some structural alterations that can protect a building, and Sheaffer (1960) has reviewed the technique more thoroughly.

Figure 11-5 Elevation of the Wren library at Trinity College, Cambridge, above flood levels on the River Cam. Trinity College is built on a terrace of the River Cam, and Sir Christopher Wren designed a set of columns to raise the library as a protection against floods and dampness.

In the early days of planning for flood control and conservation, there was much optimism about the value of *good land use* in reducing the damage caused by floods downstream. The argument was advanced that activities such as planting trees, terracing, and strip cropping that enhance infiltration would reduce the amount and speed of flood runoff and hence would reduce flood damages. Other experts countered that such methods were ineffective for controlling large floods because these events are produced by very large rainstorms or snowmelt events that exceed the storage capacity of soils in forestland and farmland alike. The controversy was reviewed by Leopold and Maddock (1954), who concluded that the value of good land use in reducing flood damage was overrated. Good land management can reduce small floods from small catchments in modest storms. It has little if any measurable impact on runoff from large storms, and especially not from large catchments. Hoyt and Langbein (1955) surveyed the evidence and concluded that "Nationwide, floods seem to roll out of forests as well as off farms." Typical Problem 11-1 considers some of the effects quantitatively.

There is still some benefit in trying to reduce flood damage by good land management. By far the largest accumulated amounts of damage on farms in upland areas are caused by floods with recurrence intervals of five years or less. In urban areas the drastic changes of land use and the high value of land require that modification of the land surface be considered along with

Explanation

1. Sump pump and drain to eject seepage
2. Properly anchored tank
3. Permanent closure of openings with masonry
4. Valve on sewer line
5. Plastic covering
6. Elevated control panel
7. Screens to prevent breakage of glass from floating debris
8. Steel bulkhead for entrance
9. Cracks sealed with hydraulic cement
10. Thoroseal coating to reduce seepage
11. Anchorage

Figure 11-6 Floodproofing involves making alterations to buildings that render them more resistant to flood damage.

other flood-control methods. Furthermore, good land management reduces soil erosion, and therefore the deposition of silt and sand within channels and downstream flood-protection structures. Good land management, therefore, helps these structures to function as they should. By far the largest benefits derived from good soil and water control, however, result from increased productivity rather than from flood control.

The construction of *engineering flood-protection works* is probably the most familiar response to flood hazard, and it is the one that has been most highly developed and heavily funded in technological societies. In recent years, some of the costs and unfortunate side effects of flood control have led many people to reassess this strategy and to try to reduce flood losses by nonstructural means wherever possible.

In some parts of valley floors, *levees* or *floodwalls* may be built along river banks or around high-value areas such as towns to keep floodwaters in the channel or out of certain areas (see Figure 11-7). Such large dykes are

Figure 11-7 Green River Valley, near Kent, Washington. The river channel has been cleared and its banks smoothed, and dykes have been built to contain the 100-year flood, as regulated by an upstream flood-control dam. Several years after the dykes were built, a drainage project was necessary to discharge water from the valley floor into the channel. The local stormwater runoff still causes extensive flooding. Industry and residential development continue to spread onto the valley floor, displacing agriculture. A firm of landscape architects has now been hired to grade the banks to a shallower angle to allow public access to the river and to improve the scenic value of the channel. (USDA–Soil Conservation Service.)

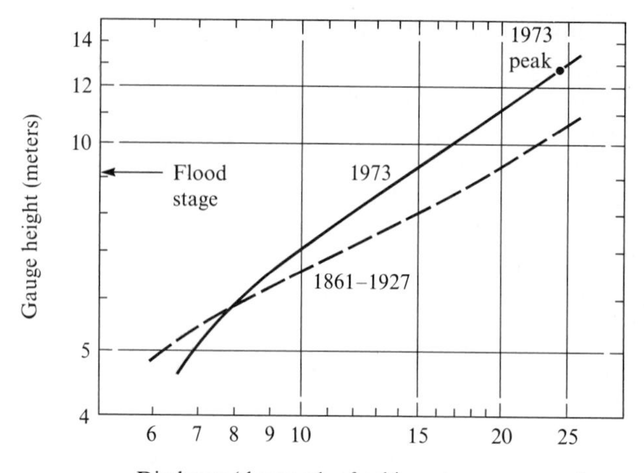

Figure 11-8 Rating curves for the Mississippi River at St. Louis, Missouri, before construction of the levee system (1861–1927) and after construction (1973). The discharge at the peak of the 1973 flood is shown; its water surface was higher than that of the same discharge before the channel was confined by levees. (From C. H. Belt, *Science,* vol. 189, pp. 681–684, 29 August 1975. Copyright 1975 by the American Association for the Advancement of Science.)

expensive, but warranted in certain areas. The vast levee system of the lower Mississippi valley contains more than 3000 miles of dykes and protects millions of acres of valuable agricultural, residential, and commercial land from inundation. The costs are increased by the necessity of providing storm drainage and sewerage from towns at times of high water when free gravity drainage into the river is not possible.

When floodwaters are confined to the channel by levees or floodwalls, instead of being allowed to spill over and be stored in the valley floor, the water level will be higher for the same discharge than it was before the river was confined. An example of particular importance is the 1973 flood of the Mississippi River which at St. Louis reached a stage of 13.18 meters (4.03 m above flood stage) when the flow was only 24,100 m^3/sec. In 1908 the same discharge was experienced, but the stage was 2.51 m lower than in 1973, owing primarily to confinement by levees and to navigation works that reduced the channel cross section by resultant deposition (Belt 1975). Stated in another way, the 1973 discharge at St. Louis is estimated to have a recurrence interval of about 30 years, but it reached a stage exceeding that of the highest flood in the 189-year record. An average rating curve (a plot of water surface elevation as a function of discharge) was constructed by Belt for the period 1861–1927 before the major levee system was constructed, and it is compared in Figure 11-8 with the rating curve applicable to 1973 conditions. The flood stage formerly was reached by 19,000 m^3/sec and after levee construction is reached by 14,200 m^3/sec.

As another example, in the Snoqualmie River valley of western Washington there are plans to protect the small flood-prone towns of the valley from spring snowmelt floods by surrounding them and some valuable agricultural land with 35 miles of levees. These walls will deny to the river an area of valley floor that it presently uses to store a portion of its floodwater during exceptionally high discharges. This extra water, which will in future remain within the confined channel, will cause higher flood levels downstream. In order to protect the downstream areas from the flooding that will be caused by the dyke system, a flood-control dam must now be built in the mountainous portion of the basin to reduce the volume of the flood. Another problem with dyking is the danger that the floodwall will be breached, as described in Chapter 1. The flood enters the protected lands quickly and is therefore more dangerous than a slow rise of the stream.

Channel improvements such as deepening, widening, lining, or straightening the channel or clearing it of vegetation and debris can increase the conveyance capacity, so that high discharges can pass without the level of the water surface being raised to dangerous heights (see Figure 11-9). Attempts at flood protection by this method are widespread on rivers ranging in size from "meander cutoffs" along the Mississippi to small streams draining a single pasture or residential neighborhood. Hoyt and Langbein report that

Figure 11-9 Sammamish Slough, Redmond, Washington. Spoils from an enlargement of the channel two weeks after grading of banks. The banks were soon vegetated by grass, but were sprayed for years to keep down brush and tree vegetation. More than a decade later a firm of landscape architects has been hired to plant trees along its banks, halt the spraying, grade the banks to a shallower angle, and encourage the development of wildlife habitat and public access. This is all possible without limiting the conveyance capacity of the channel. (USDA–Soil Conservation Service.)

straightening and deepening of the lower Mississippi channel has lowered flood levels by 2 to 12 feet for the same discharge. The U.S. Soil Conservation Service and the Corps of Engineers have "channelized" 34,240 miles of streams throughout the United States in an attempt to reduce the flood hazard (Council on Environmental Quality 1973). Although successful in some cases, the method often has undesirable side effects. Rapid passage of floodwaters in the improved channel often increases flood peaks downstream. Deepening of the channel has frequently led to the drainage of marshland on the valley floor. Although this increases the value of land for agriculture, it has eradicated thousands of acres of wetland habitat for wildfowl and large mammals. In some places, deepening and straightening have triggered further deepening and widening of the stream channel as the river seeks to reestablish an equilibrium channel, and the eroded material is deposited downstream, frequently causing a further set of problems. The channelization controversy is discussed in Chapter 18.

In some valleys where other flood-protection methods are impossible or insufficient, a portion of the valley floor is set aside for diversion of floodwaters during periods of exceptionally high water. These *channel diversions* may lead the runoff back to the channel below a town or may lead it directly to the sea or a lake. Between large floods the land can be used for agriculture, grazing, or recreation and may even be held by private owners who are compensated to leave their land in some kind of use that is not damaged by flooding. Only a few places, however, have the physiography and land use necessary for this type of flood protection.

Flood-control reservoirs are the most popular of all flood-protection measures. There are two major types of flood-control reservoir; *flood-detention reservoirs* and *multipurpose storage reservoirs.* The former are usually kept empty. When a flood occurs, a portion of the water is impounded behind the dam, and the remainder flows downstream through a pipe or over a spillway in the dam. As indicated by the worked example on reservoir flood routing in Chapter 10 (Figure 10-45), this storage can markedly reduce the peak of a flood. The degree of attenuation of the flood wave is determined by the volume of storage in the impoundment and the storage–discharge relation for the outlet, as discussed in relation to Tables 10-19 and 10-20. The floodwater drains out slowly after the storm, and between periods of heavy runoff the stream regimen is not affected because water simply drains through the reservoir as quickly as it enters. Where appropriate, old gravel quarries or opencast coal mines can be used as detention basins. If local water quality permits, the detained stormwater can be allowed to recharge the groundwater. Flood-detention structures of this kind are usually located in headwater regions and are usually small.

Multipurpose storage reservoirs are often constructed to impound water for municipal supply, irrigation, recreation, or power generation. During times of greatest flood potential, the level of the reservoir can be lowered to provide storage volume for flood runoff. These reservoirs vary from large bodies of water such as Lake Mead, storing 30 million acre-feet of water on the Colorado River, to small mountain reservoirs similar to those built solely for flood detention. The management of multipurpose reservoirs for flood control is a very complicated enterprise. Hydrologic forecasts of possible and probable floods must be weighed against the goal of maximizing water storage for irrigation, power, or municipal supply. If the storage capacity is not available when a flood strikes, there will be little attenuation of the flood wave, and the dam itself could even be overtopped and breached if the capacity of its spillway is exceeded. On the other hand, if the water level is lowered and a threatened flood does not materialize, valuable water supply for the next season is lost.

There has been a great deal of discussion about the best policy for locating flood-control dams. Some groups favor many small upstream dams, while others recommend large, downstream reservoirs. The controversy was reviewed by Leopold and Maddock (1954), who pointed out that the two kinds of reservoirs provide different degrees and locations of flood control. They gave the following example to illustrate the hydrologic aspects of the problem. Consider the 600-square-mile catchment shown in Figure 11-10, which is composed of ten 60-square-mile tributary basins, each in turn comprising three basins of 10 square miles and an area of 30 square miles that is not drained by streams large enough to put a dam on. We wish to examine the effects of dams in reducing the inundated area of valley floor resulting from design floods of different magnitudes. We will compare the effect of

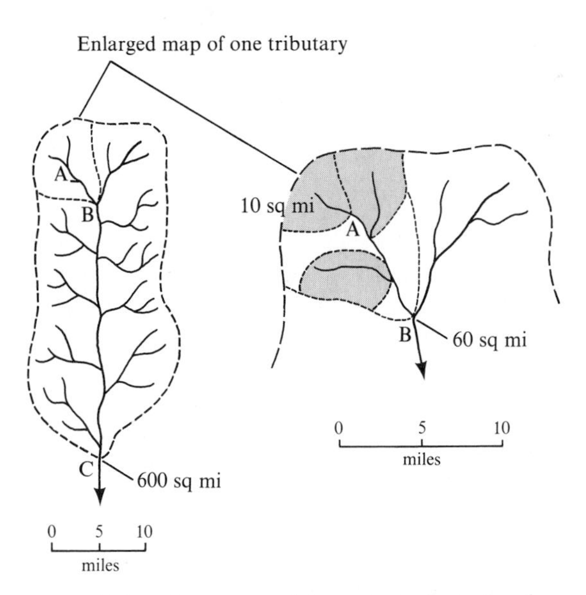

Figure 11-10 Hypothetical drainage basin for testing the effect of a series of dams on the extent of flooding. (From Luna B. Leopold and Thomas Maddock, Jr., *The Flood Control Controversy,* The Ronald Press Company, New York, 1954.)

building 30 small flood-detention structures on the thirty 10-square-mile upstream tributaries with that of building a single large dam immediately above the outlet from the basin.

For each flood the runoff from tributary basins can be routed downstream by the Muskingum method described in Chapter 10. Successive flood waves from the tributaries can be added along the channel and at each point the discharge can be compared with the capacity of the channel (see Chapter 16 for a discussion of bankfull discharge and its relationship to drainage basin area). Any point at which the bankfull discharge of the channel is exceeded by the flood discharge is considered to have experienced a flood. The bankfull channel capacity increases downstream, and the peak flood discharge may increase or decrease downstream, depending on the extent of the storm, the rate of runoff, and the storage action of the channel and artificial reservoirs.

Under the conditions specified by Leopold and Maddock, Figure 11-11(a) shows that a storm centered over a 10-square-mile tributary producing one inch of runoff in 4 hours would inundate the valley floor in the lower part of the 10-square-mile basin and for a short distance downstream. Further downstream, the flood peak would be reduced by the storage action of the channel until it could be confined within the channel whose capacity increases downstream. A small flood-detention structure designed to store two inches of runoff at station A would remove the flood hazard below that point.

If a flood produced one inch of runoff in 4 hours from one of the 60-square-mile subbasins, the situation would be as described in Figure 11-11(b). Without dams, flooding would extend further downstream than in Figure 11-11(a), but small dams on the three 10-square-mile tributaries in the subbasin would confine flooding to small areas upstream of the dams.

Figure 11-11(c) portrays results of a similar calculation made for a flood of one inch in 8 hours over the whole 600-square-mile basin. Such a storm would generate floods exceeding the channel capacity throughout the whole valley. With 30 dams detaining the runoff from the 10-square-mile tributaries, only one-half of the 600-square-mile drainage area would be controlled, and runoff from the remainder would still overtax the channel capacity throughout most of the system. Only a short stretch of the valley below each dam would be protected, because the effectiveness of any dam decreases rapidly downstream as runoff from uncontrolled parts of the basin enters the channel. If a town or other important installation were located at C in these examples, it would not be protected from the large flood even by the 30 small upstream impoundments, and it would not need protection from the smaller floods. Site C could, however, be protected by a single large impoundment immediately upstream. On the other hand, such a flood-control reservoir obviously could not provide protection upstream.

The point that emerges from these examples is that small upstream reservoirs and large downstream impoundments cannot be substituted for one another in a basin-wide flood-control scheme. They provide protection for different parts of a valley.

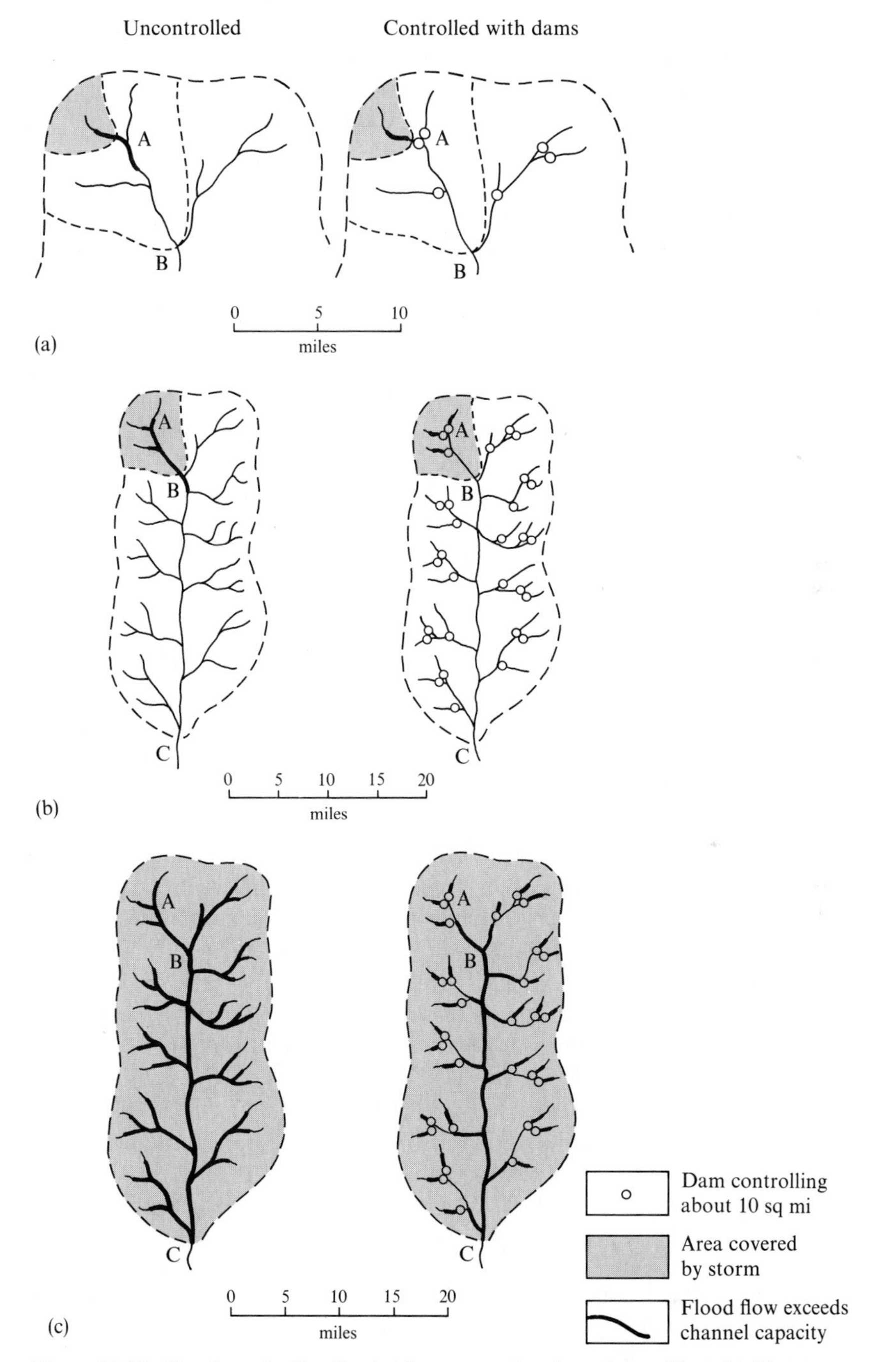

Figure 11-11 Reaches of valley flooded by storms of various sizes, with and without upstream dams. (a) Storm covering 10 square miles and generating 1 inch of runoff in 4 hours. (b) Storm covering 60 square miles and generating 1 inch of runoff in 4 hours. (c) Storm covering 600 square miles and generating 1 inch of runoff in 8 hours. (From Luna B. Leopold and Thomas Maddock, Jr., *The Flood Control Controversy,* The Ronald Press Company, New York, 1954.)

There are many other considerations that must be weighed in the location of dam sites, including economics, social policy, suitability of foundations, and values such as wildlife and scenic attractions. But the hydrologic principles are clear and should remain so in the mind of the planner.

Although of proven value in many valleys, the impoundment of rivers for flood control and other purposes has caused unfortunate side effects in some cases. Large downstream reservoirs in particular occupy valuable agricultural and residential land, and upstream impoundments frequently flood sites of great natural beauty, obliterating their scenic value and altering their recreational potential. Other dams have blocked the migration routes of anadromous fish, such as salmon and some trout. Although fish ladders can be built into a dam to allow the fish to circumvent such obstacles, this is not done in most structures because of the expense.

Silting of reservoirs reduces their economic life, and in some areas, small flood-control reservoirs have been completely filled. Sedimentation will not be a threat to the largest reservoirs for several hundred years, even on a muddy river like the Colorado. For smaller impoundments great damage is done by this process. Even partial reduction of the storage volume of a flood-control reservoir can markedly reduce its effectiveness, and in areas where sedimentation rates are significant, we can expect the flood hazard to increase steadily below some dams. In many regions of the world, including the southwestern United States, Japan, and parts of Africa, planning agencies are already considering the technical and economic aspects of removing sediment from reservoirs; these considerations will become more common in the next 25 years. Sedimentation rates will be treated quantitatively in Chapter 17.

In the Nile valley, the impoundment of silt in Lake Nasser behind the Aswan Dam and the reduction of overbank flooding has reduced the deposition of nutrient-rich silt on the Nile valley. This silt used to fertilize the fields along the river. In the absence of the silt-laden flood, farmers must buy commercial fertilizer or suffer a reduction of crop yields. In other cases, the removal of sediment from the water flowing through reservoirs has caused intense erosion downstream. After the closure of Hoover Dam on the Colorado River, for example, the clear water emerging from the dam caused extensive degradation of the channel for more than 100 miles downstream as the river picked up large amounts of sediment from the bed and banks of the channel. Even 350 miles downstream, at Yuma, Arizona, the riverbed was lowered nine feet. Such changes may render absolete many irrigation ditch headings and pumping plant intakes or may undermine bridge supports and similar structures. On other rivers the opposite effect has been observed. Before impoundment, floods carried away the sediment brought into the main stream by steep tributaries. Reduction of the peak discharges leaves the river unable to scour away the sediment that accumulates as large fans of sand or gravel below each tributary mouth. The bed of the main stream is raised, and if water intakes, towns, or other structures lie alongside the river they can be threatened again by flooding or channel shifting across the accumulating wedge of sediment.

Other effects of dam building include both harmful and useful changes of water quality (see Chapter 20) and the eradication or establishment of major wildlife habitats. The closure of Bennett Dam on the Peace River, Alberta, in 1967 reduced the large snowmelt floods on the river. This runoff formerly flooded the Peace delta and filled many delta lakes as well as Lake Athabasca. It thus annually recharged a vast habitat for waterfowl, fish, and large mammals who inhabited the surrounding swamps. Once the river floods were reduced, the lakes and swamps began to drain and became mudflats. One of the most spectacular and important wildlife habitats in North America had been disrupted by bad planning and lack of foresight.

Other mistakes continue to be made in choosing dam sites which are unsafe. Some of them have failed catastrophically, and behind others the reservoirs cannot be operated at design levels for fear of collapse. The San Fernando earthquake of 1971 triggered massive failures of the Lower Van Norman Dam in southern California. Fortunately the dam did not fail completely, but 80,000 people had to be evacuated from their homes below the dam while the level of the reservoir was lowered over a 4-day period. The 1971 earthquake alone caused $36.5 million worth of damage to dams in this tectonically active area. The Vaiont Reservoir disaster described in Chapter 15 had less fortunate results. A landslide-caused wave breached the dam and killed 3000 people in the valley below. Not every dam site is hazardous, of course, but many are, and chances continue to be taken with the money and lives of the public. The dam being built at Libby, Montana, is threatened by a huge rockslide, a portion of which failed during the early stages of construction. Although signs of hillslope instability were available in the valley before the site was chosen, they were overlooked and it is not easy to find anyone willing to admit responsibility for the choice. The hillslope subject to failure is being drained and a $2 million warning system has been installed. If the failure is a rapid one, however, the warning system will be of little use. The costs and the uncertainty continue to escalate and of course were not included in the original estimates of the cost effectiveness of the project.

While we were writing this chapter, the earthfill Teton Dam in Idaho collapsed, sending a flood wave down the valley and causing 11 deaths, the loss of 16,000 head of livestock and an estimated one billion dollars worth of property damage. Investigations by a congressional committee and by an independent panel of experts have since revealed that the dam building agency ignored many obvious geological signs of danger. The momentum to continue construction was so great that warnings by geologists were ignored, and even the agency's own personnel who believed the site unsuitable were so intimidated that they did not voice their misgivings. The inquiry found the builders guilty of "exaggerated overconfidence" in their engineering capability. Bad judgments made under the pressure to complete the structure led to its catastrophic failure. The congressional committee also found that many other dam sites in the United States constitute public hazards because of poor siting and bad design.

Even more common are dam sites that are unsuitable but do not threaten failure. Permeable foundation rocks may allow large amounts of seepage under and around the dam. The reservoir behind Hansen Dam on the Cedar River in Washington, for example, cannot be operated at its design level because of the leakiness of its foundation. There may be some excuse in that the dam was built decades ago when little experience had been accumulated. But there is a current proposal to build an even larger flood-control dam on the nearby Snoqualmie River at a location with the same geologic conditions as the Hansen Dam. There was no mention of this problem in the report on which the proposal was based. A team of graduate students from the Department of Geological Sciences at the University of Washington investigated the site and reported a high probability of leakage around the dam. Such leakage might not pose a serious threat to the structure, but it would need to be corrected for efficient operation. The geological team stressed that *before* any fair judgment could be made of the probable cost of the project, extensive drilling and design studies should be made to establish the long-term costs of construction and maintenance of the dam and of operation of the reservoir. Since the cost-effectiveness of the whole project was dubious anyway, these costs could make an important difference to the final decision. The response of the construction agency was that they would do the studies *after* the project had been approved and funded. Then, of course, it would be too late to stop if the costs escalated, as they almost certainly would. For other reasons, various groups opposed the project, the governor vetoed it, and another plan was developed. But at the time of writing, the governor has announced that he will not run for reelection, and the local newspaper reported that in response to a question from a local business group, a representative of the flood-control agency stated that "the plans [for the original dam] are always on the shelf!"

All these factors and many social, economic, and political considerations should be weighed by the planner reviewing proposals for the building of dams. There are strong pressures in favor of even larger numbers of dams than now exist, and in many countries dams are usually the only strategy considered for abating flood damage. The planner should be wary of these pressures and should make a careful review of the proposals, taking special care to pick out contradictions. In many proposals, for instance, the only way to justify the cost of building a flood-control dam is to claim that it will provide recreation benefits or major fish and wildlife habitats. Yet the operation of a reservoir primarily for flood control is difficult to harmonize with these benefits. The great depth of these lakes and the extent to which their elevation must fluctuate provide neither good feeding and nesting grounds for wildfowl, nor good feeding and spawning grounds for many fish that need shallow water. The fluctuation of levels often reduces the recreation potential of the lake by causing extensive landsliding around the shoreline (see Chapter 15), reducing accessibility for boats and lowering the overall esthetic quality of the lake.

There are many well-managed large and small reservoirs that provide flood control, and they are important and useful components of a basin-wide strategy for reducing flood damage. They are costly, however, in both economic and nonmonetary terms, and recognizing the many unsuccessful dams built in the past and the vigor of the present advocacy for more dams, the planner would be well advised to review proposals for flood-control dams very critically. It will often be found that one or more of the other strategies outlined in this section are preferable in the long run.

The realization that structural solutions to flood control often do not reduce flood losses has led various authorities concerned with flood hazard to reassess their approaches to reducing flood losses. Too often, investment in dams and other flood-control structures has simply encouraged downstream development in flood-prone areas, some of it protected and some unprotected. In many cases, the costs of such structures are justified on the basis of future development in flood-prone lands that are presently unoccupied and where no danger exists until the dam is built! Disillusionment has led many concerned people to stress the potential for controlling the land use of valley floors instead of encouraging further exposure to risk.

The general aim of *land-use control* is to limit occupance of flood-prone areas by activities that take up natural flood storage or that suffer heavily from flooding. The regulation can be accomplished by forbidding certain activities or by specifying standards of construction for buildings occupying the flood-prone area. Some administrative units, for example, have statutes forbidding the constriction of river channels. In other areas zoning laws specify types of land use that are allowed in areas with different degrees of flood hazard. The general idea is not to let owners put their land to vulnerable use and then obtain protection or relief at the public expense. Zoning can also be designed to promote uses that are productive but not subject to damage by floods. It allows maintenance of socially desirable land uses such as open space in the heart of congested urban areas. Pastureland, woods, and certain types of recreation land are not damaged significantly by heavy flooding, and even cropping may be possible if the season of flooding is favorable. Residential developments and services should obviously be kept out of the way of floods, but unfortunately this has often not been done. The central districts of many towns lie in a zone that is flooded by the 50-year or 100-year event (see Figure 11-12). Urban and industrial land use is also the most resistant to alterations, and existing development of this kind is one of the most common reasons why flood-control structures are required.

Land-use control for reducing flood losses can take many other forms depending upon local hydrology, laws, customs, and values. It is tied very closely to land-use control for other purposes and to other strategies for minimizing flood losses, such as flood insurance and flood-warning systems. In some cases the most extreme form of land-use control is government acquisition of land. This is done with the compensation of landowners in particularly hazardous areas but is costly. It allows the development of greenbelts, wildlife refuges, and other desirable breaks in the urban environment (see Figure 11-13). Still another variation upon the theme of public acquisition of land occurs when a government agency buys the development

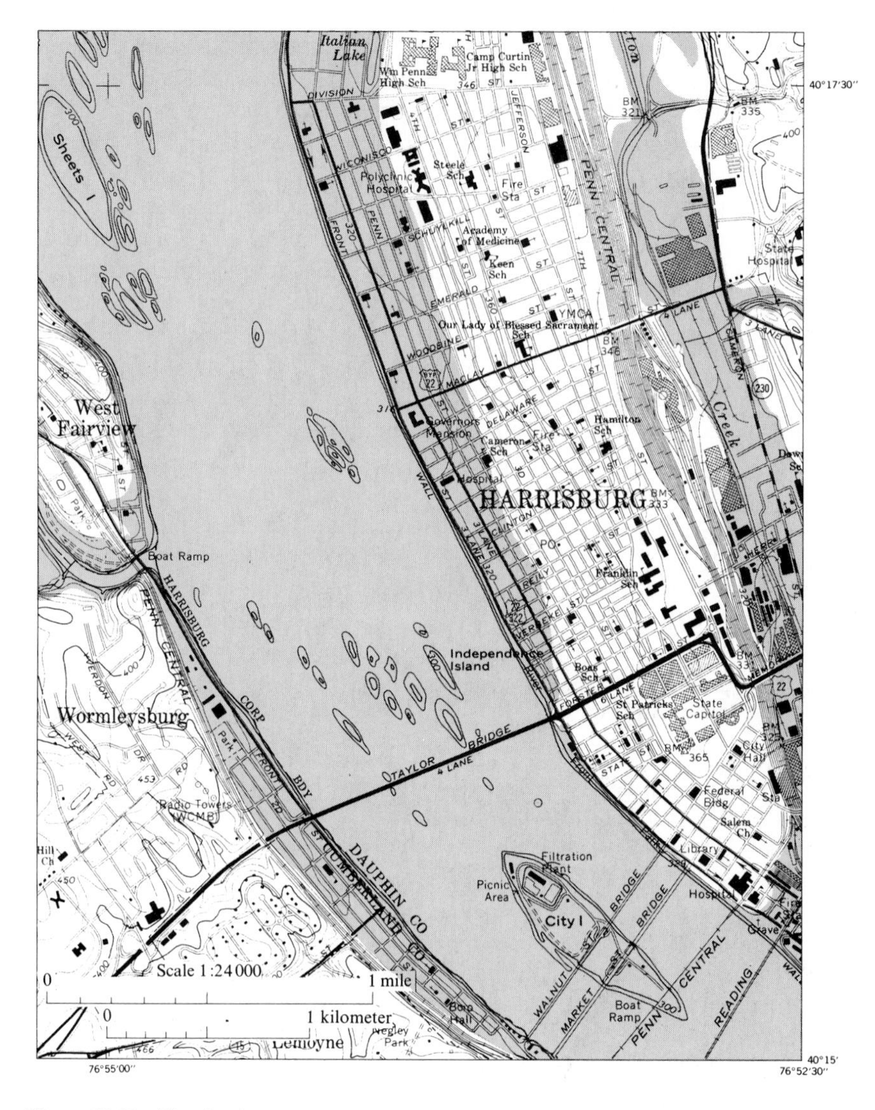

Figure 11-12 The flood-prone area (shaded) of Harrisburg, Pennsylvania. (From U.S. Geological Survey.)

rights to a piece of land. The original owner retains his property and may continue to use it as a farm, for example. But the land cannot be used for commercial, residential, or other flood-prone activities. Some of the legal aspects of land-use control are discussed by Hogan (1963), and White (1964) has made case studies of six American towns that have mixed a variety of damage-reduction strategies including land-use control.

Flood insurance, with premiums subsidized by the government, is a means of compensating for losses to activities already situated in flood-prone areas,

Figure 11-13 Urban park in the flood-prone area of Harrisburg, Pennsylvania. (U.S. Geological Survey.)

or for activities that must be located there in the future for some socially desirable end. Such insurance removes the necessity of expensive structural solutions to the flood-control problem. Moreover, if premiums are set in relation to the magnitude of the hazard, occupants of the valley floor are warned of the risk they are taking, and others are discouraged from moving into flood-prone areas.

For those who already occupy dangerous areas, an efficient *flood-warning* system can reduce losses. Damageable property, machinery, and people can be moved to upper floors or higher ground; waterproofing of fixed equipment, sand bagging, and other emergency measures can also be taken if warning is given in time. This is relatively easy to accomplish on large streams, which rise slowly, but is difficult on small streams, which rise and fall within an hour of an intense thunderstorm in the Midwest or within several hours of a rainstorm on melting snow in New England. In the United States, the regional River Forecasting Centers of the National Weather Service are charged with providing warning of floods, and they disseminate the information through a complex network of public officials and news media. Even with good forecasting, however, the final link with the people living in the flood-prone area may break down, as described in Chapter 1. This is an area of flood damage reduction that could profit from attention by planners and social scientists.

Flood Control in Small Urban Catchments

We have already reviewed the processes by which floods are increased by the drastic changes of hillslopes and channels during urbanization. The computations of these effects in Chapter 10 on urbanization show that the relative increase in discharge can be very large indeed. Faced with this problem, planners and engineers need to be aware of the range of solutions that have been applied to the reduction and control of floods from small urbanized catchments. Because of the increasing need for solutions, this problem has been the subject of a great deal of research within the last few years, and new techniques are continually being introduced in the literature of civil engineering, planning, and landscape architecture. A useful summary is provided by Hittman Associates (1974).

The conditions to be managed are portrayed schematically in Figure 11-14, which reviews the situation as we have previously described it, and points out the problems that have to be solved.

The best solution to the urban runoff problem is to detain the stormwater in small volumes as near to its source as possible, and then to release it slowly to natural stream channels or to the groundwater system. Because most of the storm runoff generated in urban areas originates on impervious surfaces such as rooftops and parking lots, one obvious method of runoff

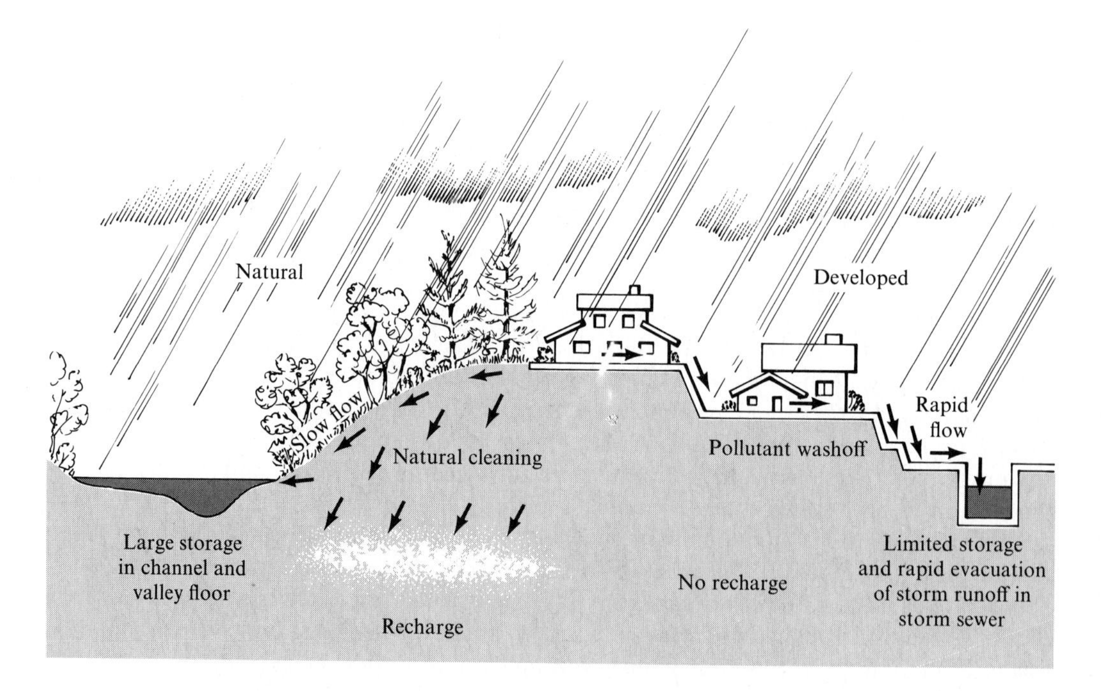

Figure 11-14 Schematic representation of problems to be managed in the control of urban storm runoff.

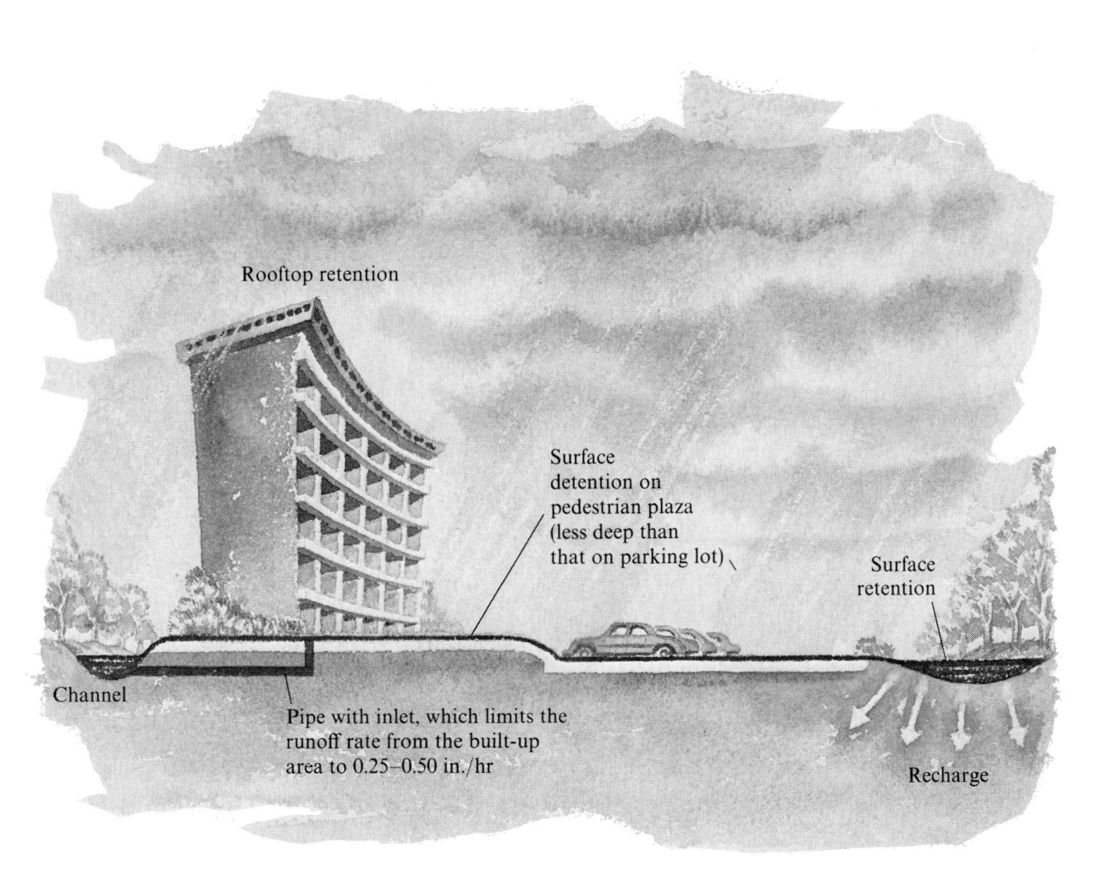

Figure 11-15 Storage of urban storm runoff.

control is the storage of the water in these very areas. Many new flat-topped buildings in industrial and commercial areas are designed to hold up to three inches of water on their roofs. The drains are designed to release the water at 0.25–0.50 in/hr to a storm sewer, natural river channel, detention pond, or recharge pit (see Figure 11-15). In some developed areas, roofs occupy as much as 50 percent of the built-up area, so rooftop storage can control a major portion of the storm runoff. Large parking lots can also be designed to store conveniently as much as three inches of water, and even pedestrian malls and plazas can be designed to provide one inch of ponding.

Detention storage may also be provided underground, as shown in Figure 11-16. Metal or concrete tanks are installed below ground, and overflow from storm sewers is diverted into them. The water may drain back to the sewer after the storm, or if heavily polluted, it may be diverted to the sanitary sewer system for treatment. If the subsoil and rock of the area are permeable, stormwater can be allowed to recharge the groundwater, which augments dry weather streamflow. Careful exploration of the local geology is necessary in some areas before recharge is allowed, however, to ensure that the stormwater does not pollute a shallow aquifer being used for water

Figure 11-16 Methods of detaining storm runoff below ground and of using it to recharge the groundwater.

supply. In other areas, there is a danger that the seepage may cause landslides on nearby hillslopes (see Chapter 15).

If local geological and hydrologic conditions are suitable, urban design may include major plans for inducing infiltration. The advantages of this strategy lie in reducing the length and capacity of storm sewers necessary to drain an area, and in increasing groundwater recharge to supply dry-weather flows. Surface recharge basins (see Figure 11-17) are excavated in permeable soils with deep water tables, which should be at least 10 feet below the floor of the pit. Storm runoff is diverted into the basin and allowed to infiltrate.

On Long Island, for example, which is underlain by deep, permeable fluvioglacial sands, more than 2000 recharge basins intercept storm runoff from small suburban neighborhoods. The basins range in size from 0.1 to 30 acres and their average depth is 10 feet. In order to maintain good conditions for infiltration, fine sediment must be dredged from the basin periodically and the floor must be tilled. Rather than use open pits, some designers employ infiltration wells, which may be shallow (up to 200 feet deep) and of large diameter (perhaps 5–15 feet) wth distributary pipes feeding water out over a wide area underground. Or they may be small diameter (12–24 inches) wells, several hundred feet deep. Stormwater is pumped

Figure 11-17 Recharge basin in Nassau County, New York. (G. E. Seaburn, U.S. Geological Survey.)

Figure 11-18 Porous pavement being installed.

down the pipe. Wells of this kind still require a storage basin at the surface because they cannot handle typical rates of storm runoff. They are also susceptible to blockage by fine sediment.

Another way of increasing infiltration now being used extensively in Europe and beginning to gain acceptance in the United States is the use of porous pavements, made of open-grated, asphalt concrete (see Figure 11-18). These again reduce storm runoff and augment low flows where soils are permeable and remain unfrozen. They can be introduced rapidly in a new development, where their variety of colors gives them great potential in design. In established urban areas, they can be introduced gradually as streets are repaved.

Where the topography is suitable or where excavation and dyking are possible, small floodwater detention basins can be located. These commonly have surface areas of 0.1 to 20 acres and storage volumes of a few thousand to 20 million cubic feet. As with the larger flood-detention structures described earlier, their purpose is to store a sufficient volume of storm runoff to reduce the peak discharge approximately to pre-urbanization levels (see Figure 11-19). They must be equipped with a spillway to convey water safely over the dam if the capacity of the basin is overtaxed. A pipe or

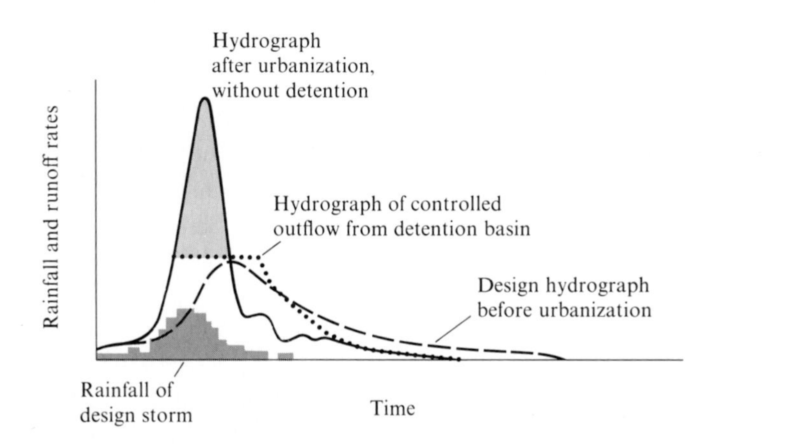

Figure 11-19 Storage requirements for a detention basin designed to keep flood peaks from an urbanizing catchment down to rural levels. The lightly shaded area represents the volume of storage required.

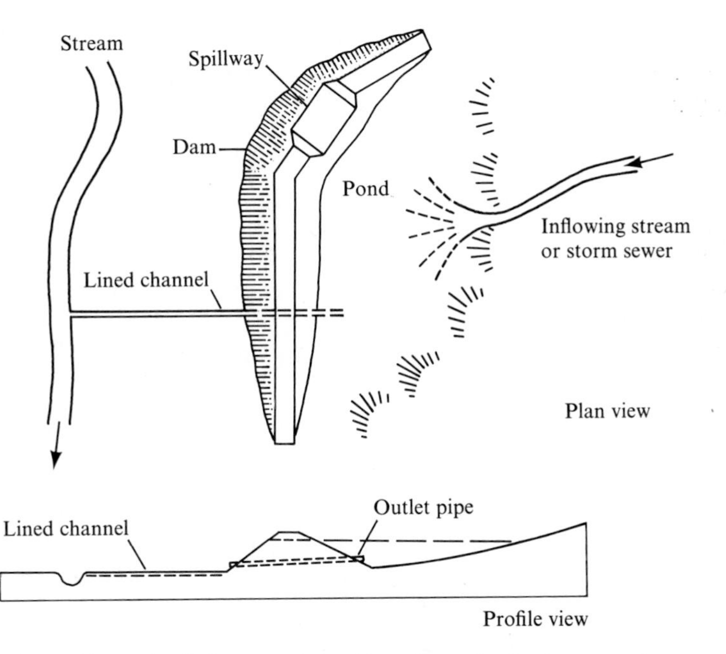

Figure 11-20 Small detention dam with a pipe outlet for the slow release of stormwater, and a spillway for passing extreme flood peaks safely.

narrow outlet weir allows water to drain out slowly during and after the storm (see Figure 11-20).

Many detention basins are allowed to drain completely after the storm and are landscaped for use as tobogganing, snowshoeing, and skating areas in winter or as picnic areas and playfields in summer (see Figure 11-21).

Figure 11-21 An urban stormwater detention basin landscaped and used for recreation. (Photograph by J. L. Hart, Midwestern Consulting, Inc.)

Activities of this sort are not pursued during heavy rain or snowmelt. One such basin in Chicago (Landscape Architecture 1974) is 11.5 acres in extent, 21 feet deep, and provides over 7 million cubic feet of storage for runoff from a 4-square-mile residential area in the 10-year, 6-hour design storm. It cost $1 million, which was less than the cost of enlarging the storm sewers of the area; and it also provided a recreational facility. Such installations fit well into city-wide designs for open "green space."

Sometimes the design of detention basins takes advantage of natural water bodies or of small artificial lakes impounded to enhance the design of a major development project. Such water bodies, however, may suffer from water-quality problems, and they require very careful design and management. First of all the drainage area they control must be large enough to supply sufficient low flow for good through-drainage in summer. This is not easy to ensure in an urban area where rapid storm runoff reduces groundwater recharge. The water must be kept deep enough to discourage the growth of weeds in shallow areas, and yet there must be sufficient storage volume for storm runoff, which could be very large from the extensive catchment required to maintain low flows. The substrate must be impermeable enough to minimize seepage losses and have enough strength to support a firm, steep bank.

Erosion during construction within the catchment (see Chapter 15) can produce rapid sedimentation in the lake, reducing its flood-control effectiveness and hastening eutrophication. Careful management of construction sites can reduce this sedimentation, and the lake can be excavated periodically with a dragline so long as some thought is given to access for such machinery when the lake is being designed. The most difficult management problems are the chemical and biological pollutants that can be washed into the lake (see Chapter 20). Urban storm drainage, especially during the first few minutes of a storm, is very heavily polluted. This "first flush" may be diverted into a sanitary sewer instead of entering the lake. Gasoline and oil in the runoff may be removed by passing the runoff through an interceptor (Landscape Architecture 1974:382).

The drainage of septic-tank effluent into these lakes, even after filtering through the soil, should be avoided by installing sanitary sewers in the catchment. Otherwise, rapid eutrophication of the lake is probable. The design and management aspects of small lakes in urban areas are reviewed by the U.S. Soil Conservation Service (1969), Rickert and Spieker (1971), Beasley (1972), and Britton et al. (1975), as well as by frequent articles in professional journals such as *Civil Engineering* and *Landscape Architecture.* The good examples of such flood-control lakes, blended into a careful basin-wide plan for the management of land and water quality demonstrate that the lakes can provide protection and enhance the esthetic value of a site. The frequent bad examples, however, highlight the many problems that can be encountered with this method of flood control.

Finally, as with large-scale flood control described earlier, there are a great many advantages to designing urban areas to take maximum advantage of the potential for flood storage within the natural landscape. Especially if the valley floor is wide, and the stream channel is very sinuous, the natural drainage system provides a storage volume that far exceeds the capacity of even a large and expensive storm sewer. Use of this storage also enhances the recharge of groundwater in valley alluvium and deep aquifers under favorable geological circumstances. In some cases, the storage and recharge can be increased by building small check dams across the stream, although these are not practical on streams that carry a heavy sediment load. As the storm water travels through the channel and valley floor it is subject to the natural cleansing action of the river and of microorganisms in the water and alluvium. The flood-prone area can be preserved as a recreation area or wooded open space, providing natural boundaries between clusters of buildings.

Taking advantage of the flood-control benefits of natural stream systems in urban areas requires imaginative design, but it can often be accomplished with greater success, less cost, and greater esthetic value than the traditional responses, which have made eyesores out of many urban areas. In one design for a 2000-acre residential development in Texas, for example, the use of the natural drainage system required runoff-control costs of $4.2 million. The computed cost of conventional methods of site drainage would have been $18.7 million (Landscape Architecture 1974). Some of the methods and design philosophy behind this project are described by Everhart (1973) (see Figures 11-22 and 11-23).

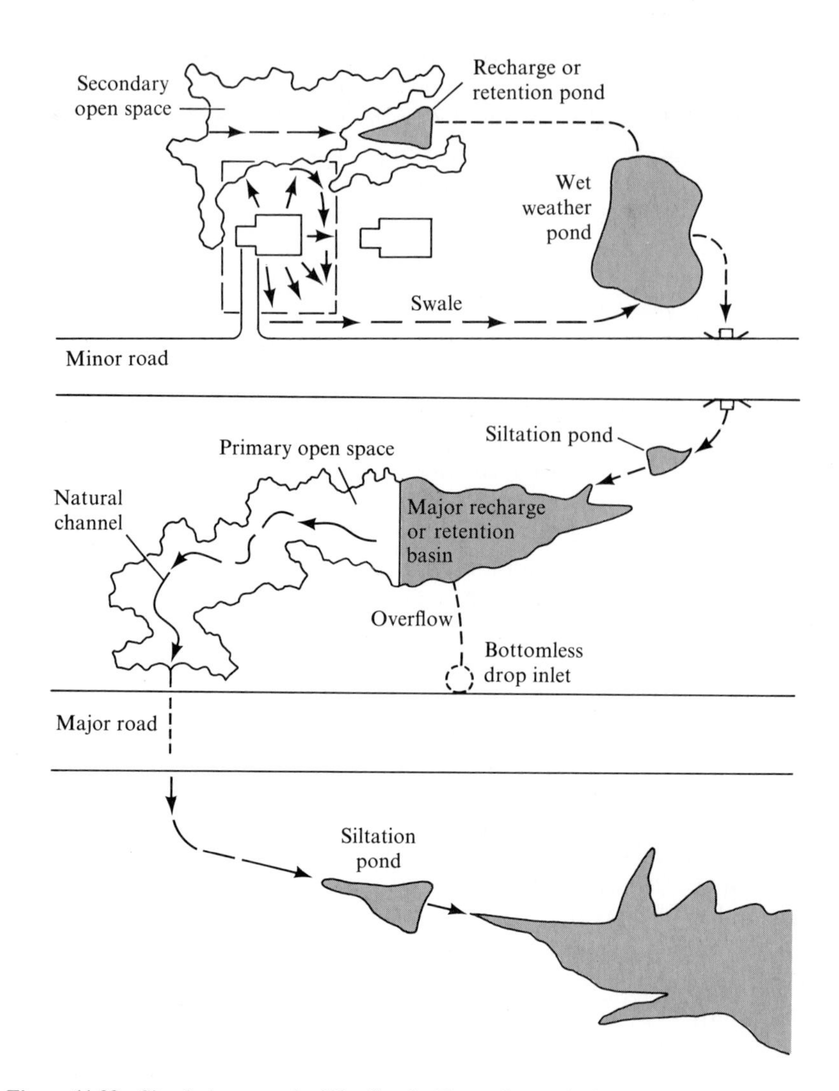

Figure 11-22 Site drainage at the Woodlands, Texas. Storm drainage is detained in small quantities close to where it is generated. It is diverted toward natural topographic depressions with permeable soils where natural vegetation and natural channels are maintained as much as possible. (Modified from Everhart 1973.)

The acceptance of the economic and esthetic benefits of exploiting the flood-control potential of natural stream systems as part of a basin-wide plan for storm runoff control in urban regions requires education and increased awareness of the flood hazard and of the potential for successful flood control among planners, engineers, administrators, and homeowners. Untermann (1973) has provided a manual which planners can use for designing stormwater control facilities on small urban catchments.

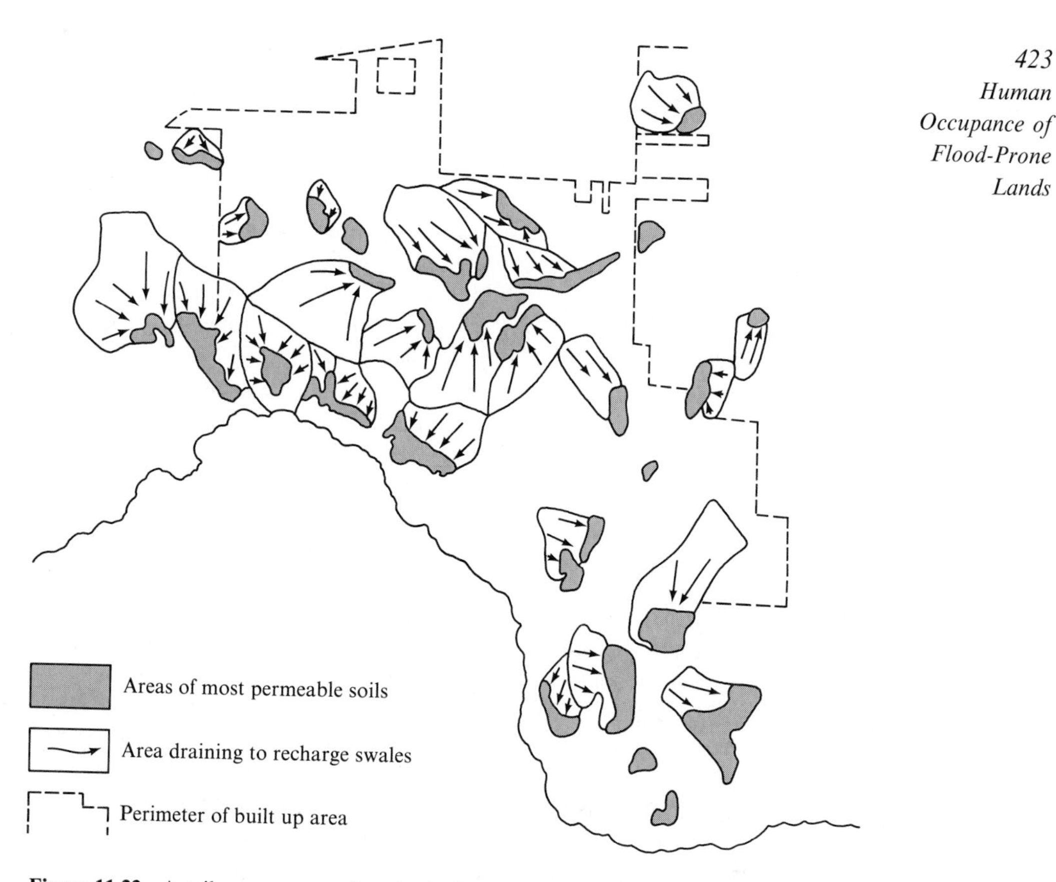

Figure 11-23 A soils map was used as the basis for delineating groundwater recharge zones into which stormwater could be diverted at the Woodlands, Texas. The low-lying areas of permeable soils in which floodwaters are detained are left under cover of hardwood trees, which provide good wildlife habitat and an attractive, varied landscape. (Modified from Everhart 1973.)

Hydrologic Information for Flood Damage Reduction

The type of hydrologic information needed for reducing flood losses depends upon the strategy employed. For flood-warning systems in small drainage basins, simple rainfall-runoff or snowmelt-runoff relations as described in Chapters 10 and 13 are needed. On larger rivers the effects of channel or lake storage must be taken into account by flood routing, and on the largest rivers that respond very slowly, simple correlations between gauge heights at upstream and downstream points are sufficient. In the United States the responsibility for developing such warning systems and for disseminating the information rests with the National Weather Service and with various state organizations. A network of stations maintained by cooperative observers relays information to River Forecasting Centers in each region. Flood warnings are issued by teletype to press, police, and local administrative agencies.

Structural solutions require a variety of hydrologic data, depending on the size and type of channel modification or detention reservoir. Each agency concerned with such designs usually has its own manual of procedures of the type described in Chapter 10. For details of the engineering philosophy and computations used in the design of structures ranging from storm sewer systems to the world's largest dams, the reader should consult Ogrosky (1964), Snyder (1964), and Viessman et al. (1972).

Many local government authorities now require that any development that involves the clearing of vegetation and the grading and paving of land should be accompanied by measures to detain storm runoff so that the peak discharge rate into stream channels does not exceed that from a similar storm before construction. The engineering departments of towns and counties are distributing their own instructions for estimating the volume of floodwater-detention structures necessary to control runoff rates. Each authority has its own modification of a procedure that is usually based on the Rational Runoff Formula. The methodology is described by Rice (1971) and Curry (1974). In some towns it is the 50-year flood peak that must be controlled to pre-development levels; in others the 2-year or 10-year flood may be chosen and written into grading ordinances or building permit regulations. Planners should be aware of local regulations. Depending on the local authority the instructions vary from clear to unintelligible, but we hope that this book will help planners understand them and perhaps even rewrite them clearly for their colleagues.

The kinds of information described above refer to individual sites. Nonstructural techniques of flood damage reduction usually require information about flood conditions throughout a valley system or administrative unit. This, in turn, requires the *mapping of flood hazard,* and in recent years the attention of hydrologists has been turning slowly to the problem of providing a spatial view of the hazard. The methodology of mapping flood hazard is still being developed and much remains to be done, though it is doubtful whether any great precision can be attained for large areas. Foster (1959) reviewed many of the problems in his discussion of flood insurance, and his words are still appropriate. It is important that the users of flood hazard maps be aware of these uncertainties and not misled by the deceptively sharp black lines drawn on the maps. Furthermore, whenever a flood hazard map is being used for planning or design, the map should be checked on the ground to see whether the delimitation is realistic.

One technique of flood hazard evaluation is by field mapping immediately after a major flood. The elevation of measured gauge heights, watermarks, flood debris, and sediment can be surveyed to known bench marks. Eyewitness accounts are also less misleading when memories are fresh. On large rivers, aerial photographs of the flood peak may be taken by various government agencies concerned with flood mapping. If an estimate of the recurrence interval of the flood can then be made, either by analysis of stream- or rain-gauge records, one has a map of the T-year flood.

The mapping of great historical floods can also provide useful information. Deposits (Costa 1974), eyewitness accounts, newspaper reports and pictures, and flood marks carved into walls or fence posts provide fragments of information that can be pieced together carefully and cross-checked for reliability. The United States Geological Survey in its series of Hydrologic Investigations Atlases has published many maps of urban areas at a scale of 1:24,000 showing the maximum flood of record.

Gauge records and a detailed topographic map can also be used to outline the flood hazard along valley floors that are heavily developed. Data on annual maximum floods or gauge heights are collected from all gauges along the reach of river under study. Supplementary flood peaks from crest stage gauges (see Chapter 16) are also used if available. A frequency analysis is made to determine the height of the 100-year flood, the 50-year flood, or any other flood of interest. The profile of the flood is then constructed by plotting the gauge heights on graph paper and interpolating the flood profile between gauges. Where no tributaries enter and valley constrictions, bridges, or steps in the valley gradient do not exist, the profile will be approximately straight. If any of these complications do exist, their effect on the flood profile must be computed (Wiitala et al. 1961) or defined by setting out additional crest-stage gauges. The backing up of tributaries by a flooding mainstream must also be considered. The flood profile so defined, as shown for example in Figure 11-24, can then be transferred to a detailed topographic map, with

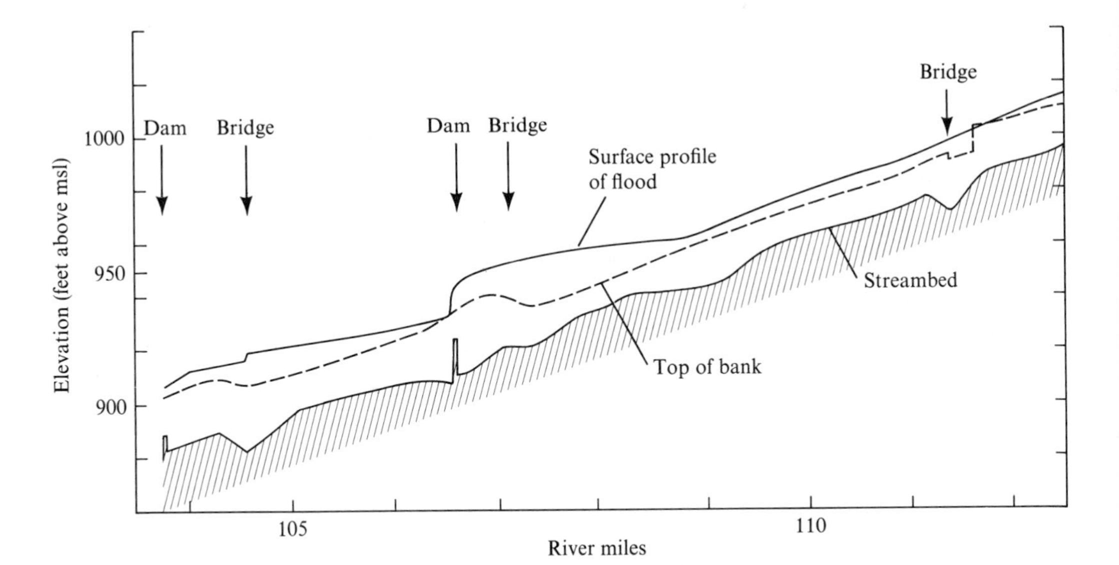

Figure 11-24 Water surface profile of a large design flood on the Yakima River at Yakima, Washington. (Data from U.S. Army Corps of Engineers.)

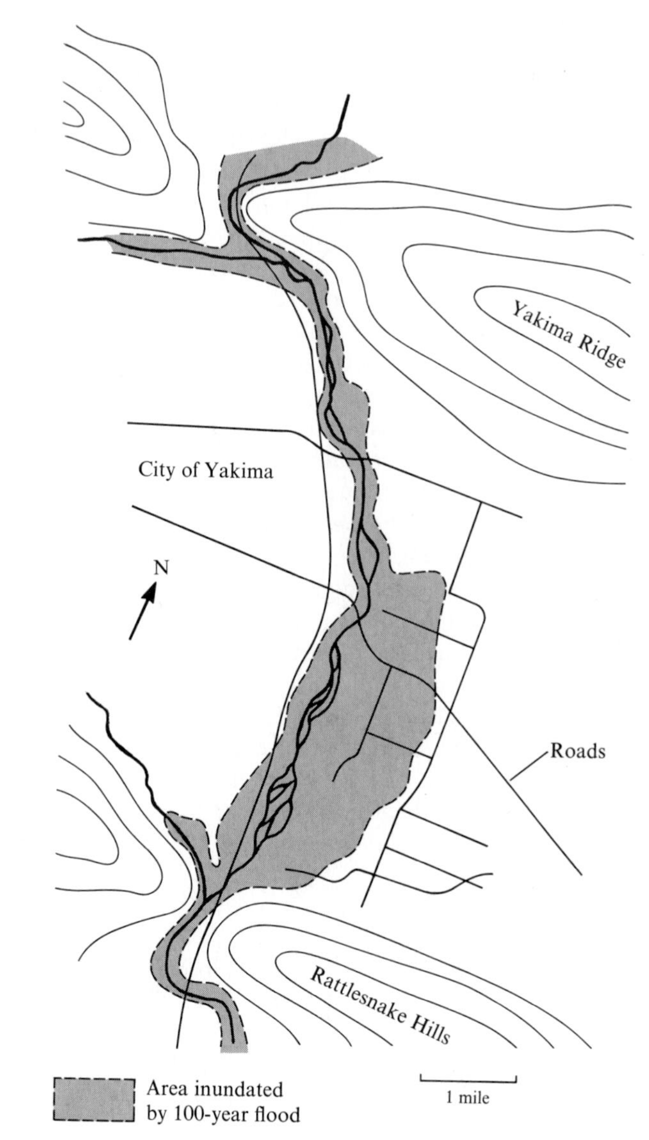

Figure 11-25 Map of the area inundated by the 100-year flood on the Yakima River at Yakima, Washington. (Data from U.S. Army Corps of Engineers.)

contour intervals on the valley floor ranging from 1 to 10 feet depending on the importance of the reach. The result is a flood map like the one shown in Figure 11-25.

Several agencies are now using flood-routing techniques to compute the flood profile. The river channel and valley floor are divided into sections and each section is assigned hydraulic parameters. The height and rate of passage of floods are then computed for each section of the reach, and the water surface elevation can be mapped at each point along the stream. The techniques have been programmed for rapid computer processing.

Each district office of the U.S. Geological Survey now publishes profiles of floodwater elevations and maps of areas that will be inundated by various floods. The information is usually released as a separate report in the *Water-Resources Investigations* series for each river valley. The use of the maps and profiles is illustrated in Typical Problems 11-4 through 11-6. Planners should familiarize themselves with these and similar publications. If no such data exist for a particular valley, the regional flood-height frequency curves described in Chapter 10 could be developed, or the dimensionless rating curve introduced in Chapter 16 could be applied.

In the United States many flood-prone areas are now being partitioned into two zones: the *flood fringe* and the *floodway.* The former is the area that would be inundated by the 100-year discharge. In this area (which can be recognized by the survey techniques just described or by methods to be introduced later) new buildings must be floodproofed and their lower floors must be at a level that will provide protection against immersion and against damage from floating ice or debris. Federally subsidized flood insurance is available for buildings in this zone. The floodway, which is outlined by engineering surveys and hydraulic calculations is the zone that could theoretically convey the 100-year flood with only a one-foot rise of water level above the height of the unconstricted flood. In other words, if the flood fringe outside the floodway were filled, the latter could pass the floodwaters with only a slight rise of stage. In the floodway, building or filling is usually forbidden and the area is maintained as green space (Figure 11-26). Because of the uncertainties inherent in the techniques for defining the floodway, some towns have challenged the positioning of the floodway through their central business districts. Federal development loans are not available for floodway land, and what makes good hydrologic sense may not be so compelling to businessmen and officials seeking to preserve their livelihoods and tax base.

Large-scale engineering delineation of flood hazard with the use of flood routing, detailed hydraulic computations, and precise topographic survey provides the most precise identification of probable flood limits and depths, and is often necessary for administration of valley floors, design of water

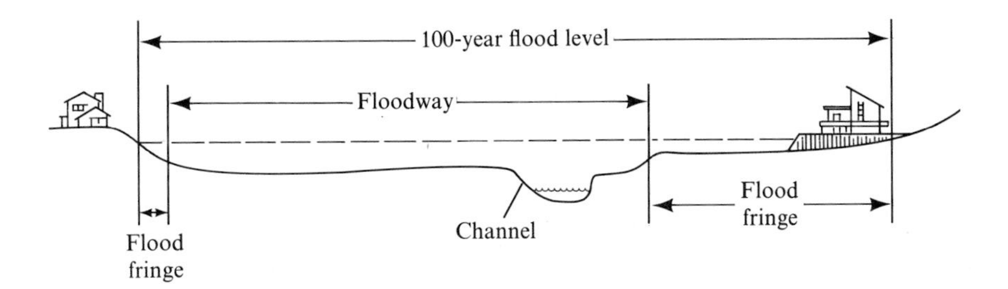

Figure 11-26 Division of the flood-prone area of a valley floor into the flood fringe and the floodway. The floodway is usually reserved for agriculture, recreation, or other uses that are not susceptible to heavy flood damage. In the flood fringe, houses can be built if special protection is installed.

intakes or drainage systems, or for settling legal disputes. In light of what we have said previously about the uncertainty in computing the sizes of floods with chosen recurrence intervals, the accuracy implied by these maps may frequently be misleading. Even under favorable conditions the mapped boundary of the 100-year flood may be in error by a hundred to several thousand feet in a horizontal direction.

A major problem with these techniques of flood mapping is that they are expensive because they require intensive inputs of skilled manpower to make engineering surveys or to characterize the hydraulics of a channel or valley floor, and they require extensive facilities for reducing data and making hydraulic computations. An even bigger problem is that these techniques are so time-consuming that the agencies charged with flood mapping cannot keep pace with the rapid suburban sprawl which is engulfing flood-prone areas in many countries. Even in rural regions in nations with land redistribution programs, the problem is an urgent one. Large estates that could withstand the flooding of their low-lying lands are being split into small holdings of a few acres. The flood hazard becomes vital to a farmer whose entire holding on a settlement scheme is within the flood-prone area. Yet in such countries there is neither time nor capital for major engineering surveys.

Lawrence Dingman of the University of New Hampshire, who has considerable experience with flood mapping, has outlined the magnitude of the problem in the Connecticut River basin. There are at least 17,000 miles of valley floor along the river. Detailed surveys and hydraulic calculations to establish 100-year flood levels cost about $5000 per mile. Detailed (2-foot contour interval) topographic mapping adds $3000 per mile, for a total required cost of $136 million. With foreseeable rates of funding and of development, the flood-prone lands will be developed long before the hazard is defined by these slow methods.

M. Gordon Wolman (1971) has pointed out that for most areas, fast, cheap ways of mapping flood hazard are available, which give a level of precision that is adequate for rural, suburban, and recreational development. Because of uncertainties in assessing the probability of floods or of computing the flood profile, even with gauge data, the slow, expensive techniques cannot be very precise, and for only a small relaxation of precision, great savings of time and money can be made by use of the more rapid techniques.

Wolman was referring to the mapping, by a combination of fieldwork and aerial-photograph interpretation, of various topographic, pedologic, and botanical features that can be correlated with floods of various recurrence intervals. As indicated in Chapter 16, an alluvial river channel is bordered by a *floodplain** that is inundated on the average once every one or two

*In this text, the term *floodplain* is reserved for a geologic feature that is being formed by the river in its present condition, and in the present climate as we will describe in Chapter 16. The reader should be aware, however, that the term is also used by engineers and planners to refer to *any* area that is subject to flooding. Thus the "100-year floodplain" is that area inundated by the 100-year flood. In this terminology there is an infinite number of floodplains, including the 50-year floodplain, the 20-year floodplain, and so on. In our terminology, there is only one. We refer to the area subject to flooding as "flood-prone land."

years. If the floodplain can be recognized and mapped, an approximate map of the 1.5-year flood can be made rapidly. In some river valleys, it is even possible to correlate river terraces (see Chapter 16) or deposits with particular flood events (Jahns 1947). Wolman suggests that if such a correlation can be established for a valley, the extent of these large floods of known frequency can be mapped from the alluvial-topographic features. In Chapter 16, we describe a method of delineating flood heights of various frequencies from measurements of the channel geometry and the use of a *dimensionless rating curve.*

In several areas, it has been shown that the extent of some soil types correlates well with the extent of major floods such as the 50-year or 100-year event. Cain and Beatty (1968), for example, compared the area of alluvial, poorly drained soils with the computed extent of flooding by events of various frequencies. The extent of flooding was computed from detailed engineering surveys, aerial photographs, rainfall-runoff and hydrograph analysis, historical flood records, and gauge measurements. Alluvial, poorly drained soils were good indicators of the extent of large floods, although very rare discharges ($>$ 100-year events) inundated small areas of terrace soils and lower footslopes with colluvial soils. Witwer (1966), in a different region, shows that the boundaries of alluvial soils outline only floods of 50-year recurrence intervals or less. Figure 11-27, however, shows that in a Vermont catchment the area of alluvium and soils with a high water table coincide quite closely with the limit of the 100-year flood.

The cost of the procedure is much lower than most other methods when good soils maps are available, but the cost rises rapidly when field checking is necessary. The technique needs to be "calibrated" in each area with a distinctive set of soil types. The use of vegetation maps is subject to similar limitations and is probably not as precise a technique as delimitation by soil type. The case of recognizing clues from vegetation, both in the field and on aerial photographs, however, makes the technique attractive where it can be used. Hack and Goodlett (1960) and Sigafoos (1964) have illustrated the use of vegetation for recognizing the extent and depth of major floods in the recent past.

Wolman has discussed the errors and costs involved in various types of flood mapping and some of the legal and administrative problems that have to be considered. As with the detailed engineering surveys, the mapping should be repeated if the flood hydrology of the catchment is significantly changed by urbanization, flood control, or other measures. The U.S. Soil Conservation Service (1972) also describes the mapping of sediment damage from major floods, for assessing flood damage due to this cause.

The possibility of catastrophic inundation due to nonhydrologic causes must also be added to flood hazard maps in some regions. In the Pacific Northwest of the United States, for example, there is a possibility that volcanic activity in the Cascade Mountains could initiate major landslides or mudflows, which could fill one of the many reservoirs in the mountain valleys, causing tremendous floods that would overtop and breach a dam (Crandell 1973), threatening communities in the mountain valleys and the

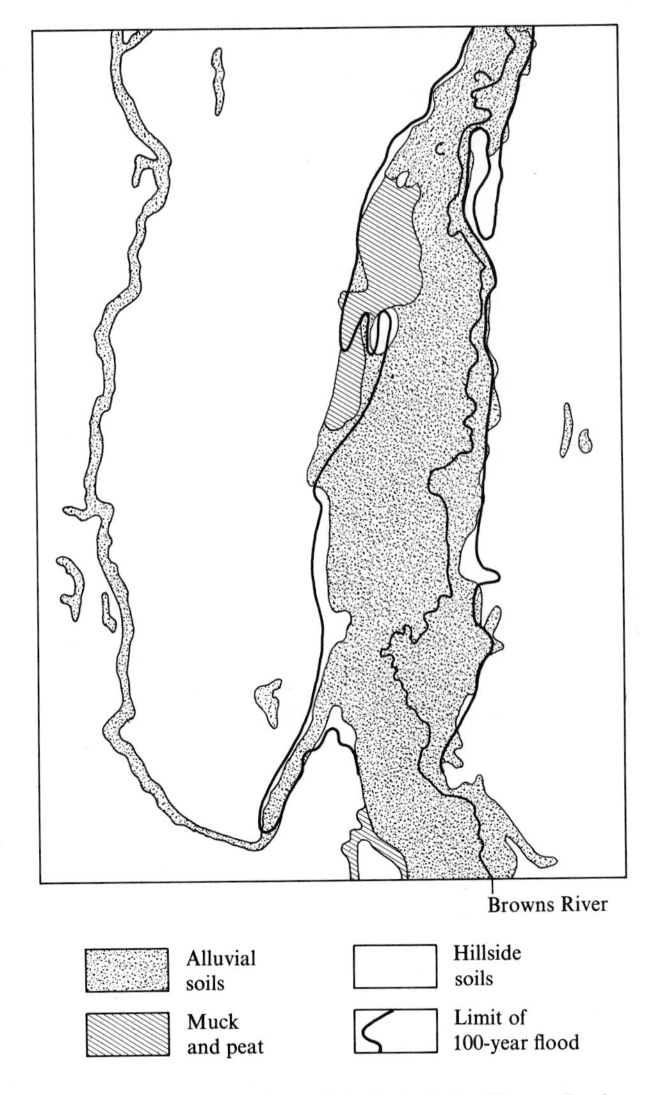

Figure 11-27 Comparison of the limit of the 100-year flood with the extent of alluvial and organic soils in the Browns River valley in Vermont. (From Vermont Department of Water Resources.)

nearby Puget Lowland. Another hazard in the region is the possibility of "glacier outburst floods" (Richardson 1968). Such flood hazards are not usually considered in standard hydrologic analyses, but although the probability of their occurrence is not high, they should be recognized in the siting of particularly important facilities such as power plants, new industrial developments, and new towns. In coastal regions, tsunamis pose a similar threat (State of Hawaii 1971).

Once planners understand the extent of the flooding problem and are aware of all sources of flood information for their region, they can help in

Figure 11-28 Corridor of land use that is not susceptible to flood damage in the midst of an urban area. Along the River Lea in northern London, the flood-prone land is reserved for parks with sports fields, boating lakes, and other recreational facilities. (The Civic Trust, London.)

the task of reducing the massive burden of flood damage to society. McHarg (1969) has discussed several imaginative plans for urban development that take into account the flood hazard. Wise development policies, even on a small scale, can concentrate settlement away from low-lying areas (Figure 11-28). The location of a sewer line, the improvement of a road, or the location of public buildings can all channel development toward high ground.

Bibliography

ACKERMANN, W. C., WHITE, G. E. AND WORTHINGTON, E. B. (1973) *Man-made lakes: their problems and environmental effects;* American Geophysical Union, Washington, DC, 847 pp.

ANTOINE, L. H. (1964) Drainage and best use of urban land; *Public Works,* vol. 95, pp. 88–90.

ASSOCIATION OF ENGINEERING GEOLOGISTS (1968) *Symposium on reservoir leakage and groundwater control;* Seattle, WA, 104 pp.

BEASLEY, R. P. (1972) *Erosion and sediment pollution control;* Iowa State University Press, Ames, IA, 320 pp.

BELT, C. B. (1975) The 1973 flood and man's constriction of the Mississippi River; *Science,* vol. 189, pp. 681–684.

BRITTON, L. J., AVERETT, R. C., AND FERREIRA, R. F. (1975) An introduction to the processes, problems, and management of urban lakes; *U.S. Geological Survey Circular 601-K.*

CAIN, J. M. AND BEATTY, M. T. (1968) The use of soil maps in the delineation of floodplains; *Water Resources Research,* vol. 4, pp. 173–182.

CIVIL ENGINEERING (1960) *Storm water control for a shopping center;* vol. 30, p. 63.

CIVIL ENGINEERING (1973) *Swampy site turned into prize winner*; vol. 43, no. 6, p. 82.

COOK, H. L. (1949) The effects of land management upon runoff and groundwater; *U.N. Scientific Conference on the Conservation and Utilization of Natural Resources,* vol. 4, pp. 193–202.

COSTA, J. E. (1974) Response and recovery of a Piedmont watershed from tropical storm Agnes, June, 1972; *Water Resources Research,* vol. 10, pp. 106–112.

COUNCIL ON ENVIRONMENTAL QUALITY (1973) Report on channel modifications: prepared by Arthur D. Little Inc.; U.S. Superintendent of Documents, Washington, DC, vol. 1, 394 pp.

CRANDELL, D. R. (1973) Map showing potential hazards from future eruptions of Mount Rainier, Washington; *U.S. Geological Survey Miscellaneous Geologic Investigations, Map I-836.*

CURRY, L. (1974) Relationship of rational and unit graph methods in retention basin design; in *Proceedings of the national symposium on urban rainfall, runoff and sediment control,* University of Kentucky, College of Engineering, pp. 33–40.

DAILY, E. J. (1961) Stormwater detention in urban areas; *Public Works,* vol. 92, pp. 146–147.

DINGMAN, S. L. (1975) An expedient approach to community-wide floodplain delineation and regulation (abstract); *EOS, American Geophysical Union, Transactions,* vol. 56, p. 360.

EARL, J. D. (1971) Where is urban hydrology practice today? *American Society of Civil Engineers, Journal of the Hydraulics Division,* vol. 97, HY2, pp. 256–264.

EDELEN, G. W. AND COBB, E. D. (1969) Floodplain mapping; *U.S. Geological Survey Techniques of Water-Resources Investigations,* Book 4, Chapter E1.

EVERHART, R. C. (1973) New town planned around environmental aspects; *Civil Engineering,* vol. 43, no. 9, pp. 69–73.

FELTON, P. N. AND LULL, H. W. (1963) Suburban hydrology can improve watershed conditions; *Public Works,* vol. 94, pp. 93–94.

FOSTER, H. A. (1959) Technical problems of flood insurance; *American Society of Civil Engineers, Transactions,* vol. 124, pp. 366–380.

GRIGG, N. S. AND HELMEG, O. J. (1975) State-of-the-art in estimating flood damage in urban areas; *American Water Resources Association, Water Resources Bulletin,* vol. 11, no. 2, pp. 379–390.

HACK, J. T. AND GOODLETT, J. C. (1960) Geomorphology and forest ecology of a mountain region in the Central Appalachians; *U.S. Geological Survey Professional Paper 347.*

HART, J. L. (1973) Swampy site turned into prize-winner; *Civil Engineering,* vol. 43, no. 6, pp. 82–84.

HITTMAN ASSOCIATES, INC. (1968) Reusing storm water runoff; *Environmental Science and Technology,* vol. 2, pp. 1001–1005.

HITTMAN ASSOCIATES, INC. (1974) Approaches to stormwater management; *Report to the Office of Water Resources Research, U.S. Department of Interior, Contract No. 14-31-001-9025,* 258 pp.

HOGAN, T. M. (1963) State floodplain zoning; *De Paul Law Review,* Chicago, vol. 12, no. 2, pp. 246–262.

HOYT, W. G. AND LANGBEIN, W. B. (1955) *Floods;* Princeton University Press, Princeton, NJ, 468 pp.

JAHNS, R. H. (1947) Geologic features of the Connecticut valley, Massachusetts, as related to recent floods; *U.S. Geological survey Water Supply Paper 996.*

KATES, R. W. (1962) Hazard and choice perception in floodplain management; *University of Chicago, Department of Geography, Research Paper No. 78,* 157 pp.

LANDSCAPE ARCHITECTURE (1974) Zero runoff; *Landscape Architecture,* vol. 65, pp. 381–395.

LEOPOLD, L. B. AND MADDOCK, T. (1953) The hydraulic geometry of stream channels and some physiographic implications; *U.S. Geological Survey Professional Paper 282.*

LEOPOLD, L. B. AND MADDOCK, T. M. (1954) *The flood control controversy;* Ronald Press, New York, 278 pp.

MCHARG, I. L. (1969) *Design with nature;* Natural History Press, New York, 198 pp.

MURPHY, F. C. (1958) Regulating floodplain development; *University of Chicago, Department of Geography, Research Paper No. 56,* 204 pp.

OBRIST, A. (1974) Ponding against the storm; *Landscape Architecture,* vol. 65, pp. 388–390.

OGROSKY, H. O. (1964) Hydrology of spillway design: small structures: limited data; *American Society of Civil Engineers, Journal of the Hydraulics Division,* vol. 90, HY3, pp. 295–310.

PARKER, D. E., LEE, G. B. AND MILFRED, C. J. (1970) Floodplain delineation with pan and color; *Photogrammetric Engineering,* vol. 26, no. 10, pp. 1059–1063.

POERTNER, H. G. (1974) Drainage plans with environmental benefit; *Landscape Architecture,* vol. 65, pp. 391–392.

RAHN, P. H. (1975) Lessons learned from the June 9, 1972, flood in Rapid City, South Dakota; *Bulletin of the Association of Engineering Geologists,* vol. 12, pp. 83–97.

RICE, L. (1971) Reduction of urban runoff-peak flows by ponding; *American Society of Civil Engineers, Journal of the Irrigation and Drainage Division,* vol. 97, IR3, pp. 469–481.

RICHARDSON, D. (1968) Glacial outburst floods in the Pacific Northwest, *U.S. Geological Survey Professional Paper 600-D,* pp. D79–D86.

RICKERT, D. A. AND SPIEKER, A. M. (1971) Real-estate lakes; *U.S. Geological Survey Circular 601-G.*

SCHWARZ, F. K., HUGHES, L. A., HANSON, E. M., PETERSEN, M. S. AND KELLEY, D. B. (1975) The Black Hills: Rapid City flood of June 9–10, 1972; *U.S. Geological Survey Professional Paper 877.*

SEABURN, G. E. (1970) Preliminary results of hydrologic studies of two recharge basins on Long Island, New York; *U.S. Geological Survey Professional Paper 627-C.*

SHEAFFER, J. R. (1960) Flood proofing: an element in a flood damage reduction program; *University of Chicago, Department of Geography, Research Paper No. 65,* 198 pp.

SHEAFFER, J. R., ELLIS, D. W. AND SPIEKER, A. M. (1970) Flood-hazard mapping in metropolitan Chicago; *U.S. Geological Survey Circular 601-C.*

SIGAFOOS, R. S. (1964) Botanical evidence of floods and floodplain deposition; *U.S. Geological Survey Professional Paper 485-A*.

SNYDER, F. F. (1964) Hydrology of spillway design: large structures: adequate data; *American Society of Civil Engineers, Journal of the Hydraulics Division,* vol. 90, HY3, pp. 239–259.

STATE OF HAWAII (1971) Flood hazard information, Island of Maui; *Report R3a, Department of Land and Natural Resources,* Honolulu.

STOREY, H. C., HOBBA, R. L., AND ROSA, J. M. (1964) Hydrology of forest lands and range-lands, Sect. 22, in *Handbook of applied hydrology* (ed. V. T. Chow), McGraw-Hill, New York.

TENNESSEE VALLEY AUTHORITY (1974) *TVA tames the river,* Knoxville, 22 pp.

UNITED NATIONS, ECONOMIC COMMISSION FOR ASIA AND THE FAR EAST (1953) *The sediment problem;* Bangkok, 92 pp.

UNITED STATES CONGRESS (1976) Teton dam disaster; *Thirtieth Report by the Committee on Government Operations,* 94th Congress, 2nd Session, Washington, DC, 37 pp.

UNITED STATES ENVIRONMENTAL PROTECTION AGENCY (1972) Investigation of porous pavements for urban runoff control; *Water Pollution Control Research Series No. 11034 DUY 03/72.*

UNITED STATES GEOLOGICAL SURVEY (1969) Floodplain mapping; *Techniques of water-resources Investigations,* book 4, chapter E1.

U.S. SOIL CONSERVATION SERVICE (1969) *Engineering field manual for conservation practices,* Washington, DC.

U.S. SOIL CONSERVATION SERVICE (1972) Sedimentation; *National Engineering Handbook No. 3,* Washington, DC.

UNTERMANN, R. K. (1973) *Grade easy: an introductory course in the principles and practices of grading and drainage;* American Society of Landscape Architects Foundation, McLean, VA.

VIESSMAN, W., HARBAUGH, T. E. AND KNAPP, J. W. (1972) *Introduction to hydrology;* Intext Educational Publishers, New York, 415 pp.

WHITE, G. F. (1945) Human adjustments to floods; a geographical approach to the flood problem in the United States; *University of Chicago, Department of Geography, Research Paper No. 29,* 225 pp.

WHITE, G. F. (1964) Choice of adjustment to floods; *University of Chicago, Department of Geography, Research Paper No. 93,* 150 pp.

WHITE, G. F., CALEF, W. C., HUDSON, J. W., MAYER, H. M., SHEAFFER, J. R. AND VOLK, D. J. (1958) Changes in urban occupance of floodplains in the United States; *University of Chicago, Department of Geography, Research Paper No. 57,* 235 pp.

WIITALA, S. W., JETTER, K. R. AND SOMERVILLE, A. J. (1961) Hydraulic and hydrologic aspects of flood-plain planning; *U.S. Geological Survey Water Supply Paper 1526.*

WISWALL, K. C. AND SHUMATE, K. S. (1976) Stormwater management design: a manual of procedures and guidelines; Prepared for the Maryland Department of Natural Resources by Roy F. Weston Inc., Environmental Consultants-Designers, West Chester, PA.

WITWER, D. B. (1966) Soils and their role in planning a suburban county; p. 15–30 in *Soil surveys and land use planning* (eds. L. Bartelli et al.), Soil Science Society of America, Madison, Wis.

WOLMAN, M. G. (1971) Evaluating alternative techniques of floodplain mapping; *Water Resources Research,* vol. 7, pp. 1383–1392.

ZINGG, A. W. (1946) Flood control aspects of farm ponds; *Agricultural Engineering,* vol. 27, no. 1, pp. 9–13.

11-1 Effects of Land Use on Flood Runoff

Using the Soil Conservation Service procedure described in Chapter 10, evaluate the effects on the volume of flood runoff from a small drainage basin originally planted to contoured row crops in good condition, if 25 percent of the land is converted to woodland. Make the calculations first by assuming that all the soils are in hydrologic group A in Table 10-3, and compute the percentage reduction in storm runoff for rainstorms of one, two, three, and five inches. Plot a graph of the percentage reduction versus storm size for group A soils. Repeat the analysis by assuming that all the soils are in group C. Assume average antecedent moisture conditions (AMC II). Write a brief interpretation of your findings.

Other data for problems of this type can be found in reviews of hydrologic evaluation of land treatment by Cook (1949) and by Storey et al. (1964:22–47) and in a handbook by the U.S. Soil Conservation Service (1972, Section 9).

11-2 Review of a Proposal for a Flood-Control Dam

Obtain a copy of a proposal for a flood-control dam in your local region. Subject it to a detailed analysis as a group project. Use the questions raised in this chapter and the techniques of flood prediction described in Chapter 10 as starting points for the discussion. Compile an annotated bibliography on the positive and negative aspects of dam construction as a strategy for minimizing flood damage.

11-3 Stormwater Detention Basins for Building Sites

From the engineering or community services department of your local government agency, obtain a copy of the regulations specifying the size of detention basins for storm runoff from building sites. Take the document and this book out to a catchment containing a residential development. Apply the regulations to the site, and calculate the size of the necessary detention basin.

Consider possible locations for the basin, how you would maintain it to keep it clean, and what other methods of reducing storm runoff are possible on the site.

11-4 Use of Flood Hazard Maps, I

Suppose that you are considering building the Garden Apartments on the valley floor of the Black River in North Springfield, Vermont. You are concerned about whether the site will be flooded. If it is subject to flooding, you would like to know the probable depth of flooding so that you could consider some floodproofing strategy, such as using the ground floor as a parking garage.

Solution

Enquiries at the Vermont Department of Water Resources have yielded the flood hazard map

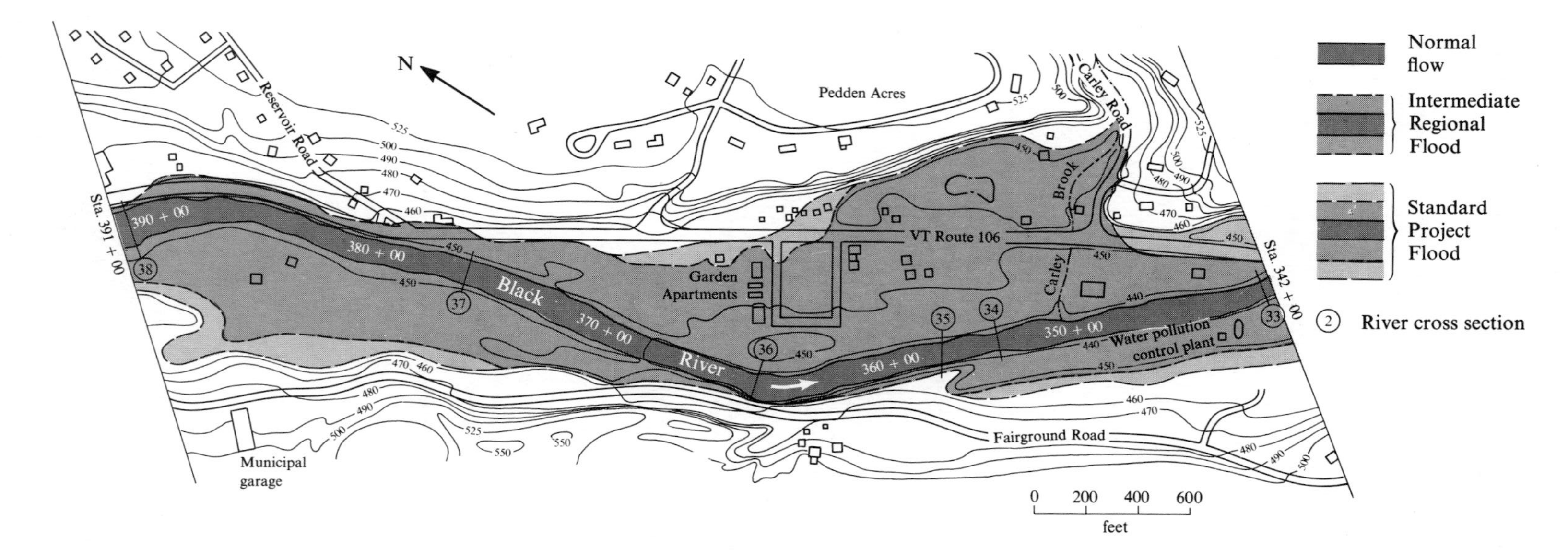

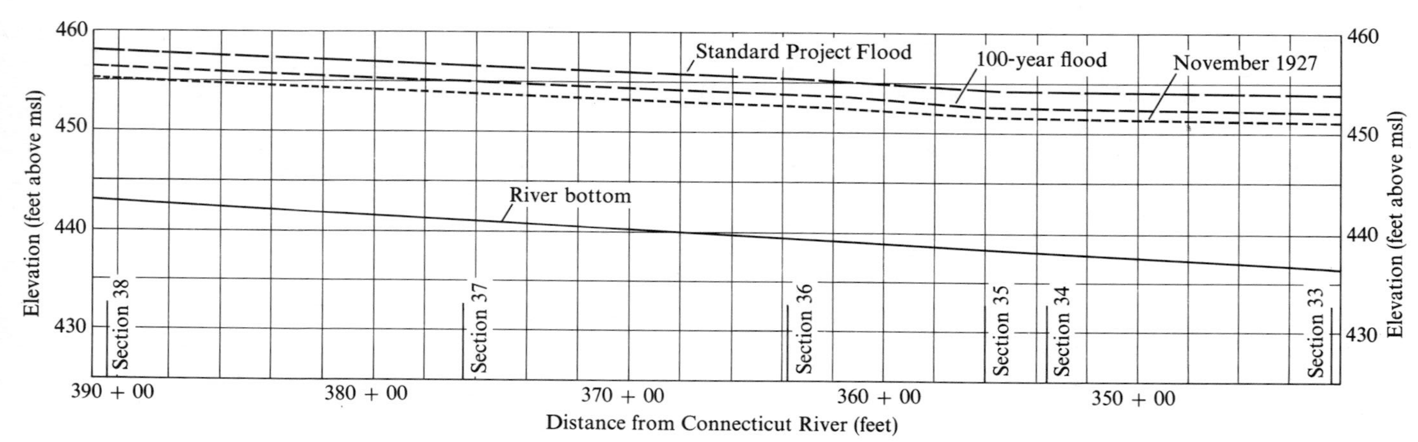

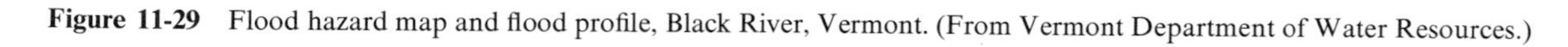

Figure 11-29 Flood hazard map and flood profile, Black River, Vermont. (From Vermont Department of Water Resources.)

and flood profile for the Black River valley in Figure 11-29. In other places, the U.S. Army Corps of Engineers, U.S. Geological Survey, or U.S. Soil Conservation Service are helpful sources of information.

Locate the site of the proposed Garden Apartments, and find that it lies 36,000 to 36,300 feet from the confluence with the Connecticut River, according to the numbers along the main channel. Also note that the site lies at an elevation of about 450 feet above mean sea level.

Transfer the distance reading to the flood profile below the map, and read the following elevations. The Intermediate Regional Flood (recurrence interval of approximately 100 years) has an elevation of about 453.5 feet, and the Standard Project Flood (a very large computed flood whose recurrence interval is not known but is probably several hundred years) has an elevation of 455 feet.

Using these elevations and the contours on the map, a rough topographic cross section of the valley can be constructed as shown in Figure 11-30.

The cross-sectional profile shows that the natural ground surface at the site will be inundated to a depth of about three feet by the 100-year flood and about five feet by the Standard Project Flood.

Many flood hazard maps have a 2-foot contour interval rather than the 10-foot interval in the example. This makes the definition of the hazard more precise. If this preliminary investigation of flooding encourages the planner to pursue the project further, an engineering level survey would then be done along a line perpendicular to the river and across the site. This would define even more precisely the elevation of the ground relative to the flood heights. But the published map and profile are useful for the early stages of the investigation.

They are also very useful, of course, if a person is considering buying the apartment house. The potential buyer can read the frequency and depth of flooding and decide whether to take the risk. Such a check should be made at the local water resources agency before buying any property along a valley floor. Other uses of the map and profile that are suggested by the present example include an indication that public facilities such as the water pollution control plant on the floor of the Black River valley should be protected with an embankment or floodwall to a height of at least 452 feet (against about 6 feet of inundation during the 100-year flood).

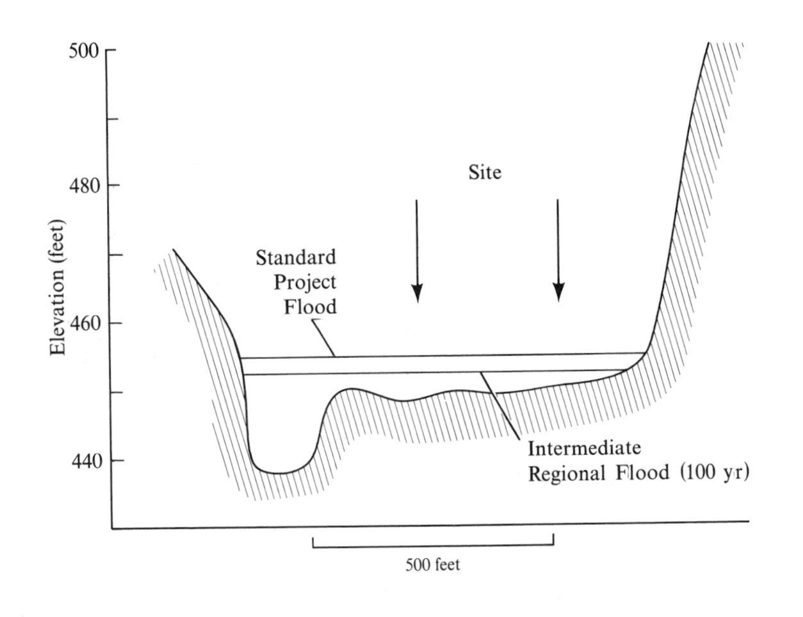

Figure 11-30 Valley cross section derived from topographic map.

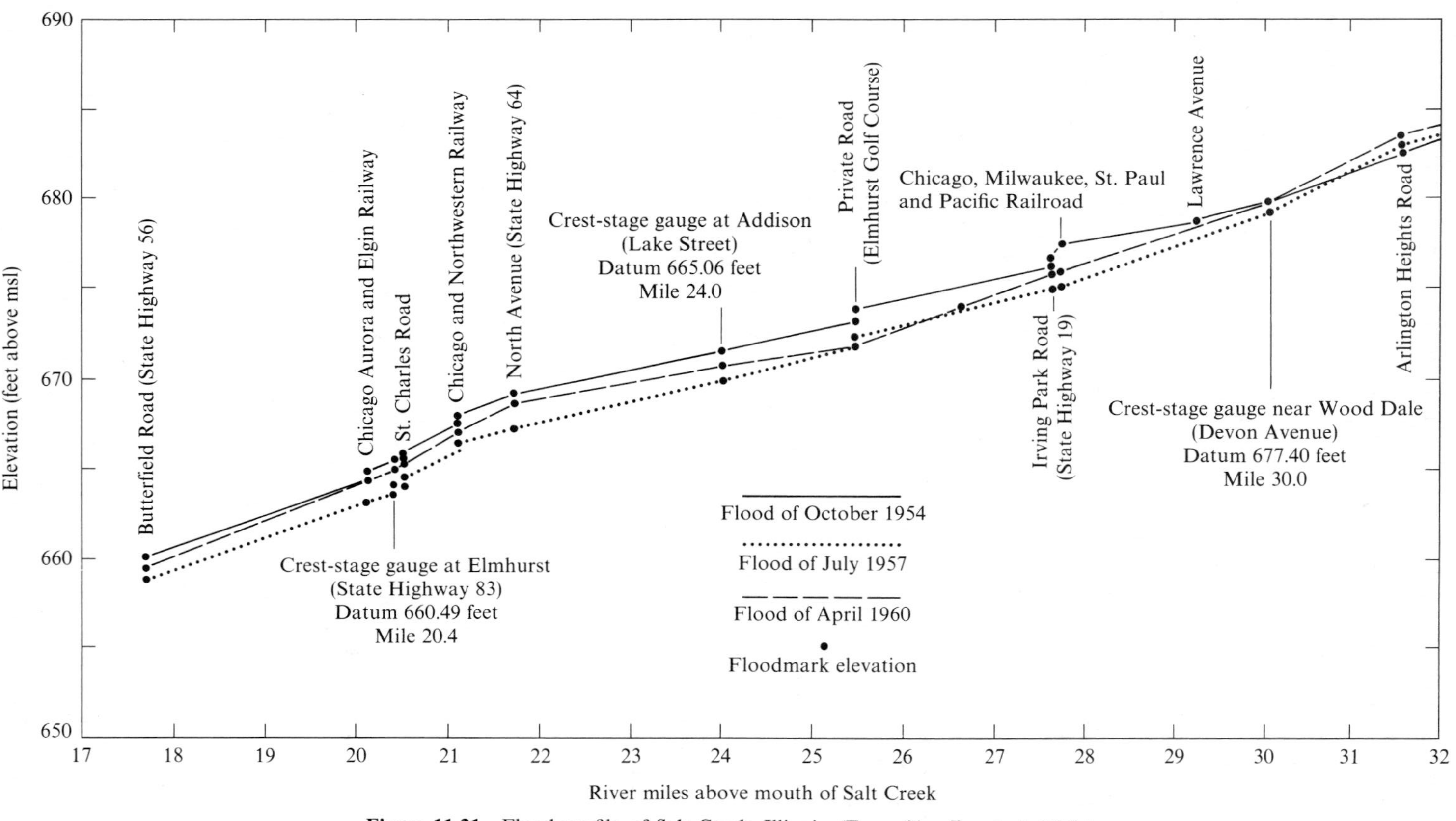

Figure 11-31 Flood profile of Salt Creek, Illinois. (From Sheaffer et al. 1970.)

The following typical problem is modified from the publication by Sheaffer et al. (1970). Suppose that you are developing a stream-side park and need to know the frequency of flooding for some site at which you plan to install facilities for visitors.

Obtain a copy of the flood map for the valley and also the flood profile, as shown in Figure 11-31. From the map, you measure that the site in which you are interested lies 23.5 river miles above the mouth of Salt Creek. In the flood hazard report accompanying the profile there is the flood-height frequency curve for the crest-stage gauge at Lake Street, 24 miles above the mouth of Salt Creek. The profile shows that the flood of October 1954 crested at 671.5 feet, while at the site of your interest (mile 23.5) the peak reached only 671 feet.

The flood-height frequency curve (Figure 11-32) indicates that the 1954 flood had a recurrence interval of 8.3 years. Thus at mile 23.5, the 8-year flood has an elevation of 671 feet. If the planner installs the structure at the 671-foot level, the chances of its being flooded in any given year are one in eight. He or she may decide to move to higher ground.

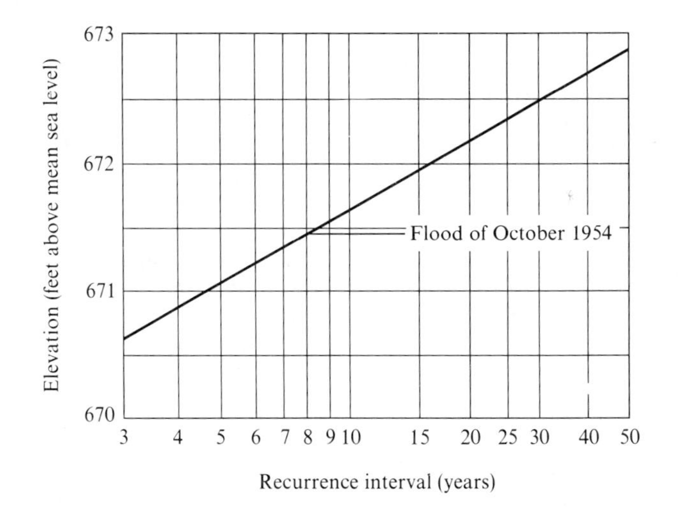

Figure 11-32 Flood-frequency graph for Salt Creek at Addison (Lake Street), Illinois. (From Sheaffer et al. 1970.)

11-6 Use of Flood Hazard Maps, III

Let us suppose that the planner in Typical Problem 11-5 is willing to accept a flood risk of 1 in 25. Find the ground elevation at which the building should be situated to enjoy that much security.

Solution

The flood-height frequency curve for the Lake Street gauge indicates that the 25-year flood has an elevation of 672.3 feet. The flood profile indicates that the water level at river mile 23.5 should be 0.5 feet below this elevation, and the building should therefore be situated at an elevation of 671.8 feet.

11-7 The Planetarium Proposal for Jewel Creek

Assume that the planner for the Regional Park Commission has made a general proposal that public-use facilities in Tilden Regional Park, Berkeley, California be expanded by the construction of a planetarium at the southeast corner of the Jewel Creek basin. Access would be provided by a side road off Nimitz Way, all traffic coming into the new complex from that road, at the far eastern edge of the basin. Parking would be provided for 200 cars. The building complex itself would cover 15 acres in addition to the parking area.

He also proposed to create a small recreational lake by building a dam 700 feet downstream of the present road crossing of Jewel Creek. No details of this dam and lake are given except that the top of the dam would be approximately level with the valley flat at the location of the dam. A picnic area would be developed on the valley flat in the vicinity of the present rock shelter. Access to the picnic area would be along the present dirt road. No road would be built connecting the planetarium complex with the picnic area.

The commission feels there may be some hydrologic considerations not mentioned by the planner. They have turned the problem over to you at this early stage with the request that you pose questions of importance and provide answers to or comments on those questions.

Write a report to the commission and include such sketches, maps, sections, computations, and profiles as are needed to explain your analysis. You do not have to provide a definite recommendation that the proposal be accepted or rejected, but your questions and analyses may affect the final decision.

Whether some other location would be more acceptable is not within the purview of your report. You are asked to comment on the proposal as it was made, to build the facilities in the Jewel Creek basin.

One commissioner asked whether the lake would fill with water and remain filled, noting the exceptionally dry winter of 1975–76. Another member raised a question about silt in such a lake.

12

Water Supply and Use

Water Use

Water supply is an important constraint in both local and regional planning for urban and rural areas. A planner often needs to know the total amount and seasonal timing of water availability at a site, or what size of reservoir is required to ensure a certain supply. These considerations affect first of all the amount of water that can be used for drinking, irrigation, or industrial use. Water supply also limits the acceptable discharge of pollutants such as heat, sewage effluent, or industrial wastes. Because the quantity and quality of water are so closely linked, the dry-season supply is a particularly severe constraint on water use for recreation and fish production. Navigation is an important consideration on some large rivers, and on rivers such as the Missouri, streamflow is artificially regulated to serve transportation in addition to uses referred to above. The generation of hydroelectric power is another important use of the water supply in some regions.

Table 12-1 lists the major uses of water in the United States and distinguishes between *gross use* and *consumptive use*. Gross use refers to the total amount of water required for some purpose. Much of this water is usable many times over so long as it is purfied. Consumptive use is that part of the water intake that becomes unavailable for further use because it is evaporated. The table shows that the largest gross users are not responsible for much consumptive water loss. The largest consumptive user is irrigated agriculture, whereas hydroelectric power generation and industrial cooling, including thermoelectric power plants, are the largest gross users. Most of the consumptive use from irrigated fields and reservoirs occurs in the dry western states. It is a heavy drain on the streamflow and groundwater resources of the region, and in some areas recent exploitation has produced an imbalance that cannot be sustained by the regional hydrologic cycle. To redress the balance, various schemes have been proposed for large-scale transfers of water from basins like that of the Columbia River, with an abundant water supply, to areas of deficiency.

Planners need to understand some of the hydrologic issues that affect water supply and some of the simpler computational methods used in design. We will cover these matters in the earlier part of the chapter and then discuss some typical problems the planner might encounter in making an analysis of water supply for a small development.

Table 12-1 Major uses of water in the continental United States. (From U.S. Geological Survey.)

USER	GROSS USE (MILLIONS OF GALLONS/DAY)	CONSUMPTIVE USE (MILLIONS OF GALLONS/DAY)	NOTES
Navigation, recreation, maintenance of aquatic ecosystems, dilution and purification of wastes	Very large, but undefinable	Low?	
Generation of hydroelectric power	2,300,000	> 11,000	Main consumptive use is evaporation from reservoirs in arid West. Unpolluted.
Industry: self-supplied	120,000	3,500	Almost all for cooling of power plants; 25% saline. Results in thermal pollution.
Industry: from public supplies	7,000	?	Process water heavily polluted.
Agriculture: irrigation	120,000	97,000	Polluted by fertilizers, pesticides, and high concentrations of dissolved salts.
Agriculture: rural domestic and livestock	4,000	3,000	Largely from local wells and springs, Heavily polluted with organic wastes.
Municipal	23,000	5,000	Heavily polluted

In earlier chapters we have reviewed the mechanisms by which water is stored and transmitted through the hydrologic cycle. The water is held temporarily in, and can therefore be drawn from, the groundwater system, ponds, lakes, and stream channels. During dry weather the drainage of these subsystems of the hydrologic cycle provides the supply upon which all human endeavor depends. Because we discussed groundwater supplies and their development in Chapter 7, the present chapter will be confined to surface supplies. It needs to be stressed that surface and groundwater supplies are related. Water can flow from one subsystem to the other, and use or disruption of one component can impair the value of the other.

Because surface water supplies fluctuate over periods of days or even years, there have been attempts throughout history to store water for use during periods of more limited supply. The topic of water supply, therefore, includes the whole field of artificial impoundments, their design, management, and environmental impact, most of which are included in manuals of engineering hydrology, sewerage design, and hydraulics rather than in the present volume.

Volumes of Runoff

The average annual volume of streamflow from the conterminous United States is equivalent to 9 inches of runoff over the 3,023,000 square miles of the country. This represents the residual after the average evapotranspiration (21 inches) is subtracted from the average precipitation.

An appreciable part of the total water discharged to the ocean is carried by a few large rivers. In the United States the Mississippi River alone discharges one-third of all the nation's runoff from a drainage area that comprises 41 percent of the total area of the conterminous states. A similar relation is true for the land areas of the whole earth. The largest river, the Amazon, carries about 23 percent of all the water draining off the continents and its drainage area represents 6 percent of the runoff-contributing area of the world.

From the tabulation compiled by Nace (1970) and using his method of analysis, Table 12-2 lists the 30 largest rivers of the United States arranged in the order of average discharge. The St. Lawrence River is included because it drains an appreciable area of the United States, although its mouth is in Canada. The list includes only rivers draining directly to the oceans. The 30 rivers discharge a total of 1706×10^3 cfs and have an aggregate drainage area of 2280×10^3 square miles, 75 percent of the total area of the conterminous states. To estimate the percentage of the total runoff represented, the cumulative values of drainage area and average discharge are plotted in Figure 12-1 against rank order from Table 12-2. Extrapolating the cumulative area to the 100-percent point and the cumulative discharge to the same position, the total runoff is read from the graph to be about 2000×10^3 cfs.

Table 12-2 Largest rivers in the conterminous United States in order of their average discharge. (From Nace 1970.)

RANK ORDER	RIVER BASIN	DRAINAGE AREA (SQ MI $\times 10^{-3}$)	CUMULATIVE AREA (SQ MI $\times 10^{-3}$)	% OF TOTAL AREA	AVERAGE DISCHARGE (CFS $\times 10^{-3}$)	CUMULATIVE DISCHARGE (CFS $\times 10^{-3}$)	ANNUAL RUNOFF (CFS/MI^2)*
1	Mississippi	1240	1240	41.0	650	650	0.52
2	St. Lawrence	396	1636	54.1	348	998	0.88
3	Columbia	258	1894	62.6	281	1279	1.09
4	Mobile	44.4	1932	63.9	63.1	1342	1.42
5	Susquehanna	28.9	1967	65.1	40.2	1382	1.39
6	Appalachicola	20.0	1987	65.7	26.7	1409	1.33
7	Sacramento	27.0	2014	66.6	23.0	1432	0.85
8	Hudson	13.4	2027	67.0	21.3	1453	1.59
9	Connecticut	11.2	2038	67.4	19.3	1472	1.72
10	Delaware	11.4	2050	67.8	18.8	1491	1.65
11	Klamath	12.1	2062	68.2	17.1	1508	1.41
12	Penobscot	9.5	2071	68.5	16.7	1525	1.76
13	Santee	15.7	2087	69.0	15.4	1540	0.98
14	Pee Dee	16.3	2103	69.5	15.2	1556	0.93
15	Potomac	13.7	2117	70.0	14.0	1569	1.02
16	Altamaha	14.2	2132	70.5	12.1	1582	0.85
17	Savannah	10.6	2142	70.8	12.0	1593	1.13
18	James	10.0	2152	71.2	10.7	1604	1.07
19	Kennebec	5.7	2158	71.3	9.9	1614	1.74
20	Pascagoula	6.6	2164	71.6	9.6	1624	1.45
21	Roanoke	9.6	2174	71.9	9.6	1633	1.00
22	Cape Fear	9.1	2183	72.2	9.5	1643	1.04
23	Pearl	6.6	2190	72.4	8.9	1652	1.35
24	Sabine	9.3	2199	72.7	8.7	1660	0.93
25	St. Johns	8.7	2208	73.0	8.4	1669	0.96
26	Merrimac	4.8	2212	73.1	7.9	1677	1.64
27	Umpqua	3.7	2216	73.3	7.4	1684	2.00
28	Brazos	44.0	2260	74.7	7.4	1691	0.17
29	Trinity	17.2	2278	75.3	7.3	1699	0.42
30	Eel	3.1	2280	75.4	7.0	1706	2.25

*1 cfs/sq mi = 13.57 in/yr

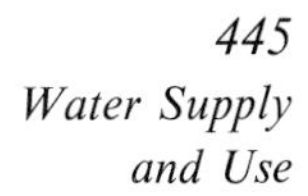

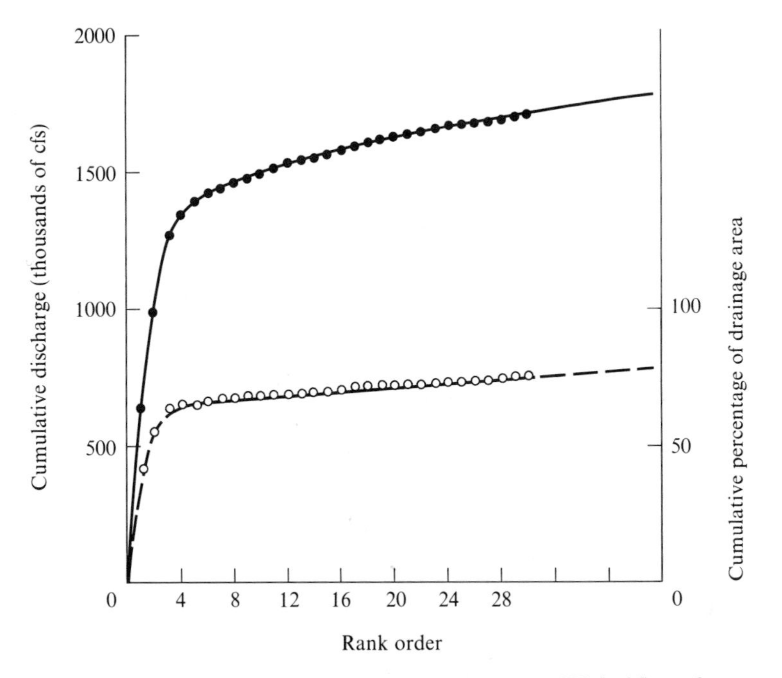

Figure 12-1 Cumulative discharge and percent of area of United States for rivers flowing to the ocean.

This independent estimate is close to the figure quoted earlier of 9 inches of runoff as the average for the country. The 2000 $\times$ 10^3 cfs expressed as inches per year is 8.98 inches from 3023 $\times$ 10^3 square miles. There are large geographical variations of runoff, however, as indicated by Figure 12-2.

For a particular site the average annual discharge of a river varies with the size of the drainage basin if the climate is approximately uniform through the catchment. For a homogeneous hydrologic region, therefore, the average annual discharge at ungauged sites can be estimated by plotting on double logarithmic paper the average annual discharge versus the drainage area at a few gauging stations. Figure 12-3 shows such a graph for several regions. For rivers not subject to large artificial withdrawals or additions, there is usually a linear relationship in such diagrams, as would be expected. The discharge from a specified area, indicated by the vertical position of each line on the graph, depends on the annual budget of precipitation and evapotranspiration. The streamflow data can be obtained from publications of the U.S. Geological Survey or similar agencies, as described at the beginning of Chapter 10.

The long-term average annual runoff, so far considered, is little more than an index of water supply. Streamflow varies from year to year and over much shorter periods as well, and these fluctuations may be major factors determining the value of the resource. To study the variability of annual

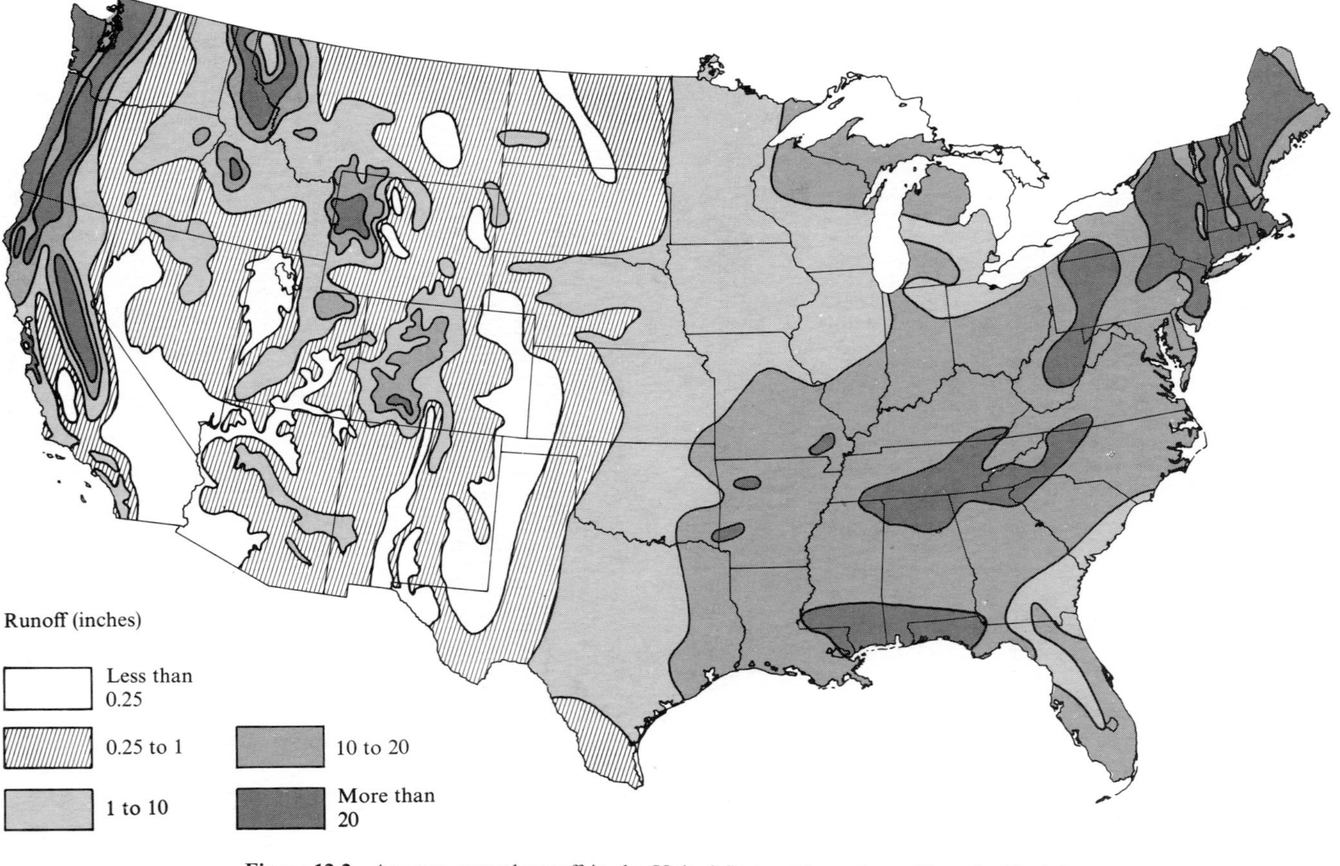

Figure 12-2 Average annual runoff in the United States. (From Leopold et al., *Fluvial Processes in Geomorphology,* W. H. Freeman and Company. Copyright © 1964 by W. H. Freeman and Company.)

amounts, the techniques of frequency analysis introduced for precipitation totals in Chapter 2 are used. Annual discharge is usually fitted quite closely by a normal distribution, and Figure 12-4 was compiled for annual runoff in the Colorado River with the aid of Equation 2-3. The discussion of Figure 2-9 is appropriate to Figure 12-4, which can be used to read, for example, that in 10 percent of all years the Colorado River at Lee's Ferry, Arizona, will experience an annual streamflow volume of less than 9.5 million acre-feet.

So far the discussion has referred only to long-term averages. Examination of Figure 12-5, however, indicates another important property of annual streamflow as well as of many other hydrologic variables. The annual discharge of the Colorado River at Lee's Ferry, Arizona, has shown major long-term fluctuations since the early part of this century. Runs of years with abundant runoff have been followed by decades with more limited supplies. The average for the 35-year period 1896–1930 was approximately 17 million acre-feet, whereas the mean annual flow for the 35-year period 1931–1965 was only about 13 million acre-feet. The legal agreement distributing

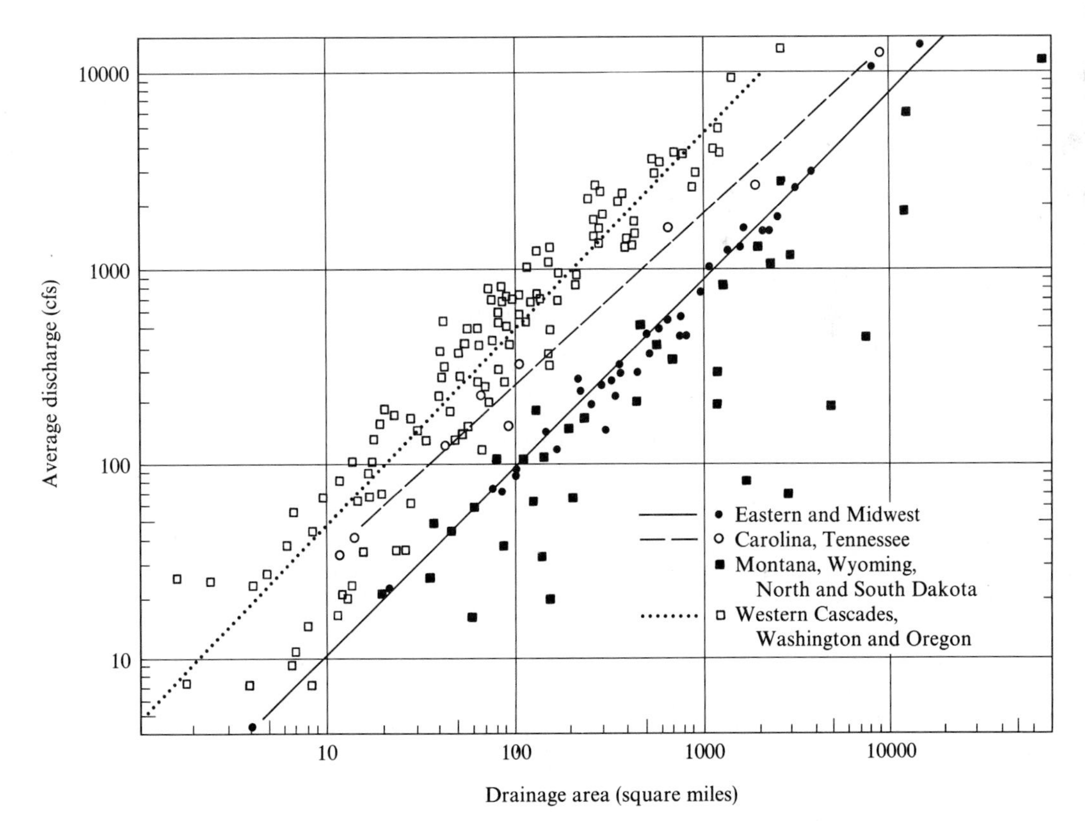

Figure 12-3 Average annual discharge as a function of drainage area for four regions in the United States.

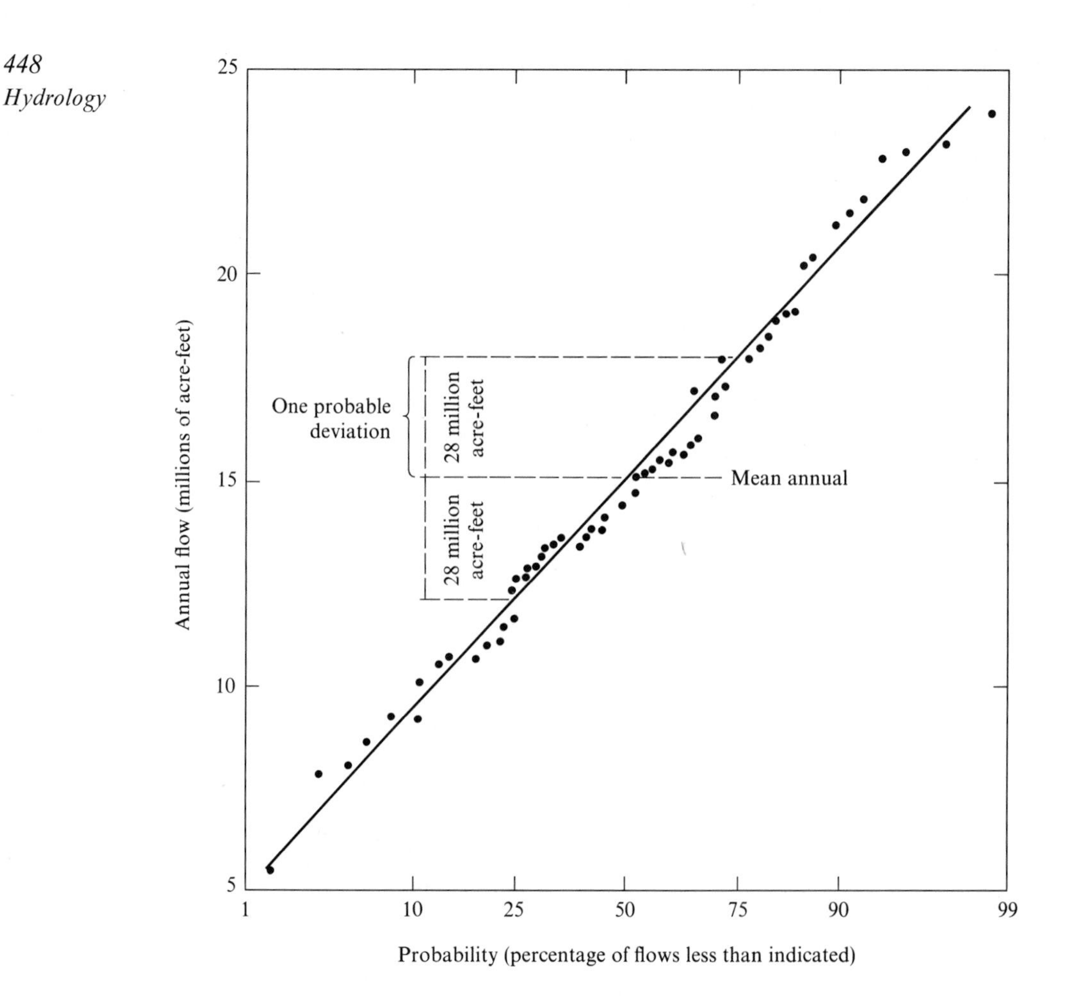

Figure 12-4 Cumulative distribution curve of annual runoff values, Colorado River at Lee's Ferry, 1896–1956. One probable deviation is an interval on either side of the mean within which 50 percent of all values should lie. The probable deviation is equal to two-thirds of the standard deviation. (From Leopold 1959.)

water rights among the riparian states was drawn up in 1922, when the 10-year moving average in Figure 12-5 was at its highest point. The legal document apportioned more water than has been available on a sustained basis at any time since.

This phenomenon by which runs of wet and dry years alternate is known as *persistence,* an effect first described by Hurst (1950) from his study of the 1000-year record of stage on the Nile. The persistence effect increases the overall variability of the supply and compounds the problem of prediction from a short record. The issue is discussed in relation to the problem of water supply by Leopold (1959). There is much long-term variability of streamflow in most rivers of the United States, with large and nonsimultaneous fluctuations. Many rivers experienced a decline some time between

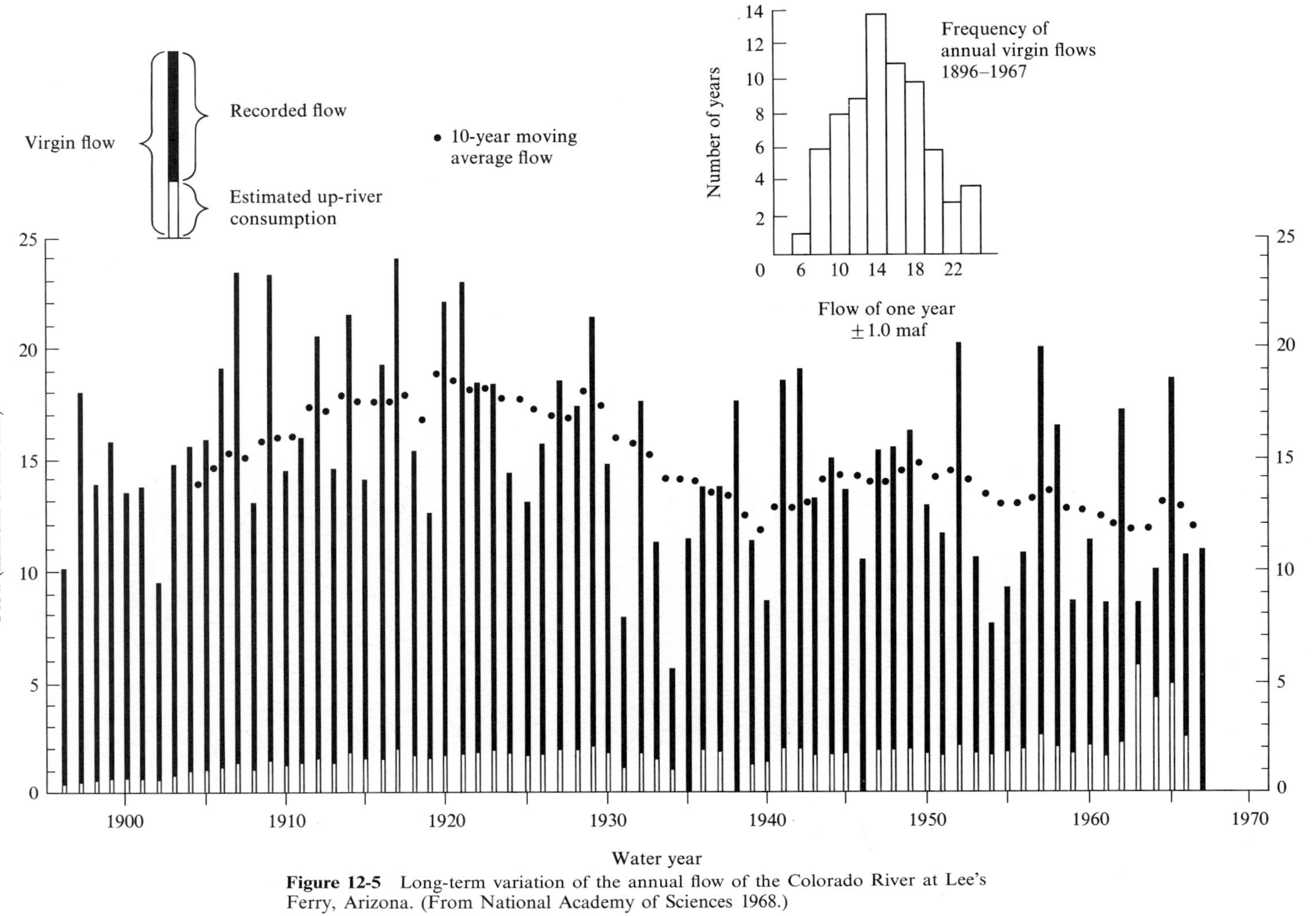

Figure 12-5 Long-term variation of the annual flow of the Colorado River at Lee's Ferry, Arizona. (From National Academy of Sciences 1968.)

1920 and 1940, and a secondary decline during the 1950's. In other countries similar fluctuations have been recognized. These variations can mean the difference between success and failure of vast undertakings in water-resource and land development, or can cause problems at the level of individual ranches and small irrigation schemes. Planners should be aware of the persistence effect, and should examine records and historical evidence in a region of interest. At present, we have no significant ability to predict these fluctuations, which do not appear to follow a regular cyclic pattern.

Timing of Runoff

The seasonal variation of flow can be at least as constraining as the total annual basin yield. A common way of expressing the seasonal regime of a river at a station is by means of an annual probability hydrograph like the one shown in Figure 12-6. Such a diagram, which is a component of most reports on water supply, is constructed in the same way as Figure 2-11. Runoff volumes for each week or month of all the years of record are subjected to a probability analysis, and discharge values exceeded in 50 percent, 75 percent, 90 percent, etc., of the years are read from the fitted cumulative frequency curves and plotted on Figure 12-6. The diagram allows the planner to see immediately the months of most severe water limitations or the season of most dependable high water. Such a diagram should be consulted before making plans for navigation, waste disposal, irrigation, or municipal supply. It can also be used for reviewing the effects of artificial

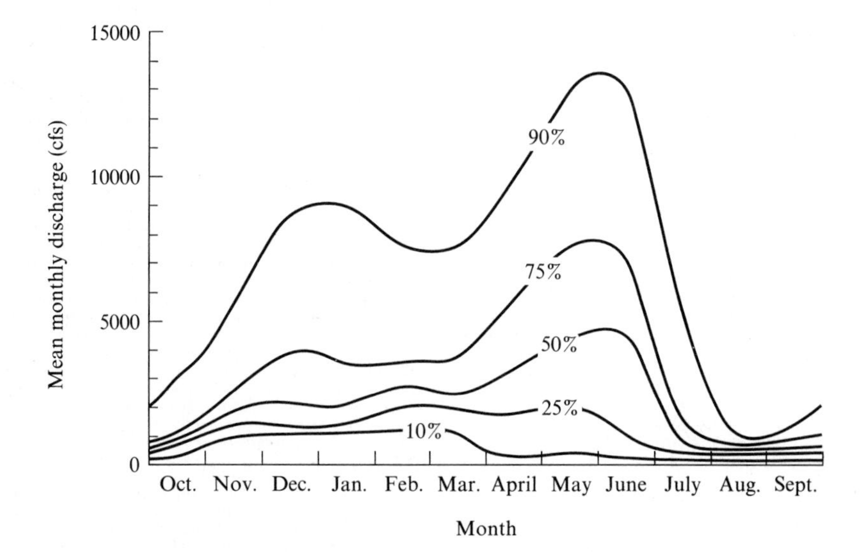

Figure 12-6 Monthly probability that the mean discharge will be less than the values indicated on the ordinate, Yakima River near Parker, Washington. (Data from U.S. Army Corps of Engineers.)

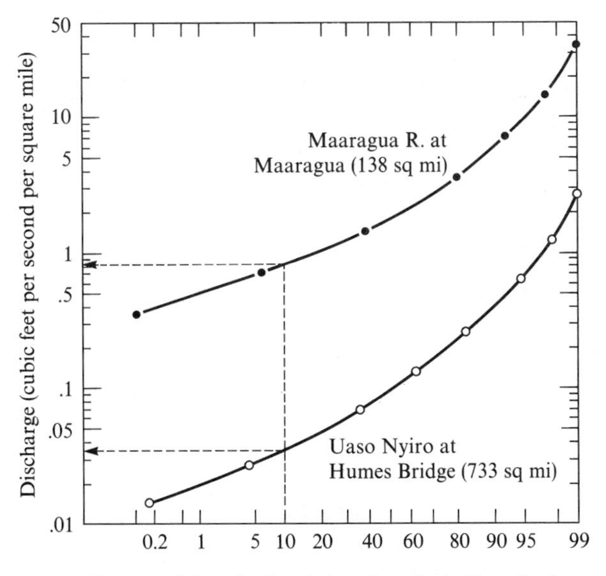

Figure 12-7 Flow duration curves for the River Maaragua in humid, central Kenya (mean annual rainfall 60 inches) and for the Uaso (River) Nyiro in semi-arid, north-central Kenya (mean annual rainfall 35 inches). The dashed lines indicate the flow values below which discharge declines for 10 percent of the time. The curves were constructed from records for the period 1956–1970.

impoundments on the water supply. The form of the annual probability hydrograph will obviously vary with the seasonality and dependability of rainfall and snowfall, with the number and sizes of natural and artificial lakes in the catchment, and to a lesser extent with the size, specific yield, and permeability of aquifers, which can store groundwater and release it slowly to supply streamflow during dry seasons.

To estimate the frequency of flows at a station, a *flow duration curve* may also be used, as illustrated in Figure 12-7, which shows the proportion of the time that discharge is less than various values for two rivers draining regions of Kenya with widely differing climates. As in other frequency analyses, the cumulative distribution of days with less than various amounts of flow is plotted. The resulting curves indicate the duration of flows less than some critical value required for fish survival, waste disposal, or municipal supply. They do not necessarily indicate that in Figure 12-7, for example, the flow of the Maaragua River falls below 113 cfs (0.82 cfs/sq mi from 138 sq mi) on 36 days of every year. It can only be said that 10 percent of all days during the period of record had discharges below that value. These days might have all occurred during one 5-year period. The annual hydrograph should be examined to find out how the sequences of low flows occur. Long

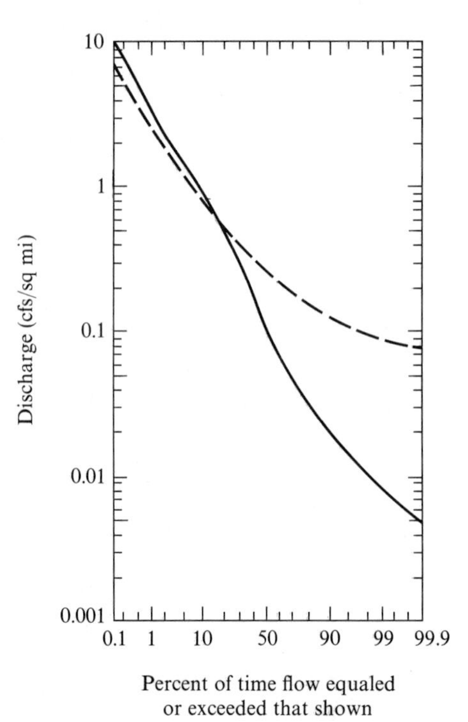

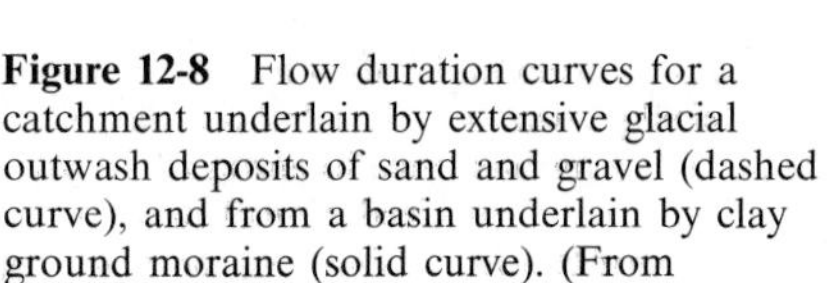

Figure 12-8 Flow duration curves for a catchment underlain by extensive glacial outwash deposits of sand and gravel (dashed curve), and from a basin underlain by clay ground moraine (solid curve). (From Schneider 1957.)

runs of relatively dry years can be much more damaging to water-resource developments than the regular occurrence of a few days with extreme low flows in each year.

In addition to the obvious effects of the size of the drainage basin, flow duration curves reflect differences in the amount and variability of rainfall excess over evapotranspiration. In cold regions the snowmelt season, another climatic factor, is important. Geology is another important regulator of runoff timing and is, therefore, reflected in the flow duration curve. Figure 12-8 shows that in a basin underlain by dense, clayey glacial till, the dry-weather flows are much lower than those from a catchment on sandy and gravelly outwash deposits. In general, rivers draining rocks with good aquifer characteristics have larger dry-weather flows. Rivers draining glaciers and permanent snowfields have low flows like those from a large ground-water supply.

Land use and water regulation alter the flow duration curve in a variety of ways. At the end of Chapter 5, we described how alterations of the vegetation cover can cause increases or decreases of runoff. These effects are most marked during low-flow periods, and several of the studies referred to earlier show that deforestation can increase summertime low flows tenfold. Rivers that are impounded in water-supply reservoirs have their flow duration curves flattened as the discharges after regulation are confined to a narrower range, with the very largest and smallest flows being avoided. Streams receiving irrigation return flow or municipal sewage effluent originally drawn from another drainage basin experience similar large additions to their summer low flows. In extreme cases, massive increases of discharge in small streams can cause serious bank erosion problems, as described by Maddock (1960).

Low-Flow Prediction

For some design purposes extreme low flows must be analyzed, and for this, one uses the techniques described in Chapter 10 for statistical analysis and prediction of floods. An *annual-minimum series* is constructed from the lowest flow of each year and is plotted on Gumbel or other probability paper with the aid of Equation 2-4 (see Figure 12-9). The resulting graph indicates the recurrence interval for various minimum discharges; the 10-year daily minimum in Figure 12-9, for example, is about 18 cfs.

For some purposes, such as fish survival or recreation, the planner may be concerned not with the lowest single day of flow in each year but with average discharges over the driest week or month of the year. Average flows for periods of various lengths can also be subjected to probability analysis. They are ranked and plotted on probability paper in the same way

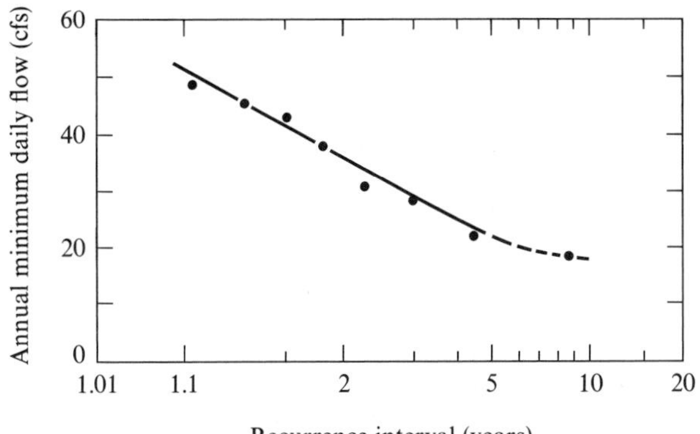

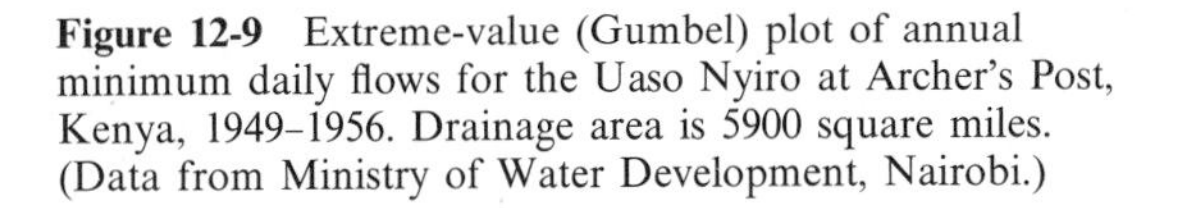
Figure 12-9 Extreme-value (Gumbel) plot of annual minimum daily flows for the Uaso Nyiro at Archer's Post, Kenya, 1949–1956. Drainage area is 5900 square miles. (Data from Ministry of Water Development, Nairobi.)

as annual daily extremes. A curve can be sketched through the plotted points for each duration, as shown in Figure 12-10, which can then be used to ascertain that, for example, the 5-year, 7-day minimum flow in the Pago River averages 0.27 cfs, or that the 10-year, 30-day minimum averages 0.155 cfs. These figures, which should be interpreted in the same manner as the recurrence intervals of floods, are often used as the basis for setting effluent standards for various wastes.

Finally, low flows can be predicted over the short term from water-balance calculations as described in Chapter 8. If a long weather record is used to calculate past runoff by this method, a statistical analysis of the predicted discharge by the methods described in this chapter can be a useful long-term prediction tool for planning.

Once the planner has obtained values of certain important low discharges (such as the 50-year minimum daily flow, or the 10-year, 7-day minimum) from a few gauging stations, the data can be regionalized and used to predict similar dry-weather flows at ungauged sites. The discharge values can be plotted against the drainage area, as shown in Figure 12-11, or the drainage

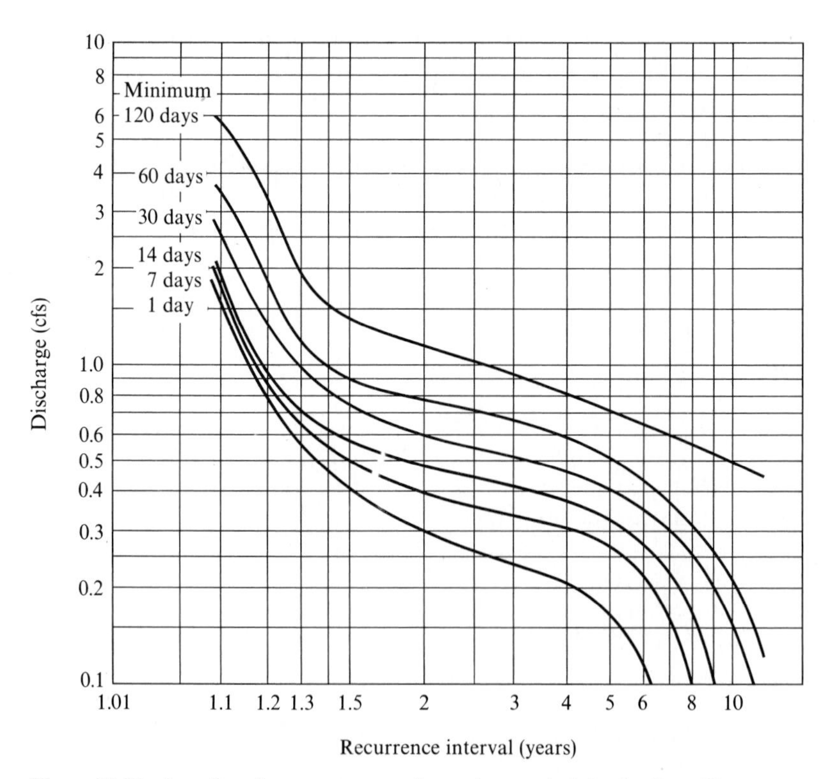

Figure 12-10 Low-flow frequency curves for various periods in the Pago River; based on analysis of data for period of October 1, 1952, to September 30, 1962. (From Ward et al. 1965.)

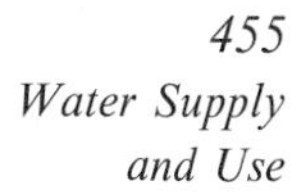

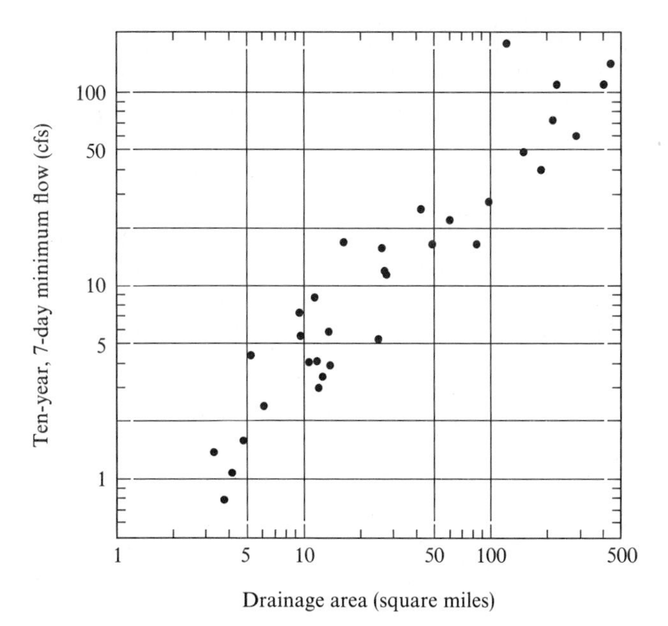

Figure 12-11 Relation of 10-year, 7-day minimum flow to drainage area for the Green and the Cedar River basins, western Washington. (Data from Hidaka 1973.)

area and some index of precipitation such as mean annual rainfall or average dry-season rainfall, can be incorporated into a multiple regression analysis to develop predictive equations for low flows of various recurrence intervals and durations.

Flow Regulation

At some sites a water supply is obtained directly from a stream. Such a supply is termed "run-of-river." If the stream is large relative to the water need, the supply is assured with no risk of shortage. In the more common case there is some possibility that the river flow will be insufficient at some time, and the chance of such an occurrence must be evaluated by the techniques outlined above and must be carefully weighed before a decision is made about waste disposal schedules, water supply, and related matters. If, for example, a coffee factory at Maaragua needs a flow of 200 cfs for processing water and into which its wastes can be discharged without drastically lowering the dissolved oxygen content of the stream, Figure 12-7 could be used to indicate the frequency of flows less than this value (1.45 cfs/sq mi). In this particular case the supply would be inadequate 40 percent of the time.

If the dry-weather flow is much smaller than the required supply, a reservoir must be built to store water during brief periods of high flow, for release during the season of lowest flow or heaviest demand. The storage volume of the reservoir should be large enough to hold and then release a volume of water equal to the maximum accumulated difference between supply (unregulated river discharge) and demand during the dry season. Because of inter-annual variations in this deficit, however, the problem of design reservoir storage is quite complicated, and again must be couched in terms of chance. The techniques used involve an analysis of the *mass curve* of stream discharge during one critical year, during several dry years, or for the entire measured or calculated record of runoff. Water-supply failures often occur during a run of dry years, so the second of these options is commonly employed.

Suppose we have isolated a period of several years with critically low runoff at a site. The stream discharge (inflow to the proposed reservoir) is plotted cumulatively, as shown in Figure 12-12. For most problems monthly

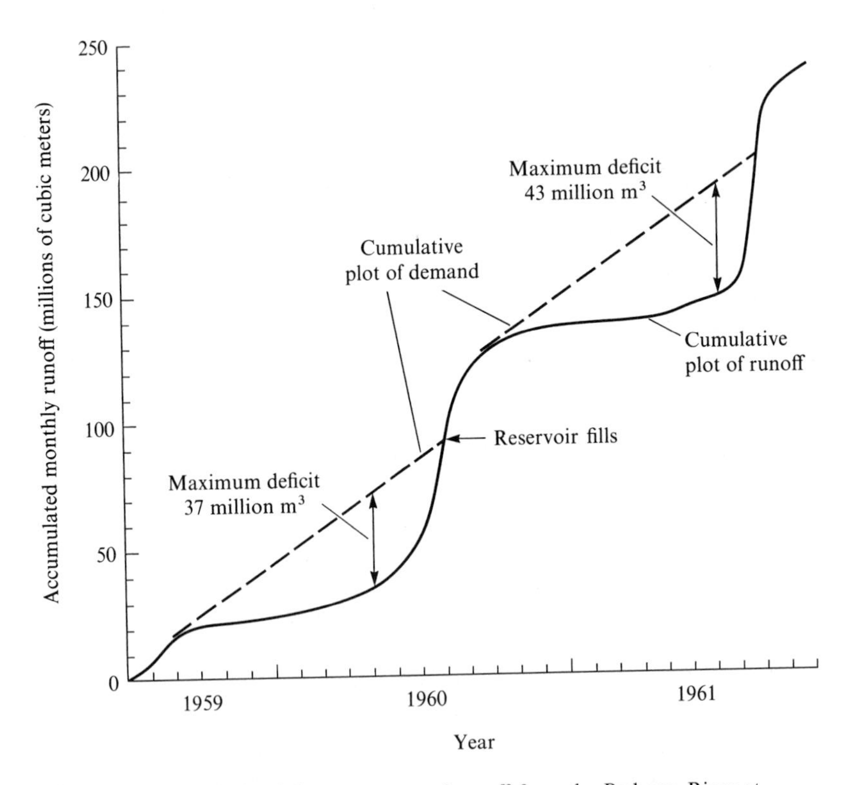

Figure 12-12 Analysis of the mass curve of runoff from the Perkerra River at Marigat, Kenya, for 2 years of low flow. A hypothetical regional water development scheme would involve diverting a constant supply of 80 million cubic meters per year out of the river for irrigation, transmission losses, industrial uses, and municipal supply. Reservoir storage volumes of 37 and 43 million cubic meters would have been necessary to maintain the demand in the 2 years considered. Drainage area is 1310 square kilometers.

runoff values are used, though for reservoirs with a small storage volume, daily values are preferable. If other small streams enter the reservoir, their measured or calculated discharge should be added and the effects of rainfall onto the lake surface, of evaporation, and of groundwater into and out of the reservoir should be estimated if they are likely to be important.

Starting from some time at which the demand rate first exceeds the rate of supply, a mass curve of demand is also drawn. The demand may vary with time, but for purposes of illustration, we will consider a uniform rate of demand. A straight line with a slope equal to the average demand is thus drawn tangent to the supply curve near the end of periods with large inflows, where supply first declines below the demand. The vertical distance between the supply and demand at any time represents the accumulated deficit. In order to ensure that the demand is met, the maximum deficit is the storage capacity necessary to maintain the required supply. Referring to Figure 12-12, if the reservoir was filled during the years of ample runoff before September 1959, water would be released from storage beginning that month, and releases would continue until July 1960 when again the reservoir would be filled to capacity. It would remain full until October 1960, from which date it would again release storage water until August 1961. While it was full, water would pass over the dam without changing the volume in storage.

If a long runoff record is analyzed, a value of storage needed to meet some specified demand can be obtained for each year. The annual storage values can be plotted on arithmetic probability paper, as shown in Figure 12-13.

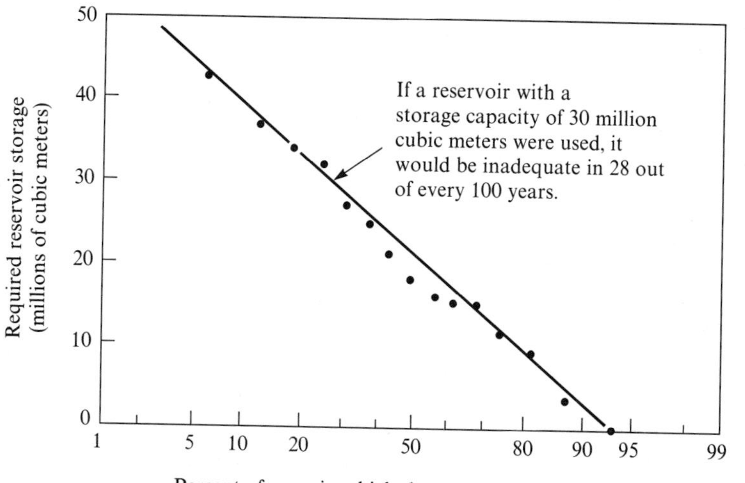

Figure 12-13 Hypothetical probability analysis of annual storage requirements for the Perkerra River at Marigat with a demand of 80 million cubic meters per year.

The cumulative frequency curve can then be used to examine the probability that the supply will fail if a certain volume of storage is provided. The amount of storage chosen for the provision of a dependable water supply depends on the economic and social impact of a deficiency. Other analyses of storage requirements for a water supply are reviewed by Riggs and Hardison (1973).

Design of Small Water-Supply Systems

In the design of a water-supply system, then, the sequence of computation begins with an estimate of the need, followed by an analysis of the water available as a function of probability. A choice of the acceptable risk is then made, and a reservoir of such a size is chosen to assure the supply for the chosen probability.

It must be recognized that the sizes of the pipes for distributing water within a city are chosen by different criteria than the size of the storage reservoir. The pipe sizes are determined not by the need merely for the home demand at the peak hour, but to supply the larger demand of fire hydrant withdrawal at a time of a big fire.

But reservoir capacity is determined by the withdrawal demand during a long dry period, usually of several weeks. The demand is ordinarily computed by use of the following average values. The usual home use is computed to be about 100 gallons per person per day, including water used for lawn sprinkling. But because most cities also sell water to industrial firms, the city water need is often computed on the basis of about 200 gal/person/day.

To judge the relative amount of water used for domestic purposes within the house compared to that used for lawn sprinkling, average figures for the Leopold household in Berkeley, California, are illuminating. In this area the rainy season is in winter and practically no rain falls in summer. The average water used during February–March is 56 gal/person/day, but in June–July it is 191 gal/person/day. The average yearly use is 109 gal/person/day. So for this family, garden sprinkling accounts for nearly three times the amount actually used in the house during the dry months. In the annual average, the garden use is nearly equal to the use within the house. The cost in Berkeley is \$0.46 per thousand gallons, or the cost per person per year is \$18.00. In the Dunne household in Seattle, Washington, where gardens require less watering, the annual use is 93 gal/person/day, the cost is \$0.35 per thousand gallons, and the annual cost is \$12.00 per person per year. During the drought in 1977, the Leopold family reduced water use by simple conservation methods to 35 gal/person/day including garden use.

In a city of 10,000 population, then, the water demand would be 2 million gallons per day, or 266,000 cubic feet, or 6 acre-feet per day. This is equivalent to a flow of 3 cfs.

If the water supply were obtained from a small river, one could compute the size of the drainage basin necessary to supply the required water. Usually, it is unwise to draw all the water out of any river because the maintenance of fish and river amenities requires that a river never be completely dried up. Assume that, even in periods of low flow, no more than half the water is to be withdrawn. Then it can be reasoned that during the driest month in summer, once in 10 years (probability .10) the flow must equal or exceed 6 cfs. A duration curve would be drawn to express the probability of 30-day average flows. A typical result would be that the 30-day, low-flow average might be a tenth of the average flow, though this ratio is highly variable and depends on the size of the river and the region. But if the ratio were 0.1, then the mean flow would have to be 10×6, or 60 cfs, and to produce this mean flow might require about 50 sq mi of basin area in the particular region under study. Thus, without storage, if the run-of-the-river diversion were to supply the demand of 3 cfs during a 30-day period of low flow, a sizable basin area would be required.

The crude computation at least indicates some of the considerations that go into a computation of water supply. But there are others as well. In the western states a permit for withdrawal of water is required in order to be assured that the downstream rights of other users are protected. The state engineer in most western states must review the application and issue the permit. Water quality is a consideration as well as quantity. Though water treatment will reduce the content of pathogens to acceptable limits, dissolved substances are not removed by treatment, and so the content of heavy metals, nitrates, total dissolved solids, and other constituents must be measured to assure that public health standards of quality are met.

Because small streams have large fluctuations in discharge and may have low-flow discharges much less than the average flow, storage is usually required even to supply towns of moderate size. This is one reason that groundwater sources are used to supply about three-fourths of the municipal supplies for cities in the United States.

Planners and environmental analysts would be well advised to know the general principles behind the hydrologic computations of a municipal water-supply design, but it is unlikely that they would be called upon to make the actual computation. The planner's problem, in practice, would ordinarily involve posing the appropriate hydrologic questions and reviewing the general assumptions behind a hydrologic design prepared by a specialist.

Bibliography

HIDAKA, F. T. (1973) Low-flow characteristics of streams in the Puget Sound region, Washington; *U.S. Geological Survey Open File Report,* Tacoma, WA.

HOWE, C. W. AND EASTER, K. W. (1972) *Interbasin transfers of water: economic issues and impacts;* John Hopkins Univ. Press, 212 pp.

HURST, H. E. (1950) Long-term storage capacity of reservoirs; *American Society of Civil Engineers, Transactions,* vol. 116, pp. 770–808.

LEOPOLD, L. B. (1959) Probability analysis applied to a water-supply problem; *U.S. Geological Survey Circular 410.*

LEOPOLD, L. B., WOLMAN, M. G. AND MILLER, J. P. (1964) *Fluvial processes in geomorphology;* W. H. Freeman and Company, San Francisco, 522 pp.

MADDOCK, T. (1960) Erosion control on Five Mile Creek, Wyoming; *International Association of Scientific Hydrology, Symposium of Helsinki, Publication No. 53,* pp. 170–181.

NACE, R. L. (1970) World hydrology: status and prospects; *International Association of Scientific Hydrology, Symposium of Reading, Publication No. 92,* pp. 1–10.

NATIONAL ACADEMY OF SCIENCES (1968) *Water and choice in the Colorado Basin: an example of alternatives in water management;* Publication 1689, Washington, DC, 107 pp.

RIGGS, H. C. (1965) Effect of land use on low flows in Rappahannock County, Virginia; *U.S. Geological Survey Professional Paper 525-C,* pp. C196–C198.

RIGGS, H. C. (1967) Some statistical tools in hydrology; *U.S. Geological Survey Techniques of Water-Resources Investigations,* book 4, chapter A1, pp. 1–39.

RIGGS, H. C. (1972) Low-flow investigations; *U.S. Geological Survey Techniques of Water-Resources Investigations,* book 4, chapter B1, pp. 1–18.

RIGGS, H. C. AND HARDISON, C. H. (1973) Storage analyses for water supply; *U.S. Geological Survey Techniques of Water-Resources Investigations,* book 4, chapter B2, pp. 1–20.

RODDA, J. C. (1965) A drought study in southeast England; *Water and Water Waste Engineering,* vol. 71, pp. 316–318.

SCHNEIDER, W. J. (1957) Relation of geology to streamflow in the upper Little Miami Basin; *The Ohio Journal of Science,* vol. 57, no. 1, pp. 11–14.

THOMAS, D. M. AND BENSON, M. A. (1970) Generalization of streamflow characteristics from drainage-basin characteristics; *U.S. Geological Survey Water Supply Paper 1975.*

THOMAS, M. P. (1966) Effect of glacial geology upon the time distribution of streamflow in eastern and southern Connecticut; *U.S. Geological Survey Professional Paper 600-B,* pp. B209–B212.

WARD, P. E., HOFFARD, S. H. AND HATHAWAY, J. C. (1965) Hydrology of Guam; *U.S. Geological Survey Professional Paper 403-H.*

12-1 Mean Annual Runoff versus Drainage Area

Obtain from your local library a copy of the publication for your state entitled *Water Resources Data for (Washington), Part 1: Surface Water Records.* In other countries similar publications or water-agency records are available. Choose a homogeneous hydrologic region, and for all the gauging stations in that region, plot the mean annual discharge versus the drainage area. Calculate the average discharge per unit area and the mean annual rate of evapotranspiration by subtracting the annual runoff from the average annual precipitation. How does the result of this calculation compare with the calculated potential evapotranspiration for the same region?

12-2 Reliability of Annual Streamflow Totals

Your interest in a stream as a water supply for a small town requires an investigation of the reliability of annual runoff totals. You want to find its mean annual runoff, the variability around this mean, the probability that the runoff in any one year will be less than some critical value, and the reliability of the mean of your sample record as an estimate of the true long-term mean discharge.

Solution

Obtain a list of the annual flows for all years of record, arrange them in order of their magnitude, and use Equation 2-3 to plot the cumulative frequency distribution, as shown in Figure 12-4.

Read the value below which the discharge will decline in 10 percent of the years.

Read the standard deviation from your graph. Sixty-seven percent of all years will have discharges within 1 standard deviation of the mean. Fifty percent of all years will have discharges within 0.68 times the standard deviation of the mean. This value is called the *probable deviation.*

The reliability of the sample mean as an estimate of the true long-term mean is expressed by the *standard error of the mean,* $S_{\bar{x}}$, where

$$S_{\bar{x}} = \frac{\text{standard deviation}}{\sqrt{\text{number of years of record}}}$$

There is a 68-percent chance that the true mean will lie within 1 standard error of the sample mean.

Annual runoff totals for individual years can be obtained from the U.S. Geological Survey publications *Water Resources Data for (Ohio).* A compliation of annual totals up to 1950 for a large number of stations in the United States can be found in Water Supply Papers 1301 through 1319.

12-3 Seasonality of Water Supply and Demand

For your local region plot a graph using average monthly values to show the monthly pattern of runoff from one river and the pattern of water demand. Write a short essay on the hydrologic reasons for the imbalances at various seasons and on the strategies employed in your region for ensuring that the demand is met.

Also obtain runoff values at the gauging station for each month of the year over a period of at least 10 years from one of the sources referred to in Typical Problem 12-2. Conduct a separate frequency analysis on the data for each month, and sketch a probability graph like Figure 12-4. The curve through the points will not usually be straight for monthly data. Read from the curve the values below which the montly discharge will fall in the lowest 10, 25, and 50 percent of all years, and plot an annual probability hydrograph like Figure 12-6.

12-4 Flow Duration Curves

From the U.S. Geological Survey publication *Water Resources Data for (Washington), Part 1: Surface Water Records* or a similar compilation, construct flow duration curves for one year at two stream-gauging stations with strikingly different climates.

Choose classes of discharge, such as < 50 cfs, 50–99.9 cfs, 100–149.9 cfs, etc. Tabulate these and count the number of daily discharge values falling into each class. Calculate the percentage of days in the year in each discharge class and then cumulate these percentage frequencies. If 36 days had discharges less than 50 cfs, and another 36 had flows between 50 and 99.9, then (36 + 36)/365, or 19.7 percent had discharges less than 100 cfs. These last two numbers are plotted against one another and the process is repeated cumulatively for each class.

Now suppose that a water development proposal calls for the construction of a water-supply reservoir and for heavy consumptive use of the stream for irrigation, but promises to maintain the discharge above 0.01 cubic feet per second per square mile (csm). For comparison, read from the flow duration curve the frequency with which the stream in its natural state falls below this level.

In a class discussion, review what other kinds of questions concerned with water you would ask the proponents of the scheme if you were asked to review the plans.

12-5 Low-Flow Analysis

Obtain a list of annual minimum daily flows at five gauging stations in your region. The data can be obtained from annual volumes of the U.S. Geological Survey publication *Water Resources Data for (Alabama), Part 1: Surface Water Records,* and their predecessors in the U.S. Geological Survey Water Supply Paper series. The stations chosen should monitor a wide range of drainage areas.

Make a frequency analysis of the annual-minimum series for each station, and plot a frequency curve on Gumbel paper. Read from each curve the 5-year, 10-year, and 25-year annual minimum flows, and plot them against drainage area. Join the points with smooth curves which (with the addition of a few more points) could be used for estimating this one aspect of water supply for ungauged basins in the region.

12-6 Water Supply for Suburban Developments

A developer proposes to build a subdivision consisting of 50 single-family homes at the edge of a growing community. What problems of water supply for these homes are involved? Assume that the development is in a western state, New Mexico, where the annual precipitation is 14 inches, so that irrigation or lawn water will be required during part of the year.

What is the law regarding water development and use? The applicable requirements are administered by the state engineer who must review any petition for water use and determine the appropriate priority under the law. The priority for use is determined by priority in time, that is, an earlier user has priority over a new user.

In this case the water would be supplied by the city because the planned development is within the city limits, and the city has already received a permit from the state engineer. The developer, then, must arrange with the city to provide water, and the city engineer must approve the developer's plan for water distribution to the new homes.

Are there restrictions on water use applicable to the people who buy the homes in the development? The city engineer has the power to restrict use for lawn irrigation if necessary. There is seldom any restriction placed on use within the house.

Assume the development discussed in Typical Problem 12-6 was outside the city limits.

This places the developer in a quite different position, for the city is not bound to furnish water to the new development. An application must be made to the state engineer for a permit to develop an independent water-supply system. Granting of the permit depends on several things. The first is whether the planned development is in a "closed basin," that is, a basin declared by the state engineer closed to further water development. He so declares where he has found that the total developable water in the basin has already been equaled by use. In such basins a present use of water must be eliminated before additional use elsewhere is begun. Elimination of a present use means purchase of a water right somewhere in the basin and cessation of that use.

Further, the source to be developed must be specified whether ground or surface, and the general engineering plans for the water development must be submitted to the state engineer for approval. If a dam is to be built, its location, size, spillway specifications, and method of construction must be furnished. If groundwater is to be developed, the number of wells, location, depth, depth of perforation, and well diameter must be furnished.

In this problem where a new source of water is to be developed, the planner should seek the professional assistance of a registered engineer to draw up the specifications and prepare the design. It is necessary to make preliminary tests of water quality both for bacteriological and physical characteristics. For groundwater one or more test wells may be required and pumping tests made to determine the quantity produced for different amounts of drawdown.

Water treatment would be required for a supply to serve 50 families. The treatment may be as simple as chlorination, but depending on the source and type of storage, a more complicated treatment may be required. The State Department of Health usually sets the standards to be met by water-supply developments.

Maintenance is one of the most commonly overlooked aspects of such a water-supply system. Whether a ground or surface source is used, there is a continuing need for maintenance of dams, spillways, pumps, valves, pipes, or treatment facilities. Most small independent sources experience greater difficulty with maintenance than with failure of the source to supply the necessary quantity and quality. In a city this service is provided by professional people under the city engineer or by the municipal utility district, but maintenance can be a headache for a group of homeowners without such an organization. Pump maintenance in groundwater systems, spillway and inlet maintenance in surface reservoirs, and water treatment facilities in both are the features most in need of constant monitoring and maintenance.

12-8 Water Supply for a Household

An individual landowner plans to build a house on a property of 10 acres.

Whether in eastern or western states, the first priority for water use applies to a home or domestic supply. In other words, a landowner may develop a water supply on his own land for home use. This priority does not apply to irrigation use. However, in most states he must file with the state engineer a plan for any dam, or in the case of a drilled well, must file a log of the well. In the latter case a professional well-driller will ordinarily keep a record of the materials encountered and thus provide the log needed for filing.

The usual situation is for the landowner to call in a professional well-driller and contract for a well. He pays by the foot drilled whether or not water is found, but an experienced driller can give an experienced guess on the depth to water and the most likely place for placing the well.

It is at this point in the procedure that a landowner needs geologic or hydrologic advice. Whereas an experienced driller will, in many instances, be able to furnish such advice on the basis of experience, many combinations of geology, terrain, and climate offer no obvious clues and a more professional analysis is required.

Unfortunately, there is no organized service of geologists and groundwater hydrologists to provide at modest cost the technical analysis needed. In most states the office of the state geologist would like to be in a position to offer such service, but lack of personnel and funds precludes doing so on a regular basis. Nevertheless, it is to that office that the ordinary landowner should turn for advice, even if the geologic fieldwork cannot be offered.

There are many professionals in the consulting business furnishing field mapping and office analysis, but these are usually so costly that the person needing a single well cannot afford the service. Consulting offices cater to clients who need a large enough unit of work to afford a substantial fee. The consultant cannot set up an office similar to that of a physician in which a fairly large number of clients, each paying a small fee, can support a viable business. The reason is that each case requires more work, often fieldwork, than can be done for the price of an office visit.

Lacking the professional service a landowner feels he can afford, he may be tempted to turn to a water dowser and the forked stick, an alternative not recommended here. A more rational explanation and analysis are possible; however, the problem is that the organizational infrastructure may not exist to bring available knowledge to bear on the case.

13

Snow Hydrology

Importance of Snow Hydrology

At temperatures of about 0°C or less, precipitation occurs as snow, the accumulation and melt of which are of great economic and social significance in cool temperate and subarctic regions. Few people in such regions have neutral feelings about snow. To some it represents several months of skiing, snowshoeing, and breathtaking scenery after each storm. To others, the snow season is a terrible time of shoveling, traffic hazards, and isolation, with the esthetic value of snow increasing with the third power of its distance from the observer. Planners must consider snow from both positive and negative points of view, and the particular problem with which they are concerned determines the kind of information they need about snow hydrology.

The total seasonal water content of snowfall is of great concern to water managers because the resulting meltwater fills reservoirs and groundwater systems and controls the amount of water available for irrigation, municipal use, and power generation. In northern Vermont, for example, 25–35 percent of the annual precipitation occurs as snow. It melts from late March to early May and generates about one-half of the annual runoff within a 30-day period. At this time the limited groundwater bodies of the region are recharged with meltwater, which therefore also supplies a portion of the summer runoff. During the rest of the year, 65–75 percent of the precipitation produces less than one-half of the annual runoff because it is subject to evapotranspirative losses. Snow is even more important to the water supply in mountainous regions of the western United States. A thin winter snowpack in the Cascade Mountains of Washington forces the planners for power companies and municipalities to develop contingency plans for curtailing water supply or power generation in the following summer and autumn. Unusually light winter snowfall causes unemployment and economic losses in ski areas from California to the Alps. The supply of soil water from snowmelt is vital in the vast wheat regions of interior North America and the U.S.S.R. A deep snowpack not only insulates winter wheat against low temperatures but supplies the moisture for early growth in spring. Annual totals of snow depth or of water equivalent are analyzed and predicted with the aid of the statistical tools described in Chapter 2 for liquid precipitation.

The planner is also concerned with the rates at which snow melts, for herein lies a major environmental hazard, the snowmelt flood. To understand the rates at which runoff is produced by this process, we will have to consider again the energy-balance concepts developed in Chapters 4 and 5. Most of this chapter will be concerned with methods of predicting snowmelt.

Snow Measurement

Snow can be caught in standard rain gauges, but wind eddies around the top of the gauge reduce the catch and the resulting information is suspect even if the gauge is shielded to reduce the eddying problem. *Snowfall* is expressed in terms of its *water equivalent* (i.e., mm of water). A better measure of fall is obtained by laying out a *snow board* and measuring the melted volume of snow that accumulates on the board during each storm. Often, only the depth of accumulation is measured and is converted to water equivalent by assuming a snow density of 0.10 (i.e., 1 m of snow contains 0.10 m of water equivalent). Large errors can result from this calculation, however, since the density of newly fallen snow ranges from about 0.05 to 0.20 depending on the air temperature during the storm.

Many processes affect the amount of snow that remains on the ground, and a great deal of effort is put into the measurement of *snow accumulation.* The amount of snow on the ground depends not only on snowfall but on the intensity of drifting by wind, and therefore on topographic and vegetative patterns. The amount of snow that is intercepted by the vegetation cover can vary from zero to more than 30 percent of the gross precipitation under a dense coniferous forest (see Chapter 3). Midwinter melting also reduces the snow cover in a pattern that varies with altitude, topography, and vegetation type. For these reasons measurements of snow accumulation, which

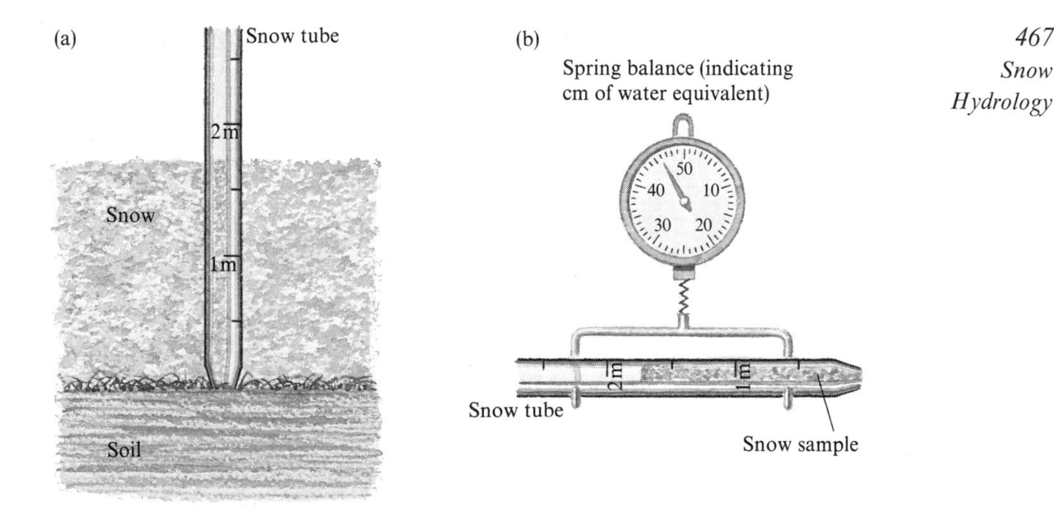

Figure 13-1 (a) Snow tube with sharp cutting edge is forced through the snowpack to the ground surface. The depth of snow is read from a scale on the outside of the tube. (b) The tube containing the snow sample is lifted out of the pack, vegetation and soil are cleaned out of the end of the tube, and it is then placed in a cradle suspended from a spring balance. The balance is graduated to display the amount of snow in the tube in terms of centimeters of water equivalent.

are of great significance for estimating the potential water yield of a catchment, must be carefully designed.

To measure snow accumulation within a river basin, a number of *snow courses* are laid out. Each course consists of perhaps ten* stations marked by stakes to which the observer can return periodically. The courses should be spread through the catchment to sample various altitude zones, vegetation types, and topographic conditions. At each measuring station a *snow sampler* is used. The sampler (see Figure 13-1) is a tube with a sharp cutting edge that can be forced down through the pack to the ground surface. The depth of the snow is measured with a scale on the outside of the tube, and a core of snow can be lifted out of the pack. The tube and snow are then weighed on a spring balance, calibrated in millimeters of water equivalent. The procedure is not quite as easy as it sounds, particularly in deep snowpacks during a long day of snowshoeing or skiing. One has to have a strong interest and a keen sense of humor.

*The number of stations per course and the number of courses necessary to define the snowpack of a basin to within acceptable levels of precision depend on the spatial variability of the pack and can only be estimated by a pilot study of the variability and a calculation of required sample sizes, which can be found in any elementary text on statistics.

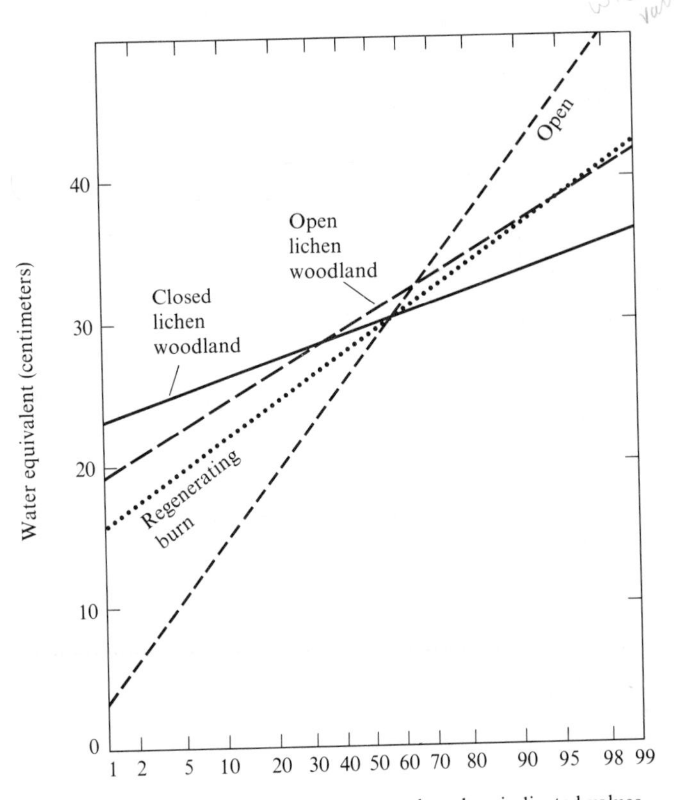

Figure 13-2 Cumulative frequency curves for normal distributions of snowpack water equivalent under four types of vegetation cover at Schefferville, Quebec. Tree canopy densities were closed lichen woodland, 25 percent; open lichen woodland, 15 percent; regenerating burn, 11 percent; open, 6 percent. (Data collected by John Fitzgibbon.)

A single snow-course survey will usually yield information of the kind shown in Figure 13-2, which illustrates the differences in the cumulative frequency distribution, mean and standard deviation between sample measurements of water equivalent under different types of vegetation cover. In this example from the Canadian subarctic, 500 measurements of the peak snow accumulation were made just before the melt season began along transects across a 35-square-kilometer drainage basin. Results from another year were almost identical. The water equivalents were normally distributed under each cover type, but the mean was lowest in open areas and greatest in the denser boreal forest. The values had a large standard deviation in the open where the snow drifted vigorously in the fierce winds of the tundra. Under such circumstances, deep drifts accumulate in hollows, and ridge

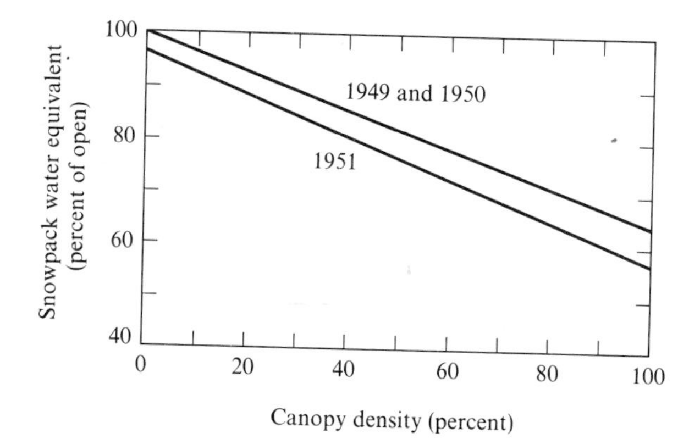

Figure 13-3 Effect of forest canopy density on the water equivalent of the snowpack during the accumulation season in the upper Columbia River basin. (From U.S. Army Corps of Engineers 1956.)

tops are bare. Under the closed lichen woodland (canopy density, 25 percent) there is much less drifting, and the water equivalents are more uniform. Small differences between the mean values result from sampling errors, different evaporation rates, and large-scale wind transport of snow from the bare tundra ridges to the intervening forested valleys. In a less extreme environment such as the western United States, results shown in Figure 13-3 are more common. Denser forest canopies intercept more snow, and so there is less available for melt in the springtime.

Repeated measurements at snow courses throughout the season allow the hydrologist to quantify the accumulation, modification, and melting of the snow. Because of the variability of snow depth, density, and water equivalent, and the difficulty of measurement, even a careful snow survey gives only an approximate figure or an index of the amount of potential runoff stored in the pack. In some basins, ground snow surveys are supplemented by aerial photography to indicate the percentage of the area that has a snow cover.

Several methods of monitoring the water content of the snowpack automatically and of transmitting the data to planning centers are now being developed. It is only possible to make such measurements at a few points in a basin, and they can only provide a rough index of the basin-wide snowpack.

Snow-course surveys are made by many agencies and are frequently published. In the United States the Soil Conservation Service, the Geological Survey, and the Forest Service collect data at snow courses, and the National Weather Service monitors snowfall at shielded rain gauges. Many hydropower and water-supply agencies also maintain snow courses. In Canada and Europe there is a similar mixture of data sources.

Evolution of the Snowpack

The density of new snow (the depth of water obtained by melting a unit depth of snow) varies from about 0.05, if the air temperature during the storm is around −10°C, to 0.20 at 0°C to +1°C. Soon after the snow falls, however, its density begins to increase. Gravitational settling, wind packing, melting, and recrystallization are the processes responsible for the change. Midwinter melts and rainstorms form ice crusts on the snow surface. The density increases from an average of 0.10 to 0.30–0.50 by springtime. For this reason, measurements of snow depth are unreliable indicators of water equivalent.

The melting and recrystallization are particularly important agents of snow metamorphosis because they prepare the pack to release water during snowmelt periods. After a cold period, all or a part of the snowpack has a temperature below 0°C. With the onset of warmer conditions, some melting occurs at the surface, and water percolates into the pack where it encounters ice below 0°C and freezes. This process has two effects. First, the recrystallizing films of water increase the size of ice crystals in the pack. Second, the freezing process liberates latent heat, which warms the snowpack. This process may continue until the temperature is 0°C throughout the pack. The snow then consists of rounded ice crystals, usually two to five millimeters in diameter, and has water-holding and transmitting properties very much like those of a coarse sand. Skiers know this as "corn snow." Percolating meltwater brings the liquid water content of the snowpack up to field capacity, which is usually between 2 and 8 percent by volume. Under these conditions of temperature, texture, and moisture, the snowpack is said to be *ripe*. Further additions of meltwater will generate runoff.

In addition to understanding what is happening under his skis, of course, the planner needs to know how much runoff will be released by the snow under various critical meteorologic conditions. For this we need to consider separately the melt and the processes by which snowmelt runoff is generated.

Snowmelt

Snowmelt is expressed in terms of the depth of water released from the pack. To melt one gram of ice at 0°C, 80 calories of heat must be transferred to the snowpack. This *latent heat of fusion* is supplied by several hydrometeorologic processes. Since water held in the pores of the snowpack is already in the liquid state, the amount of energy needed to release one gram of water from the pack is 80 × (1 − fractional water content by weight) calories. The ratio of the weight of ice to the total weight of a unit volume of snow is called the *quality of the snowpack*, and it usually lies between 0.90 and 1.00 for melting snow. The measurement of this parameter is described by the U.S. Army Corps of Engineers (1956). Errors in estimating this value are generally less than those inherent in the estimation of the various energy sources, and so we will ignore the correction for snowpack quality in the following discussion.

The Energy Budget Method

As we have described in Chapters 4 and 5 for a water body and a vegetation cover, the energy balance of a snowpack can be expressed as

$$Q_s(1 - \alpha) + Q_{lw} + Q_h + Q_e + Q_p - Q_m = Q_\theta \tag{13-1}$$

in which most of the terms are defined with Equation 4-2 and are expressed in cal/cm²/unit time. Q_p is the energy advected to the snowpack by rainfall, and Q_m is the energy used to melt snow. Minor conduction of heat from the ground below the snowpack can be ignored in the calculation of snowmelt floods, although it can contribute up to 2.0 cm/mo of baseflow runoff in midlatitudes.

Before a melting snowpack can release runoff, it must become isothermal at 0°C, as we described in the previous section. There can be no change of heat storage, therefore, and Q_θ becomes zero. This is a good approximation unless the surface of the pack cools to very low temperatures at night, when melting stops. Either Q_m or Q_θ must be zero at all times. When the snow is melting, therefore, Equation 13-1 reduces to

$$Q_m = Q_s(1 - \alpha) + Q_{lw} + Q_h + Q_e + Q_p \tag{13-2}$$

Note that we have reversed our use of signs in this equation from Equation 4-2. Energy flows *into* the snowpack are considered positive; those out of the pack are negative. Net solar radiation and the heat advected by precipitation are always positive. The other heatflows on the right of Equation 13-2 can be positive or negative depending on whether they are directed into or out of the pack. The term Q_e, for example, is negative when energy is lost from the pack through evaporation and is positive when the condensation of water vapor on the pack releases latent heat which is then available for snowmelt.

Calculating snowmelt from the energy balance depends upon evaluating the terms on the right side of Equation 13-2, and using the equation

$$M = \frac{Q_m}{80\rho} \tag{13-3}$$

where M is the rate of release of meltwater in centimeters per day, ρ is the density of water (gm/cm³), and 80 cal/g is the latent heat of fusion of ice.

Solar radiation can be measured directly or estimated by methods described in Chapter 4. It varies with latitude, season, time of day, cloud cover, and vegetation cover. Under a coniferous forest canopy, the insolation can be approximated from Figure 13-4. We have much less information under

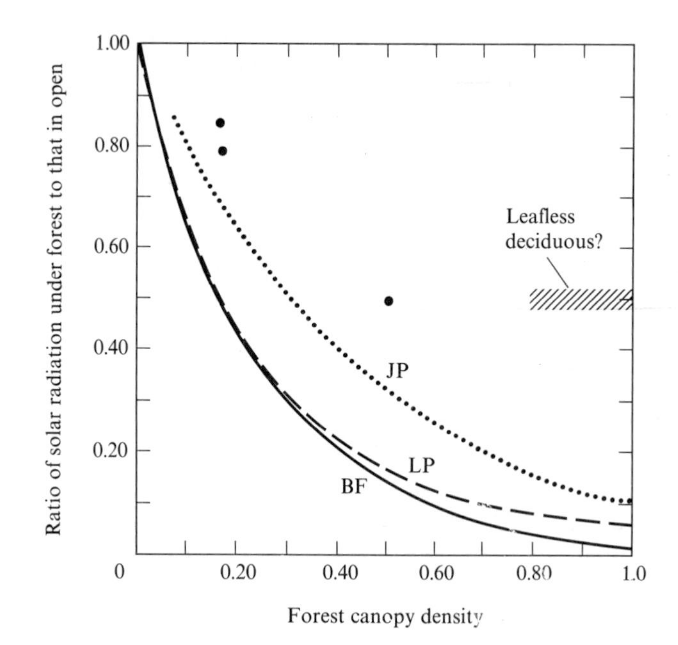

Figure 13-4 Ratio of solar radiation (Q_s) under a forest to that in the open for a range of canopy density (expressed in terms of the fractional area of land covered by the vertical projection of the crown). The canopy density can be measured quickly from an aerial photograph. Curves BF and JP are for balsam fir and jack pine from Vezina and Pech (1964); LP represents lodgepole pine (U.S. Army Corps of Engineers 1956); and the solid circles are for open boreal spruce forest (Petzold 1974).

leafless deciduous forests, but the assumption that about 50 percent of the solar radiation penetrates a full cover is probably close to reality. The gradient and orientation of the land surface also affects the amount of insolation on different hillsides, but this effect is not large when averaged over a drainage basin with a variety of hillslope shading.

The *reflectivity* (or albedo) of the snowpack, α, is high and variable. It can be measured with two pyrheliometers set up as shown in Figure 4-5(a). Figure 13-5 shows a graph that is commonly used to estimate albedo. These curves often underestimate the reflectivity during the melt season in Vermont and Quebec, however. Values of 0.70–0.80 can persist for a few weeks over a thick, clean snowpack, and only in the late states of the melt, when vegetation debris and dust are becoming concentrated on the surface of the pack, does the albedo fall to 0.50–0.60. Where the snow is thin and dirty, of course, the lower value can be reached very early in the melt, because the dark snow absorbs more radiation.

Net longwave radiation can be estimated from the Brunt Equation (4-11) with the signs reversed, though, as described earlier, this equation can lead to gross errors for periods as short as one day, particularly if cloud conditions are not specified. Because the snow surface temperature cannot exceed 0°C (273°K), the outgoing longwave radiation can reach a maximum of 661 cal/cm²/day. The proportion of this radiation that escapes to space will depend on the cloud cover. Longwave radiation from the atmosphere depends on the air temperature and vapor pressure. Since vapor pressures vary only slightly over a snowpack, the U.S. Army Corps of Engineers (1956) proposes the following simplification of the Brunt Equation over a melting snowpack under clear skies:

$$Q_{lw} = (0.76\sigma T_a^4 - 661)(1 - aC) \qquad (13\text{-}4)$$

where σ is the Boltzmann constant (1.19×10^{-7} cal/cm²/°K⁴/day), T_a is air temperature (°K), and a and C are defined in relation to Equation 4-11. Under the cold atmospheric conditions that prevail during snowy seasons, the net longwave radiation is usually negative even on melt days. The loss is often large by comparison with the other components in Equation 13-2, although it is reduced by clouds.

A forest cover has a similar effect upon the longwave radiation balance, which can be calculated from

$$Q_{lwf} = F(\sigma T_f^4 - \sigma T_s^4) + (1 - F)(0.76\sigma T_a^4 - \sigma T_s^4)(1 - aC) \qquad (13\text{-}5)$$

where F is the fractional forest cover (as defined in the caption of Figure 13-4) and T_f is the temperature of the trees (usually taken as the air temperature). In Figure 13-6 we have calculated some values with the aid of this equation to show the general nature of the forest effect. On a cold day the net longwave radiation is strongly negative in the open, but the loss is reduced as the forest canopy thickens (Figure 13-6(a)). Net radiation is negative, however, throughout most of the range of cover density. With an air temperature of 10°C, the net longwave radiation is slightly negative

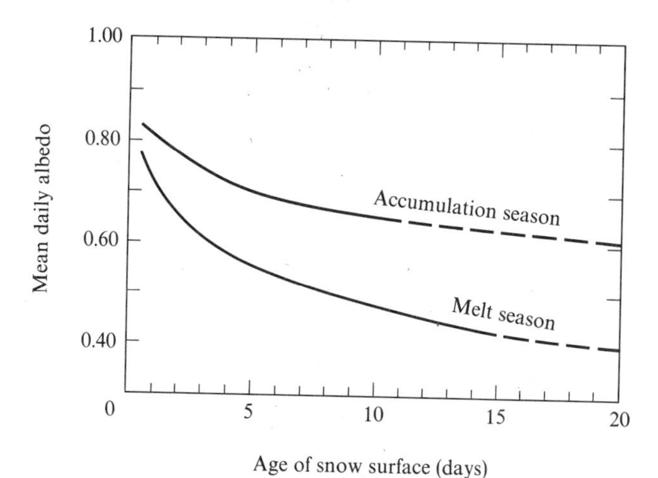

Figure 13-5 Variation of the mean daily albedo of a snowpack with time. See text for qualification. (From U.S. Army Corps of Engineers 1956.)

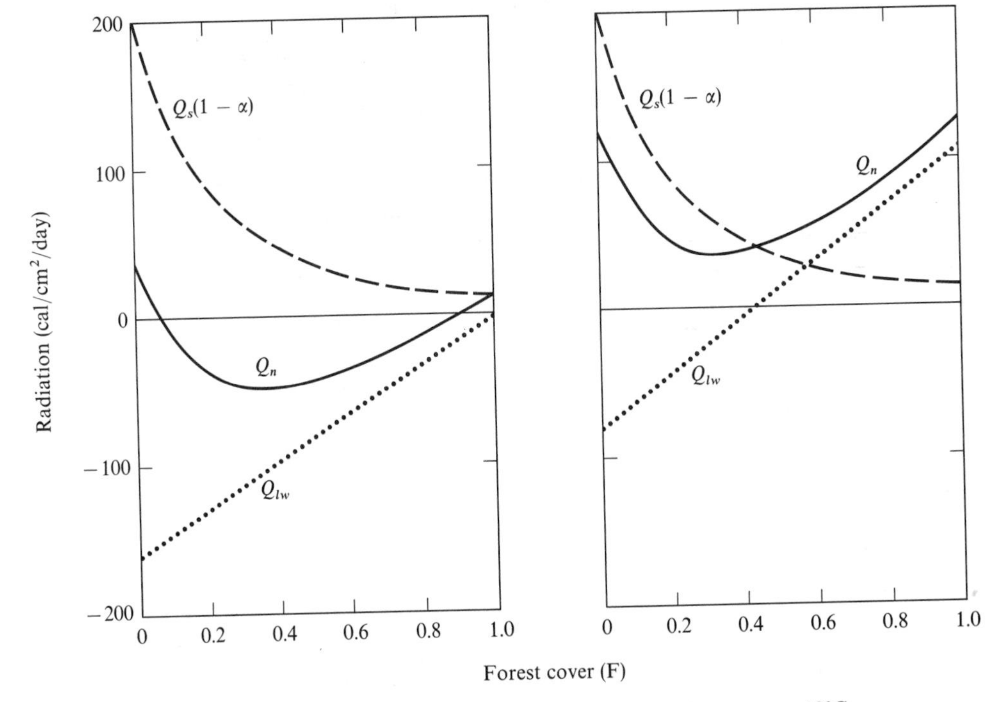

Figure 13-6 The radiation balance under coniferous forest covers of varying density. The calculations were made using a solar radiation of 500 cal/cm²/day, an albedo of 0.60, Figure 13-4, and Equation 13-5 for a clear sky.

under the thin canopies and positive under the denser cover, and net radiation is positive throughout the range of the data. The net radiation available for melting is lowest under thin vegetation canopies, rather than under a dense forest as might be first thought. As another example, Table 13-1 presents radiation budgets under a clear sky at noon during the snowmelt season at Durham, New Hampshire.

Table 13-1 Radiation budgets at noon under a clear sky during the spring snowmelt season at Durham, New Hampshire. Units are cal/cm²/min. (From Federer and Leonard 1971.)

COVER TYPE	SOLAR RADIATION REACHING THE SNOW SURFACE	$Q_s(1-\alpha)$	NET LONGWAVE RADIATION	NET ALLWAVE RADIATION
Open	1.00	0.50	−0.06	0.44
Leafless hardwoods	0.55	0.28	0	0.28
Conifers	0.10	0.05	+0.06	0.11

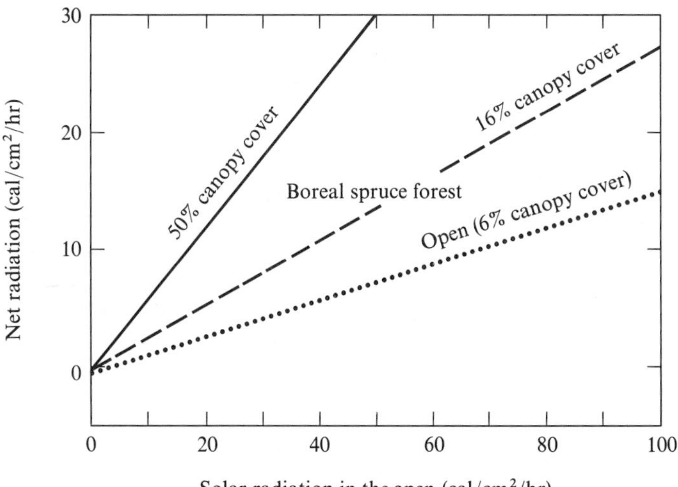

Figure 13-7 Relation of net and solar radiation over snow for an open spruce forest of two densities (16- and 50-percent canopy cover) and for open areas during the spring snowmelt season at Schefferville, Quebec. (From Petzold 1974.)

Net allwave radiation can also be estimated directly from solar radiation as shown in Figure 13-7. The relationship varies with the type of vegetation and the air temperature, but it is sufficiently consistent for use in snowmelt calculations if defined for the melt season.

In summary, it is not easy to estimate the radiation balance of a snowpack very precisely without expensive instrumentation. Anthony Price of the University of Toronto and Dunne have used the expressions given above in Vermont and in the Canadian subarctic. On most days the expressions gave reasonable results, but then occasionally the calculated net longwave radiation varied from measured values by very large amounts. In the next section we will introduce simplified expressions for the radiation balance, which were used with some success. We will continue for the moment our discussion of the physics of snowmelt.

Sensible and latent heat fluxes into the snow pack occur when turbulent eddies in the windstream carry small parcels of warm or moist air down to the surface. Cooler, slow moving parcels of air are lifted away from the surface by the eddies. Turbulent transfer of sensible heat to the snowpack can be calculated from this equation:

$$Q_h = 3.84 \frac{u_a(T_a - T_s)}{\left[\ln\left(\frac{z_a}{z_0}\right)\right]^2} \qquad (13\text{-}6)$$

where the symbols are the same as used in previous chapters, the energy units are cal/cm²/day, and u_a (km/day) is measured at height z_a (m) above the snow surface. If wind measurements are only available for open areas but must be applied to forests, the measured values are usually corrected by means of this equation:

$$u_f = (1 - 0.8F)u_{\text{open}} \tag{13-7}$$

Correction Values

The roughness of the snowpack surface, z_0, is difficult to measure or estimate. For a flat snow surface, values from 0.005 to 0.025 m are commonly used. Where the surface is characterized by drifts, however, it is the hummocks that are the important roughness elements. In this case, z_o can be estimated by the method shown in Figure 13-8. The parameter becomes more difficult to estimate in dense forests and during the later stages of the melt season when ground vegetation protrudes through the snow surface.

If water vapor from moist air condenses on a snowpack, 590 calories of heat are released by each gram of condensate. This is enough energy to melt approximately 7.5 gm of ice, which when added to the condensate yields a total of 8.5 gm of potential runoff. The transfer of latent heat for melting can be calculated from the equation

$$Q_e = \frac{7592\ u_a(e_a - e_s)}{p\left[\ln\left(\frac{z_a}{z_0}\right)\right]^2} \tag{13-8}$$

where p is the atmospheric pressure (mb) and u_a may be corrected for the effect of forest density by the use of Equation 13-7.

These last two terms, Q_h and Q_e, dealing with the turbulent exchange of energy between the atmosphere and the snowpack, should also be corrected for the stability of the atmosphere. If the snowpack is colder than the atmosphere, the air near the surface is cooled, becomes denser and resists being lifted away from the snowpack surface, and cannot be replaced quickly by warmer or moist air. The turbulent exchange of energy is thereby reduced. When this occurs (i.e., when $T_a > T_s$), the right-hand side of Equations 13-6 and 13-8 should be divided by

$$K = 1 + \frac{732310 z_a(T_a - T_s)}{(T_a + 273)u_a^2} \tag{13-9}$$

where z_a (m) is the height of wind measurement u_a (km/day).

Energy advected by rain can be evaluated from

$$Q_p = PT_a c_w \rho_w \tag{13-10}$$

where P is the depth of rainfall (cm), T_a is the air temperature, which is assumed to characterize the falling rain, c_w is the specific heat of water (cal/gm/°C), and ρ_w is the density of water (gm/cm^3). These last two quantities are equal to 1.0. Although the heat supplied from this source is not negligible in a large, warm rainstorm, the water contributed directly by rainfall far outweighs the amount of meltwater generated by Q_p, since even low rainfall intensities are much higher than extreme rates of snowmelt.

A more detailed description of the application of these equations to the prediction of hourly and daily snowmelt is given by Price and Dunne (1976). The relative magnitudes of the three major energy contributions (net radiation, sensible heat transfer, and latent heat transfer) vary through the day and between days with changes of weather (see Figure 13-9). Net radiation is usually negative at night, whereas the other two components can be large

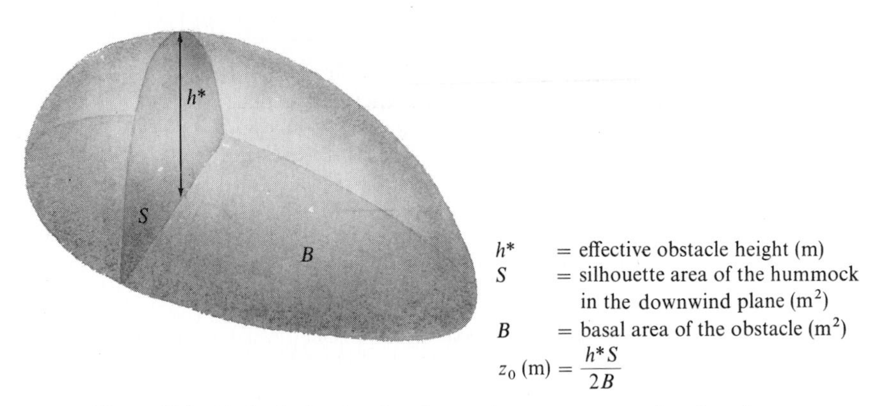

Figure 13-8 Method of estimating the roughness parameter of surface forms.

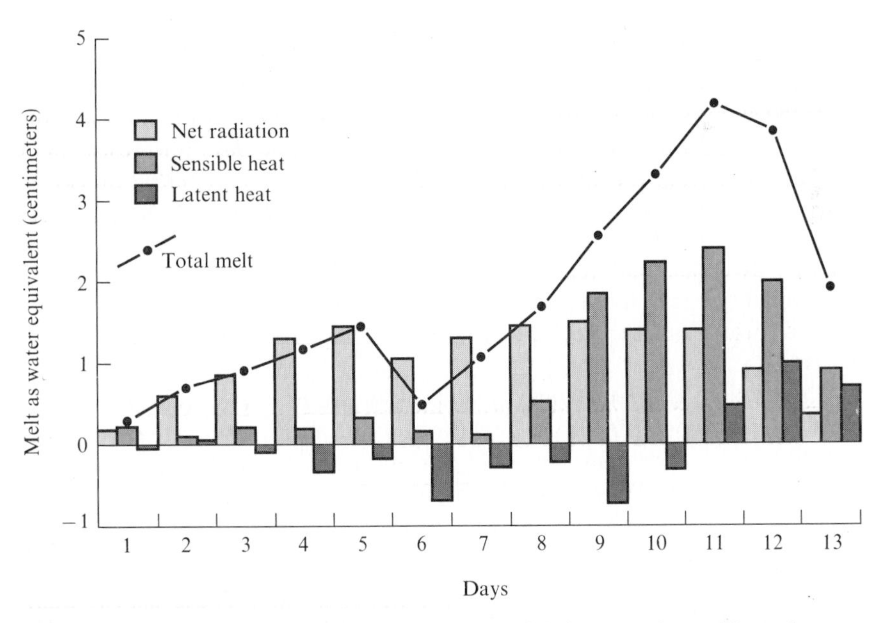

Figure 13-9 Daily totals of melt due to radiation, sensible heat transfer, and latent heat transfer at Schefferville, Quebec, during May 1973. The highest melt rates were associated with the warm sector of a large weather disturbance, which moved over Schefferville from the eighth to the eleventh day, and drew a strong flow of warm moist air over the snowpack. (From Price 1975.)

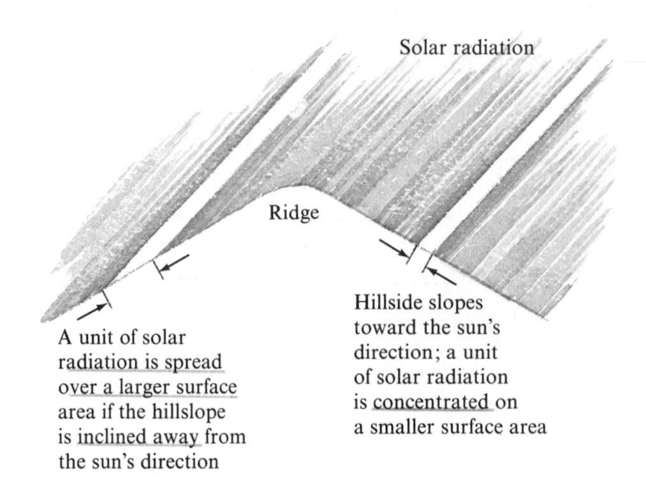

Figure 13-10 Effect of topographic gradient and orientation on the intensity of solar radiation per unit area of ground surface.

and positive. During sunny weather, radiant energy is usually the dominant control of snowmelt. Hillslopes with different orientations and gradients are irradiated differentially (see Figure 13-10), however, and this causes south-facing slopes to lose their snow cover early in the melt season by which time north-facing areas are barely contributing to runoff. By the time melting on the north-facing slopes is rapid, the southerly hillslopes have often lost their snow cover. Areas with different vegetation also experience differences of melt rate. In a landscape with a variety of hillslope gradients, orientations, and vegetation, therefore, the melting due to radiation falling on various parts of the catchment will be out of phase and the average rate of melting will usually be less than one would expect on a uniform surface. Melt rates from a varied landscape will be out of phase on both a daily and a seasonal basis.

The sensible and latent heat contributions are not subject to such modification by topography and vegetation, apart from a reduction of wind velocity in the forest and possible variations of windspeed due to topographic exposure. These energy inputs are the same on all parts of the landscape, which therefore shed meltwater uniformly. In many regions, such as subarctic Quebec, northern New England, and the Cascade Mountains, the highest melt rates and the most damaging snowmelt floods occur not on sunny days but at times when a strong flow of warm, moist air invades the region in the warm sector of a cyclonic weather disturbance. The Snohomish flood described in Chapter 1 is an example. Such conditions are usually associated with cloudy, rainy conditions with low solar radiation.

The turbulent transport of sensible and latent heat is high, however, and is simultaneous in all parts of the landscape. Melting from these heat sources can also continue through the night.

The U.S. Army Corps of Engineers have provided two equations for the sensible and latent heat contributions based on a different theory of turbulent exchange between the atmosphere and the snowpack. The expressions are

$$Q_h = 0.04 \frac{p}{p_0} (z_a z_b)^{-0.167} (T_a - T_s) u_b \tag{13-11}$$

$$Q_e = 0.19 (z_a z_b)^{-0.167} (e_a - e_s) u_b \tag{13-12}$$

where p and p_0 are the pressures at the site and at sea level (mb), z_a and z_b are the heights (m) above the snowpack at which T_a (°C) and u_b (km/day) are measured.

Basin Snowmelt Equations

A valuable summary of the physical theory of snowmelt has been provided by hydrologists of the U.S. Army Corps of Engineers (1956, 1960). These publications should be read by every planner with a major interest in snowmelt. For rapid calculations of melt rates over a catchment, however, the theoretical equations (which resemble those in the previous section) were simplified, as described below. This work was done in mountainous regions of the western United States and should be used only with caution in regions with a different climate. John Fitzgibbon of the University of Saskatchewan and Dunne checked the predictions against field measurements of melt in subarctic Quebec, however, and there at least the simplified equations gave results of sufficient accuracy for most applications in planning. Robert Hendrick of the Agricultural Research Service also found them suitable for conditions in northern Vermont (Hendrick et al. 1971). Such a check is advisable if accurate predictions are necessary.

Snowmelt days are divided into *rain periods* and *rain-free periods*. During rain a complete cloud cover is assumed, and this allows several simplifying assumptions to be made. First, the solar radiation input will be small. If measurements of Q_s are available they can be used, but if they are not available, an estimate of this quantity will not lead to gross error. In many regions, the net solar radiation term, $Q_s(1 - \alpha)$, is usually equivalent in the open to about 0.25 cm of melt (M_s) per day during rain. Net longwave radiation is usually controlled by the temperature of the cloud base, but for a low, complete cloud cover this can be approximated as the air temperature at a lower level in the empirical equation

$$M_{lw} = 0.142 T_2 \tag{13-13}$$

where the melt is given in cm/day and the air temperature (°C) is measured

2 m above the snowpack. This equation would also be appropriate for a complete forest canopy.

The largest melt contributions during rain result from the sensible and latent heat fluxes and can be estimated from

$$M_h + M_e = 0.00059 T_2 u_2 \tag{13-14}$$

where windspeed (km/day) varies with vegetation cover according to Equation 13-7. The amount of rain and its resulting melt (Equation 13-10) must be added to the four melt totals listed above.

The radiation balance is a more important component of the energy balance during rain-free periods than during rainstorms and must be estimated with some care. Vegetation cover and cloud cover must also be considered. Direct measurements of solar radiation are desirable or may be obtained from published records, as described in Chapter 4. If measurements are not available, Equation 4-8 or 4-9 may be used. The amount of this radiation reaching the snowpack should be reduced by the proportion F, which indicates the degree of shading under forest vegetation. Net longwave radiation is also affected by the proportion of forest cover, or the proportion of cloud cover, C. The turbulent heat fluxes are obtained from simplified versions of Equations 13-11 and 13-12. The simplified equations for total basin snowmelt under these conditions are as follows:

Open areas (< 10% tree cover):

$$M = 0.0125 Q_s(1 - \alpha) + (1 - 0.1C)(0.104 T_2 - 2.13) + 0.013 C T_a + 0.00078 u_2 (0.42 T_2 + 1.51 T_d) \tag{13-15}$$

Lightly forested areas (10–60% cover):

$$M = 0.01(1 - F) Q_s (1 - \alpha) + 0.00078 u_2 (1 - 0.8F)(0.42 T_2 + 1.51 T_d) + 0.14 F T_2 \tag{13-16}$$

Heavily forested areas (60–80% cover):

$$M = 0.00078 u_2 (1 - 0.8F)(0.42 T_2 + 1.51 T_d) + 0.14 T_2 \tag{13-17}$$

Densely forested areas (> 80% cover):

$$M = (0.19 T_2 + 0.17 T_d) \tag{13-18}$$

In these equations all the melt rates are in cm/day, insolation (Q_s) is in cal/cm²/day, cloud cover (C) is expressed in tenths of the sky, T_2 and T_d (°C) are the temperatures of the air and dewpoint* 2 m above the snowpack, u_2 (km/day) is the windspeed, and F is the forest canopy cover expressed as a decimal fraction.

---

*The dewpoint temperature (T_d) is the temperature at which the air would become saturated if it were cooled at constant pressure without any change of moisture content. It therefore reflects the vapor pressure. The dewpoint temperature is often published along with other weather records. If not, the dewpoint temperature can be obtained from Figure 4-6(a) using measured air temperature and vapor pressure or relative humidity.

From the foregoing discussion of the energy balance it should now be obvious to the reader what each of the terms in Equations 13-15 to 13-18 is describing. The expressions involve several major approximations and empiricisms which we cannot pursue here, however. The equations work surprisingly well in a wide range of snowy climates, though of course the results should be checked at least roughly before being applied in important design studies. Checks can be made against snow-course measurements of melt over one season or against measured or estimated runoff in a small drainage basin. At least the predicted melt rates should be checked to see whether they agree reasonably well with those measured in neighboring regions. After many devastating snowmelt floods, detailed hydrologic and hydrometeorologic studies have been made and the results published by agencies such as the U.S. Geological Survey and the Canadian Atmospheric Environment Service.

Hendrick et al. (1971) used the simplified Corps of Engineers equations to predict snowmelt from diverse catchments in northern New England. After first checking the predictions against snow stake measurements, they divided up the Sleepers River Experimental Watershed (110 sq km) into 96 uniform environments on the basis of elevation, hillslope gradient, aspect, and vegetation cover, and computed snowmelt in each environment. They also showed that during seasons with high radiation inputs, the diversity of the landscape reduced the overall basin average rate of snowmelt. In the White Mountains of New Hampshire, with a great range of hillslope gradient, elevation, and forest density, melt rates in the various environments are strongly out of phase and the result is a generally low average rate of water release. In the Champlain Valley of northwestern Vermont, on the other hand, there is a limited range of environments, and the snow leaves the whole area almost simultaneously creating a large potential for runoff during radiation melts. During seasons dominated by turbulent transfers of latent and sensible heat, however, when differential irradiation is not so important, the differences are much less marked. At these times, the more extensive snow cover at high elevations, the heavy rainfalls, shallow mountain soils, and steep channels can provide optimum conditions for rapid runoff, and some very large and damaging floods have been generated by rain on melting snow in the mountains of New England and elsewhere.

Temperature Index Method

The energy-balance methods outlined in the previous two sections are useful for understanding the melting process and for design purposes where the effects of a hypothetical set of weather conditions are considered. For other purposes, such as flood warning or reservoir operation, a heat index is often used to provide a rapid estimate of melt.

The most obvious and widely available index of the energy available for snowmelt is air temperature. If a number of daily measurements of snowmelt are made and plotted against mean air temperature for the day, a linear relationship usually results, as shown in Figure 13-11. The measurement of melt may be a direct one with the use of a snow sampler or simply the total daily snowmelt runoff, and the user of the results should be aware of which measure was used. The general form of such a relationship is expressed as

$$M = a + kT_a \tag{13-19}$$

where M is the total melt (cm) for the day or other period, T_a is the mean air temperature for the period (°C), and k is known as the *degree-day factor* or *degree-day index*. Alternatively, the equation describing the line in Figure 13-11 can be described by an equation of the form

$$M = k(T_a - T_b) \tag{13-20}$$

where T_b is some nonzero base temperature that can be obtained by trial. The average daily temperature is usually taken as the mean of the maximum and minimum values. In mountainous terrain a correction of 5.5°C per 1000 m must be allowed if there is a difference of elevation between the meteorologic station and the zone for which melt is measured or calculated.

The procedure is simple, but it must be calibrated for each region and vegetation type (see Figure 13-12). Values of k appropriate to direct measurements of snowmelt generally lie between 0.15 and 0.45 cm/day/°C, though they may rise as high as 0.70 for short periods. The subarctic data give smaller coefficients. The value of 0.40 cm/day/°C is commonly quoted in the literature of temperate Europe and North America. Figure 13-11 illustrates the main problem with the technique, however. The scatter of points around the best-fit regression line is great, and therefore Equations 13-19 and 13-20 have low predictive ability. The problem can be partially offset by increasing the number of measurements. In the Canadian sub-

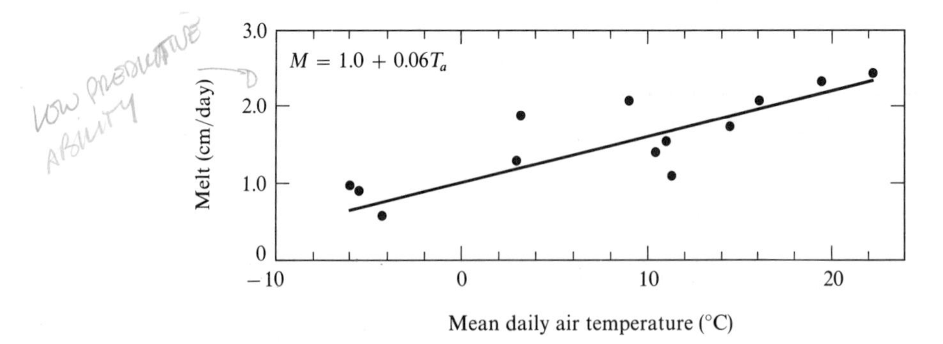

Figure 13-11 Correlation of measured snowmelt and mean daily air temperature under closed lichen woodland (spruce forest with a canopy cover of 25 percent) at Schefferville, Quebec, 1972 and 1973. (Measurements by John Fitzgibbon.)

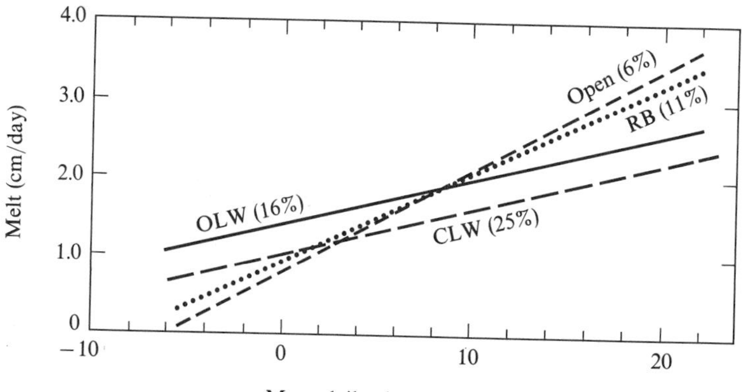

Figure 13-12 Correlation of measured daily snowmelt and mean air temperature under closed lichen woodland (CLW), open lichen woodland (OLW), regenerating woodland in burnt areas (RB), and in the open at Schefferville, Quebec, 1973. The figures in parentheses represent the canopy cover density. (Measurements by John Fitzgibbon.)

arctic, Fitzgibbon and Dunne have used expressions developed from one season's measurements to predict melts in another season and compared the predictions to measured values. Average errors of prediction ranged from 0.45 to 0.55 cm/day for melt rates averaging 1.3 to 1.6 cm/day. The method is not very precise, therefore, but for rapid, approximate predictions of melt, or in the absence of energy-balance data, it yields usable results.

Snowmelt Runoff

Early in this chapter, we discussed the ripening of a snowpack and pointed out that no runoff will be generated until sufficient heat has been transmitted to the pack to raise its temperature to 0°C and then to melt 2 to 8 percent (by weight) of the pack. The heat can be supplied by snowmelt or by rain. For a more complete discussion of the heat transfer during ripening and the timing of water release, the reader should consult the U.S. Army Corps of Engineers reports (1956).

The heat necessary to ripen a snowpack can be a very large component of the energy budget. A cold, deep mountain snowpack can absorb 10 cm or more of rainfall and can mean the difference between safety and catastrophe to the inhabitants of a nearby lowland. For short-period forecasting of runoff, therefore, mountainous drainage basins are usually divided into zones on the basis of altitude and vegetation cover. Because snow depth, pack temperature, rainfall, and melt rates vary between these environments, the onset of runoff will vary. Furthermore, the snowpack melts at different rates in these environments, and so the area contributing runoff must be adjusted

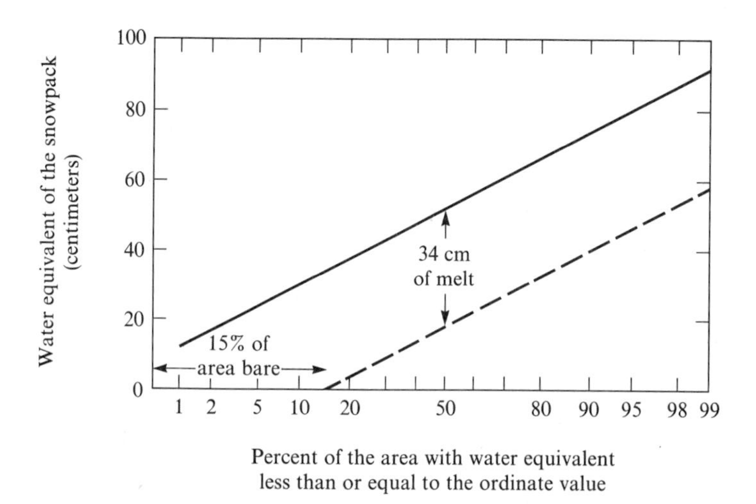

Figure 13-13 Estimating the area of snow cover by subtracting snowmelt from snow survey data on accumulation. After 34 cm of melt is subtracted from the peak snow accumulation survey, only 85 percent of the area is covered by snow.

throughout the melt season. This can be done by snow sampling through the melt season, from sequences of aerial photographs, or by subtracting calculated values of melt from each curve in Figure 13-2, which then indicates roughly the area which becomes bare, as shown in Figure 13-13.

Once runoff has begun, all melting or rainfall releases water from the pack. The paths that meltwater takes to a stream after leaving the snowpack are illustrated in Figure 13-14 and have been discussed already in Chapter 9. In very cold regions with shallow snowpacks and thin vegetation cover, the topsoil is invaded during autumn by a dense layer of ice lenses known as "concrete frost," which does not thaw out during the winter. The infiltration capacity can be lowered to 0.02 cm/hr or less by this ice, and rapid surface runoff occurs as shown in Figure 13-14(b) (Dunne and Black 1971; Dunne et al. 1976). Not all regions experience concrete frost, however. Under thick snow packs or dense vegetation in temperate regions, the soil may remain unfrozen, or may be occupied by "porous frost" or needle ice, which does not decrease the infiltration capacity. Stephenson and Freeze (1974) have described the runoff process illustrated in Figure 13-14(a), and the mechanism illustrated in Figure 13-14(c) has been observed during snowmelt on the Sleepers River Experimental Watershed in Vermont. Post and Dreibelbis (1942) and Trimble et al. (1958) have demonstrated how soil frost and infiltration capacity are related to snow depth and vegetation cover. Dunne et al. (1975) have discussed the recognition of zones that will produce saturation overland flow during snowmelt.

The details of the surface runoff process are not of interest in the usual

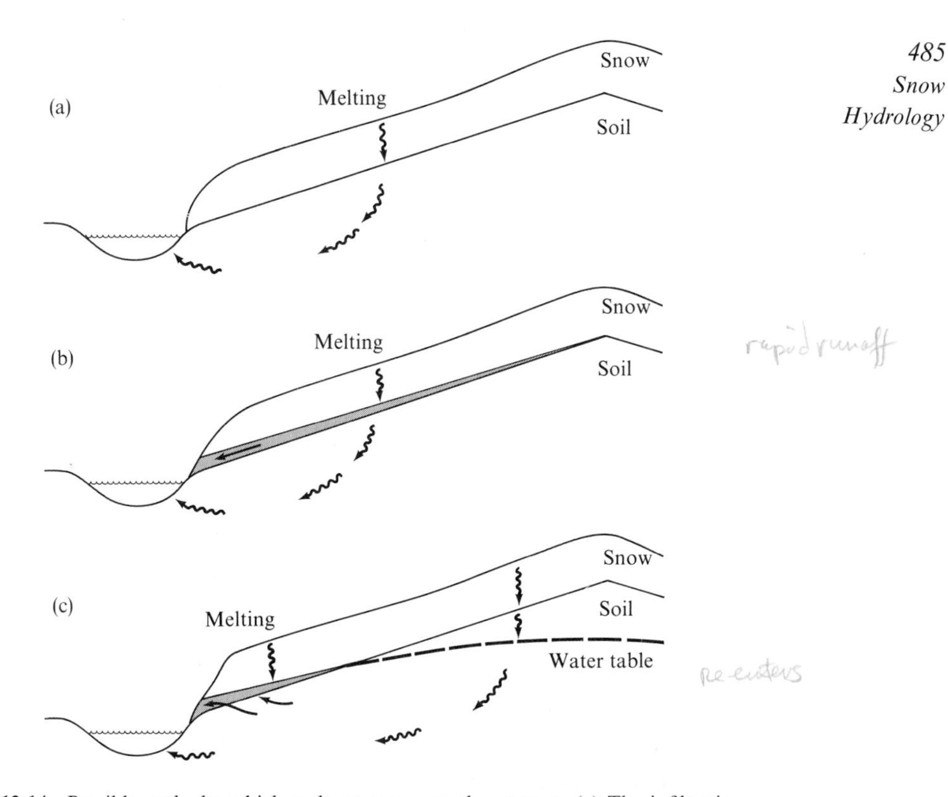

Figure 13-14 Possible paths by which meltwater may reach a stream. (a) The infiltration capacity of the soil exceeds the rate of meltwater percolation; all water enters the soil and moves to the channel as subsurface flow. (b) The rate of meltwater percolation through the pack exceeds the infiltration capacity of the soil usually because of the presence of concrete frost. A saturated layer develops in the lower few centimeters of the snow, and water moves downslope as saturated flow. (c) The infiltration capacity exceeds the meltwater percolation rate, and water enters the soil. It drains downward to raise the water table, which then intersects the ground surface near the base of the hillside. Saturation overland flow is produced by the return of subsurface water to the ground surface and by the direct runoff of meltwater, which percolates down through the snowpack onto the expanding saturated zone.

planning problem, although in some applications for small catchments it is necessary to make a rough estimate of the lag between melt and runoff. Although the speed of downward percolation varies with the rate of melting and the characteristics of the snow, it is possible to estimate the lag between melting and the time at which water reaches the ground surface from the depth of the pack, as shown in Figure 13-15. Ice lenses in the snow may complicate such a relationship, however, and increase the lag. If some of the meltwater forms a saturated layer at the ground surface, as shown in Figure 13-14(b) and (c), its rate of movement downslope varies from about 4 m/hr on a 10° slope to 10 m/hr on a 25° slope. During this percolation in the saturated layer, the diurnal wave of snowmelt is attenuated, but unless the

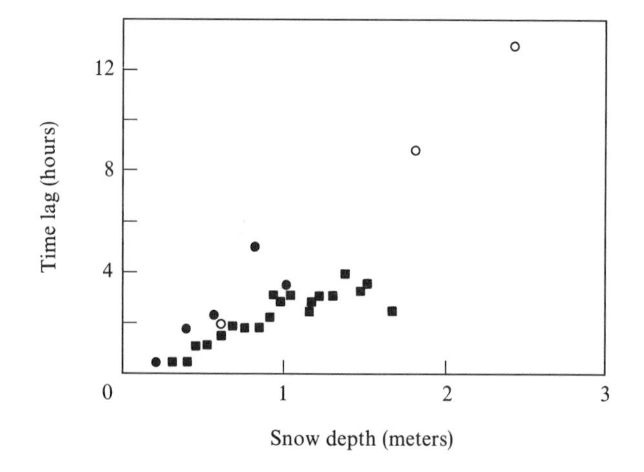

Figure 13-15 Approximate time lag between the peak rate of snowmelt at the surface of the pack and the peak rate of percolation of unsaturated flow from the bottom of the pack. The fate of the water after it reaches the soil surface is determined by the infiltration capacity of the soil and the presence of a saturated area, as shown in Figure 13-14. (Data from Anderson 1968 (■), Dunne et al. 1976 (●), and Sharp 1951 (○).)

hillslope is very long and gentle and the drainage basin is large or occupied by lakes, the daily melt cycle is important in controlling the form of the snowmelt hydrograph.

Predicting streamflow hydrographs from snowmelt or rain on melting snow is a complicated process which uses the techniques described in Chapter 10 but with uncertainties added by our limited knowledge of the processes that generate runoff under a snowcover. It is first necessary to estimate how much runoff will be generated on the hillslopes by the calculated snowmelt, and this can be done using the methods for assessing infiltration rates, average loss rates, or runoff volume described in Chapters 6 and 10. In basins with an adequate record of streamflow, the daily meltwater runoff can be predicted from a degree-day index equation, such as the one shown in Equation 13-19 or 13-20, where the dependent variable is the daily volume of snowmelt runoff.

Once the rate or volume of runoff is known, standard techniques of hydrograph prediction can be used. The unit hydrograph and the *S*-hydrograph are favorite tools for this purpose. The application is complicated considerably by the varying rates of melting and by the thinning of the snowpack if percolation through the pack is an important component of the lag time. Unit hydrographs work best, therefore, in large catchments. Other techniques treat the whole snowpack and drainage basin as a reservoir and use reservoir flood routing as described in Chapter 10. Muskingum routing is used for the same purpose. The subject is too complicated for discussion in this text; the interested reader should consult the U.S. Army Corps of Engineers manuals (1956, 1960). For a 600-square-kilometer basin in the Sierra Nevada, Rantz (1964) used a daily water balance to compute the seasonal snowmelt runoff hydrograph. A recent development, pioneered by hydrologists at Stanford University, is the use of computer-based mathematical models of the snowmelt and runoff processes. Anderson and Crawford (1964) have incorporated these processes into the Stanford Watershed Model referred to in Chapter 10. Although there is some uncertainty about whether the model accurately describes the runoff processes in many cases, it has been "calibrated" and used with success for design problems in many snowmelt regions. Planners may eventually use or at least review the results of this model or its variants at some time in their work. The application of computer models of snowmelt runoff to large design problems will increase rapidly, and although the topic is beyond the scope of this text, planners should become acquainted with the kind of information computer models use and produce.

Bibliography

ANDERSON, E. A. (1968) Development and testing of snowpack energy balance equations; *Water Resources Research,* vol. 4, pp. 19–37.

ANDERSON, E. A. AND CRAWFORD, N. H. (1964) The synthesis of continuous snowmelt runoff hydrographs on a digital computer; *Technical Report No. 36, Department of Civil Engineering, Stanford University.*

COLBECK, S. C. (1972) A theory of water percolation in snow; *Journal of Glaciology,* vol. 11, pp. 369–385.

COLBECK, S. C. (1974) Water flow through snow overlying an impermeable boundary; *Water Resources Research,* vol. 10, pp. 119–123.

DUNNE, T. (1978) Field studies of hillslope flow processes; Chapter 7 in *Hillslope hydrology* (ed. M. J. Kirkby), John Wiley & Sons, London.

DUNNE, T. AND BLACK, R. D. (1971) Runoff processes during snowmelt; *Water Resources Research,* vol. 7, pp. 1160–1172.

DUNNE, T., MOORE, T. R. AND TAYLOR, C. H. (1975) Recognition and prediction of runoff-producing zones in humid regions; *Hydrological Sciences Bulletin,* vol. 20, pp. 305–327.

DUNNE, T. AND PRICE, A. G. (1976) Estimating net radiation over a snowpack; *Climatology Bulletin,* McGill University, vol. 18, pp. 40–48.

DUNNE, T., PRICE, A. G. AND COLBECK, S. C. (1976) The generation of runoff from subarctic snowpacks; *Water Resources Research,* vol. 12, pp. 677–685.

FEDERER, C. A. AND LEONARD, R. E. (1971) Snowmelt in hardwood forests; *Proceedings of the Eastern Snow Conference,* vol. 29, pp. 95–109.

FEDERER, C. A., PIERCE, R. S. AND HORNBECK, J. W. (1972) Snow management seems unlikely in the Northeast; *National Symposium on Watersheds in Transition,* American Water Resources Association, pp. 212–219.

FÖHN, P. M. B. (1973) Short-term snowmelt and ablation derived from heat- and mass-balance measurements; *Journal of Glaciology,* vol. 12, pp. 275–289.

HENDRICK, R. L., FILGATE, B. D. AND ADAMS, W. M. (1971) Application of environmental analysis to watershed snowmelt; *Journal of Applied Meteorology,* vol. 10, pp. 418–429.

KUZMIN, P. O. (1972) *The melting of snow covers;* Israeli Program for Scientific Translations.

LETTAU, H. (1969) Note on roughness-parameter estimates on the basis of roughness-element description; *Journal of Applied Meteorology,* vol. 18, pp. 828–832.

LIGHT, P. (1941) Analysis of high rates of snow melting; *EOS, American Geophysical Union Transactions,* vol. 22, pp. 195–205.

MARTINEC, J. (1965) A representative watershed for the research of snowmelt-runoff relations; *International Association of Scientific Hydrology, Symposium of Budapest, Publication No. 66,* pp. 494–501.

PETZOLD, D. E. (1974) Solar and net radiation over snow; *Climatological Research Series No. 9,* McGill University, Department of Geography, 77 pp.

PETZOLD, D. E. AND WILSON, R. G. (1973) Solar and net radiation over melting snow in the subarctic; *Proceedings of the Eastern Snow Conference,* vol. 31, pp. 51–59.

POST, F. A. AND DREIBELBIS, F. R. (1942) Some influences of frost penetration and microclimate on the water relationships of woodland, pasture, and cultivated soils; *Soil Science Society of America Proceedings,* vol. 7, pp. 95–104.

PRICE, A. G. (1975) "Snowmelt runoff processes in a subarctic area"; Ph.D. dissertation, McGill University, 185 pp.

PRICE, A. G. AND DUNNE, T. (1976) Energy balance computations of snowmelt in a subarctic area; *Water Resources Research,* vol. 12, pp. 686–694.

RANTZ, S. E. (1964) Snowmelt hydrology of a Sierra Nevada stream; *U.S. Geological Survey Water Supply Paper 1779-R.*

SHARP, R. P. (1951) Meltwater behavior in firn on upper Seward Glacier, *International Association of Scientific Hydrology, Symposium of Brussels.*

STEPHENSON, G. R. AND FREEZE, R. A. (1974) Mathematical simulation of subsurface flow contributions to snowmelt runoff, Reynolds Creek, Idaho; *Water Resources Research,* vol. 10, pp. 284–294.

TRIMBLE, G. R., SARTZ, R. S. AND PIERCE R. S. (1958) How type of soil frost affects infiltration; *Journal of Soil and Water Conservation,* vol. 13, pp. 81–82.

U.S. ARMY CORPS OF ENGINEERS (1956) *Snow hydrology: Summary report of snow investigations;* North Pacific Division, Portland, OR, 437 pp.

U.S. ARMY CORPS OF ENGINEERS (1960) *Runoff from snowmelt;* Engineering Manual 1110-2-1406, 59 pp.

VEZINA, P. E. AND PECH, G. (1964) Solar radiation beneath conifer canopies in relation to crown closure; *Forest Science,* vol. 10, pp. 443–451.

13-1 Measurement of Volumes of Snowmelt and Runoff

Find a small rural drainage basin with an area of about 50 hectares conveniently near your home or college. Place a staff gauge in the stream and develop a rating curve, as described in Chapter 16, so that the discharge can be measured.

When the snowpack is just beginning to release meltwater, use a graduated pole to measure the depth of the snow at 20 to 50 locations along lines across the basin. The snow should have a density of about 0.35 to 0.40 at this time. Use this value and the mean snow depth to calculate the volume of water available in the pack at the beginning of the melt season. Set up a can as a rain gauge and read it each day.

Measure the stream stage at several times each day and compute the total runoff during the melt period. Estimate the evaporation loss using Equation 4-24 and data from your local meteorological station. This loss will be small.

For the whole melt season, evaluate each term in the water balance equation:

$$\text{Melt} + \text{Rainfall} = \text{Runoff} + \text{Evaporation} + \text{Groundwater recharge}$$

The last term can be evaluated by subtraction when the others are known.

13-2 Energy-Balance Computation of Snowmelt

Calculate the amount of water that would be released from a snowpack in the open and under a coniferous forest with a canopy density of 50 percent on a day with the following average weather conditions: $Q_s = 450$ cal/cm^2/day, $T_a = 8°\text{C}$, $u_a = 300$ km/day, relative humidity $= 60$ percent with clouds covering 40 percent of the sky. Wind and temperature data were collected 2 m above the snow surface.

Assume that the snowpack is isothermal at 0°C and that the albedo is an average value for a rural site early in the melt season. The snowpack surface is free of drifts but has developed small ripples during the melt.

Solution

In Equation 13-2, a value of $\alpha = 0.75$ is a reasonable estimate and $Q_s(1 - \alpha) = 112$ cal/cm^2/day in the open.

From Equation 13-4 for the open,

$$Q_{lw} = [0.76 \times (1.19 \times 10^{-7}) \times 281^4 - 661] \times (0.1 + 0.9C) = -45 \text{ cal/cm}^2\text{/day}$$

where the last term in parentheses is the substitution for $(1 - aC)$ suggested in the discussion of Equation 4-11. There is a net loss of longwave radiation from the snowpack.

The contribution of sensible heat is calculated from Equation 13-6, using a value of $z_o = 0.025$:

$$Q_h = \frac{3.84 \times 300 \times 8}{\left[\ln\left(\frac{200}{0.025}\right)\right]^2} = 114 \text{ cal/cm}^2\text{/day}$$

Latent heat flux is calculated from Equation 13-8. An average value of $p = 1000$ mb can be used in the absence of a measurement. The vapor pressure is calculated from relative humidity $= 60 = 100(e_a/e_s)$ where, from Figure 4-6(a), $e_s = 10$ mb at an air temperature of 8°C. Therefore, $e_a = 6$ mb. These values give

$$Q_e = \frac{7592 \times 300(6.0 - 6.11)}{1000 \times \left[\ln\left(\frac{200}{0.025}\right)\right]^2} = -3 \text{ cal/cm}^2\text{/day}$$

which indicates that there is a slight cooling of the snowpack by evaporation because the atmospheric vapor pressure is slightly less than the saturated vapor pressure of the snow surface.

The values of Q_h and Q_e must be corrected for the stability of the atmosphere flowing over a cold surface by dividing them by

$$K = 1 + \frac{732{,}310 \times 2 \times 8}{(8 + 273) \times 300^2} = 1.46$$

The values so obtained are inserted into Equations 13-2 and 13-3 to yield

$$Q_m = 112 - 45 + 78 - 3 = 142 \text{ cal/cm}^2\text{/day}$$

$$M = \frac{142}{80} = 1.8 \text{ cm/day}$$

For a forest with $F = 0.5$, Figure 13-4 is used to alter Q_s, Equation 13-5 is used to calculate the net longwave loss, and Equation 13-7 is used to calculate windspeed under the forest for insertion into Equations 13-6, 13-8, and 13-9. Strictly speaking, Equations 13-6 and 13-8 should not be used under a forest canopy, but they are adequate for obtaining estimates of snowmelt for planning purposes. The answer should be about 0.9 cm/day, depending on some judgments that must be made in some steps of the calculations.

To investigate this method further, the interested reader might choose a range of forest covers and calculate snowmelt under the given meteorologic conditions, plotting melt against canopy density. Another useful exercise involves altering one characteristic of the snowpack, such as α or z_o and plotting the calculated snowmelt for a range of each value if all other data are held constant.

13-3 Calculation of Snowmelt with Simplified Basin Equations, I

Calculate the total amount of water that would be released from a ripe snowpack on a day with 1.6 cm of rain, an air temperature of 10°C, and a windspeed of 250 km/day. The basin of interest has a canopy cover of 25 percent.

Solution

$M_s = 0.25 \times 0.5 = 0.12$ cm/day, from the text and Figure 13-4.

$M_{lw} = 0.142 \times 10 = 1.42$ cm/day, from Equation 13-13.

$M_h + M_e = 0.00059 \times 10 \times 250 \times (1 - 0.20) = 1.18$ cm/day, from Equations 13-7 and 13-14.

The energy advected by the rain is given by Equations 13-3 and 13-10 and is equivalent to 0.20 cm of melt.

The total amount of water released by melt and by rain, therefore, is 4.5 cm/day.

13-4 Calculation of Snowmelt with Simplified Basin Equations, II

Calculate the melt from an isothermal, ripe snowpack under a 70-percent coniferous canopy cover on a day with the following meteorologic conditions: $Q_s = 550$ cal/cm²/day, $u_2 = 200$ km/day, $T_2 = 10$°C, relative humidity = 80 percent.

Solution

First, Figure 4-6(a) must be used to evaluate the dewpoint temperature of the atmosphere, as follows. At a temperature of 10°C, $e_s = 12$ mb, and since the relative himidity is 80 percent, $e_a = 9.6$ mb. The same graph can be used to show that 9.6 mb is e_s at a temperature of about 7°C, which therefore is the dewpoint temperature.

Equation 13-17 then gives

$$\begin{aligned} M &= 0.00078 \times 200 \times (1 - 0.8 \times 0.7) \\ &\quad \times (0.42 \times 10 + 1.51 \times 7) + 0.14 \times 10 \\ &= 2.4 \text{ cm} \end{aligned}$$

III
Geomorphology

14

Drainage Basins

Geomorphology and Planning

Geomorphology is the study of surface forms on the earth. It is therefore the science of landscape, the processes and history of development of land forms, the relation of geologic material to surface features, and by logical extension, the extrapolation of such knowledge toward forecasting the effects of different actions on the landscape. Planning should include in its domain the forecasting of effects, but this has not always been the case. Experience has shown that planning without knowledge of geomorphic processes can produce sins of both omission and commission. Even if a planner is not expert in geomorphic subjects, an awareness of natural processes can be of great assistance.

In this book emphasis is placed on the geomorphic processes associated with the action of rainfall and runoff on the landscape. These are the agents that form and alter most landscapes, even in arid regions. Consequently, they affect most frequently the landscapes and installations with which the planner is concerned.

Geomorphic Effects of Climatic Change

While we stress current processes, it is valuable to recall what was said in Chapters 2 and 9 about climatic change and climatic fluctuations. The earth's surface reflects the net result of a long history during which continents have been moved, mountains have been uplifted and worn away, plant and animal communities have developed, flourished, and disappeared. But perhaps even more important to modern man has been the effect of the very recent geologic past, the events of the Pleistocene during which time great climatic changes occurred and ice sheets modified large areas of the earth. It was only about 10,000 to 15,000 years ago that the last continental glaciers retreated from middle latitudes. Both our present climate and our present landscapes reflect those events. The earth is still in a general climatic regime only slightly removed from the glacial epoch and certainly far from the moderate, uniform, warm conditions that prevailed throughout most of geologic time. We are closer to an ice age than to the geologic norm in climate.

There is a pragmatic reason why changes in the recent geologic past are of importance. The climatic variations that punctuated the Pleistocene and that are still continuing on a smaller scale today affect the landscape through alterations in the operation of the hydrologic cycle, especially the interaction of precipitation, vegetation, and soils. Similar alterations in the hydrologic cycle are also one of the effects of human activities. Thus, effects of humans and effects of climatic change operate through the same mechanisms, have similar effects on the land, and are therefore difficult to separate. This becomes important when observed changes in river regimen, in erosion conditions, in sediment production, and in some vegetative changes are to be interpreted. It represents one justification for providing land planners, landscape architects, geographers, and urban planners with some basic knowledge of geomorphology and the earth's surficial processes.

Drainage Basins in Planning

Often, carefully laid schemes are disrupted by something that occurs many miles away from the site. A brush fire or careless logging in nearby hills may release large amounts of sediment, which alters stream channels and flooding conditions for many miles downstream. The sprawl of urbanization or the location of a power plant may ruin vistas. Conflicts can arise between upstream and downstream users of a water supply who wish to take advantage of the stream for irrigation, drinking, waste disposal, or cooling, or merely for looking at. The degree to which these effects are felt over large areas often surprises and perplexes the users, but that is because it is easy to forget that hillslopes, gullies, rivers, groundwater bodies, urban storm drains, industrial cooling systems, and irrigated fields are linked as components of drainage basins.

A drainage basin is the area of land that drains water, sediment, and dissolved materials to a common outlet at some point along a stream channel. The term is synonymous with *watershed* in American usage and with *catchment* in most other countries. The boundary of a drainage basin is known as the *drainage divide* in the United States and as the *watershed* in other countries. Thus the term *watershed* can mean an area or a line. The drainage basin can vary in size from that of the Amazon River to one of a few square meters draining into the head of a gully. Any number of drainage basins can be defined in a landscape (see Figure 14-1) depending on the location of the drainage outlet on some watercourse. Because the hydrologic and geomorphic effects of natural and human processes within a catchment are focused at its outlet, the drainage basin of interest to planners is often defined as the area draining to some critical point at which they intend to install something. Planners should be equally aware, however, that the drainage basin they have defined in order to make some design calculation is a portion of some larger drainage basin whose downstream portion may suffer from the effects of the design unless they are careful.

In some landscapes, the topographic limit of the drainage basin may not coincide with the boundary between subsurface drainage systems. Chapter 7

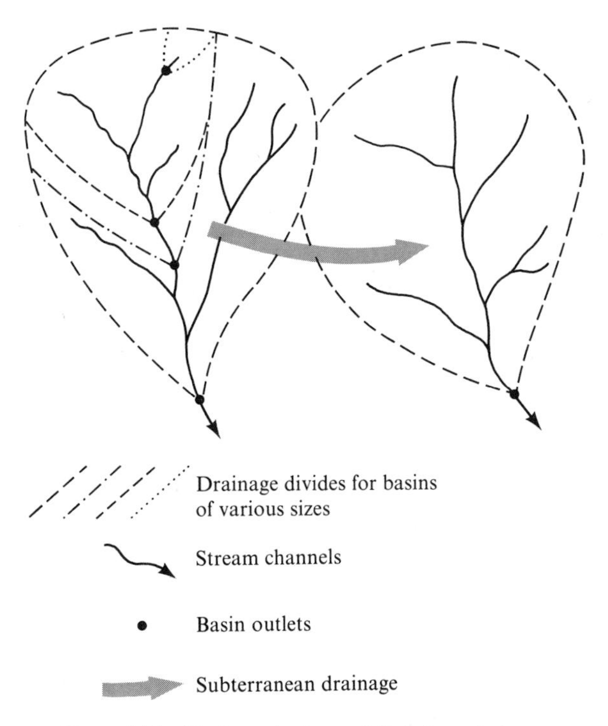

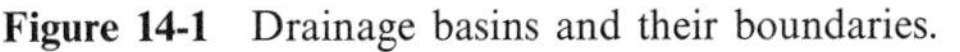

Figure 14-1 Drainage basins and their boundaries.

mentioned some of the geological conditions that could control groundwater movement. Planners should be aware of such conditions, because pollution or disruption of a groundwater supply by their actions could cause extensive damage and suffering elsewhere and could involve them in litigation.

Not all the problems or constraints of physical planning can be examined solely within the context of the drainage basin, therefore, and many of them can even be removed by solutions such as inter-basin transfers of water or waste. However, there is much to be gained from examining the drainage basin as a convenient unit for understanding the action of hydrologic and geomorphic processes and for appreciating the spatial linkages between different areas that can affect both regional and site planning. As more and more planning is based on an understanding of natural processes, this is increasingly being done. There are now many planners who see the drainage basin not only as a constraint but as an opportunity to be exploited with imaginative and beautiful designs (McHarg 1969).

Drainage-Basin Form

The form of a drainage basin provides the planner with various constraints and opportunities. Steep slopes with intricate drainage patterns may limit access, confining roads to drainage divides. They may provide secluded building sites with attractive vistas, but require careful planning of waste disposal because the thin soils and steep slopes are unsuitable for septic tanks. Steep slopes also need to be examined carefully for landslide hazard. In very gentle topography the subdivision of land, construction, and the provision of facilities may be easy and cheap. The problems there may be the avoidance of flooding, ensuring sufficient grade for the operation of storm and sanitary sewers, or simply making the site visually interesting. These are just a few examples of the effects of drainage-basin form on the planning process. Other indirect effects include the manner in which hillslope gradient, channel densities, channel gradients, and valley-floor widths control the generation of floods. Here we will review a few of the more useful measures of drainage-basin form, mostly developed by Robert E. Horton and Arthur Strahler, who laid the foundations for the quantitative description of drainage basins.

A catchment is drained by a hierarchical network of channels (Figure 14-2), whose size increases downstream from small rills through gullies to small and large river channels as the amount of water and sediment they must carry increases. The streams in these channels have cut the valleys in which they lie and are therefore responsible for the overall orientations of valleys and hillslopes in the area. In regions that have been glaciated, the valleys may have been significantly widened and deepened by ice, but in most cases the glaciers followed preexisting river valleys and in many cases caused little modification.

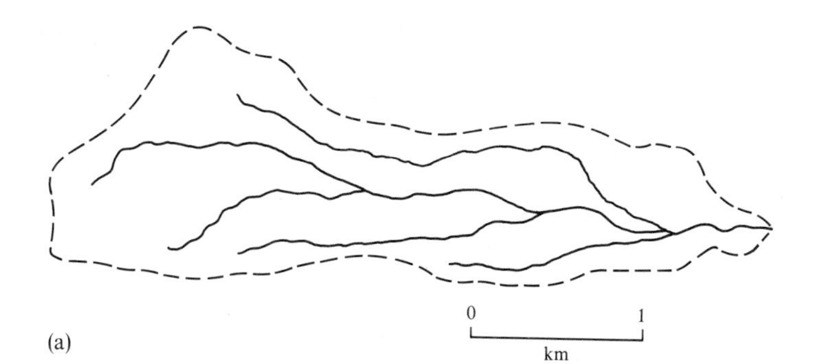

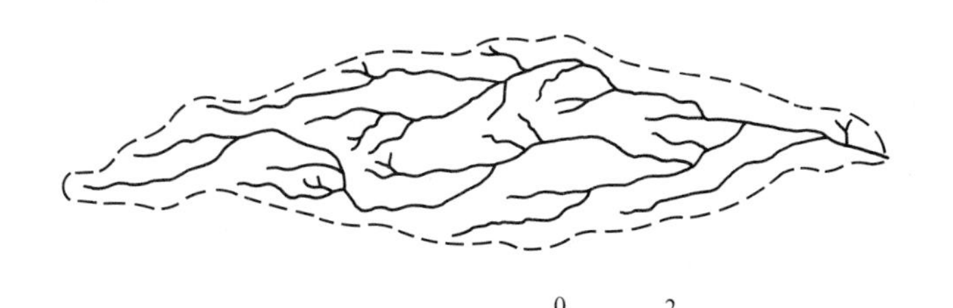

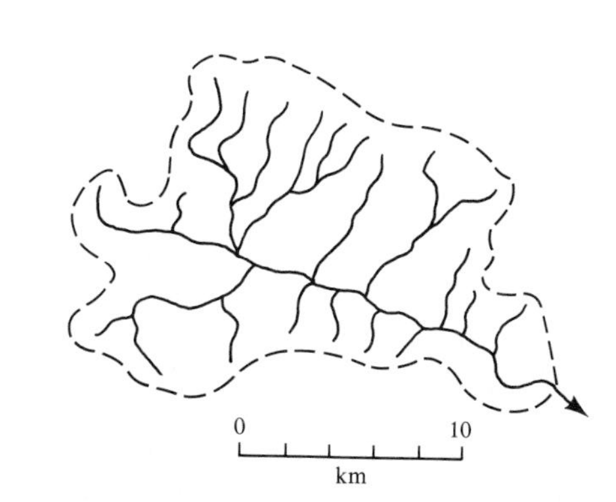

Figure 14-2 Examples of drainage networks. (a) Unnamed tributary of the East Fork River at New Fork, Wyoming, showing a palmate, dendritic pattern. (b) Arroyo Frijoles, near Santa Fe, New Mexico, showing a dendritic pattern. (c) Dog River, near Northfield, Vermont, having a palmate, dendritic pattern where three tributaries merge at the upper end of the basin, and a pinnate network lower down the valley, where parallel tributaries flow from the steep mountain slopes that flank the drainage basin.

In fairly uniform rocks the plan view of the network of river channels is usually dendritic, as shown in Figure 14-2(a) and (b). In mountain valleys many small tributaries may feed a larger stream like the pinnae of a pinnate leaf, as shown in Figure 14-2(c), while at valley heads or valley junctions the network may be strongly palmate. This last situation is a very hazardous one for town sites because simultaneous flooding in the tributary valleys often concentrates heavy runoff at the junction. Montpelier, the capital city of Vermont, is in just such a location and has frequently suffered damaging floods.

Rock structures such as major bedding planes, joints, or faults are exploited by streams in some areas to form oriented dendritic networks. Where alternating beds of weak and resistant rocks occur, the drainage network becomes trellised as the weaker formations are carved into valleys between ridges of the resistant members (see Figure 14-3). If the uplands are high and steep enough, transportation routes are confined to the river gaps, which are therefore chosen for settlement sites, commercial centers, and other purposes. The patterns of transportation and settlement in many parts of southeastern England are developed to take advantage of a landscape drained by a trellised stream network.

As a means of comparing rivers of different size or importance within or between networks, Robert Horton proposed a system of stream ordering, which is illustrated in Figure 14-4, as modified by Strahler. The smallest streams of the network, which have no tributaries, are called first-order streams. When two of these first-order streams coalesce they form a second-order stream, and further along its course this stream may join another second-order channel to form one of the third order, and so on. A low-order

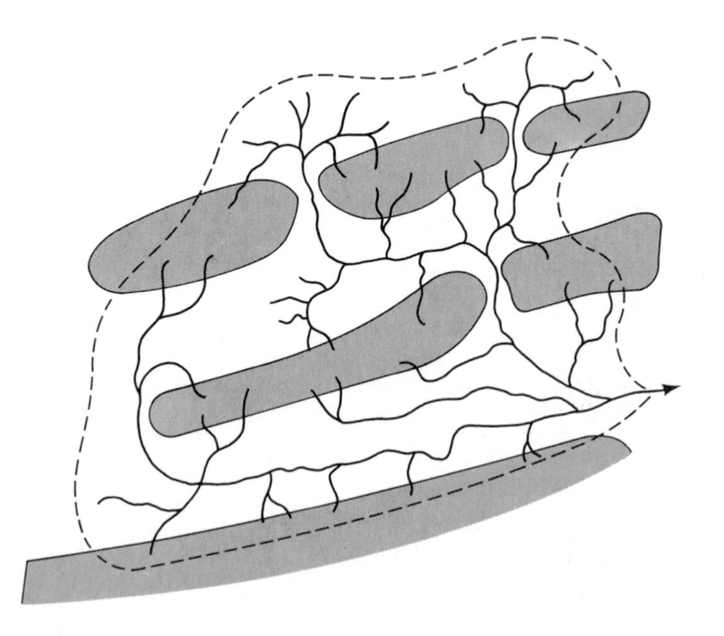

Figure 14-3 A trellised drainage pattern. Rivers have breached the resistant rock (shaded) in a few, narrow gaps. On the intervening weaker rocks (unshaded), many river channels drain the valleys between the more resistant highlands.

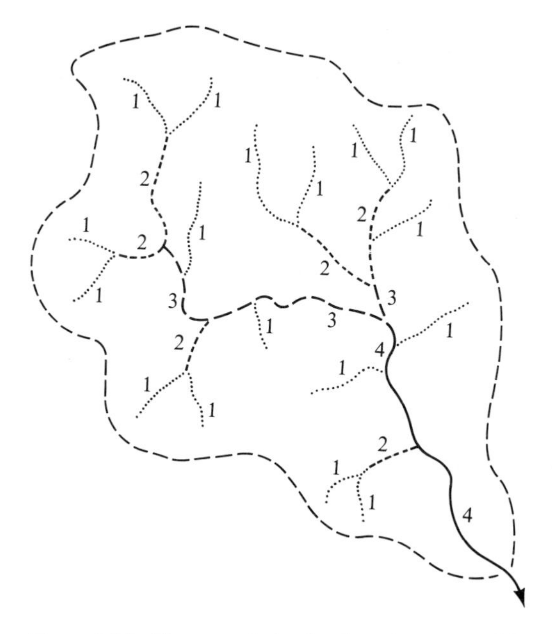

Figure 14-4 Stream ordering by the rules proposed by Strahler (1964).

stream, such as one of first order joining another of higher order, does not alter the rank of the latter. The rules are illustrated in the diagram.

A description of stream orders is usually made by tracing the network from a topographic map or aerial photograph. The stream segments of each order can then be coded by color. One limitation is the size of the smallest stream that can be recognized from the map or photograph. On many standard topographic map series, the blue lines representing streams often refer only to perennial watercourses or to those seasonal streams that have well-defined channels. During storms, however, the stream network will expand into formerly dry gullies and unchanneled swales and marshlands. This expansion can be foreshadowed by extending the traced channel lines upstream into contour crenulations on the map or swales visible through a stereoscope on the photograph. Leopold and Miller (1956) showed by example that the size or drainage area of the smallest or first-order stream depends on the map scale. When a basin was mapped at large scale, what had appeared to be a first-order stream on a smaller-scale map became a fifth-order stream. Thus all the orders determined from the small-scale map were increased by four on the large-scale map.

The density of channels in the landscape is an easily obtained measure of the dissection of the terrain. The *drainage density* is defined as the length of all channels in the drainage basin divided by the basin area. When the channels are extended on the map into swales, the values of drainage density for the three maps shown in Figure 14-2 are as shown in Table 14-1. Because of the limitation on the recognition of small channels, the map scale and

Table 14-1 Drainage density for each drainage basin shown in Figure 14-2.

	AREA (SQ KM)	TOTAL CHANNEL LENGTH (KM)	DRAINAGE DENSITY (KM/SQ KM)
Tributary to East Fork, WY	3.4	11.1	3.27
Arroyo Frijoles, NM	32.6	99.8	3.06
Dog River, VT	245.2	121.9	2.01

details of whether channels were extended should always be given with values of drainage density.

This measure of the texture of dissection is associated with various geomorphic and hydrologic conditions of interest to the planner. Areas with high drainage density are associated with high flood peaks, high sediment production, steep hillslopes, general difficulty of access, relatively low suitability for agriculture, and high development costs for the construction of buildings and the installation of bridges, roads, and other facilities.

The relief and slope characteristics of drainage basins can be expressed in a variety of ways, but probably the simplest is the *relief ratio,* defined as the difference in elevation between the highest and lowest points of the basin divided by the length of the basin in a line roughly parallel to the major drainage. A second measure is the gradient of the stream channel. Within a region of roughly uniform climate and geology, these parameters are useful indices of sediment production and flood peaks (Schumm and Hadley, 1961; see also Figures 10-26 and 10-43). Because they are usually correlated with average hillslope gradient in the basin, the relief ratio or channel gradient can be used to indicate areas possessing the advantages and disadvantages of various types of terrain described earlier.

As river channels join downstream, the resulting stream drains larger and larger catchments. One surprising result of these random confluences is that in most river basins the drainage area increases with distance downstream at a predictable rate. A double-logarithmic plot of drainage area against distance along the main channel from the drainage divide to the outlet of a catchment yields a straight line. John T. Hack of the U.S. Geological Survey, whose many quantitative studies of landscape form have generated much of our understanding of the way drainage basins function, obtained the following equation from the graph in Figure 14-5 for drainage basins in the Shenandoah River basin of Virginia:

$$A = 0.57L^{1.67} \tag{14-1}$$

where A is the drainage area in square miles and L is the channel length in miles. Many other studies of drainage basins of all sizes in different parts of the world plot remarkably close to Hack's original line. The diagram includes points from some of the world's largest rivers for comparison. Equation 14-1

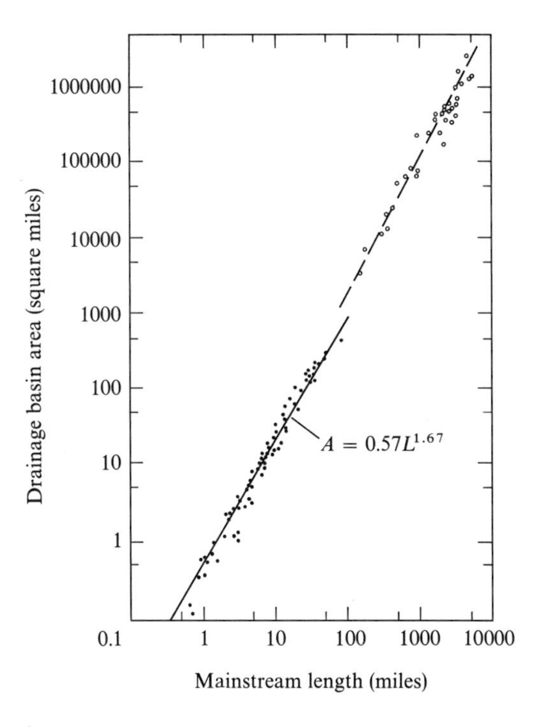

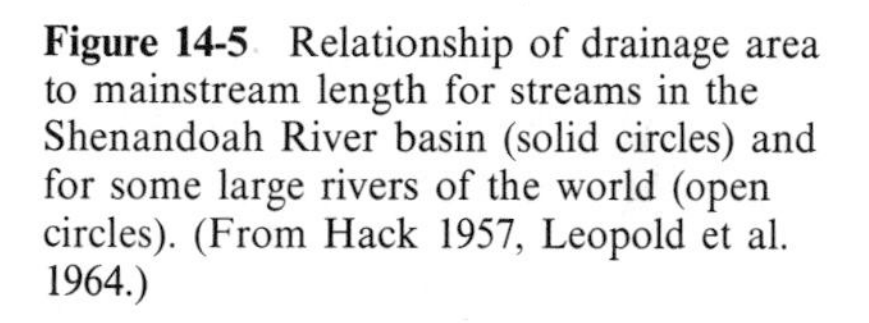
Figure 14-5 Relationship of drainage area to mainstream length for streams in the Shenandoah River basin (solid circles) and for some large rivers of the world (open circles). (From Hack 1957, Leopold et al. 1964.)

indicates the rate at which drainage area increases with distance along the main channel, although it should be recognized that for a single catchment the curve would be broken by several steps as large increments of drainage area are added at tributary junctions.

The rate at which drainage area increases with distance downstream is the dominant control of the rate at which runoff increases downstream, and we have already discussed in earlier chapters the relationship between drainage area, annual yield, low flows, and flood discharges. We will see in later chapters how the rate of increase of drainage area affects sediment yields, channel characteristics, and water quality. The downstream increase of catchment size is also related to many of the morphometric variables described above. There is usually an inverse relationship between drainage area and relief ratio, average hillslope gradient, channel slope, and sometimes drainage density.

The problems faced by the planner will therefore change in emphasis downstream. The lower the location in a drainage basin, the more likely are problems posed by upstream water use, flood sizes, land drainage, and flat, uninteresting terrain. Adequate water supply, trafficability, and construction difficulties are less likely. Upstream, he is often constrained by limited water supplies from a single stream or from groundwater, by steep terrain, and by many of the problems and possibilities referred to above.

Bibliography

GREGORY, K. J. AND WELLING, D. E. (1973) *Drainage basin form and process;* Halsted Press, New York, 456 pp.

HACK, J. T. (1957) Studies of longitudinal stream profiles in Virginia and Maryland; *U.S. Geological Survey Professional Paper 294-B,* pp. 45–97.

HORTON, R. E. (1945) Erosional development of streams and their drainage basins: hydrophysical approach to quantitative morphology; *Geological Society of America Bulletin,* vol. 56, pp. 275–370.

LEOPOLD, L. B. AND MILLER, J. P. (1956) Ephemeral streams: hydraulic factors and their relation to the drainage net; *U.S. Geological Survey Professional Paper 282-A.*

LEOPOLD, L. B., WOLMAN, M. G. AND MILLER, J. P. (1964) *Fluvial processes in geomorphology;* W. H. Freeman and Company, San Francisco, 504 pp.

MCHARG, I. L. (1969) *Design with nature;* Natural History Press, New York, 198 pp.

SCHUMM, S. A. (1954) The relation of drainage basin relief to sediment loss; *International Association of Scientific Hydrology Publication No. 36,* pp. 216–219.

SCHUMM, S. A. AND HADLEY, R. F. (1961) Progress in the application of landform analysis in studies of semi-arid erosion; *U.S. Geological Survey Circular 437.*

STRAHLER, A. N. (1964) Quantitative geomorphology of drainage basins and channel networks; Section 4-2 in *Handbook of applied hydrology* (ed. Ven te Chow), McGraw-Hill, New York.

Typical Problems

14-1 Construction of a Topographic Map by Pace, Compass, and Estimation

Experience indicates that until one has actually made a topographic map, one will overlook many map details. In practice, maps that are more detailed than published quadrangle maps are often needed for description, whereas field surveying by plane table may not be justified. Thus, there is justification for getting minimal experience in constructing such a map by eye. The following instructions are used for an exercise at the University of California, and so they refer to the Berkeley region. The reader can substitute details appropriate to his or her own area.

Location

The mapped area will encompass about 23 acres (or 1000 ft × 1000 ft) of which the center is latitude 37°54′47″, longitude 122°14′40″. It is located in Tilden Regional Park and appears on the U.S. Geological Survey quadrangle, "Richmond, California."

Show your computation of position by latitude and longitude to locate the position on the topographic quadrangle sheet. Include this computation sheet in the folder with your final map.

Map, General Instructions

A map will be drawn at a scale of 1 inch = 100 feet and contour interval of 10 feet. An unnamed draw or tributary will traverse the center of the map. We will call this draw "Jewel Creek" because it enters Wildcat Creek near Jewel Lake. The U.S. Geological Survey topographic map will be helpful and should be used in any way that assists the fieldwork, but the field map will be far more detailed. Key elevations, however, can be read directly from the published map and used to control the contours on the field map.

Suggested Procedure

1. Gather materials and equipment.

Obtain a copy of the Richmond quadrangle and locate the point given by latitude and longitude. Carry to the field a piece of mapping paper about 12″ × 12″ or 14″ × 14″, preferably a Strathmore drawing paper. The paper might best be placed on a light-weight plane table board or other hard board usable in the field. It would be wise to have a Brunton or Forester's compass, or if either of these is not available, any simple compass. For fieldwork, highly recommended is a field notebook that has cross-section paper in it, in a leather notebook case. An engineer's scale is necessary (10 divisions to the inch). A pocket-type slide rule is useful. Carry a couple of sheets of graph paper (5 or 10 lines to the inch) or a notebook containing graph paper. Have a good eraser.

2. Obtain length of stride.

Either before you go to the field or after you get there, lay out a known distance of between 500 and 1000 feet. This can be done with a surveyor's tape or it can be measured on a good map where individual streets are clearly shown. Walk the distance between the known points at your normal walk—do not stretch out the length of step. Walk it in each direction, counting as you go.

A pace is the distance from right footprint to left footprint. A stride is two paces or the distance from right to right footprints. Count strides as you walk the known distance. Walk in the two directions, point A to B and back to A. Divide the known distance by the number of strides to get an average stride length.

Keep a record of your average stride. It is a very useful field tool.

3. Lay out several identified points.

With your mapping paper in hand, locate yourself in the field within sight of the point known from latitude and longitude.

By far the easiest points to identify in the field are on a trail, a road, or a culvert. Bends in the road may be another.

Place a mark in the center of your map sheet representing the center of the area where the latitude–longitude point is located. With respect to that point, place one of the identified places, such as a culvert. Pace from that first point to the second identified point, and get the compass bearing from the first to the second.

Once you know the distance and compass bearing, and orienting your map sheet so magnetic north is at the top of your sheet, plot the position of the second point in its proper relation to the first known point, using your compass as a crude protractor and your engineer's scale to measure the distance.

Plot the location of several other identifiable points in a similar manner.

From the published topographic quadrangle, read as closely as possible an elevation of one of the identified points, and write that elevation near that point on your map. Now, by looking at the points in the field, estimate their respective elevations, and enter the estimated elevation for each identified point.

4. Sketch the main linear features.

The main linear features on any piece of ground are the streams, gullies, roads, trails, electric lines, and fences. To the extent of your map, sketch the location of all such features. At this stage, paced distances are more useful than compass bearings. For example, in sketching the course of the stream, pace a few distances to make sure the position sketched relative to the identifiable points is correct. Any long straight line, such as a fence or center line of a road, should be plotted in a direction taken with the compass.

Remember that in using a Brunton compass, always have the mirror toward your body, not away from it. Then you can read the bearing of your line of sight by noting the angle on the engraved cardinal direction.

Scale paced distances or estimated distances

with the engineer's scale. Check some of your estimated distances by pacing. It is a good idea to estimate a distance by eye, then to pace it. This increases your ability to estimate as your experience grows.

5. Tentatively sketch the contours.

Controlled by the few elevations noted at the identifiable points, sketch in a tentative fashion a few contours, perhaps 50 feet different in elevation, or at the least 20 feet different. Later, intermediate contours will be drawn to the required contour interval. This preliminary sketching of contours will begin to lay out the general form of the topography.

The most important instrument you will own is an eraser. Use it.

Use of Sketched Profiles

At this stage your map will have on it the principal linear features, main road, creek, and draws or swales. A few key elevations have been entered, and a few rudimentary contours sketched.

Now locate a line across the topography, usually orthogonal to the main valley, the location of which can be quite accurately shown. Locate the ends of this line on your field map by the usual symbol for the ends of a cross section and label it A – A′.

Choose a horizontal distance scale on the cross-section paper. It will be most convenient to use your map scale, 1 inch = 100 feet. Choose an elevation scale without great vertical exaggeration, say 2 × (100 feet = 1 inch horiz., 50 feet = 1 inch vert.).

As is usual in river work, make the cross section from left to right looking downstream or downhill.

Looking at the ground across the line of the chosen cross section, and estimating the elevations of at least a couple of points along it, sketch the cross section on your cross-section paper. This forces you to make the elevations along that section consistent and in general agreement with the appearance of the whole topography along that line. It also makes you divide distances in a way that conforms to what you see.

Now plot the positions on the map where the contours cross your cross-section lines on the planimetric map, reading distances and elevations from the cross-section paper.

Draw sketch contours through the points of the cross sections. To obtain consistency with other parts of the topography, you may have to move the contours slightly from their plotted positions on the cross sections. As a check, plot on the cross section the adjusted elevations taken from the contour lines. These plotted points should agree in general with the original cross section.

Your map should show those features most prominent on the ground, roads, trails, vegetation masses, buildings, and whatever else is important for the work at hand.

Trace the rough map onto a clean piece of tracing paper, smooth the contours, and decide on the placement of lettering. Labels for contours should follow a straight or a smoothly curved path, for the sake of appearance. Trial positions for such labels can be put on the traced map.

A map should include the usual features: north arrow, date, name, location, symbols, contour interval, scale, declination and arrow to true north.

Ink the final map on tracing paper.

Fold the sheets for inclusion in a manila folder using three brass staples so that the sheet may be unfolded after it is bound. The folded sheet should be exactly page size, $8\frac{1}{2}'' \times 11''$. Include in the folder all your work sheets including the original map sheet made in the field.

A word of caution: Do not shortcut the fieldwork. Sketch enough cross sections or make extra notes. Another few minutes in the field can save you hours in the office trying to remember what it looked like.

14-2 Measurement of Drainage Area

Assume it is necessary to determine the discharge from a storm of given frequency expected in a channel at a designated point. The first information necessary is the drainage area. The present problem concerns outlining the drainage area and its measurement.

Solution

The drainage area is needed above a point located in the NE$\frac{1}{4}$ SE$\frac{1}{4}$ Sec. 5 T 31 N R 108 W shown on the 1:24000 quadrangle sheet, titled "Boulder, Wyoming." (Substitute your local area.)

The location is the mouth of an unnamed tributary to New Fork River that includes Vible Reservoir, a small stock pond. The mouth of the basin for which the drainage area is needed may also be located at latitude 42°41′00″, longitude 109°44′15″, which is an alternative and more specific location than the township and range description given above.

There are several features of a modern topographic map in the United States that assist in location. The sections (one square mile in area) are outlined in red and are at this scale 2.64 inches on a side, for the scale is 2000 feet to the inch. The red number in the center of the red square is the section number. If the section is divided into four parts and each in turn is divided into four parts, it can be seen that the first of the division gave four quarters that can be designated as NW, NE, SW, or SE. The same is true for the further subdivision. So the description given may be read as "The desired point is in the northeast quarter of the southeast quarter of section five."

Note that there are four black crosses located on a quadrangle sheet marking the intersections of $2\frac{1}{2}$ minutes of both latitude and longitude. These crosses line up with the black ticks at the margin marked respectively 40′ or 42′30″. The connection of the black tick at the margin with a cross makes it easier to interpolate the desired position, remembering that from a cross to the edge of the sheet is always 2′30″ or 150″ (150 seconds) of arc on the earth's surface.

Also note the small black numbers near blue ticks along the map margin. These are locations in metric units from base lines. On the sheet in question at the southeast corner appears a latitude designation 47 20000 m N or 4720 kilometers north of the equator. At the top margin appears the designation 6 03000 m E or 603 kilometers east of the nearest main meridian. The location desired can be read from these as 4726.230 kilometers north and 603.240 kilometers east of the respective base lines.

Having found the desired spot on the map, begin drawing a line uphill perpendicular to each contour, through the bulges or topographic noses and over the saddles. When nearing the upper part of the basin, begin again at the spot on the stream but on the opposite side from the previous line. The lines drawn will meet at some point on the drainage divide.

Compute the area outlined. This is most easily done by constructing a rectangular grid and counting squares, estimating partial squares. Usually it is easiest to trace the basin outline on to a sheet of graph paper, and count the squares on that grid.

In this basin there will be found 37.6 square inches in the area, and because the map scale is 2000 feet to the inch, each square inch is 4×10^6 square feet. A square mile is $(5280)^2$ or 27.9×10^6 square feet, so 1 square inch is .143 square miles and the whole basin has a drainage area of $.143 \times 37.6$, or 5.4 square miles.

15

Hillslope Processes

Hillslopes cover virtually the whole landscape. They are the land to be managed, the sites upon which forestry, agriculture, urban construction, and other human activities must be carried on in harmony with natural processes. The response of hillslopes to use by humans controls the long-term productivity of the land for cultivation, or its long-term suitability for residential development or other use. If unsuitable use causes accelerated soil erosion, hillslope failure, or other disruption, the land may become virtually worthless. Furthermore, sediment released by such activities is carried downstream and may have other damaging consequences, which we shall examine in Chapters 17 and 18.

The planner needs to be aware of hillslope processes and of the ways in which they can be disrupted. In particular, if the planner can recognize at an early stage in the planning process those areas that are most susceptible to damage, heavy costs and great inconvenience can be avoided. If a variety of options is considered, the land might even be put to better use.

Hillslopes vary in scale from a few meters in height and length to several kilometers long and more than a kilometer high. Their gradients vary from almost horizontal to vertical, and the dominant hill-forming processes vary from the imperceptibly slow to the catastrophic. Almost all geomorphic processes on hillslopes are concerned with the action of water, and therefore with the way in which runoff is produced. For this reason it is useful to outline the set of hillslope-forming processes by referring to the summary diagram of runoff processes outlined in Figure 9-1.

If, in subhumid regions or on hillslopes disturbed by human use, water does not infiltrate the soil, it remains on the ground surface and is responsible for a set of processes such as *rainsplash, sheetwash,* and *gullying* that are associated with Horton overland flow. Although saturation overland flow occurs on restricted areas of humid, thickly vegetated landscapes, the gradients of those areas are generally low, the surface is well vegetated, and the amount of surface erosion accomplished is negligible. In such regions, most water infiltrates the soil and rock, and causes a lowering of the strength of these hillslope materials. Where the topography is steep, the materials may be weakened to a point at which they can no longer resist the downslope component of gravity, and they will fail as catastrophic *mass movements,* such as landslides. Where the gradient is gentle, large gravitational stresses cannot be generated and the soil moves downhill by slow mass movement, or *soil creep,* as individual soil particles are jostled by frost action or wetting and drying. Figure 15-1 indicates some of the conditions in which each of these erosive processes is dominant. It is not meant to indicate that they cannot occur elsewhere. Soil creep, for example, occurs on virtually all soil-covered hillslopes. We will now review each of the major processes of hillslope erosion and consider how they affect human activity and how humans can take action to avoid or control the worst consequences of hillslope processes.

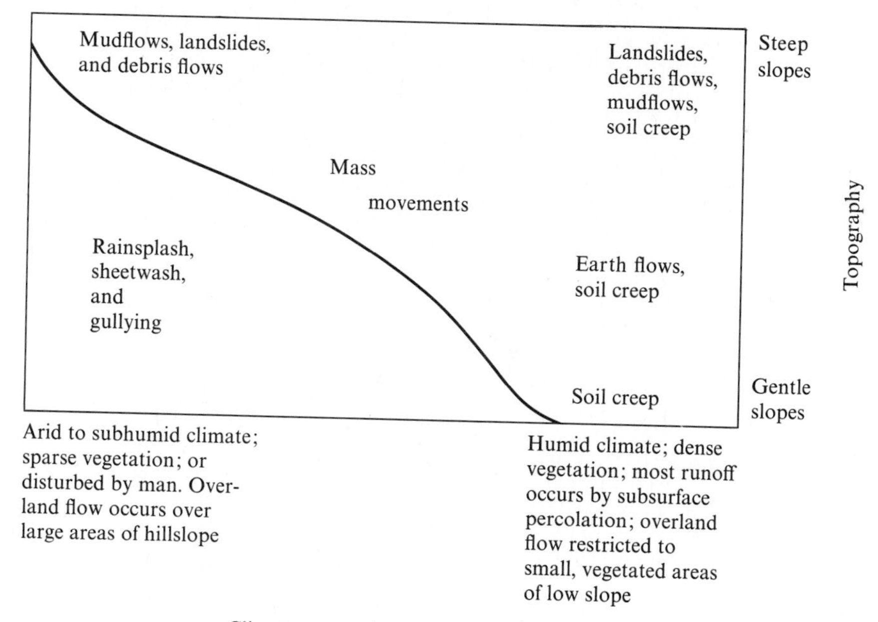

Figure 15-1 Conditions of climate, vegetation, land use, and topography under which various geomorphic processes dominate hillslope erosion, although they may not be restricted to the areas shown. Compare with Figure 9-16.

Geologically Normal and Accelerated Rates of Erosion

In its natural state, the soil covering a hillslope remains in a state of approximate balance for very long periods of time. The physical and geochemical processes by which rock minerals are broken up and decomposed are known collectively as *weathering.* They produce a layer of sand, silt, and clay, which mantles the solid bedrock and provides the inorganic skeleton for the soil (Figure 15-2). The clay particles are particularly important, because they are able to hold loosely on their surfaces plant nutrient ions, such as

Figure 15-2 Soil weathering from bedrock near Molo, Kenya. Chemical decomposition penetrates the rock along joints and other avenues of water percolation. Blocks of rock are isolated and slowly weathered to pebbles and eventually to sand, silt, and clay. The weathering sequence can be observed along a line from the lower part of the photograph to the soil surface.

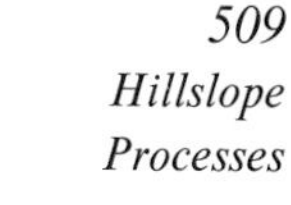

calcium, potassium, and phosphorus. The soil is also able to hold water and air in its pores, and these together with the nutrients provide a medium for plants and soil fauna that are able to store, decompose, and circulate organic matter composed of hydrogen, carbon, nitrogen, and sulfur. The organic matter (humus) is also capable of holding the nutrients in a form that makes them available for plant growth. A large number of soil-forming processes interact to develop distinctive layers, or horizons, with different physical and chemical characteristics (Fitzpatrick 1971).

Inputs of minerals by weathering and of organic matter by leaf fall and litter decomposition are balanced by other processes of export. Some of the organic materials are decomposed and returned to the atmosphere. Nutrients in solution are cycled through the vegetation or leached into streams by subsurface flow. The solid mineral particles and some of the organic material are removed by erosion of the surface or by mass movement (Figure 15-3). In the natural state, the depth, horizonation, and chemical characteristics of the soil attain a balance between these inputs and outputs of material. The soil achieves an equilibrium state that depends on the

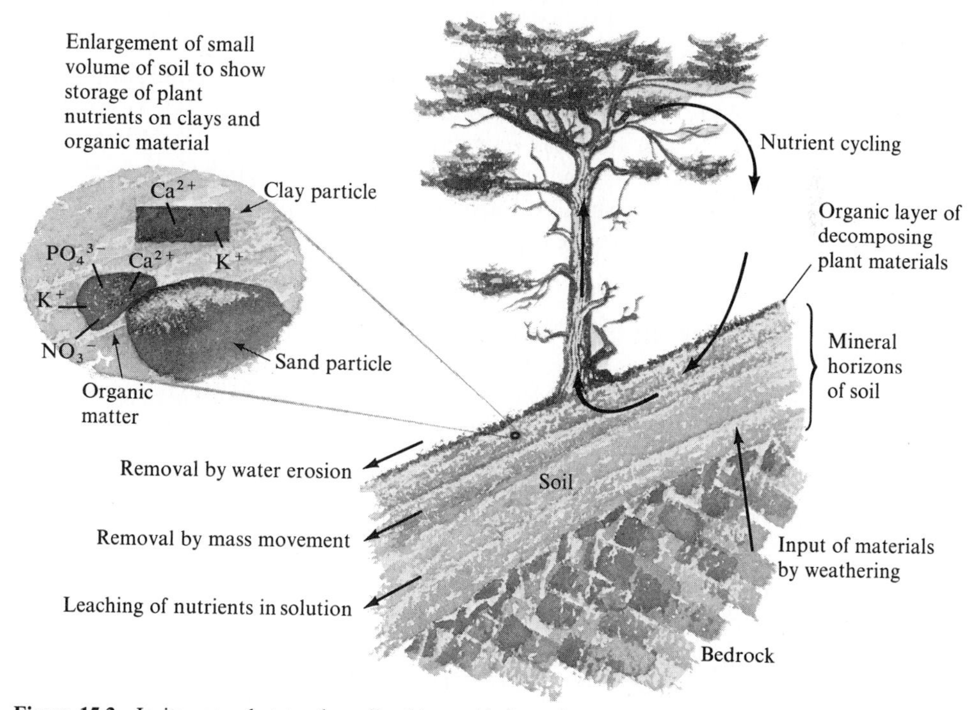

Figure 15-3 In its natural state, the soil achieves a balance between the input and output of materials.

local climate, geology, vegetation, and topographic conditions. This equilibrium is more or less stable for very long periods of time by comparison with the span of human history. The rate of erosion during this state of balance is usually referred to as the *geologically normal* rate of erosion, though it can fluctuate from time to time as climate varies. It is difficult for us to estimate what this rate was during the recent past in most areas, because few measurements have been made. However, later in this section we shall introduce some figures on rates of erosion under forest and dense grass, which approximate the geological norm for humid regions.

Human occupance of land almost always increases the rate of hillslope erosion by significant and sometimes catastrophic amounts. The term *accelerated erosion* is often used to emphasize this increase, and in regions where Horton overland flow is the dominant erosive agent, this term is used synonymously with *soil erosion,* although we have pointed out above that soil erosion is a natural process whose intensity varies from place to place even in the absence of humans.

Good management of soil for long-term productivity requires minimizing the acceleration of soil erosion so that the soil is not degraded of its store of nitrogen and other plant nutrients. In many rural regions this is a difficult task, expensive of both money and labor, and in many areas the battle is being rapidly lost. In such areas the bad effects of accelerated erosion are felt both on-site and downstream where the eroded soil comes to rest. On urban construction sites, roads, and mining spoil heaps, where erosion is very rapid, there is no question of maintaining the soil in balance; it has been destroyed. In these cases, as well as below severely eroded rural areas, the bad effects of accelerated erosion are felt downstream, as we shall see in Chapters 17 and 18. Mass movements are not usually considered under the term "accelerated soil erosion," but they can also be accelerated by human activity, as we shall describe later. They also have disruptive effects both on-site and downstream.

Hillslope Erosion by Water

In the early 1940's, experimental work by W. D. Ellison demonstrated the role of raindrops in splashing soil particles into motion. Earlier work in Europe and the United States had pointed out that the raindrops composing a storm of moderate size and intensity possess tremendous amounts of kinetic energy by virtue of their mass and their velocity of fall (kinetic energy $= 0.5 \times$ mass $\times$ velocity^2). Ellison showed that a significant part of this energy mobilizes soil, and is the major process initiating soil erosion at both the geologically normal and accelerated rates in areas subject to Horton runoff. He took high-speed photographs of the rainsplash erosion, as shown in Figure 15-4. It is possible to see clearly from the photographs that the impact of the drop generates a small explosion of soil and water. Large soil aggregates are dispersed and the smaller particles are splashed

Figure 15-4 High-speed photograph of a raindrop impact on a soil surface. Soil particles are splashed in all directions, but those traveling downhill move farther than those splashed uphill. (USDA–Soil Conservation Service.)

over several feet. The soil particles splashed downhill travel farther than those that move uphill, causing a net downslope transport of soil. Repeated billions of times per rainstorm, this downslope transport is what we measure as *rainsplash erosion.* After centuries of concern the fundamental cause of soil erosion finally became apparent from Ellison's work.

To appreciate this process, you may recall occasions when you have walked in the desert or on a construction site during a heavy rainstorm. Your trouser legs record a profile of the heights to which soil particles can be splashed by raindrops. After the rainstorm each area of soil protected by a pebble stands up on a small earth pillar, indicating the depth of soil moved by rainsplash.

Rainsplash erosion is particularly important on steep slopes devoid of vegetation, just the conditions created during parts of the year on agricultural fields, road cuts, and mining and construction sites. A thick vegetation cover intercepts virtually all the kinetic energy of rainfall (see Figure 15-5) and herein lies the critical and dominant role of vegetation in reducing soil erosion. Many cultivated crops leave the soil exposed to raindrop impacts (Figure 15-6). In general, the denser the vegetative cover, the lower the rate of erosion; this factor, which is controlled in turn by climate and land use, dominates the effects of all other controls such as rainfall energy and hillslope gradient. In fact, the area of the United States that experiences the heaviest rainfalls, namely the southeast, also has one of the lowest geologic rates of erosion because of the thick natural forest cover. When the forests of the region were cleared for planting corn, tobacco, and cotton in the late

Figure 15-5 Thick ground cover of grass and clover. A thick vegetation cover protects the soil very efficiently from the kinetic energy of raindrops. Few if any drops reach the soil directly but rather drop from a small height, often onto organic litter. (USDA–Soil Conservation Service.)

Figure 15-6 Under a corn crop bare ground is exposed to raindrop impact and sheetwash. In this picture overland flow is running off a hillslope on a Maryland farm. (USDA–Soil Conservation Service.)

eighteenth and early nineteenth centuries, however, the result was catastrophic accelerated erosion. Parts of Brazil, the Malagasy Republic, Malàwi, and many other tropical and subtropical regions have suffered the same fate.

Another important factor controlling the intensity of rainsplash erosion is the resistance of soil to dispersal. High organic content and moderate amounts of clay and calcium seem to be the factors promoting the development of stable soil aggregates. Silt and sand are more easily splashed, and many tillage operations break down the aggregates to the detriment of soil resistance. The factors controlling resistance to splash are reviewed by Bryan (1968).

As described in Chapter 9, in areas where rainfall intensity exceeds the infiltration capacity of the soil, water accumulates on the land surface and runs downslope as an irregular sheet. The depth and velocity of this water increase downslope as more water is generated by precipitation excess, until the force applied to the soil by the water is sufficient to exceed the resistance of soil to erosion. The intensity of erosion depends on the product of water depth and hillslope gradient. It may increase or decrease with distance downslope depending on the combination of water depth and hillslope gradient along the hillslope profile. A common situation is for erosion to increase with distance as the gradient steepens downslope, and then for erosion to decrease, or even for some deposition to take place on the gentle footslope. These relationships are summarized in Figure 15-7. Our picture

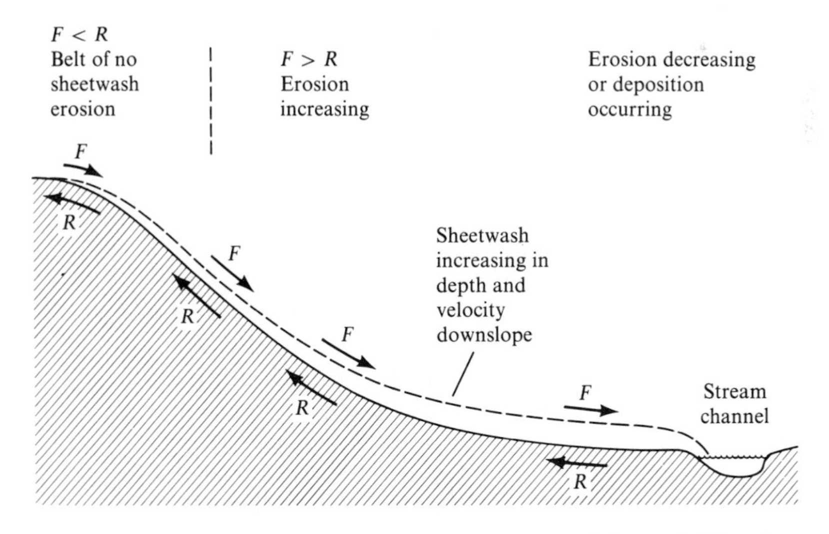

Figure 15-7 The relative magnitudes of the eroding force of sheetwash (F) and the resistance (R) of the soil to erosion determine the width of a belt of no sheetwash erosion near the top of a hillslope. Below this belt the intensity of sheetwash erosion depends on the erosive force F (which is proportional to the product of hillslope gradient and water depth). Note that rainsplash erosion does occur in the belt of no sheetwash erosion.

Figure 15-8 A hillslope undergoing rapid rainsplash and sheetwash erosion as a result of heavy grazing, cutting, and burning. The pedestals of soil surviving under each tree indicate that the hillslope, which has a gradient of only two percent, has lost 60 cm of topsoil during the last 50 years. In the foreground, the soil surface shows signs of vigorous surface wash.

of a uniform sheet of water is an idealization. In fact, the sheet contains tiny streams and threads of water that are slightly deeper and faster than the average. They move back and forth across the hillslope during the rainstorm and together with rainsplash are responsible for most of the erosion and transportation of soil from the hillslope. We will refer to this erosion process, however, as *sheetwash erosion,* because the net effect of the movement of these threads through many rainstorms is to remove a more or less uniform depth of soil from the hillslope. Often the term sheetwash erosion encompasses rainsplash erosion because the two are difficult to separate in the field, but we have chosen to discuss them separately here. The effects of combined sheetwash and rainsplash erosion are apparent in Figure 15-8.

Some of the controls of rainsplash erosion also affect sheetwash erosion. Rainfall intensity determines the amount and rate of precipitation excess and therefore of the amount and depth of runoff. Vegetation supplies organic matter to the soil, provides favorable conditions for soil microfauna, which with plant roots promote loose topsoil with a high infiltration capacity and low rates of sheetwash. Soil texture is another determinant of infiltration capacity, as we discussed in Chapter 6.

The resistance of soil to erosion is another important factor, though one that is not clearly understood. Large soil particles are more difficult to move than finer ones, but the very finest silt and clay particles are more difficult to move than sand. A good soil structure in which particles are held together by organic material and cations into a porous aggregate is known to resist sheetwash erosion as it resists dispersal by rainsplash. A moderate amount of clay in the soil binds sand particles together, but high clay content lowers the infiltration capacity and promotes runoff.

Hillslope gradient affects sheetwash erosion by increasing the erosive force the runoff applies to the soil surface, and hillslope length is important because, as indicated in Figure 15-7, the depth of overland flow increases with the distance downslope and its ability to erode is proportional to its depth. At the top of a hillslope, the depth of the sheet is not sufficient for the eroding force to overcome the resistance of the soil to erosion. There will be a zone near the top of the slope in which sheetwash erosion cannot occur, though rainsplash erosion can. The relative importance of rainsplash and the force applied by sheetwash changes with the distance downslope. On very long hillslopes with low infiltration rates, sheetwash usually becomes the dominant process, although rainsplash is still exceedingly important in mobilizing soil which is then transported by the sheetwash. The combination of the two processes can be appreciated by closely observing surface runoff during a lull in a storm when abundant water is moving over the surface, but only an occasional large raindrop falls. The drop passes through the sheet of overland flow and splashes soil up into the flow. A small, faint plume of sediment can be seen drifting downslope with the sheetwash.

If the minute streams of water cut separate channels, as shown in Figure 15-9, the process is known as *rill erosion.* The concentration of runoff in

Figure 15-9 Rill erosion on an agricultural field in Illinois. (USDA–Soil Conservation Service.)

these small channels causes an increase in the efficiency and intensity of soil removal. On agricultural fields, tillage operations may obliterate the rills each year, causing their significance to be overlooked.

If the rills become engraved into the land surface to depths of more than about one foot, and especially if they are not obliterated by cultivation and do not migrate back and forth across the hillslope, they are generally classified as gullies (Figure 15-10). *Gully erosion* produces incisions ranging in size from a foot in depth and width and a few feet long to several tens of feet deep, hundreds of feet wide and miles long. Geomorphologists do not yet understand exactly how rills and gullies form and remain stable, but they are generally associated with very intense water erosion under the most severe combinations of the controlling factors listed above.

The amount of soil exported by gully erosion is usually small by comparison with that removed by rainsplash and sheetwash erosion, but the process can do great damage to an individual piece of land. In a heavily cut and grazed region in northern Kenya, gully erosion measured by Dunne has mobilized only 1 percent as much sediment as sheet erosion. Leopold et al. (1966) found that gully growth supplied only 1.4 percent of the sediment yield of a small grazed catchment in a semi-arid area near Santa Fe. Rainsplash and sheetwash erosion supplied 97.8 percent and soil creep 0.7

Figure 15-10 Gully erosion after land was cleared and graded for housing construction. In a single winter the main gully was cut to a depth of 8 feet and an average width of 50 feet, and was more than 500 feet long. Juanita Creek basin, Kirkland, Washington. (USDA–Soil Conservation Service.)

Figure 15-11 Construction sites undergo intense sheetwash and gully erosion because the vegetative cover is removed and the surface is continually steepened, churned, and compacted. (USDA–Soil Conservation Service.)

percent. Brune (1950) measured gully contributions of 3.4 to 4.0 percent in an agricultural basin in Illinois; sheet erosion supplied 79–96 percent, with the remainder coming from erosion in the valley bottom, and probably from soil creep. Glymph (1957) compared gully and sheetwash erosion at many localities in the agricultural lands of the east and central United States. Gullies supplied from 0 to 89 percent of the total sediment yield, but most of the high values came from one region of Mississippi where the process is particularly severe. Seventy-five percent of Glymph's values were less than 30 percent.

We do not yet have any quantitative estimates of the contributions of rill and gully erosion on urban construction sites, road cuts, or mined areas and spoil heaps, where conditions favor all types of water erosion. As we will show in the next section, rates of soil loss from such areas are very high because the vegetative cover is removed, the permeable topsoil is removed, and the dense subsoil is exposed, compacted, and graded into steep slopes and is frequently disturbed by heavy machinery. Some of the conditions encountered on these sites are shown in Figure 15-11. Mined areas and their spoil heaps suffer rapid water erosion for the same reasons.

Even in forested and agricultural areas, roads are important sources of sediment and need constant care to minimize damage to stream systems (Figure 15-12). More will be said in Chapter 17 about the effect of urbanization and other types of land use on the sediment yields of drainage basins.

Figure 15-12 Sediment washing from a rural road in eastern Kenya.

Rates of Water Erosion

Because of the difficulties of separating rainsplash, sheetwash, and rill erosion, their effects are usually measured together by monitoring the amount of sediment lost from hillside plots like those shown in Figure 15-13, or from small catchments. Another method of obtaining the same data is by repeated measurement of the exposure of stakes or nails (Figure 15-14). A detailed discussion of techniques for evaluating various kinds of erosion is provided by Leopold et al. (1966) and by Dunne (1977).

Thousands of measurements of rates of soil loss have been made from hillslopes with different gradients, length, vegetative covers, and soil conservation techniques. We make no attempt to summarize this literature, except to present a few representative measurements under different types of land use. They simply emphasize the overriding importance of vegetative cover and therefore of land use. Figure 15-15(a) is perhaps the most widely quoted summary of soil loss from agricultural plots in the midwestern United States. Figure 15-15(b) shows a similar compilation from studies in Tanzania. The results speak for themselves in view of our foregoing discussion. A sample of data is listed in Table 15-1.

Figure 15-13 Experimental plots for measuring runoff and soil erosion under various crops and cultivation practices at Marlboro, New Jersey. The large tanks trap the water and sediment caught in the trough at the bottom of each plot. (USDA–Soil Conservation Service.)

Figure 15-14 Erosion pins on a hillside near Santa Fe, New Mexico. Nails and washers are placed in lines on the slope, and the height from the top of the nail to the washer is measured annually with a millimeter scale.

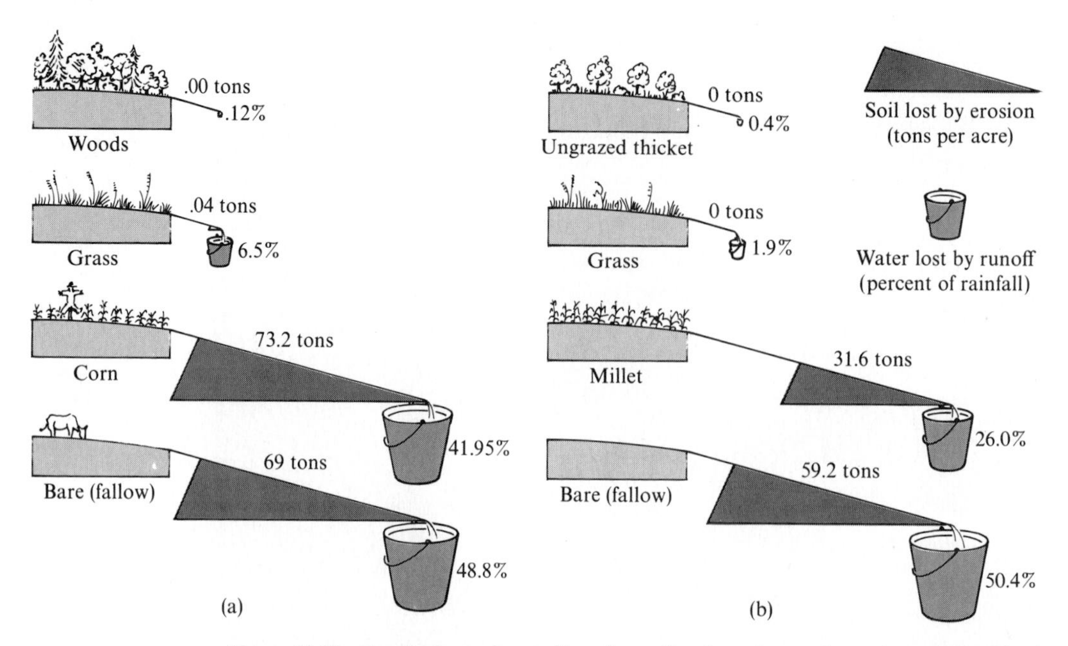

Figure 15-15 Results from plot studies of runoff and erosion under various types of land use. (a) Midwestern United States. (Soil Conservation Service.) (b) Mpwapwa, Tanzania. (From Rapp et al. 1972.)

Table 15-1 Some measurements of soil loss from hillside plots.

LAND USE	LOCATION	SOIL LOSS (TONS/ACRE/YR) [1 TONNE/HECTARE = 0.4474 TONS/ACRE]	SOURCE
FOREST			
Primeval	Oklahoma	0.01	Smith and Stamey (1965)
Burned annually	Oklahoma	0.11	Smith and Stamey (1965)
Primeval	North Carolina	0.002	Smith and Stamey (1965)
Primeval	Kenya	0.09	Dunne, unpublished
Burned semiannually	North Carolina	3.08	Smith and Stamey (1965)
Second growth	Ivory Coast	0.40	Nye and Greenland (1960)
Woodland, protected	Texas	0.05	Smith and Stamey (1965)
Woodland, burned annually	Texas	0.36	Smith and Stamey (1965)
Woodland, protected	Ohio	0.01	Smith and Stamey (1965)
Woodland, protected	North Carolina	0.08	Dils (1953)

continued

Table 15-1, *continued*

LAND USE	LOCATION	SOIL LOSS (TONS/ACRE/YR) [1 TONNE/HECTARE = 0.4474 TONS/ACRE]	SOURCE
AGRICULTURE, CULTIVATED GRASSLANDS			
Bluegrass	Midwestern U.S.	0.02–0.34	Smith and Stamey (1965)
Alfalfa	Midwestern U.S.	0.03–0.15	Smith and Stamey (1965)
Clover and grass	Virginia	0.01–0.07	Smith and Stamey (1965)
Bermuda grass	S.W. United States	0.00–0.10	Smith and Stamey (1965)
Fescue grass	Georgia	0.20	Barnett (1965)
Hayland	Washington	0.01–0.08	Smith and Stamey (1965)
Hayland	North Carolina	0.31	Smith and Stamey (1965)
Tropical perennial grasses	Puerto Rico	1.20	Smith and Stamey (1965)
Tropical kudzu	Puerto Rico	0.18	Smith and Stamey (1965)
Hayland	India	1.05–8.30	Vasudevaiah et al. (1965)
Tropical pasture grasses	India	0.20	Vasudevaiah et al. (1965)
Dubgrass	India	0.47	Battawar and Rao (1969)
Grass	India	1.4–1.8	United Nations (1951)
AGRICULTURE, CROPLANDS			
Bare fallow	Georgia	100	Barnett (1965)
Bare fallow	Tanzania	59.2	Rapp et al. (1972)
Bare fallow	Ivory Coast	45	Dabin (1959)
Bare fallow	Midwestern U.S.	69	Bennett (1939)
Bare fallow	India	14.6–15.4	United Nations (1951)
Maize (corn)	India	4.7	Battawar and Rao (1969)
Maize	India	10.54	Vasudevaiah et al. (1965)
Maize	Midwestern U.S.	17.86	Jamison et al. (1968)
Maize	Rhodesia	2.1–4.5	Hudson and Jackson (1959)
Maize	Midwestern U.S.	73.2	Bennett (1939)
Millet	Tanzania	31.6	Rapp et al. (1972)
Hill rice	India	1.0	Battawar and Rao (1969)
Hill rice	India	9.4	Vasudevaiah et al. (1965)
Hill rice	Java	11.2	Nye and Greenland (1960)
Hill rice	Guinea	0.4–1.8	Nye and Greenland (1960)
Arhar grain	India	5.5	Battawar and Rao (1969)
Cowpeas	India	1.9	Battawar and Rao (1969)
Urid grain	India	21.0	Vasudevaiah et al. (1965)
Peanuts	India	6.2	Vasudevaiah et al. (1965)

continued

Table 15-1, *continued*

LAND USE	LOCATION	SOIL LOSS (TONS/ACRE/YR) [1 TONNE/HECTARE = 0.4474 TONS/ACRE]	SOURCE
RANGELAND			
Dry woodland and rangeland	Southern California	2.7	Krammes (1960)
Dry woodland and rangeland, after fire	Southern California	24.7	Krammes (1960)
Dry woodland and rangeland	New Mexico	21.2	Leopold et al. (1966)
Sparse grassland	Alberta	7.7	Campbell (1970)
Dry woodland and rangeland, heavily cut and grazed	Kenya	13–76	Dunne, measurement
Grass and scrub	India	1.5–1.8	United Nations (1951)
URBAN			
Road cuts	Georgia	79–237	Diseker and Richardson (1961)
Building sites	Maryland	125–219	Wolman and Schick (1967)
Building sites	Maryland	189	Guy (1965)
MINING			
Land devegetated by smelter fumes	Ontario	26.1	Pearce (1973)
Spoil bank	Ohio	87	*Geotimes* (1971, Dec)
RURAL ROADS			
Forest roads in a jammer unit	Idaho	29.7	Megahan and Kidd (1972)
Forest roads	Idaho	7.9	Copeland (1965)
Rural road	N. Kenya	53.7	M. Norton-Griffiths,

Prediction of Soil Erosion

In the planning of agricultural land use, rural development schemes, and small water-resource developments, it is frequently necessary to estimate the rate at which soil is being lost from hillslopes, or the rate at which it will be lost if a new crop is introduced, if soil conservation programs are begun, or if there is a fluctuation of the weather, such as a run of very wet years.

This is being done with increasing frequency, care, and success as we accumulate unfortunate experiences about the potential impact of soil erosion upon agricultural production and engineering works, and as the techniques for making predictions of erosion improve.

The best method of predicting soil loss is to have some local field data that are representative of the range of conditions found in the area of interest. Most planners are not in a position to monitor soil erosion in the field. They should be aware, however, that some agricultural engineers, geomorphologists, and hydrologists do monitor soil loss, and that data are available in some areas. This is particularly true in agricultural regions where the Soil Conservation Service, and the Agricultural Research Service in the United States, and similar organizations elsewhere have accumulated a great deal of information about soil loss from cultivated fields. Planners are often in a position to encourage further collection of such data to document both the impact of human activities and of natural fluctuations of weather. The necessary fieldwork is not expensive and can yield valuable information for planning.

A second method of predicting erosion is by the use of multivariate equations developed from data collected at large numbers of experimental sites. Many examples of multiple regression equations are available in the literature, but they are only useful in the area for which they were developed. The prediction–equation approach has been most highly and usefully developed by the Soil Conservation Service and the Agricultural Research Service over the past 35 years, and many predictive equations have evolved. The most comprehensive and useful of these has become known as the *Universal Soil-Loss Equation* (Wischmeier and Smith 1965) and continues to undergo refinement and testing both within the United States and in other countries (Battawar and Rao 1969, Hudson 1971). The technique can also be used to predict soil loss from construction sites (Wischmeier and Meyer 1973).

The equation predicts *only* the amount of soil moved from its original position on a field or hillside. It does not estimate the net soil erosion resulting from the difference between erosion and deposition. Nor can its results be compared with the sediment yield of a river basin. The following computations provide an index of net soil erosion that allows the planner to recognize sensitive sites and to plan strategies to minimize soil loss.

The Universal Soil-Loss Equation is

$$A = RKLSCP \tag{15-1}$$

where A = soil loss (tons per acre)
R = the rainfall erosivity index
K = the soil erodibility index
L = the hillslope-length factor
S = the hillslope-gradient factor
C = the cropping-management factor
P = the erosion-control practice factor.

The rainfall erosivity index is calculated from the kinetic energy (E) of each rainstorm multiplied by the maximum 30-minute intensity of the storm

(I_{30}). The kinetic energy varies with the rainfall intensity (I) according to the relationship

$$E = 916 + 331 \log_{10} I \tag{15-2}$$

where the energy is expressed in foot-tons per acre per inch of rainfall and the rainfall intensity in inches per hour. The rainstorm is divided into portions of uniform intensity for the application of this equation. The energy per inch of rain is calculated from Equation 15-2 and is multiplied by the amount of rain falling during the time interval. Increments of energy are then summed to obtain E for the whole storm.

The rainfall erosivity index can then be calculated from the sum of the $E \times I_{30}$ product for each storm during the period of interest, such as a year or a season, according to the equation

$$R = \frac{\sum_{i=1}^{n} E_i I_{30i}}{100} \tag{15-3}$$

The factor can be calculated from the rainfall intensity values obtained with a recording gauge. For the eastern and central United States, annual erosivity values have been mapped (see Figure 15-16), and graphs of seasonal distribution and probability of occurrence of various amounts of rainfall energy are given by Wischmeier and Smith.

The soil erodibility factor, K, is the average soil loss, in tons per acre, per 100 foot-tons per acre of rainfall erosivity (i.e., per unit value of the erosivity index given in Equation 15-3) when the soil is exposed as cultivated bare fallow under specified conditions of hillslope length and gradient. Values of K were obtained by direct field measurement on a few agricultural soils in the United States (see Figure 15-17), and K values for other soils can be estimated from these by comparing their physical properties. For most of the major soil series in the United States such erodibility factors can be obtained from the Agricultural Research Service or from the Soil Conservation Service. Wischmeier et al. (1971) have extended the method by the development of Figure 15-18, which can be used to estimate K values for both agricultural soils and for the dense subsoils exposed on construction sites. The soil properties needed for the newer technique are the percentage of soil particles between 0.002 and 0.1 mm, the percentage greater than 0.1 mm, the organic matter content, the soil structure, and the permeability. The use of the nomograph is illustrated by the path of the dashed line. The authors of the work claim that about 95 percent of all K values estimated in this way should lie within ± 0.04 of the true value.

The length and slope factors are usually evaluated together from Figure 15-19. If convex or concave hillslopes are being studied, the length and slope values used should be modified. If the lower end of the hillslope is much steeper than the upper portion, the gradient of the steeper portion should be used with the overall length. On a concave hillslope the gradient and length of the upper, steeper segment only should be used.

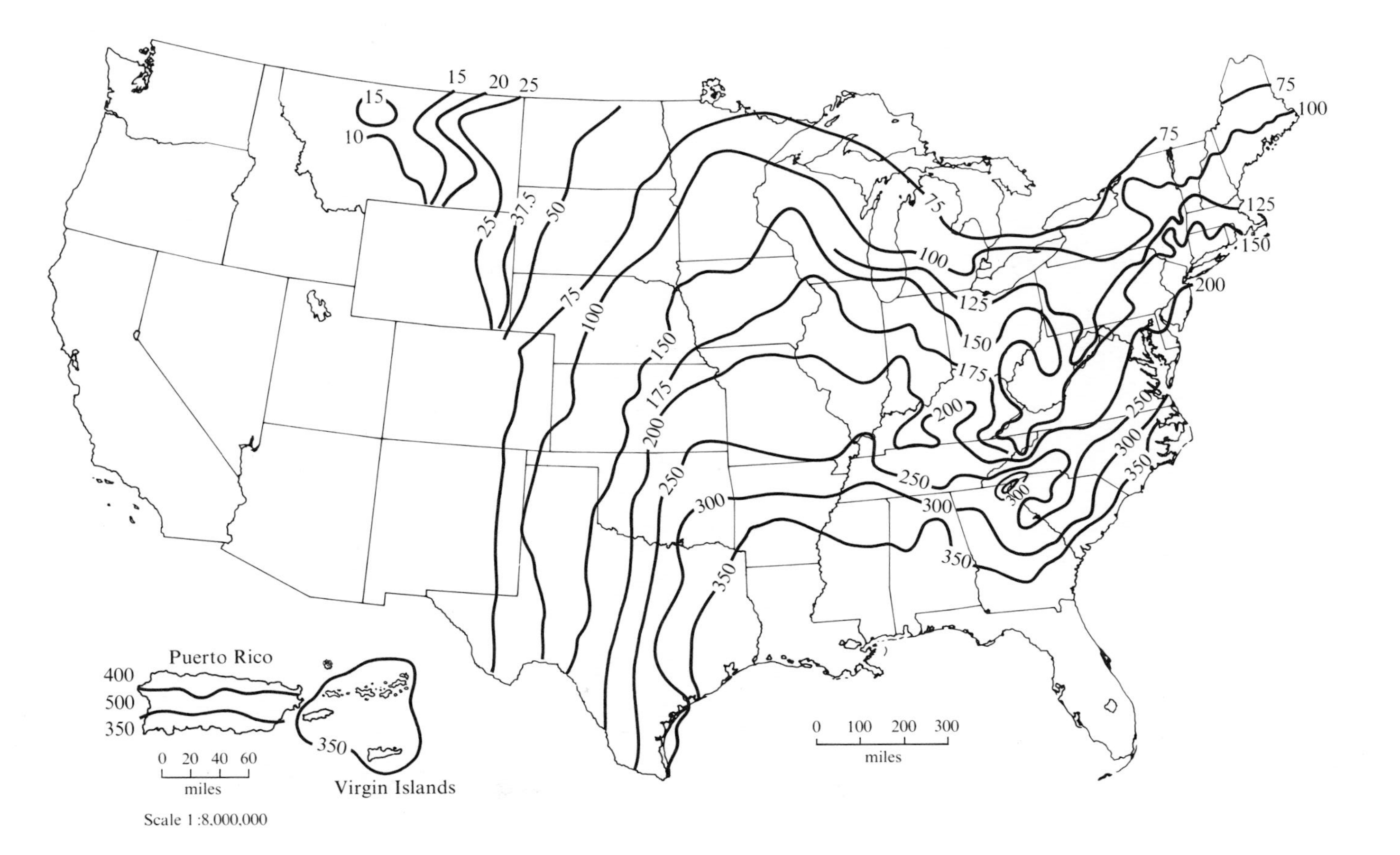

Figure 15-16 Average annual totals of erosivity in hundreds of foot-tons per acre for the eastern and central United States. Note that the values on this map differ from those given by Wischmeier and Smith (1965). The Agricultural Research Service has determined that the higher values shown in *Agricultural Handbook 282* are unrealistically high. (From U.S. Soil Conservation Service 1975b.)

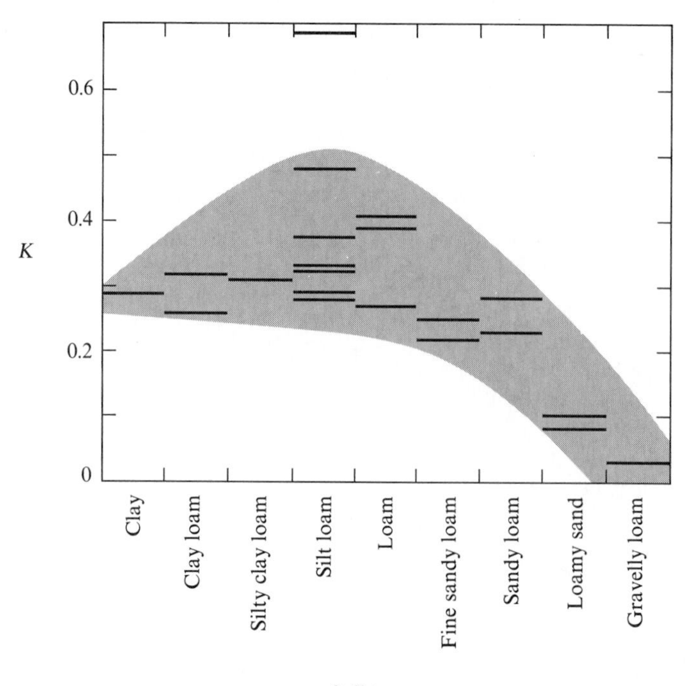

Figure 15-17 Summary of measured K values for a range of soils in the eastern and central United States. Each horizontal bar indicates one plot measurement. The shaded area represents the range of the data available so far, apart from a single measurement on a silt loam. (Data from Wischmeier and Smith 1965.)

The cropping-management factor, C, includes the effect of vegetative cover, the sequence of crops in a rotation, the stage of the crop, tillage practices, and residue management. Values of C are given in the Wischmeier and Smith report for many different crop rotations and for seven field conditions covering four crop stages and the fallow period. The C value for each crop stage in Table 15-2 can be weighted by the proportion of the annual total erosivity for the location (also given for the United States by Wischmeier and Smith) to calculate a weighted average value of the crop-management factor, C_w.

Thus,

$$C_w = \frac{\sum_{i=1}^{n} C_i \cdot \%R_i}{100} \tag{15-4}$$

where $\%R_i$ is the percentage of rainfall erosivity units occurring in the ith crop stage. The U.S. Soil Conservation Service (1975c) also releases C values for other agricultural regions, and for woodland and pasture, as illustrated in Tables 15-3 and 15-4. Wischmeier and Meyer (1973) provide some guidance on C values for construction sites.

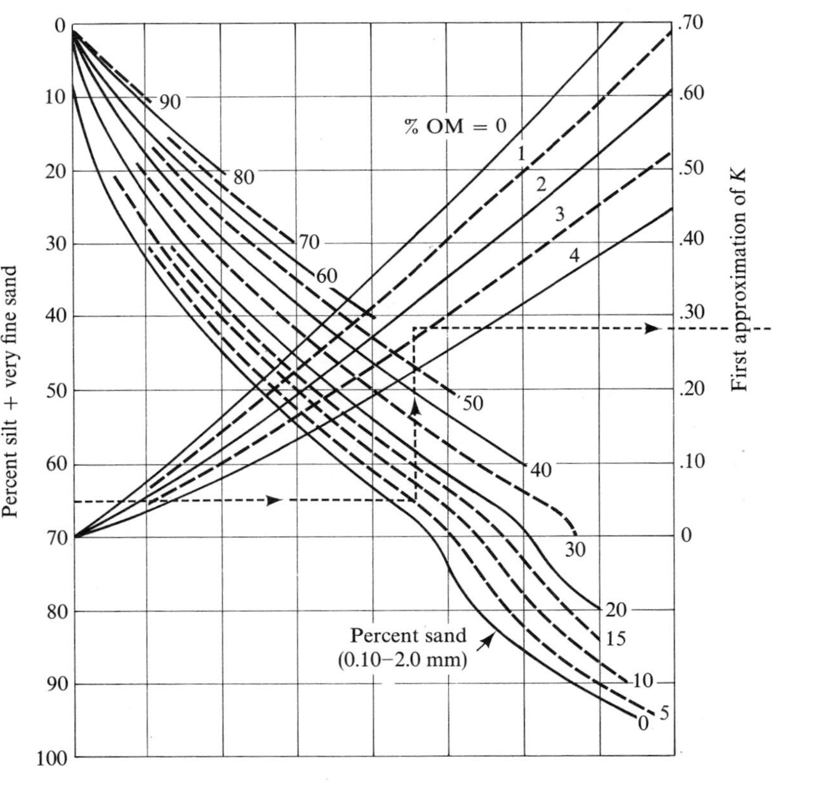

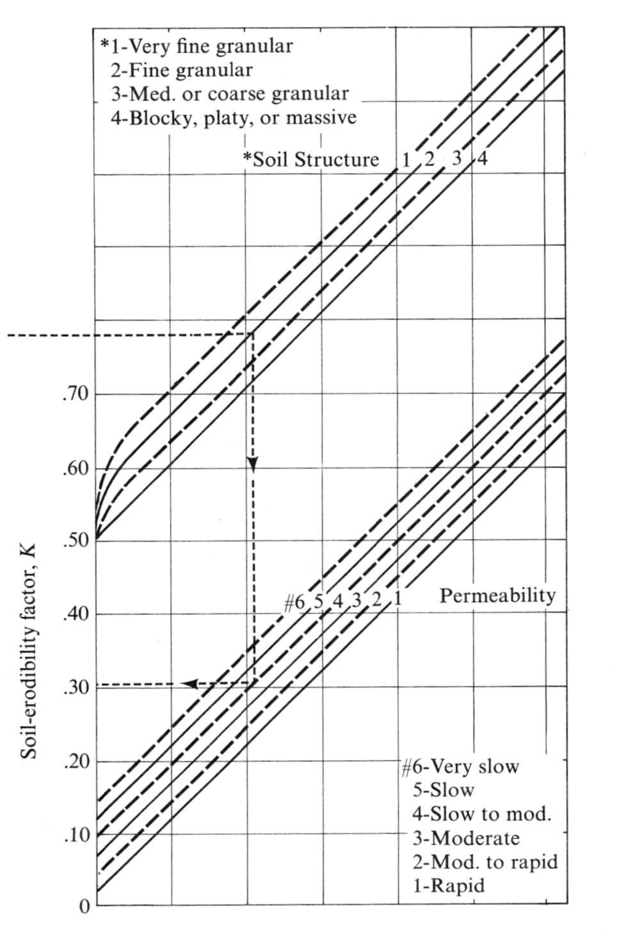

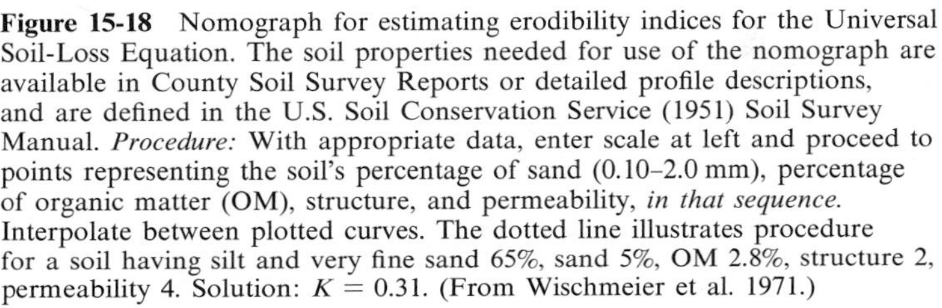

Figure 15-18 Nomograph for estimating erodibility indices for the Universal Soil-Loss Equation. The soil properties needed for use of the nomograph are available in County Soil Survey Reports or detailed profile descriptions, and are defined in the U.S. Soil Conservation Service (1951) Soil Survey Manual. *Procedure:* With appropriate data, enter scale at left and proceed to points representing the soil's percentage of sand (0.10–2.0 mm), percentage of organic matter (OM), structure, and permeability, *in that sequence.* Interpolate between plotted curves. The dotted line illustrates procedure for a soil having silt and very fine sand 65%, sand 5%, OM 2.8%, structure 2, permeability 4. Solution: $K = 0.31$. (From Wischmeier et al. 1971.)

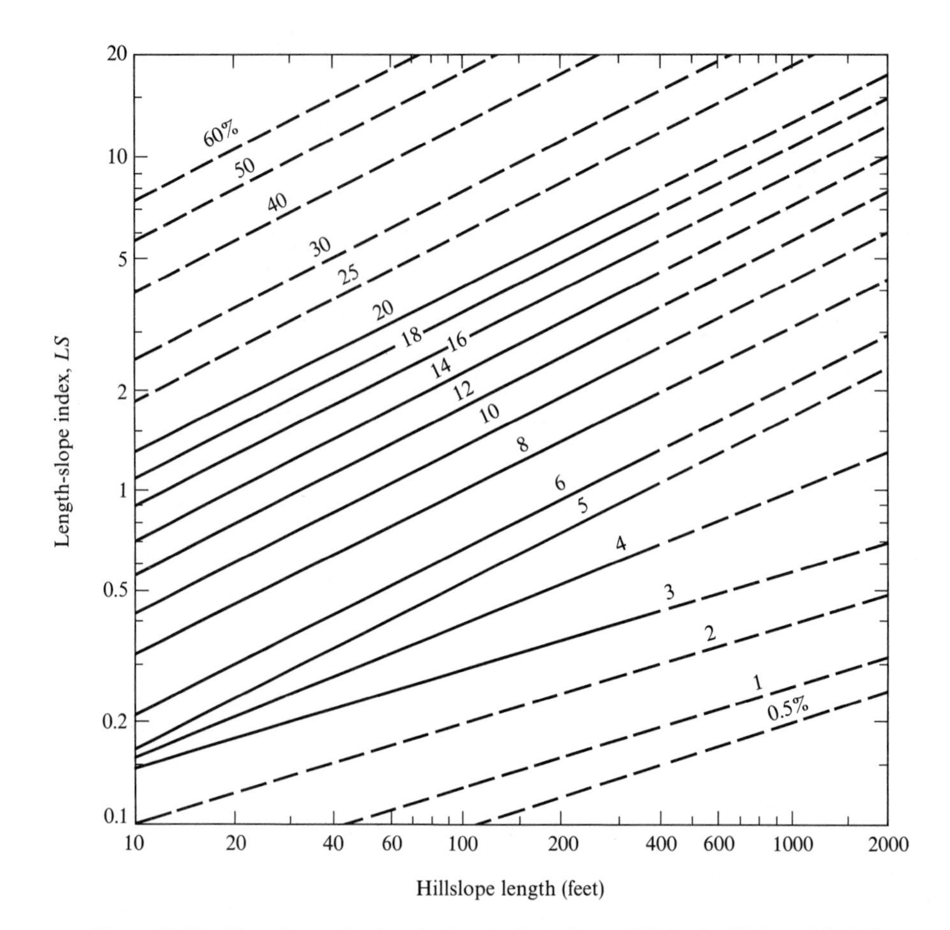

Figure 15-19 Chart for evaluating the length–slope factor, *LS*, in the Universal Soil-Loss Equation. The solid lines represent conditions within the range of data from which the curves were derived. The dashed lines are based on extrapolations and should be used with care. (From U.S. Soil Conservation Service 1975b.)

Table 15-2 Examples of cropping-management factors, *C* in the Universal Soil-Loss Equation, for agricultural land. Many other crop combinations are listed in the original reference as well as in technical releases for particular regions. The crop stages are also defined in the original source. (From Wischmeier and Smith 1965.)

	CROP STAGE				
CROP	FALLOW	SEEDLING	ESTABLISHMENT	MAJOR GROWTH	RESIDUE
Corn with conventional tillage in rotation with small grains and grass	0.36	0.63	0.50	0.26	0.30
Alfalfa					0.02

Table 15-3 Cropping-management factors, *C* in the Universal Soil-Loss Equation, for woodland. (From U.S. Soil Conservation Service 1975b.)

TREE CANOPY (% OF AREA)	% OF AREA COVERED BY > 2 INCHES OF FOREST LITTER	UNDERGROWTH	*C*
100–75	100–90	Grazing and burning controlled	0.001
		Heavily grazed and burned	0.003–0.011
70–40	80–75	Grazing and burning controlled	0.002–0.004
		Heavily grazed and burned	0.01–0.04
35–20	70–40	Grazing and burning controlled	0.003–0.009
		Heavily grazed and burned	0.02–0.09
<20	Treated as grassland or cropland		

Table 15-4 Cropping-management factors, *C* in the Universal Soil-Loss Equation, for pasture, rangeland, and idle land. (From U.S. Soil Conservation Service 1975b.)

TYPE OF CANOPY AND AVERAGE FALL HEIGHT OF WATER DROPS	CANOPY COVER (%)	GROUND COVER*	PERCENT GROUND COVER					
			0	20	40	60	80	95–100
No appreciable canopy		G	.45	.20	.10	.042	.013	.003
		W	.45	.24	.15	.090	.043	.011
Canopy of tall weeds or short brush (0.5 m fall ht)	25	G	.36	.17	.09	.038	.012	.003
		W	.36	.20	.13	.082	.041	.011
	50	G	.26	.13	.07	.035	.012	.003
		W	.26	.16	.11	.075	.039	.011
	75	G	.17	.10	.06	.031	.011	.003
Appreciable brush or brushes (2 m fall ht)	25	G	.40	.18	.09	.040	.013	.003
		W	.40	.22	.14	.085	.042	.011
	50	G	.34	.16	.085	.038	.012	.003
		W	.34	.19	.13	.081	.041	.011
	75	G	.28	.14	.08	.036	.012	.003
		W	.28	.17	.12	.077	.040	.011
Trees but no appreciable low brush (4 m fall ht)	25	G	.42	.19	.10	.041	.013	.003
		W	.42	.23	.14	.087	.042	.011
	50	G	.39	.18	.09	.040	.013	.003
		W	.39	.21	.14	.085	.042	.011
	75	G	.36	.17	.09	.039	.012	.003
		W	.36	.20	.13	.083	.041	.011

*G = Cover at surface is grass, grasslike plants, decaying compacted duff, or litter at least 2 inches deep.
W = Cover at surface is mostly broadleaf herbaceous plants (as weeds with little lateral-root network near the surface, undecayed residue, or both).

Table 15-5 Erosion-control practice factor (*P*) for the Universal Soil-Loss Equation. (From U.S. Soil Conservation Service 1975b.)

LAND SLOPE (%)	*P* VALUES			
	CONTOURING	CONTOUR STRIP CROPPING*	CONTOUR IRRIGATED FURROWS	TERRACING†
2.0–7	0.50	0.25	0.25	0.10
8.0–12	0.60	0.30	0.30	0.12
13.0–18	0.80	0.40	0.40	0.16
19.0–24	0.90	0.45	0.45	0.18

*Using a 4-year rotation of maize, small grain, meadow, and meadow.
For prediction of contribution to off-field sediment load.

The erosion-control practice factor, *P*, varies with such techniques as contour cultivation, strip cropping, and terracing. Recommended values are given in Table 15-5.

The Universal Soil-Loss Equation can be extended from individual fields and hillslopes to small drainage basins by dividing the basin map into areas of uniform soil type, topography, and agronomic conditions, and computing the soil loss for each combination.

In the United States, a concept has been developed of a "tolerable soil-loss rate," which is the erosion rate below which land productivity can be maintained. The values of tolerable erosion estimated for different soils in the United States vary from about one to six tons per acre per year. In planning agricultural development, therefore, this rate can be substituted for *A* in Equation 15-1, and for the fixed values of *R*, *K*, *L*, and *S* for a field, the values of *C* and *P* can be manipulated to balance the equation. This exercise indicates the management options available for keeping the erosion rate below the tolerable level, and policies can then be followed to encourage the adoption of these practices among farmers. It should be realized, however, that the values decided upon for the agricultural areas of the United States refer mostly to good conditions of soil and climate, a high level of management skills on the farm, and a situation in which fertilizers and energy inputs have been very cheap. These last two conditions are not likely to last much longer, even in North America. In many other countries with a soil-erosion problem, conditions are even less encouraging for all the factors referred to. It is likely, therefore, that tolerable rates of soil erosion in some areas are lower than one ton per acre per year.

On urban construction sites, road cuts, and rural roads, the Universal Soil-Loss Equation can be used to predict rates of sediment production and their impact upon turbidity levels of receiving streams, as well as to design strategies for sediment control within the disturbed area (Wischmeier and Meyer 1973).

The approach incorporated in the equation has many possibilities for application and refinement. Its application to other countries is limited by lack of field data, but this might be remedied by a survey of the few such data that exist in a country (see Fournier 1967, Temple 1972, and Hudson 1971 for valuable summaries of past work on soil erosion in tropical countries). A relatively few field measurements and some educated guesses by experienced soil scientists, agricultural engineers, geomorphologists, and hydrologists could also be used to extend the approach. Valuable information of this kind is now beginning to accumulate in several countries and is published in such outlets as the Journal of Soil and Water Conservation in India, the Interafrican Soils Conference, and the Soils Bulletins of FAO. Battawar and Rao (1969), for example, have evaluated the trend of C values through the cropping stages of several important crops in India.

Even where it is not possible to make such quantitative estimates of soil loss, it is still possible to map the distribution of erosion intensity in a region, to gain insights into the major controls, and to make predictions of the relative soil-loss rates to be expected from some change of land use. This can be done by means of an *erosion survey,* in which the degree of erosion is estimated visually on an ordinal scale, as indicated in Table 15-6. The scale could, of course, be altered to suit any regional conditions. Such a survey can be carried out rapidly over large areas with individual fields or hillslopes, or over areas of up to 100 acres classified together. At each site, other factors can be noted, such as average land gradient, drainage density, rock type, soil type, vegetation, land use, and climate. By means of a graphical portrayal or statistical analysis, the intensity of erosion in a region can be related to its controls and the results can be used to make semi-quantitative predictions of what to expect under various management alternatives. An early example of this methodology was given by Renner (1936), and in the rush to practice multiple regression analysis on two or three years of plot measurements of soil loss, the potentialities of the erosion survey for use in planning over larger areas has been neglected.

Significance of Soil Erosion

The total cost of accelerated soil erosion, either in monetary terms or in human suffering, has never been calculated and probably never could be. In the United States during the 1930's, soil erosion was recognized as a major threat to the continued productivity of the land, and a great deal of research and investment went into soil conservation. Important advances were made and the productivity of most of the nation's agricultural and grazing lands was stabilized and often increased. This was accomplished, however, with the aid of vast amounts of cheap energy and fertilizer. The effects of the rising costs of these commodities on the soil conservation program are uncertain. More recently, concern about soil erosion has shifted to

Table 15-6 Ordinal scale measurements of soil erosion intensity for regional erosion surveys.

TYPE OF EROSION	EROSION CLASS	DESCRIPTION	SOURCE
Sheet erosion*	1	No apparent erosion	Soil Conservation Service (1951)
	2	Slight erosion; less than 25% of the A horizon removed	
	3	Moderate erosion; between 25% and 75% of the A horizon removed	
	4	Severe erosion; more than 75% of the A horizon removed	
	5	Very severe erosion; removal of the entire A horizon, and perhaps parts of the B or C horizon	
Gully erosion	1	Occasional; less than 3 gullies per acre or gullies more than 100 feet apart	Soil Conservation Service (1951)
	2	Frequent; 4 or more gullies per acre and gullies less than 100 feet apart, but more than 75% of the land area is drained by gullies	
	3	Very frequent or destructively large gullies	
Water erosion along road sides	1	Sheet erosion; minute rills occasionally present	DeBelle (1971)
	2	Rill erosion; rills up to 6 inches deep	
	3	Initial gully erosion; numerous small gullies 6–12 inches deep	
	4	Marked gully erosion; numerous gullies 12–24 inches deep	
	5	Advanced erosion; gullies or depressions over 24 inches deep	

*Can also be used for surveying wind erosion. The original source suggests a classification for describing wind deposition.

viewing the process as a source of water pollution (see Chapter 19). In the 1970's concern has been growing again about the limited acceptance of soil conservation programs throughout the United States.

In many other regions of the world the situation is much more disconcerting. Growing populations are creating land pressure, and every available piece of land must be used to produce food or the cash crops needed to generate foreign exchange. In some countries the sight of maize growing on steep, rilled road cuts indicates this pressure on the land. Limited remnants of forest on the steep slopes of important water catchments are being invaded by agriculture, which soon generates rapid sheet and gully erosion. Because the cultivators of these lands cannot cover up their sins with cheap fertilizers, crop yields soon decline drastically and the pressure to clear new land increases.

Conservationists, hydrologists, agronomists, and others often wring their hands about the supposed stupidity of such people, but many of the cultivators are fully aware of the consequences. They simply have no option: families must be fed. Until the concept of population control gains popularity and surplus rural labor is diverted into other types of production, such lands will continue to be used. We can only view the fertility of these lands as an asset to be used up within a short time as a temporary solution to the problem. In the meantime, we can work on ways of minimizing the damage and of reclaiming exhausted, eroded lands for productive forestry or grazing when they can no longer support agriculture. Particularly on lands that do have a long-term agricultural potential, we can search for methods of soil conservation that are practicable and acceptable to subsistence agriculturalists as well as commercial operators. In the past many mistakes have been made in forcing people to use soil conservation practices that not only were unpopular but did not work (Berry and Townshend 1972). As more and more research stations and educational facilities are founded in these areas, it can only be hoped that both the soil conservationists and the agriculturalists will learn enough to combat the problem with locally developed expertise.

The effects of soil erosion are felt both on-site and downstream. On the hillside, erosion selectively removes the smaller and less dense constituents of the topsoil. These are clay particles and organic material, which, as we outlined previously, are the constituents that store nutrients in a form available to plants. A review of the removal of each of the major nutrients is given by Barrows and Kilmer (1963), and the effect on soil fertility is illustrated by Peterson (1964). In addition to the removal of plant nutrients, the removal of loose topsoil with its good infiltration and water holding and rooting characteristics is damaging. In some subhumid tropical regions the productivity of severely eroded lands seems to be limited not so much by nutrient deficiencies as by the availability of water. The effects of decreased soil fertility on crop yields can be seen in Figure 15-20 for a severely eroded field recently cleared of forest in the southern Appalachian Mountains.

Gully erosion does not usually account for a large proportion of the soil removed from an area, but it can have great local significance. Gullies increase the cost of road maintenance and may disrupt fields so that they are impossible to cultivate in an efficient way. In other places, gully formation lowers the local water table, drying out soils and bringing on the demise of some plants. Bennett (1939) has presented impressive pictures of the effects of sheet erosion and gully erosion in many parts of the world.

The downstream effects of accelerated soil erosion can also be severe. Deposition of eroded sediment in reservoirs and harbors (Gottschalk 1945) incurs heavy costs for new investments or maintenance. Sedimentation within the stream channel damages fish and can ruin their spawning habitats. For this reason, there is much concern about the large volumes of sediment deposited in channels below logged areas (Platt 1971) and mining

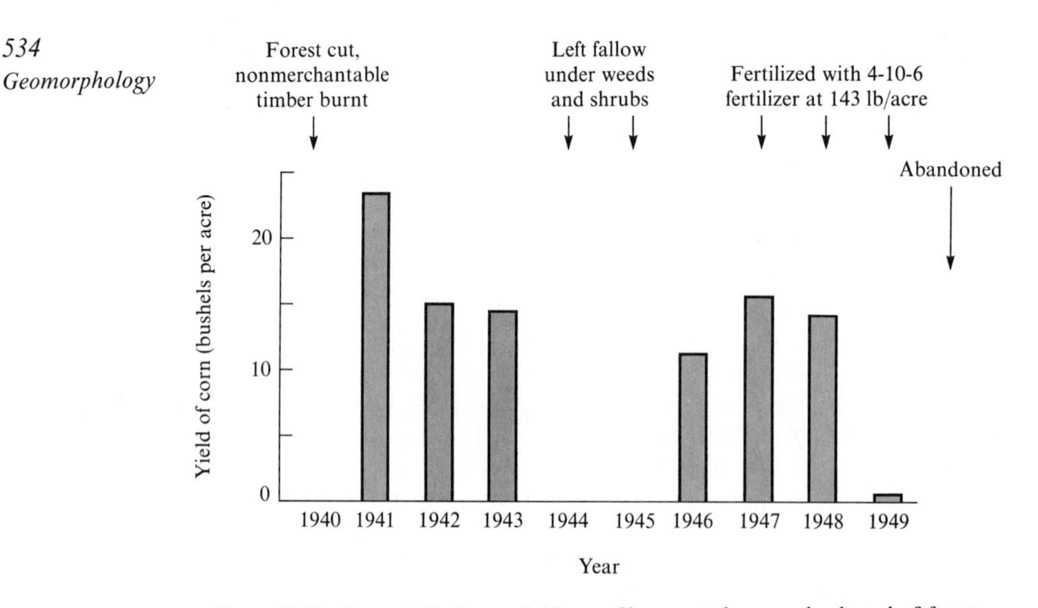

Figure 15-20 Corn yields from a field on a 50-percent slope newly cleared of forest on a mountain farm in the southern Appalachian Mountains. (Data from Dils 1953.)

areas (Smith 1940). The conveyance capacity of the channel can also be reduced temporarily by this deposition, with increased frequency of overbank flooding (see Chapter 18). This is a frequent problem in urban regions where construction releases large amounts of sediment, which may be deposited in storm drains or natural channels just at the time the storm runoff from these areas is increasing (Leopold 1972). Sediment is deposited on the valley floor by the resulting floods, and in rural regions the newly created overbank deposits can be advantageous or disadvantageous according to their characteristics. Until the closing of the Aswan High Dam, the fabled Nile flood maintained the fertility of Lower Egypt for thousands of years by bringing nutrient-laden silts and clays weathered from the volcanic highlands of Ethiopia. If the sediment is sandy or gravelly and devoid of nutrients, however, it may lower the fertility of fields on the valley floor (Happ 1944). The plant nutrients adsorbed on eroded soil particles are often deposited in a lake or reservoir and cause eutrophication, a subject of increasing concern, as we will describe in Chapter 20. Because it originates on large areas of a river catchment, rather than from a few obvious outlets, eroded sediment is referred to as a "non-point" source of pollution.

Control of Soil Erosion and Sedimentation

In this section, we will give a brief, introductory review of the various strategies used to control soil loss from disturbed hillsides in both rural and urban areas. For more complete information, the reader can refer to one of the many excellent textbooks on agricultural engineering, or to the increasing number of handbooks on controlling urban soil loss or on the construction of good forest roads. We have listed some useful books in the bibliography.

Soil conservation on croplands can be accomplished by agronomic means or by mechanical means. The first set of methods includes the choice of crops and ways of plowing, planting, and tilling. From the foregoing discussion, it is obvious that maintaining a thick *vegetative cover* is the best means of reducing soil losses (see Figure 15-21). Choice of a *crop rotation* that includes plants that provide a good cover and improve the structure of the soil is therefore useful. Where rotation is not possible, it is usually necessary to encourage rapid, heavy growth of the crop by *fertilizing* with chemicals or manure. Sometimes it is possible to grow a *cover crop* (usually a legume) between cash crops and to plow in the cover crop later as a "green manure." Many crops are now being planted without previous plowing of the land. This "no-till" method reduces runoff and soil loss dramatically.

If the cropland is gently sloping and smooth with a fairly stable topsoil and moderately high infiltration capacity, losses of soil and water can be reduced by a factor of up to four by carrying out all plowing, planting, and tillage operations in rows parallel to the contour. This *contour cultivation* increases surface roughness and holds up overland flow, enhancing infiltration and reducing the velocity of runoff to nonerosive levels.

Another means of reducing soil loss is by *strip cropping*, which involves planting crops in strips along the contour and alternating those that provide a good cover with grains or inter-tilled crops that do not (see Figure 15-22).

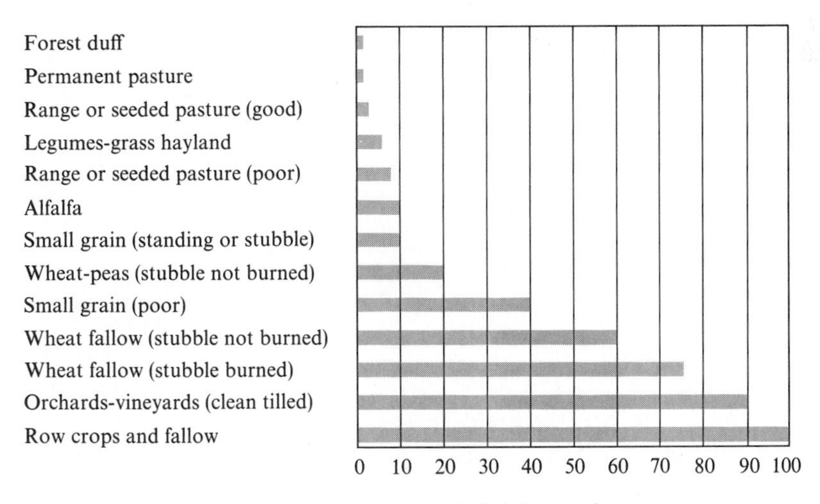

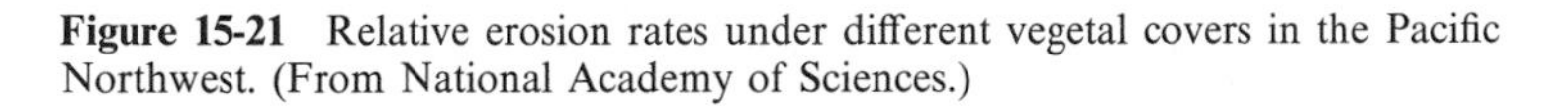

Figure 15-21 Relative erosion rates under different vegetal covers in the Pacific Northwest. (From National Academy of Sciences.)

The good cover, such as grass, encourages a high infiltration rate, generates little or no runoff, and slows down the runoff from the alternating strips, filtering out some of the eroded soil.

Careful tillage can improve soil structure and infiltration, and *residue management* can also provide important soil conservation benefits if crop residues are spread over or plowed back into the land as a mulch. This is a complicated matter on large commercial farms, however, and is still the subject of much research. On small subsistence holdings in many areas, crop residues are frequently burnt, and although this provides a degree of pest control, important soil conservation possibilities are lost.

Leopold and Maddock (1954) reviewed the results of studies on the efficacy of various land management practices. The data showed that the most efficient method of reducing soil erosion and increasing infiltration was the practice of *mulching*, that is, covering the ground surface with organic debris such as manure, hay, straw, or compost. Because of its simplicity, the mulching of small areas cleared for urban construction can also be an important tool for the planner and the developer to use for sediment control (Wischmeier and Meyer 1973).

Figure 15-22 Strip cropping and contour cultivation in Wisconsin. (USDA–Soil Conservation Service.)

Figure 15-23 Terraces: (a) Hand-dug bench terraces on a steep hillside in central Kenya. (b) Large-scale terraces on the gentle hillsides of a commercial farm. (b, USDA–Soil Conservation Service.)

The most common mechanical means of combating soil erosion in agricultural lands is by *terracing* (see Figure 15-23). The general strategy is to divide the hillslope into short lengths with low gradients so that overland flow cannot develop erosive depths and velocities. Runoff may be ponded on the terrace to encourage infiltration and conserve moisture in regions of dry farming. Alternatively, the runoff may be intercepted by a shallow, vegetated channel and conducted away across the slope on very low gradients so that erosive velocities do not develop. On large, commercial farms, terraces can be constructed by machinery on hillslopes with gradients less than 10 percent. Bench terraces can be built by hand on much steeper slopes, as the spectacular cultivated hillsides of the Rhine Valley, Southeast Asia, and Italy show. Irrigation systems are often incorporated into these bench terraces. The system requires massive investments of human labor, but with careful planning and maintenance and sometimes with heavy fertilization, they have remained productive for centuries. One of the disadvantages of terracing is that if the works fail, they usually concentrate runoff and cause gullying. When innovations such as terracing are being introduced into a region, therefore, it is important to educate farmers about the importance of maintaining any structure designed for the control of stormwater, whether it be a terrace or an earthen detention dam.

Gully control and reclamation is another form of soil conservation to which a variety of vegetative and mechanical techniques have been applied. They involve such techniques as the planting of trees, shrubs, and vines in the floor and walls of the gully; diverting stormwater around the gully; or building various kinds of dams to retain sediment and eventually fill the gully up. The techniques are well reviewed by texts on soil and water conservation listed in the bibliography.

On rangelands, soil loss can be minimized by *controlling the numbers of stock* so that the cover is not reduced too much. In most grazing regions, general guidelines have been established for the carrying capacity of the range. These have often been made rather casually in some pastoral nomadic regions, however, because of the transfer of attitudes and values from commercial ranching areas. In nomadic areas, the whole concept of carrying capacity and the importance of various levels of soil loss needs to be reviewed. For large, commercial ranches, there are several ways of reclaiming eroded areas and improving pastures. They include seeding and mechanical alterations of the soil surface. But most of the world's grazing lands do not benefit from these techniques because of the shortage of capital, expertise, or value. In such regions, the careful management of grazing, as already practiced by some herdsmen, is the only feasible way of reducing soil loss.

On most forest lands the rates of soil erosion are low. In some of the drier woodlands, the most common agent accelerating soil loss is fire, especially if the plant litter and organic duff on the forest floor are consumed. Krammes (1960) provides an example from southern California where a fire led to a tenfold acceleration of soil loss. In dense forests in humid regions most fires are relatively cool and do not destroy the organic duff. Consequently, burning produces only a slight increase in soil-loss rates.

Logging also causes only a slight increase of erosion so long as the duff layer is not destroyed. In some dry forests, however, clear felling of trees on steep slopes has generated rapid sheet erosion, requiring expensive operations to control soil loss and runoff during replanting (Maragaropoulos 1967). The most important control of soil loss from commercial forests is the extent and nature of road construction. Many studies have now shown that skid trails and haul roads are the primary sources of sediment during logging. The roads are compacted and bare, and often have steep gradients and cross shallow drainage lines. They are also disturbed by heavy traffic during the frequent heavy rainstorms experienced in many mountainous, forested areas. Since roads and trails may cover 5 to 35 percent of a logging area, depending on the topography and the logging method used, the rates of soil loss quoted for rural roads in Table 15-1 can be very high relative to normal rates of erosion in forested regions.

Because of the damage caused by sedimentation from logging areas in scenic and valuable mountain streams, a great deal of research has been done on minimizing soil losses from logging roads. All roads should be laid out on the gentlest gradient possible, and preferably on ridge tops where they will not cross drainage channels. Unstable rock and soil formations should be avoided, and care should be taken not to undercut or overload unstable slopes. Runoff from the road should be controlled and spread safely over the undisturbed forest floor. Drains and culverts should be carefully designed, maintained, and cleared, especially during periods of heavy rain. Logging operations should be kept out of stream channels. After logging is finished, the roads should be closed, scarified, and planted with vegetation if they are not absolutely necessary for fire control or recreation in future plans for the regenerating forest. These and other techniques are discussed in detail by Packer and Christensen (1973).

In mining areas, both surface mine workings and spoil heaps remain sources of sediment for very long periods of time, while their steep slopes are devoid of vegetation. In some areas the acid nature of the wastes discourages revegetation for more than a century, as in northern England and parts of Appalachia. Revegetation in arid regions is also very slow, while on calcareous coal mine wastes in western Washington, the forest can regenerate within 40 years. The high rates of soil erosion on mined areas listed in Table 15-1 are being combated in some areas by modifications of the same techniques as those described above. *Restoring a vegetative cover* is difficult (Davis, 1971), but careful species selection, mulching, fertilization, and irrigation are now being used successfully. Wastes from animals, agricultural processing plants, and sewage treatment plants can be used as mulches or applied by irrigation (Sopper 1970, Lejcher 1972, Gemmell 1972).

The major difficulty seems to be in deciding to reclaim these areas and how to pay for the work. Both soil stability and revegetation can be enhanced by backfilling the mine (Griffith et al. 1966), grading of spoil heaps (Lindstrom 1952) and construction of terraces on the spoil heaps. These terraces, which cost only about $10 per acre at present prices (Curtis 1971) can reduce the loss of sediment severalfold and conserve water to encourage plant growth. Some kinds of surface mines can be employed as landfills and recreation areas with proper grading, water control, and revegetation (Taylor 1966). In areas where strong legislation exists, strip-mined land can be returned to agricultural production or used for residential development without disrupting the economics of the mining operation (Striffler 1967).

It is in urban areas where most new developments in soil control are occurring, and several important publications now cover the details of the various techniques available (Diseker and Richardson 1961, Johnson 1961, Powell et al. 1970, Kao 1974, U.S. Soil Conservation Service 1976, Dallaire 1976). The first method of controlling soil losses from construction sites is to *limit the time during which the sites are open.* Because the major problem is caused by the clearing of construction sites and road cuts, one obvious method of control is to *maintain as much of the original vegetation and topsoil as possible.* Good site planning and architecture that takes advantage of the natural features of the site rather than obliterating them will reduce soil loss. Revegetation of the stripped areas should be initiated as quickly as possible after construction is finished. Various types of effective vegetation have now been tested, and they are often planted at the same time that the land surface is treated by mulching with sawdust or bark, spraying with a surface stabilizer and fertilizer, or even covering temporarily with matting, netting, or polyethylene sheets. Figure 15-24 shows some results of mulching to control soil loss on construction sites.

Grading operations on the site can sometimes be designed to produce flat *benches* or other surface configurations that will minimize soil loss. On various parts of the site, *detention basins* store sediment and storm runoff from the site. These basins catch coarse sediment very effectively, but finer particles take hours to settle through the water, and it is very difficult to detain these without using a catch basin that can store all the runoff from a storm. This is expensive.

Previously in this section, we have been concerned with combating soil loss after it has become a problem. Wise planning can often avoid causing the problem entirely. Land characteristics that favor soil erosion under various kinds of development can be recognized early in the planning process. Such information can be used to plan land use that is compatible with the natural limitations and potentialities of each area. In the particular case of minimizing soil loss, it is possible before planning decisions are taken to isolate those areas with steep, long slopes, erosive soils, or generally infertile conditions, or to recognize many other combinations of conditions that will affect the success of the planned venture.

The U.S. Department of Agriculture developed a method of classifying lands according to their agricultural capability. The classification is listed in Table 15-7 and is based on such characteristics as soil depth and texture, land gradient, evidence of previous erosion, stoniness, poor drainage, and soil chemistry. For details of the procedure the reader should consult

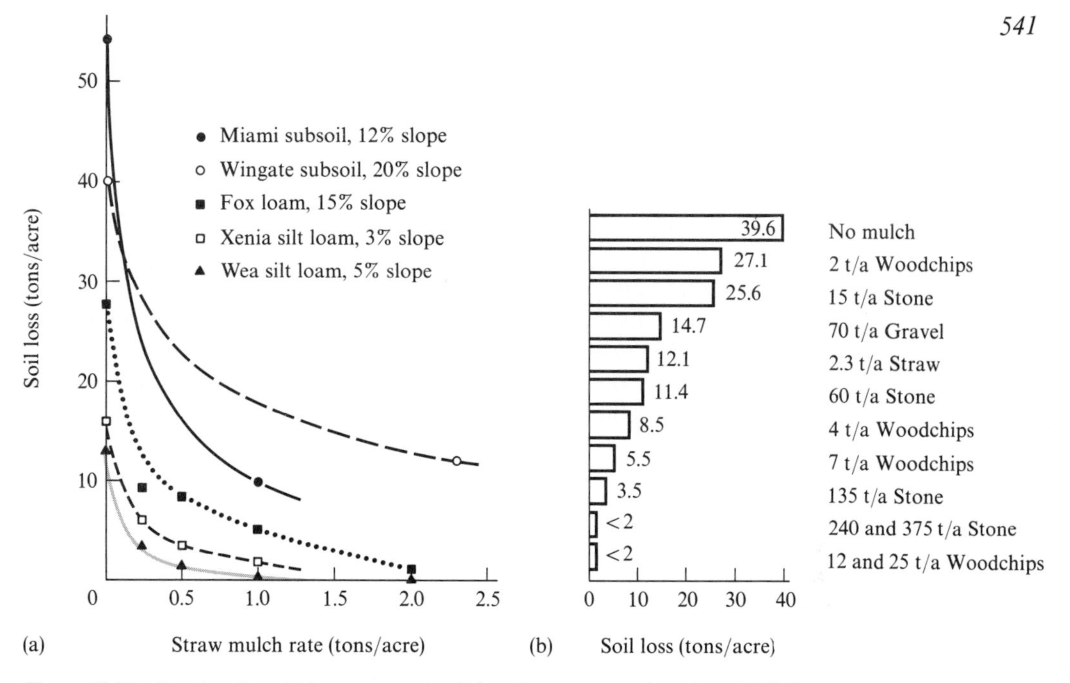

Figure 15-24 Results of mulching to control soil loss from construction sites. (a) Soil losses from 5 inches of simulated rain in 2 hours on a 35-foot-long hillslope as affected by the amount of straw used as a mulch. (b) Soil losses from 5 inches of simulated rain in 2 hours on a construction site with a gradient of 0.20 and a hillslope length of 35 feet as affected by various types of mulch. The abbreviation t/a indicates the amount of mulch in tons/acre. (From Wischmeier and Meyer 1973.)

Klingebiel and Montgomery (1961) or U.S. Soil Conservation Service (1951). Hudson (1971) also provides a review of land capability classification systems in several other countries. A planning unit and its surroundings can be mapped onto aerial photographs according to such a classification scheme. Figure 15-25 shows how the various classes might be distributed in a typical landscape. The classification then indicates which zones should be avoided by various types of land use and where conservation practices are needed. Through the best use of each area and through the encouragement of necessary conservation practices, it is often possible to forestall damage to the land and ensure development in harmony with natural processes. Much remains to be done with this tool, especially for rural planning in developing nations.

The U.S. Department of Agriculture is now turning its attention to developing land capability classifications for use in urban planning (Bartelli et al. 1966). Most of the factors taken into account are those affecting such things as septic-tank performance, the capacity of the soil to bear loads, and flood hazard, as we have described in Chapters 6 and 11. The portrayal of areas with erosive soils and steep gradients, however, is also useful for indicating where the most severe sedimentation problems are likely to occur during construction. Urban planners are increasingly taking advantage of these facilities. The Southeastern Wisconsin Regional Planning Commission, for example, contracted with the Soil Conservation Service to map the soils of a 2689-square-mile region undergoing urbanization. The commission claims that the soil survey, which costs about $200 per square mile, will save approximately $300 million in the cost of residential land development in the area during the next 25 years. Some of these savings will be made by averting problems of erosion and sedimentation. Similar efforts are being made in forestlands to produce maps showing the susceptibility of soils to damage from erosion. It is possible to excerpt a map of one char-

Table 15-7 Land capability classification developed by the U.S. Department of Agriculture. Classes I–IV are suitable for crop cultivation; classes V–VIII are suitable only for grazing, forestry, or wildlife habitat. The classification can be applied to other land uses, such as urbanization, with only minor modification. (From Klingebiel and Montgomery 1961.)

CLASS	DESCRIPTION
I	Few limitations on land use. Low gradients; soils deep, well drained, fertile, and easily worked. Overland flow absent or very rare.
II	Moderate limitations on use and moderate risks of damage; needs careful soil management to prevent deterioration, but practices are easy to apply. Limitations due to one or a combination of: gentle gradients, moderate susceptibility to erosion, less than ideal soil depth, slightly unfavorable soil structure, slight salinity problems, occasional damaging overland flow, soil wetness.
III	Severe limitations on use or risks of damage. Regular cultivation requires careful soil conservation practices. Limitations due to one or a combination of: moderately steep slopes, high soil erodibility, frequent overland flow, very slow permeability of subsoil, waterlogging, shallow soil, low water-holding capacity, low fertility, moderate salinity.
IV	Very severe limitations on use or risks of damage that restrict the choice of plants and require very careful soil conservation practices which are difficult to apply and maintain. Have the limitations of class III but in a more extreme form.
V	No erosion hazard but have limitations of wetness, stoniness, flooding.
VI	Severe limitations due to gradient, erosion hazard, effects of past erosion, stoniness, shallow depth, wetness, low water-holding capacity, or salinity.
VII	Very severe limitations even for grazing and forestry. Same characteristics as class VI but in a more extreme form.
VIII	Extreme limitations on any kind of land use except wildlife habitat and wilderness areas. Characteristics of class VI in extreme form.

Figure 15-25 Illustration of several land-capability classes defined by the U.S. Department of Agriculture. In moving from class I to class VIII the planner's choices usually become fewer and the risks of causing damage to the land become greater. (USDA–Soil Conservation Service.)

acteristic, such as erosion hazard or limitations on septic tanks, from the information gathered in a general land capability survey. An example of a capability map of soil limitations for urban development in a small drainage basin is given in Figure 15-26. Some of the uses of such information and the savings they can bring are discussed by Klingebiel (1966), Koch and Clay (1966), and Wohletz (1966).

Mass Wasting

It is difficult to appreciate the awesome power and danger of a landslide or other catastrophic mass movement until you see one. Whether you see the motion or the resulting deposit, it is hard to remain unimpressed. You can walk over the deposit of the slide that blocked Madison Canyon, Montana, during the 1959 earthquake and gain some feeling for the 42 million cubic yards of rock that crashed from as high as 1400 feet and filled a 4000-foot section of the canyon floor to a depth of 200 feet (see Figure 15-27). Some of the blocks are larger than a house. The geologic conditions in which the landslide occurred are sketched in Figure 15-28. Deeply weathered and

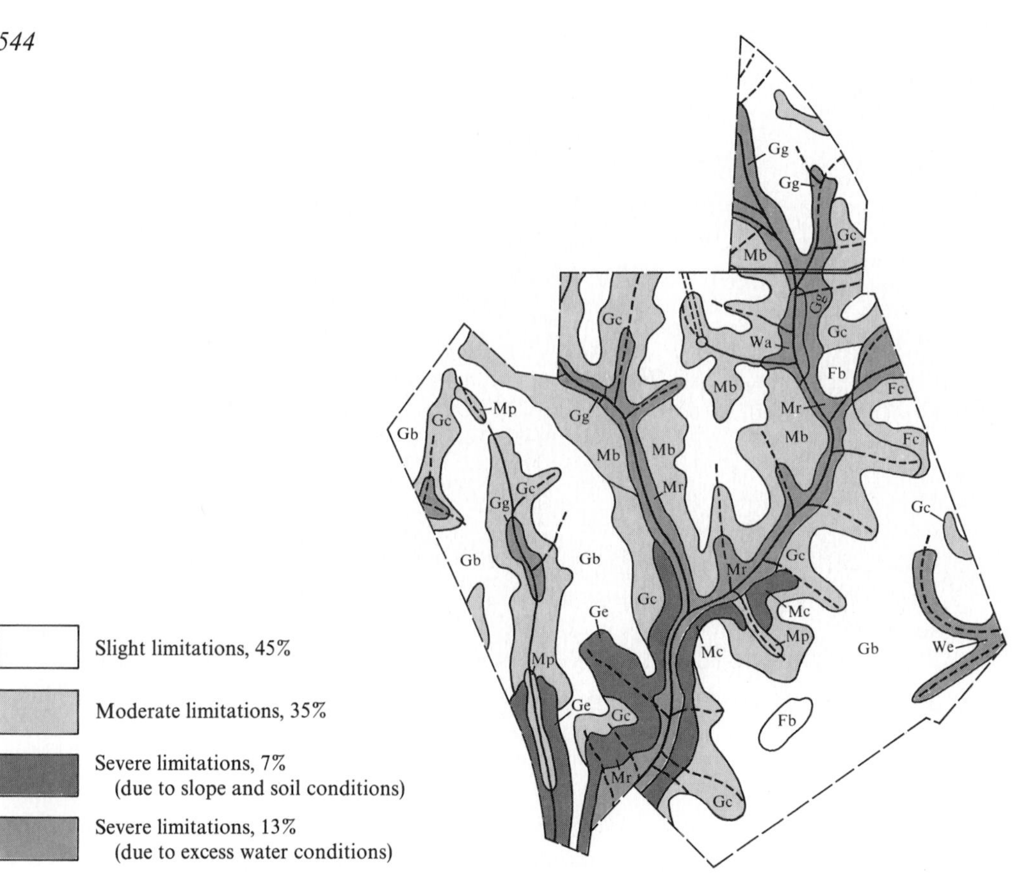

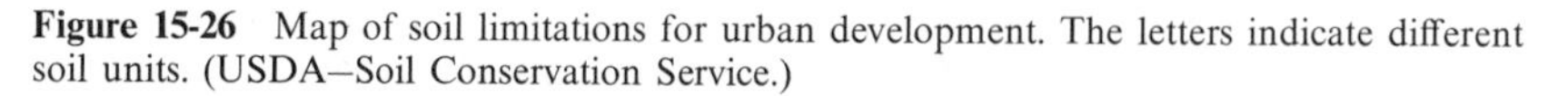

Figure 15-26 Map of soil limitations for urban development. The letters indicate different soil units. (USDA–Soil Conservation Service.)

mechanically weak schist and gneiss were held high on a steep mountain slope by a bed of hard dolomite. Vibrations caused by the earthquake shattered the dolomite, which then could not support the metamorphic rocks above, and a thick layer of the mountainside slid onto the valley floor and 400 feet up the opposite wall of the canyon. Twenty-eight campers along the river were engulfed, a section of road was destroyed, and the Madison River was blocked by a 200-foot-high rock dam. A lake is still impounded behind this dam and drains over the landslide debris.

A view of such a landslide will undoubtedly impress upon all those concerned with planning the need to understand mass wasting processes and their controls. The example from Seattle described in Chapter 1 illustrates how even much smaller hillslope failures need to be understood to save the planner embarrassment and heavy costs. In particular, the Madison Canyon slide raises several questions that are pertinent in all areas experiencing mass movements, questions we will address in this chapter. Why did the

Figure 15-27 Madison Canyon landslide, Montana. In the foreground is the lake that was impounded by the slide. (J. R. Stacy, U.S. Geological Survey.)

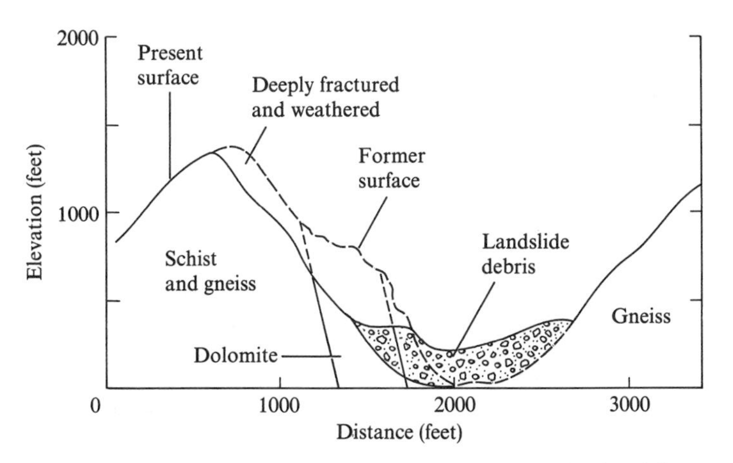

Figure 15-28 Sketched geologic cross section through the site of the Madison Canyon landslide.

landslide occur at that place and time? Could the potential hazard have been recognized before the slide occurred? Could anything have been done to stabilize the source of the slide?

In addition to the direct danger from landslides in terms of loss of life and destruction of buildings, some examples of their economic costs are given by Sheng (1966) for Taiwan. Landslides triggered by a single typhoon led to the deposition of 19.5 million cubic meters of sediment in a $100 million reservoir only one year after completion. This sediment occupied one-third of the dead storage volume provided for 80 years of reservoir operation. In another area, sediment mobilized by landslides was deposited 20 m thick in the valley below, completely burying a power plant. Elsewhere on the island, landslides have destroyed villages and roads, formed dams that broke and caused disastrous flooding, and supplied sand and gravel, which is spread over agricultural lands during floods. Similar disasters have been described recently from Brazil (Jones 1973), Hong Kong (So 1971), and the southeastern United States (Williams and Guy 1971).

Even small failures, involving a few tons of material or a foot of movement, can damage structures and disrupt service along highways, railroads, sewers, water mains, or electrical cables. The breakage of electrical cables by even a very small slide can create a fire hazard. In small landslides, the greatest damage to structures is often done by falling trees rather than by the geological materials.

Types of Hillslope Failures

Several classifications of mass movements are available in the geologic and engineering literature, the most complete being the one developed by Varnes (1958). We will discuss only the simpler classification of the most common landslides shown in Table 15-8 and illustrated in Figure 15-29, which is abstracted from Varnes' original illustration. The differences between these

Table 15-8 Classification of hillslope failures.

Falls
- Rockfalls

Slides
- Planar failures: rock slides, debris slides
- Rotational failures: slumps

Flows
- Debris avalanches
- Debris flows
- Earthflows
- Mudflows (including lahars)
- Solifluction*

Soil creep

*Solifluction is the slow, viscous, downslope flow of waterlogged soil underlain by an impervious layer, most commonly frozen ground. Although this process is gaining considerable importance in planning because of the development of arctic and subarctic regions, we will not treat solifluction in this text. We suggest that those concerned with northern regions consult Washburn (1973).

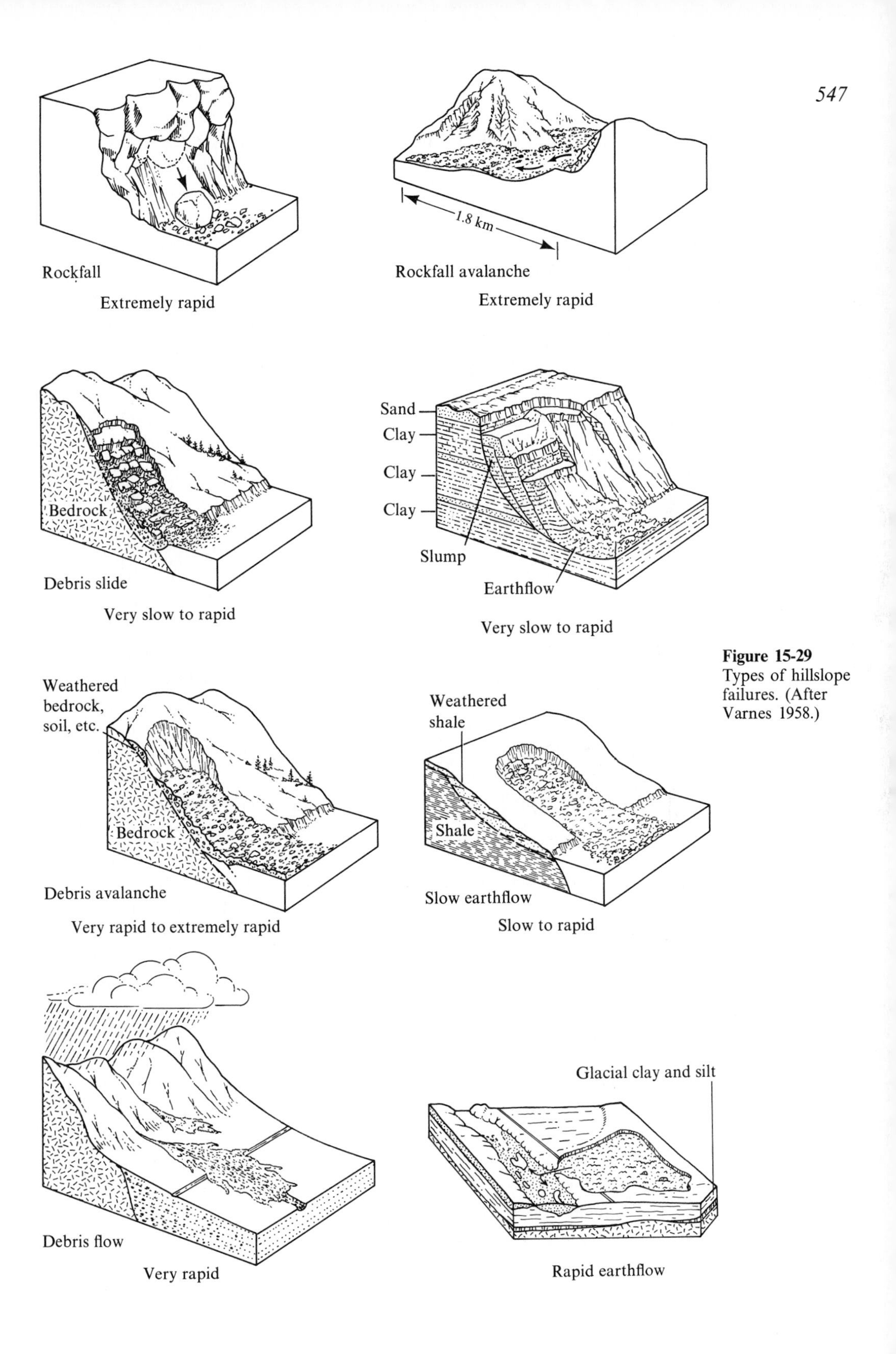

Figure 15-29 Types of hillslope failures. (After Varnes 1958.)

processes are not sharply defined, and the types of movement grade into one another, so that there is often a difference of opinion among experts about the classification of a particular hillslope failure. This does not affect the planning process, however, unless there is an important difference of opinion about the cause, future development, or most appropriate control method. The planner should be aware that mass wasting features are difficult to study and to predict and that any prognosis about them is subject to at least as much uncertainty as is the case with floods.

Rockfalls occur along steep cliffs where material breaks loose and reaches the valley floor by free fall, bounding, and rolling. The movement is very fast and can vary in size from single rocks to millions of cubic yards of material. The rockfall that occurred at Elm, Switzerland, in 1881, for example, involved 13 million cubic yards of rock. The rockfall from a peak in Mount Rainier National Park, Washington, in 1963 consisted of 14 million cubic yards. Most rockfalls, however, are much smaller.

The fallen material may be removed by a river at the base of the cliff, or it may accumulate as *talus slopes,* or *screes,* like those shown in Figure 15-30. The foot of such a cliff or talus slope is a very dangerous place, especially during heavy rains or snowmelt. The cost of maintaining transportation routes through such terrain can be appreciated by traveling through the Fraser Canyon in British Columbia, the "Million Dollar Highway" in the San Juan Mountains of Colorado, or several routes through the Alps. In addition to the danger from rockfalls there is also a chance that the talus slope itself will fail as a debris slide or debris flow (described later). Not all cliffs and talus slopes are active in the present climatic regime; some have been inactive since shortly after the last advance of ice in the Pleistocene. The level of activity can be gauged roughly from the degree of weathering, lichen colonization, and vegetation on the talus, or from sequences of aerial photographs on which individual blocks or debris trains may be visible. Excavation of the toe of a stable talus slope may induce debris slides within it, however.

In *slides* the failure occurs along a surface or within a narrow zone of deformation. The failure surface may be planar or arcuate, and the slide material may be greatly deformed or largely undeformed. The rate of movement can range from feet per year to feet per second depending on the hillslope gradient, the water content, and the existence of special conditions, such as the presence of a lubricating cushion of compressed air, which is thought to have enabled some slide blocks to travel long distances on low gradients (e.g., the Blackhawk landslide in southern California described by Shreve 1965). The size of relatively undeformed blocks involved in landslides may range from a few cubic yards to several cubic miles and may travel distances ranging from feet to miles.

According to its constituent material, a planar hillslope failure will usually be termed a *rockslide* or *debris* (*weathered rock and soil*) *slide.* The plane along which sliding occurs is usually determined by some structural feature, such as a bedding plane or the base of the weathering zone. Rockslides

Figure 15-30 Cliff and talus slope subject to frequent rockfalls along the Pinedale–Jackson road, Wyoming.

generated by a large storm in 1967 caused much property damage and loss of life in the densely settled urban area of Rio de Janeiro (Jones 1973). Debris slides are similarly destructive, and much more common (see Figure 15-31). Where the geologic materials are deep, uniform, and cohesive, rotational slumps are common (Figures 15-29, 15-32). A slump block does not usually move far but causes significant ground breakage, destruction of installations at its foot, and undermining of those hilltop sites with vistas

Figure 15-31 Debris slides covering access road to a tunnel leading to the Nilo Peçanha powerplant near Rio de Janeiro. (F. O. Jones, U.S. Geological Survey.)

that are so popular with architects and homeowners. Although most of the material in a slump is usually undeformed, the leading edge is frequently deformed during sliding and forms a frontal lobe or rampart of hummocky topography as the material spreads out at the toe of the slide. In other examples, this toe is eroded by a river. The upper rim of the slump is usually broken by scarps and back-tilted blocks (see Figure 15-33). Strong differential movements on and around the slump block destroy buildings.

In *flows* the movement of debris resembles that of a viscous fluid; there is more or less continuous internal deformation of the material rather than a narrow zone of shearing. High proportions of water, air, and fine, granular material, especially clay, favor flow. Dry flows include the destructive air-

Figure 15-32 Upper end of a slump that carried several houses over the edge of a cliff at Golden Gardens, Seattle, in May 1974. The movement began very slowly and accelerated to a rate of about 10 feet/day as the houses were destroyed. The trees on the right of the photograph are on the landslide bench shown in Figures 1-10 and 1-11.

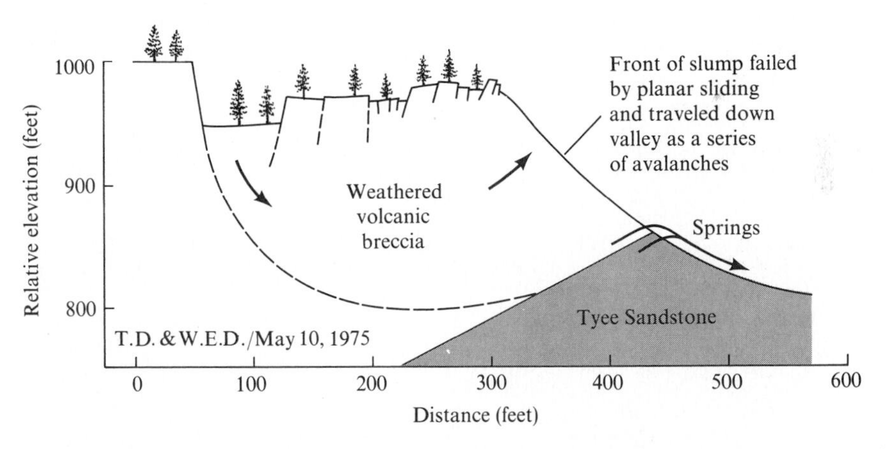

Figure 15-33 Cross section of Castle Creek Slump, Coast Range, Oregon. The large failure involving 20 million cubic yards of deeply weathered volcanic breccia began over the contact with the underlying sandstone where a line of springs emerged.

lubricated avalanches that occur when some large rockfalls, such as those mentioned above at Elm, Switzerland, and Mount Rainier, incorporate compressed air as they reach the ground. In these two cases, it is claimed that the compressed air, together with the high velocity impacts between particles, produced *rockfall avalanches* that traveled down valley at speeds over 100 miles per hour for distances of 1.1 miles in the Swiss case, and 4.3 miles at Mount Rainier. The Elm failure took 115 lives; the Washington event occurred during winter in an uninhabited portion of the National Park. It reached within half a mile of one of the most popular campgrounds in the region (see Figure 15-34) and at another time of year, loss of life could have occurred (Crandell and Fahnestock 1965).

There is a continuum in the style of movement from *debris slide* through *debris avalanche* to *debris flow* as the moving mass breaks into smaller and smaller fragments and incorporates more water as it moves down a hillslope or valley (see Figure 15-35).

Figure 15-34 A rockfall and avalanche, originating from Little Tahoma Peak on Mount Rainier in December 1963, spread large blocks of rock for a distance of more than 4 miles over the Emmons Glacier and the floor of the White River valley. The terminus of the rockfall avalanche lies only one-half mile from the heavily used White River campground in Mount Rainier National Park. (Austin S. Post, U.S. Geological Survey.)

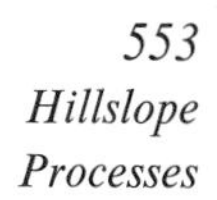

Figure 15-35 A debris avalanche in the Wallowa Mountains of Oregon. (Photograph by David P. Dethier.)

Williams and Guy (1971) described debris avalanches generated by a hurricane in central Virginia. The failures typically left scars 200 to 800 feet long, 27 to 75 feet wide, and 1 to 3 feet deep, and moved an average of 75,000 cubic feet of sediment to streams. The addition of such large amounts of debris to the rivers caused many of them to become streams of mud and boulders, which increased the destructiveness of the hurricane floods.

In the Coast Range of Oregon William Dietrich, a graduate student of the University of Washington, found a debris slide that occurred in 1972 at the head of a small valley and released approximately 250 cubic yards of deeply weathered volcanic rock. The material slid down a steep gully on a 20° slope and continued down the valley of Cape Creek for a distance of one-half mile. As it traveled, it incorporated large amounts of water from the stream, which was in flood at the time. The movement developed into a flow, which increased in thickness and width downstream as it scoured more weathered material from the valley sides and valley floor. At constrictions in the valley, the flow increased in thickness to 18 feet and then thinned again in wider sections. It flowed around two right-angled bends, riding up on the outside of the bend and eventually coming to rest in a series of ramparts. The original 250-cubic-yard slide had grown to approximately 250,000 cubic yards by the time it came to rest. This example and the previously quoted rockfall avalanches illustrate how landslides that develop into flows can affect areas far from the source of the slide. The density and viscosity of the flowing mass give it the transporting power to move large boulders, houses, bridges, and most other structures.

Earthflows, whether slow (moving at rates of feet per year) or rapid (feet per minute) consist mainly of fine-grained materials, especially clays either from sedimentary clay, shales, volcanic deposits, or other rocks that weather to a clayey soil. They usually occur when a slump or other failure mobilizes clays that have a very high water content. Once in motion, the mass begins to flow and it develops a surging and spreading toe of dense, viscous material, which can flow over gentle gradients, especially if confined in a river valley.

Small, slow earthflows can occur beneath grass or forest vegetation, and are often first recognized when their bulging toes disrupt a communications line during wet weather. Crandell and Varnes (1961) measured the rate of movement of a slow earthflow that had originated in clay-rich, deeply weathered volcanic deposits in the San Juan Mountains of Colorado. The toe of the slide moved only 30 feet in 12 years. During its 350-year history, however, the earthflow had moved 12,000 feet, indicating that these flows can move much more rapidly and dangerously at some times.

A particularly dangerous type of rapid earthflow occurs in "sensitive clays" most commonly found in areas such as the Saint Lawrence Valley of Canada, which were inundated by saline or brackish water during the land- and sea-level fluctuations of the Pleistocene epoch. The structure and water content of these clays are such that once disturbed (by an earthquake or a minor hillslope failure) they lose almost all their strength. A fluidlike clay stream with unbroken blocks of clay then moves rapidly downhill or down a valley, picking up everything in its path. On April 20, 1971, for example, a small bank failure (200 ft × 500 ft) occurred on the edge of the Petit-Bras River at the town of Saint-Jean Vianney, Quebec. The townsite is underlain by sensitive marine clay, and behind the town is a large marshy area that feeds subsurface water down through the clay. Four days after the initial failure, a much larger earthflow began. Most of the movement took place within a few minutes, and a nine million cubic yard crater was formed within one and a half hours. A mixture of sand and clay surged into the river and flowed 1.8 miles downstream at a speed of 16 miles per hour. The stream of mud reached 60 feet up the valley walls, and scoured the floor of the river valley as much as 20 feet, eroding adjacent hillsides and reducing their stability.

The scarp of the major failure retreated several hundred feet through a new residential development (see Figure 15-36), and people trapped in the houses were carried to their deaths in the mudstream. As we shall describe later, the recognition of some simple local geologic and hydrologic indicators would have suggested that the residential development be located elsewhere.

Mudflows develop where mass failures in fine-grained material mix with streams. Mudflows also develop in mountainous deserts, where intense sheetwash and rainsplash erosion can wash vast quantities of mud into canyons. As the mud flows downstream, it can pick up more sediment or lose water, thickening as it flows. Because of its high density and viscosity, the mudflow has the power to transport buildings and other structures in its path. The steep hillsides behind Los Angeles, California, are the source of frequent mudflows of this type, especially where fire and urbanization have left hillslopes devoid of vegetation. The mudflows travel down canyons and spread over gently sloping fans of debris at the mountain front. Unfortunately, these fans are now the sites of suburban developments (see Figure 15-37). But the signs of the hazard were there before the development was planned.

A devastating, though rare, type of mudflow or debris flow called a *lahar* occurs in volcanic regions. The steep sides of many volcanoes are clothed with thick layers of ash, or lava that has been chemically altered by weathering or by volcanic gases. This mechanically weak material is often saturated by water from heavy rainstorms or snowmelt, which may be accelerated by volcanic heat. These conditions sometimes give rise to large debris slides,

Figure 15-36 Oblique aerial photograph of the scar left by a rapid earthflow at Saint-Jean Vianney, Quebec in 1971. The failure has undermined a portion of the housing development in the upper central part of the photograph. (Geological Survey of Canada.)

Figure 15-37 Mudflow in Glendora, southern California, following a heavy rainstorm on a steep, sparsely vegetated basin at whose mouth the suburbs lie. (Los Angeles Times Photo.)

which move down valleys as tremendous debris flows or mudflows. Many such events have occurred around Mount Rainier and Mount St. Helens in Washington (Crandell and Mullineaux 1975). The Osceola mudflow, for example, occurred from one side of Mount Rainier about 5000 years ago and involved nearly half a cubic mile of clay-rich debris. Its depth of flow exceeded 500 feet in places as it traveled 40 miles down a mountain valley before spreading over 65 square miles of the Puget Lowland. Although this is the largest such debris flow in the region, several other large ones have occurred more recently, including one of 50 million cubic yards in 1947, and there is no reason to suspect that such activity has ended. The deposits of former lahars cross the heavily used National Park on Mount Rainier as well as the sites of several towns. There is concern in the Pacific Northwest that one of these mudflows could invade one of the many reservoirs in the area, creating a large wave that could overtop and breach the dam (see Figure 15-38).

Soil creep is the slow downhill movement of debris that results from disturbance of the soil by freezing and thawing, wetting or drying, or slow plastic deformation under the soil's own weight. It occurs on virtually all hillsides, and though it is slow it is responsible for most of the downslope transport of debris to stream channels in heavily vegetated regions. Even in areas where landslides are an important hazard, their relatively low frequency and limited extent restrict their contribution to the sediment budget of the catchment below that of slow, ubiquitous soil creep. In a small basin in the Coast Range of Oregon, for example, Dietrich (1975) estimated that soil creep down the steep, forested mountain slopes supplied 99 percent of the sediment lost from the six-square-mile catchment, while debris avalanches and debris flows transported the remainder. In most of the densely vegetated regions where creep is the dominant erosive agent, sediment yields are generally low and do not pose planning problems. In such regions soil creep may disrupt a building foundation or a fence but is not a major hazard. In a few places, however, where the underlying rocks weather to clay-rich plastic soils, the rate of soil creep may average up to several centimeters per year (Kojan 1969) in soil mantles several meters deep. Such rates pose obvious problems for construction and the early recognition of such mobile areas on certain rocks can avert problems in development.

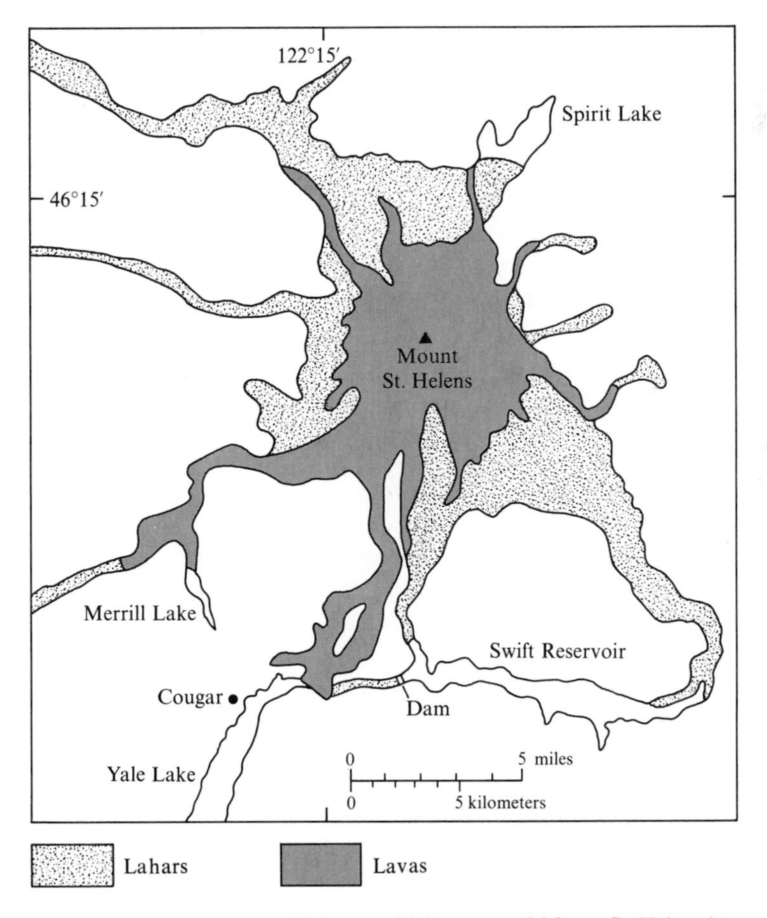

Figure 15-38 Distribution of lavas and lahars around Mount St. Helens in the Cascade Range of Washington. In the last several thousand years, lahars have reached the present sites of reservoirs around this and other volcanoes in the same region. (From Crandell and Mullineaux 1975.)

Within a particular region, very few types of landslides are encountered, and more emphasis may be placed on the agents causing landslides so that planning can be done for future control or avoidance of the landslide hazard. For these purposes, other classifications may be needed. Sheng (1966), for example, used the scheme shown in Table 15-9 for classifying landslides in Taiwan. This or many kinds of classifications could be used for specific purposes and regions. However, as far as possible, it is better to follow established groupings so as not to confuse the literature any further.

Table 15-9 Classification of the immediate causes of landslides. (From Sheng 1966.)

1. Runoff concentration
2. Seepage
3. Stream cutting
4. Road excavation
5. Cultivation
6. "Geologic" processes
7. Combinations of the above

Factors Controlling Mass Wasting

Planners are not concerned with detailed engineering mechanics of landslides, and so we will not present such analyses here. However, it is useful for planners to appreciate the general nature of the forces controlling hillslope stability and to understand the role of water in hillslope failures, as well as the effects of human activity, volcanism, and earthquakes. A discussion of controls will also enable the planner to appreciate the nature of work that must be done by engineers and geologists to recognize areas of landslide hazard and in some cases to design methods of stabilization.

Figure 15-39 represents a dry, uniform soil mantle overlying much stronger bedrock. The soil mantle is subject to two major opposing forces: the downslope component of the soil weight ($W \sin \alpha$), which acts to shear the soil along a potential failure plane parallel to the hillslope, and the resistance of the soil to shearing (sometimes called its shear strength). If the downslope force exceeds the shear resistance, the hillslope will fail.

The shear strength can vary widely between various geologic materials. It is controlled by such properties as the cohesive forces due to electrical forces between clay particles or to the binding action of tree roots and the frictional forces resulting from surface friction and interlocking between grains of soil or blocks of rock. The frictional forces increase with the size of the force acting at right angles to the potential failure plane (usually called the normal stress, $N = W \cos \alpha$, in Figure 15-39).

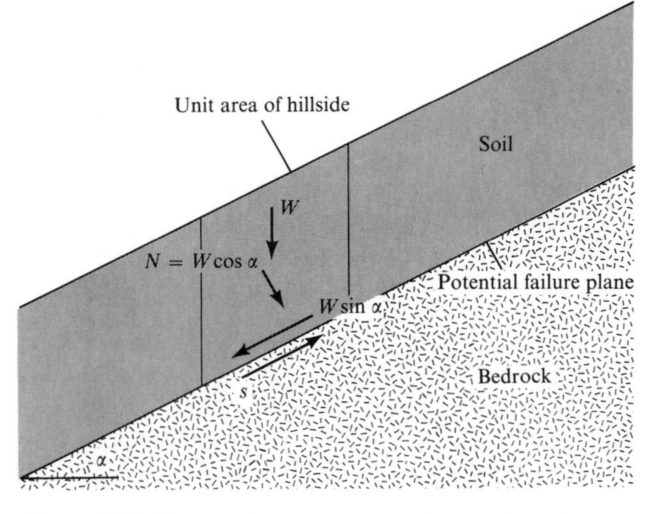

Figure 15-39 Forces acting upon a dry, uniform soil mantle overlying bedrock. W = weight of soil per unit area of hillside; α = angle of slope of the hillside; $W \sin\alpha$ = downslope component of the soil's weight per unit area of hillside; $N = W \cos \alpha$ = component of the soil's weight per unit area acting normal to the potential failure plane (the normal stress); s = resistance of the soil to shear, a force per unit area of hillside (the shear strength of the soil).

In soils or rocks containing water, the shear strength is strongly affected by the water pressure in voids between the grains or blocks. This pressure supports a portion of the soil's weight and therefore reduces the normal stress that is effective in producing friction. The relationship of shear strength to its controls is summarized by the equation

$$s = c + (N - p) \tan\phi \qquad (15\text{-}5)$$

where c = cohesion
N = component of the weight of the saturated soil acting normal to the failure plane ($= W\cos\alpha$)
p = pressure of water in the voids of the soil or rock
ϕ = angle of internal friction, which reflects the degree of interlocking and surface friction within the material and whose measurement is described in standard soil mechanics texts (Lambe and Whitman 1969).

The water pressure can be measured, as described in Chapter 7, by inserting a piezometer into the material and measuring the height to which water stands in the pipe (Figure 15-40). The height (in feet) multiplied by the specific weight of water (62.4 lb/ft^3) gives the water pressure at the base of the piezometer. The water pressure within the material on a hillside is ex-

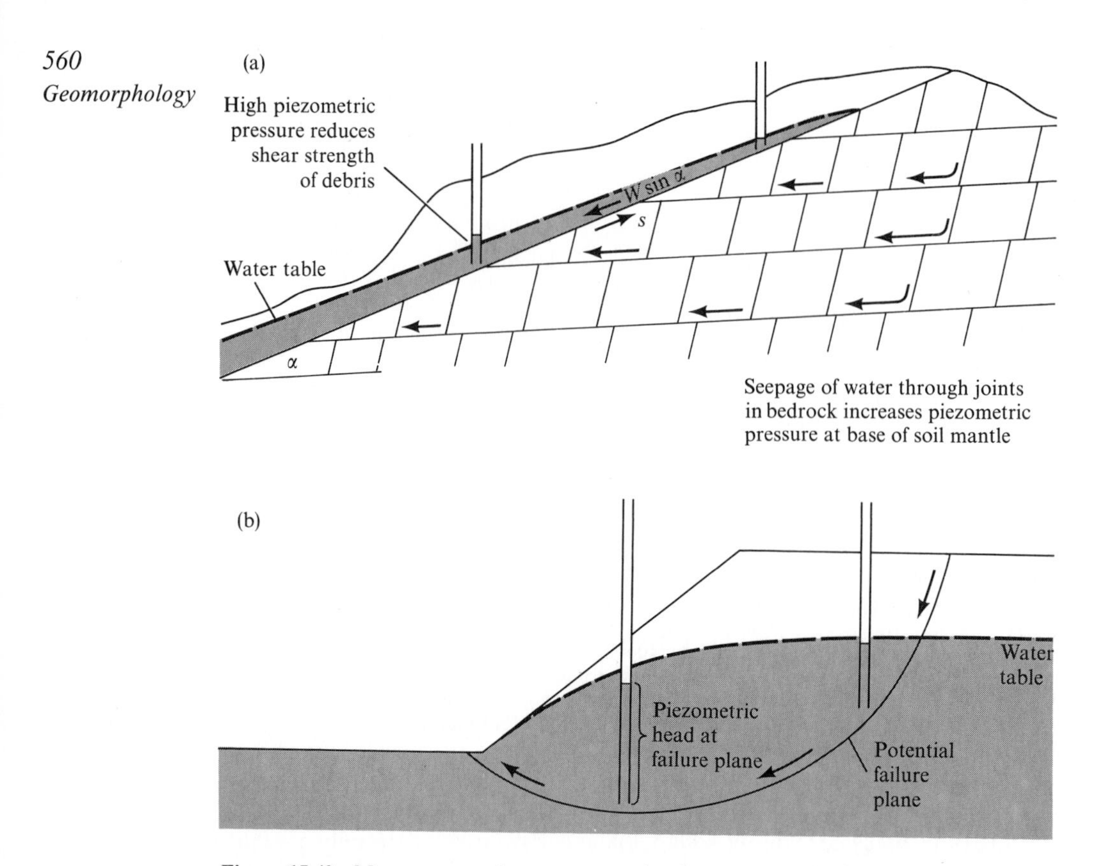

Figure 15-40 Measurement of water pressure in piezometers at the base of typical hillslope failures. (a) Planar failure. (b) Rotational failure.

tremely important in determining its strength, and a large majority of hillslope failures can be traced to the action of water. The higher this piezometric pressure is in a material, the lower is its shear strength and the greater the susceptibility to failure.

In Figure 15-39 we considered a case in which the probable failure plane was obvious. The same is true in the saturated soil shown in Figure 15-40(a). In other cases the probable failure surface is not obvious before a landslide occurs and need not be planar. If there is no distinct break in weathering characteristics in a glacial till or a deeply weathered granite in the tropics, for example, the shear strength may vary gradually with depth. Also the permeability of the material may decrease uniformly with depth so that there is no obvious horizon to impede percolation and build up water pressures. In such cases the probable failure surface and therefore the degree of stability of the hillslope can only be assessed by extensive calculation, or in some cases by making a field survey of the depths of failure of landslides in the area. In deep, uniform materials that possess some cohesion, such as marine clays and shales, the failure surface is arcuate, as shown in Figure 15-40(b), and again can only be predicted by extensive, trial-and-error calculation. In the cases shown in Figure 15-40, the downslope force ($W \sin \alpha$), the normal stress ($W \cos \alpha$), the water pressure, and therefore the shear strength vary along the failure surface.

The general controls of hillslope stability can be summarized as follows: Landslides occur when the gravitational forces acting along the potential failure surface exceed the resistance of geologic materials to failure. Some processes influence the downslope disturbing force by affecting the weight and distribution of material on the hillside, or by altering the angle of slope. Other processes affect the shear strength of the material by changing the cohesion (c), by changing the frictional properties of the rock ($\tan \phi$), or by increasing the piezometric pressure at the failure surface. These controls and some of the ways in which they can be altered are summarized in Table 15-10.

Steep hillslopes are obviously favorable zones for landslides. Table 15-11, for example, shows the density of small landslides on slopes with different gradients in an area of southern California. Any steep slope should be considered carefully as a possible site for hillslope failures, and in some geologic conditions, landslides can occur on quite gentle gradients. In mountainous regions, intense degradation of valley floors by rivers or by glaciers in the recent past has produced high steep hillslopes along which failure is common. In tectonically active areas, slow *tilting* of the ground by mountain building forces creates an additional hazard by steepening some slopes.

Table 15-10 Controls of hillslope stability.

CONTROLS OF THE DOWNSLOPE FORCE
Hillslope gradient
Steepening of the slope by tectonic tilting
Undercutting of the slope by geomorphic processes or human interference
Loading of the upper end of the slope
Short-term downslope stresses generated by earthquakes
CONTROLS OF THE SHEAR STRENGTH
Nature of the geologic materials: rock type and structure (joints, faults, angle of dip); nature of weathering products
Water-pressure changes due to fluctuations of rainfall and snowmelt; diversion of stormwater; submergence; fluctuation of reservoir levels; leakage from canals, irrigated fields, septic tanks, sewerage lines, and water pipes; reduction of evapotranspiration following change of vegetation. Concentration of groundwater flow by geologic structures, such as joints, or by the sequence of geologic materials
Earthquake vibrations, which can reduce the strength of weakly cemented sands or silts
Tree roots, which can increase the cohesion of soils; this cohesion is lost when the roots decay after logging or burning.

Table 15-11 Density of landslides as a function of hillslope gradient under two vegetative covers in southern California. (From Corbett and Rice 1966.)

HILLSLOPE GRADIENT (%)	NUMBER OF LANDSLIDES PER ACRE	
	UNDER BRUSH	UNDER GRASS
40–54	0	0.15
55–69	0.25	2.02
>70	0.88	4.09

Undercutting by a stream or spring at the hillslope base and the removal of previously failed debris maintain a steep slope. Undercutting is also common during construction of highways and railroads, buildings and parking lots. Robinson et al. (1972), describe how construction of an interstate highway in Colorado caused movement of a 770,000 cubic yard landslide. The disturbed material consisted of glacial deposits and deeply fractured and weathered granite and sedimentary rocks which covered a high slope that had been steepened by glacial erosion during the Pleistocene. In the same area there is evidence that several other major ancient slides have occurred in similar geologic circumstances.

Overloading often occurs during road construction, especially along many forest roads and residential developments in steep areas, where material derived from undercutting the upper hillside is side cast onto the lower hillslope as a relatively loose fill to widen the road bed (see Figure 15-41). Stormwater is then often diverted from the road onto the loose fill. The effect of road construction on the location of landslides in logging areas is illus-

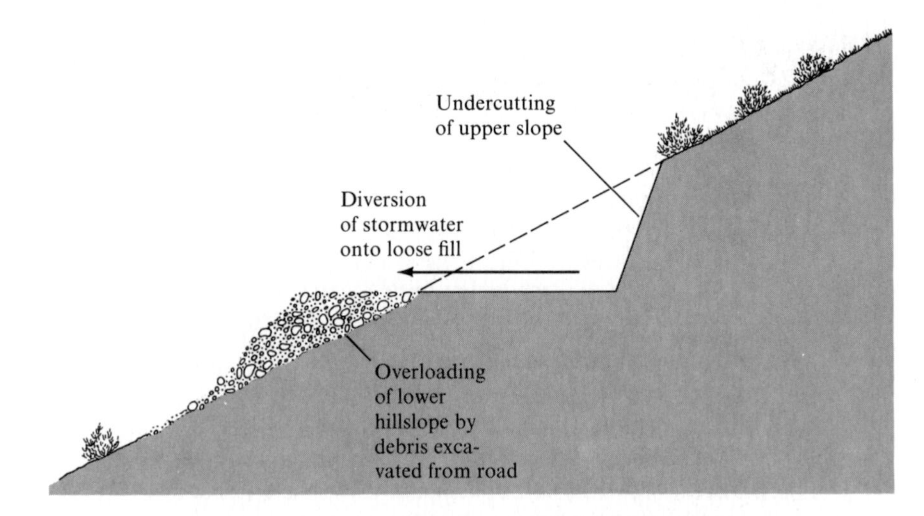

Figure 15-41 Alteration of hillslope stability by road construction.

Table 15-12 Effects of roads on the occurrence of landslides.

LOCATION	IMPACT	SOURCE
Cascade Mountains, Oregon	During heavy rains and snowmelt of 1964–65 winter 0.4 slides/1000 acres in undisturbed areas 3.9 slides/1000 acres on logged areas 126 slides/1000 acres in vicinity of roads	Dyrness (1967)
S. Fork Salmon River, Idaho	89 failures during one storm, 90% along roads; mostly due to fill failures and obstruction of drainage	Swanston (1970)
Coast Mts., southwest British Columbia	48 failures examined, 29% associated with road construction	O'Loughlin (1972)

trated in Table 15-12. In view of the costs of road maintenance, the esthetic impact of failures, and the ecologic and economic effects of downstream sedimentation caused by landslides, much more attention needs to be paid to the alignment, construction, maintenance, and reclamation of roads, whether they are for logging, recreation, urbanization, or mining in steep terrain. Several good guides are available (e.g., Packer and Christensen 1973) but planners, conservationists, and resource managers need to give more thought to the planning stage and to the education and supervision of builders and users. Whatever care is taken, it is likely that roads in most mountainous regions will always promote landslide activity and accelerated sedimentation in stream channels and lakes. In urban areas, the overloading of slopes is occurring more frequently as buildings are sited on artificial fills on steep hillsides that were formerly avoided. Increases in land-use pressure, advances in building technique, and the desire for spectacular views all contribute to this problem.

Overloading and the creation of unstable slopes frequently occur during the deposition of spoil from mining operations. Figure 15-42 shows a common situation below many open-pit mining operations in Appalachia. A catastrophic example of a debris flow from a mine spoil occurred in the Welsh coal-mining village of Aberfan in 1966. Shale mining-waste had been deposited in a large tip on a 30-percent slope in well-jointed sandstone. The loosely packed spoil accumulated over a spring in the sandstone. The tip had occasionally moved in previous years; the spring emerging from the toe of the deposit had caused intermittent failures of increasing size. The base of the tip had been failing slowly along a well-defined failure surface as the shear strength of the material was lowered by the water pressure. Eventually, heavy rains raised the piezometric pressure enough to cause a large slump. This acted as a trigger, breaking the weak interparticle frictional contacts in the loosely packed material. The weight of the overburden was suddenly transferred to the pore water, and the whole mass behaved like

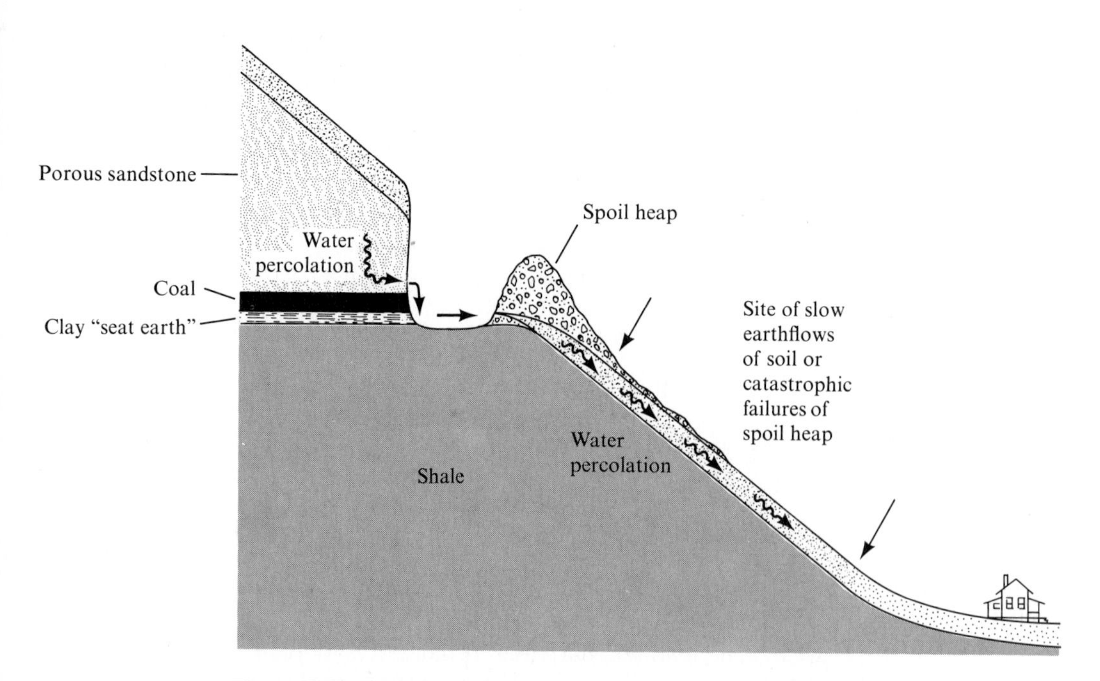

Figure 15-42 Typical situation below many open-pit mining operations in Appalachia. Overloading and water percolation cause slow earthflows, debris slides, and debris flows.

a viscous liquid. A huge debris flow traveled 2000 feet down a 22-percent slope at a speed of 20 miles per hour. Before coming to rest, the debris flow razed several homes and a school, and killed 144 people, including 116 children in the school. At the time the tip was deposited, very little was known about geologic and hydrologic conditions that could affect its stability, and even less was known about debris flows that might result from a failure. Some signs of instability and impending failure were evident for several years before 1966, however, and if those responsible had been able to interpret the signs correctly and had considered the probable path of a resulting debris flow, a disaster could have been avoided.

In areas prone to *earthquakes,* landslides often constitute a major component of the earthquake hazard. Earth motions cause short-period increases in the downslope stress within hillside material. Simonett (1967) described a large number of landslides generated by a large earthquake in New Guinea. The number of slides per unit area decreased away from the epicenter.

The shear strength of hillslope materials depends on their *geology.* This is especially true of bedrock, whether it is fresh or slightly weathered. The dip of the rocks and the degree of orientation of joints, faults, and other lines of weakness are very important in determining the stable angle of slope. The collapse of 240 million cubic meters of rock into the Vaiont Reservoir in the Italian Alps in 1963 demonstrates these effects. A weak limestone with interbedded clays dipping toward the reservoir (Figure 15-43) had been further weakened by solution of the rock and by the post-Pleistocene development of joints, as shown in Figure 15-44. These weaknesses and the rise in groundwater pressures as the reservoir level was raised resulted in a landslide that sent a 100-meter-high wave of water over the dam, killing nearly 3000 people farther down the valley. Again valuable clues were ignored about the general instability of the site and the possible catastrophic nature of the failure. Kiersch (1964) quotes an Italian govern-

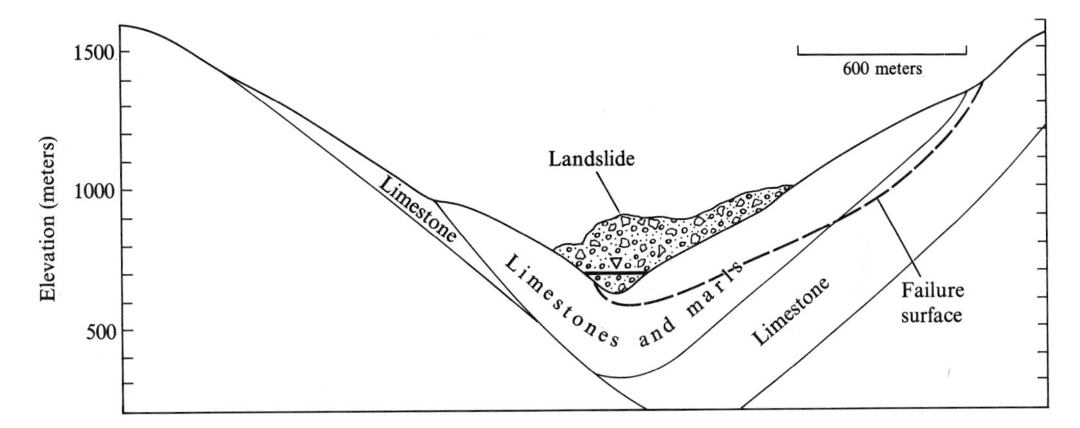

Figure 15-43 Geologic cross section of the landslide at Vaiont Reservoir. (Modified after Kiersch 1964.)

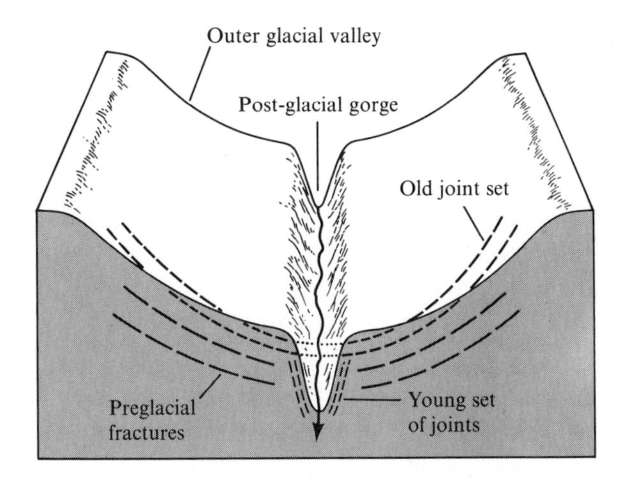

Figure 15-44 Joint systems in the rocks of the Vaiont Valley. The old set developed after the glacier left the valley, and the young set developed as the inner gorge was cut by streams after glaciation. (From Kiersch 1964.)

ment report citing the "lack of coordination between technical and governmental officials."

Terzaghi (1962) reviewed the effects of joint patterns on the stability of rock slopes. The intensity of fracturing and the mineralogy of a rock have a strong influence upon its shear strength. Some rocks weather to debris with a particularly low strength and a high water-holding capacity. In some cases, the strength of a bedrock hillside may depend not on the properties of the majority of the rock but on the nature of inconspicuous zones of weathered material along joint surfaces. The influence of rock type on slope stability is illustrated by the Madison Canyon slide shown in Figure 15-28. O'Loughlin (1972) examined 14 major slides associated with road construction in the Coast Mountains of British Columbia. Nine of the slides occurred in deeply weathered and mechanically weak sedimentary and metasedimentary rocks that covered 20 percent of the region; on the 80 percent of the area covered by coherent dolerites, only five such failures occurred.

Among unconsolidated sediments such as sands, silts, and clays, the shear strength is much lower than in coherent bedrock. Shear strength generally decreases as the particle size and degree of packing (density) of the material decrease. In some rocks, particularly clays, the shear strength can deteriorate so that an artificial slope that was stable when it was formed might fail 10 to 50 years later as the clay is weakened by fissuring, water penetration, and weathering.

In weakly cemented sand and loess (wind-blown silt), *earthquake vibrations* can break the intergranular bonds and the material can become a fluid mass like the huge dry flows of loess that destroyed villages and killed thousands of people in Kansu Province, China, during the earthquake of 1920.

In most landslides, water is the critical factor determining where and when the strength of geologic materials will be lowered to the failure point. Equation 15-5 indicates that as the water pressure in the material increases, the shear strength decreases. There are many ways in which this can happen by both natural and man-made causes. We will illustrate a few, but in landslide-prone areas anyone concerned with natural processes or the environmental impact of development should always be aware of the danger of water concentrating in geologic materials.

The sequence of rocks is perhaps the most common reason for the impedance of drainage and the build-up of water pressure. We have already shown examples of saturation developing above bedrock (Figure 15-40) and above a silty clay beneath sand (Figure 1-10). Any other features, such as concentrations of joints or topographic depressions, which focus subsurface water, also localize failures. These therefore occur most often in topographic hollows and swales. O'Loughlin, for example, found that 65 percent of the landslides he examined originated in drainage depressions or seepage hollows. The role of water also explains why most landslides occur during wet seasons, severe rainstorms, or periods of snowmelt.

Figure 15-45 Embayments caused by slumping into Lake Roosevelt, Washington. (F. O. Jones, U.S. Geological Survey.)

People often cause or contribute to failures by altering the hydrology of hillsides. Diversion of stormwater from roads or roofs, leakage of water from canals, irrigated fields, septic tanks, water pipes, and sewerage lines are frequent causes of small but damaging landslides. The submergence of unconsolidated sediments by the impoundment of reservoirs has been responsible for large landslides in some areas. Excellent examples of such failures occur around Lake Roosevelt, Washington, behind Grand Coulee Dam (see Figure 15-45). Jones et al. (1961) described huge failures involving both slumping and flowage during and since the filling of Lake Roosevelt. The unconsolidated silts and sands are weakened by saturation and fail below the lake surface. The material flows outward onto the lake bed, producing slumps in the high terraces along the shore. These failures have undermined houses, farms, communications links, and valuable farmland. If a reservoir level is lowered rapidly, the drainage of water from the banks often cannot keep pace with the drawdown. This situation produces high piezometric pressures in the bank material relative to the reservoir level. This is the reason for the large number of unsightly landslides around some artificial reservoirs, whereas natural lakes in the same region have stable banks. The process lowers the recreational value of many flood-control and water-supply reservoirs and is most common in silty bank materials.

Another type of land use that causes some increase of landslide frequency is clear felling of forests, although the logging itself does not cause as many failures as the associated road system. During the first few years after cutting, the decay of roots decreases the cohesive strength of the soil and

Table 15-13 Summary of studies of increased landsliding in clearcut logging areas.

LOCATION	IMPACT	SOURCE
Southeast Alaska	During 100 years (approx.) before logging, 13 slides in a basin; during 10 years after logging began, 116 slides in same area	Bishop and Stevens (1964)
Cascades Mts., Oregon	During heavy rains and snowmelt of 1964–65 winter, 0.4 slides/1000 acres in undisturbed areas and 3.9 slides/1000 acres in logged areas	Dyrness (1967)
Southern California	Conversion of brush to grass led to a sevenfold increase in volume of soil slips in a 9-year storm and to a 2.8-fold increase in a 32-year storm	Rice and Krammes (1970)

makes failure more likely. Table 15-13 summarizes some of the experimental studies of the effects of this process in logging areas. Even less dramatic changes of vegetation, such as replacing brush with grassland, can trigger extensive landsliding if other conditions favor failure (see Tables 15-11 and 15-13).

Recognition and Avoidance of Landslide Hazard

We have reviewed the interplay of forces controlling the occurrence of landslides and have pointed out how humans can disrupt the balance. Planners need to be aware of how their actions might produce such disruption with expensive consequences both on-site and downstream. To avoid such consequences, planners should be aware of the possibilities for recognizing zones of potential landslide hazard. Delineation of hazardous zones at an early stage of planning can save large amounts of time and money in road alignment, building location, and esthetic, ecologic, and economic values downstream. Where the landslide-prone area cannot be avoided, measures for stabilizing hillslopes can be undertaken at an early stage in the development, but only if the hazard can be recognized. Too often recognition occurs only after a hillslope fails and causes damage. Stabilization then may be very expensive or impossible.

We do not suggest that the planner try to acquire the expertise of the engineering geologist or the soils engineer, and the suggestions that we present below should not be viewed as substitutes for detailed site analysis. We recommend only that the planner be able to recognize clues that suggest that hillslope stability may be a problem and that the services of an expert may be needed. *Detailed site analysis is usually undertaken only if there is some reason to believe that a problem exists.* Student planners in our classes have become proficient at gathering information on problem areas both in the field and from other sources. They are generally able to decide: (1) whether a site is in a stable area; (2) whether there is enough uncertainty to warrant a rapid reconnaissance or a detailed site investigation by an expert; or (3) whether the site is so obviously unstable that it should be avoided unless extensive stabilization works can be justified.

Techniques for evaluating landslide hazard involve at least one of the following: (1) recognition of past hillslope failures; (2) recognition of conditions that are conducive to landsliding; (3) recognition of the de-stabilizing effect that some planned development might have.

Recognition of past hillslope failures usually depends on a variety of procedures in the field in the office. Some landslides can be recognized easily by their scars, either in the field or on aerial photographs. For this reason it is easiest to locate them soon after catastrophic wet periods, which generate large numbers of slides. This is particularly true in tropical and subtropical regions where red or yellow soils exposed in landslide scars contrast strongly with vegetation. Within a few years the scars may be re-vegetated in humid regions, and they are then more difficult to detect. Discontinuities in the vegetation cover may help locate older slides. Some pioneer species, such as blackberry or alder, are useful, for example, in the Puget Lowland and Cascade Mountains of Washington. Patches of young trees of similar age amidst a forest of mixed ages sometimes indicate former landslides, though other processes can produce a similar situation.

Landslide deposits yield another set of clues. Some deposits form obvious bulging, hummocky lobes, on which vegetation may lean in all directions. The upper part of a slide or slump may have one or more scarps separating narrow flat steps. Talus slopes produced by rockfall are generally free of vegetation and are conspicuous. In other places, the deposit may have been removed by river or wave action, or the slide mass may have come to rest on a very gentle gradient and spread out with very little topographic expression. There are countless examples of inconspicuous old landslides that have been remobilized when their lower ends were excavated, and it is common to see houses located on the toe of a slide where a slight rise of land offers better drainage or a view. If the geologic material can be seen in cross section, such as in a river bank, road cut, or building excavation, jumbled blocks of loose material with contorted bedding or none at all might indicate landslide debris, although again the absence of bedding is not by itself diagnostic. Comparison with the other geologic formations in the vicinity should indicate whether such material has been deformed or moved from its original stratigraphic position by landsliding.

Landslide hazard in some areas may be suggested by the juxtaposition of a steep cliff and a bench, like that shown in Figure 1-10. Such a bench might have resulted from the retreat of the upper cliff, and a wise planner should consider what geomorphic process could be responsible for such a bench before being carried away with the view. Cliff retreat is not the only possible reason for such a bench, but if landslide debris is found on the bench, the danger should be recognized.

Slow, continuing movement can also be recognized in the field by watching for minute details. The cracking of soil or of buildings, the deformation of fence lines, and the breakage or patching of pavement surfaces are useful signs. Spaces between sidewalks, curbstones, and road beds are particularly useful, as is the tilting of a sidewalk that can be recognized if it has been resurfaced with thicker material on one side than on the other.

Other landslides in urban areas can be located from files in city engineering offices, highway departments, newspaper offices, and from federal and state organizations involved in disaster relief. Such records are rarely compiled in a useful form, although this is slowly being done in some cities. Some large, old landslides are plotted on published geologic maps, although this information is frequently not utilized by homeowners and authorities who issue building permits. The house shown in Figure 15-46 is built entirely on a deep, old landslide deposit that was indicated on a U.S. Geological Survey map published 10 years before construction.

Whether located by fieldwork or from historical records, hillslope failures can be plotted onto topographic maps and aerial photographs. Once the observer is familiar with what a landslide and its deposit look like on the aerial photograph, failures can be mapped rapidly over large areas from the photographs with the aid of a stereoscopic viewer. The features that may be used to map landslides from photographs are described in several publica-

Figure 15-46 House situated on an old landslide deposit, portrayed on a U.S. Geological Survey map for 10 years before construction.

tions (Liang and Belcher 1958, American Society of Photogrammetry 1960, Breitag 1968, Kojan et al. 1972). If each site can be visited, the observer can identify various factors responsible for the slide, such as the geologic conditions, hillslope gradient, and type oı human interference; this may also be possible from the aerial photographs. Identification of controlling factors allows the observer to extrapolate nis or her findings to other sites where failure is probable but has not yet occurred.

Recognition of conditions conducive to landsliding may be the only warning of instability if the topographic expression of landslides is not clear in heavily vegetated terrain, or if the deposits have been removed by stream action or grading operations. In some areas, changes in the conditions of stability are still taking place, especially in regions affected by Pleistocene glaciation. Landslides may not have occurred yet at a particular site but might be expected at any time, especially if there is disturbance by an outside agent. Figure 15-44 shows a good example of such conditions at the Vaiont Reservoir. A set of joints had developed as stresses in the bedrock and were released when a Pleistocene glacier melted from the valley. A second set of joints had developed parallel to the inner canyon walls during intense post-Pleistocene downcutting by the stream. These joint sets and the solution of the limestone by circulating groundwater had greatly weakened the rock but in the relatively short period of geologic time available had not caused failure. Only a relatively slight disturbance of the groundwater conditions was necessary to initiate the catastrophe. Even if there were no other signs of landsliding in the vicinity, these geologic conditions suggested the possible landslide hazard, as some experts had already warned.

Perhaps the most obvious feature to look for in an area when considering landslide hazard is topography. Steep slopes are always suspect, although gradient alone is not a very good criterion. Furthermore, some very gentle slopes may be highly unstable; the bench in Figure 1-10 is a good example of this. For this reason, land-capability maps or erosion-hazard maps that indicate the water erosion hazard due to gradient can be very misleading indicators of mass wasting. Other topographic conditions that should be viewed with suspicion are topographic hollows and the mouths of canyons.

Rock type and structure are particularly useful indicators of landslide hazard. Formations that are mechanically weak, deeply weathered, intensely fractured or steeply dipping should be investigated carefully. Strong cliff-forming rocks such as some lava flows and well-cemented sandstones overlying weak rocks such as volcanic breccias and shales provide favorable situations for catastrophic rockfalls. Stratigraphic relationships that impede groundwater and build up high piezometric pressures should also be recognized. Figure 15-47, for example, shows the geologic conditions in which a 50-million-cubic-yard rockslide occurred in the Gros Ventre Canyon of Wyoming. A permeable sandstone overlies impermeable, plastic shale on a steeply dipping (20°) contact. The Gros Ventre River had eroded through the sandstone. Heavy rainfall and snowmelt recharge the groundwater each

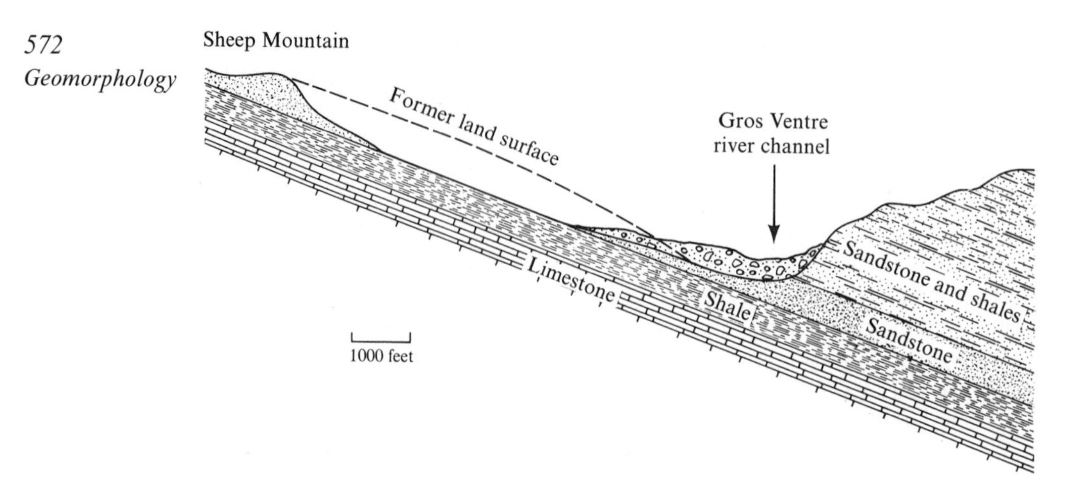

Figure 15-47 Sketched geologic cross section of geologic conditions at the site of a 50-million-cubic-yard landslide in the canyon of the Gros Ventre River, Wyoming. (From Alden 1928.)

year and abundant seepage occurs along the base of the slope on which the landslide occurred. Abundant slides occur in the area. With good geologic mapping it is to be hoped that such a hazard would be recognized and avoided today. But disasters still occur in similar situations.

Geologic reports, drilling logs, and field observations can be used as sources of information on stratigraphy and the mechanical characteristics of rocks. Topographic features may indicate the stratigraphy, even if geologic contacts are hidden by dense vegetation, thick soil, younger deposits, or landslide debris; the bench shown in Figure 1-11 is an example of this.

Other geologic hazards, such as seismicity, volcanism, or the continuing rise of the land during tectonism, produce special landslide hazards in some regions, and the planner should be aware of these. Figure 15-38, for example, is a map of ancient debris flows around Mount St. Helens, Washington. It indicates the path such flows might take in the likely event of a recurrence.

Hydrologic conditions may also indicate potential landslide sites, and certain signs warrant further investigation even if no landslides have been reported there. Seepage into or out of a rock or soil mass in the vicinity of a steep slope should always be considered suspect. Ponds, marshes, or streams may drain into a potential slide mass on top of a hill. Seepage can also be observed at the foot or in the middle of a hill. A search for seepage should be conducted during the wettest part of the year. If the area is heavily vegetated or urbanized, the field investigator would be well-advised to stop periodically and listen; this may enable him to trace shallow subsurface flow. Increasing use is being made of color and infrared aerial photographs for recognizing zones of infiltration, subsurface flow, and seepage lines

(Tanguay and Chagnon 1972). Botanical evidence of seepage can also be useful, either from aerial photographs or in the field. Some plant species, such as horsetails (Equisetum), nettles, and various tree species, are frequently associated with either wet or disturbed areas. The best indicator species vary from region to region, but an awareness of vegetation is valuable in most areas and most planning problems.

Recognition of the potential destabilizing effect of some planned development usually requires some thought, a careful look at the field conditions, and an evaluation of the environmental impact of the proposed development by experts in engineering geology and soil mechanics, if there is some initial cause for concern. But the experts would not usually be called in for such a consultation *unless the planner senses the danger first.* This is one of the reasons for which we reviewed some controls of hillslope failure. Although not trained in engineering, planners must appreciate that proposals they develop–deciding road alignments, locating residential areas, encouraging strip mining or clearcutting, recommending dam construction, and so on–can disrupt the stability of a hillslope and put the project and human lives in jeopardy. Even major engineering structures are destroyed by hillslope failures (Kiersch 1964). Planners should be aware of the possible mass-wasting problems and should know what questions to ask of engineers and engineering geologists *at an early stage* in the planning process.

Landslide Hazard Maps

As with the flood-hazard problem, described in Chapter 11, losses due to landslides are increasing as residences, highways, recreational developments, and major structures spread into unstable regions. This spread is occurring so rapidly, especially in developing countries, that it outstrips the capabilities of existing engineering services for detailed stability analyses on a site-by-site basis. Many developments occur piecemeal and no single investment, such as a residence, is considered sufficiently large to warrant a careful engineering study. For these reasons, potential hillslope failure is never considered in most development, even in areas prone to landsliding.

To counteract this trend, and perhaps to reduce landslide damage if the information is used by planners, an effort is now being made to map the susceptibility of hillslopes to failure. This technique usually involves mapping landslides and studying their association with such factors as hillslope gradient, stratigraphy, or undercutting by streams. The associations can then be used to define classes of landslide hazard, as shown by the example in Table 15-14. Different schemes will be needed in areas with different controls of hillslope stability, and each local classification should be designed to give the best resolution between classes with and without landslides. An area can then be mapped according to this classification, as shown in Figure 15-48.

Table 15-14 Example of a classification scheme for slope stability. *Note:* This classification scheme is given only for illustration. It is useful in the Puget Lowland of western Washington (Tubbs 1975) but not necessarily elsewhere. For other regions, a new ranking, appropriate to local controls of slope stability, must be developed.

Class 4:	All sand-over-clay contacts
Class 3:	Gradients >15% underlain by clay or silt*
Class 2:	Gradients >15% underlain by sandy or gravelly materials
Class 1:	Gradients <15%, except on a sand-over-clay contact

*The choice of 15% was based on a consideration of the mechanics of failure and the shear strength of the silt-clay bed.

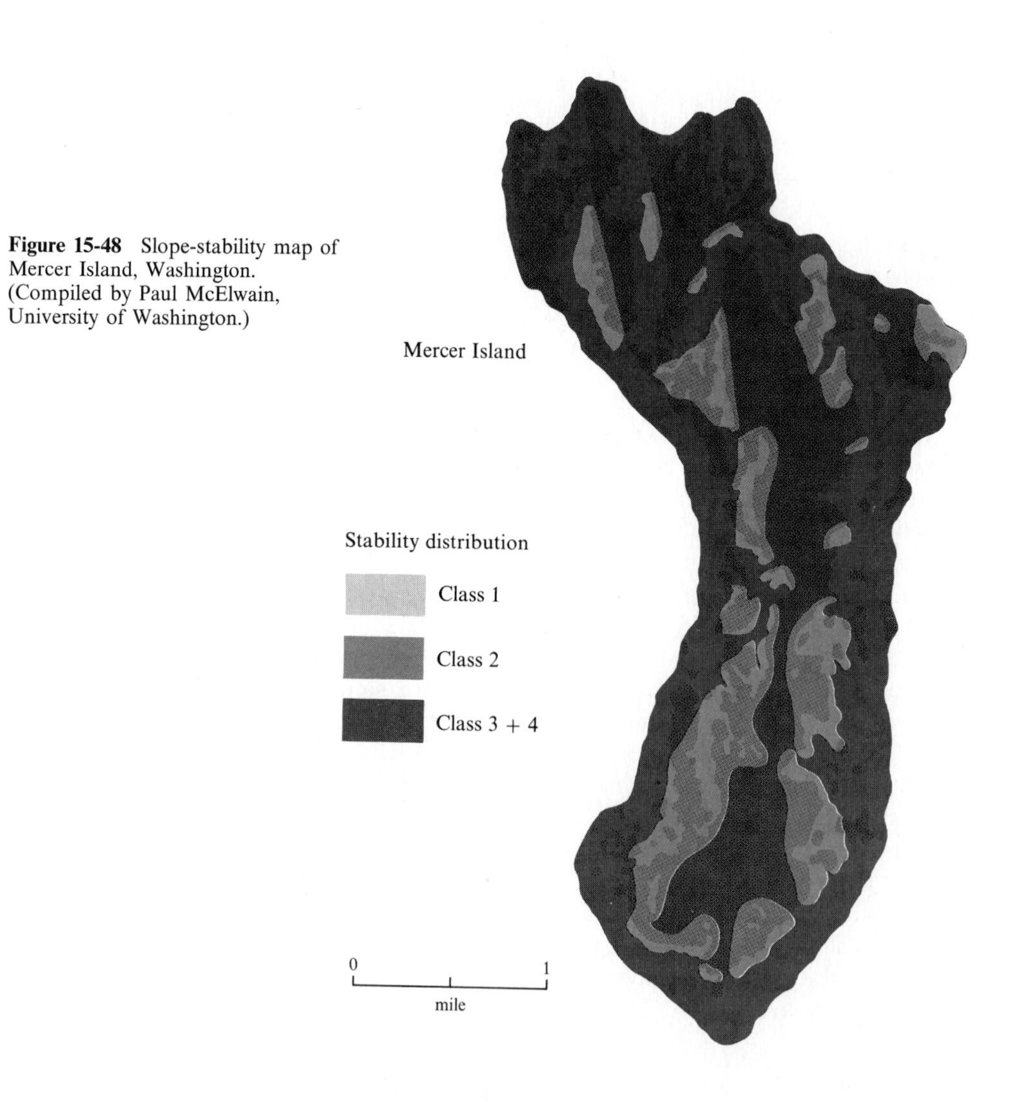

Figure 15-48 Slope-stability map of Mercer Island, Washington. (Compiled by Paul McElwain, University of Washington.)

The hillslope gradient can, be determined directly from the topographic map by making a scale on the edge of a piece of cardboard. For example, on a map with a scale of 1:24,000 and a contour interval of 25 feet, a 30-percent slope is indicated by a contour spacing of 24 contours per inch. Marks spaced 24 to the inch are marked on the edge of the cardboard, which is then oriented perpendicular to the contours at many points on the map. If the contours on the map are closer than the marks on the card, the hillslope gradient exceeds 30 percent and can be so labeled. Later, all the points with gradients exceeding 30 percent can be surrounded by a boundary. The process can be repeated for gradients exceeding 15 percent or any other value with the aid of a new scale. Other factors such as stratigraphy can then be added to the map. For example, the gradient map can be laid over a geologic map, and at each sampling point the stratigraphy can be identified (e.g., sand, glacial till, sand-clay contact). The materials or contacts that the earlier analysis has shown to be associated with landsliding can be mapped onto the hillslope gradient overlay. Finally, the various classes of hillslope stability, such as those listed in Table 15-14, can be identified and shaded on the overlay to produce a result like that in Figure 15-48. Tubbs (1975) has produced a similar map of Seattle using the classification scheme listed in Table 15-14. The density of new slides occurring in each of his four stability classes during the winter of 1972 is shown in Table 15-15.

Table 15-15 Density of landslides occurring in each of the slope-stability classes defined in Table 15-14 during the winter of 1972, Seattle. (From Tubbs 1975.)

	SLOPE-STABILITY CLASS			
	1	2	3	4
Density of slides per sq mi	0.034	0.77	4.7	12.4
Damage costs ($ per sq mi)	50	9,000	25,000	161,000

In California, many other techniques of mapping landslide hazard have been developed in a search for methods of presenting such information in a form that is useful to administrators, planners, and landowners. Some maps have been published showing simply the location of landslides recognized on aerial photographs (e.g., Brabb and Pampeyan 1972). Others show landslide deposits with the isopleths (contours) of landslide density superimposed (e.g., Campbell 1973). The most ambitious maps so far produced show seven classes of landslide susceptibility based on combinations of gradient and rock type. The ranking of landslide susceptibility in each of these classes was based on the measured areal frequency of landslides (Brabb et al. 1972).

The use of slope stability and landslide susceptibility maps in planning is growing rapidly, and we think that this development is salutary. The trend is not without its critics, however. Landowners, civic boosters, and taxation agencies who find a part of the land in which they are interested classified as unstable often resist the development and publication of such information for understandable reasons. The situation here is the same as in the case of flood mapping, where a town will even seek legal recourse to have its central business district reclassified "out" of a floodway that has been recognized as such by a team of hydrologists and engineers. Such resistance will probably continue and can only be combated by a general awareness of natural processes and their danger among the general public and those concerned with wise land-use planning.

More serious from a methodological point of view are the criticisms of soils engineers and engineering geologists who rightly claim that it is impossible to evaluate accurately the stability of an individual hillside without a thorough engineering investigation involving drilling, the measurement of shear strength and groundwater pressures, and a consideration of any stabilization work that may be necessary. To our minds, this is not a criticism of the value of slope stability maps but merely reflects a misunderstanding of their purpose.

A slope-stability map is a statement of the probability of hillslope failure. Donald Tubbs puts the point succinctly when he says, "In some areas, the association between landsliding and certain easily recognized geologic and topographic features is so strong that those areas should be presumed guilty until they are proven innocent." On the other hand, a slope-stability map indicates to planners those areas where they should encounter no difficulties —unless they do something imprudent. Mapped zones of greatest hazard should immediately attract the planner's attention. Through consultation with an expert, he or she can decide whether the zone can be stabilized at a reasonable cost and chance of success, or whether it should be avoided altogether. Even in a zone of generally unstable slopes, detailed investigation may indicate that a particular site is relatively stable or can be stabilized.

For example, Richard Untermann, a landscape architect at the University of Washington, was interested in a residential lot overlooking Lake Union in Seattle in an area known for its landslide hazard. He investigated the site and neighboring lots where houses had been damaged by failure, and he decided that the hillslope materials only became unstable on sites with bad subsurface drainage. Richard drained his land carefully, paying particular attention to the control of storm runoff, and built his house there. The house has not moved—so far! This is an example of a qualified person making a decision to live in a place with full awareness of the potential hazard, and taking precautions to minimize the problem. Other homeowners might not want to take the chance. Or a planner may decide that although the risk is acceptable for his or her own residence, it is too great to justify locating a larger investment there, or for building large numbers of cheaper residences or apartments without adequate drainage. Another example of development in a zone of high landslide hazard occurred during the construction of Interstate Highway 5 through Seattle. The road follows the sand-clay contact that localizes most of the landslides in the city. Extensive buttresses and drainage systems were designed, but many failures nevertheless occurred during construction. It was decided, however, that the resulting cost would be smaller than the cost of a different road route, which would cause greater economic and social dislocation.

There are several limitations inherent in slope-stability maps, however, and the user should be aware of these. The first limitation is imposed by the scale of the map. At the most frequently used scale of 1:24,000, for example, a line 0.01 inch thick covers 20 feet on the ground. This is the absolute limit of resolution for most maps. Uncertainties in placing the line correctly during field mapping, aerial-photograph interpretation, and final drafting at least double this uncertainty. It is simply not possible to locate the edge of an unstable slope to within 50 feet on a map of 1:24,000 scale, even if the other uncertainties discussed below were not present. Many landslide susceptibility maps are published at scales of 1:50,000 or 1:62,500, and on these the uncertainty in the locations of lines or points is increased. If the contour interval is large relative to the heights of the hillslopes, low, steep slopes will be overlooked as potential failure sites.

Minor, but important, errors in geologic mapping are more common than nongeologists believe. These may occur through inaccurate placing of boundaries between formations, misinterpretation of a stratigraphic sequence, or overlooking an inconspicuous but important formation. If the slope-stability map has been constructed on the basis of geological associations, these errors could lead to incorrect classification of some areas. The best defenses against such errors are to use good geologic maps from a reputable agency, to check some of the contacts in the field, and to visit a number of landslides and check the underlying geology. If a large number of landslides occurs in a zone for which there is no obvious reason on the geologic map, a search can be made for unmapped formations, perhaps covered by younger deposits.

Slope-stability maps of the kind described above cannot be used for assessing the stability of artificial slopes produced by cutting or filling. The stability of each case depends on the way it was cut, drained, compacted, or buttressed.

Finally, the most important limitation of these maps is that they portray slope stability *rather than landslide hazard.* In Figure 15-49, a hillslope failure threatens the buildings both above and below. Yet they are both in areas that would be classified as stable according to a scheme such as that shown in Table 15-14. At the top of the hill there is a danger from undermining; at the bottom of the hill, the slide debris might bury or severely damage structures. On either side of an unstable zone there lies another, more diffuse, zone subject to landslide hazard, which declines from some high value at the sites of potential failures to a low value at some distance above or below the failure zone. The width of the zone of highest landslide hazard depends on the topography above and below the failure site, the size and nature of the probable failure, and the dynamics of the failed mass after it has been set in motion. Even detailed engineering analyses do not usually treat these factors; such analyses are usually concerned only with the initial failure.

We know of no attempt to map landslide hazard per se, but the user of landslide susceptibility maps should be aware of the problem, and we can offer a few suggestions of what to consider in loosely defining the hazard. Because the top of the steep hillside in Figure 15-49 is retreating, the chances of some point being undermined must be expressed in relation to a chosen time period and to the rate of retreat of the hillslope. The best indicators of the rate of hillslope retreat are accounts of the recent geologic history of the area, a comparison of old and new maps, and a comparison of sequences of aerial photographs. If these are not available, an examination of the size of individual slump blocks might indicate how far the cliff could retreat in a single event.

The problem of estimating landslide hazard is even more difficult at the lower end of the hill. If the failure moves as a slump, it will not usually travel far, and the landslide hazard will decrease rapidly away from the zone indicated on the slope-stability map. Avalanches and flows, on the other hand, can move miles, and in these cases the landslide hazard is extensive and decreases slowly with distance from the initial failure zone. The extreme cases of this situation are the valleys on Mount St. Helens subject to lahars (Figure 15-38) and the area inundated by the wave from the Vaiont Reservoir. A rough idea of the extent of landslide hazard can be obtained by observing the type and distance of movement of past failures in the region, either from aerial photographs or in the field. Landslides that might occur

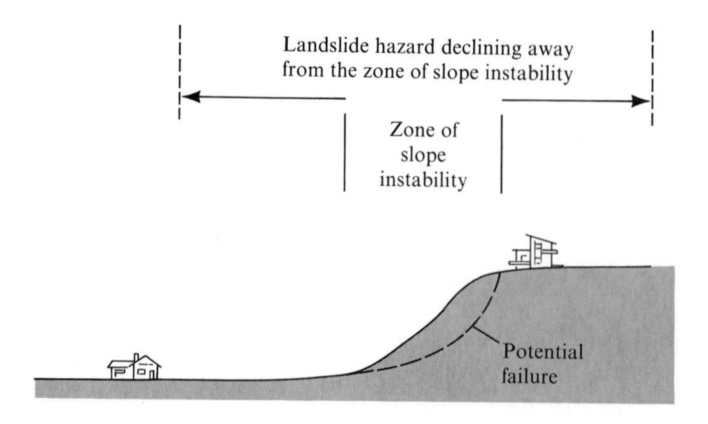

Figure 15-49 The extent of slope instability and of landslide hazard.

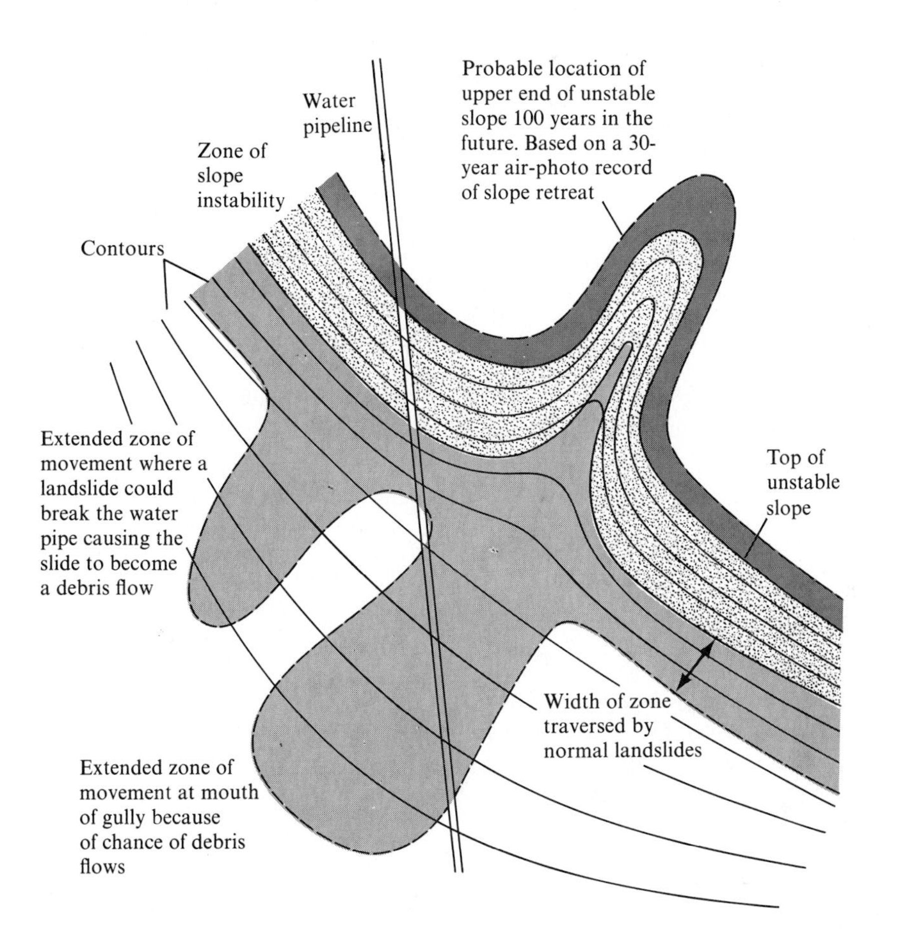

Figure 15-50 One possible form of a landslide hazard map for a site.

near stream channels deserve particular attention because they might extend the hazard for great distances downstream. Figure 15-50 illustrates a map of landslide hazard constructed according to these suggestions. The boundaries on such maps, however, will always be defined only vaguely, and all users of such maps should be aware of these limitations.

Control of Landslides

The stabilization of areas that have failed, or that show the first signs of impending failure, or that have a high probability of failure is the responsibility of soils engineers and engineering geologists. The planner should, however, know that stabilization techniques exist and that the technique employed varies with the nature of the failure and local site conditions. We have listed some techniques of stabilization in Table 15-16. These and some

Table 15-16 Examples of methods for controlling mass movements. (From Baker and Marshall 1958.)

TYPE OF MOVEMENT	METHOD FOR CONTROL
Falls	Flattening the slope
	Benching the slope
	Drainage
	Reinforcement of rock walls by grouting with cement, anchor bolts
	Covering of wall with steel mash
Slides and flows	Grading or benching to flatten the slope
	Drainage of surface water with ditches
	Sealing surface cracks to prevent infiltration
	Subsurface drainage
	Rock or earth buttresses at foot
	Retaining walls at foot
	Pilings through the potential slide mass

more specialized techniques are described by Root (1958) and Baker and Marshall (1958). The techniques are usually expensive, and their cost is best avoided by early recognition of the landslide hazard and by planning appropriate land use. In many geological situations, there is no feasible method of stabilizing hillslopes.

Bibliography

ALDEN, W. C. (1928) Landslide and flood at Gros Ventre, Wyoming; *American Institute of Mining and Metals Engineering, Transactions,* vol. 76, pp. 347–361.

AMEMIYA, M. (1970) Land and water management for minimizing sediment, in *Agricultural Practices and water quality* (eds. T. L. Willrich and G. E. Smith), Iowa State University Press, Ames, pp. 33–45.

AMERICAN SOCIETY OF PHOTOGAMMETRY (1960) *Mannual of photographic interpretation;* Washington DC, 833 pp.

BAKER, R. F. AND MARSHALL H. E. (1958) Control and Correction, Chapter 8 in *Landslides and engineering practice* (ed. E. B. Eckel), Highway Research Board Special Report 29, National Academy of Sciences, Washington, DC, pp. 150–188.

BARNETT, A. P. (1965) Using perennial grasses and legumes to control runoff and erosion; *Journal of Soil and Water Conservation,* vol. 20, pp. 212–215.

BARROWS, H. L. AND KILMER, V. J. (1963) Plant nutrient losses from soils by water erosion; *Advances in Agronomy,* vol. 15, Academic Press, New York, pp. 303–316.

Bartelli, L. J., ed. (1966) *Soil surveys and land use planning;* Soil Science Society of America, Madison, WI, 196 pp.

Battawar, H. B. and Rao, V. P. (1969) Effectiveness of crop-cover for reducing runoff and soil losses; *Journal of Soil and Water Conservation in India,* vol. 17, no. 3,4, pp. 39–50.

Beaseley, R. P. (1972) *Erosion and sediment pollution control;* Iowa State University Press, Ames, 320 pp.

Beer, C. E., Farnham, C. W. and Heinemann, H. G. (1966) Evaluating sedimentation prediction techniques in western Iowa; *American Society of Agricultural Engineers, Transactions,* vol. 9, pp. 828–833.

Bennett, H. H. (1939) *Soil Conservation;* McGraw-Hill, New York, 993 pp.

Berry, L. and Townshend, J. (1972) Soil conservation policies in the semi-arid regions of Tanzania: a historical perspective; *Geografiska Annaler,* vol. 54A, pp. 241–254.

Bishop, D. M. and Stevens, M. E. (1964) Landslides on logged areas in southeast Alaska; *U.S. Forest Service Northern Forest Experiment Station Research Paper NOR-1.*

Brabb, E. E. and Pampeyan, E. H. (1972) Preliminary map of landslide deposits in San Mateo County, California; *U.S. Geological Survey Miscellaneous Field Studies Map MF-344.*

Brabb, E. E., Pampeyan, E. H. and Bonilla, M. G. (1972) Landslide susceptibility in San Mateo County, California; *U.S. Geological Survey Miscellaneous Field Studies Map MF-360.*

Breitag, E. (1968) Landslide identification; *Photo Interpretation,* vol. 7, pp. 15–21.

Briggs, R. P., Pomeroy, J. S. and Davies, W. E. (1975) Landsliding in Allegheny County, Pennsylvania; *U.S. Geological Survey Circular 728,* 18 pp.

Brown, C. B. (1950) Effects of soil conservation; in *Applied sedimentation* (ed. P. D. Trask), pp. 380–406, John Wiley, New York.

Brune, G. M. (1950) The dynamic concept of sediment sources; *American Geophysical Union Transactions,* vol. 31, pp. 587–594.

Bryan, R. R. (1968) The development, use and efficiency of indices of soil erodibility; *Geoderma,* vol. 2, no. 1, pp. 5–26.

Campbell, I. A. (1970) Erosion rates in the Steveville Badlands, Alberta; *Canadian Geographer,* vol. 14, pp. 202–216.

Campbell, R. H. (1973) Isopleth map of landslide deposits, Point Dume Quadrangle, Los Angeles County, California; an experiment in generalizing and quantifying areal distribution of landslides; *U.S. Geological Survey Miscellaneous Field Studies Map MF-535.*

Campbell, R. H. (1975) Soil slips, debris flows, and rainstorms in the Santa Monica Mountains and Vicinity, Southern California; *U.S. Geological Survey Professional Paper 851.* 51 pp.

Chittenden, D. B. (1973) Prevention and control of soil erosion: the state of the art; in *Soil erosion: causes and mechanisms, prevention and control,* Highway Research Board Special Report 135, pp. 129–140.

Copeland, O. L. (1965), Land use and ecological factors in relation to sediment yields; *U.S. Department of Agriculture Miscellaneous Publication 970,* pp. 72–84.

Corbett, E. S. and Rice, R. M. (1966) Soil slippage increased by brush conversion; *U.S. Forest Service Pacific Southwest Forest and Range Experiment Station Research Note PSW-128.*

Crandell, D. R. and Fahnestock, R. K. (1965) Rockfalls and avalanches from Little Tahoma Peak on Mount Rainier, Washington; *U.S. Geological Survey Bulletin 1221-A.*

Crandell, D. R. and Mullineaux, D. R. (1975) Technique and rationale of volcanic hazards appraisal in the Cascade Range, Northwestern Unites States; *Environmental Geology,* vol. 1 no. 1, pp. 23–32.

Crandell, D. R. and Varnes, D. J. (1961) Movement of the Slumgullion earthflow near Lake City, Colorado; *U.S. Geological Survey Professional Paper 424-B,* pp. 136–139.

Curtis, W. R. (1971) Terraces reduce runoff and erosion on surface-mine benches; *Journal of Soil and Water Conservation,* vol. 26, no. 5, pp. 198–199.

Dabin, B. (1959) Schedule of three years of erosion (1956–57–58) at the Adiopodoumé station, Ivory Coast; *Proceedings of the 3rd Interafrican Soils Conference,* Dalaba.

DALLAIRE, G. (1976) Controlling erosion and sedimentation at construction sites; *Civil Engineering,* vol. 46, pp. 73–77.

DAVIS, G., ed. (1971) A guide for revegetating bituminous strip-mine spoils in Pennsylvania; *Research Committee on Coal Mine Spoil Revegetation in Pennsylvania,* 46 pp.

DEBELLE, G. (1971) Roadside erosion and resource implications in Prince Edward Island; *Geographical Paper No. 48, Policy Research and Coordination Branch, Department of Energy, Mines and Resources.* Ottawa.

DIETRICH, W. E. (1975) Sediment production in a mountainous basaltic terrain in central Coastal Oregon; M.S. thesis, University of Washington.

DILS, R. E. (1953) Influence of forest cutting and mountain farming on some vegetation, surface soil, and surface runoff characteristics; *U.S. Forest Service Southeastern Forest Experiment Station; Station Paper 24* Asheville, NC, 55 pp.

DISEKER, E. G. AND RICHARDSON, E. C. (1961) Roadside sediment production and control; *American Society of Agricultural Engineers Transactions,* vol. 4, no. 1, pp. 62–64, 68.

DUNNE, T. (1977) Evaluation of erosion conditions and trends; in *Guidelines for watershed management, FAO Conservation Guide 1,* pp. 53–83 (ed. S. H. Kunkle), United Nations Food and Agriculture Organization, Rome.

DYRNESS, C. T. (1967) Mass-soil movements in the H. J. Andrews Experimental Forest; *U.S. Forest Service Pacific Northwest Forest Experimental Station Research Paper PNW-42.*

ELLISON, W. D. (1947) Soil erosion studies, Parts I to VII; *Agricultural Engineering,* vol. 28, pp. 28, 145, 197, 245, 297, 349, 402, 442.

ELLISON, W. D. (1948) Erosion of soil by raindrops; *Scientific American,* vol. 180, pp. 40–45. Offprint 817.

FITZPATRICK, E. A. (1971) *Pedology: a systematic approach to soil sciences;* Oliver and Boyd, Edinburgh, 306 pp.

FOSTER, A. R. (1973) *Approved practices in soil conservation;* Interstate Printers and Publishers, Danville, IL, 497 pp.

FOURNIER, F. (1967) Research on soil erosion and soil conservation in Africa; *African Soils,* vol. 12, pp. 53–96.

FREDRIKSEN, R. L. (1963) A case history of a mud and rock slide on an experimental watershed; *U.S. Forest Service Pacific Northwest Forest Experiment Station Research Note PNW-1.*

GEMMELL, R. P. (1972) Use of waste materials for revegetation of chromite smelter waste; *Nature,* vol. 240, pp. 569–571.

GILLULY, J., WATERS, A. C. AND WOODFORD, A. O. (1960) *Principles of geology;* W. H. Freeman and Company, San Francisco.

GLYMPH, L. M. (1957) Importance of sheet erosion as a source of sediment; *American Geophysical Union Transactions,* vol. 38, pp. 903–907.

GOTTSCHALK, L. C. (1945) Effects of soil erosion on navigation in Upper Chesapeake Bay; *Geographical Review,* vol. 35, pp. 219–238.

GRIFFITH, F. E., MAGNUSON, M. O. AND KIMBALL, R. L. (1966) Demonstration and evaluation of five methods of secondary backfilling of strip-mine areas; *U.S. Bureau of Mines, Report of Investigations 6772,* Washington, DC.

GUY, H. P. (1965) Residential construction and sedimentation at Kensington, Maryland; *Proceedings of the Federal Interagency Sedimentation Conference, Miscellaneous Publication No. 970,* U.S. Department of Agriculture, Washington, DC, pp. 30–37.

HAGGETT, P. (1961) Land use and sediment yield in an old plantation tract of the Serra do Mar, Brazil; *Geographical Journal,* vol. 172, pp. 50–62.

HAPP, S. C. (1944) Effect of sedimentation on floods in the Kickapoo Valley, Wisconsin; *Journal of Geology,* vol. 52, no. 1, pp. 53–68.

HAUPT, H. F. (1959) Road and slope characteristics affecting sediment movement from logging roads; *Journal of Forestry,* vol. 57, pp. 329–331.

HORTON, R. E. (1945) Erosional development of streams and their drainage basins; *Bulletin of the Geological Society of America,* vol. 56, pp. 275–370.

HUDSON, N. (1971) *Soil conservation;* Cornell University Press, Ithaca, NY, 320 pp.

HUDSON, N. AND JACKSON, D. C. (1959) Results achieved in the measurement of erosion and runoff in Southern Rhodesia; *Proceedings of the 3rd InterAfrican Soils Converence,* Dalaba.

HYDE, J. H. (1975) Upper Pleistocene pyroclastic-flow deposits and lahars south of Mount St. Helens Volcano, Washington; *U.S. Geological Survey Bulletin 1383-B,* 20 pp.

JAMISON, V. C., SMITH, D. D. AND THORNTON, J. F. (1968) Soil and water research on clay pan soil; *Technical Bulletin 1379,* U.S. Department of Agriculture, Washington, DC, 107 pp.

JENNY, H. (1941) *Factors of soil formation;* McGraw-Hill, New York, 281 pp.

JOHNSON, A. W. (1961) Highway erosion control; *Transactions American Society of Agricultural Engineers,* vol. 4, no. 1, pp. 140–152.

JONES, F. O. (1973) Landslides of Rio de Janeiro and the Serra das Araras Escarpment, Brazil; *U.S. Geological Survey Professional Paper 697.*

JONES, F. O. EMBODY, D. R. AND PETERSON, W. L. (1961) Landslides along the Columbia River valley, northeastern Washington; *U.S. Geological Survey Professional Paper 367.*

KAISER, V. G. (1965) Appraisal of soil erosion and sedimentation damage on grain farmlands of the Pacific Northwest, in *Report of meeting on erosion and sedimentation, 1964–65 flood season,* Columbia Basin Inter-agency Committee, pp. 5–8.

KAO, D. T. Y., ed. (1974) *Proceedings of the National Symposium on Urban Rainfall and Runoff and Sediment Control;* College of Engineering, University of Kentucky, Lexington, 246 pp.

KIERSCH, G. A. (1964) Vaiont reservoir disaster; *Civil Engineering,* vol. 34, pp. 32–39.

KLINGEBIEL, A. A. (1966) Costs and returns of soil surveys; *Soil Conservation,* vol. 32, no. 1, pp. 3–6.

KLINGEBIEL, A. A. AND MONTGOMERY, P. H. (1961) Land capability classification; *Agricultural Handbook No. 210,* Soil Conservation Service, U.S. Department of Agriculture, Washington, DC, 21 pp.

KOCH, C. J. AND CLAY, J. W. (1966) Guide for urban expansion; *Soil Conservation,* vol. 32, no. 1, pp. 6–8.

KOJAN, E. (1969) Mechanics and rates of natural soil creep; *Proceedings of International Association of Engineering Geologists,* Prague, First Session, pp. 122–154.

KOJAN, E., FOGGIN, G. T. AND RICE, R. M. (1972) Prediction and analysis of debris slide incidence by photogrammetry, Santa Ynez–San Raphael Mts, California; *Proceedings of the 24th Geological Congress,* section 13, pp. 124–131.

KRAMMES, J. S. (1960) Erosion from mountain sideslopes after fire in southern California; *U.S. Forest Service Pacific Southwest Forest and Range Experiment Station Research Note 171,* Berkeley, CA.

LAMBE, T. W. AND WHITMAN, R. V. (1969) *Soil mechanics;* John Wiley & Sons, New York, 553 pp.

LARSE, R. W. (1971) Prevention and control of erosion and stream sedimentation from forest roads; *Proceedings of the Symposium on Forest Land Use and the Stream Environment,* Oregon State University, pp. 76–83.

LEJCHER, T. R. (1972) Strip-mine reclaimation utilizing treated municipal wastes; *Proceedings of the National Sympossium on Watersheds in Transition,* American Water Resources Association, pp. 371–376.

LEOPOLD, L. B. (1951) Vegetation of Southwestern watersheds in the 19th century; *Geographical Review,* vol. 41, pp. 295–316.

LEOPOLD, L. B. (1972) River channel change with time: an example; *Geological Society of America Bulletin,* vol. 84, pp. 1845–1860.

LEOPOLD, L. B., EMMETT, W. W. AND MYRICK, R. M. (1966) Channel and hillslope processes in a semi-arid area, New Mexico; *Professional Paper 352-G, U.S. Geological Survey,* Washington, DC.

Leopold, L. B. and Maddock, T. M. (1954) *The flood control controversy: big dams, little dams, and land management;* Ronald Press, New York, 278 pp.

Liang, Ta and Belcher, D. J. (1958) Airphoto interpretation; Chapter 5 in *Landslides and engineering practice* (ed. E. B. Eckel), Highway Research Board Special Report 29, National Academy of Sciences, pp. 69–92.

Lindstrom, G. (1952) Effects of grading stripmine lands on the early survival and growth of planted trees; *U.S. Forest Service Central States Forest Experiment Station Technical Paper 130.*

Maragaropoulos, P. (1967) Woody vegetation as a pioneer action action towards restoring of totally eroded slopes in mountainous watersheds; in *Forest hydrology* (eds. W. E. Sopper and H. W. Lull), pp. 613–623, Pergamon Press, Oxford.

Megahan, W. F. and Kidd, W. J. (1972) Effects of logging and logging roads on erosion and sediment deposition from steep terrain; *Journal of Forestry,* vol. 70, no. 3, pp. 136–141.

Nilsen, T. H. and Turner, B. L. (1975) Influence of rainfall and ancient landslide deposits on recent landslides (1950–71) in urban areas of Contra Costa County, California; *U.S. Geological Survey Bulletin 1388.*

Nobel, E. L. and Lundeen, L. J. (1971) Analysis of rehabilitation treatment for sediment control; *Proceedings of the Symposium on Forest Land Use and the Stream Environment,* Oregon State University, pp. 86–96.

Nye, P. H. and Greenland, D. J. (1960) The soil under shifting cultivation; *Technical Communication 51,* Commonwealth Bureau of Soils, Harpenden, England, 156 pp.

O'Loughlin, C. L. (1972) A preliminary study of landslides in the Coast Mountains of Southwestern British Columbia; Chapters 3–4, in *Mountain geomorphology* (eds. H. O. Slaymaker and H. J. McPherson), Tantalus, Vancouver, pp. 101–112.

Packer, P. E. (1953) Effects of trampling disturbance on watershed condition, runoff and erosion; *Journal of Forestry,* vol. 51, pp. 28–31.

Packer, P. E. and Christensen, G. F. (1973) Guides for controlling sediment from secondary logging roads; *U.S. Forest Service Intermountain Forest and Range Experiment Station,* Ogden, UT, 42 pp.

Pearce, A. (1973) Mass and energy flux in physical denudation, defoliated areas. Sudbury, Ontario; Ph.D. dissertation, McGill University, 235 pp.

Peterson, J. B. (1964) The relation of soil fertility to soil erosion; *Journal of Soil and Water Conservation,* vol. 19, pp. 15–19.

Platt, W. S. (1971) The effects of logging and road construction on the aquatic habitat of the South Fork Salmon River, Idaho; Testimony at U.S. Senate hearings on *Clearcutting practices on national timberlands,* pp. 113–115.

Powell, M. D., Winte:•, W. C. and Bodwitch, W. P. (1970) *Community action guidebook for soil erosion and sediment control;* National Association of Counties Research Foundation, Washington, DC, 63 pp.

Rapp, A., Murray-Rust, D. H., Christiansson, C. and Berry, L. (1972) Soil erosion and sedimentation in four catchments near Dodoma, Tanzania; *Geografiska Annaler,* vol. 54A, pp. 255–318.

Renner, F. G. (1936) Conditions influencing erosion on the Boise River Watershed; *Technical Bulletin 528,* U.S. Department of Agriculture.

Rice, R. M. and Krammes, J. S. (1970) Mass wasting processes in watershed management; *Proceedings of the Symposium on Interdisciplinary Aspects of Watershed Management;* American Society of Civil Engineers, pp. 231–259.

Robinson, C. S., Lee, F. T., Moore, R. W., Carroll, R. D., Scott, H., Post, J. D. and Bohman, R. A. (1972) Geological, geophysical and engineering investigations of the Loveland Basin landslide, Clear Creek County, Colorado; *U.S. Geological Survey Professional Paper 673.*

Root, A. W. (1958) Prevention of landslides; Chapter 7 in *Landslides and engineering practice* (ed. E. B. Eckel), Highway Research Board Special Report 29, National Academy of Sciences, pp. 113–149.

SCHWAB, G. O., FREVERT, R. K., EDMINSTER, T. W. AND BARNES, K. K. (1966) *Soil and water conservation engineering;* 2nd edition, John Wiley & Sons, New York, 683 pp.

SHARPE, C. F. S. (1938) *Landslides and related phenomena;* Columbia University Press, New York, 137 pp.

SHENG, T. C. (1966) Landslide classification and studies in Taiwan; *Chinese-American joint commission on rural reconstruction, Forestry Series No. 10,* 97 pp.

SHREVE, R. L. (1965) The Blackhawk landslide; *Geological Society of America Special Paper 108,* 47 pp.

SIMONETT, D. S. (1967) Landslide distribution and earthquakes in the Bewani and Torricelli mountains, New Guinea; in *Landform studies from Australia and New Guinea* (eds. J. N. Jennings and J. A. Mabbutt), pp. 64–84.

SMITH, O. R. (1940) Placer mining silt and its relation to salmon and trout on the Pacific Coast; *Transactions of American Fisheries Society,* vol. 69, pp. 225–230.

SMITH, R. M. AND STAMEY, W. L. (1965) Determining the range of tolerable erosion; *Soil Sciences,* vol. 100, no. 6, pp. 414–424.

SO, C. L. (1971) Mass movements associated with the rainstorm of June 1966 in Hong Kong; *Transactions of the Institute of British Geographers,* vol. 53, pp. 55–65.

SOPPER, W. E. (1970) Revegetation of stripmine spoil banks through irrigation with effluent from municipal sewage; *Compost Science,* vol. 11, no. 6, pp. 6–11.

STRIFFLER, W. D. (1967) Restoration of open cast coal sites in Great Britain; *Journal of Soil and Water Conservation,* vol. 22, pp. 101–103.

SWANSTON, D. N. (1970) Mechanics of debris avalanching in shallow till soils in southeastern Alaska; *U.S. Forest Service Pacific Northwest Forest Experiment Station Research Note PNW-103.*

TANGUAY, M. G. AND CHAGNON, J-Y. (1972) Thermal infrared imagery at the St. Jean-Vianney landslide; *1st Canadian Symposium on Remote Sensing, Centre for Remote Sensing, Department of Energy, Mines and Resources, Ottawa,* pp. 387–402.

TAYLOR, R. N. (1966) Gravel pit to landfill to park with rolling hills; *Public Works,* vol. 97, pp. 105–106.

TEMPLE, P. H. (1972) Measurements of runoff and soil erosion at an erosion plot scale with particular reference to Tanzania; *Geografiska Annaler,* vol. 54A, pp. 203–220.

TERZAGHI, K. (1962) Stability of steep slopes on hard, unweathered rock; *Geotechnique,* vol. 12, pp. 251–270.

TRIMBLE, S. W. (1974) *Man-induced soil erosion on the southern Peidmont, 1700–1970;* Soil Conservation Society of America, 180 pp.

TUBBS, D. W. (1975) Causes, mechanisms and prediction of landsliding in Seattle; Ph.D. dissertation, University of Washington, 88 pp.

UNITED NATIONS (1951) *Methods and problems of flood control in Asia and the Far East;* Bureau of Flood Control of the Economic Commission for Asia and the Far East, Bangkok, 45 pp.

U.S. SOIL CONSERVATION SERVICE (1951) Soil survey manual; *U.S. Department of Agriculture Handbook No. 18.*

U.S. SOIL CONSERVATION SERVICE (1975a) Guidelines for the use of the universal soil-loss equation in Hawaii; *Technical Note, Conservation Planning No. 1,* Honolulu, 40 pp.

U.S. SOIL CONSERVATION SERVICE (1975b) Procedure for computing sheet and rill erosion on project areas; *Technical Release No. 51,* Washington, DC, 15 pp.

U.S. SOIL CONSERVATION SERVICE (1975c) Universal soil-loss equation; *Technical Note, Conservation Agronomy No. 32,* Portland, OR, 34 pp.

U.S. SOIL CONSERVATION SERVICE (1976) *Erosion and sediment control guide for urbanizing areas in Hawaii;* Honolulu.

VARNES, D. J. (1958) Landslide types and processes; Chapter 3 in *Landslides and engineering practice* (ed. E. B. Eckel), Highway Research Board Special Report 29, National Academy of Sciences, pp. 20–47.

VASUDEVAIAH, R. D. SINGH TEOTIA, S. P. AND GUHA, D. P. (1965) Runoff-soil loss determination studies at Deochanda Experiment Station: II. Effect of annual cultivated grain crops and perennial grasses on 5 percent slope; *Journal of Soil and Water Conservation in India,* vol. 13, nos. 3 and 4, p. 36.

WASHBURN, A. L. (1973) *Periglacial processes and environments;* St. Martin's Press, New York, 320 pp.

WILLIAMS, G. P. AND GUY, H. P. (1971) Debris avalanches: a geomorphic hazard; in *Environmental geomorphology* (ed. D. R. Coates), Department of Geology, State University of New York, Binghampton, pp. 25–46.

WISCHMEIER, W. H., JOHNSON, C. B. AND CROSS, B. V. (1971) A soil erodibility nomograph for farmland and construction sites; *Journal of Soil and Water Conservation,* vol. 26, no. 5, pp. 189–192.

WISCHMEIER, W. H. AND MEYER L. D. (1973) Soil erodibility on construction areas; pp. 20–29 in *Soil erosion: causes, mechanisms, prevention and control;* Highway Research Board Special Report 135, Washington, DC.

WISCHMEIER, W. H. AND SMITH, D. D. (1965) Predicting rainfall-erosion losses from cropland east of the Rocky Mountains; *Agriculture Handbook No. 282,* U.S. Department of Agriculture.

WOHLETZ, L. R. (1966) Soil maps in land planning; *Soil Conservation,* vol. 32, no. 1, pp. 8–9, 18–19.

WOLMAN, M. G. AND SCHICK, A. P. (1967) Effects of construction on fluvial sediment, urban, and suburban areas of Maryland; *Water Resources Research,* vol. 3, no. 2, pp. 451–464.

Typical Problems

15-1 Universal Soil-Loss Equation for Farmland

Estimate the average annual rate of soil loss from a 50-acre hillside with the following attributes:

Rainfall erosivity: 20,000 foot-tons per acre per year (i.e., $R = 200$)

Soil: fine, sandy loam; $K = 0.2$ to 0.4

Hillslope length: 3000 feet

Hillslope gradient: 5 percent

Therefore, using Figure 15-19, $LS = 0.9$.

Cover: Continuous corn, residues left on field. A weighted average value of C was obtained using values for each cropping stage and a curve showing the seasonal distribution of rainfall erosivity, both from Agricultural Handbook 282. Using Equation 15-4, $C_w = 0.43$.

Conservation practice: contour cultivation, $P = 0.5$ (from Table 15-5).

Solution

For $K = 0.2$:

$$A = 200 \times 0.2 \times 0.9 \times 0.43 \times 0.5$$
$$= 7.7 \text{ t/acre/yr}$$

For $K = 0.4$:

$$A = 15.5 \text{ t/acre/yr}$$

For the 50-acre hillside, therefore, the soil loss would range between 387 and 774 t/yr.

Suppose that you want to install a catch basin below a construction site to prevent eroded sediment from entering a stream. You want to predict the volumes of sediment to be handled in a year to assess how big the basin needs to be or whether it will have to be dredged periodically while the site is open.

Solution

The site has the following characteristics:

Area: 4 acres

Average hillslope length: 200 feet

Average hillslope gradient: 10 percent

The *LS* index in Equation 15-1 is 1.93 (from Figure 15-19).

Average annual erosivity index: 300 (from an analysis of rainfall intensity records, or from Figure 15-16).

Construction plans call for removing the topsoil to a depth of about 2.5 feet, which will expose a subsoil with the following characteristics (obtained from the county soil survey report and defined in the U.S. Soil Conservation Service Soil Survey Manual): 60% silt and very fine sand; 35% sand; 0.5% organic matter; fine granular structure, and slow to moderate permeability. Use of Figure 15-18 in the manner shown by the dotted line on that nomograph produces a soil erodibility index, *K*, of 0.53.

If the graded site is left without cover or conservation practice, Equation 15-1 predicts that the annual soil los per acre will be

$$A = 300 \times 0.53 \times 1.93 \times 1.0 \times 1.0$$
$$= 307 \text{ t/acre/yr}$$

or 1,228 t/yr from the whole site. Average densities of reservoir sediments range from 60 to 80 lb/ft^3 for clay sizes, 80 to 100 lb/ft^3 for mixtures of clay, silt, and sand, and 95–130 lb/ft^3 for sand and gravel. A figure of 100 lb/ft^3 seems appropriate for the soil in our problem, so the annual sediment accumulation in the catch basin would occupy a volume of 24,560 ft^3.

If the catch basin is also designed to act as a storm runoff detention structure, as described in Chapter 11, the designer would have to take into account the reduction of storage volume due to sediment accumulation, which might necessitate increasing the size of the basin.

It is important that the designer realize that the erosivity values shown in Figure 15-16 or similar compilations are long-term averages only. Some extra storage in the catch basin should be designed to store sediment during a year with highly erosive rainstorms. Wischmeier and Smith (1965) tabulate probabilities of obtaining annual *R* values exceeding various levels for stations throughout the eastern United States. Probabilities for other regions can be obtained by calculating annual *R* values from rain gauge records with the aid of Equations 15-2 and 15-3, and subjecting the *R* values to a frequency analysis, as illustrated for annual rainfall totals in Chapter 2. The values often exhibit a log-normal frequency distribution. If, for example, such an analysis indicates that in any one year there is a 10-percent probability of an *R* value of at least 450, the previously calculated soil loss should be increased by 50 percent. If the site is only to be opened for a part of the year, seasonal *R* values can be obtained from the handbook by Wischmeier and Smith, or from an analysis of rain gauge records.

Depending upon local grading and sediment-control ordinances, it may also be necessary for the planner to consider the chances that a large storm might erode a large amount of soil when the catch basin is almost full. To judge the probabilities of rainstorms with various erosivities, it is possible to compute values of EI_{30}, and therefore of *R*, for single large storms with the aid of Equations 15-2 and 15-3. The erosivity values can then be subjected to an extreme-value analysis, as described in Chapter 10, to yield estimates of single- storm *R* values with recurrence intervals of 10 years, 50 years, and so on. The handbook by Wischmeier and Smith contains probability tables for single-storm *R* values in various regions of the United States.

Suppose the planner finds, from one of these two sources, that there is a 10-percent chance of an *R* value equal to or greater than 200 for a single storm at his site in any year. What would be the soil loss during that storm?

After considering these rates of soil loss, the costs of installing and maintaining a large catch basin, and the possibility of it being overloaded, the designer might wish to choose some treatment of the construction site to reduce soil erosion. Such treatments would affect the values of C and P in Equation 15-1. Some of the possibilities are indicated in Table 15-5, Figure 15-24, and various references on urban sediment control listed in the bibliography. The data in Figure 15-24(a), for example, suggest that applying one ton of straw mulch per acre of the site would result in a C value of about 0.2, reducing the average annual soil loss from the 4-acre site to 246 t/yr.

15-3 Universal Soil-Loss Equation

Suppose that you wanted to control the average rate of soil loss in Typical Problem 15-1 to 2 t/acre/yr or less. What options would you have for manipulating the cover and the erosion-control practice?

Solution

Using the maximum value of $K = 0.4$,

$$A_{max} = 2 = 200 \times 0.4 \times 0.9 \times C \times P$$

Therefore, $C \times P = 0.028$. Possible erosion-control practices that might be more effective than simple contour cultivation are strip cropping and terracing. From Table 15-5, for a 5-percent slope, the P values for these would be 0.25 and 0.10, respectively.

With strip cropping, the following value would be necessary:

$$C = \frac{0.028}{0.25} = 0.11$$

With terracing, the following value would suffice:

$$C = \frac{0.028}{0.10} = 0.28$$

In the case of terracing, the erosion-control practice factor would be 0.10, but the LS factor would also be changed by the terracing. The planner would have to consult local agricultural engineers to find out what types and spacing of terraces work well in the area of interest. Suppose a spacing of 100 feet is recommended, without change of slope. The new LS factor, from Figure 15-19 then becomes 0.54 and a value of

$$C = \frac{2}{200 \times 0.4 \times 0.54 \times 0.10} = 0.46$$

would suffice to keep the soil-loss rate below 2.0 t/acre/yr.

The table of C values for various combinations of cover, cropping sequence, and residue management must then be examined in Table 15-2, or in similar publications for other regions (e.g., Soil Conservation Service, 1975a, b). A realistic combination of crops and agronomic practices must then be chosen, taking into account local climate, soils, economics, traditions, and tastes.

15-4 Definition of Landslide Hazard

You are involved in the planning of a large recreational complex in a mountainous region. The development will include skiing facilities, major access roads, a dense network of secondary roads, and many buildings. On your first trip to the area you notice a number of landslide scars, talus slopes, and some areas of gentle but hummocky terrain below the outlets of a few canyons. You begin to suspect landslide hazard, and after carefully questioning local inhabitants and government officials, you decide that the hazard is present, that it will be a major threat to the safety of people and buildings, and that it is likely to cause large costs for highway maintenance.

Outline how you would plan and carry out a program of investigation to define the spatial and temporal patterns of the hazard. At what stage, and for what purposes would you call in an expert in engineering geology?

Solution

Spatial

1. Map landslide scars and deposits (ground and aerial photographs). If there are important differences in the type of movement, such as

slumps and debris flows, identify the various processes.

2. Study their relation to major spatial controls (rock type, stratigraphic contacts, faults, joint patterns, and other structures in the rock, steep slopes, topographic hollows, undercutting by streams, road construction, and so on).

3. List major landslide-prone situations and rank them according to frequency of landslides per unit area covered by the landslide-prone condition.

4. Map the landslide susceptibility of the area on the basis of this ranking. Need geologic and topographic maps, geologic reports.

5. By examining the style of past movements in the various zones, consider the approximate extension of the landslide hazard beyond the margins of the zone of unstable slopes, and add this to your map.

6. If earthquake activity has been intense (or is potentially great) in the region, the pattern of activity should be examined in reports on earthquake hazard. Major faults should be recognized from the geologic maps of the area. If volcanic activity is a potential initiator of mass movements, the probable paths of the movements should be mapped from old deposits and from topographic maps.

7. Call in an engineering geologist to examine the area and the maps for any special problems that may arise, and to advise on the location of all installations, including roads.

Temporal

1. Examine files of highway maintenance departments or railroads in the region, files of newspapers, and any other available records of the timing of landslides.

2. If only a rough seasonal picture is necessary, the average number of slides per month can be plotted on a graph.

3. If a short-term, early warning system is needed for closing highways or alerting road maintenance crews, the number of recorded slides per day or per week could be correlated with an antecedent precipitation index, or rainfall totals during the preceding one-day, two-day, or seven-day period. Such a correlation, together with weather forecasts of rainfall or snowmelt, could be used to predict short-term landslide hazard, and the accuracy of the correlation should improve as experience is gained by studying landslides in the region.

4. If earthquake or volcanic activity are thought to have potential for initiating mass movements, advice should be sought from experts in geophysics and geology about the possibility of forecasting such activity.

16

River Channels

Importance in Planning

Whenever humans make changes on the land surface or its vegetation, they alter some aspect of the hydrologic cycle with some concomitant effect on the water collecting in the channel system. These effects may include the amount, timing, and location of water reaching the channels. The drainage network, being an interconnected system, can pass on a variety of effects to places far distant from the location where the change was made. It is the continuity within the system that leads to off-site consequences, and the essence of planning includes the anticipation of such consequences.

The ability to recognize the possibilities of such effects depends on knowledge of the normal characteristics of channels, their internal processes, and thus their probable reaction to imposed change. Beyond mere recognition of possibilities, quantitative estimation requires still greater sophistication in such knowledge.

Changes made on the landscape alter the timing and amount of water flowing, especially peak or flood flows and low-flow conditions. Over time these also affect channel shape and stability. Especially because changes of the latter kind are delayed and off-site, they may have unwanted often costly results. Increase in magnitude and frequency of flooding is a typical result of urbanization. The planner should know, for example, that this effect is greatest immediately downstream of the urbanized area but decreases or fades away with distance downstream.

Construction often requires the physical relocation of channels. If in the construction of a new reach to carry the water, insufficient attention is paid to dimensions and form, the flowing water will not remain passively within its newly assigned place, but will by erosion or deposition alter its bed and banks with possible adverse or unexpected results. Therefore, environmental planners need tools of analysis and design to assist in such decisions.

Because human use of the land usually alters channel factors, planners, land managers, environmentalists, and engineers have, or should have, a special interest in how channels react to changes on the basin upstream. The difficulty is that even when the principles are well understood, forecasting the specific nature of the river's reaction to a basin alteration is tenuous at best. Therefore, the more widespread the understanding of those facts that are firmly established, the more likely that handling the channel system will at least be given some thoughtful consideration.

Discharge, Velocity, and Flow Resistance

Water flowing in a channel is being pulled downhill by gravity, the gravitational force of the weight of the water having a downhill or forward component, just as an automobile parked on a hill might move forward because of the component of its weight in a downhill direction. Thus, the stream gradient is an essential hydraulic factor. Counteracting the downhill force is the drag or resistance of the banks and bed, tending to retard the flow. Because, in general, the water neither accelerates nor decelerates but maintains an approximately constant velocity, the downhill component of gravitational force is equal to and opposite to the resisting force.

Discharge is the product of cross-sectional area of flowing water and its velocity. Figure 16-1(a) shows a rectangular flume one foot wide in which

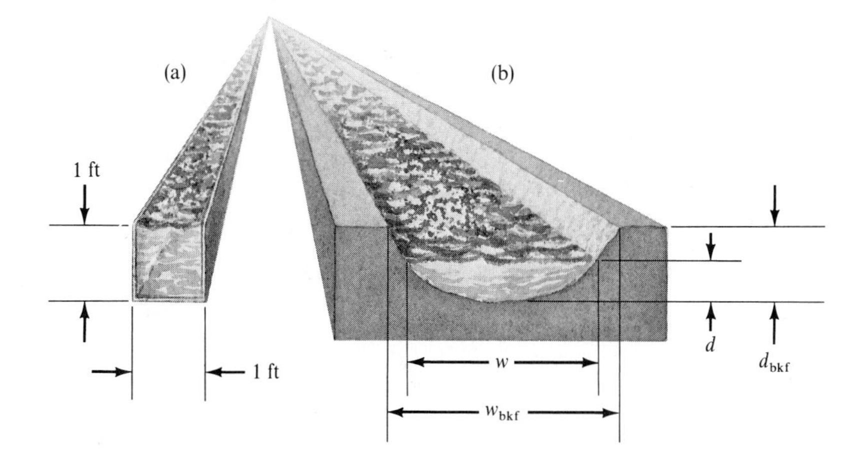

Figure 16-1 Definition diagrams of water flowing in conduits: (a) in a rectangular trough, (b) in a stream channel where d is depth and w is width.

the water flows one foot deep. The cross-sectional area of the flowing water is then $1 \times 1 = 1$ sq ft. Each foot of length of the flume then contains one cubic foot of water. If the discharge out of the flume is one cubic foot each second, or one cfs, then the velocity of the water along the flume length must be one foot per second. Thus the discharge, Q, which stands for quantity expressed as cfs, is the product of cross-sectional area times velocity or

$$Q = Au = wdu \tag{16-1}$$

where Q = discharge in cfs
A = area in sq ft
u = velocity in ft/sec
w = width in ft
d = depth in ft.

The same nomenclature can be applied to a natural stream channel sketched in Figure 16-1(b), in which the depth and width of water at the time are d and w. But the flow pictured does not fill the whole channel, which has a bankfull depth of d_{bkf} and a bankfull width of w_{bkf}. The velocity would be somewhat greater at bankfull than at the stage depicted, so the discharge at bankfull or when the channel was at capacity would be

$$Q_{bkf} = w_{bkf} \times u_{bkf} \times d_{bkf} \tag{16-2}$$

Velocity as well as the depth and width increases as discharge increases.

If the velocity increases with rising discharge, what factors are responsible? The velocity depends on the depth, the slope, or the water surface gradient, and inversely on the boundary resistance. The relation can be expressed by the well-known Chezy Formula:

$$u = C\sqrt{RS} \tag{16-3}$$

where u = water velocity
C = a resistance factor that is large for smooth boundaries and small for rough boundaries offering much resistance
R = hydraulic radius, the ratio of cross-sectional area of flowing water to wetted perimeter, A/wp; R is about equal to the mean depth for wide channels
S = the energy gradient, closely approximated by the slope of the water surface.

In simplest terms velocity increases as the square root of depth and the square root of slope, and with the smoothness of the boundaries.

The comparable formula widely used in American engineering practice is the Manning relation:

$$u = \frac{1.49 R^{2/3} S^{1/2}}{n} \tag{16-4}$$

in which R and u, defined as before, are in feet and feet per second, and n, called the Manning resistance coefficient, has been experimentally de-

Table 16-1 Manning roughness coefficients for various boundaries. (From Ven te Chow, ed. 1964:7–25.)

BOUNDARY	MANNING ROUGHNESS, n ($FT^{1/6}$)
Smooth concrete	0.012
Ordinary concrete lining	0.013
Vitrified clay	0.015
Shot concrete, untroweled, and earth channels in best condition	0.017
Straight unlined earth canals in good condition	0.020
Rivers and earth canals in fair condition—some growth	0.025
Winding natural streams and canals in poor condition—considerable moss growth	0.035
Mountain streams with rocky beds and rivers with variable sections and some vegetation along banks	0.040–0.050
Alluvial channels, sand bed, no vegetation	
1. Lower regime	
Ripples	0.017–0.028
Dunes	0.018–0.035
2. Washed-out dunes or transition	0.014–0.024
3. Upper regime	
Plane bed	0.011–0.015
Standing waves	0.012–0.016
Antidunes	0.012–0.020

termined for a variety of boundaries. Table 16-1 lists some values of the Manning roughness factor. Note that, different from the Chezy friction coefficient, C, the value of n increases as the boundary is rougher, causing increased friction.

A practical tool of great assistance is the series of photographs of rivers published by the U.S. Geological Survey in which the value of the Manning coefficient has been computed from field observations of depth, velocity, and slope. One can compare the river under consideration with the published photographs and use the value of n for the illustration most like the river under study. The publication is *Water Supply Paper 1849* by Barnes (1967).

There are other ways of expressing the resistance factor that are common in experimental and theoretical work, especially useful because they are dimensionless, whereas n varies slightly with depth. Because they are often used in the literature of hydraulics, the following expressions for resistance factor are listed here for reference. They are all related to the Chezy coefficient, C, discussed earlier:

$$\mathbf{f} = \frac{8gds}{u^2} \tag{16-5}$$

This is the Darcy-Weisbach coefficient in which g is gravitational acceleration; d, s, and u are depth, slope, and velocity; $\mathbf{f}$ increases with increased resistance or friction.

$$\frac{u}{u_*} = \frac{u}{\sqrt{gds}} = \frac{C}{\sqrt{g}} = \frac{\sqrt{8}}{\sqrt{\mathbf{f}}} \tag{16-6}$$

The definitions are the same as before except u_*, the shear velocity, is $\sqrt{\tau/\rho} = \sqrt{gds}$. The ratio of velocity to shear velocity is a friction factor that increases with a decrease in frictional resistance.

The roughness coefficient is dependent on the size distribution of bed particles, as well as channel irregularities such as bars, dunes, bends, and vegetation. In straight reaches of natural channels in which the bed material contains some gravel, the hydraulic roughness can be estimated from the empirical equation of Leopold and Wolman (1957) that has been verified by Limerinos (1970) in an independent study.

$$\frac{1}{\sqrt{\mathbf{f}}} = 2\log\frac{d}{D_{84}} + 1.0 \tag{16-7}$$

where $\mathbf{f}$ is defined as above, d is the mean depth of flow, and D_{84} is the size of bed material that 84 percent is finer than, d and D_{84} both being in the same units.

Discharge is measured by the observation of water velocity and the cross-sectional area of the flowing water. At a gauging station the factor continually recorded is water surface elevation (river stage), and at the same location, periodic velocity measurements are made as described below. About 11,000 gauging stations are operated in the United States. The discharge data are published by the U.S. Geological Survey in the Water Supply Papers or in the annual reports for each state entitled, for example, "Surface Water Records for California."

Water Stage and Rating Curve

In river investigations the *stage* refers to the elevation of the water surface, usually above some arbitrary datum. Stage is recorded at a gauging station or river measurement station shown and explained in Leopold (1974:40). Stage is often abbreviated O.S. for "outside stage."

In brief, the gauging station consists of a pit or a well connected by a pipe to the water in the river, thus keeping the water surface in the stilling well always at the same level as that of the river. A float rides on the water surface in the stilling well and is connected by a wire to a recorder in the gauge house. The recorder is either a pen drawing a line on a chart or an instrument that punches holes in a tape at intervals that are a digital record of stage as a function of time. Periodically, usually once a month, a hydrographer visits a station to collect the record and to make a discharge measurement by current meter. Such a measurement is made by wading at low flow and from a cable over the stream or from a bridge at high flow. The measurement consists of counting the rate of revolutions of the cups of the meter, which is convertible into velocity. Such a direct measure of velocity is made at about 30 places across the stream cross section, and at each the depth is also measured. Thus, cross-sectional area of flowing water and the velocity are measured, the product of which is discharge.

From the standpoint of the environmental planner, the available gauging station data usually apply to drainage areas considerably larger than the areas of concern, and therefore it is useful to supplement the published gauging station records with direct observation at or near the place where information is needed. The gauging station nearest to the area under consideration may be several miles away and on a stream where the drainage area is 15 square miles, whereas the information needed is on a much smaller stream draining half a square mile. Much can be learned by a few simple observations at the location in question. This involves the installation of a staff gauge and observations of actual flow during at least a few storms. A staff gauge, a cross section, and a relation of the reading of water surface elevation on the staff gauge to simultaneous discharge are needed. The procedures for choosing a section, installing a gauge plate, measuring a cross section, and making a discharge measurement by timing of floats are described in Typical Problem 16-3. In connection with these procedures, Typical Problems 16-1 and 16-2 are also worthy of study.

A gauging station and a measuring bridge are shown in Figure 16-2, and a typical staff gauge installation on a small channel is illustrated in Figure 16-3.

The discharge determined by the methods referred to above is plotted against the corresponding elevation of the water surface; this graph is called the rating curve, an example of which is shown in Figure 16-4. To establish a rudimentary rating curve for a staff gauge, only a few storm observations are needed, but a complete curve for a gauging station may take months or even a few years, the time being required to experience a range of discharges.

In addition to the record of stage in punched tape or pen trace in the gauge house, stage can also be read visually as the water surface elevation on an enamel plate marked off in hundredths of feet and attached firmly to a post or other stable object. The datum or actual elevation above mean sea level of the zero reading of the plate must be established by leveling from a bench mark, but this is not absolutely required as long as the datum of the gauge plate remains fixed.

The stage, gauge height, or level of the water surface goes up and down from day to day, reflecting the precipitation of the recent past upstream

Figure 16-2 A stream gauging station. In the small building there is a well connected to the stream. The height of the stream surface is recorded continuously by monitoring the water level in the well. (Photograph by Richard F. Hadley.)

Figure 16-3 Staff gauge installed in a small channel.

from the particular river location. Thus, the channel is more full at some times than at other times. Most of the time it has only a moderate amount of water, and a few days a year it is full or nearly so.

A typical cross section is shown in Figure 16-5, and this is the same location to which the rating curve of Figure 16-4 applies. To visualize this loca-

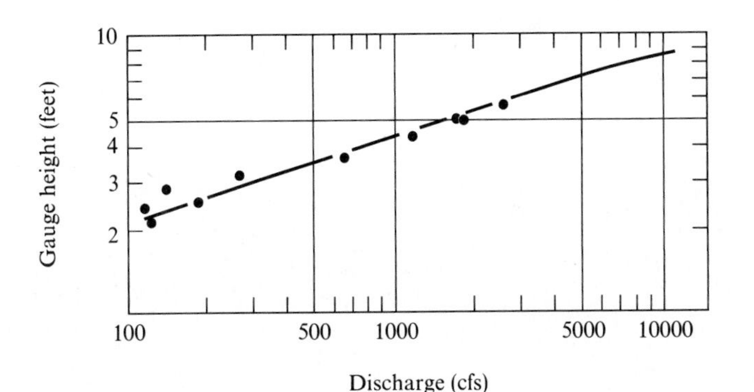

Figure 16-4 Rating curve for the New Fork River at Boulder, Wyoming.

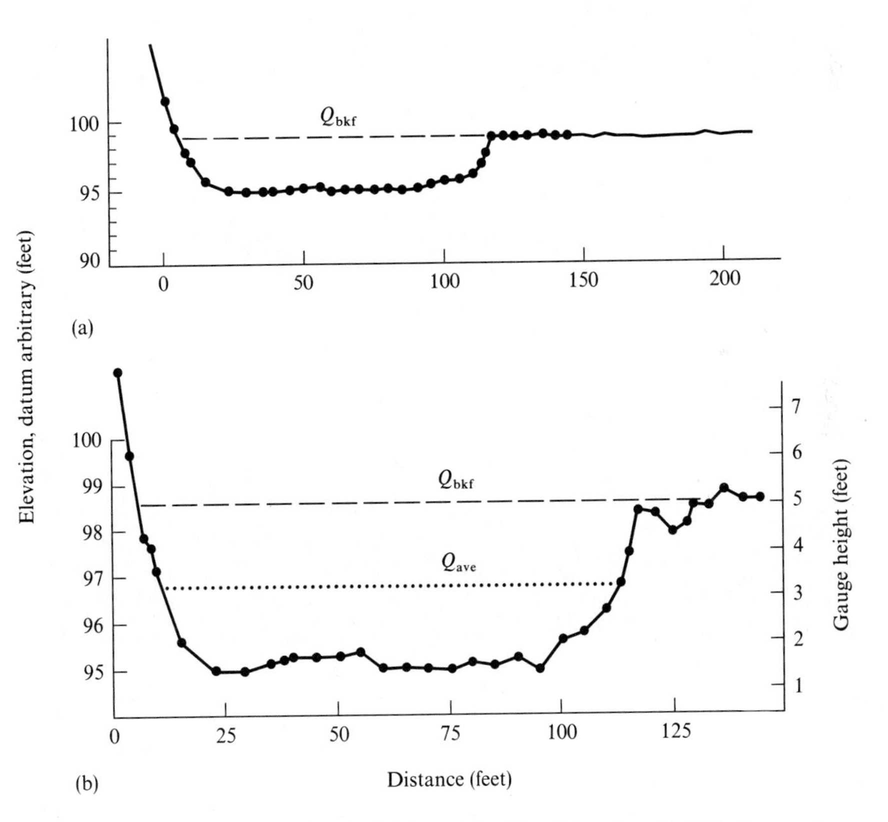

Figure 16-5 Cross section of New Fork River at Boulder, Wyoming. (a) This diagram is at a small scale to show the cliff on left and a broad flat on the right. (b) This enlarged view shows water level of mean annual discharge and at bankfull. Drainage area 552 sq mi, slope .0012, bed material D_{50} 40 mm.

(a)

(b)

Figure 16-6 Reach of the New Fork River at Boulder Wyoming, near gauging station: (a) view upstream, (b) view downstream.

tion on the ground, Figure 16-6 has photographs of the same location looking upstream and downstream. The surveyed cross section is just upstream from the camera location and goes from the steep bank seen on the left bank to the flat floodplain covered with willow and cottonwoods on the right bank (right and left bank always refer to a stream looking in the direction of water flow).

Channel Shape

The shape of the stream or river channel is highly organized and is similar for rivers of the same size in a comparable climate. The shape is a complex result of many interacting factors of which there are two general classes: factors related to the debris load, its size, lithology, amount, and depositional forms; and factors related to water flow, that is, hydraulic factors.

The channel is self-formed in that the water in the channel and the debris it carries result in the channel. The water carves and maintains the conduit containing it. Also, the channel is self-adjusting, for if the time and volume characteristics of its water or debris are altered by man, by climatic change, or by alterations of the protective vegetal cover on the land of the basin, the channel system adjusts to the new set of conditions.

A stream channel is gently rounded in cross section, roughly parabolic, though sometimes trapezoidal with straight sloping sides. Large rivers are very much wider than deep, whereas small streams may be only as wide as several times the depth.

The channel is formed and maintained by the flow it carries but is never large enough to carry without overflow even discharges of rather frequent occurrence. Unusual or infrequent floods will invariably overflow the ordinary channel banks. Channel geometry is the word we introduced to describe the physical size, shape, and characteristics in relation to the hydraulic factors of velocity, roughness, slope, and flow frequency.

The shape and size of river cross sections in a single geographic area are worthy of close attention in order to perceive the great variety that exists and yet how certain parameters are closely related and highly organized. Such a comparison is not generally available in the literature, so examples are included here.

In Figures 16-7 and 16-8, cross sections and photographs of several channels at gauging stations are presented. Each cross section is plotted at two scales: one to show the relation of the channel to the valley floor nearby; and the second at an enlarged vertical scale to show the relation of average discharge to the channel and to show the level of the bankfull stage, that is, when the channel is filled to capacity.

These cross sections have in common certain features that might be illustrated by the generalized or typical picture shown in Figure 16-9. Between the hillslopes bordering the valley is the irregular but generally horizontal valley floor or valley flat. The rounded or parabolic stream channel winds through it. But the valley flat often does not consist of only one level; rather it shows two or even several levels. Only that level or berm nearest the channel and at the lowest elevation is being constructed by the present stream. The higher berms or levels were constructed by the stream at some time in the past when it flowed at that higher elevation. These higher flats are abandoned floodplains called terraces, no longer under construction. It is only the lowest one near the stream that is the modern or present floodplain. This distinction between floodplain and terrace is highly important to the environmental planner, as will be discussed in the next section.

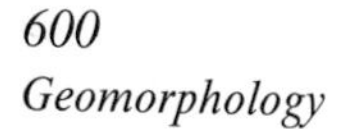

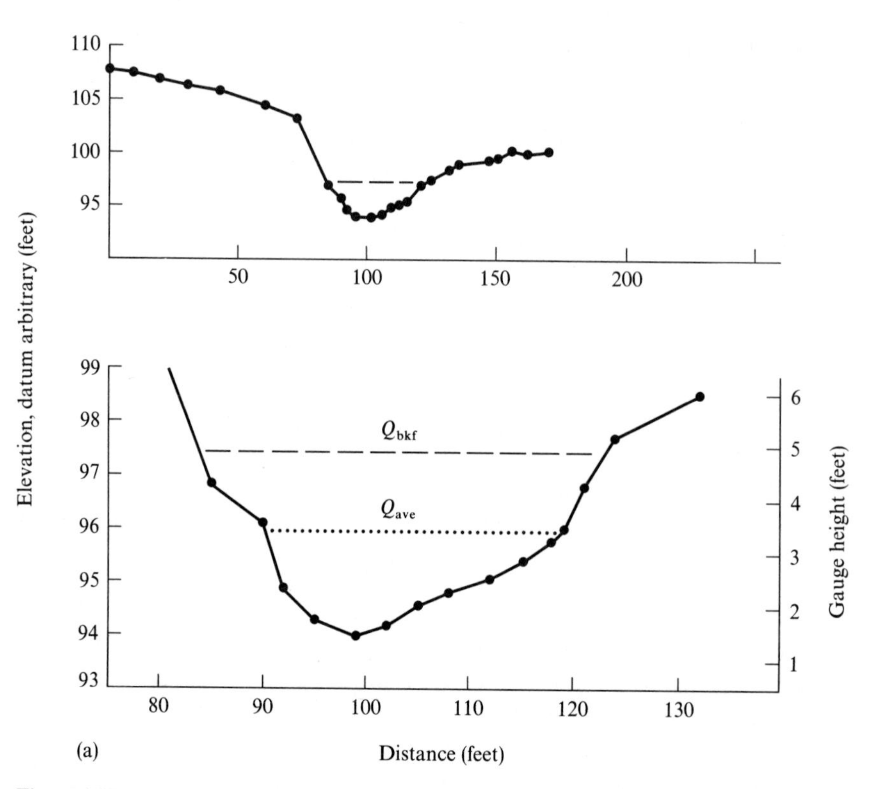

Figure 16-7 (a) New Fork River below New Fork Lake: drainage area 36.2 sq mi; slope .0051; bed material D_{50} 52 mm; $Q_{1.5}$ 355 cfs; Q_{bkf} 340 cfs; Q_{ave} 50.5 cfs.
(b) Fall Creek near Pinedale, Wyoming: drainage area 37.2 sq mi; slope .040; bed material D_{50} 210 mm; $Q_{1.5}$ 350 cfs; Q_{bkf} 315 cfs; Q_{ave} 39.4 cfs.
(c) Silver Creek near Big Sandy, Wyoming: drainage area 45.4 sq mi; slope .004; bed material D_{50} 45 mm; $Q_{1.5}$ 600 cfs; Q_{bkf} 390 cfs; Q_{ave} 43 cfs.

The Floodplain and Bankfull Stage

The cross sections of Figure 16-7, as well as general observation, illustrate the fact that most river channels are bordered by a relatively flat area or valley floor. When the water fills the channel completely or is at bankfull stage, its surface is level with the floodplain. This is a word of great importance to the environmental planner as well as to the geomorphologist, and its definition must be understood in terms of river morphology as well as in terms of flood potential. It is defined as follows*:

The floodplain is the flat area adjoining a river channel constructed by the river in the present climate and overflowed at times of high discharge. Each part of this definition is important and will be elaborated.

---

*Refer also to the footnote on p. 428 on this term as used by engineers and other specialists concerned with water resources and land-use planning.

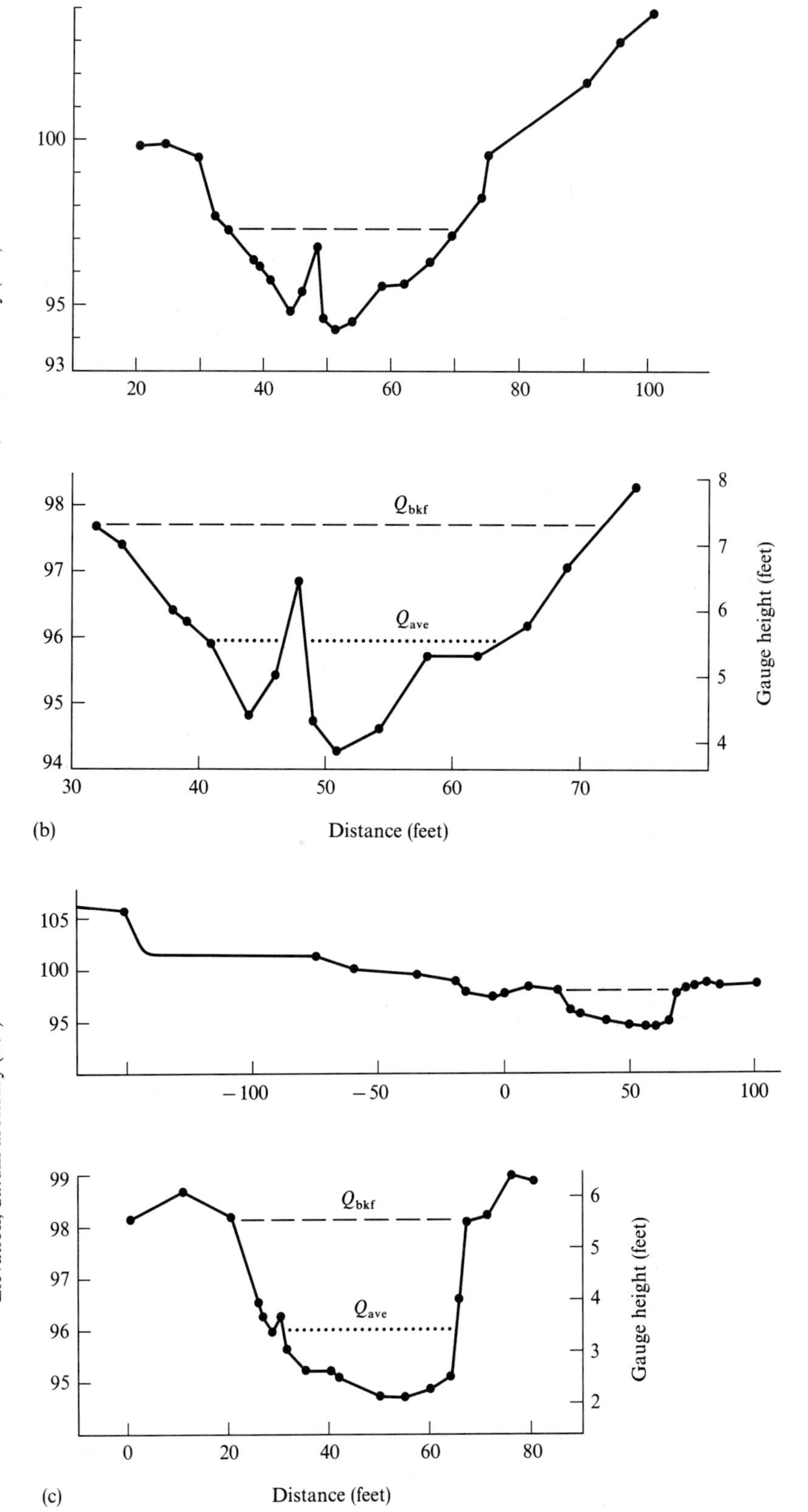

Elevation, datum arbitrary (feet)
100
95
93
20
40
60
80
100
98
97
96
95
94
30
40
50
60
70
Q_{bkf}
Q_{ave}
8
7
6
5
4
Gauge height (feet)
(b)
Distance (feet)
105
100
95
−100
−50
0
50
100
Elevation, datum arbitrary (feet)
99
98
97
96
95
0
20
40
60
80
Q_{bkf}
Q_{ave}
6
5
4
3
2
Gauge height (feet)
(c)
Distance (feet)

continued

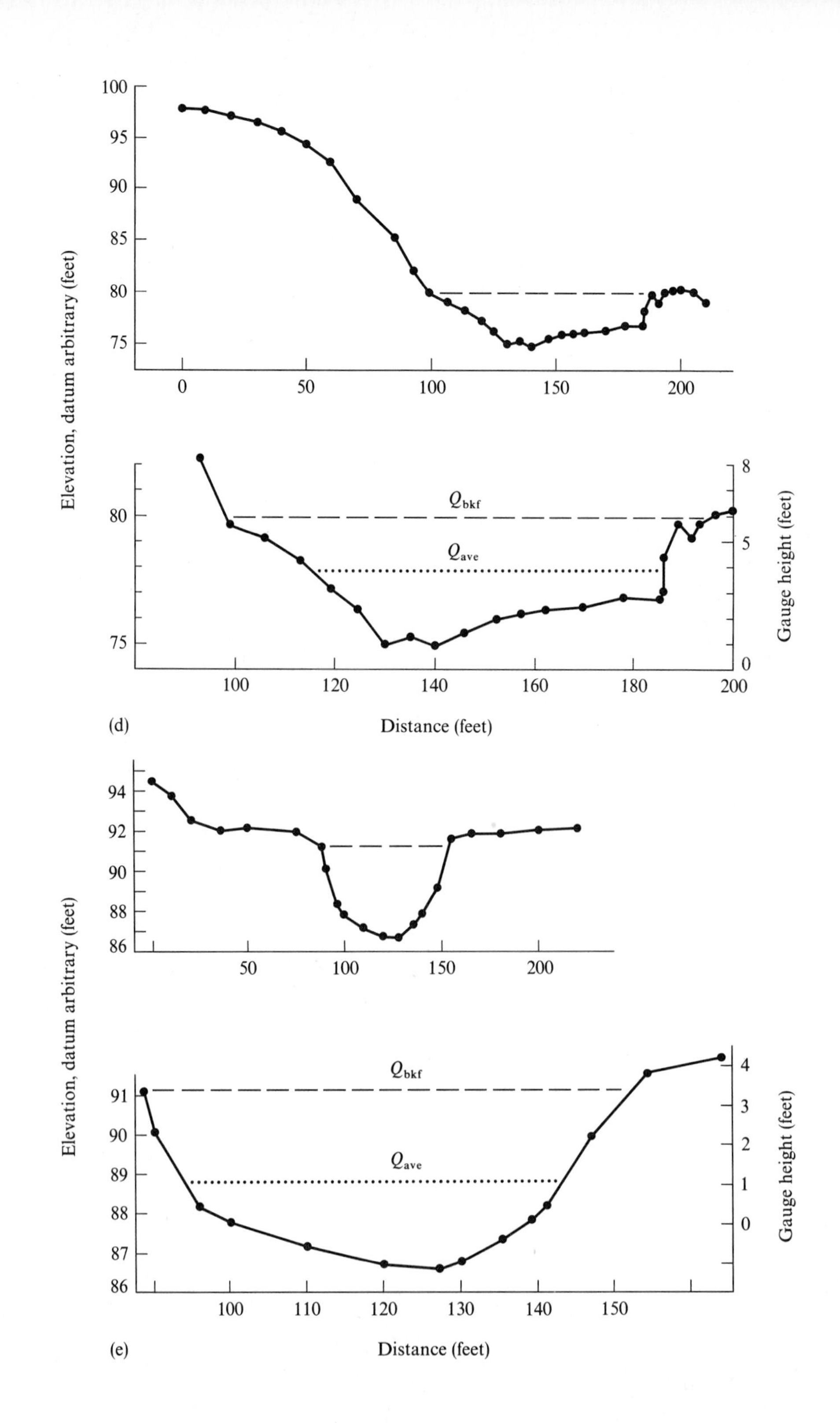

100
95
90
85
80
75
0
50
100
150
200
Elevation, datum arbitrary (feet)
80
75
100
120
140
160
180
200
Q_{bkf}
Q_{ave}
8
5
0
Gauge height (feet)
(d)
Distance (feet)
94
92
90
88
86
50
100
150
200
Elevation, datum arbitrary (feet)
91
90
89
88
87
86
100
110
120
130
140
150
Q_{bkf}
Q_{ave}
4
3
2
1
0
Gauge height (feet)
(e)
Distance (feet)

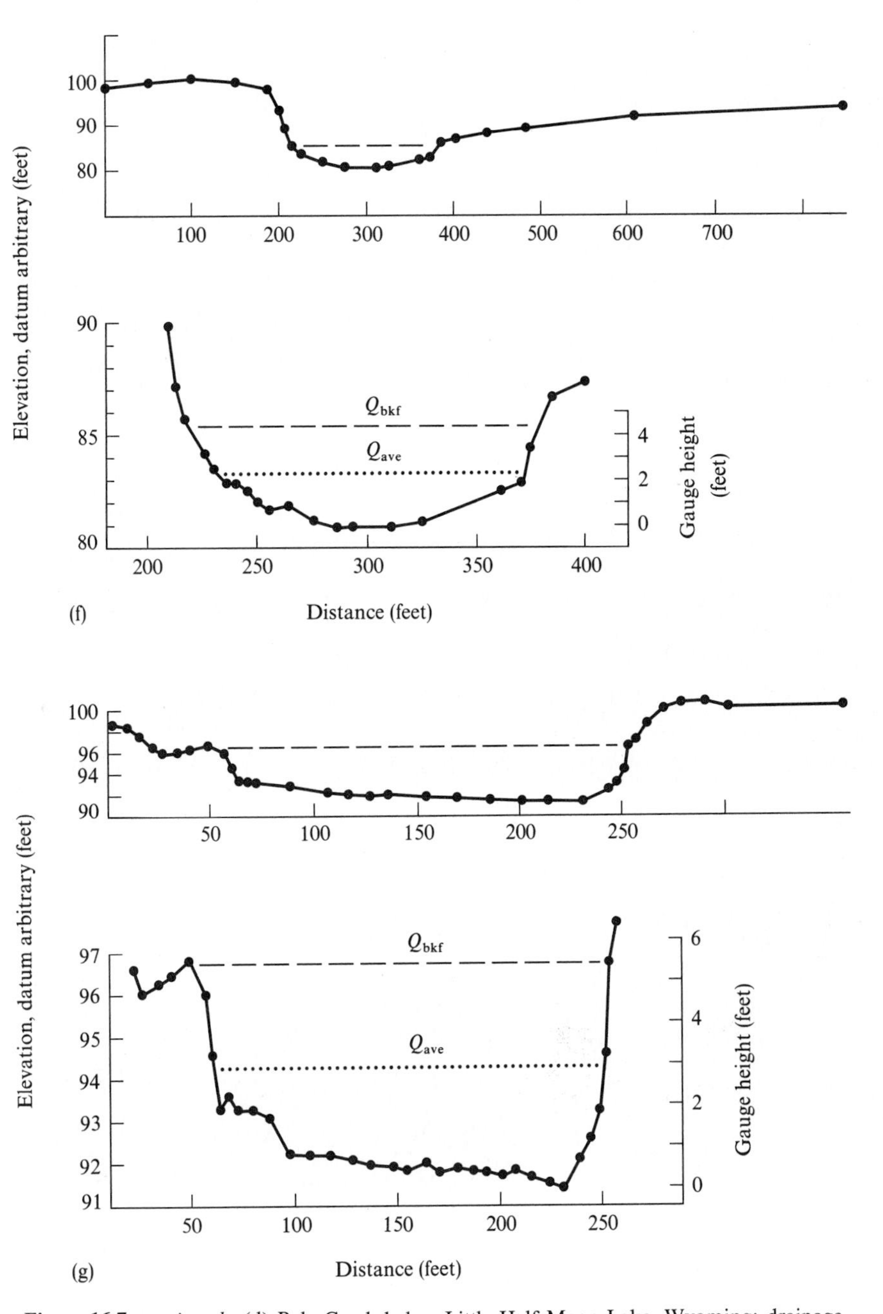

Figure 16-7, *continued* (d) Pole Creek below Little Half Moon Lake, Wyoming: drainage area 87.5 sq mi; slope .0018; bed material D_{50} 128 mm; Q_{bkf} 930 cfs; Q_{ave} 107 cfs.
(e) East Fork River at Highway 353 near Boulder, Wyoming: drainage area 102 sq mi; slope .0023; bed material D_{50} 90 mm; Q_{bkf} 770 cfs; Q_{ave} 110 cfs.
(f) Green River at Warren Bridge near Cora, Wyoming: drainage area 468 sq mi; slope .0037; bed material D_{50} 140 mm; $Q_{1.5}$ 2600 cfs; Q_{ave} 492 cfs.
(g) New Fork River near Big Piney, Wyoming: drainage area 1230 sq mi; slope .00146; bed material D_{50} 64 mm; $Q_{1.5}$ 4400 cfs; Q_{bkf} 4200 cfs; Q_{ave} 660 cfs.

(a)

(b)

(c)

(d)

(e)

(f)

(g)

Figure 16-8 (a) New Fork River below New Fork Lake near Pinedale, Wyoming, looking downstream. Bankfull stage is lower than the gravel surface seen through trees at right, but is at top of sand and silt across the stream just above the top of the staff gauge in the foreground.
(b) Fall Creek near Pinedale, Wyoming, looking upstream. Bankfull stage just covers large boulders in foreground.
(c) Silver Creek near Big Sandy, Wyoming, looking upstream. Bankfull stage is near top of gravel above man's head.
(d) Pole Creek below Little Half Moon Lake near Boulder, Wyoming, looking downstream. Forested area across stream at right is the floodplain, the surface of which is coincident with bankfull stage.
(e) East Fork River at Highway 353 near Boulder, Wyoming, looking downstream. Sand and silt point bar at right in foreground. Gravel terrace covered with willow at mid-distance across the stream. Bankfull stage is at top of point bar in right foreground.
(f) Green River at Warren Bridge near Cora, Wyoming, looking upstream. Bankfull stage poorly marked, but is at level of top of white boulder in middle foreground.
(g) New Fork River near Big Piney, Wyoming, looking upstream. Cable car of gauging station in upper left.

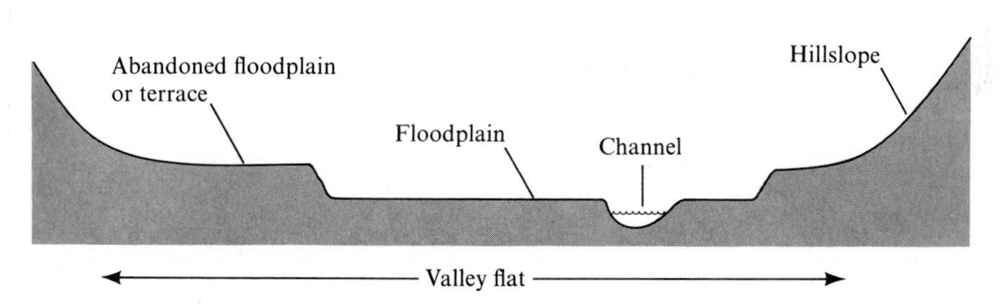

Figure 16-9 Diagrammatic cross section of a valley showing relation of present channel to the floodplain and to a terrace (abandoned floodplain).

The flat floor of a valley was constructed by the river during lateral migration and by deposition of sediment. At some time in the past the river occupied each and every position on the flat valley floor. Through time the river moves laterally by erosion on one bank and simultaneous deposition on the other. Throughout this process of lateral movement, however, the channel maintains its width and depth about as presently observed. Though the loci of erosion and deposition are closely linked to a curve in the channel and to the sequence of deeps and shallows, given time the position of these loci is sufficiently random during the long process of movement that the channel can and has occupied all positions on the observed valley flat. The river continually is building by deposition new flat land at that level as it moves laterally, and the flat valley is the most direct evidence of lateral migration.

The progressive sequence of channel migration and consequent floodplain construction is sketched in Figure 16-10, showing four hypothetical channel positions. Usually the coarsest material, gravel or cobbles, is on the channel bed, but it is often absent in the higher part of the channel banks. The coarse bed material tends to be covered with finer sand or silt, resulting in channel banks that appear finer in texture than most of the streambed.

An example of surveyed cross sections over a period of years (1953–1972) is shown in Figure 16-11. A series of cross sections, monumented at each end so resurveys could be made at identical positions, was established by Leopold on Watts Branch near Rockville, Maryland, and these have provided a graphic record of channel migration through time (Leopold 1973). From such surveys the generalized diagram in Figure 16-10 was constructed.

Because of a change in climate or a change in watershed condition by grazing, urbanization, or other influences, the river may change its level

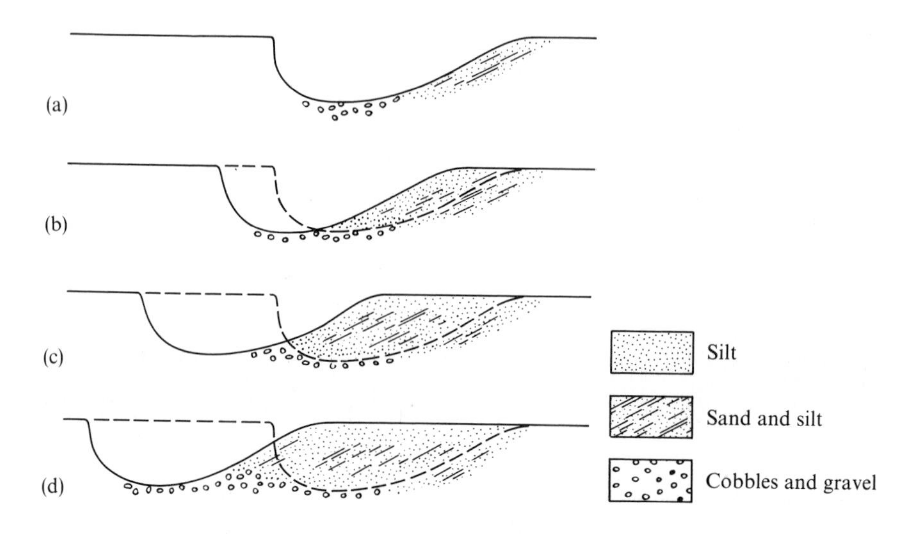

Figure 16-10 Four sequential stages in progressive construction of a floodplain as stream moves laterally, diagrammatically shown. (a) Initial stage showing gravel and sand on stream bed. (b) Stream erodes left bank as it deposits on point bar to right of diagram. (c) Later stage showing stream bed gravel covered over with sand and silt, finer material deposited near the top of the point bar. (d) Still later stage indicating how progressive lateral movement builds floodplain with cobble or coarse material at base and finest material near surface. A field example is shown in Figure 16-11.

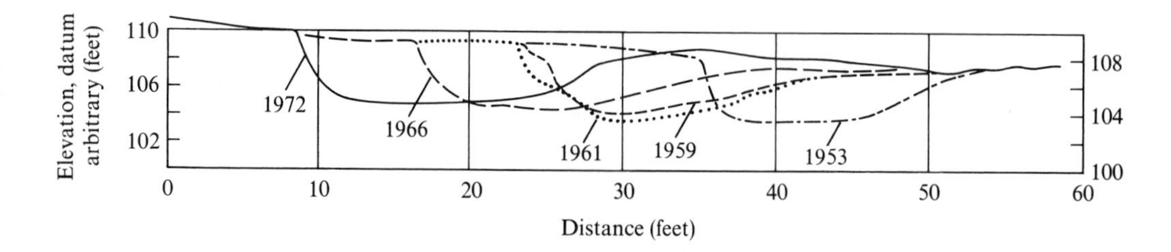

Figure 16-11 Successive cross sections of Watts Branch, near Rockville, Maryland, at section 44A–45. Five surveys over a 19-year period are pictured. The channel is eroding a low terrace on the left bank and building a floodplain about 1.5 feet lower on the right bank.

Figure 16-12 Rio Puerco del Oeste at Manuelito, New Mexico. Trench in valley floor leaves former floodplain as a terrace, too high to be flooded, 25 feet above present stream bed.

either upward (aggradation) or downward (degradation). It will build a new level of floodplain appropriate to the new bed elevation. The former floodplain it had been constructing is thus abandoned. An abandoned floodplain is called a *terrace* (Figure 16-12). The stream cuts down or builds up when watershed conditions change because there is a new relation between discharge and sediment production. But it is not possible to forecast what will be the effect of a particular alteration. The change of elevation of a river channel is the net effect of complex interactions, and its forecast is beyond present knowledge except under special circumstances.

In many areas of the country, the valley flat is not being presently constructed for it is a terrace. In the San Francisco Bay region, for example, many streams are incised in deep, narrow canyonlike trenches, and the valley flat is never flooded for it is a terrace too high to be reached by floodwater (Figure 16-13). Other terraces are flooded on occasion, but from river to river the probability of flood covering a terrace is highly variable. In contrast, the floodplain under construction is flooded frequently and at a relatively consistent recurrence interval of 1.5 years in the annual flood series.

Figure 16-13 Terraces along Walker Creek near Marshall, Marin County, California. The valley flat is the top of the terrace (right), standing 15 feet above the channel bed. A lower terrace stands eight feet above the bed (left middle ground). Floodplain is narrow and not distinguishable here.

The definition also includes the idea that the floodplain is frequently flooded as a natural attribute of rivers. The floodplain is indeed part of the river under storm conditions.

The valley flat in some regions may not be formed in the manner described above. Hack and Goodlett (1960) found that mountain valleys of the central Appalachians are subject to recurrent debris flows associated with extreme storm events. The deposits from these debris flows form the valley floor in many reaches, and definite stream channels carved in these deposits are impermanent, for the succeeding flood may dam or divert or greatly enlarge them. Where such debris flows are important, levels, berms, or terraces may be distinguished and even ascribed to particular flood years, but a floodplain, as defined here and having a constant frequency of overflow, cannot be identified or does not exist. The use of the word floodplain as here defined applies to the level or berm constructed by a combination of progressive lateral migration with some overbank deposition.

The river channel flowing full has a special significance from several standpoints. At higher flow, water spills out of the channel and overflows the flat-lying land near the river, floodplain, or terrace. The river does not construct its channel large enough to accommodate the highest discharges without overflow. The bankfull stage corresponds to the discharge at which channel maintenance is most effective, that is, the discharge at which moving sediment, forming or removing bars, forming or changing bends and meanders, and generally doing work that results in the average morphologic characteristics of channels.

It is human encroachment on the floodplains of rivers that accounts for the majority of flood damage. Because it is a natural attribute of rivers to produce flows that cannot be contained within the channel, the floodplain is indeed a part of the river during such events. It is therefore important that planners know something about these characteristic features and thus possibly counteract to some degree the emphasis placed on flood-control protection works. More logical is flood damage prevention by the restriction of floodplain use.

Why is the natural channel not large enough to carry the maximum flood? First, one can argue that whatever size of flood is visualized, there is the possibility of an even larger one. The limit is discussed in engineering hydrology as the maximum probable flood. Because the extreme conditions associated with the maximum probable flood have not been observed in one place and at the same time, the probability of the combination is impossible to estimate. Ordinarily, the computed but never experienced maximum probable flood in an area is imagined to be possible only once in 1000 years or longer. It is, however, not expressed in terms of frequency. Such an improbable event never having been experienced is unlikely to be the mechanism that determines the size of the usual river channel. Hence, the channel would be of such a size that it could not contain such an extreme event without overflow, but might contain a smaller more frequent event. What then would be the size of a flood sufficiently common to govern the channel size? This is the question considered by Wolman and Miller (1960), who argued that the very large events were too infrequent to govern channel characteristics, though when they did occur their effectiveness for channel change would be great. In contrast, low flow, common and frequent, is ineffective and thus does not contribute to shaping the channel, but merely conforms to or flows within the channel that exists. Thus, there must be some flow of intermediate size, large enough to be effective in causing change, but sufficiently frequent that the product of its frequency and effectiveness would be greater than that of any other size of flow event. Testing this idea by computing the flow size that transports the largest total amount of sediment over a period of years, Wolman and Miller concluded that the bankfull stage is the most effective or is the dominant channel-forming flow. This most effective of flows has a recurrence interval of 1.5 years in a large variety of rivers.

If a river reaches bankfull stage at a constant frequency (recurrence interval), it follows that the floodplain is flooded at a constant frequency, the floodplain being by definition the valley level corresponding to the bankfull stage. But a terrace, being at a higher level, is flooded less frequently, and terraces may be at different levels above the floodplain, depending on the past history of the individual river. Therefore, there is no consistency among rivers in the recurrence interval of flooding of the terraces that exist. In this fact lies the importance to the environmental planner of recognizing the difference between floodplain and terrace. The consistent flooding frequency of the floodplain is a useful characteristic.

Data on the Frequency of Bankfull Discharge

Data on the discharge at channel capacity or on the gauge height of the bankfull condition are not published or even determined in a systematic manner despite their importance to planners, environmentalists, and everyone interested in floods and flooding. Flood stage is designated by the National Weather Service or the Corps of Engineers for a few locations on a few rivers. Some agencies refer to it as the "zero-damage stage." Nothing is available for most rivers. There is no indication on the rating curve or other easily available portion of a gauging station record from which the bankfull condition can be recognized. The best approximation is obtained by constructing a frequency curve, reading the discharge having a recurrence interval of 1.5 years in the annual flood series, obtaining the corresponding gauge height from the rating curve, and checking the height against field observations.

An example of the survey and computation to compare the observed bankfull height and the water level corresponding to a discharge having a recurrence interval of 1.5 years is given below.

At the gauging station on the New Fork River near Boulder, Wyoming, a leveling line was surveyed across the channel and adjoining valley flat, the latter presumed to be the floodplain and not a terrace. This survey is plotted on Figure 16-5. The upper diagram on this survey shows that a level flat extends at least 250 feet away from the right bank of the stream channel at an elevation of 98.5 feet, arbitrary datum. This is assumed to be the level of the floodplain and the elevation corresponding to bankfull stage on the basis of the field conditions. A frequency analysis will now be made to compare this elevation with that of the discharge of recurrence interval 1.5 years.

The highest discharge for each year of record at the gauge was tabulated, arranged in rank order of size, plotted according to the method described in

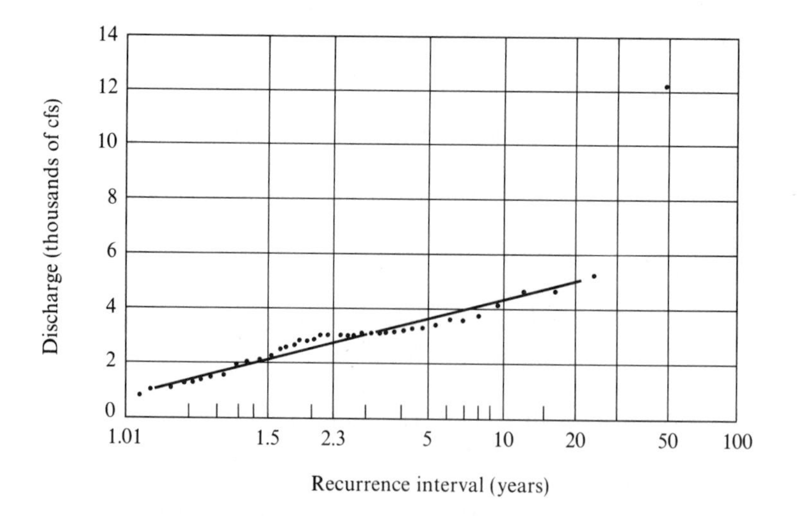

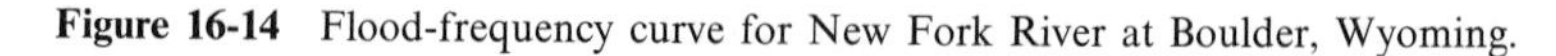
Figure 16-14 Flood-frequency curve for New Fork River at Boulder, Wyoming.

Chapter 10, and shown in Figure 16-14. From that graph the discharge corresponding to 1.5 years is read as 2200 cfs. Entering this discharge in the rating curve of the station, shown as Figure 16-4, the gauge height can be read as 5.3 feet. Comparing this value with the gauge height of bankfull stage determined in the field, 5.0 feet, it can be seen that the agreement is close, confirming for this stream the correspondence of floodplain level or bankfull level with the water level of the discharge having the usual recurrence interval, 1.5 years. Conversely, the stage of 5.0, or discharge 1750 cfs, has a recurrence interval of 1.28 years, a reasonable agreement with the expected value of 1.5 years.

For a short list of gauging stations, field surveys are available locating the bankfull stage. For comparison the discharge of 1.5-year recurrence interval was determined from station frequency curves. The discharges determined by the two methods are compared in Figure 16-15. Reasonable agreement is shown for basins whose bankfull discharges range through three orders of magnitude.

The determination of bankfull stage is also the determination of the level of the floodplain. When one or more terraces exist, some of which may differ from the floodplain by only a moderate vertical distance, it may not be obvious which is the floodplain. A frequency analysis as described is the best way of deciding, and the technique is useful in many planning problems

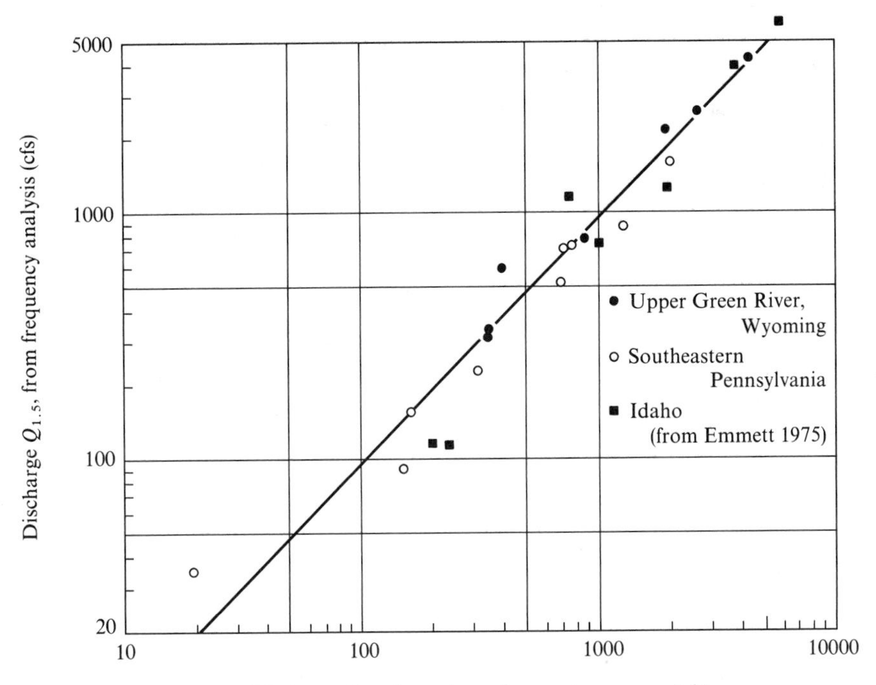

Figure 16-15 Comparison of discharge having a recurrence interval of 1.5 years with that determined from field observation of bankfull stage.

because floodplain protection is usually involved. Clearly, it is necessary to know what is liable to flooding.

The field determination of bankfull stage is difficult when the floodplain is narrow or not flat and well defined. A field procedure for making the survey and the physical indications of bankfull conditions at the channel margin are described in the Typical Problem 16-2. Another description of the field procedure is given by Emmett (1975:A35).

With these criteria in mind, an inspection of Figures 16-7 and 16-8 is useful, for we show on each surveyed cross section the level determined to be bankfull. And in Table 16-2 we give the discharge values corresponding to bankfull and the 1.5-year recurrence interval. Following also are some notes concerning the cross sections and their corresponding photographs.

Referring to the cross sections in Figure 16-7 at the New Fork River below New Fork Lake, the left side of the channel abuts a steep bank, the top of which is a glacial outwash terrace. The right bank is bordered by a rough and uneven flat, possibly also a terrace but not sufficiently smooth to provide any useful morphological evidence. The bankfull stage was identified by changes of bed material at the margin of the channel, and by change in vegetation, moss, grass, and trees above and no vegetation below.

Fall Creek near Pinedale is a typical mountain torrent with a bouldery bed. No morphologic character of the channel itself marked bankfull, but there was a clearcut margin with moss and grass above and no vegetation below that was chosen as bankfull.

Table 16-2 Channel geometry data, upper Green River basin, Wyoming.

	DRAIN AREA (SQ MI)	AVE DISCH. Q_{ave} (CFS)	DISCH. OF 1.5 RI $Q_{1.5}$ (CFS)	DISCH. BANKFULL Q_{bkf} (CFS)
1. LaBarge Creek near LaBarge Ranger Station	6.3	13.9	115	
2. New Fork below New Fork Lake	36.2	50.5	355	340
3. Fall Creek near Pinedale	37.2	39.4	350	315
4. Silver Creek near Big Sandy	45.4	43	600	390
5. North Piney Creek near Mason	58.0	56	330	
6. Pole Creek below Little Half Moon Lake	87.0	107	790	870
7. East Fork at Hwy 353	102			770
8. Boulder Creek below Boulder Lake	130	195	1700	
9. Green River at Warren Bridge	468	492	2600	2600
10. New Fork near Boulder	552	394	2200	1900
11. New Fork near Big Piney	1230	660	4400	4200
12. Green River near Fontenelle	3970	1550	9000	

Silver Creek near Big Sandy has a slight berm on the right bank and a definite break in slope on the left bank that was chosen as bankfull. This stage was one foot lower than the valley flat, a definite and wide flat area especially on the right side, considered a terrace.

Pole Creek below Little Half Moon Lake has a wide flat on the right bank and a concomitant change in vegetation that constituted good morphologic evidence of the bankfull stage.

East Fork River at Highway 353 near Boulder has a definite break in slope on the left bank at gauge height 3.4, and a concomitant change in vegetation and in deposited grain size.

Green River at Warren Bridge had no sure morphologic evidence, but a change in vegetation and the nature of deposition on the channel margin was used as evidence.

The New Fork near Big Piney had a well-marked flat on the left bank at gauge height 5.3 and a concomitant vegetation break.

We hope these notes and the photographs convey the idea that identification of the floodplain in the field is by no means simple. The difficulty is greater in foothills and mountains than in the Piedmont and Coastal Plain of the eastern United States. In the latter physiographic provinces, it is usual for a low terrace, only slightly above floodplain level, to exist on streams draining 30 square miles or less. The floodplain in these small basins is often much narrower than the valley flat, which is mostly comprised of a terrace. This is illustrated by an example discussed in relation to Figure 16-26.

DEPTH d (FT) AT		VELOCITY u (FT/SEC) AT		WIDTH w (FT) AT		BED MATERIAL SIZE D_{50} (MM)	SLOPE	GAUGE HEIGHT AT Q_{bkf} (FT)
Q_{ave}	$Q_{1.5}$	Q_{ave}	$Q_{1.5}$	Q_{ave}	$Q_{1.5}$			
0.79	1.8	1.4	3.4	12	17.2			
1.0	2.0	1.8	4.1	29	44	52	.0051	4.9
0.98	1.8	2.3	7.1	19	25	210	.04	7.3
0.80	2.8	1.6	4.3	31	48	45	.004	5.5
0.96	2.1	1.9	4.6	31	33			
1.2	3.1	1.5	3.1	59	82	128	.0018	6.0
						90	.0023	3.4
2.3	4.3	.78	2.5	112	155			
1.7	3.2	2.3	5.4	130	145	140	.0040	4.3
1.4	3.7	2.5	4.5	105	132	40	.0012	5.0
1.6	4.1	2.2	4.3	175	232	64	.0015	5.4
3.3	6.2	2.0	5.6	245	268		.00052	6.9

Average Channel Dimensions

Having recognized the fact that bankfull depth or gauge height is often not obvious in the field, two other channel characteristics will be discussed that can be very helpful in interpreting features encountered in the field. They are the average values of width, depth, and cross-sectional area at bankfull, and the average values of discharge at bankfull. Both of these are highly correlated with size of the basin (drainage area) in a given region.

For hydraulic consideration, width is usually the width of the water surface at any given discharge. Depth is the quotient of cross-sectional area of flowing water divided by width, and thus represents the depth of a rectangular channel of the same area. Though other depths that consider the elliptical or trapezoidal cross section might be used, mean depth as computed above has been found to be a useful one.

Rivers increase in size downstream as tributaries enter. Therefore, drainage area at any point is closely correlated with many size and discharge characteristics. To provide a general picture of river channel dimensions then, the bankfull width, depth, and cross-sectional area as functions of drainage area are useful. Figure 16-16 presents these dimensions for four regions in the United States. It can be seen that there is a considerable consistency even among regions. For the same drainage area, bankfull dimensions in the eastern United States and the San Francisco Bay area are very nearly the same. Channels in Wyoming, in an area of lower annual precipitation and runoff, are smaller. As usual, mean relations for rivers are drawn through points having considerable scatter, but the averages are useful tools in planning even though any individual place on a river may not agree closely with the mean.

A comparable graph could be constructed for any region by making some field measurements of local channels. The slope of any of the lines is nearly the same for various regions, and a moderate number of field observations could be used to place the vertical position of each line. Slopes comparable to those in the figure could be used as a guide.

For example, then, a basin of one square mile can be expected to have a channel 1.6 feet deep and 15 feet wide at the top, both in east central United States and the Bay region of California. In Wyoming it would be 1.3 feet deep and 8 feet wide on the average.

Within a hydrologically homogeneous region, of which eastern Pennsylvania is an example, these average values of channel dimension are sufficiently consistent that the degree of deviation from them can be interpreted as the magnitude of the effect of urbanization.

Relation of Bankfull Discharge to Drainage Area

Not only are average channel dimensions similar for streams of a given drainage area in a region, but the bankfull discharge is closely correlated with drainage area. This fact is highly useful because of all parameters

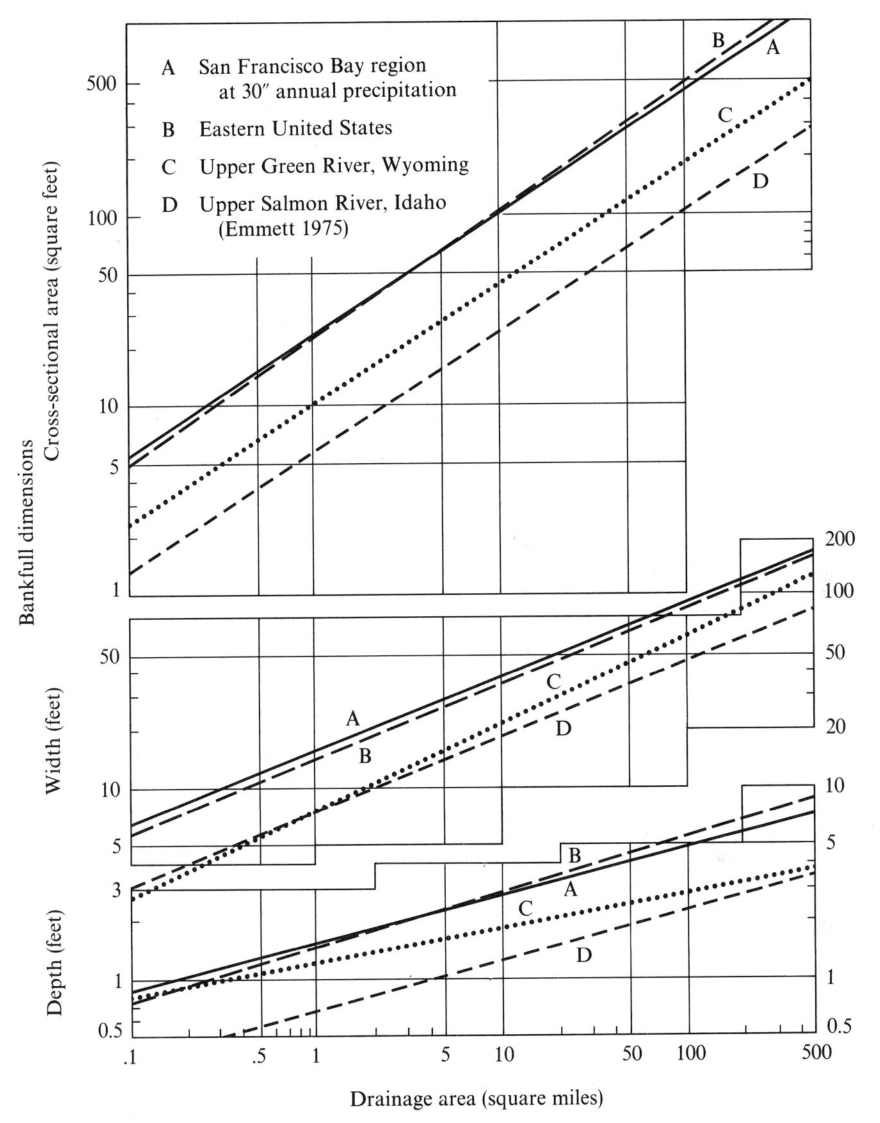

Figure 16-16 Average values of bankfull channel dimensions as functions of drainage area for four regions.

applicable to a geographic location, the easiest to determine is drainage area. This means that having measured the basin area on a map, a quantitative estimate of channel dimensions and bankfull discharge can be made with the understanding that the variability is large.

The variability as well as the general trend is illustrated by Figure 16-17. The values of bankfull discharge were obtained in one of two ways: by survey of a channel cross section in the field (a channel geometry survey) or by computing the discharge for a recurrence interval of 1.5 years. The scatter

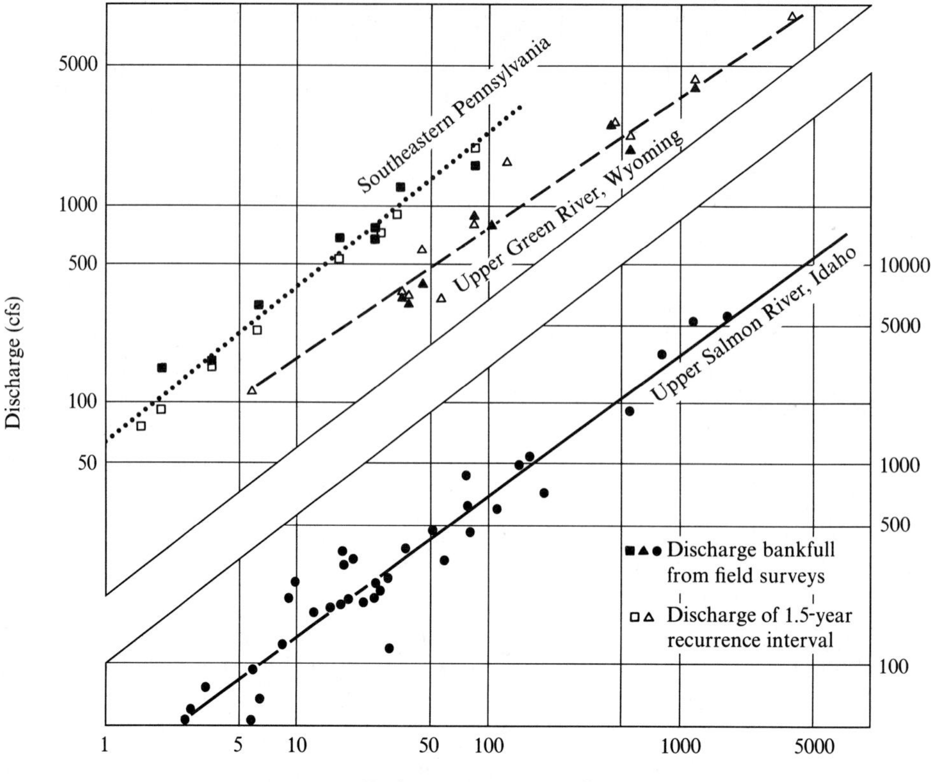

Figure 16-17 Data for individual stations plotted as discharge bankfull as a function of drainage area. (Salmon River data from Emmett 1975.)

of points about the line of best fit is of the same magnitude for both the methods by estimating.

Despite the variability, a general relationship extending through a range of drainage area of four orders of magnitude can be a useful tool. The relationship reveals well the differences from one region to another. Such a comparison is presented in Figure 16-18 for five regions in the United States.

Areas in the Piedmont and Appalachians of the eastern United States resemble those in the far northwest and the San Francisco Bay region, having annual values of runoff of 10 inches or more. The upper Green River in Wyoming and the Salmon River basin, Idaho, have much lower values of bankfull discharge, reflecting the smaller annual runoff of 1 to 10 inches.

The equations for the mean lines in Figure 16-18 are*

$$\text{West Cascades and Puget Sound: } Q_{\text{bkf}} = 55D_A^{0.93}$$

$$\text{San Francisco Bay region: } Q_{\text{bkf}} = 53D_A^{0.93}$$

*To avoid confusion with channel cross-sectional area, we are using D_A for drainage area in this chapter and Chapter 18.

$$\text{Pennsylvania: } Q_{\text{bkf}} = 61 D_A^{0.82}$$
$$\text{Upper Green River: } Q_{\text{bkf}} = 36 D_A^{0.68}$$
$$\text{Upper Salmon River: } Q_{\text{bkf}} = 28 D_A^{0.69}$$

It can be seen that bankfull discharge does not increase as fast as drainage area. If the two were directly proportional, the exponent of the regression line would be 1.0 instead of lower values. Flood discharge, or any infrequent flow of a chosen frequency of occurrence, will always increase more slowly than drainage area for the following reasons. Storms cover limited areas (see Chapter 2). Storms have the characteristic also of having one or more spots or loci of high intensity precipitation, but in the same storm the margins have lower intensity. The larger the proportion of the whole storm one considers, the smaller is the average intensity for the heavy fall in the storm centers. The spot of heavy rain is diluted, as it were, by the lower fall in the edges. Storm discharge in a river reflects this, and there is less water contributed per unit area for large than for small areas.

It is characteristic of river basins that discharge of any chosen frequency of occurrence will increase less rapidly than area drainage. The relationship can be represented by a function of the type

$$Q_F = c D_A^{\,n} \tag{16-8}$$

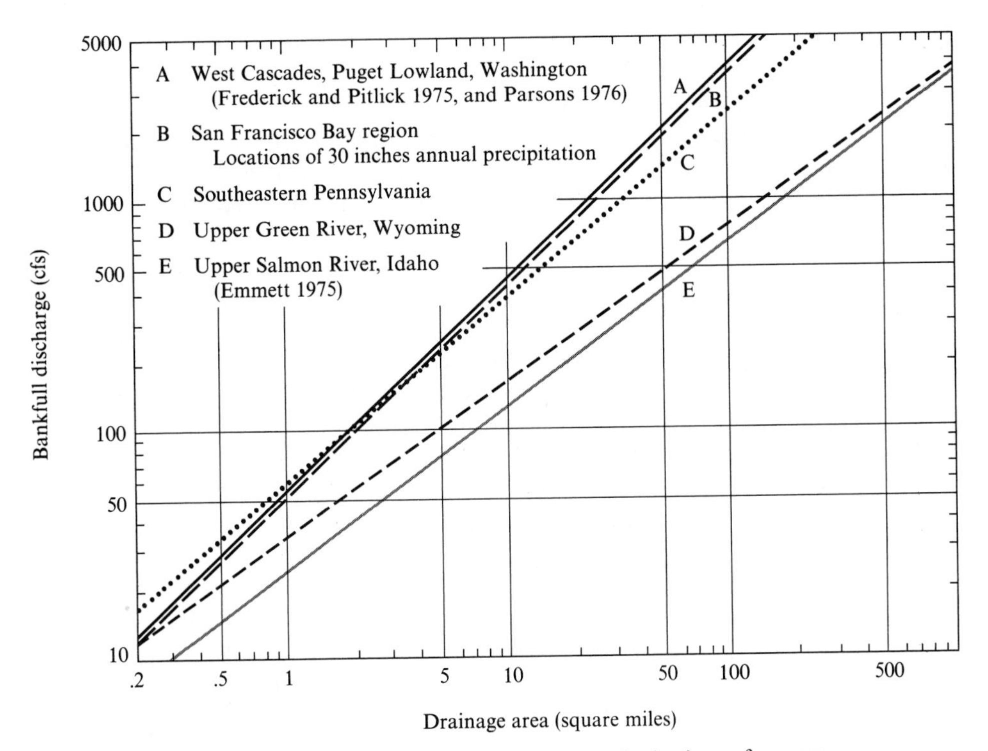

Figure 16-18 Bankfull discharge as a function of drainage area in the form of average relations for five regions.

where Q_F is the discharge from a flood of a given frequency, D_A the drainage area, c a coefficient that depends on the climate and the frequency of the flood, and n an exponent having the characteristic value less than unity, often between 0.7 and 0.8.

Mean Annual Discharge

The U.S. Geological Survey publishes annually for each state the daily discharge record of every gauging station in the state. This is the mean value for each day of the year. The year, however, is the water year, October 1 through September 30. The water year does not coincide with the calendar year because the precipitation and thus the runoff is seasonal—the climatic winter season may be considered to be from about November to March. Two successive winter seasons are more likely to be different than are successive parts of the same season, so rather than use the calendar year for the compilation of discharge data, the hydrologic or water year is used.

The mean flow for a day is the discharge that, if continued uniformly for 24 hours, would give the same volume of water as that actually observed during the day. By the same reasoning, the mean annual flow is the constant discharge value, that, if continued uniformly, would give the same volume of water as that observed in the period of record. The mean annual flow is also the arithmetic average of all the mean daily flows in the period of record.

Total annual runoff is the difference between the total precipitation over the drainage area and the total evapotranspiration. Over a large part of the eastern United States, the total precipitation varies with distance in a gradual and uniform fashion, and the total evapotranspiration does likewise. Within a district or region where the precipitation is gradually changing areally, the evapotranspiration changes in a like manner, with the result that the total or average runoff varies but little from one area to another. Such a circumstance leads to the result that each unit of area contributes to runoff about the same volume of water on the average. Thus, the average annual discharge varies directly with the drainage area in a one-to-one relation. Therefore, for a large part of the eastern United States, a plot of these parameters gives an equation of the form

$$Q_{\text{ave}} = D_A^{1.0} \qquad (16\text{-}9)$$

where Q_{ave} is the average annual discharge, D_A is the drainage area, and the exponent is nearly identically unity. Examples of the relation of average annual discharge to drainage area are given in Figure 16-19.

In mountainous regions, however, there is a large variation of annual precipitation with elevation and with exposure (windward versus leeward). The region, therefore, is not hydrologically homogeneous, and the average annual runoff is not a direct function of drainage area. Rather, within such a heterogeneous region, those parts that are hydrologically similar will, as

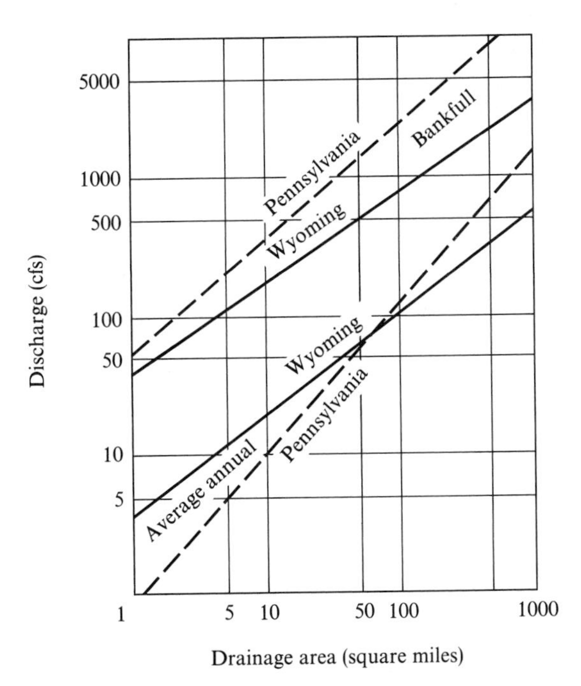

Figure 16-19 Bankfull and average annual discharges as functions of drainage area for upper east branch Brandywine Creek, Pennsylvania and for upper Green River basin, Wyoming.

before, show a direct relation of mean annual discharge to drainage area. Then, the plot of mean annual discharge as a function of drainage area will be a series of essentially straight and parallel lines, having slopes slightly smaller than unity. Mountainous and inhomogeneous areas of California are typical of this situation.

Relation of Mean Annual to Bankfull Discharge

Whereas average discharge amounts to from 1.0 to 4.0 cfs per square mile of drainage area, depending on both location and basin size, the bankfull discharge ranges from 40 to more than 100 cfs per square mile. Thus, the average discharge is contained well within the channel banks. One may ask how full the channel is when the discharge is equal to the mean annual value, and whether there is any consistency in such a ratio. Data show that there is a very consistent relation, as will be demonstrated.

A comparison of discharge values at bankfull and average annual is best visualized by graphical presentation. An example in Figure 16-19 compares typical basins in Pennsylvania where high discharges are caused by heavy rains, usually in summer, and a Wyoming area where high flows result from spring snowmelt. In the former, bankfull discharge is 40 times the mean annual for small basins and 10 times for large basins. This variation results from the fact that mean storm rainfall intensity is greater over small than over large areas.

In the mountainous area of western Wyoming where floods are derived from snowmelt, bankfull discharge is 5 to 10 times the mean annual flow for basins of all sizes, apparently because snowmelt covers a basin more uniformly than does precipitation of a single storm in the eastern states. Channel size for different sizes of basins reflects these regional characteristics.

In each channel cross section shown in Figure 16-7, a line has been drawn representing the water surface level corresponding to the mean annual discharge. Inspection of these for the several stations in the upper Green River basin, Wyoming, shows that an average flow fills the channel somewhat less than half full. In Table 16-3 for these examples, the percentage of bankfull depth represented by the mean annual flow is tabulated. Through a range of drainage areas from 6 to 1200 square miles, the figure remains in the same magnitude, about 40 percent. Because in the usual low-flow condition, a channel has much less water than half full, the average annual discharge is a high not a low flow.

For the same list of stations, the bankfull discharge is on the average seven times the average annual discharge. That the average discharge is a relatively high flow is shown by the percentage of time, indicated on a flow duration curve, that the discharge of a river equals or exceeds the mean annual value. This figure is very constant among a large variety of rivers and closely equals 25 percent. That is, on 75 percent of all days the discharge is less than the mean annual.

Relation of Recurrence Interval to Number of Floods

It is often a source of confusion that the recurrence interval computed by the use of an annual flood series does not give the actual number of flood experiences. Flood frequency may be computed from a gauging station record in two ways, by the use of the annual flood array or by a partial-duration array. Though this difference was discussed in Chapter 10, its application to the bankfull discharge is treated further here.

The bankfull stage is the stage of incipient overflow. Discharges that exceed this stage must overtop the banks, spread out over the floodplain, and potentially cause damage. A flood may be defined as discharge in excess of channel capacity. The planner should know not only how often it can be expected but how many such occurrences may be expected.

The discharges useful for a flow frequency computation are momentary peaks or the highest discharge in each storm event. This momentary or highest value is the important quantity, because if a house or factory is flooded, even for a moment, damage results. The momentary peak values of discharge and gauge height are therefore published.

Table 16-3 Relation of average discharge to bankfull discharge, upper Green River basin, Wyoming.

	DRAINAGE AREA (SQ MI)	RATIO $\frac{Q_{bkf}}{Q_{ave}}$	PERCENT OF CHANNEL DEPTH FILLED BY Q_{ave}
LaBarge Creek near LaBarge Ranger Station	6.3	8.3	44
New Fork below New Fork Lake	36.2	6.7	50
Fall Creek near Pinedale	37.2	8.0	54
Silver Creek near Big Sandy	45.0	9.1	29
Pole Creek below Little Half Moon	87	8.1	39
Green River at Warren Bridge	468	5.3	53
New Fork near Boulder	552	4.8	38
New Fork near Big Piney	1230	6.4	39
Average		7.1	43

The highest flood peak in a given year is called the annual flood. When these are tabulated for each year of record, arranged in order of magnitude, they provide the basis for a frequency curve of the annual flood series. When from this curve one reads the value of the discharge equaled or exceeded once in 1.5 years, an approximation of bankfull discharge, it has the following meaning: once in 1.5 years, or 2 years out of 3 on the average, the annual flood (highest momentary peak of the year) will equal or exceed the specified value of discharge.

But the highest momentary peak in a year does not mean there were not other flows nearly as great. If all momentary peaks, not merely the highest in each year, are included in the array, the list is known as the partial-duration series discussed in Chapter 10. These two arrays for a given station are identical for the highest flows, but diverge at lower values. The discharge having a recurrence interval of 1.5 years on the annual flood series has a recurrence interval of 0.92 in the partial-duration series. Therefore, the bankfull discharge has the following frequency characteristics: the annual flood will equal or exceed bankfull once every 1.5 years or 2 years out of 3 on the average. A discharge equal to or greater than bankfull will be experienced each 0.92 years, or 1.1 times each year, that is, slightly more often than once a year.

An example of the difference in these definitions is given in the following count of momentary peaks experienced on the Green River at Warren Bridge, near Cora, Wyoming. All peaks equal to or exceeding 2600 cfs are included. This value of discharge has a recurrence interval of 1.5 years in the annual flood series. In a sample of 26 years, there were 42 occurrences or 1.6 per year. This is in accord with theoretical characteristics of the two arrays, that the value of annual flood having a recurrence interval of 1.5 years is a discharge that occurs somewhat more often than once a year.

These frequency arrays of momentary peak discharges do not convey any information concerning the number of times the mean daily flow is equal to or larger than bankfull. In the spring snowmelt season, several days may experience a mean daily discharge greater than bankfull, while in the flood event there was only one momentary peak. The number of days the mean exceeds bankfull would be shown on the duration curve of daily values.

In a sample of eight stations in western Wyoming, the percentage of days the mean daily discharge equaled or exceeded bankfull was read from the duration curve of each station. The percentage of time varied between 1.3 and 4.5 percent, and the average was 2.1 percent. That is, on the average, these stations experienced 2.1 percent of the days or about eight days each year during which the mean flow for the day equaled or exceeded bankfull.

Channel Patterns

The natural channels of rivers increase in size downstream as tributaries enter and add to the flow. The channel is neither straight nor uniform, yet its average size characteristics change in a regular and progressive fashion. In upstream reaches the channel tends to be steeper. The gradient decreases downstream as width and depth increase. The size of the sediment debris on the bed also tends to decrease, often from boulders in the hilly or mountainous upstream portions, to cobbles or pebbles in middle sections, and sand or silty sand farther downstream. Even within a given reach, the channel is not uniform, rather the bed undulates in elevation in quite a regular repeating pattern. Shallow parts that we have called *riffles* alternate with deeps or *pools*.

These terms we drew from our vocabulary as fishermen. The riffles differ from pools in the concentration or density of larger rock sizes. The bed of a pool may be mostly sand or sand mixed with a few cobbles, whereas the riffle nearby will be mostly gravel or cobbles. The riffle is a topographic high, or a local hillock on the bed, and water leaving the pool and approaching the riffle must converge as shown in Figure 16-20; that is, bed water must actually rise upward and converge with water near the surface. Because of the restricted cross-sectional areas over the riffle, the mean velocity over the riffle must be greater than in the pool. A typical pool–riffle sequence is shown in Figure 16-21.

This convergence and increased velocity over the riffle are more pronounced at low flow than at high. With increase in discharge accompanied by increase in mean velocity and in depth, the water surface slope, which is the measure of the rate of energy expenditure, also changes. Over the pool the surface slope increases with discharge, and over the riffle it decreases. At high stage the gradients become equal, the water surface appears smooth rather than stepped, and all visual indications of the presence of pool and riffle are obliterated. In stream-gauging parlance, the riffles "drown out." This drowning out, or obliteration of the surface indications of the pool–riffle alternation, occurs when the channel is about three-fourths bankfull, or when the mean depth is about three-fourths of the bankfull depth. Also

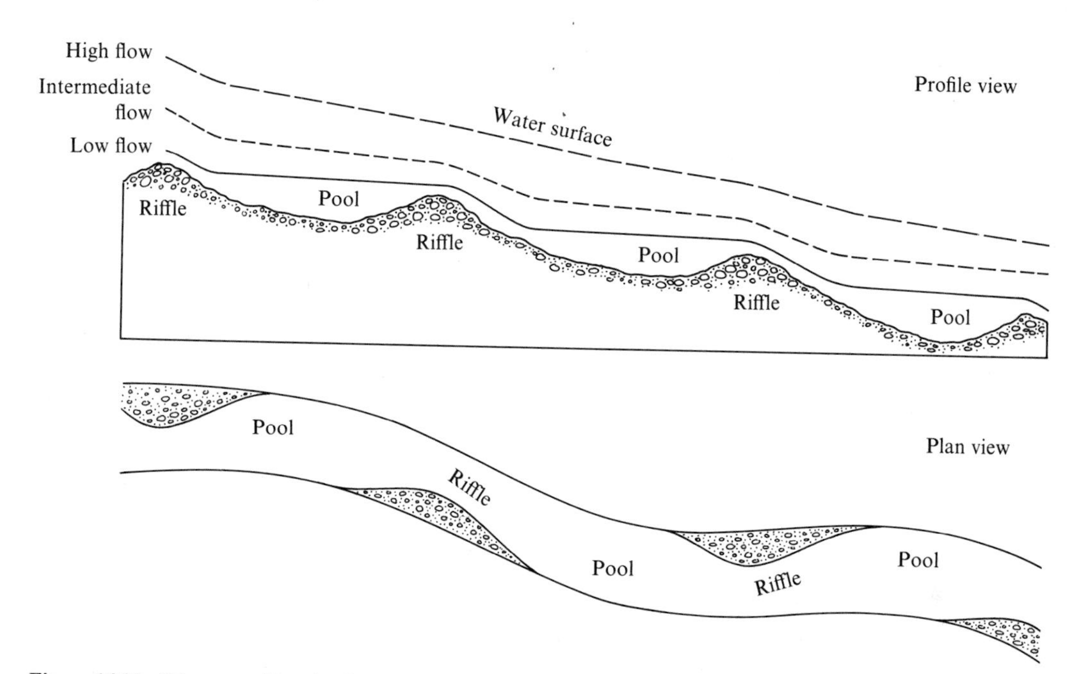

Figure 16-20 Diagram of longitudinal profile and plan view of a pool-riffle sequence. Water surface profiles in upper figure represent high, intermediate, and low flow conditions.

Figure 16-21 Pool (foreground) and riffle (middle distance) in East Fork River near Highway 353, Boulder, Wyoming.

at about that depth, the debris on the bed becomes mobile and those particles or rocks of average or median size are moving. In gravel bed streams, the layer of bed material in motion is about as thick as the diameter of the median or average particle. The layer in motion, then, is about one diameter thick.

The riffles are caused by the deposition and maintenance of a gravel bar that characteristically occurs alternately on one side and then on the other side of the channel. The distance between successive bars averages five to seven channel widths.

Sediment moving near the bed concentrates near the convex bank and tends then to be deposited (Leopold 1974:82–84). In this way the convex bank is gradually extended streamward in the growth of a *point bar,* as this deposit is known.

The point bar is the deposit formed around and against the convex bank in a channel bend. The top level of the point bar is generally flat and at the height of the floodplain. The continual streamward extension of the point bar as the channel migrates laterally is a major process of floodplain formation.

The meandering pattern refers to the plan view of a channel executing rounded curves of repetitive and uniform shape in which the ratio of channel length to down-valley distance exceeds 1.5. This is merely an arbitrary assignation, for this ratio changes in the unbroken continuum of shapes varying from straight to very looping. Really straight reaches are rare and seldom does one see a straight reach of length exceeding 10 channel widths.

Meander bends of a river are neither semicircular nor sinusoidal. The shape is of a slightly different but special kind called the sine-generated curve (Leopold and Langbein 1966). In this curve the angle of deviation from the mean down-valley direction is a sine function of the distance along the channel. Rivers meander rather than flow straight because the meandering pattern is a closer approach to minimum uniformly distributed work than are alternative shapes. Once having adopted a meandering pattern, a river will not change to a straight one as long as the climate does not change.

Channel bends cause a large amount of energy loss because work is necessary to deflect the direction of water flow. Therefore, the slope of the water surface is steeper in a meandering channel than it is in a straight channel carrying equal discharge. This fact is in direct opposition to the intuitive, but incorrect, conclusion that because the meandering channel is longer, its slope is therefore less steep than a straight channel.

In a channel that has an irregular rather than a meandering habit, or in other words a channel that is more or less straight, successive riffle bars tend to occur on alternate sides of the channel (Figure 16-22). In a channel that adopts a meandering pattern, the shallow section will tend to be at the crossover or point of inflection where the curvature changes. In a fully developed meander pattern in which the change of direction is pronounced, the deepest part of the channel is near the concave bank at the place of

Figure 16-22 Gravel bars in a nearly straight channel alternate from one bank to the other, and at low flow they force the water to take a sinuous path. Silver Creek near Big Sandy Crossing, Wyoming.

maximum curvature opposite to the point bar on the convex bank. This relation is sketched in Figure 16-23, which shows transitional forms between straight and meandering.

The profile shown on Figure 16-20 for the straight channel indicates that even at high flow the pool-and-riffle reach shows subdued but real change in water surface slope, steeper over the riffle and flatter over the pool. This alteration of steep and less steep water surface is usually not discernible by eye at high flow but can be demonstrated by survey. On the other hand, the meander has an essentially constant water surface gradient at high flow, and this is steeper than the mean gradient of a pool-and-riffle straight channel of equal size and discharge. This characteristic difference is an essential element in explaining why rivers meander and form the beautifully regular ribbonlike bends so often seen when traveling by air.

A third major channel pattern is braiding (Leopold and Wolman 1957). A *braided stream* separates around islands. The dividing and rejoining may consist merely of two channels around a single island, typical of the Mississippi River in eastern Iowa upstream of Keokuk, or it may consist of many channels around multiple islands as in proglacial streams emanating from ice-covered areas in Alaska. In the latter type of braiding, individual channels are often separated by a gravel bar of rapidly changing size and shape and an unstable and constantly varying distribution of channels and bars (Fahnestock 1963).

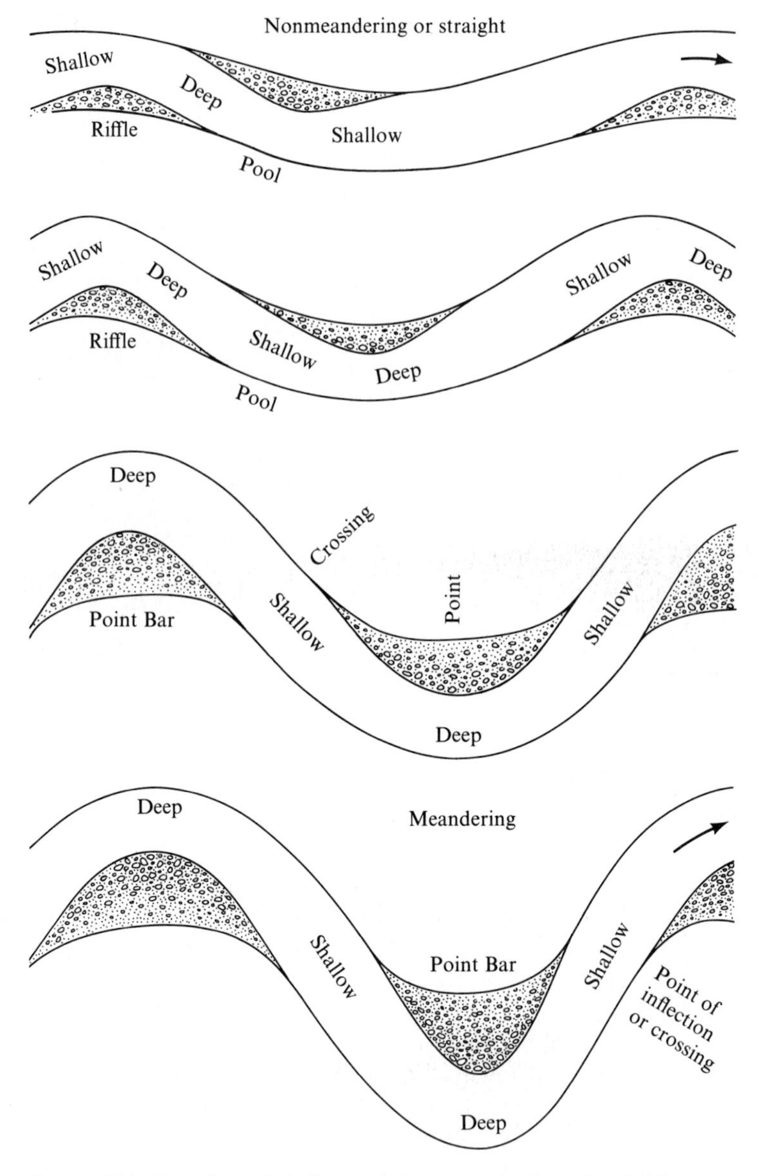

Figure 16-23 Locations of shallow and deep zones in channels of different sinuosity. Riffle bars on alternate banks characterize straight channels, but point bars on convex banks characterize meander bends.

The different patterns are widely different in stability. Straight channels migrate laterally most slowly. Meanders tend to move by erosion along concave banks, with equal deposition on convex banks. Braided channels are most unstable, especially where separated by bars of gravel rather than vegetated islands. Braided channels are subject to evulsions, that is, rapid, unpredictable abandonment of a channel segment and creation of another channel some distance away.

In the course of designing projects, it is sometimes necessary for planners to obliterate a reach of stream and move it to a different location. They should be cognizant of the form and characteristics of channels so that they can design with appropriate dimensions. More detailed discussion of channel morphology is found in material written for geomorphologists (Leopold, Wolman and Miller 1964). For present purposes, emphasis is on those characteristics most important to planning.

Reading the Riverscape for Planning Purposes

The preceding descriptions of stream channels may be made more understandable if coordinated by a few examples that indicate what features of importance should be observed in the field. Among the questions that come to mind when one looks at a channel from the standpoint of planning are the following: Is the valley floor subject to frequent flooding? Is the channel stable and likely to remain unchanged over a period of a few years, or is it liable to rapid shifts, bank cutting, and lateral migration? Is the sediment load large and the bed subject to scour?

The examples chosen for discussion are channels illustrated in this volume or descriptions easily available in publications. It is assumed in each example that the location under consideration can be identified on a map and that, therefore, the drainage area can be known approximately.

Watts Branch, a tributary to the Potomac River near Rockville, Maryland, is observed at a point where the drainage area is 3.7 sq mi. The cross sections in Figure 16-11 are at this location. The valley floor is about 300 feet wide and nearly flat. The banks are silt or clayey silt, vertical on the concave banks, and sloping toward the channel on convex banks. The bed is gravel with some sand and silt on the channel bars but mostly silt in the pools. The channel pattern is sinuous to strongly meandering. Its average width at the top of the banks is 18 feet, and the bank height (channel depth) is about 4 feet. (Photographs and cross sections are published in Leopold, Wolman and Miller 1964:87, Figure 3-26; Leopold 1973:1848–1851; Leopold 1974:40, 80).

With these observations available, what might the planner do with them? The first question is how often the valley floor might be inundated. The graph of average channel dimensions (Figure 16-16) is helpful. For a channel draining 3.7 sq mi in the eastern United States, the average bankfull width can be expected to be about 20 feet and the mean depth about 4 feet. Thus, we may conclude that the channel as observed is of normal size and that the adjacent flat areas are probably floodplain. The cross section of Figure 16-11 suggests that the point bar, being overflowed, lies slightly below, perhaps 1 foot below, another level. Because many valley floors of streams draining 5 square miles or less in that region exhibit a low terrace only slightly above the floodplain, this interpretation of the Watts Branch valley is a logical one. But this terrace is low enough to be inundated frequently, though not as often as the floodplain.

The meandering character of the channel indicates that it tends to move laterally, eroding the concave banks and extending point bars from convex banks. The silty banks and gravel stream bed suggest there is but little bed scour at high flow, and that the major process of bank erosion is connected with freeze and thaw in winter.

Seneca Creek, another tributary to the Potomac River in Maryland, is situated about 15 miles north of Watts Branch. Near Dawsonville it drains 100 square miles, is 80 feet wide, and has silty banks 5 to 6 feet in height and a bed of gravel with sand and silt in pools. It is pictured in Figures 16-24 and 16-25. Photographs and cross section are published (Leopold 1974:42, 44, 79). The photographs show that long reaches of channel are closely bordered by trees, the roots of which are exposed in the bank. The channel is sinuous but not meandering.

Figure 16-24 Gravel bar forces low flow to opposite bank, Seneca Creek near Dawsonville, Maryland.

Figure 16-25 Near bankfull flow drowns out the bar that at low flow causes a riffle, same reach as Figure 16-24, Seneca Creek near Dawsonville, Maryland.

With these data obtainable by inspection, we may again turn to Figure 16-16 and see that the expected channel width at bankfull for a basin of 100 square miles is 110 feet, and the expected depth is 5.5 feet, in reasonable agreement with observed dimensions. Because low terraces are uncommon along channels draining areas of 50 square miles or more in this region, we infer that the wide flat valley floor bordering Seneca Creek is the floodplain and, therefore, will be flooded on the average of once a year or more.

The sinuous but nonmeandering channel bordered by trees may be expected to be moving laterally but slowly, and bank erosion is inhibited by tree roots. The channel, then, is probably relatively stable and the valley flat subject to frequent flooding.

Walker Creek near Nicasio, California, a few miles north of San Francisco Bay, is pictured in Figure 16-13, where the drainage area is about 30 square miles. The channel bed is about 50 feet wide and stands 12 to 14 feet below a broad to narrow but prominent valley floor. A definite berm or level stands 5 to 6 feet above the channel bed. The stream bed is mostly gravel.

Inspection of the graph of expected channel dimensions (Figure 16-16) for 30 square miles in the Bay region gives the information that bankfull width is 60 feet and bankfull depth 3.5 feet. We may immediately conclude that the broad valley flat is a terrace and is never flooded by the main stream, though it may be wetted by runoff from bordering hillslopes. The level 5 feet above the bed is also a terrace but is flooded frequently. This low terrace is narrow, exists only as isolated spurs within the main trench, and so would ordinarily not be considered as a site for any construction. The high banks are subject to undercutting and sloughing, so one would not build any structure close to a vertical bank 12 feet high.

Indeed, this physiographic situation is typical of most of the minor valleys in the Berkeley Hills and much of the San Francisco Bay region. The valley floor stands as a terrace high above the present stream bed. The floodplain is so narrow, being developed within the trench, that it is usually not even identifiable.

Horse Creek, a tributary to the Green River draining eastward out of the Wyoming Range, joins the Green near Daniel, Wyoming. Its drainage area there is 124 square miles, its water surface width excluding the gravel islands is 70 feet (50 feet when undivided by islands), and its banks have a height of 3 feet. The channel is covered with gravel, and the banks are of similar gravel with a thin cover of silt. Elongated islands of gravel are common, so the channel continually divides and rejoins in a typical braided pattern. Standing 5 feet above the bed is a broad valley flat covered with sagebrush, occasional willow, and cottonwood. Cross sections and photographs are published (Leopold and Wolman 1957:43).

Figure 16-16 indicates that at 124 square miles in the region the expected bankfull width is 55 feet, and the depth is 3 feet. The top level of the islands corresponds then to the floodplain, and the valley flat, 2 feet above the floodplain, is flooded only infrequently. But the braided pattern suggests channel instability, frequent shifts in position, bank cutting on concave banks, and a relatively high sediment load. Field inspection shows numerous places where local ranchers have built rock revetment or rock walls to try to reduce the rate of bank cutting. Thus, field indications confirm the inferences drawn from channel pattern and texture of bed and bank.

These brief examples will, we hope, indicate that inferences important to planning may be drawn from field inspection in conjunction with some of the generalized graphical relations presented in this volume.

Use of the preceding principles for more detailed analysis is shown in the following example. The Stroud Water Laboratory of the Academy of Natural Sciences of Philadelphia is located near the channel of East Fork White Clay Creek on the road between London Grove and Chatham, Pennsylvania, and shown on West Grove quadrangle, Pennsylvania. The drainage area at that location is 2.8 square miles. The questions posed are: (1) Is the laboratory building likely to be reached by floodwaters? (2) What is the extent of area that might be covered by an extreme flood? (3) Is the stream channel stable in its present location, or is bank erosion liable to move the stream laterally and endanger the building?

The most efficient way to use the time spent in field analysis is to make a sketch map. Walking along White Clay Creek one has the impression that there exist two levels of berms, one of which probably is the floodplain. A brief reconnaissance suggests that the identifiable levels stand at 2.5 to 3 feet and 4.5 to 5 feet above the mean bed elevation. This determined the classification of mappable features.

From the topographic map and field inspection, a sketch map was drawn showing the planimetric configuration of the channel through a reach of about 3000 feet, or a distance equal to 200 channel widths. Fences, roads, and buildings were placed on the map. The configuration of the bounding hillslopes was indicated in an approximate way by hachures. The initial inspection suggested that two levels existed near the channel, the higher one constituting the main valley floor. A crosshatch symbol was adopted to indicate the area of the level standing 2.5 to 3.0 feet above the channel bed, and the number 5 was written at locations where the higher level was obvious. Walking along the channel, mapping as one went, resulted in the map shown in the upper part of Figure 16-26. The map shows that the lower level or berm was discontinuous but present along one or both sides of the channel over the whole distance.

From the completed map, locations were chosen to measure several cross sections, one of which is plotted in the lower part of Figure 16-26. The cross section was surveyed by measuring down to the ground surface from a taut horizontal string using a stretched tape for distance from origin. This method is described in Typical Problem 16-3.

The map and cross sections confirmed the initial impression that the valley floor stands 5 feet above the stream bed, and a persistent berm or bench stands 2.5 feet above the bed. One of these two must be the floodplain and it is necessary to determine which. On the cross section the base of the laboratory building is at an elevation of 105 feet. It is necessary to find whether floodwater could reach that level.

Horizontal lines were drawn on the cross section at the level of the valley floor, 100.0, and at the level of the persistent berm, 98.5. Until the floodplain is identified, both the valley floor and the inner level are called berms, that is, flat zones.

Entering Figure 16-16 for the eastern United States, one can read values of width, depth, and cross-sectional area at bankfull as functions of drainage area. From that figure, expected dimensions are compared in Table 16-4 with dimensions measured from the cross section on White Clay Creek.

Better agreement with expected values is furnished by the low than the high berm, especially in the bank height (depth) value.

On the basis of the above analysis, it is concluded that the floodplain of White Clay Creek is not the wide valley flat but the narrow zone or low berm. The upper berm, therefore, is a terrace or abandoned floodplain,

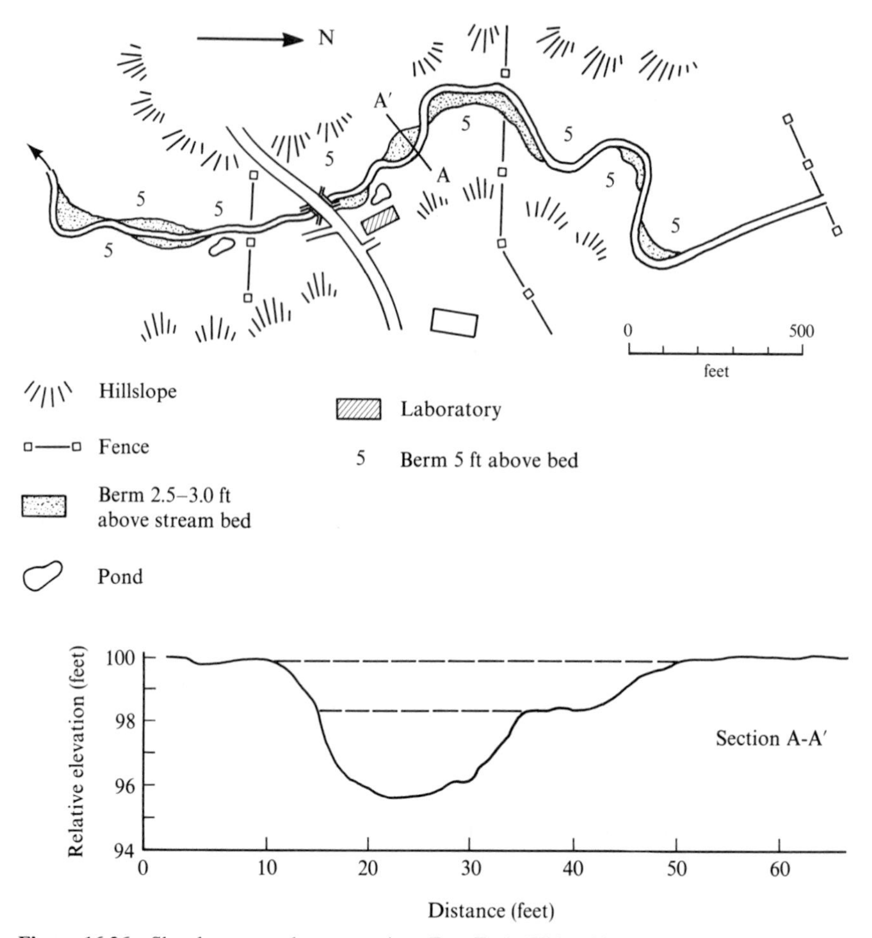

Figure 16-26 Sketch map and cross section, East Fork White Clay Creek at Stroud Water Laboratory, Pennsylvania. The areal extent of each of two levels of valley floor is shown, one 5 feet, another 2.5–3.0 feet above channel bed.

the abandonment probably the result of agriculture since colonial times in the area. The ages of the trees lining White Clay Creek indicate that the channel is relatively stable, moving laterally only on channel bends and these apparently at a normal rate. This relative stability is also indicated by the channel pattern, which is sinuous, not braided. The laboratory building is sufficiently high above the floodplain that it will not be flooded in ordinary floods. The frequency of flooding can be estimated from the dimensionless rating curve described later in this chapter.

This example reminds the reader that perfect agreement between computed and measured results cannot be expected, and the predominance of evidence is often the deciding factor. Other examples might have been chosen in which the results were definitive and unequivocal.

Table 16-4 Observed and expected dimensions of White Clay Creek, Pennsylvania.

	BANKFULL WIDTH (FT)	BANK HEIGHT (FT)	CROSS-SECTIONAL AREA (FT^2)
Observed at high berm	38	3.5	77
Observed at low berm	17	2.1	35
Expected value, Fig. 16-16	20	2.0	40

The Hydraulic Geometry

The consistent relations among width, depth, velocity, slope, roughness, and discharge remained long unobserved because the individual values of all the parameters except discharge are not published. Measurements of width, depth, and velocity are made in connection with a discharge determination, but are kept in the files of the U.S. Geological Survey. For most purposes it is the discharge that has practical value, but in the understanding of details, the unpublished part of the record is also necessary.

In the files of each district office of the Water Resources Division of the Geological Survey, usually in the capital city of a state, the current meter records are kept on a form called 9–207. Each horizontal line gives the results of a discharge measurement by current meter. The main column headings are as follows (we have inserted some explanatory comment):

Number (indicates the total number of discharge measurements made since the station was installed)

Date

Made by (names of hydrographers)

Width (of water surface)

Area (cross-sectional area of flowing water)

Mean velocity

Gauge height (reading on the staff gauge plate; datum usually arbitrary)

Discharge (in cfs)

Rating (because the rating curve shifts or changes with time at many stations, the particular curve in use is indicated)

Shift adjustment, percent difference (refers to comparison of this measurement with previous ones)

Method (0.6 means the velocity was measured 6/10 of the distance from surface to stream bed; 2.8 means two measurements of velocity, at 2/10 and 8/10 distance from surface to bed)

Number of measured sections (how many places across the stream was the velocity measured)

Gauge height change (the change in feet of water surface during the measurement)

Time (hours required to make the measurement)

Measurement rating (E excellent, G good, F fair, and P poor)

Remarks (useful information in understanding the results)

From the data on Form 9–207, the values of width, depth, and velocity, and the corresponding discharge can be plotted as shown in Figure 16-27. Figure 16-28 shows a photograph of the Green River at this location. The scatter of points is typical and always present to some degree in such data. At most stations the scatter of points is caused by the fact that discharge measurements are not made at the same place each time; for at low flow when the hydrographer is wading, he will choose a cross section that looks best at the time. At high flow when wading is impossible, he will measure from the cable at the station or from a bridge. At the Fontenelle site used as an example, wading measurements are made at a discharge of up to 800 cfs and from the cable at higher flows. The wading measurements are made at

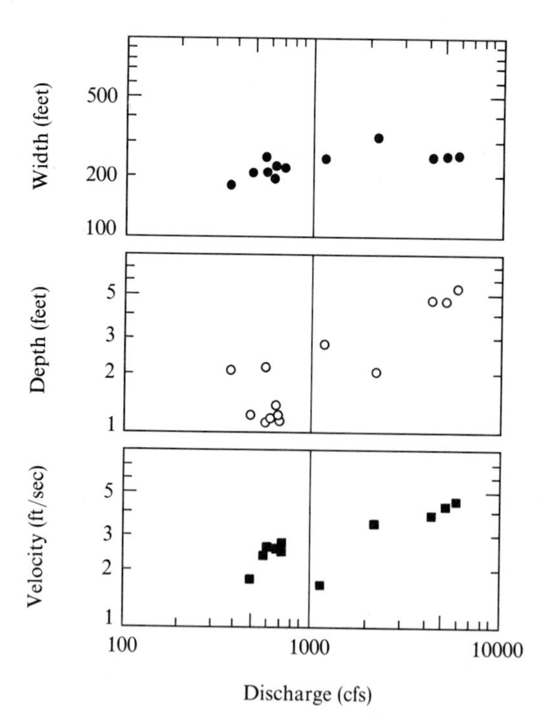

Figure 16-27 Data (1962) on width, depth, and velocity plotted against discharge without any selection of points to eliminate ice-covered or low-flow measurements, Green River near Fontenelle, Wyoming.

Figure 16-28 Gauging station on Green River near Fontenelle, Wyoming. Wire weight for measuring stage is hanging from the inclined board.

sections as far upstream as half a mile and downstream 1000 feet. The variation in cross section is understandably great.

As an indication of the necessity for use of the hydrographer's notes, at the same Fontenelle station for which the data were scattered in Figure 16-27, a selection was made of measurements made at the cable, assuring that the same cross section was measured at all the discharges. The resulting plot is shown in Figure 16-29. The scatter has been largely eliminated. Straight lines drawn through each set of points in the graphs of width, depth, and velocity give a reasonable representation of the array of data. But the plot of gauge height as a function of discharge often does not give a straight line.

It is usual to draw by eye a straight line through the points of each graph, understanding that such data are scattered and that any single line is a broad generalization. The criticism has been raised that the estimation of the line of best fit by eye is inferior to fitting a line by least squares. Another is that the actual relation may not be a straight line on log-log paper, and that a curvilinear relation may be closer to reality. Those comments have merit, but the nature of the data and the purpose to which they are to be put should be examined. It is primarily the overall consistency in pattern when such lines are drawn for a number of stations on various rivers that gives some assurance that the generalizations are not spurious. In those special investigations where measurements were consistently made at

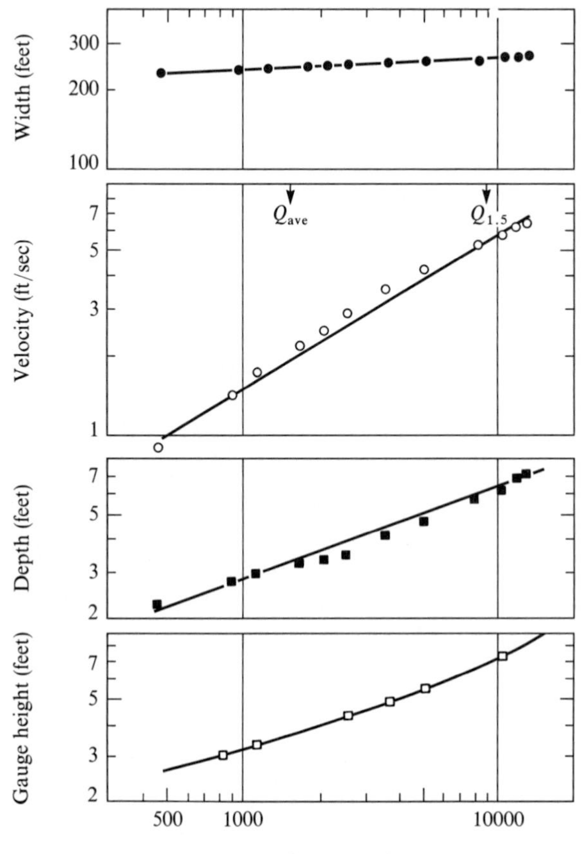

Figure 16-29 Width, depth, and velocity as functions of discharge, using only data collected from the cable and thus representing a single cross section, Green River near Fontenelle, Wyoming. Bottom graph is the rating curve.

identical sections (Wolman 1955), the scatter of points is greatly reduced and the simple procedures widely used find justification.

The computation of mean depth as area/width is a source of some concern. Because area and width are measured in the field and shown on Form 9-207, this method of computing depth is practical and consistent.

The notes of the hydrographer should be inspected in each case to eliminate points that are not representative of the cross section. At many gauging stations the high discharges are measured from a highway bridge, and in this case the width does not continue to increase with increasing discharge when the water fills the space between the bridge abutments. This also causes both velocity and depth to increase faster with discharge than would be the case if the cross section were entirely natural. Some values in the

Fontenelle record were obtained by digging holes in the ice covering; the data for such measurements will usually be aberrant compared with normal conditions because an ice cover increases the frictional resistance to flow and the velocity tends to be unusually low.

If a straight line on log-log paper represents the data, then the power function may be represented in the form

$$y = ax^n \tag{16-10}$$

and in the case of these graphs,

$$\begin{aligned} w &= aQ^b \\ d &= cQ^f \\ u &= kQ^m \end{aligned} \tag{16-11}$$

using the symbols introduced by Leopold and Maddock (1953), in which a, c, and k are coefficients and b, f, and m are the exponents. The value of b, f, and m are the slopes of the respective lines. On Figure 16-29, the values are $b = 0.05$, $f = 0.35$, and $m = 0.60$.

Because discharge is the product of area and velocity,

$$Q = Au = wdu$$

from the equations above:

$$Q = (aQ^b)(cQ^f)(kQ^m)$$

$$\begin{aligned} b + f + m &= 1 \\ a \times c \times k &= 1 \end{aligned} \tag{16-12}$$

In the case of Figure 16–29, $b + f + m = 0.05 + 0.35 + 0.60 = 1.0$ and $a \times c \times k = 170 \times 0.24 \times 0.024 = 1.0$. The coefficients represent the theoretical values of width, depth, and velocity when the discharge is unity (1.0), but on many rivers the discharge is never actually 1 cfs and so the coefficient has no physical reality. Also, the coefficient is highly dependent on the slopes of the lines and therefore is influenced in the actual construction of lines of best fit, both by the slope and the vertical or ordinate positions of the lines drawn.

At stations where there is no cable and high-water measurements are made by wading or from a bridge, it is often not possible, even with the full use of the field notes made by the hydrographer, to select data points that represent a single cross section. In such cases there remains an irreducible scatter in the plotted graphs, and a line of best fit is a broad generalization. Some examples again drawn from stations in the upper Green River are shown in Figures 16-30–16-32.

But there is, nevertheless, in a hydrologically homogeneous region, a pattern that is displayed by the curves of width, depth, and velocity plotted against discharge. This pattern provides a general description of the channels in the region. The pattern is shown by the graphs for channel sites in the upper Green River basin, Wyoming, in Figure 16-32. Among the stations the respective lines are generally parallel. The slopes of these lines, which are thus the exponents in the equations, are listed in Table 16-5. Station curves differ much more in ordinate position than in slope, as can be seen in Figure 16-33.

These curves represent the changes in hydraulic factors at a given cross section as discharge changes and are thus called at-a-station curves. The different discharges represent different frequencies of occurrence, higher discharges being less frequent.

A comparison among locations along the river channel system may be made if some particular frequency of discharge is used. A recurrence interval of 1.5 years is meaningful, or the mean annual flow is most often chosen because the data are readily available from published data.

For the upper Green River example, the plots of width, depth, and velocity at bankfull discharge, and therefore approximately a constant frequency

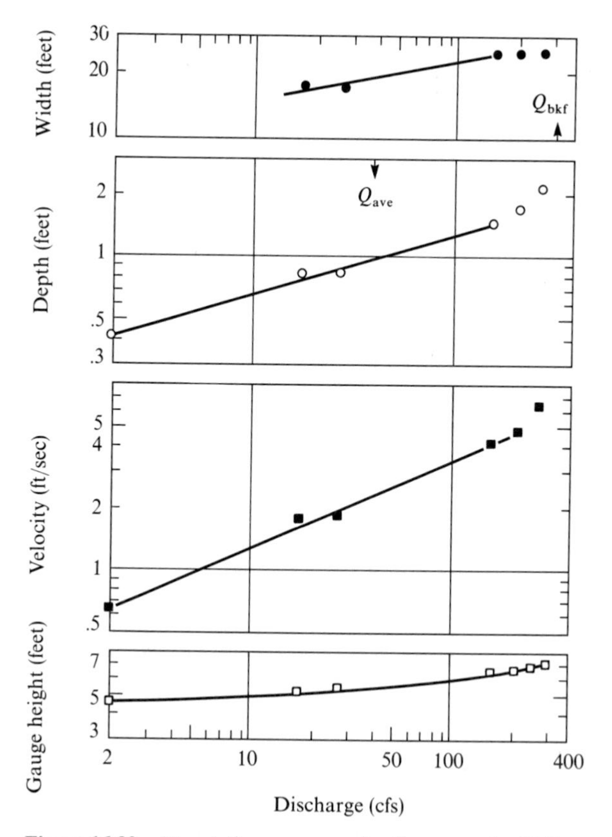

Figure 16-30 At-a-station curves and rating curve for Fall Creek near Pinedale, Wyoming.

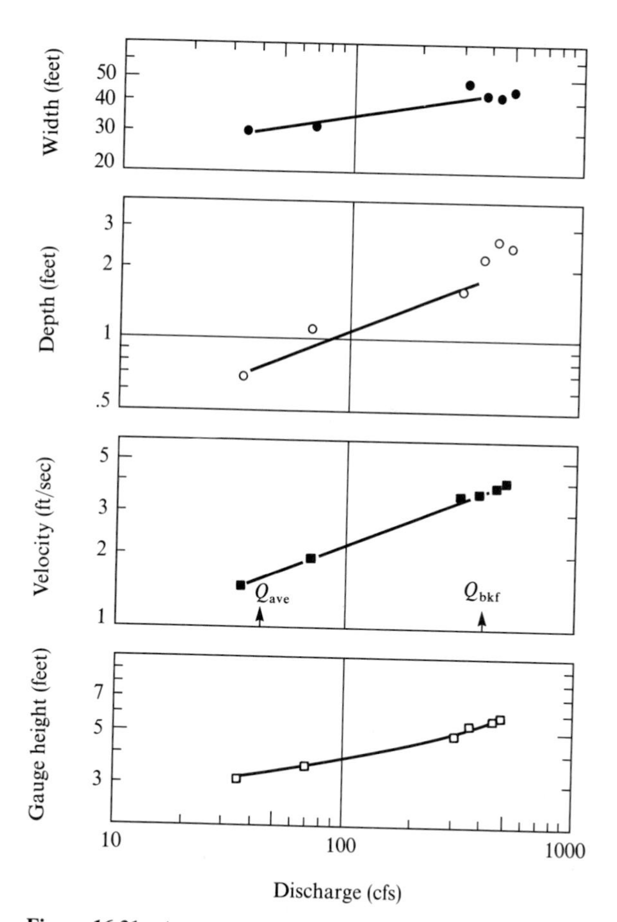

Figure 16-31 At-a-station curves and rating curve for Silver Creek near Big Sandy, Wyoming.

(1.5 years recurrence interval), are shown in Figure 16-34. Tabulated values for this plot appear in Table 16-2. The same equations listed for the at-a-station case also apply to the downstream relations; for the Green River basin the slopes of the lines are represented by the exponents

$$b = 0.55$$
$$f = 0.35$$
$$m = 0.10$$

For a large number of basins, the average exponent values have been found to be

$$b = 0.50$$
$$f = 0.40$$
$$m = 0.10$$

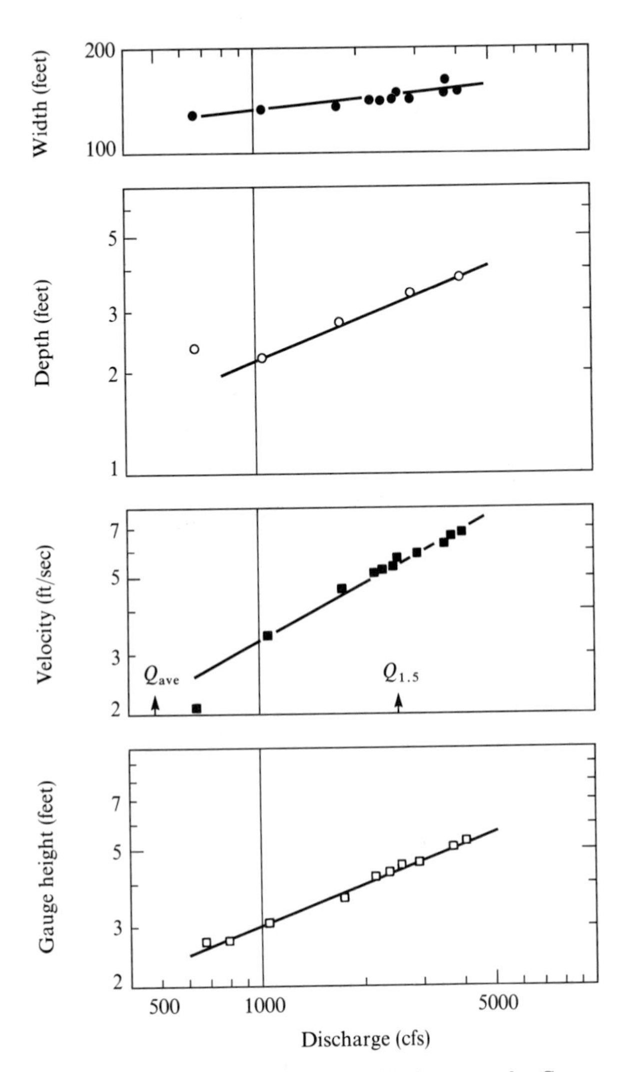

Figure 16-32 At-a-station curves and rating curve for Green River at Warren Bridge, Wyoming.

The data for the upper Green River fit the average relations about as closely as is usually found when such curves are constructed for a particular river basin. Because of the nature of river data, the scatter of points is such that the exact fitting of a single line through the plotted points is a matter of judgment. The meaning would not be greatly enhanced if the curve were fitted by least-square methods.

The sum of the exponents should equal unity, and when fitting a straight line by eye, adjustment should be made until the exponents differ from unity

Table 16-5 Exponents in the at-a-station equations of the hydraulic geometry, upper Green River, Wyoming.

STATION	EXPONENTS			
	b	f	m	SUM
LaBarge Creek near LaBarge Ranger Station	0.16	0.39	0.44	0.99
New Fork below New Fork Lake	0.22	0.34	0.43	0.99
Fall Creek near Pinedale	0.22	0.29	0.42	0.93
Silver Creek near Big Sandy	0.18	0.44	0.38	1.00
North Piney Creek near Mason	0.08	0.42	0.50	1.00
Pole Creek below Little Half Moon	0.16	0.45	0.34	0.95
Boulder Creek below Boulder Lake	0.14	0.29	0.55	0.98
Green River at Warren Bridge	0.09	0.39	0.49	0.97
New Fork near Boulder	0.19	0.45	0.33	0.97
New Fork near Big Piney	0.25	0.34	0.37	0.96
Green River near Fontenelle	0.05	0.34	0.59	0.98
Average	0.16	0.38	0.44	0.98

by 5 percent or less. Also, the product of width, depth, and velocity for any discharge should approximate that discharge within 5 percent, but further refinement is in practice probably not worthwhile, considering the variety of local conditions occurring in river channels.

Another way of visualizing the relation of the at-a-station hydraulic geometry to the downstream hydraulic geometry is to plot at-a-station curves for several stations located on basins of chosen size on the same sheet. This has been done in Figure 16-35, in which the curves are for stations located at five values of drainage area from 1 to 100 square miles in the eastern United States. The lower end of each plotted line is at an abscissa value equal to the average annual discharge. The upper end of each line is bankfull discharge. The number on each line is the drainage area in square miles.

Consider the curve of width as a function of discharge. If through the lower end of each of the plotted lines a new line were drawn, it would have a steeper slope than any on the graph. It would represent the increase of width with mean discharge as the drainage area increases. Its slope, value of b, would be close to 0.5. If a comparable line were drawn through the upper ends of the graphs, it would represent the increase of width with bankfull discharge in a downstream direction, and its slope also would be close to 0.5. The parallelograms describing at-a-station and downstream graphs gave rise to the name hydraulic geometry.

In the downstream plots the fact that the width exponent b is larger than f, the depth exponent, means that downstream along the channel system, width increases faster than depth so that the width/depth ratio increases. This can be seen in Figure 16-36, in which the cross section at locations in the upper Green River basin have been plotted at different scales, such that the width as drawn is the same for each. Note how the depth decreases relative to the width as drainage area and thus river size increases.

The hydraulic geometry exhibits the consistent manner in which natural stream channels are shaped to carry water and sediment load imposed from upstream. This consistency indicates that natural channels, self-formed and self-maintained, seek a shape and size consonant with the size of the basin upstream and its sediment yield. Alteration in this preferred natural shape and size will lead to a tendency toward erosion or deposition as the channel processes operate toward reestablishment of a quasi-equilibrium under the new conditions.

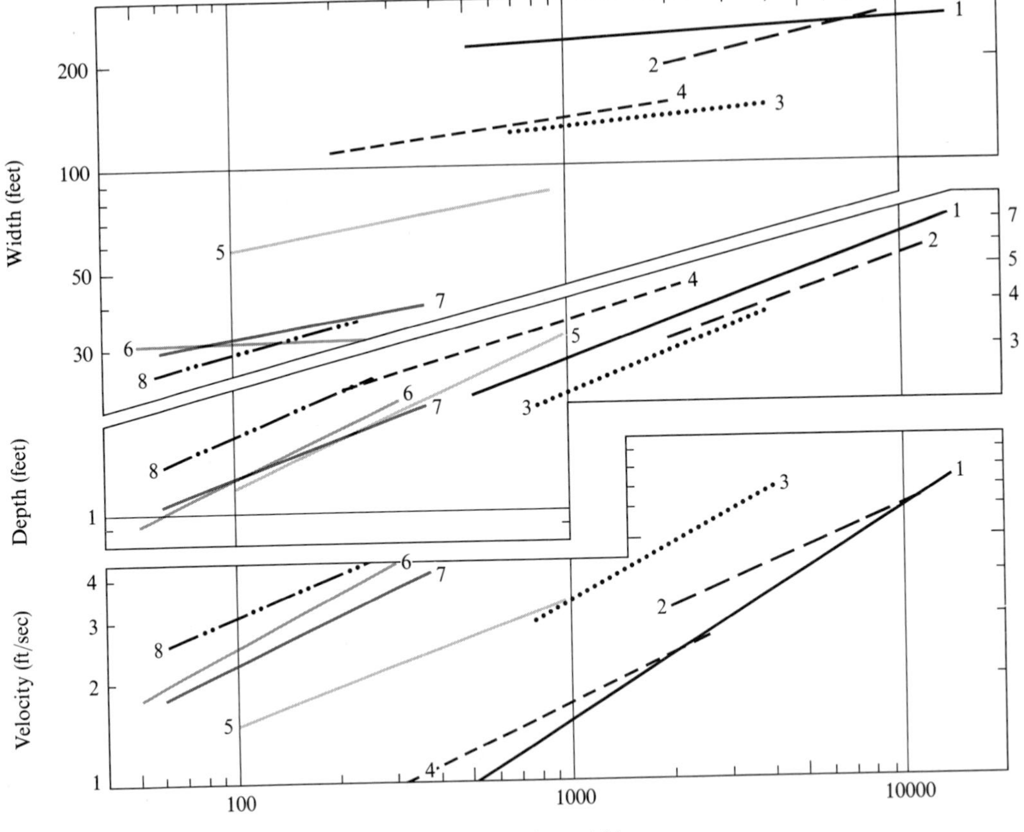

Figure 16-33 At-a-station curves for stations in upper Green River Basin, Wyoming: (1) Green River near Fontenelle; (2) New Fork at Big Piney; (3) Green River at Warren Bridge; (4) Boulder Creek below Boulder Lake; (5) Pole Creek below Little Half Moon Lake; (6) North Piney near Mason; (7) New Fork below New Fork Lake; (8) LaBarge Creek near LaBarge Meadows Ranger Station.

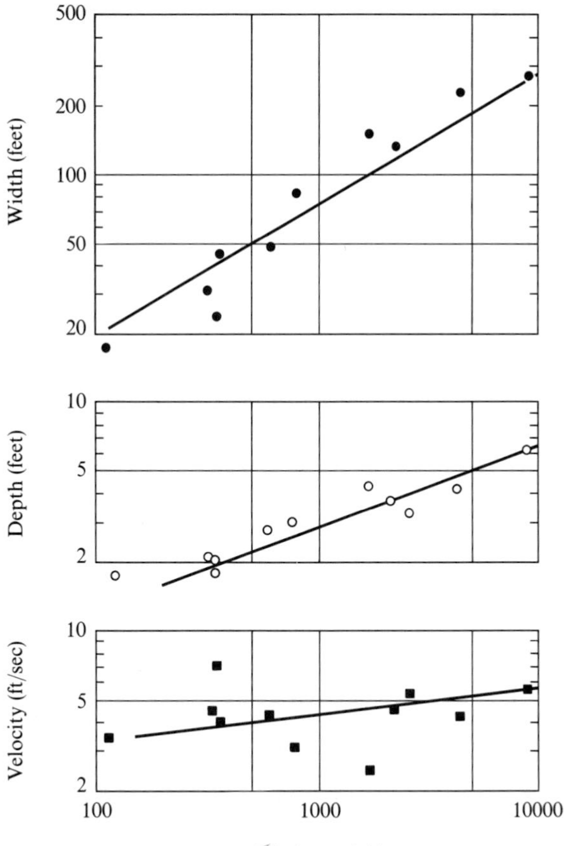

Figure 16-34 Downstream change of width, depth, and velocity with bankfull discharge, upper Green River basin, Wyoming.

Width as a Diagnostic Parameter

Following a suggestion of W. B. Langbein, the consistency in size of channels in a region exemplified by the hydraulic geometry has been exploited for estimating both water yield (average annual discharge) and flood peak expectancies. The most consistent parameter is bankfull width, for it more than any other easily measured characteristic of the channel is correlated with flow parameters such as average annual discharge and discharges having specific recurrence intervals. In many regions gauging station records are sufficiently numerous that the usual methods of flood-frequency analysis and average flow determination are applicable. But in semi-arid regions where most channels are ephemeral, flow measurements are scarce, gauging stations are difficult to operate, and any flow at all is sporadic and unpre-

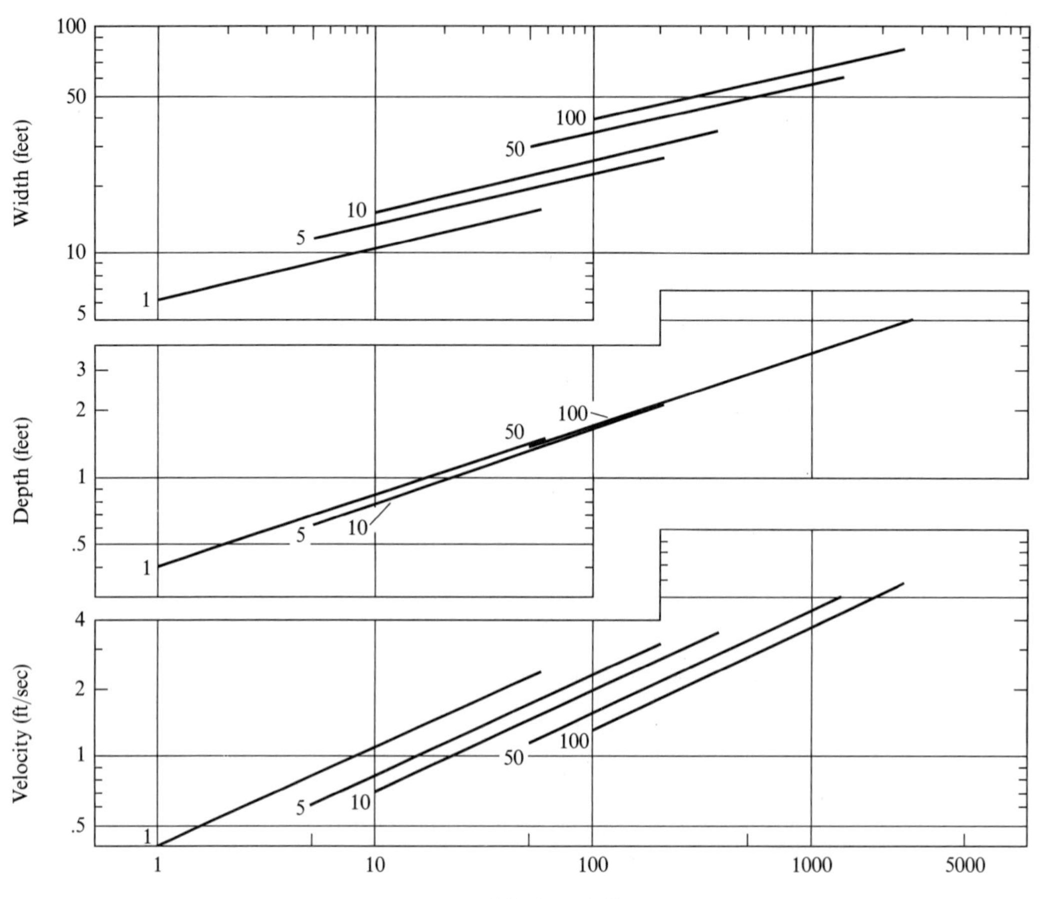

Figure 16-35 At-a-station curves for basins of drainage area 1, 5, 10, 50, and 100 square miles, eastern United States. The lower end of each curve is at average discharge. The upper end at bankfull discharge.

dictable. Under such conditions even mean annual discharge is obtained only by long periods of gauging station record, owing to the large variance from season to season and year to year.

It is logical then to ascertain whether measurable characteristics of the channel are correlated with the desired but unmeasured flow parameter.

The channel width at bankfull stage is correlated statistically with mean annual discharge and with discharges of recurrence intervals of 2, 10, 25, 50, and even 100 years. The correlation coefficients are typically 0.75 to 0.93, better correlation for frequent than for infrequent floods.

These good correlations can be of use to the environmental planner who needs a quick and easy method of estimating flood discharge from field indications. In many planning problems, however, it is less important to

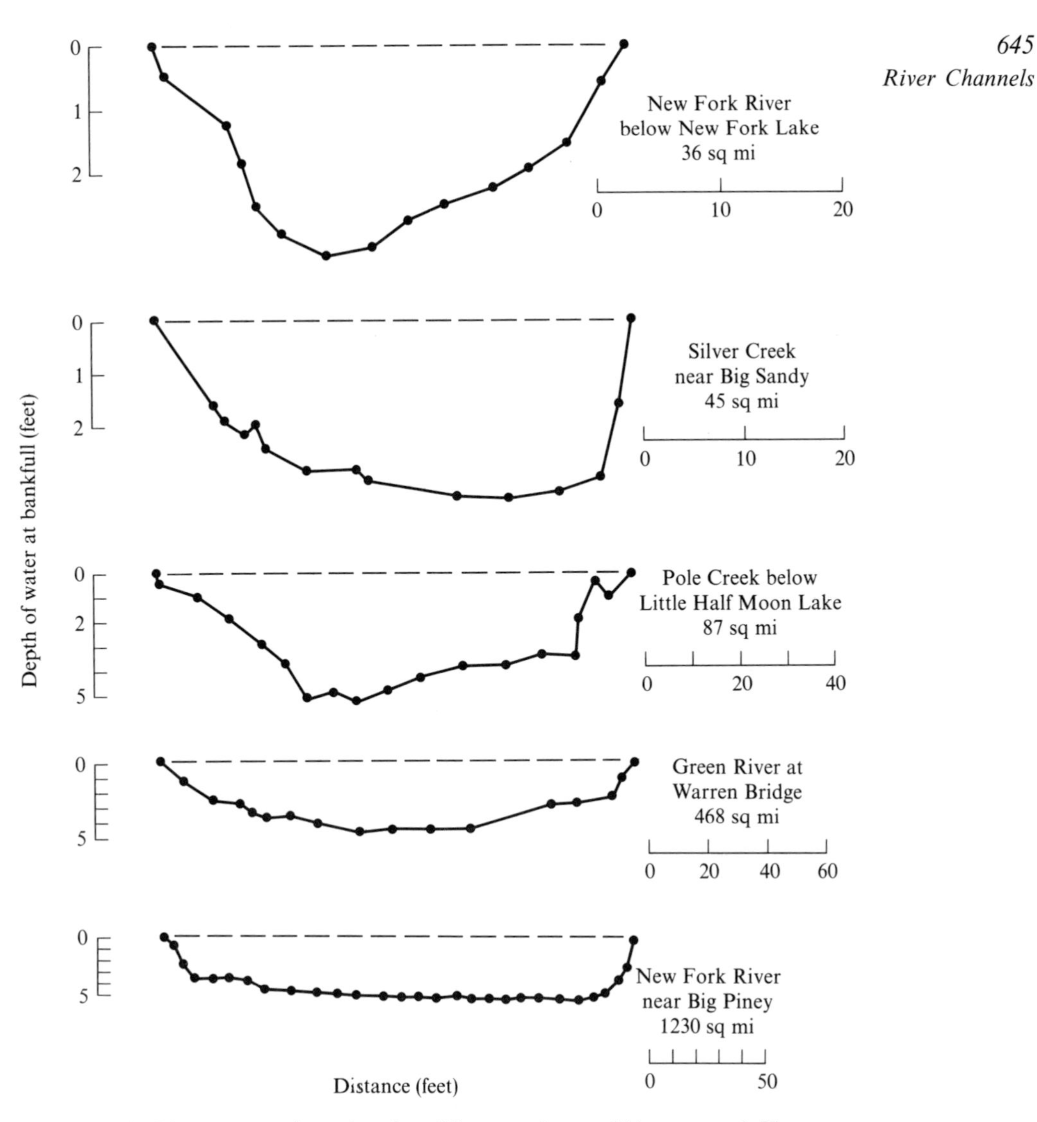

Figure 16-36 River cross sections plotted at different scales so widths are equal. The vertical exaggeration is five times in each case, upper Green River basin.

know the discharge expected at some recurrence interval than to know the height of water surface or the depth associated with that discharge.

Nevertheless, because field methods have been developed for a wide variety of areas and channel types, the general method should be known. There has grown up a disconcerting variety of criteria for specifying what width to measure in the field. The various procedures with a few exceptions use some approximation to the bankfull width as defined in this volume. In addition to or instead of channel width as the field parameter, some use basin characteristics usually including drainage area and elevation.

Because the principal method consists simply of measuring a channel width in the field and, from the nomograms presented, reading off an estimated value of mean discharge or flood discharge, the environmental planner would be well advised to find from the accompanying bibliography, publications that report on these methods for his particular region.

California: Hedman, E. R. (1970) Mean annual runoff as related to channel geometry of selected streams in California: U.S. Geological Survey Water Supply Paper 1999E, 17 pp.

Colorado: Hedman, E. R., Moore, D. O. and Livingston, R. K. (1972) Selected streamflow characteristics as related to channel geometry of perennial streams in Colorado: U.S. Geological Survey Open File Report, 14 pp., Denver, Colorado.

Kansas: Hedman, E. R., Kastner, W. M. and Hjel, H. R. (1974) Selected streamflow characteristics as related to active-channel geometry of streams in Kansas: Kansas Water Res. Board Tech. Rept. 10.

Kansas: Osterkamp, W. R. (1977) Effect of channel material on width-discharge relations for perennial streams, with special emphasis on streams in Kansas–a progress report: U.S. Geological Survey Open File Report 77–38, Lawrence, Kansas.

Missouri: Hedman, E. R. and Kastner. W. M. (1974) Progress report on streamflow characteristics are related to channel geometry of streams in the Missouri River basin: U.S. Geological Survey Open File Report.

New Mexico: Kunkler, J. L. and Scott, A. G. (1976) Flood discharges of streams in New Mexico as related to channel geometry: U.S. Geological Survey Open File Report 76–414, 29 pp., Santa Fe, New Mexico.

Wyoming: Lowham, H. W. (1976) Techniques for estimating flow characteristics of Wyoming streams: U.S. Geological Survey Water Resources Investigations 76–112, Cheyenne, Wyoming, 83 pp.

Many areas: Riggs, H. C. (1974) Flash flood potential from channel measurements: International Assoc. Sci. Hydrology, Symposium of Paris, IAHS–AISH Publ. No. 112, pp. 52–56.

Riggs, H. C. (1976) A simplified slope-area method for estimating flood discharges in natural channels: Jour. Research, U.S. Geological Survey, vol. 4, no. 3, pp. 285–291.

The Dimensionless Rating Curve

The rating curve, a plot of gauge height versus discharge, expresses both the shape and hydraulic character of a reach of channel. Obviously, a rating curve for a small stream will differ from that for a large one, but the difference in size may be eliminated if the rating curve is expressed in terms of ratios and is thus dimensionless. Partly because of the conservative nature of velocity (large rivers have only slightly larger velocity than small ones), the dimensionless rating curve represents well a surprising variety of channels.

The curve is a plot of a ratio representing depth versus a ratio representing discharge, so it is similar in character to the usual rating curve of a channel cross section. In this graph the ordinate is the ratio of a depth, d, at a given discharge, Q, to the depth at bankfull, or is d/d_{bkf}. The abscissa is the ratio of the discharge, Q, to bankfull discharge, or Q/Q_{bkf}. If the discharge under consideration, d, is identical to d_{bkf}, then the ratio on the ordinate scale is 1.0. At this condition the discharge, Q, is equal to bankfull, so the abscissa value is also 1.0. The curve, then, goes through the coordinates $x = 1$, $y = 1$ (Figure 16-37).

For the construction of the dimensionless rating curve, the bankfull discharge is obtained by the field survey procedure or approximated by reading the discharge for 1.5 years recurrence interval from a station frequency curve. For selected current meter readings listed on Form 9–207, discharge, width, and cross-sectional area are tabulated, from which depth is calculated by dividing area by width. The ratios are then computed and plotted, and a mean line is drawn through the plotted points.

Figure 16-37 presents lines representing stations in the eastern United States (Leopold, Wolman and Miller 1964:219) and stations in Idaho (Emmett 1975:A37).

It has been shown that positions on such a curve have a relation to recurrence interval. On the graph the discharge having a recurrence interval in the eastern United States of 50 years is 4.3 times larger than the bankfull discharge at the same location; the depth of the 50-year recurrence interval flood is 1.8 times the bankfull depth.

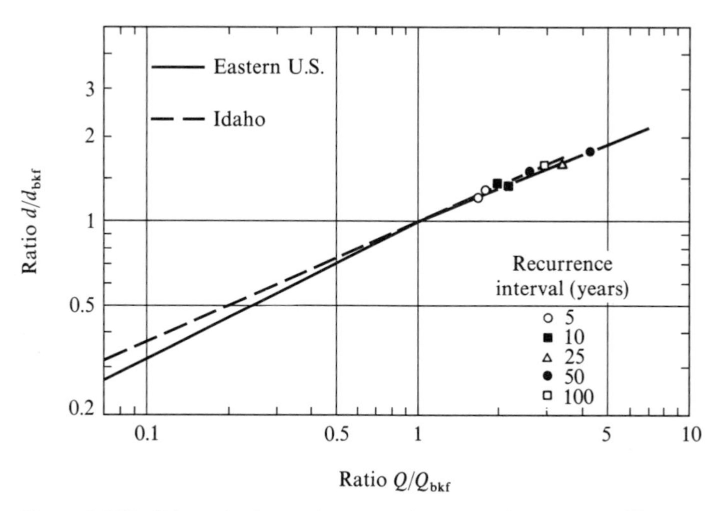

Figure 16-37 Dimensionless rating curve for two regions, eastern United States and Idaho. (Idaho data from Emmett 1975.)

The ratio of a flood discharge of any recurrence interval to bankfull discharge is larger in eastern basins than in the mountainous basins of Idaho, probably because in the latter areas spring snowmelt is responsible for the highest discharges. Excessive rainstorms in the East give higher rates.

Whereas for a flood discharge of a specific recurrence interval, the value of Q/Q_{bkf} varies from one region to another, the comparable value of d/d_{bkf} is amazingly conservative. Dimensionless rating curves have been constructed by different authors for four regions. The values of d/d_{bkf} for four recurrence intervals are shown in Table 16-6.

Table 16-6 Values of the ratio, d/d_{bkf} (depth ÷ bankfull depth), for various values of recurrence interval, from four regions in the United States. (Sources: (1) Leopold, Wolman and Miller 1964:219; (2) Emmett 1975:A37; (3) S. C. Parsons 1976; (4) J. E. Frederick and J. Pitlick 1975.)

RECURRENCE INTERVAL (YR)	EASTERN U.S. (1)	IDAHO (2)	WESTERN CASCADES, WA (3)	PUGET LOWLAND, WA (4)	AVERAGE
1.5	1.0	1.0	1.0	1.0	1.0
5.0	1.1	1.2	1.3	1.3	1.2
10.0	1.2	1.3	1.5	1.4	1.4
50.0	1.8	1.4	2.0	1.7	1.7
100.0		1.5	2.2	1.8	1.8

The consistency of these figures provides confidence that the relation may be used in one of the most common practical problems facing a planner or landscape architect. There is a ubiquitous need to determine what area on a valley floor will be flooded by a discharge of a given recurrence interval, or to what elevation the water surface of that flood would reach. The data show that the flood of 50-year recurrence interval may be expected to overflow the floodplain to a depth equal to 1.7 times the depth of water at bankfull stage. Thus, once the level of the floodplain as defined in this volume is established, the elevation above that level can be easily determined.

The gauging stations used in the four quoted analyses were not chosen to have any particular characteristics. Apparently the width of floodplain is not a crucial factor, though one can imagine that channels incised in a narrow gorge within terrace scarps will have different characteristics than those where the floodplain is wide, flat, and well defined. Apparently these differences lead to compensating differences in velocity of flow over the floodplain that result in a conservative value of d/d_{bkf} for a flood of given recurrence interval.

Other workers have added useful modifications to the procedure for constructing dimensionless curves. If the channel lies within a wide, flat floodplain, when overflow occurs, the water surface width may increase greatly with only a minor increase in the cross-sectional area of the flowing water. Then the computation of depth, as area divided by width, would show a lower mean depth than before the overflow. In such a case the depth, d, to be used in d/d_{bkf}, might better be computed as the addition of the increment of gauge height above bankfull stage to the depth computed for bankfull. The latter procedure was suggested by the Washington students in their analyses of the Western Cascade and Puget Lowland regions quoted in the preceding tabulation. Interestingly, their values of d/d_{bkf} for various recurrence intervals differ but little from those computed in the usual way.

Emmett (1975:A37) pointed out that the slope of the line in the dimensionless rating curve should be equal to the value of f (exponent in $d \propto Q^f$) in the hydraulic geometry for the stations used in its construction. By the same reasoning, if the rating curve ordinate were the ratio of cross-sectional areas of flowing water, A/A_{bkf}, then the slope of the line would be $b + f$ of the hydraulic geometry because area is merely depth times width.

Flow area ratio gives somewhat less scatter of points than depth ratio, but depth "offers a better visual indication of the amount of water . . ." (Emmett 1975).

The dimensionless rating curve can be used to estimate the recurrence interval of a flow that is overbank by noting the ratio of depth above bankfull to bankfull depth; or as mentioned, the overbank area floodable by a discharge of chosen recurrence interval can be estimated. The latter relation offers a relatively simple procedure for delineating floodable lands on the valley floor, as diagrammed in Figure 16-38.

The utility of the dimensionless rating curve depends on the identification of the floodplain or the designation of bankfull level. For this identification the high correlation of bankfull condition to the discharge of 1.5-year recurrence interval provides a simple solution where a flood-frequency graph can be constructed.

Mapping Valley Area Subject to Flooding

A procedure utilizing the dimensionless rating curve for mapping the extent of overflow for a flood discharge of chosen frequency is shown in Figure 16-38. It includes the topography of a reach of Watts Branch near Rockville, Maryland, just downstream of the gauging station. From the individual shots where elevation was determined while surveying by plane table, channel-bed elevations were plotted on the profile in the upper part of the figure. In a different symbol, elevations read at the top of the bank (elevation of floodplain) are also plotted. In this profile the abscissa is distance along the channel. Mean lines are drawn through these sets of points, the upper one representing water surface at bankfull and the lower one the mean profile of the channel bed. As expected, the two lines are parallel. The vertical distance between them is therefore depth bankfull, and in this case is 4.0 feet.

The dimensionless rating curve indicates that the ratio of depth to bankfull depth for a flood having a 10-year recurrence interval is 1.4, or the floodwater spreads out over the valley floor to a depth overbank equal to four-

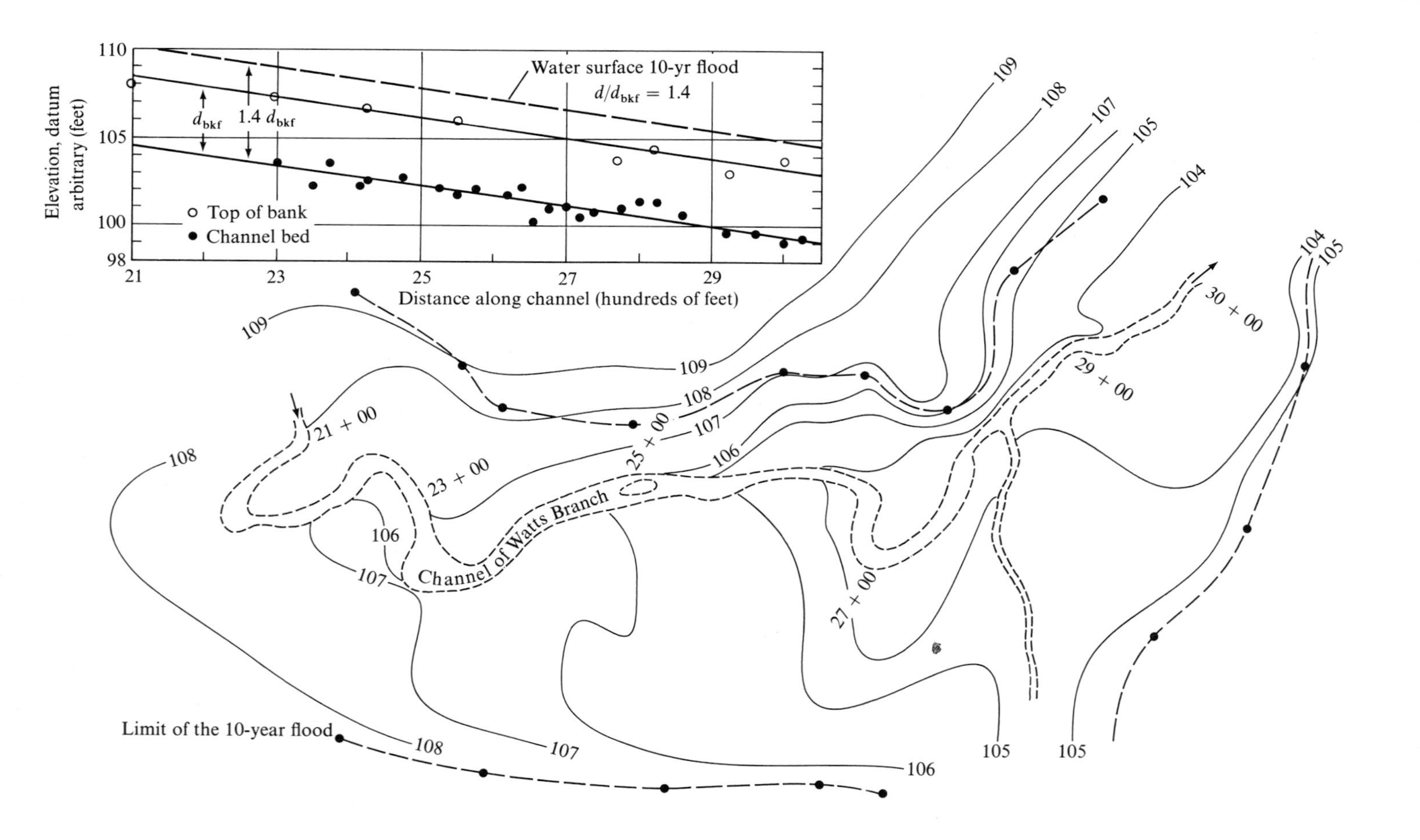

Figure 16-38 Plot of the limit of area probably flooded by a discharge of 10-year recurrence interval on a topographic map of a reach of Watts Branch near Rockville, Maryland. Elevations of water surface are read from the profile using the information from the dimensionless rating curve that the depth for a 10-year flood is 1.4 bankfull depth.

tenths the bankfull depth. A third line, dashed in the profile, is drawn parallel to the others and at an elevation such that it is higher than bankfull by four-tenths, or 1.6 feet higher than bankfull. On this profile representing the probable water surface of the 10-year flood, elevations can be read and are plotted as crosses on the topographic map. Then dashed lines connecting these crosses represent the probable extent of flooding by a flood having a recurrence interval of 10 years.

The same method may be applied to a valley for which a published topographic map is available. An approximation of bankfull depth may be read from Figure 16-16 when the drainage area has been determined. Then the extent of flooding for a flood of chosen recurrence interval can be plotted on the topographic map, using elevations determined from a profile as here demonstrated.

Bibliography

Barnes, H. H. (1967) Roughness characteristics of natural channels; *U.S. Geological Survey Water Supply Paper 1849,* 213 pp.

Chow, Ven te, ed. (1964) Handbook of applied hydrology; McGraw-Hill, New York.

Emmett, W. W. (1975) The channels and waters of the Upper Salmon River area, Idaho; *U.S. Geological Survey Professional Paper 870-A.*

Fahnestock, R. K. (1963) Morphology and hydrology of a glacial stream; *U.S. Geological Survey Professional Paper 422-A.*

Frederick, J. E. and Pitlick, J. (1975) A dimensionless rating curve for the rivers of the Puget Lowland; Unpublished report. Department of Geological Sciences, University of Washington.

Hack, J. T. and Goodlett, J. C. (1960) Geomorphology and forest ecology of a mountain region in the Central Appalachians; *U.S. Geological Survey Professional Paper 346,* 66 pp.

Langbein, W. B. and Leopold, L. B. (1966) River meanders: theory of minimum variance; *U.S. Geological Survey Professional Paper 422-H.*

Leopold, L. B. (1973) River channel change with time: an example; *Bulletin Geological Society of America,* vol. 84, pp. 1845–1860.

Leopold, L. B. (1974) *Water: a Primer;* W. H. Freeman and Company, San Francisco, 172 pp.

Leopold, L. B. and Langbein, W. B. (1966) River meanders; *Scientific American,* June, pp. 60–69.

Leopold, L. B. and Maddock, T. (1953) The hydraulic geometry of stream channels and some physiographic implications; *U.S. Geological Survey Professional Paper 252.*

Leopold, L. B. and Wolman, M. G. (1957) River channel patterns: braided, meandering and straight; *U.S. Geological Survey Professional Paper 282-B.*

Leopold, L. B., Wolman, M. G. and Miller, J. P. (1964) *Fluvial processes in geomorphology;* W. H. Freeman and Company, San Francisco, 522 pp.

Limerinos, J. T. (1970) Determination of the Manning coefficient from measured bed roughness in natural channels; *U.S. Geological Survey Water Supply Paper 1898-B,* 47 pp.

Miller, R. A., Troxell, J. and Leopold, L. B. (1971) Hydrology of two small river basins in Pennsylvania before urbanization; *U.S. Geological Survey Professional Paper 701-A.*

Parsons, S. C. (1976) A dimensionless rating curve for rivers of the Western Cascade Mountains, Washington; Unpublished report, Department of Geological Sciences, University of Washington.

Rantz, S. E. (1971) Suggested criteria for hydrologic design of storm-drainage facilities in the San Francisco Bay Region, California; U.S. Geological Survey Open File Report, Menlo Park, CA, 69 pp.

Wolman, M. G. (1955) The natural channel of Brandywine Creek, Pennsylvania; *U.S. Geological Survey Professional Paper 271.*

Wolman, M. G. and Miller, J. P. (1960) Magnitude and frequency of forces in geomorphic processes; *Journal of Geology,* vol. 68, no. 1, pp. 54–74.

Typical Problems

16-1 Cross Section and Rating Curve, New Fork River at Boulder, Wyoming; Hydrograph of Daily Flows

When a planner or landscape architect deals with a stream channel, one of the most important questions that should be asked initially concerns how high the water is under various circumstances of drought or flood. The height of water is always related to some datum, and measured relative to that datum by a gauge plate or a staff gauge. To translate water surface elevation to discharge, a rating curve is used. The rating curve and its construction then become one of the first and simplest tools of hydrologic analysis.

To construct a rating curve from published data in Water Supply Papers or Water Resources Data, use simultaneous values of discharge and gauge height. In the present problem the channel cross-section data from a level survey are furnished in Table 16-7.

Problem

1. Survey notes provide the data to plot a cross section across the channel near the gauging station. Develop a cross section and a rating curve.

2. Plot a hydrograph of daily flow for four and one-half months.

Procedure

1. Survey notes are provided. The cross section surveyed is about 100 feet upstream from the bridge and 200 feet upstream of the gauge house. The cross section is very close to the cable location. A photograph of the river at this location is shown in Figure 16-6.

Plot two cross sections at different scales, one to show just the channel at an enlarged scale, and one to show the relation of the channel to adjacent cliff and valley flat.

On each of the two cross sections, construct two ordinate scales. The principal scale shown on the left side of the page is elevation in feet, coinciding with the printed lines of the cross-section paper. The second scale placed at the right side of the sheet is a scale of gauge height in feet. It will not have the 0, 5, 10 feet levels coincident with the dark lines of the paper, but because both elevation and gauge height are in feet, the printed grid can be used for both.

The relation, of course, can be established by one simultaneous reading of gauge height and elevation, and this is given in the survey notes.

Table 16-7 Survey notes for cross section near cable, about 100 ft upstream of bridge, New Fork River near Boulder, Wyoming, July 25, 1974.

STATION (FT)	ELEVATION (FT)	STATION (FT)	ELEVATION (FT)
0 + 1.7	101.5	0 + 80	95.2
0 + 4	99.7	0 + 85	95.1
0 + 7	97.9	0 + 90	95.3
0 + 8.5	97.7	0 + 95	95.4
0 + 10	97.2	1 + 00	95.7
0 + 15	95.7	1 + 05	95.8
0 + 23	95.0	1 + 10	96.3
0 + 29	95.0	1 + 13.5	96.9
0 + 35	95.1	1 + 15	97.6
0 + 38	95.2	1 + 17	98.5
0 + 40	95.3	1 + 21	98.9
0 + 45	95.3	1 + 25	98.0
0 + 50	95.3	1 + 28	98.2
0 + 55	95.4	1 + 29.5	98.6
0 + 60	95.1	1 + 33	98.6
0 + 65	95.1	1 + 36	98.9
0 + 70	95.1	1 + 40*	98.7
0 + 75	95.0	1 + 44†	98.7

*The valley flat that begins at station 1 + 40 continues on at same level at least to station 2 + 50.

†Elevation of 100.00 feet begins at U.S.G.S. Bench Mark #5 and corresponds to 6.42 feet on the gauge plate. The bench mark is a bronze plate set in concrete near the gauging station. To left of station 0 + 1.7 is cliff about 30 feet in height.

The rating curve can be produced from simultaneous values of gauge height and discharge given in Water Supply Paper 1683, page 200. Plot the gauge height as ordinate, and the discharge as abscissa on log-log paper.

2. Choose any year during which daily discharges are published for the New Fork River at Boulder. This stream has its high flows as a result of spring snowmelt, but much water is diverted for irrigation during the same period. This delays the peak at the Boulder station. The spring runoff has its peak about June 1. This delay is obvious in nearly any year.

16-2 Channel Geometry Survey

The object of the survey is to establish the slope of the water surface at bankfull stage and to relate this grade line to the datum of the gauging station. Therefore, profiles will be run on the stream-bed center line, the water surface, top of banks, and the surface of any terrace or abandoned floodplain that can be identified. All these profiles should be essentially parallel if the surveyed reach is sufficiently long. Though terraces may have a different down-valley gradient than the present stream, in any short survey this difference in gradient is usually not detectable. The length of the survey should be no less than 30 times the river width.

Procedure

Before going into the field for actual surveying, it is highly desirable to organize flood-frequency data for gauging stations in the general area to be surveyed. Select a group of stations in the locality, choosing if possible stations that command a variety of drainage area sizes. Seven to ten stations are usually ample if they have a range of basin size from a few square miles to 500 square miles. From flood-frequency curves of the individual stations or from regional flood-frequency curves (U.S. Geological Survey Water Supply Papers, e.g., No. 1683), tabulate the discharge having a recurrence interval in the annual flood series of 1.5 years. Plot this discharge value as ordinate and the drainage area in square miles for the stations as abscissa. This graph available in the field during the survey is useful.

Equipment needed in the field includes copy of U.S. Geological Survey publication for that state entitled "Surface Water Records"; graph of 1.5-year recurrence interval discharge versus drainage area; surveyor's level and tripod; Philadelphia leveling rod; tapes (cloth or metallic), 100-foot and 150-foot lengths; some 2-foot lengths of iron bar for defining the ends of each cross section; small maul or hatchet; engineer's mining transit notebook; leather notebook carrying case; pencils; pocket slide rule; scale divided in tenths of inches; maps; camera and film.

Channel surveys can be made where there is no gauging station, but it is usual to survey at gauging stations until a body of data is acquired.

Having located the gauging station, often a task made easier by having in hand the U.S. Geological Survey publication, find the outside gauge plate. If a gauging station key is available, open the gauge house and look for a rating curve for the station. If you find it, copy enough figures of gauge height and discharge to provide a plotted rating curve when needed. Gauge height at time of survey is recorded.

Set up the surveyor's level at a place from which you can see the maximum distance upstream and down. There is usually a brass plate or bench mark on the step or in a concrete monument nearby. Take a shot on this bench mark to obtain a height of instrument. Take a shot on the top of the gauge plate in order to relate the gauge datum to the rest of the survey. Usually if the actual elevation of the bench mark (brass plate) is unknown, it is given an assumed elevation of 100. Establish the HI (height of instrument) to hundredths of feet, reading the top of the gauge plate to hundredths but the channel readings to tenths of feet. When the instrument must be moved, read the elevation of the turning point to hundredths of feet. Thus, the survey is carried to hundredths, though profiles and cross sections are only read to tenths of feet.

Having established the relation of the HI to the gauge datum, the longitudinal profiles should be run. The instrument man and the rodman should walk together along the length of the stream to be surveyed, and agree on the nature of the evidence for bankfull and terrace heights. At each shot the rodman will signal to the instrument what class of reading will be ascribed to that particular shot.

Before the survey begins, the drainage area of the gauge should be obtained from the published record. Entering the graph of drainage area and discharge for a 1.5-year recurrence interval, the latter value is read. The value is entered in the rating curve to obtain the gauge height corresponding to that discharge. This value of gauge height is then inspected at the staff gauge to obtain an approximate level near which evidence for bankfull discharge may be expected. The evidence sought is of the following types:

1. Elevation of topographic break from vertical bank to flat floodplain.

2. Topographic break from steep bank to more gentle slope.

3. Change in vegetation from bare to grass, from moss to grass, from grass to sage, from trees to grass, or from no trees to trees.

4. Change of texture of deposited material from clay to sand, or sand to pebbles, or boulders to pebbles.

5. Highest elevation below which no fine debris of needles, leaves, pine cones, or seeds occur; in some instances it is the upper limit of such fine debris.

6. Change in texture (size) of fine material lodged between cobbles or rocks. This change is often from fine sand to fine gravel.

High-water marks consisting of debris hanging in trees or on brush are deceiving. Such debris usually is hanging at a higher elevation than the mean water surface of the flow that deposited it.

Profile distances are determined by pacing or by tape. Each shot is categorized as thalweg (deepest part of the channel), water surface, high-water or bankfull indication, or terrace. When the survey is complete, the profile data are plotted in the notebook, with different symbols indicating the various types of field evidence. The several profiles should have similar slopes. If they do not, the survey should continue a longer distance.

The elevation at which the smoothed profile of the thalweg passes the gauge plate scale is an estimate of the gauge height of zero flow. Where the profile of the bankfull indicators passes the gauge plate scale is the gauge height of bankfull determined in the field. This gauge height is then entered into the rating curve and the bankfull discharge determined. For the drainage area this discharge is plotted and compared with the discharge of 1.5-year recurrence interval estimated from drainage area alone.

Where several elevations of possible bankfull stage were determined in the field, one may be chosen as most likely from comparison with the expected elevation of the flow of 1.5-year recurrence interval.

One or more cross sections are chosen that contain the best evidence of the topographic and flow levels seen in this profile survey. A cross section is surveyed with enough shots to represent it fairly when the shot elevations are connected by straight lines. Usually 20 shots are needed in the channel itself.

You have measured the location of each cross section along the profile, and therefore from the profile you can compute the elevation of the bankfull stage for each cross section and plot it on the cross section. This should also be done in the field. Along the cross section, notes should be kept on changes in vegetation and in surface texture.

A pebble count is made on a riffle bar. The b axis (intermediate axis) is measured for 100 rocks chosen at random. The size classes used differ by the square root of two, beginning in mm 4, 5.6, 8, 11, 16, 22, etc.

Experienced persons may reduce the work to a profile of water surface, a cross section, and a pebble count. But even when the process is abbreviated, the steps outlined are in principle considered.

Plotting the profile and the cross section in the notebook while still at the site is insurance that the main features have been noted and that the data are consistent.

A photograph, one facing upstream and one down, is useful in later recollection of the main features.

16-3 Making a Discharge Measurement without a Current Meter by the Use of Floats

Because so many planning problems involve flow in channels where no gauging station exists, it is necessary to develop an approximate rating curve by cross-sectional survey and one or more direct measurements of discharge. But the environmental planner may not have easy access to a Price current meter, the standard device for measuring discharge in the United States. Measurement by float gives good results when done with care. Float measurements of velocity are often the only way of obtaining a velocity during a flood, even if a meter is available.

Problem

Measure in the field the discharge in a channel using a cross-sectional survey and float velocity.

Procedure

Equipment needed includes a watch with a second hand, surveyor's rod, sharp knife, notebook, pencil, 100-foot or 30-meter tape, staff gauge consisting of marked lumber, meter stick, enamel plate or other scale for installation, light maul, hammer, and nails.

Choose a straight reach, preferably with an obvious or well-developed pool-and-riffle sequence, and channel banks relatively uniform. Install the staff gauge in a vertical position near the foot of the pool or just upstream of the riffle. The easiest way to put in a staff gauge is to drive vertically a steel fencepost and tie with wire the gauge plate to the steel. The lower end of the gauge is driven an inch or more into the stream bed so that even at the lowest flow the water level can be read on the scale of the plate. Place the gauge plate near the bank but facing the shore from which it will be read. A vertical wooden stake carrying the plate must be braced by diagonals to the bank, but a steel post driven well into the stream bed will stand without braces (Figure 16-3).

A cross section will be measured at or very close to the staff gauge. Stretch a tape across the stream so that it is taut and more or less horizontal. The zero end of the tape is on the left bank (left, facing downstream). The tape is used to measure horizontal position but not elevation.

Very near the tape and above it a cord is stretched. Carpenter's string is best because it is strong and can be stretched taut. The cord is adjusted until it is horizontal at any arbitrary distance above the ground. A string level or hand level is useful to assure that the string is horizontal.

Using the taut string as a datum, measure downward from the string to the ground or stream bed at such horizontal intervals that the cross section is well represented by straight lines drawn from point to point in the plot. The vertical distance of the ground surface below the string is measured with a surveyor's rod held vertically at the chosen location, a reading taken where the string passes the scale on the rod. About 50 points are needed. In a channel 20 feet wide, horizontal spacing of points might be 1 foot. Continue the cross section for one channel width on each side of the stream, so if the channel were 10 feet wide, the cross section should be at least 30 feet long. Note the position of left edge and right edge of water for the present flow. Plot the cross section in the field notebook, and compute the cross-sectional area of the water.

Stretch the tape along the channel bank for a distance equal to at least 10 times the channel width in the straightest reach near the cross section. Cut floats from wooden matches, tree branches, or sticks. There should be about eight pieces of wood large enough to be seen as they float down the channel. Near the upper end of the taped distance, throw in a float a few feet above the beginning of the measurement reach. As the float passes the zero point, note the time on the second hand of the watch. Record in seconds the time the float required to go the measured distance. Record the position of the float relative to the channel centerline, the time of travel, and the distance.

Repeat, using another float, and in total make five to eight float measurements at mid-channel, one-third of the width from the left bank, one-third of the width from the right bank, or other positions. Discard data for any float that becomes stalled in a swirl or hangs up on a rock.

Average the float times, and get an average speed in feet per second or meters per second. Multiply the float velocity by 0.8 to convert surface velocity to mean flow velocity. Multiply mean velocity by cross-sectional area to obtain discharge.

Be sure to note the water surface reading (gauge height) on the gauge plate. This gauge reading and the corresponding discharge give one point on the rating curve.

16-4 Flood-Frequency Analysis for Determination of Bankfull Discharge

Problem

From published records, determine the discharge expected to be equaled or exceeded in 1.5 years, 10 years, and 50 years for the New Fork River at Boulder, Wyoming. The determination will be made in two ways. The water level of the 1.5-year flood will be placed on the channel cross section plotted in Typical Problem 16-1.

Procedure

1. The procedure is explained in Chapter 10. Use flood-frequency plotting paper. If it is not available, you can use arithmetic or logarithmic probability paper. Records for plotting the station flood-frequency curve are published in U.S. Geological Survey Water Supply Paper 1683.

Draw a smooth curve through the plotted points, and read the discharge values for recurrence intervals of 1.5, 10, and 50 years from the smooth curve.

2. For any location, even where there is no gauging station, similar computations are possible using regional flood-frequency curves, except in the semi-arid southwestern United States. The procedure is discussed in Chapter 10. The needed hydrologic region may be determined from the map in the back envelope in Water Supply Paper 1683. The appropriate curves are Figures 2 and 3 in Water Supply Paper 1683.

The mean annual flood is determined by the drainage area. The ratio to the mean annual flood is read off the appropriate curve for the recurrence intervals of 1.5, 10, 25, and 50 years. Multiplying the ratio by the mean annual flood gives the discharge equaled or exceeded at the chosen recurrence interval. Plot on the same graph as that of Step 1 the discharge for each recurrence interval.

Results from the regional curve will not be identical to those from the station data, but the order of magnitude will be the same.

Note that the regional curve procedure can be applied at any spot, but the curves do not extend down to drainage areas less than 30 square miles. This is an example of how engineering hydrology does not specifically treat the problems often encountered by the planner or landscape architect.

3. Draw a horizontal line on the enlarged plot of the channel cross section at the New Fork gauge (Typical Problem 16-1) or for a gauge in your own region, representing the water level of the 1.5-year recurrence interval discharge, an approximation of bankfull. This is done by choosing between the two values of 1.5-year discharge determined from station and regional analyses, a value of discharge to be used. Enter the chosen value in the rating curve, read the gauge height, and plot on the cross section a horizontal line representing the water surface.

16-5 Dimensionless Rating Curve

Empirical studies have shown that a rating curve can be developed in dimensionless terms that represent an average stream channel in a region. Also, regional differences are small so that one general curve will give a good approximation of the relation of discharge to gauge height for an ungauged area. With this tool the recurrence interval of various overbank flows can be estimated.

Problem

A cross section was surveyed at the gauge plate on North Fork Strawberry Creek between Haviland and Giannini Halls on the Berkeley campus. Records of a single storm are available. Estimate the recurrence interval of the peak flow.

Procedure

1. Plot the cross section—data are provided in Table 16-8. Use a vertical scale of 1″ = 1′ elevation, and 1″ = 4′ horizontal distance.

2. Make an additional scale of gauge height in the same cross-section graph. The relation of gauge height to elevation is given.

Table 16-8 Surveyed cross section of North Fork Strawberry Creek 50 ft southwest of wooden bridge (near Giannini and Haviland Halls).
Note: Gauge height 0.50 ft corresponds to elevation 7.25 ft. Location on campus of University of California, Berkeley.

DISTANCE (FT)	ELEVATION (FT) (DATUM ASSUMED)	DISTANCE (FT)	ELEVATION (FT) (DATUM ASSUMED)
0 + 1.0	9.65	0 + 13.0	7.30
2.0	9.50	13.5	7.15
3.0	9.65	14.0	7.10
4.0	9.60	14.5	7.00
5.0	9.70	15.0	7.02
6.0	9.35	15.5	7.05
7.0	9.25	16.0	7.00
8.0	9.10	16.5	6.92
9.0	9.02	16.8	7.00
10.0	8.88	17.4	8.65
11.0	8.70	18.0	8.65
11.5	8.65	18.5	8.90
12.0	8.35	19.0	9.12
12.4	8.20	20.0	9.50
12.5	7.45	21.0	9.78
12.8	7.30		

3. Prepare a tabulation in which gauge-height values represent the first column. By counting squares on the cross-section graph, compute the cross-sectional area for each value of gauge height. Plot a graph of cross-sectional area in square feet as a function of gauge height.

4. Flood-frequency curves can be constructed from the Rantz (1971:14) equations applicable to the San Francisco Bay region. The discharge values can be adjusted for the present urbanization by multiplying the values from the equations by factors read from Rantz (1971:16, Figure 3), and a new flood-frequency curve representing present urbanization drawn.

From the latter curve, estimate the bankfull discharge approximated by the 1.5-year recurrence interval. This value may be designated Q_{bkf}.

5. Now a rating curve is needed to determine at what gauge height this discharge would occur. Several observations of velocity are available. For each the cross-sectional area can be determined, and the product of velocity times area is the discharge.

The float and estimated velocities are surface velocities and must be multiplied by 0.8 to approximate mean velocity.

By current meter, February 25, 1976,

$gh = 0.38$ $\quad u = 0.14$ ft/sec

$A = 0.54$ ft^2 $\quad Q = 0.08$ cfs

On January 2, 1977.

gh 0.60, surface $u = 0.85$ ft/sec by floats
gh 0.88, surface $u = 2.0$ ft/sec by floats
gh 2.40, surface $u = 3.5$ ft/sec, estimated
gh 2.55, surface $u = 5$ ft/sec, estimated
gh 2.65, surface $u = 6$ ft/sec, estimated

For each, compute a mean velocity, and plot velocity as a function of gauge height. Draw a smooth line through the points. For several values of gauge height, read a value of mean velocity and multiply by the corresponding cross-sectional area to get a discharge. Plot discharge against gauge height to obtain a rating curve.

6. The peak stage on January 2, 1977, was 2.75 feet. From the rating curve find this discharge. Let this be called Q. Divide Q by Q_{bkf} to obtain a value of Q/Q_{bkf}. Enter this value on the abscissa scale in the dimensionless rating curve, Figure 16-37, and read a recurrence interval.

7. The above determination of recurrence interval will be of the same magnitude as that obtained by entering Q in the flood-frequency graph obtained in the previous problem.

16-6 Average Channel Dimensions

One of the remarkable characteristics of stream channels is the consistency of dimensions among examples within a region if the drainage area is the same. Because urbanization and other human-caused influences tend to change channel size, it is often useful to know the usual or mean dimensions of a natural channel in a region.

Problem

Data have been collected in the field on channel size for a few regions. Plot graphs of width, depth, and cross-sectional area as functions of drainage area for streams in a region.

Procedure

Data for some streams in western Wyoming are shown in Table 16-2. Similar data are published for the following regions:

Pennsylvania: Miller, Troxell and Leopold (1971) U.S. Geological Survey Professional Paper 701-A, Table 1, p. A8.

Idaho: Emmett (1975) U.S. Geological Survey Professional Paper 870-A, Table 12, p. A32, and Table 9, p. A-26 for drainage area.

For a given region on log-log paper, plot three graphs of width, depth, and area (width × depth) as functions of drainage area, for a discharge of 1.5-year recurrence interval or discharge bankfull, whichever is tabulated. Plot the graph for each region on a separate sheet.

The river channel is subject to rapid fluctuations in the water it receives from upstream. These changes result from storm runoff or snowmelt To accommodate these changes the flow might increase in depth alone, other factors remaining constant. Or velocity could increase alone. These do not occur. Rather, all variables change simultaneously but at different rates.

Problem

Using the current meter measurement data in Table 16-9, develop graphically the changes in width, depth, and velocity with change of discharge at a given river cross section.

Procedure

Furnished is a summary of many current meter measurements made on the New Fork at Boulder, Wyoming. These are not published but, for any station, are kept on file at the District Office of the U.S. Geological Survey. To request such data from an office of the U.S. Geological Survey, ask for data on Form 9–207.

On log-log paper, plot three graphs. On each

Table 16-9 Current meter data, New Fork River near Boulder, Wyoming, from U.S.G.S. Form 9–207, 1963–1965. Data includes days only when there is no ice cover.

DATE	WIDTH (FT)	AREA (SQ FT)	MEAN VELOCITY (FT/SEC)	GAUGE HEIGHT (FT)	DISCHARGE (CFS)
9/14/65	96.0	113	2.35	3.15	266
10/14/65	95.0	157	1.77	3.15	278
4/23/66	85.0	124	1.64	2.75	203
5/10/66	136	526	2.21	4.38	1160
6/14/66	102	214	3.00	3.71	642
7/21/66	87	87.2	3.96	2.90	346
9/8/66	78	106	1.09	2.41	116
9/14/64	78	61.3	1.57	2.02	96
10/10/64	74	53.5	1.43	1.97	76.7
11/15/64	84	77.2	1.97	2.39	152
4/20/65	100	143	2.90	2.99	414
6/2/65	97.5	183	3.08	3.35	564
6/17/65	82.0	675	3.71	5.62	2610
7/28/65	82.0	603	2.87	4.96	1730
9/14/65	96.0	113	2.35	3.15	266
10/14/65	95.0	157	1.77	3.15	278
8/27/63	63.0	76.4	1.61	2.16	123
10/3/63	100	125	2.82	2.87	352
11/4/63	87.0	111	1.68	2.44	187
5/31/64	135	259	3.83	3.96	992
7/11/64	136	636	2.81	4.87	1790
8/17/64	124	119	2.23	2.66	266
9/14/64	78.0	61.3	1.57	2.02	96
10/10/64	74.0	53.5	1.43	1.97	77

the abscissa is discharge in cfs. The respective ordinates are width, depth, and velocity. Width, velocity, and discharge are tabulated for each measurement. Compute and tabulate values of depth, the ratio of area/width. Points for times of ice cover have not been included in the table.

Measure the slopes of the lines computed as the tangent (height/distance) of the mean line through the plotted points.

At high discharge, apparently the width is constructed by the bridge below the cable. Therefore, the mean line should be drawn through points of discharge 1800 cfs or less.

The scatter is large, as usual, at a gauging station where measurements are taken at different cross sections.

16-8 The Downstream Hydraulic Geometry

As drainage area increases when tributaries join a river, discharge increases. The increased discharge could be accommodated by an increase in width, depth, or velocity. But the river in quasi-equilibrium does not allow one factor to change while others remain constant. Rather, it divides the change among all the dependent factors. As a result, all the factors change in a remarkably consistent fashion from one river to the next.

Problem

Plot the relations of channel width, depth, and velocity as functions of discharge of a given frequency, as discharge increases downstream as a result of tributary entrance.

Procedure

At-a-station curves have been plotted for a number of stations, comparable to the plot already prepared for the New Fork River at Boulder. The values of width, depth, velocity, and discharge equal to bankfull (1.5-year recurrence interval) have been read off these curves and are tabulated in Table 16-2.

The stations are all in the upper Green River basin in Wyoming, so the streams are of comparable character.

Plot the three graphs, width, depth, and velocity versus discharge, the three dependent factors being ordinate values and discharge as abscissa. Plot on log-log paper.

Draw a line of best fit by eye for each graph. Measure the slope of each of the mean lines.

These slopes will be of the order of 0.5 for width, 0.4 for depth, and 0.1 for velocity. The sum of the slope should be unity or very close to unity. If yours do not add up to 1.0, readjust one or more of your lines so that the sum is close to unity.

For example, see Leopold and Maddock (1953) U.S. Geological Survey Professional Paper 252; or Leopold, Wolman and Miller (1964:242–245).

17

Sediment Production and Transport

The Load of Rivers

Precipitation falling on the landscape, together with the action of biological agents, breaks down rocks by weathering, forming soil and rock debris, and dissolving some of it. Precipitation feeds the rivers, carries the dissolved load, and moves the weathered debris. These various actions, both by unconcentrated water and water gathered into rills or channels, gradually move the rock debris toward the oceans, ultimately lowering the continents and depositing the materials in the sea. Successive periods of uplift have over geologic time raised land masses out of the sea so that the leveling process never becomes complete. But the downcutting or denudation of the land masses proceeds on all continents inexorably.

The rate of denudation seems slow when viewed in a human perspective, but the amount of debris moved is immense. The denudation may be expressed in various ways, each conveying a somewhat different impression. The number of centimeters or feet of lowering of a continental area per 1000 years is one form of expression often used. Another is the number of tons of debris per square mile per year. Yet another is the tons per day of debris carried by a river past a particular point. Because the rate of transport and of degradation varies widely in time and place, each form of expression is useful for particular purposes. For example, it is worthy of note how climate affects the rate of denudation, and how important are uncommonly great floods as compared with more frequent but smaller ones, and how the clastic or detritus load compares with the dissolved load.

Table 17-1 Dissolved and suspended load in selected rivers in different climatic regions of the United States. (From Leopold, Wolman, and Miller 1964.)

RIVER AND LOCATION	ELEVATION (FT)	DRAINAGE AREA (SQ MI)	AVERAGE DISCHARGE, Q (CFS)	DISCHARGE ÷ DRAINAGE AREA (CFS/SQ MI)
Little Colorado at Woodruff, AZ	5,129	8,100	63.3	.0078
Canadian River near Amarillo, TX	2,989	19,445	621	.032
Colorado River near San Saba, TX	1,096	30,600	1,449	.047
Bighorn River at Kane, WY	3,609	15,900	2,391	.150
Green River at Green River, UT	4,040	40,600	6,737	.166
Colorado River near Cisco, UT	4,090	24,100	8,457	.351
Iowa River at Iowa City, IA	627	3,271	1,517	.464
Mississippi River at Red River Landing, LA		1,144,500†	569,500†	.497
Sacramento River at Sacramento, CA	0	27,000‡	25,000‡	.926
Flint River near Montezuma, GA	256	2,900	3,528	1.22
Juniata River near New Port, PA	364	3,354	4,329	1.29
Delaware River at Trenton, NJ	8	6,780	11,730	1.73

*Computation of load, dissolved or suspended, depends on discharge for same period. Years of record pertain to number of years used for related values of discharge and of suspended and dissolved load. Where two figures are shown, the first is for suspended load and the second is for dissolved load.

Table 17-1 provides some idea of the order of magnitude of clastic and dissolved load for selected basins. Table 17-2 shows the range of sediment yield for various regions in the United States.

Considering the materials constituting a floodplain, colluvial slope deposit, point bar, river bed, or other feature, it is implicit that the flowing water transports mineral grains of various sizes and deposits them. They were eroded from some place upstream and are seen deposited, perhaps temporarily, in one or another kind of sedimentary feature. This leads logically to a consideration of the manner in which clastic debris is carried by moving water, the relation of transportation to deposition and the sources of the deposited material.

Sources and Composition

Rocks are agglomerations of minerals, each having a specific chemical composition. The chemical and mechanical disintegration of rocks reduces over time even the hardest and most solid rock material. This is the result of the

YEARS OF RECORD IN SAMPLE*	AVERAGE SUSPENDED LOAD	AVERAGE DISSOLVED LOAD	TOTAL AVERAGE SUSPENDED AND DISSOLVED·LOAD	TOTAL AVERAGE LOAD ÷ DRAINAGE AREA (TONS/SQ MI/YR)	DISSOLVED LOAD AS PERCENT OF TOTAL LOAD (%)
	←(MILLIONS OF TONS/YR)→				
6	1.6	.02	1.62	199	1.2
1	6.41	.124	6.53	336	1.9
5	3.02	.208	3.23	105	6.4
1	1.60	.217	1.82	114	12
26–20	19	2.5	21.5	530	12
25–20	15	4.4	19.4	808	23
3	1.184	.485	1.67	510	29
3	284	101.8	385.8	337	26
3	2.85	2.29	5.14	190	44
1	.400	.132	.53	183	25
7	.322	.566	.89	265	64
9–4	1.003	.830	1.83	270	45

†From U.S.G.S. records for Vicksburg, Mississippi station.
‡Estimated.

action of water and the materials dissolved in it. Even apparently pure water carries minor amounts of chemical substances, some of which are acids. Even in dilute form, these substances gradually attack the margins of mineral grains making up a rock and, given sufficient time, not only cause mineral grains to separate but cause changes in the chemical nature of the grains, forming new mineral species.

One of the most simple and ubiquitous chemical reactions concerns carbon dioxide, which constitutes about two percent of the air in the earth's atmosphere. In contact with water, some goes into solution and the dissociated ions recombine to form minor amounts of H_2CO_3, carbonic acid. Though hardly a powerful acid, in constant contact with rock it gradually effects chemical changes in some rock minerals. Similarly, the small amounts of chloride and sulfate in the air derived from ocean water spray and other sources combine with water to form dilute concentrations of even more effective acids: hydrochloric and sulfuric acids.

Chemical change in a mineral caused by the slow but effective action of acids and other chemicals in water usually results in the expansion of the physical size of the mineral suffering decomposition. A slight increase in

Table 17-2 Sediment yield from drainage areas of 100 square miles or less of the United States. (From U.S. Water Resources Council 1968, Part 5, Chap. 5, p. 4.)

REGION	ESTIMATED SEDIMENT YIELD (TONS/SQ MI/YR)		
	HIGH	LOW	AVERAGE
North Atlantic	1,210	30	250
South Atlantic-Gulf	1,850	100	800
Great Lakes	800	10	100
Ohio	2,110	160	850
Tennessee	1,560	460	700
Upper Mississippi	3,900	10	800
Lower Mississippi	8,210	1,560	5,200
Souris-Red-Rainy	470	10	50
Missouri	6,700	10	1,500
Arkansas-White-Red	8,210	260	2,200
Texas-Gulf	3,180	90	1,800
Rio Grande	3,340	150	1,300
Upper Colorado	3,340	150	1,800
Lower Colorado	1,620	150	600
Great Basin	1,780	100	400
Columbia-North Pacific	1,100	30	400
California	5,570	80	1,300

volume experienced by a large number of individual minerals making up a once solid rock promotes their separation into discrete mineral particles. The additional mechanical effects of alternate wetting and drying, of freeze and thaw, and the growth of plant roots into incipient cracks in the rock, all promote the process of weathering and breakup. Thus, a rock once solid becomes comminuted into grains or particles of varying size.

The study of chemical interactions among rocks, minerals, water, and the atmosphere constitutes a special branch of geology and merges with chemical hydrology. The development of the soil profile, the zonal differentiation of surficial rock material under the influence of climate, is part of another whole science: pedology, or soil science. Though soil characteristics and soil movement are of immense importance both in hydrology and in land use and management, for present purposes we will restrict the consideration to those aspects of soil and rock materials related to fluvial processes.

Weathering obviously affects some minerals more and faster than others. Grains of feldspar and silica constitute by far the largest percentage of mineral grains making up soil and debris moved by water. Feldspars constitute 30 percent and silica 28 percent of all minerals. Silica is a hard and resistant material, relatively insoluble and immune to chemical decomposition except by solution.

Though there are several mineral species composed entirely of silica, by far the most common is quartz, which occurs in a variety of igneous rocks, such as granite, in which grains of quartz solidified or crystallized out of the original molten magma. The decomposition of other minerals in the rock frees individual quartz grains. Being relatively resistant to decomposition, quartz grains are moved and concentrated by flowing water and thus constitute the bulk of many sedimentary rocks, especially sandstone. Thus, grains or particles of quartz make up most of the sand carried by rivers. Cobbles or gravel are composed of many kinds of rock, but those composed of quartz or high in quartz are especially common. The lithology of gravel usually has a high representation of vein quartz, quartzite and chalcedony, each of which is similar in chemical composition but varies in mode of origin and texture or crystalline character.

It is these products of rock weathering that make up soil materials and the clastic load of streams. By the process of downhill and downriver movement, the materials making up the highlands of continents are over time moved toward the oceans, and the land surface is reduced in elevation. The sand, gravel, boulders, silt, and clay seen on the beds of streams are in the process of transport, though when the water is low enough for the bed to be easily seen, the rate of transport is small and the downstream movement usually not discernible. At high flow the amount of material in motion is surprisingly large though usually unseen because then the water is cloudy. These materials, whether in motion or temporarily at rest, are called the debris load of the river. Debris load, sediment load, or clastic load are synonymous terms.

Size Distribution

Sizes of the particles are described by the terms clay, silt, sand, cobble, boulder, and the size categories applicable to each are shown in the following tabulation:

	SIZE RANGE (MM)
Clay	Smaller than 0.0039
Silt	0.0039–0.0625
Sand	0.0625–2.0
Gravel	2.0–64.0
Cobble	64.0–256.0
Boulder	256.0–4096.0

The lithology of these particles is described by usual rock or mineral names, quartzite, sandstone, basalt being examples of the former and quartz, feldspar, magnetite examples of the latter.

The debris load of a stream is distributed in the channel in various topographic bed forms, bars, pools, riffles, and point bars, as previously described. Each form may have one or several units within which the size distribution of particle size is characteristic. Any unit consists not merely of one size but of an assortment of sizes.

The size of sediment particles is usually expressed as a distribution graph, showing the percentage by weight in the sample represented by each size class, determined by sieving through a nest of sieves.

Drying, weighing, and sieving must be used for materials in the sand size with only small amounts of material of silt or clay size. For the smaller fraction, dry sieves are not satisfactory, and more elaborate wet procedures are used that involve filtering or settling in water.

Many geomorphic and hydraulic problems involve particle sizes coarser than sand, and the samples required are both too large and too heavy to be carried into the laboratory for weighing. In such problems some quantitative expression of sediment size is needed to express material at the surface, on a hillslope, or on a gravel bar in a river. The surface particle size is a determinant of the surface roughness and, therefore, is influential in determining the velocity of water flowing in a channel of over a hillslope.

For this reason the scheme we use involves the sampling of the surface. If the surface particles are primarily gravel, cobbles, or boulders, we make a count of 100 particles picked up at random from the surface, measuring and recording the size of each. The usual situation involves the determination of the size of gravel making up the surface of a location on a hillslope.

The procedure is to select a zone or area considered homogeneous. As the researcher walks over the selected area, he reaches over the toe of his boot with eyes closed or averted and touches with an extended finger a rock. The rock is picked up and measured with a scale along its intermediate, or b, axis, being neither the longest nor the shortest axis. The measurement is made in millimeters and recorded as the lower limit of the size class into which the rock falls. The size classes vary, not as the doubling of the lower class, but as $\sqrt{2}$, so that the classes go in a series: 2, 2.6, 4, 5.6, 8, 11, 16, 22, 32, 45 mm, etc. The procedure is explained in detail in Leopold (1970) but is briefly described here.

Pebbles or material less than 2 mm in diameter cannot be counted by this method, and when a finger touches such fine material, it is recorded as < 2 mm. The size of each rock picked up is tabulated and when about 100 have been measured, the basic data are comparable to columns 1 and 2 in Table 17-3.

A large number of rocks have been measured and weighed, and the average weight in each size class is shown in column 3. This tabulation is applicable to most field sites. When the numbers in column 2 are multiplied by average weights in column 3, the estimated total weight of sample in each size class is obtained, column 4.

Because large rocks present larger areas on the surface, there is a high probability of picking up the large ones, so a correction is made by dividing

Table 17-3 Size distribution data, gravel bar, right bank, Pole Creek below Hoot Owl Bridge, at Station 37 + 50 near Pinedale, Wyoming 1967. (From Leopold 1970.)

	RANDOM PEBBLE COUNT IN A 10 × 20 FOOT AREA, DATA							
SIEVE OPENING HELD ON (MM)	NO. OF ROCKS	AVERAGE WEIGHT OF ONE ROCK (G)	ESTIMATED TOTAL WEIGHT (G)	MEAN DIAMETER SQUARED (SQ MM)	$\Sigma wt/d_2$ (G/MM^2)	PERCENT OF TOTAL	$\frac{\text{PERCENT}}{\text{LOG }(2)^{1/2}}$	MEAN SIZE SIEVE OPENING (MM)
256		40,000		92,000				303
180		15,000		45,800				214
128		5,600		23,100				152
90		2,100		11,500				107
64	2	700	1400	5,770	0.243	5.7	38	76
45	11	255	2800	2,920	0.959	22.6	150	54
32	20	94	1880	1,445	1.300	30.6	204	38
22.6	18	34	612	718	0.854	20.1	134	26.8
16	12	12	144	360	0.400	9.4	63	19.0
11.3	15	4.5	68	179	0.380	9.0	60	13.4
8	5	1.6	8	90	0.089	2.1	14	9.5
5.6	1	0.52	1	44	0.023	0.5	3	6.6
4		0.21		22				4.7
Total	84		6913		4.248	100.0		

each total weight (column 4) by the square of the mean diameter of the particle size, those values shown in column 5. The quotient is shown in column 6. The figures in column 6 are added, and the sum is divided into each figure in the column, yielding the percentage of the total represented by each size class. These percentages are shown in column 7.

In order to make the final graph of percent by weight of each size class independent of the particular sieve sizes used, values in column 7 are divided by the log of the diameter interval of the size categories. We use intervals that differ by the factor $\sqrt{2}$. Thus, values of column 7 are divided by $\log \sqrt{2} = 0.150$ to give column 8. Values in column 8 are plotted on log-log paper against the geometric mean size of the interval. These geometric means are shown in column 9, so column 8 is plotted against column 9, resulting in the plot shown in Figure 17-1.

The advantage of the scheme is that it yields a dominant particle size, the size class that represents the largest percentage of the total sample weight, and it approximates the result that would have been obtained by collecting the surface rocks, sieving them for size, and weighing and plotting the data in a similar way.

The method is practical and, once learned, easy to use in the field. There is no reason for guessing about the dominant size of surface particles or for passing off the grain size with descriptive and elusive adjectives.

In the example shown in Figure 17-1, we would say that the dominant size was 48 mm. Several other characteristics about the material could also

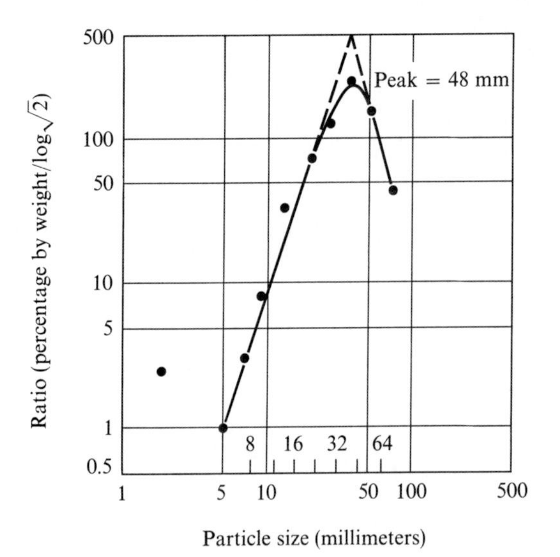

Figure 17-1 Size distribution of sample of stream-bed gravel, Pole Creek below Hoot Owl Reach at station 37 + 50, near Pinedale, Wyoming. (From Leopold 1970.)

be derived from the graph, as explained by Leopold (1970). But a statement of the dominant size tells immediately what part of the range from sand through gravel to cobbles the sample lies, and in a quantitative and reproducible way.

Modes of Debris Transport

The velocity is greater in the main body of flow than near the boundaries, bed, and bank. There is therefore a shear in the flow, that is, faster water flowing past slower, described as the velocity gradient. In turbulent flow the momentum of the faster water is transmitted to the slower by eddies or swirls that begin and end in a random manner. These eddies, of course, dissipate their kinetic energy of motion into heat, which for the most part represents a loss of energy without doing any work, a loss to frictional dissipation. But some of the energy is put to mechanical work by setting into motion rocks or debris on the stream bed.

Sketches in Figure 17-2 show some aspects of this process. In (a) the velocity at each of two levels is represented by the length of the arrows. The higher velocity tends to roll up the intervening water in a spiral motion; and the roll becomes, in fact, a local eddy which causes faster water to be mixed with slower and, as a net result, slows the mean velocity of the whole by frictional resistance.

This same shearing motion, clockwise as seen from the right, applied to a rock on the stream bed, is shown diagrammatically in sketch (b). A rock subjected to such shear tends to be rolled and can thus be dragged along, or if it is deflected upward by hitting another rock, can be bounced up into the higher velocity flow above. The skipping motion just described is called *saltation,* in which a particle is dragged up into the flow only to fall back to the bed because of its own weight. Its trajectory is generally parabolic, but the downstream distance moved is much larger than the vertical, like the path of a bullet fired at only a slight angle above the horizontal.

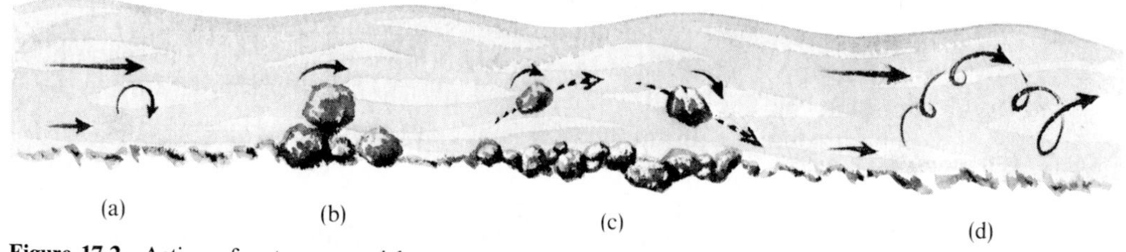

Figure 17-2 Action of water on particles near stream bed: (a) Direction of shear due to decrease of velocity toward bed. (b) Tendency of velocity to roll an exposed grain. (c) Diagram of saltating grains. (d) Suggested motion of a grain thrown up into turbulent eddies in the flow.

Sketch (c) suggests the trajectories shown by dashed lines of two particles in saltation, with the rolling or rotational motion shown by the solid curved arrows. The particle in motion is then subject to four components of force in addition to its rotational motion. It is pulled downstream by the drag of the faster moving water in which it is embedded. Its drag or resistance to this motion is a resisting force directed upstream. It is subjected to a lift vertically upward representing the upward motion it acquired as it left the stream bed. It is being pulled downward vertically by its own immersed weight.

If the particle is small, it can be carried upward into the main body of flow by an eddy because its own immersed weight gives it a fall velocity smaller than the upward velocity in the eddy. As it falls it may be entrained in another upward moving eddy and be given an additional upward lift in a manner similar to the soaring of a hawk or to a glider, both of which move from one upward moving air current to another, and thus, be sustained aloft against the gravitational downward pull of its own weight. This eddying upward and fall downward is suggested by the spiraling arrow indicating a particle trajectory in sketch (d).

These trajectories and their relation to the main-flow velocity at different levels above the bed lead to the definition of the two important modes of debris transport by moving water, the transport by suspension and by traction. The particles lifted up by eddies in the main flow and moving long distances downstream before settling to the bed constitute the *suspended load.* The particles, usually much larger, that are rolled, dragged, skipping, or saltating constitute the *bed load.*

The distinction between suspended and bed load is generally made on the basis of this difference in motion or path, but a given particle may alternate from one to the other; and in any event, its trajectory cannot be observed except by sophisticated photography in a laboratory. It is physically and theoretically more satisfying to define the two segments of load in terms of the manner in which their respective immersed weight is carried on the stream bed. Bed load is defined here as that segment of the load the weight of which is supported by the solid or grain bed of the stream. The suspended load is defined as that part of the weight carried by the fluid and, thus, in turn exerts its weight not on the stream-bed surface but through the bed-particle interstices to the fluid within the porous bed surface.

The support of bed load by the solid grain bed is similar to the relation between the floor and a bouncing ball. Though the ball is only in intermittent contact with the floor, it is obvious that the weight of the bouncing ball is carried by the floor and specifically at the point of contact between the floor and the ball.

Suspended load, however, is akin to the glider or hawk. The upward support to the wing is given by the fluid in which it is immersed. The weight is transmitted through the fluid not as a single vertical line but through a cone of the fluid, the base of which is on the ground and the apex of which is at the soaring object. In the case of a suspended particle, its weight is carried by the fluid, which in turn is not supported by the grain surface as much as by the fluid in the interstices between grains. Thus, the ultimate support of suspended load is not by surface grains but by the fluid well below the stream-bed surface.

Methods of Measurement

In practical matters the distinction between modes of transport arises in the differences in technique of measurement. The suspended load can be satisfactorily sampled by obtaining a sample of water and the grains of sediment dispersed within it. A sampler was devised by personnel of the federal agencies and has been widely adopted.

The suspended load sampler consists of an aluminum fish within which can be placed a common one-pint milk bottle. Immersed in the stream, water enters a nose nozzle while displaced air is emitted through a vent. The design of the nozzle and vent is such that the velocity of entering water is the same as that of the ambient flow, and so there is no tendency to differentially measure water and its suspended sediment.

At a sediment measuring station, samples are obtained once a day or week, with greater frequency during high or unusual flows. Stage is simultaneously observed. In the laboratory the sample is filtered and dried; the sediment is weighed and expressed as a percentage of the weight of the water-sediment mixture. This concentration is published for sites observed by the U.S. Geological Survey in the Water Supply Paper series. For any state, the information on location, dates, and places of publication of data on suspended sediment may be obtained from the district office of the U.S. Geological Survey, usually in the capital city of the state. Suspended sediment has been measured at several hundred sites in the country.

But it is sometimes necessary in planning problems to ascertain by direct measurement the sediment load of a stream because published data do not appear applicable. A suspended-load sampler, small and light enough to be operated by hand (without a winch) and held by a person who is wading, is widely used and might be inspected, if not borrowed, in a local office of the Water Resources Division of the U.S. Geological Survey. To substitute for a long record, measurements several times during a single hydrograph rise will roughly define a relation of suspended sediment load as a function of discharge, as demonstrated by Leopold and Miller (1956:12).

In principle, making several field measurements during the rise and fall of a single flood hydrograph is, as in the measurement of lag time, a practical way to determine how the local stream fits in the general field of such data.

Data on bed load are scarce. No organized network of measurement is extant, despite the fact that the bed load constitutes usually 5 to 50 percent of the total debris load of a river. Thus, as much as one-half of the total load is essentially unmeasured, and indeed, little is known of the range of the percentage represented by the unmeasured bed load.

There are a few studies in which the total load carried over a period of time has been determined. The results have been derived from a comparison of successive maps of the deposit of sediment accumulated in a reservoir or lake. But such a method requires first that a reservoir exists on the river, and that sufficient time has elapsed between two surveys so that a measurable difference can be observed, a time represented by years not weeks. Then the amount contributed by bed load must be surmised from the size distribution of accumulated debris, or there has to be a concurrent suspended load station from which the total suspended load could be computed. This could, theoretically at least, be subtracted from the total load measured, and the difference would represent the part contributed by bed load.

In cases where it has been of importance to know the bed load, it has ordinarily been computed by one or more of the existing formulas; but this has the disadvantage of not knowing, in the end, whether the result is correct. But at least an order of magnitude can be obtained in this manner.

In recent years the need to know the amount of debris carried as bed load has stimulated a renewed effort to develop a practical sampler. Both the National Research Council of Canada in collaboration with the University of Alberta and the U.S. Geological Survey have experimented with new sampling procedures. The most promising appears to be the Helley-Smith sampler of the U.S. Geological Survey, an improved version of the Arnhem sampler of Zurich, Switzerland. This sampler, shown in Figure 17-3, has a flaring throat that tends to equalize entry velocity and ambient velocity. The sampler has a nylon net bag, which traps the sediment reaching it but allows the water to pass through. The sampler has been made in two versions, an 80-pound type that is lowered into the stream by a winch and cable and a light-weight version that can be used by a person wading in the stream. Plans and characteristics have been published (Helley and Smith 1971), and there is no patent or copyright, so the design is freely available to all. Details on availability of finished products can be obtained from the U.S. Geological Survey, Denver, Colorado.

Transport Mechanics

Recognizing at the outset that this is one of the most complicated and mathematical parts of the subject of fluvial processes, we cannot hope to answer satisfactorily even the majority of principal questions arising in the reader's mind. Rather, our purpose in this discussion is to present in essentially nonmathematical form the main considerations and the nature of both reasoning and evidence concerning the transport of debris by flowing water. The treatment follows the development of the subject by Ralph A. Bagnold,

(a)

(b)

Figure 17-3 (a) Bed-load sampler of the Helley-Smith type being lowered from a bridge by Dr. W. W. Emmett. (b) Sampler will orient in flow direction when it is submerged.

F.R.S. (1973), who stands uniquely in this field as the most innovative theoretician concerned with understanding the basic mechanisms, and not diverted by the elusive goal of providing a universal formula that would satisfy engineering requirements for computed results under specific circumstances. Similarly, we attempt to transmit an understanding of the principles rather than provide a formula for specific computations.

The flow of granular solids propelled by a medium, whether fluid or gas, is widely observed in nature and used in commerce. The movement of sand grains by wind in the desert, the transport of coal as a slurry in pipes, the practice of hydraulic dredging, and the movement of sand or gravel on the river bed are a few examples that come to mind. These all partake of the following essential features:

1. The motion of the grains is in the form of a shearing motion in which various layers of solid grains slide over one another.
2. An impelling force or tractive force exerted by the fluid in the direction of motion is required to maintain the motion.
3. The solids are heavier than the fluid and, therefore, tend to sink or are pulled downward by gravity.

4. If the motion is persistent and approaches a steady state, the forces acting on each layer of the moving solids must be in equilibrium in a statistical or time-averaged sense.

These may seem overly basic or unduly complicated statements when trying to get at the flow of gravel grains on a river bed. But by thinking about a bowl full of round marbles, and gradually tipping it until the marbles begin to roll off, it can be visualized that the particles (marbles) must roll upward relative to their neighbors in order to start motion. This upward motion relative to the grain layer below dilates that part of the whole. The mean separation distance must increase before motion begins. The bulk must dilate.

Now if the motion were caused not by increasing the gradient of the surface but by the drag of a moving fluid across the surface, the dilation before motion would also be required. But the dilation is upward and must therefore be working against or must overcome the force of gravity pulling the grains downward.

If the motion of grains is continuous, as when sand particles are being transported by a river, then also there must be a continuous supporting stress acting upward, and it must be equal to the immersed weight of the particles.

There are only two mechanisms by which such an upward supporting stress is transmitted: the first is by a continuous or intermittent grain to grain contact; the second is the transfer of momentum from the fluid to the supported solids. These two possible mechanisms define the difference between bed load and suspended load. Bed load is, then, that part of the grain load whose weight is supported by grain-to-grain intermittent contact and thus by the grains of the unmoving bed. Suspended load is that part of the grain load whose weight is supported by the upward transfer of momentum in the turbulent eddies and is thus supported by the column of fluid.

Many individual particles are supported by both mechanisms simultaneously, but statistically the load is divided into these two discrete parts.

The transport rate is the product of the weight times the mean forward velocity of the grains. The weight here involved is the excess weight or dry weight less the buoyant weight of the water displaced. Force times distance is work. Velocity is distance per unit time. Work per unit time is power. Thus, weight $\times$ distance $\div$ time $=$ power. The transport rate of sediment, expressed as the immersed weight passing in a unit time over a unit width of the bed, is thus expressible in terms of work rate or power.

The dynamic transport rate expressed above is not the same as the work rate until corrected by a dynamic coefficient of friction that relates the normal or downward stress due to the immersed weight to the forward or tractive stress needed to maintain the load.

The work rate of transport then is the product of the immersed weight of solids times the forward velocity times the dynamic friction coefficient.

The river carrying a load can be considered a transporting machine that, like all other machines, can be characterized by the relation

$$\text{Rate of doing work} = \text{Available power} \times \text{Efficiency}$$

The available power in a river derives from the movement of a mass of water from higher to lower elevation. The energy of position, or the potential energy represented by the height above base level or sea level, was, of course, provided by the sun's energy, which evaporated the water from some low elevation and moved it as atmospheric vapor to some place where as precipitation it fell on a landscape having an elevation or position higher than sea level. This stored energy of position is given up as sensible heat and work enroute as the water flows to lower elevation.

Because water flowing in a river does not accelerate or has a nearly constant downstream velocity, the fall of elevation in any given distance represents the potential energy changed into kinetic energy and then into heat or used in doing work. The major part of the energy is used up in friction associated with the motion of flow and is turned into heat, lost mostly by radiation and convection to the ambient air. A small part of the energy is used to transport sediment or alter the channel.

The available power over a unit area of stream bed is proportional to the product of the discharge rate times the gradient or slope of the water surface. Both of these terms are measurable in rivers. Since the available power is readily computed, the principal problem is in ascertaining the efficiency factor. A computed value of the rate of load transport would then be possible. A full explanation is given by Bagnold (1966, 1973).

The above discussion conveys a few fundamental concepts:

1. Bed load and suspended load can be defined in theoretically acceptable terms, reflecting the mechanics of the support forces.

2. Sediment transport concerns the use of available power or rate of doing work, and in this sense, a river can be viewed as a transporting machine.

3. Stream power or rate of doing work is the theoretically correct parameter expressing transporting ability, and is a simple product of two terms possible of measurement in the field: discharge and water surface slope, and multiplied by a constant, the specific weight of water.

The relation of stream power to bed-load transport rate presented in Figure 17-4 is a generalization of the few data in the published literature and in unpublished form representing the actual measurements of bed load in rivers, and for which the observed values of necessary parameters are available. The plot follows the general theory just discussed in that the dependent variable is the bed-load transport rate, i_b, in kilograms per meter of channel width per second. The independent variable is stream power per unit width of channel, ω, in kilograms per meter of width per second, computed as $\gamma \bar{u} \bar{d} s$, where γ is the unit weight of water, $\bar{u}$ the mean velocity

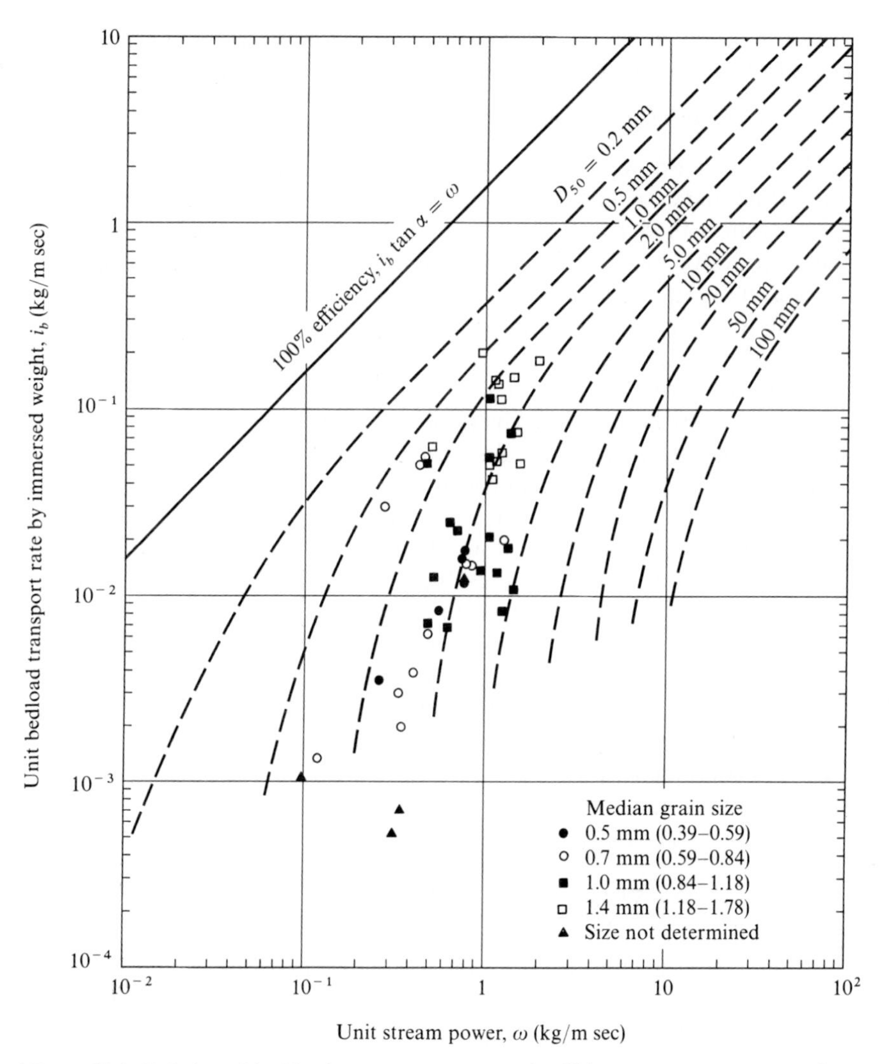

Figure 17-4 Relation of bed-load transport rate per unit width to stream power per unit width, the family of curves being the D_{50} for the moving load. Plotted points are bed-load measurements at the East Fork bed-load trap, Wyoming. (From Leopold and Emmett 1976.)

in meters per second, $\bar{d}$ the mean depth in meters, and s the river slope. Thus, the plot is dimensionally consistent.

The data used to derive the generalized lines represent various rivers in which the median size of bed particles differs. The generalized lines represent the relation of the two parameters, i_b and ω, for different values of D_{50}, the median grain size in mm. Individual measurements in river data do not align on the appropriate curves until the mean particle size of the actual bed grains in motion begins to approach the median bed particle size. That is, when motion begins especially on gravel bed streams, the first particles that move are usually smaller in size than the mean. With increased discharge and thus load, the dominant size of the moving particles tends to increase and may approach the median size of bed particles in general. At low stages or soon after initial bed motion, plotted points would not be expected to coincide with the drawn curves.

The graph is not meant to be an engineering tool for design. It is useful, however, in indicating the order of magnitude of the bed load transported. Because the power expended in transport cannot exceed the power available, data cannot exist in the region above the line labeled "limit." This also may serve as a rough guide for order of magnitude (Leopold and Emmett 1976).

The transport rate of suspended sediment usually is greater than that of bed load in the same stream, providing that there is fine material available for suspension. Ordinarily, the bed load portion consists of some 20 to 40 percent of the total load, but there are many exceptions to this general statement.

Grains that appear in the suspended load are nearly always less than 0.5 mm in diameter. Though the range of suspended load size is from clay through silt to medium sand, the debris moved as bed load may vary from 1 mm to 1000 mm and the transport rate depends greatly on the debris size.

Even after a sediment rating curve is developed or chosen, there are complications in its application. Basically, one must compute the total time the river is flowing at each of several ranges of discharge, multiply the time by the rate of transport applicable, and sum to give the total load passing the measuring point during the time considered. This means that it is necessary to have a record of stage or discharge during the period in question, or a flow duration curve for the period. This may be obtained by direct observation over a period, as is provided by a gauging station record, or can be approximated from nearby stations when a record is not available.

In addition to the problem of obtaining a stage or discharge record, some other decisions must be made, and for these, some suggestions are offered here. The graph applies to load per unit of width. The width used should be not the water surface width but the width of the active bed transport. This is approximated by bed width but is often somewhat less than bed width, perhaps 80 percent of the bed.

The particle size is that of the debris in motion, which, except for a few hours or days of highest flow, is smaller than the median size observed on the stream bed. The size is best determined by an actual bed-load sample obtained with a sampling device.

Finally, the transport rate is variable both in time and space, even under constant conditions of water discharge. Bed load in a natural river appears to move in narrow streamers or bands between which at the same time no appreciable movement is occurring. Because in natural streams of any depth, direct observation of bed-load motion is usually impossible, what we know is inferred from the hundreds of samples taken with a Helley-Smith sampler and in our bed-load trap, an experiment being conducted on the East Fork in Sublette County, Wyoming (Leopold and Emmett 1976). In many repetitions and on several rivers of different size, it is a common occurrence that the sampler repetitively placed at the same place only minutes apart in time may catch different amounts of debris, amounts that may be different tenfold. The variability inherent in eddies of all sizes must ultimately be reflected in an erratic and random movement or a change in the location and intensity of bed motion.

Moreover, for the same discharge, the falling limb of a flood hydrograph usually carries less sediment load than the rising side. It is presumed that the explanation for this difference involves availability of debris for transport, but the full explanation is far from clear. It is recognized that sediment moved during passage of a flood does not come from the erosion of stream banks, for whatever erosion does take place tends to be balanced by deposition. Material making up the excess of transport over deposition is mostly prepared anew for transport by bank sloughing; tributary entrance either of ephemeral or perennial streams; small cones or debris fans; and other sources prepared by freeze and thaw, bank caving, and other discontinuous processes.

When a flood rise occurs, this prepared material is moved more easily than that constituting the firm or undisturbed stream margins. As the water rises in the channel on the rising limb of a hydrograph, the prepared material is moved, and at the same discharge on the falling stage the easily moved material is gone, so that the sediment transport rate is smaller. The load-discharge graph typically has, then, a hysteresis loop, an example of which can be seen in Figure 17-5.

Sediment Yield

Sediment production or sediment yield refers to the rate at which sediment passes a particular point in the drainage system, usually expressed as volume or weight per unit of area per unit of time. The usual values are in tons per square mile per year.

Sediment yield is determined best by a survey of deposits in reservoirs. Alternatively, it can be computed by integration from a hydrograph of flow and a sediment rating curve, that is, a curve relating sediment transport rate as a function of discharge. Separate sediment rating curves must be developed by field measurement for suspended load and bed load.

The sediment rating curve used together with a flow hydrograph will allow the estimation of the total sediment moved past the point in question during the flood event or time period represented by the hydrograph. In other cases the problem of concern is the total sediment yielded over a period of years by a unit of drainage area. Long-term average sediment yield may in any instance be merely a rough approximation, but data are extant over a wide range of climatic and topographic conditions. In Figure

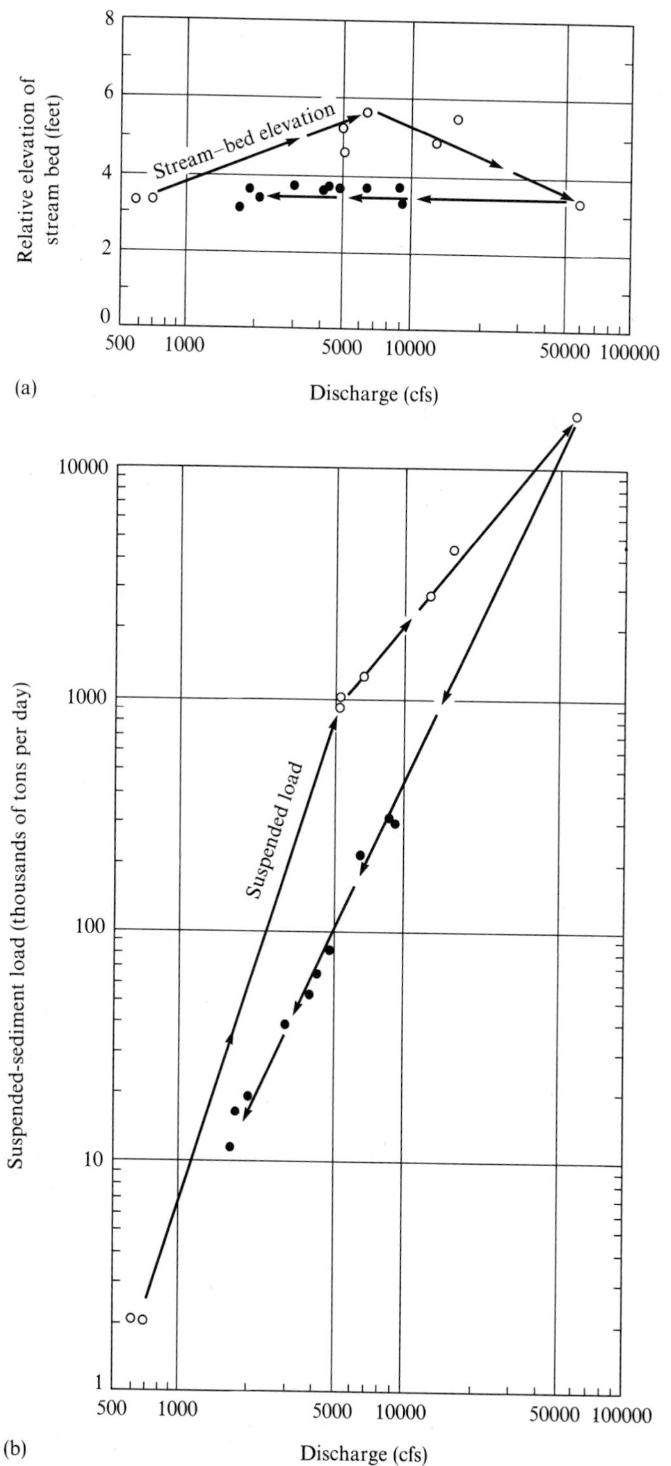

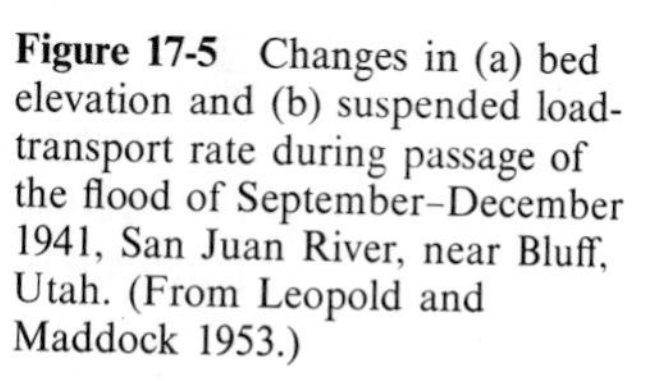

Figure 17-5 Changes in (a) bed elevation and (b) suspended load-transport rate during passage of the flood of September–December 1941, San Juan River, near Bluff, Utah. (From Leopold and Maddock 1953.)

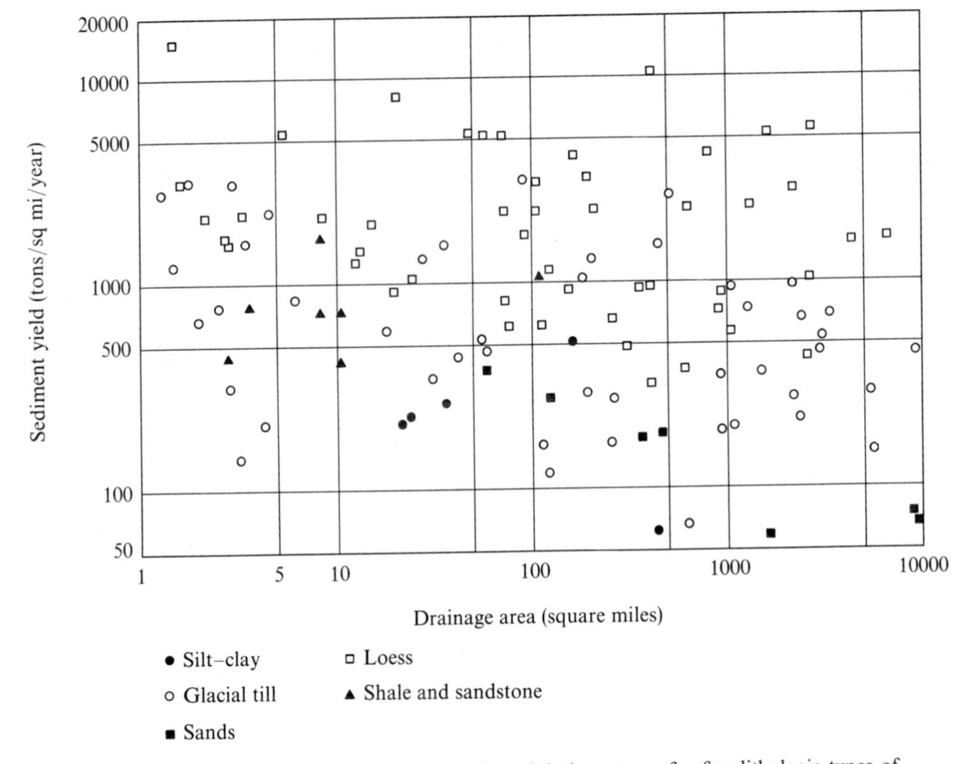

Figure 17-6 Sediment yield as a function of drainage area for five lithologic types of bedrock in midwestern United States. The larger values of yield result from clear-tilled and small grain cropland. (Data from Brune 1951.)

17-6 are some of these data organized on a basis of average annual yield per square mile of basin. The average yield is usually larger per square mile from small basins than from large, but the main impression given by the data is that of extreme variability.

The data in the figure were compiled by Brune (1951), mostly from rates of accumulation in reservoirs. There was in the 1930's an extensive program in the Soil Conservation Service to make reservoir surveys for sedimentation-rate determinations. The method consists of spudding from a boat, which involves dropping a long iron rod vertically to the lake floor at many places. The spud or rod has concentric grooves in which the bottom mud lodges as the spud penetrates the sediment. In the thin deposits the spud reached the original soil surface, the soil of which is much darker than newly deposited material, and the change of color and texture of the sediment caught in the grooves of the rod indicated the bottom of the sediment deposit in the reservoir. The top of the sediment is shown by the level at which no mud sticks in the grooves. The boat position is recorded by triangulation, and in this way a map is constructed of sediment thickness from which the total volume is computed. Knowing the date of construction of the reservoir, the average rate of sediment accumulation is obtained.

To convert the volume of sediment into weight, it is necessary to know the average density of sediment in lake deposits. Deposited sediment has a large void space indicated by the value often used as the average specific weight for this conversion: 100 pounds per cubic foot.

Inspection of Figure 17-6 indicates first that a large number of measurements have been made, yet the graph does not include all of Brune's data. Plotted here are those for which a lithology of source area could be specified, which we put in five classes as indicated by the symbols. The variance in sediment yield is large and only slightly reduced by singling out one lithologic type. Yet there is a difference among lithologies, loess clearly yielding a larger number of tons per square mile per year than other types. Glacial till appears to be an intermediate producer of sediment, and the classes of sands and silt-clay the lowest producers of sediment.

Brune notes the variance and states that land use, especially the degree to which the watershed is cultivated in clear-tilled and small-grain crops, importantly influences sediment yield. Figure 17-6 suggests that within any one lithology and for a given basin size, annual sediment yield is increased tenfold by the cultivation of crops that give little protection against erosion.

This is made more specific in a nomogram constructed by Brune (1948) that gives an estimated annual yield of sediment as a function of annual runoff in the region, drainage area, and the percentage of the basin that is in cultivation (Figure 17-7). Sediment yield per square mile increases with decreasing drainage area, with increasing runoff, and with increasing percentage of land in cultivation. The reasons for the influence of annual runoff and percentage cultivated are obvious, for the more bare soil and the more runoff water, the greater the probable erosion. But the effect of drainage basin size is less clear. The lower sediment yield per unit area from large than from small basins probably lies in the tendency for some eroded sediment to lodge or be trapped in wide valley floors, extensive point bars, and broad colluvial areas present in large drainage basins.

Regional variations in sediment yield are indicated in Table 17-2, compiled by the Water Resources Council. Orders of magnitude agree with those shown in Figure 17-6.

Urbanization has effects on sediment yield similar to those of clear-tilled agriculture. During construction when a building site is laid bare, erosion can be serious during the rainy season. After urbanization is more or less complete, vegetation is established and lawns developed, sediment production decreases, but data show that an area considered urbanized still produces more sediment than it did under the original vegetative cover in undisturbed conditions. Even when only a few lots are bare in an apparently developed or urbanized area, they can produce enough sediment to make the local streams cloudy or silty during storms. An example is Rock Creek

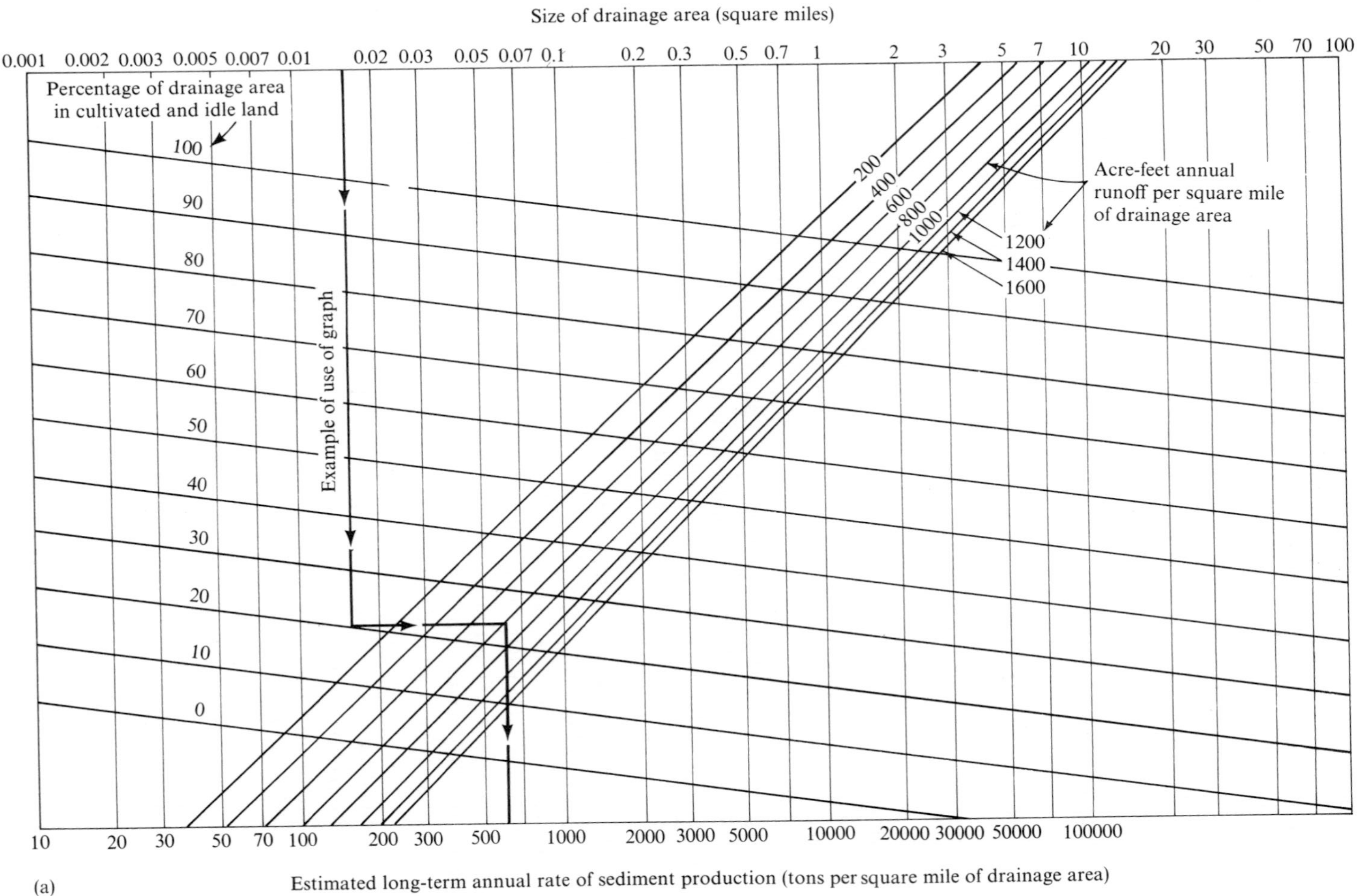
Size of drainage area (square miles)
0.001 0.002 0.003 0.005 0.007 0.01 0.02 0.03 0.05 0.07 0.1 0.2 0.3 0.5 0.7 1 2 3 5 7 10 20 30 50 70 100
Percentage of drainage area in cultivated and idle land
100
90
80
70
60
50
40
30
20
10
0
Example of use of graph
200
400
600
800
1000
1200
1400
1600
Acre-feet annual runoff per square mile of drainage area
10 20 30 50 70 100 200 300 500 1000 2000 3000 5000 10000 20000 30000 50000 100000
Estimated long-term annual rate of sediment production (tons per square mile of drainage area)

(a)

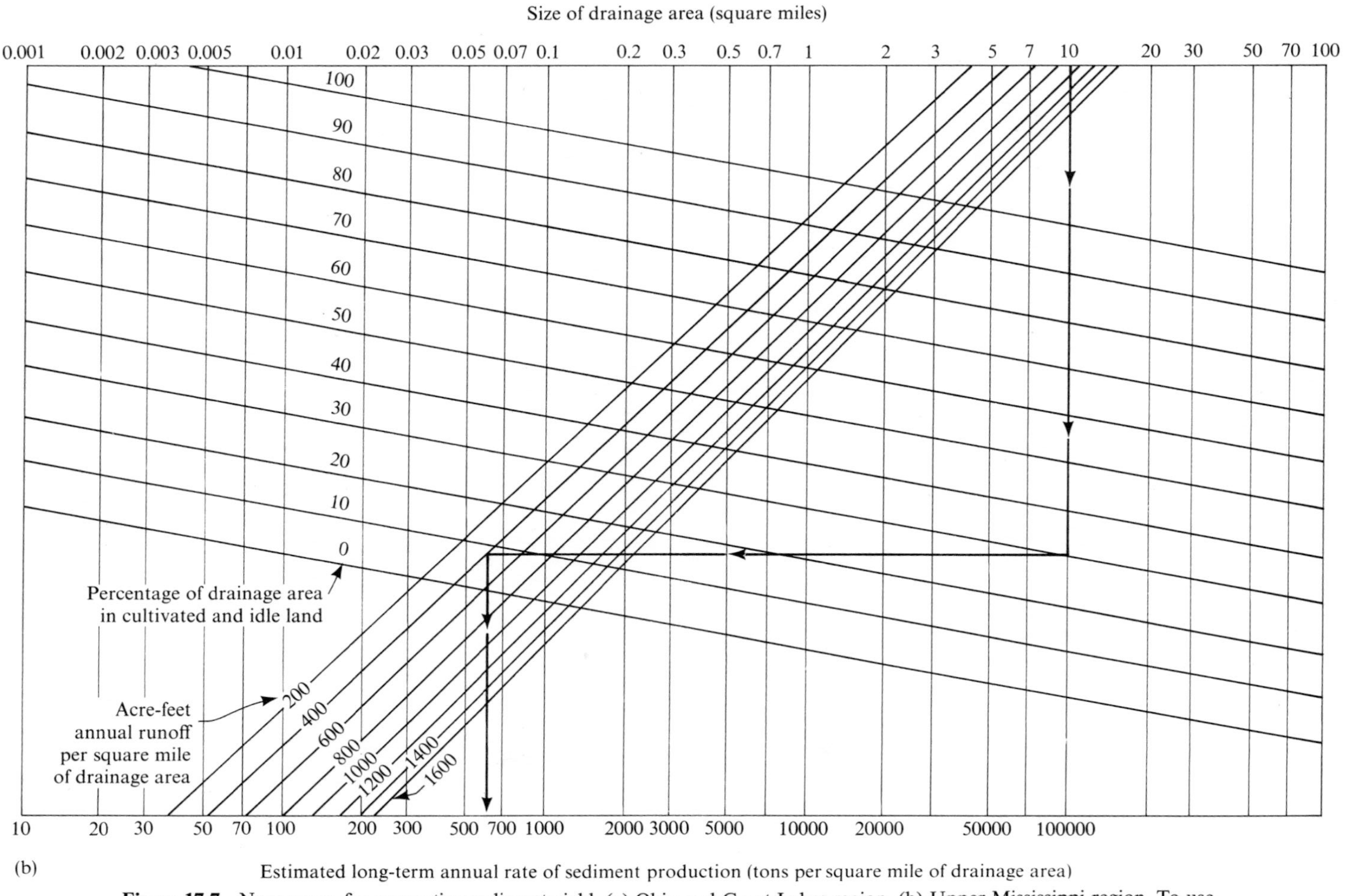

Figure 17-7 Nomogram for computing sediment yield. (a) Ohio and Great Lakes region. (b) Upper Mississippi region. To use, enter drainage area, move vertically to percentage of basin cultivated, move horizontally to annual runoff, move vertically to read sediment yield. (From Brune 1951.)

flowing from suburban Maryland through the District of Columbia. In what seems to be a developed basin and where little new construction is evident, the creek is brown with silt in nearly every winter storm.

Wolman and Schick (1967) studied sediment records and made direct measurements in natural, urbanizing, and urban areas in Maryland and constructed the graph shown in Figure 17-8 that shows the differences mentioned above. Areas under construction produce 10 to 100 times more sediment than is found in otherwise comparable rural or natural areas.

To give an example of reasoning that might be used in the estimation of sediment yield, consider the following problem. A suburban development is planned on a 40-acre tract located in eastern Nebraska. A short distance downstream is a small reservoir, and the question is to what extent will the planned development affect the life of the reservoir. The reservoir controls a drainage area of seven square miles, underlain by a friable sandstone generally covered by a silty colluvium containing some medium to fine gravel. The swales and valleys are filled with silty alluvium.

Figure 17-6 suggests a sediment yield of 1000 to 2000 tons/sq mi/yr. Turning to Table 17-2, the regions listed that appear most applicable are the Missouri River and Arkansas-White-Red River basins, showing

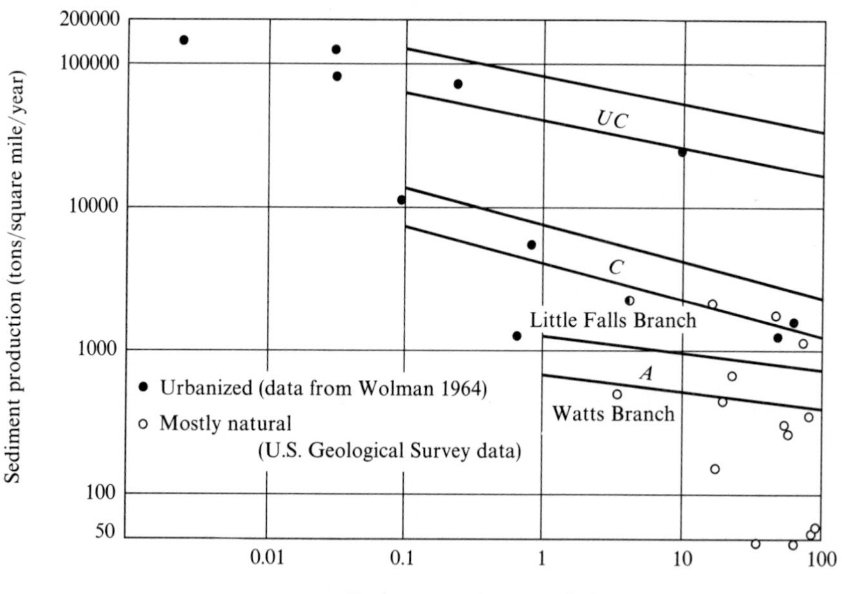

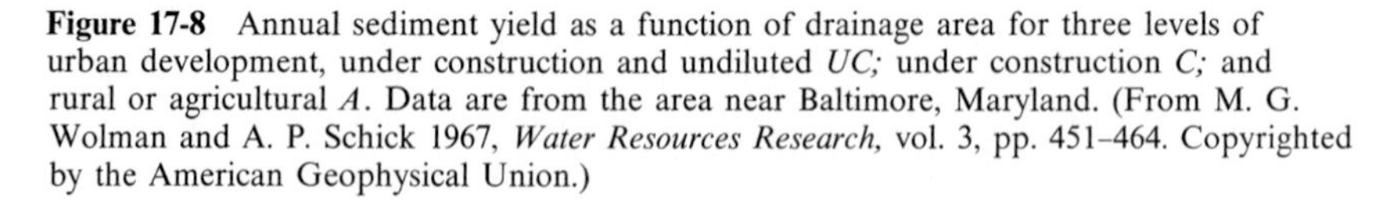
Figure 17-8 Annual sediment yield as a function of drainage area for three levels of urban development, under construction and undiluted *UC;* under construction *C;* and rural or agricultural *A*. Data are from the area near Baltimore, Maryland. (From M. G. Wolman and A. P. Schick 1967, *Water Resources Research,* vol. 3, pp. 451–464. Copyrighted by the American Geophysical Union.)

average values of sediment yield of 1500 and 2200 tons per square mile per year, respectively.

The next check is Figure 17-7, using the nomogram for the upper Mississippi basin. The annual runoff for western Nebraska can be read from the runoff map published in most complete form in the National Atlas of the United States (U.S. Geological Survey 1970), page 119; or in condensed form in Leopold, Wolman and Miller (1964), page 59; and in Langbein (1949). From this map it is ascertained that the annual runoff is probably about 2.0 inches, or 106 acre-feet per square mile.

Estimate first the sediment production rate before the projected development. For the basin above the reservoir, enter Figure 17-7 at seven square miles, move downward to an estimate of the percentage of basin in cultivation, which from general observation must be about 50 percent for Nebraska. Then move horizontally to an annual runoff of 100 acre-feet, and move down to sediment production to read about 1500 tons per square mile per year.

From Figure 17-8, during construction the sediment yield can be expected to be 50 times the normal rate. From 40 acres of development, the additional sediment yield to the reservoir would then be

$$50 \times 1500 \times 40 \div 640 = 4700 \text{ tons per year}$$

The volume of reservoir capacity allotted to sedimentation would be obtained as follows. Over a period of 20 years, for design purposes, the sediment to be stored with no development is

$$7 \times 20 \times 1500 \times \frac{2000}{100} = 4.2 \times 10^6 \text{ cu ft} = 96 \text{ acre-feet}$$

in which the specific weight of deposited sediment is 100 pounds per cubic foot. Added to this is sediment produced during a 2-year period of construction from the 40 acres:

$$2 \times 4700 \times \frac{2000}{100} = 188{,}000 \text{ cu ft} = 4.3 \text{ acre-feet}$$

Thus the sediment volume to be provided in the reservoir during a 20-year life is 100 acre-feet.

Bibliography

BAGNOLD, R. A. (1966) An approach to the sediment transport problem from general physics; *U.S. Geological Survey Professional Paper 422-1.*

BAGNOLD, R. A. (1973) The nature of saltation and of "bedload" transport in water; *Proceedings of the Royal Society of London,* Series A, vol. 332, pp. 473–504.

BRUNE, G. M. (1948) Rates of sediment production in midwestern United States; *U.S. Soil Conservation Service, SCS-TP-65,* 40 pp.

BRUNE, G. M. (1951) Sediment records in midwestern United States; *International Association of Science Hydrology,* Tome III, Brussels Symposium, pp. 29–38.

HELLEY, E. J. AND SMITH, W. (1971) Development and calibration of a pressure-difference bedload sampler; *U.S. Geological Survey Open File Report,* Menlo Park, CA, 18 pp.

LANGBEIN, W. B. et al. (1949) Annual runoff in the United States; *U.S. Geological Survey Circular 52,* 14 pp.

LEOPOLD, L. B. (1970) An improved method for size distribution of streambed gravel; *Water Resources Research* vol. 6, no. 5, pp. 1357–1366.

LEOPOLD, L. B. AND EMMETT, W. W. (1976) Bedload measurements, East Fork River, Wyoming; *Proceedings of the National Academy of Science, USA,* vol 73, no. 4, pp. 1000–1004.

LEOPOLD, L. B. AND MADDOCK, T. (1953) The hydraulic geometry of stream channels and some physiographic implications; *U.S. Geological Survey Professional Paper 252.*

LEOPOLD, L. B. AND MILLER, J. P. (1956) Ephemeral streams: hydraulic factors and their relation to the drainage net; *U.S. Geological Survey Professional Paper 282-A,* 36 pp.

LEOPOLD, L. B., WOLMAN, M. G. AND MILLER, J. P. (1964) *Fluvial processes in geomorphology;* W. H. Freeman and Company, San Francisco.

U. S. DEPARTMENT OF THE INTERIOR, GEOLOGICAL SURVEY (1970) *The national atlas of the United States;* Washington, DC, 417 pp.

U.S. WATER RESOURCES COUNCIL (1968) *The nation's water resources;* Washington, DC.

WOLMAN, M. G. AND SCHICK, A. (1967) Effects of construction on fluvial sediment, urban and suburban areas of Maryland; *Water Resources Research* vol. 3, no. 2, pp. 451–464.

18

Channel Changes

Consideration of the hydrograph leads logically to the conclusion that alteration of the ground surface of a basin may change both the infiltration rate (and thence the volume of runoff from a given storm) and the lag time of the basin. Usually these operate in the same direction–increase of runoff volume and shortening of lag time. Both tend to increase peak flow from a given storm. Such increase may be expected to result from a variety of land-use alterations, such as urbanization, grazing, agriculture, forest clearcutting, and others.

The same factors affecting surface runoff will also tend to change sediment load. It can be expected then that the stream channels will react to a new relation of discharge and sediment and to the altered flow regimen, especially the frequency-magnitude characteristics of high discharge.

Channels change by erosion and deposition. These take time. Therefore, a change in discharge and sediment parameters does not produce an immediate change in the stream channel but rather initiates a change that may extend over a period of time. The channel is altered both by the change in flow frequency and by the change in sediment yield, and these two alterations may not be coincident in time.

Thus, the channel may exhibit different responses to changes in surface conditions of the basin depending on the relative magnitude of the change in flow and the change in sediment yield from the original or undisturbed state.

It is perhaps the most important hydrologic problem in the present state of knowledge to determine quantitatively how much alteration in the rainfall-runoff-erosion relation is necessary to make a given type and amount of channel change. This does not sound like a difficult problem considering the mass of information available on the effect of varying agricultural crop and tillage practices on runoff volumes, peaks, and sediment production. But there is a void in translating these established relations to stream channel response. For example, there are masses of data from plot studies to show that the change from grass cover to corn increases runoff volumes, runoff peaks, and sediment. But what such a change will do to the channels downstream is not documented. The channel reaction will depend not only on the type of change of plant cover and soil condition but on the percentage of the basin affected, the climatic region, the topography and slope, and presumably many other interconnected basin characteristics.

Furthermore, the types of observations required to accumulate such knowledge, though in themselves simple enough, require time—for, as mentioned above, the channel reaction to change is not immediate. Time is a plentiful commodity in the geologic setting, but at the human scale it is dear. Nor will the substitution of space for time be entirely satisfactory in such a search. To some extent it will be necessary to instigate long-term observational programs in which causal factors and their effects will be monitored. But pending the acquisition of such data, the only course of action is to analyze those cases that appear capable of yielding some generalizations; however tentative they may be, we will begin here with a discussion of channel changes known to have taken place in that span of geologic time about which most is known and the duration of which was sufficiently long to have produced significant changes: the Holocene period. Then the changes observed to have occurred within the span of recent observation will be discussed, though the final results of the channel reaction may not yet have been fully achieved.

Climatic Change

The retreat of the ice sheets was not followed by a progressive climatic warming. Fluctuations from warmer than present to colder have punctuated the Holocene over much of the Northern Hemisphere at least, and the changes were not areally uniform either. There appears to have been a more or less general warm period about 5000 years before present referred to in North America as the Altithermal, and it had a counterpart in Europe. Thereafter the climate became colder and wetter, and mountain glaciers advanced, interrupted in the southwestern United States by further fluctuations, including what appears to have been a warm and dry period lasting a century or more in immediately pre-Columbian time. This picture of climatic changes in the Holocene has been put together by a large number of scientists in different fields using evidence from pedology, stratigraphy, archeology, palynology, and river morphology.

The climatic changes left their imprint on river systems; in fact, rivers offer the most obvious evidence of these events of the recent geologic past, the most prominent of which is the existence of terraces bordering rivers nearly everywhere. Stratigraphic evidence of successive periods of erosion followed by deposition implies that there have been changes in the production of water and sediment. The dating of these events depended for a long time on soils and archeology, but the development of dating techniques by radiocarbon and by vegetation changes indicated by pollen profiles have greatly expanded knowledge of the dates and character of these episodes. That the successive periods of aggradation alternating with degradation before the nineteenth century were climatic in origin is in accord with general knowledge of hydrology; alternative explanations such as tectonic uplift, eustatic variations in base level provided by the ocean, or the effects of grazing animals on vegetal cover can in most cases be eliminated as possible causes.

Probably the sequence most intensively studied and best understood is that of the semi-arid southwestern United States, where under the leadership of Dr. Kirk Bryan and later expanded by his students and followers, a general chronology of events has been established, differing in detail from one area to another, but applicable generally to a swath of the continent from Montana to the border of Mexico. In simplest outline the observed morphologic character of alluvial valleys and the inferred climatic character are illustrated in Figure 18-1 and Table 18-1.

The four generalized valley cross sections in Figure 18-1 deserve some comment. The erosion preceding Deposition I cut down to bedrock. The sediments of Deposition I contain a well-developed caliche horizon and in some places contain extinct fauna indicating a late Pleistocene age. Most of that alluvial fill was eroded away before the subsequent Deposition II. Some of the oldest portions of Deposition II are marked at the top by a pedocalic paleosol attributed to the more arid climate of the Altithermal period, about 5000 years before present (BP). The younger alluvium of Deposition II contains paleoindian artifacts and pottery dated in the interval AD 950–1300.

All the sections have certain characteristics in common: the wide valley floor or most extensive level is underlain by Deposition II material. This is true in the California section as well as in the others, where W. W. Haible, whose observations furnished the section shown, obtained wood near the base of the major alluvial fill dated at 3000–4000 years BP.

The modern gully as well as most previous periods of downcutting eroded down to bedrock in many places. Both the deep and the steep-walled or boxlike character typified previous gullies as well as the modern gully.

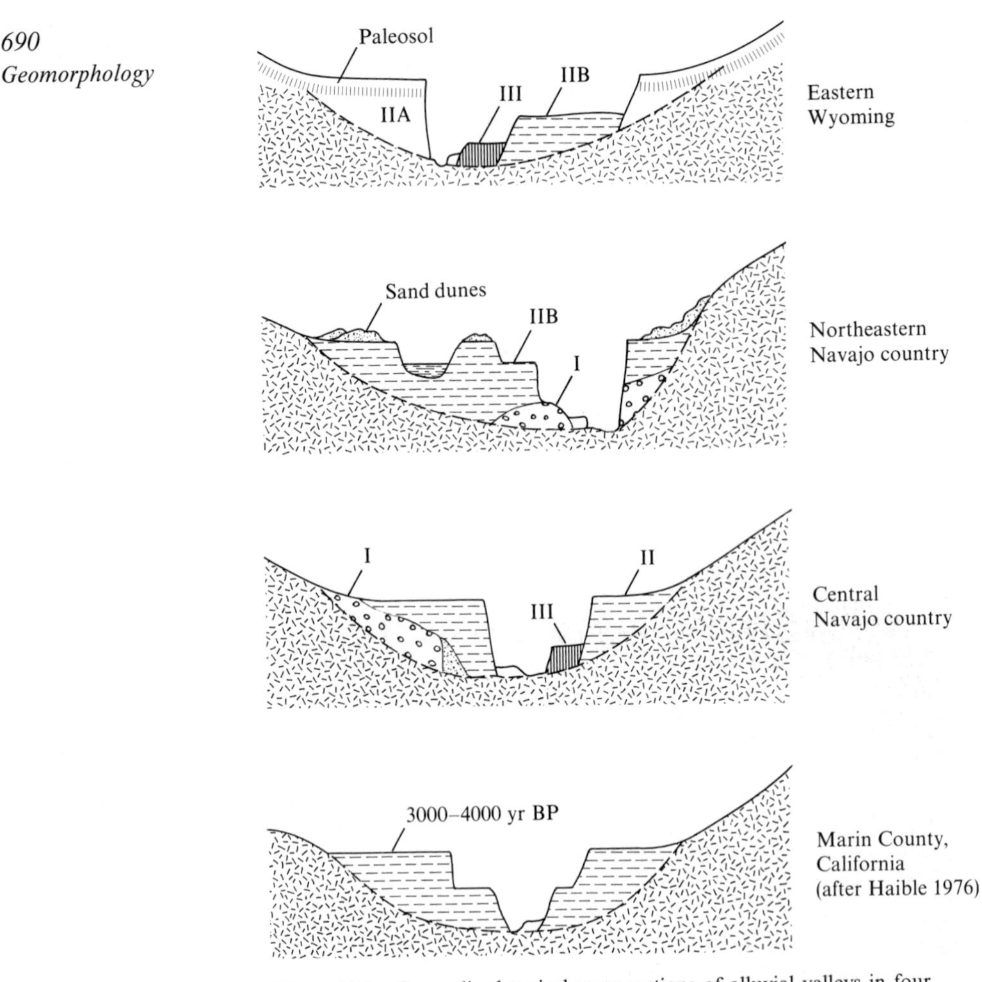

Figure 18-1 Generalized typical cross sections of alluvial valleys in four areas of western United States. In each, previous epicycles of erosion had cut down to bedrock as has the modern gully. Stratigraphic units are briefly described in Table 18-1.

The high terrace, constituting in many places the valley flat, stands 15 to 30 feet above the present or recent channel bed, and a middle terrace, where it exists, stands 6 to 10 feet above the stream bed.

In all cases the modern gully began downcutting in the period 1880–1900 and was presumably the result of the combination of climatic factors, especially an increase in intensity of summer storms and exceptionally heavy grazing by stock (Leopold 1951).

It is now well established (Broecker 1975) that in about 1945 a climatic shift began in the United States, and probably elsewhere, toward what appears to be cooler and probably wetter conditions. Even before this was

Table 18-1 Alluvial chronology of western valleys, United States.

PERIOD AND NAME (AFTER J. T. HACK)	CHARACTER AND DATE	INFERRED CLIMATE
Deposition I, Jeddito	Includes extinct animals—Pleistocene	More arid at end of period
Erosion	Sand dunes locally; paleosol of more arid type	More arid
Deposition II, A and B, Tsegi	Often subdivided into parts; paleosols at end of first phase; second phase contains artifacts dated as late as AD 1200	More humid
Erosion	AD 1200–1400	Warmer, more arid
Deposition III, Naha	Ended with nineteenth century erosion	More humid, colder near end
Modern gully erosion beginning about AD 1880	Generally intensified and in places caused by overgrazing	Summer rainfalls exceptionally intense
Initial aggradation or cessation of gully extension	Began 1940–1960	Trend toward cooler and more precipitation in most but not all regions

clearly evident in climatic records, we had observed a slow aggradation of arroyo channels at cross sections established as part of the Vigil Network, a program of repeated resurvey of selected channels. Some data on the observed deposition have been published (Leopold, Emmett and Myrick 1966; Leopold and Emmett 1972). There is also visual evidence in many places in Wyoming, Colorado, and New Mexico that gully floors that were bare of vegetation in the 1940's were green and often moist in the 1970's, as illustrated in Figure 18-2. Gullies eroding during the first half of the century have begun to aggrade or fill.

The totality of these observations shows clearly that the relations among rainfall, runoff, vegetation, and sediment altered, creating at times a condition in which more sediment was eroded from the landscape than could be carried through the valley by the available water. This condition alternated with one in which flood flows were capable of eroding the valley alluvium. How much and what type of climatic change would be required to provide such a reversal in river action? The answer must be deduced and must be qualitative at best. It can be argued that a shift toward greater aridity would decrease the floods capable of eroding the valley fill, would decrease the density of vegetation, and thus would promote erosion from the hillslopes, but with insufficient flow to carry the sediment out of the valley, with the result that the valleys would aggrade.

Figure 18-2 Tributary to Chinle Wash near Dinnehotso, Arizona, looking upstream. Herbaceous vegetation is becoming established on the point bars and on the bed in some places as deposition has begun to replace erosion since the 1950's.

The logic of that reasoning, however, is countered by direct evidence of three kinds. First, the occurrence of sand dunes is stratigraphically associated with gully formation. Dune sand lies stratigraphically on the eroded remnant of Deposition I and in the channel eroded in that alluvium. Sand dunes, now stabilized, lie on top of the alluvium of Deposition IIB, indicating that the deposition phase ended with the onset of more arid conditions.

Second, the cessation of Deposition IIA in Wyoming was followed by a period of soil formation resulting in the pedocalic paleosol containing strong accumulation of calcium carbonate in the B horizon, an indication of semi-arid conditions. Similarly, the eroded remnants of Deposition I are strongly calichified, also implying aridity. Aggradation after 1950 was coincident with the trend toward cooler and presumably wetter conditions of the third quarter of the present century (Leopold 1976).

The difference between localities of higher versus lower annual precipitation in the semi-arid zone is in the number or frequency of low-intensity and small rains, which contribute to the growth of vegetation, high annual totals accumulating from a large number of small rains (Leopold 1951).

It is not possible, however, to specify quantitatively how much change in what climatic parameters are necessary for reversal from valley aggradation to erosion. In the present century the observed change toward aggradation cannot yet be correlated with a specific change in annual, seasonal, or intensity-level of precipitation at any specific location (Leopold, Emmett and Myrick 1966:240). Reliance so far must be placed on general regional tendencies.

The desirability of pinpointing the amount and type of change in rainfall-runoff conditions necessary to cause specific channel changes should be obvious when one considers the environmental consequences of the many landscape alterations caused by man's activities. These alterations affect the hydrologic cycle through the effect on infiltration capacity, amount, and effectiveness of vegetal cover, changes in concentration or lag time of basins, and changes in hydraulic roughness and thus water velocity in channels and in overland flows. These result in changes in the volume of storm runoff, peak discharge, rate of erosion, and transport rate of sediment–in other words, in all the factors that determine channel stability and the tendencies for channel change.

It is in this respect that the effects of human activities on the hydrologic cycle are similar to the effects of climatic change, the main difference being merely that humans have not yet had a major or regional effect on precipitation itself. But changes due to climatic fluctuation and human activity markedly and similarly affect the land phase of the cycle.

From the channel changes of the Holocene, a few generalizations may be drawn that have application to understanding or forecasting the effects of urbanization, grazing, vegetation clearing, and agriculture. Though they are inferences and for the most part qualitative, they are of some value for forecasting or interpreting channel change.

Changes in frequency and magnitude of high flows probably is the factor most conducive to channel change. To the extent that this is true, the shortening of lag time of small basins by sewerage, channel straightening, and smoothing is most likely to cause channel erosion downstream.

Channel slope changes slowly and only slightly. It is highly conservative and is likely to maintain close to its original value whether the channel is aggrading, degrading, or stable.

Urbanization

When an area is developed for housing or other urban purposes, the immediate hydrologic effect is to increase the area of low or zero infiltration capacity and to increase the efficiency or speed of water transmission in channels or conduits (Leopold 1968; Rantz 1971). Collateral effects are an increased sediment yield at least temporarily and, in many instances, a decrease in drainage density or number of actual channels to carry the increased sediment load. The subsequent effect on stream channels is considered here. These latter effects as in the case of climatic change, depend on the relative efficacy of the flows to carry away the sediment produced from the basin, and on the ability of the channels to remain in equilibrium with the newly imposed conditions.

In urban development the smallest rills, swales, or incipient channels tend to be looked upon by the construction crew as being insignificant, unimportant, or a nuisance. Many a house is built over a buried pipe intended to carry runoff toward some obvious channel; so, in fact, first-order channels, ordinarily ephemeral, are either eliminated by grading or put in a pipe or conduit. The net result is the virtual expurgation of many first- or even second-order channels that under natural conditions played a role in keeping both sediment and runoff distributed or divided among many small channels, each of which played its part in delaying movement of flood peaks, providing channel storage and slowing the average speed at which water was delivered to the larger stream channels.

An example is provided by the urban stream Rock Creek that drains much of the area of Washington, DC. Figure 18-3 shows the drainage net of this stream in 1913 before modern suburban expansion and in 1964. A large percentage of small channels identifiable in early aerial photographs are eliminated or buried underground in pipes.

Though further urban development in that basin is no longer visually obvious because so little open land other than parks remains to be devel-

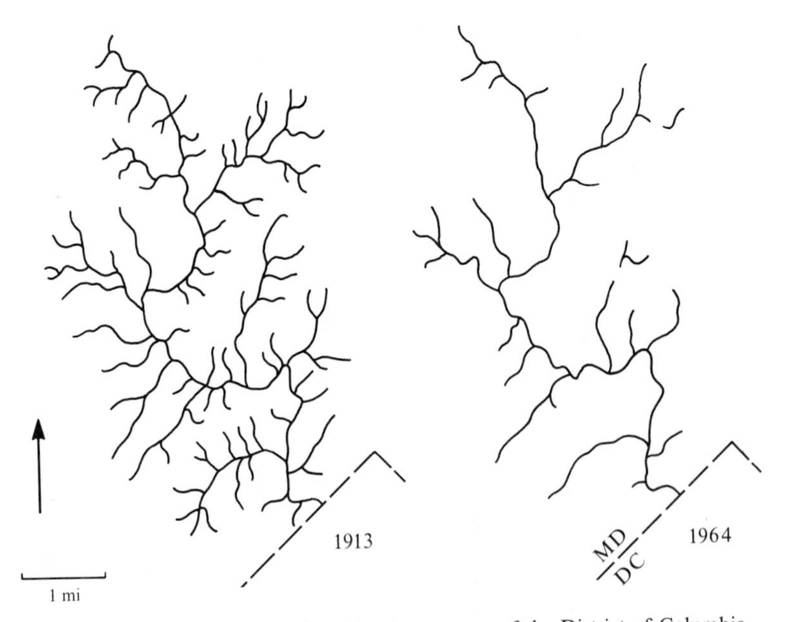

Figure 18-3 Drainage net of Rock Creek upstream of the District of Columbia–Maryland line in 1913, before modern urbanization and again in 1964. (From U.S. Geological Survey in Dept. Interior 1964.)

oped, the creek is muddy with silt in every freshet, and the number of times each year the discharge reaches or exceeds bankfull has increased 10 to 20 times over its previous frequency.

For basins of drainage area less than 10 square miles, urbanization apparently tends to increase, on the average, the cross-sectional area of a channel at bankfull level. Though this enlargement may lag in time the changes of basin surface that cause it, the implication is that for small drainage areas the increase in flood peaks overcompensates for concurrent increases in sediment yield. This conclusion follows from the fact that enlargement means more channel erosion than deposition. The amount of enlargement must be related to some average value of channel size in a particular region.

The channel size expected in the usual, normal, or unurbanized state might best be expressed as a function of drainage area. It is established that on the average the bankfull width and depth increase downstream with the increase of discharge according to the power functions:

$$w \propto Q^{0.5}$$
$$d \propto Q^{0.4}$$

It follows then that the bankfull cross-sectional area, $A = wd$, varies with bankfull discharge downstream as

$$A \propto Q^{0.9}$$

Using also the relation of bankfull discharge to drainage area, that for the humid-temperature region of the eastern United States is on the average

$$Q \propto D_A^{0.75}$$

Then in terms of drainage area, the cross-sectional area of the channel should be

$$A \propto (D_A^{0.75})^{0.9} = D_A^{0.67}$$

By the same reasoning, bankfull width and depth would vary with basin area as

$$w \propto Q^{0.5} = (D_A^{0.75})^{0.5} = D_A^{0.37}$$

and

$$d \propto Q^{0.4} = (D_A^{0.75})^{0.4} = D_A^{0.30}$$

Figure 16-16 presents a plot of stream cross-sectional area as a function of drainage area for rural or unurbanized channels, the lines representing the mean relations for a large number of examples. The slopes of the lines on Figure 16-16, respectively 0.66, 0.39, and 0.28, are very close to the exponents derived above.

Despite considerable differences in climate and especially in mean annual precipitation, data show that for the same drainage area the channel size is similar. The scatter of points for individual locations is as great as are the regional differences among the average values. Such graphs can be con-

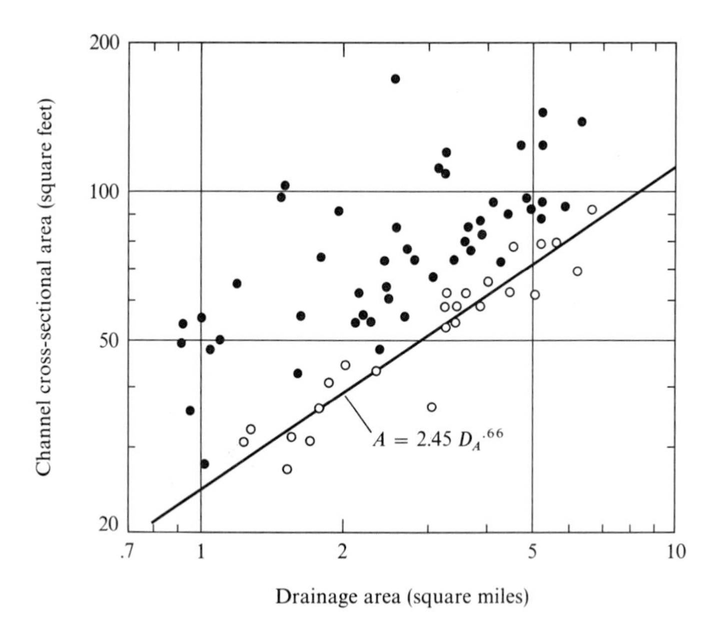

Figure 18-4 Channel cross-sectional area at bankfull as a function of drainage area in Pennsylvania. Open circles represent channels in rural or unurbanized condition; solid circles are channels draining areas urbanized to various extent. (From Hammer 1971.)

structed for any region by plotting at-a-station graphs for each gauging station, and from the published record compiling a flood-frequency curve for each station. Reading the value of discharge for a recurrence interval of 1.5 years, this discharge may be entered into the at-a-station curves to read values of width and depth, which then can be plotted against drainage area. They will represent an approximation to bankfull conditions. Because slightly different average lines would be obtained from a different set of gauging stations in the same region, it is urged that these curves be considered general guides, and that for a particular problem, station data for the vicinity be compiled as mentioned above as a check against the curves here presented.

These average channel dimensions can be used to estimate the enlargement of channel size resulting from urbanization. At the suggestion of Leopold, Thomas Hammer (1972) measured actual channel size downstream of areas having different degrees and types of urbanization, and from the analysis he determined by correlation techniques the relative importance of various factors influencing channel enlargement.

In Figure 18-4 Hammer's data are plotted in the form of channel area at bankfull as a function of drainage area, separating the unurbanized or natural basins from those affected by urban development. The average relation for the natural basins gives the equation

$$A = 24.5D_A^{0.66}$$

which is a close approximation to the generally expected relation described above. Unfortunately, Hammer does not list the width and depth that made up his measured cross-sectional area of channel at bankfull, but from the other data and the equations shown earlier, a probable relation of width and depth to drainage area that would agree with his cross-sectional area values might be

$$w = 15.5D_A^{0.38}$$

and

$$d = 1.57D_A^{0.28}$$

These lines compare favorably with the dimensions of natural channels unaffected by land alteration, Figure 16-16.

The cross-sectional areas of channel draining basins that have been altered by urbanization, nearly without exception lie above the mean line for natural or unaltered basins, meaning that either the width or depth or both have increased as a result of the increase of flood magnitude and frequency and the concomitant increase in sediment load.

Hammer analyzed his data by evaluating for each basin the parameters in the urbanization process that he thought should change flood and sediment discharge. By multiple correlation he evaluated the relative influence of each parameter in promoting channel change. These influential parameters, arranged in their order of importance, are listed in Table 18-2.

It is interesting that the age of the urban development has a strong influence. It seems probable, though not so interpreted by Hammer, that age correlated with channel enlargement because it takes a certain amount of time for the increased flood discharge from a storm of given magnitude to exert its effect on the receiving channel. One can hardly visualize any hydrologic parameter of a developed area, say a housing project, that in itself

Table 18-2 Ratio of enlarged channel area to natural channel area in a basin of one- to five-square-mile area if all the basin area were in use as specified. (From Hammer 1972.)

LAND USE	RATIO
Wooded	0.75
Previous developed land	1.08
Impervious area less than 4 years old; unsewered streets and houses	1.08
Cultivation	1.29
Houses more than 4 years old fronting on sewered streets	2.19
Sewered streets more than 4 years old	5.95
Impervious areas more than 4 years old	6.79

changes with time, but it is quite possible that the effect of a land surface alteration becomes noticeable in the channel downstream only after some time has elapsed for the channel to adjust.

It will be noted that wooded areas have slightly smaller channels than the average, which presumably included a mix of woods and open or non-wooded areas.

Only a few sets of data are available to define the relative contribution of width and depth to the change in channel cross-sectional area. The most complete are the surveyed cross sections across the channel of Watts Branch, a tributary to the Potomac River near Rockville, Maryland. At a reach where the drainage area is 3.7 sq mi, Leopold (1973) surveyed 14 cross sections more or less annually over a 20-year period (1953–1972).

For these cross sections the mean depth increased 23 percent during the observation period, and the mean width at top of channel banks (bankfull width) decreased 35 percent. The sections become more rectangular, smaller, deeper, and narrower primarily by deposition within the channel, and to a lesser extent by overbank deposition that raised the elevation of the stream bank. The mean bed elevation rose slightly in 10 of the 14 cross sections. Typical changes are illustrated in Figure 18-5.

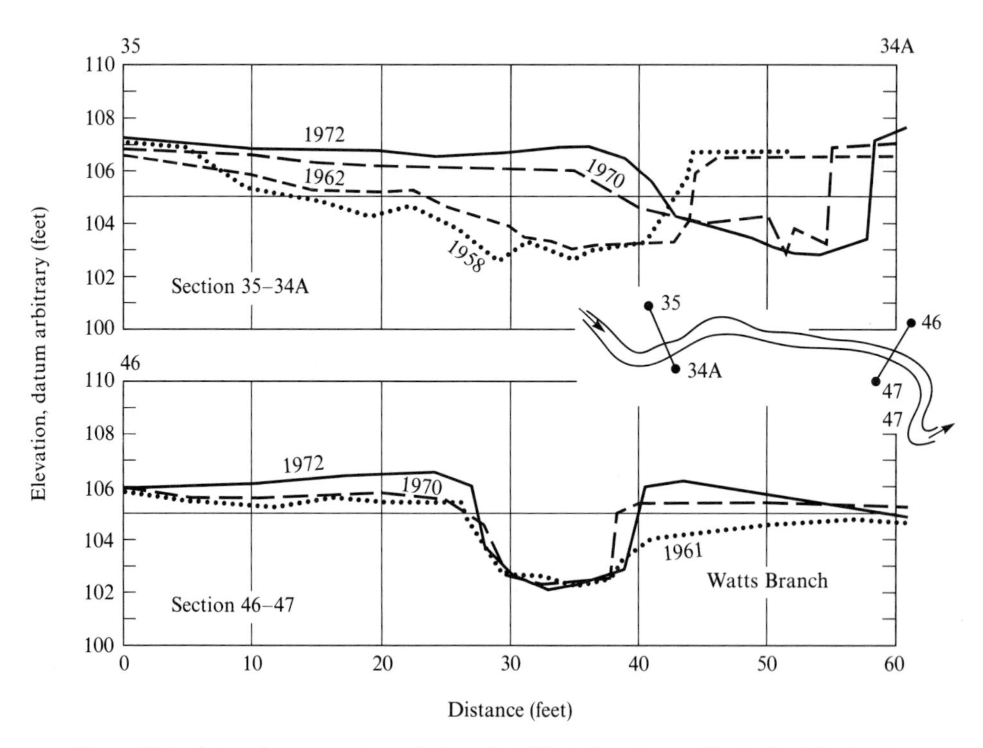

Figure 18-5 Selected cross sections of channel of Watts Branch near Rockville, Maryland, drainage area 3.7 sq mi, over a period of years. Changes are attributed to urbanization. Map at right center shows position of cross sections 35-34A (upper) and 46-47 (lower) relative to local channel bends. (From Leopold 1973.)

Though this decrease in channel area seems to run counter to the observation of Hammer, the difference might best be attributed to the time sequence at which observations were made. Though urbanization, expressed as the total number of houses or structures within the basin, increased linearly with time during the two decades of observation, the channel changes did not follow in the same linear fashion. The number of occurrences per year of discharges exceeding bankfull varied from zero to three for the first 12 years, but in the next 5 years this number varied from four to eleven. That is, a decade of urbanization seemed not to have appreciably changed the frequency of high flows, but thereafter the number of high flows increased rapidly, progressively, and massively.

Various measures of sediment deposition seemed to follow a similar time sequence. In the first decade in-bank channel deposits gradually narrowed the cross section, but in the second decade overbank deposition rapidly increased the height of the banks.

The progressive enlargement of channel area began about 1967, coincident with the marked increase in number of high discharges. In two representative cross sections, the change in channel area between 1968 and 1972 was respectively 40 and 62 percent. Thus, in the second decade, progressive channel enlargement had begun and probably had not achieved a stable channel area.

It is conjectured that in the next decade, with continued urbanization, the channel cross section will expand by erosion because of the high floods and that the channel will then exhibit the change seen by Hammer in the Pennsylvania streams, the net enlargement resulting from urbanization.

One other set of surveyed cross sections is available that shows the effects of urbanization. Santa Fe, New Mexico, annual precipitation 14 inches, has been expanding like most other cities. In 1967 a new development north of the city was about to begin, and plat maps became available of the projected roads and building sites. We monumented and surveyed some cross sections of the arroyo channels in this development chosen to represent, we hoped, some that would be unaffected and some that would definitely be downstream of projected roads and houses. These were resurveyed in 1974, an elapsed time of 7 years and after the roads and some houses were built. Presumably the sediment observed to have been deposited in the channels was derived from the driveways, averaging 20 feet in width and 150 feet in length, at an average gradient of 20 percent. The houses or disturbed areas averaged 60 by 50 feet. Thus, ground made less pervious to infiltration and tending toward increased runoff averaged 6000 sq ft in each lot of 150,000 sq ft, or 4 percent of the lot area.

The total length of street area in the 0.23 sq mi subbasin is about 5000 lineal feet, which at 30 feet in width takes up an area of 3.4 acres. Thus, the area comprising houses and driveways is 2.9 acres, and the roads constitute 3.4 acres or 6.3 acres in a subbasin of 148 acres, or 4 percent of the total area.

The percentage of total subbasin area is not large, but the alterations are characterized by numerous steep cut banks along roads and driveways, rendering the disturbed areas less permeable than originally. Piñon and juniper woodland in this area is nearly devoid of grass, so the infiltration rate even under natural conditions is low and the soil is easily eroded. The silt, when it occurs, is in the near-surface skeletal soil profile. The channel bottom is loose, fine gravel mixed with coarse sand. Vegetation is sufficiently sparse to have essentially no soil-binding effect. The channels are ephemeral, carrying water only a few days each year during the thunderstorm rains of summer. No winter discharge occurs.

Sections 2 and 1 (Figure 18-6), which receive runoff from 11.5 and 16.5 acres of basin area respectively, were chosen to represent upstream channels of order 2 that would presumably be unaffected by roads or building sites. If these partake of the general trend of channel change in those draining areas of 0.1 to 5 square miles in the region, they would be expected to be very slowly aggrading or stable. The annual aggradation rate for similar channels in the Santa Fe area is 0.05 feet. The expected was observed. The channel bed at Section 2 did not change elevation and that of Section 1 aggraded about 0.1 foot in the seven years.

In contrast, Sections 3 and 4 (Figure 18-7), receiving runoff from 132 and 146 acres respectively and draining an area where roads and 18 houses were built in the period between the surveys, showed marked changes. At Section 3, pictured in Figure 18-8, the channel degraded 2 feet and narrowed from an average width of 19 feet to 12 feet. At Section 4 the channel bed degraded or cut 1.8 feet and narrowed from 31 to 21 feet.

These observed results of urbanization are in keeping with other experiences in that the increase in flood frequency and magnitude tending toward channel downcutting or enlargement eventually overcompensates for the presumed concomitant increase in sediment yield. In Watts Branch, Maryland, the first few years of urban development led to deposition and slight channel contraction followed by a marked increase in channel size. This channel enlargement coincided with markedly increased frequency and size of flood discharges.

Channel changes resulting from urbanization have not been extensively documented, but those mentioned in the literature appear to be comparable to the examples described quantitatively above. Ian Douglas (1974) has published exceptionally good data on changes in sediment load in New South Wales and Malaysia, and he provides qualitative information on channel changes of two types. Dumaresq Creek has been widened in the immediate vicinity of new bridges built in Armidale, New South Wales, and immediately downstream from the widened sections, "Sedimentation occurs due to loss of velocity in the wider channel sections." The experience shows the necessity of maintaining continuity of slope and channel cross section if channel changes are contemplated.

The Anak Ayer Batu in the Kuala Lumpur area has been impacted by urbanization, changing progressively "from a deep winding channel . . . to create a straighter, shallower, wider and steeper channel, much to the cost

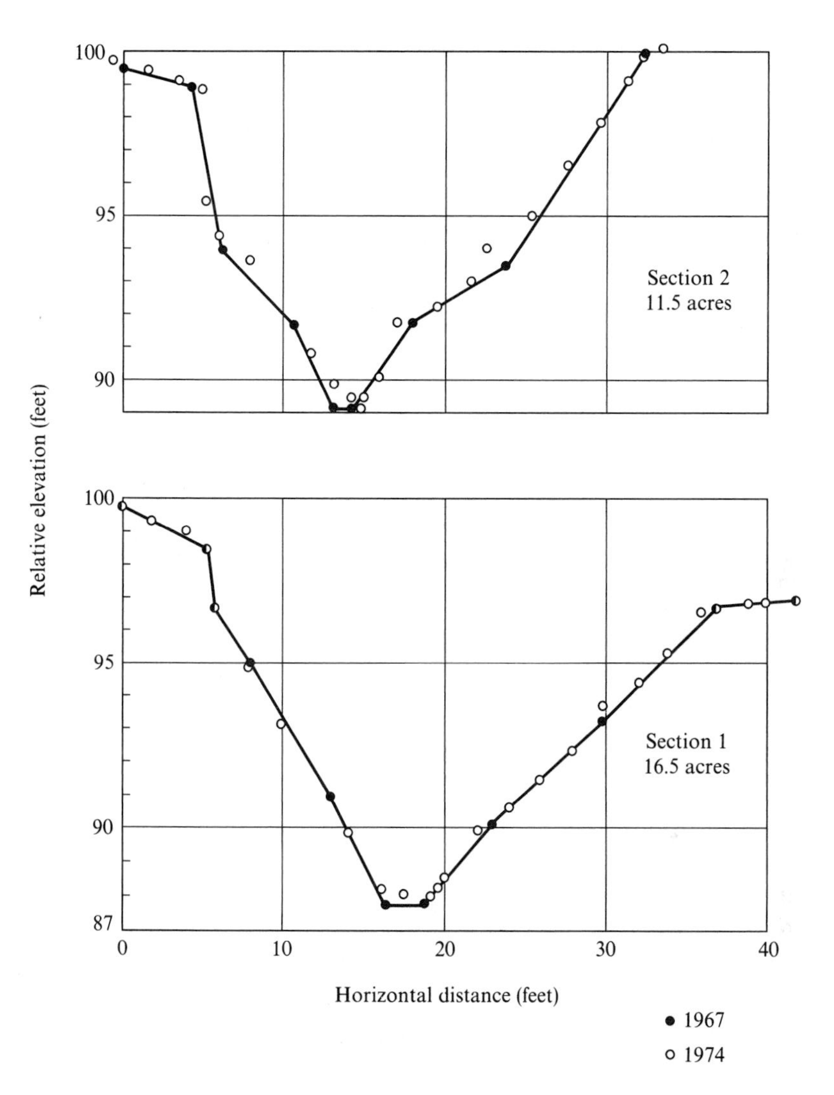

Figure 18-6 Channel cross sections of tributary to Cerro del Piedra Arroyo, Santa Fe, New Mexico, in 1967 (solid circles and line) and in 1974 (open circles). These sections are in a channel not affected by the local urbanization. Channel filling (aggradation) has taken place in the 7 years between surveys.

of the University of Malaya which has had to install gabions along the river . . ." (Douglas 1974:316).

Another qualitative observation sufficiently general to be worthy of mention concerns the bank erosion immediately downstream from the end of a concrete urban channel. Where the trapezoidal or rectangular artificial section ends and a natural channel begins, the banks of the latter tend to erode back in bulbous, caving reentrants, making the channel exceptionally

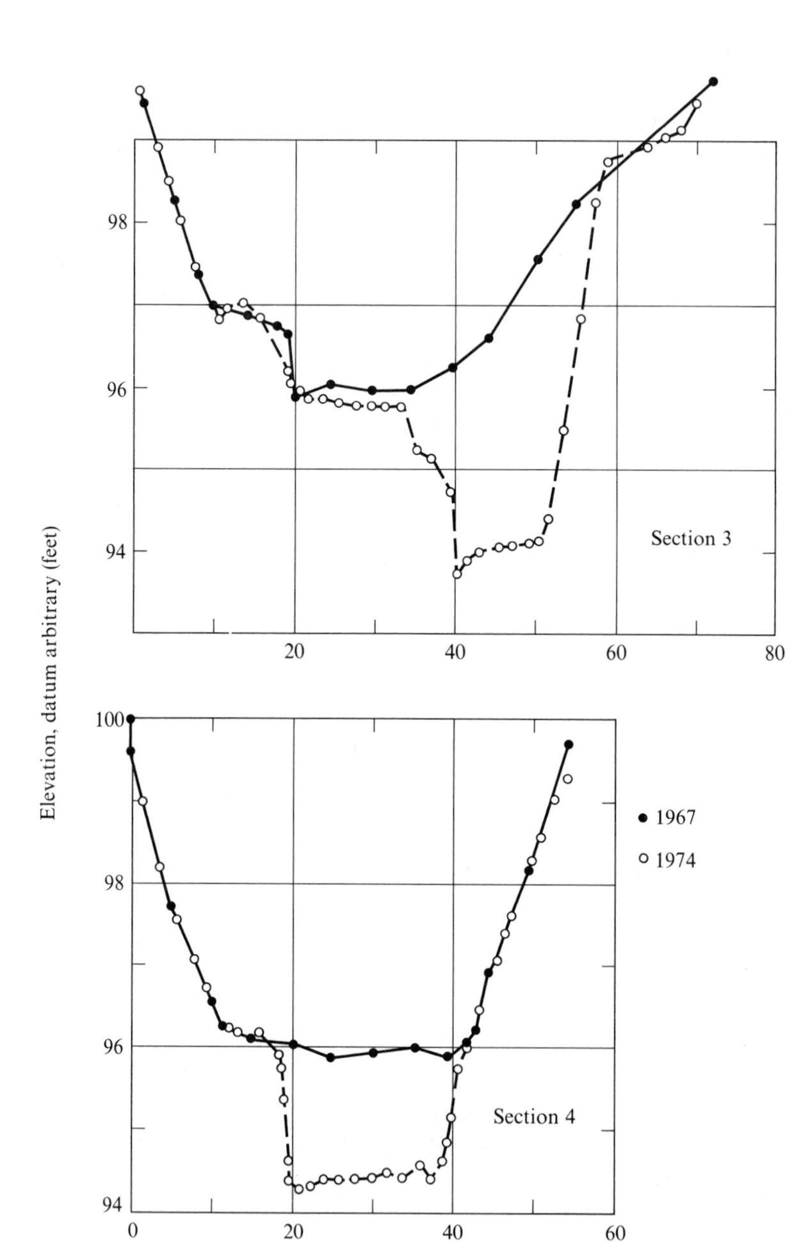

Figure 18-7 Channel cross sections being affected by local urbanization. The channel is Cerro del Piedra Arroyo, Santa Fe, New Mexico, surveyed in 1967 (solid circles and line) before urbanization and in 1974 (open circles) after construction of roads and houses.

Figure 18-8 Channel of Cerro del Piedra Arroyo, Santa Fe, New Mexico, after urbanization had begun. The stream bed had formerly been level with the flat area seen at the right, but it eroded downward 2 feet and simultaneously narrowed the low water channel.

wide for a short distance. The danger in this local bank recession is that the lower end of the concreted section will be undermined laterally or at the toe. Of course, the original purpose of making a concrete channel is to decrease the flooding and bank erosion caused by the urbanization, increasing the hydraulic efficiency by straightening and thus providing a smooth artificial channel. This speeds the runoff, but also increases the peak discharge, often necessitating further downstream extension of the artificial channel section. And so each action creates the need for further construction and more concrete.

Channel changes in urban streams cause costly construction and maintenance and unsightly, dangerous conditions. Many a small stream through an urban area will be seen encased in concrete, lined with a high, strong wire fence to keep children from falling into the fast-moving water during storm flow.

An alternative design might well have been a stream-side park with bicycle paths, picnic places, attractive vegetation, and hydrologically desirable flood capacity that will tend to decrease rather than increase flooding downstream. Where this latter alternative is chosen, the urban creek can be an esthetic amenity and a social asset.

The principles of hydraulic design discussed in Chapters 10 and 16 can be applied to determine the discharge capacity required. The low-flow channel should be kept as close as possible to its original depth and width, maintaining curves, pools, and riffles appropriate to the size of the drainage area. Overbank flood capacity can be provided by the park or green-space area bordered if necessary by low, flat levees to contain the design discharge.

Even if the final hydraulic design is computed by a hydrologic engineer, it is the responsibility of the environmental planner to examine the possible design alternatives and to recommend some appropriate combination of land-use control and hydrologic design that tends to maintain and enhance the esthetic and hydrologic potential of the urban stream.

Channel Change–Effect of Logging, Grazing, and Clearing

Far more quantitative information is available on sediment production and flood peak change resulting from various land uses than on channel change. The hydrologic principles are clear, but how specific channels will react to particular land treatments is difficult to forecast. At present the best guide that can be offered is a few examples that indicate the nature of channel reaction.

The most completely documented picture of channel change associated with clearcut logging is the study by Janda et al. (1975) on Redwood Creek, California. Commercial logging of redwood (Figure 18-9) has removed practically all the trees from a major portion of the drainage basin. The combination of steep slopes, erosive and landslide-prone soil, high precipitation of an intense character, and the extent of vegetation removal creates a potential for greatly increased sediment load delivered to the channel system. Janda demonstrated that the channel has generally aggraded, by amounts varying from near zero in one reach to more than 10 feet over a long distance. Because the river is large, the extent of this aggradation is immense.

Clearcut logging as practiced in the redwood country leaves the soil bare and the tractor trails, landing areas, and skid trails provide large volumes of disturbed soil to easy movement downslope. The marked reduction in evapotranspiration leaves the soil more moist and thus more susceptible to gravitational downslope movement into stream channels both by mass movement and by landslide.

The inflow of masses of sediment into the river causes the channel to erode laterally, undermining steep slopes that are incipient landslides, and

Figure 18-9 Soil churning by tractor logging in clearcut redwoods, basin of Redwood Creek, California.

the resulting mass movement into the channel exacerbates the channel instability.

The most extensive channel changes resulting from grazing are those in New Zealand. Sheep raising is an important commercial use of land, especially on the North Island, and pasture rotation, fertilization, and grass planting are extensively utilized so that in most areas the soil is effectively protected from even the intense concentration of animals. But as in the case of logging, some combinations of steep slopes, heavy rains, and unstable soil create conditions of active gullying that cannot be reversed even when the initial stimulus is removed. This combination is especially severe in the upper portions of the Waikato Basin, with the result that through a long reach the river bed has aggraded as much as 30 feet. The channel in this aggrading zone is composed of fine gravel with sand, a wide unstable river bed without definite banks and an appearance of active braiding.

The intense erosion of unstable hillslopes in New Zealand caused channel aggradation, whereas in the southwestern United States overgrazing and a concomitant climatic shift caused gully and sheet erosion on hillslopes but deep gullying of valley alluvium. There is no theory that will forecast accurately how a channel system will react to severe alteration of the vegetation–soil complex of a drainage basin. However, a few general observations might be offered.

Subtle channel changes may be expected from changes in the vegetation–soil complex that result in increased sediment load and flood peaks, but these changes will ordinarily be slow and not obviously related to the basic cause. The subtle changes usually will take the form of increased channel shifting, bank erosion, and instability of bars.

The most obvious and serious problems arise where steep slopes are underlain by clayey material. Observation of the extent and nature of past and present landslides and soil creep is one way a planner may anticipate the extent of the potential erosion problem.

Areas where fine gravel and sand are available in quantity as potential sediment load of the streams are likely to experience channel aggradation, widening, and rapid shifting of channels if the load is materially increased. Where valley alluvium and hillslope material are mostly silt or clayey silt, the probable consequence of disturbance of vegetal cover on the basin will be channel erosion, deepening, and gullying.

Alteration of Stream Channels–Channelization

Under legislation authorizing federal assistance to achieve land drainage and flood damage reduction, a large amount of work has been done to alter stream channels. On large rivers the "downstream" programs of three federal agencies, the Corps of Engineers, the Bureau of Reclamation, and the Tennessee Valley Authority, have led to the construction of 8300 miles of levees and floodwalls, and 5000 miles of "channel improvement." Under the "upstream" program of the Soil Conservation Service, there has been an additional 3200 miles of channel "improvement." These figures do not include the construction of dams for storage of floodwaters. A summary that includes analyses of the results of a large sample of individual projects has been prepared for the Council on Environmental Quality under the title "Report on Channel Modifications" (1973).

The procedures used for project evaluation have been discussed not only in handbooks of the constructing agencies but elsewhere. The above-mentioned report presents for the projects used as a sample, a summary of the costs and benefits anticipated when each project was authorized. Aspects of the program related to flood damage reduction were analyzed in depth in "The Flood Control Controversy" (Leopold and Maddock 1954), which, though written some years ago, describes procedures, advantages, and disadvantages applicable today. The principal change that has occurred since that book has been in the increased number of projects and the increased cost. The average annual flood damage nationwide, which was increasing in 1954, has continued to increase to the present and for the same reason. The use of flood-prone land continues to rise faster than the application of measures to reduce flood damages. This continues to be one of the foremost challenges to land planners–finding ways to control the use of flood-prone areas, and ways of requiring those who seek the advantages of use of floodable areas to assume a fair proportion of the financial risk involved in such use.

There is a large amount of literature dealing with most aspects of channel modification to achieve increased agricultural production and decreased losses from flooding. But several effects, all of which fall in the category of project costs, tend to be overlooked or unevaluated. These are, however, of concern to the planner and environmentalist and will, for that reason, be mentioned here.

Among the potential costs or disadvantages accruing from channel modification are

1. Channel instability or effects of channel readjustment to the imposed conditions;

2. Downstream effects especially increased bank erosion, bed degradation or aggradation;

3. Esthetic degradation, especially the change in stream biota and the visual alteration of riparian vegetation, of stream banks and channel pattern or form.

The last of these, esthetics, constitutes a subject sufficiently broad to demand a chapter of the present book, Chapter 22.

Studies of the interrelations among width, depth, velocity, slope, roughness, and sediment load through the technique of hydraulic geometry bring out a conclusion of singular significance. A channel out of equilibrium or in the process of adjustment to new conditions cannot be differentiated from those considered to be in equilibrium. That is, the interrelations of the hydraulic factors to imposed discharge and load tend to approximate a mean or average. There is sufficient variance in these relationships among streams considered to be stable or adjusted that those thought to be out of adjustment and in the process of change are not sufficiently different to be identified on the basis of their hydraulic characteristics. Stating this conclusion in another way, a stream subject to disruption or change by humans immediately responds by adjustments among hydraulic variables in such a way that an approximation to the original or unaltered condition is approached. The readjustment process may take time, for it is accomplished by erosion and deposition, but in the meantime internal relations are maintained within the constraints of unalterable or not yet altered parameters.

An example is provided by the Pequest River, New Jersey. In the middle of the basin is an area of flat, fertile land near Great Meadows that had originally been subject to flooding. This reach of the channel was dredged into a deep trapezoidal flume in about 1950 for the purpose of eliminating overbank flow. The original channel is estimated to have had a bankfull discharge of about 550 cfs and, judging by unaltered streams in the region, probably had a mean width of 44 feet, a depth at bankfull of 3.3 feet, and a cross-sectional area of 140 square feet.

The dredged cross section was about 10.5 feet deep, top width 90 feet, bottom width 36 feet. In the sketch of Figure 18-10, the presumed cross

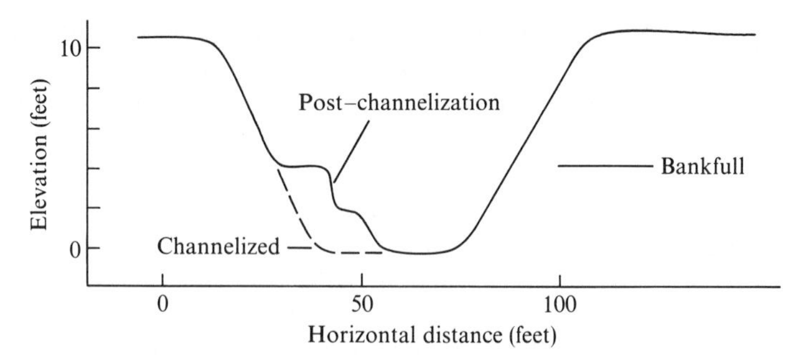

Figure 18-10 Result of channelization of Pequest River, New Jersey at Big Meadow, where drainage area is about 70 square miles. The original channel was about 40 feet wide and 3 to 4 feet deep. Deposition within the dredged cut has rebuilt a channel approximately the original size. The full line shows 1969 cross section, 19 years after the channelization.

section after channelization is shown by dashed line. That observed in 1969 is plotted in solid line.

In 1969 (post-channelization condition) there was deposition on the left side of the dredged trapezoid with a flat top or berm 4 ft above the bed. This narrowing of the channel by deposition created a new cross section considerably smaller than that originally dredged, and more in keeping with the expected dimensions of natural channels. The top of the berm on the left bank is presumed to be coincident with the bankfull stage. The dimensions of the original dredged channel are compared in Table 18-3 with those expected for a natural channel and those resulting from deposition.

This example illustrates how a channelized reach will alter its cross section toward a form and size commensurate with the discharge experienced in the reach, or toward dimensions approximating the original or unaltered stream of its drainage area.

Emerson (1971) studied the channelized Blackwater River, Missouri, and showed that river shortening by the elimination of meanders caused the river to deepen progressively for 60 years with no sign of abatement. Bridges

Table 18-3 Comparison of dimensions of Pequest River, New Jersey, before and after dredging.

	EXPECTED IN NATURAL CHANNEL	DREDGED CHANNEL AT 4-FT DEPTH	CHANNEL AFTER SELF-ADJUSTMENT
Bankfull discharge (cfs)	550		550
Top width (ft)	44	60	35
Mean depth (ft)	3.4	4.0	4.0
Cross-sectional area (ft^2)	140	240	140

were undermined and had to be rebuilt. From this a conclusion of far-reaching significance can be drawn. An imposed change of river slope can cause an instability quite irreversible in any short period of time, and is the most difficult change to which a stream must adjust. In contrast to the example of the Pequest River mentioned above, mere deposition on the channel margin is not sufficient to reestablish quasi-equilibrium. One may conclude that of all the imposed channel changes possible, the one most likely to cause continuing problems is severe shortening with its consequent change in channel gradient.

Equally severe but probably more seldom met as a field example is the radical change in drainage area. Dzurisin (1975) discussed a diversion of ephemeral Furnace Creek wash into the much smaller Gower Gulch, increasing the area contributing to the latter from 5 km^2 to 440 km^2. Even 33 years after, the resulting instability had not yet been arrested. At the diversion, vertical erosion lowered a knickpoint, and Furnace Creek wash responded by headward dissection. Dzurisin estimates that during the next few hundred years there can be expected "(1) increasing depth and extent of headward dissection in Furnace Creek wash and its tributaries upstream from the diversion point, and (2) readjustment of the Gower fan profile in response to accumulation of coarse diverted debris" (p. 309).

Looking generally at the present practices of channelization, Keller (1975) argues the same points made previously here, that the design of projects might easily include the maintenance of principal morphologic features of natural streams. The design should include the construction of a pool–riffle sequence with an average spacing of six times the channel width, the maintenance of natural width–depth ratio appropriate to stream size, a high-water channel of a size in accordance with flood potential, and a low-water normal channel having those natural characteristics that maintain biologic productivity.

Bibliography

Broecker, W. (1975) Climatic change: are we on the brink of a pronounced global warming? *Science*, vol. 189, pp. 460–463.

Council on Environmental Quality (1973) *Report on channel modifications;* Washington, DC, 3 vol.

Douglas, Ian (1974) The impact of urbanization on river systems; *Proceedings of the International Geographical Union Regional Conference*, New Zealand Geographical Society, pp. 307–317.

Dzurisin, D. (1975) Channel responses to artificial stream capture, Death Valley, California; *Geology*, June, pp. 309–312.

Emerson, J. W. (1971) Channelization: a case study; *Science*, vol. 173, pp. 325–326.

Hack, J. T. (1942) The changing physical environment of the Hopi Indians of Arizona; *Papers of the Peabody Museum of American Archeology and Ethnology*, Harvard University, vol. XXXV, no. 1.

Haible, W. W. (1976) Holocene profile changes along a California coastal stream; M.S. thesis, Department of Geology, University of California, Berkeley, 74 pp.

Hammer, T. (1972) Stream channel enlargement due to urbanization; *Water Resources Research*, vol. 8, no. 6, pp. 1530–1540.

Janda, R. J., Nolan, M. K., Harden, D. R. and Colman S. M. (1975) Watershed conditions in the drainage basin of Redwood Creek, Humboldt Co., California, as of 1973; *U.S. Geological Survey Open File Report* no. 75.568, 260 pp.

Keller, E. A. (1975) Channelization: a search for a better way; *Geology*, May, pp. 246–248.

Leopold, L. B. (1951) Rainfall frequency—an aspect of climatic variation; *Transactions, American Geophysical Union*, vol. 32, no. 3, pp. 347–357.

Leopold, L. B. (1968) Hydrology for urban land planning—a guidebook on the hydrologic effects of urban land use; *U.S. Geological Survey Circular 554.*

Leopold, L. B. (1973) River channel change with time—an example; *Bulletin of Geological Society of America*, vol. 84, pp. 1845–1860.

Leopold, L. B. (1976) Reversal of erosion cycle and climatic change; *Quaternary Research*, vol. 6, pp. 557–562.

Leopold, L. B. and Emmett, W. W. (1972) Some rates of geomorphological processes; *Geographia Polonica*, vol. 23, pp. 27–35.

Leopold, L. B., Emmett, W. W. and Myrick, R. M. (1966) Channel and hillslope processes in a semi-arid area, New Mexico; *U.S. Geological Survey Professional Paper 352-G*, pp. 193–253.

Leopold, L. B. and Maddock, T. (1954) *The flood control controversy*, Ronald Press, New York, 278 pp.

Leopold, L. B. and Miller, J. P. (1954) A post-glacial chronology for some alluvial valleys in Wyoming: *U.S. Geological Survey Water Supply Paper 1261*, 89 pp.

Rantz, S. E. (1971) Suggested criteria for hydrologic design of storm-drainage facilities in the San Francisco Bay Region of California: *U.S. Geological Survey Open File Report*, 69 pp.

U.S. Department of Interior (1964) *The nation's rivers*, Government Printing Office, Washington, DC.

IV

River Quality

19

Physical Characteristics of Water

Planners are increasingly concerned with the maintenance of water quality. By the quality of water, we mean the physical, chemical, and biological attributes that affect the suitability of water for agriculture, industry, drinking, recreation, and other uses. These attributes are obviously linked to the available quantity of water, which was treated in Chapter 12. The most important physical characteristics affecting the usefulness of water are its sediment concentration and its temperature. In this chapter, therefore, we look at sediment and heat as pollutants.

Effects of Sediment

The debris load carried by rivers is a natural attribute of the river system, necessary for the maintenance of relative stability among bed and banks, erosion and deposition. To deprive the river of its load will lead to downstream adjustments, either bed erosion or bank cutting. After the construction of Boulder Dam on the Colorado River, the clear water released from the reservoir eroded the river bed for miles downstream, degrading the channel as much as 10 m vertically. Below Fort Peck Dam on the Missouri River the effects included bank erosion and bed armoring as the finer material in the bed was winnowed away. On the other hand, erosion in a drainage basin resulting from logging, grazing, urbanization, or other human-induced changes can overload a river, causing aggradation of the bed, change of channel pattern, and sedimentation of lakes and reservoirs.

Depending on climate, geology, and vegetation, each part of every river system had, before the advent of human influence, a natural and appropriate sediment load. The Colorado and Missouri rivers had generally high silt contents and the Mississippi was called "Big Muddy." In contrast, there are streams famous for water clarity, reflected in their names; the Clearwater of Idaho and the Upper Green in Wyoming. Other names conjure up the vision of fine fishing: the Salmon River of Idaho, the Brule River of Wisconsin, and the Madison River of Montana.

Under what conditions, then, sediment should be considered a pollutant is a matter of definition. Even though the Rio Grande has always had a high sediment concentration and therefore is naturally muddy, to a water-supply engineer the suspended sediment load is a nuisance and might be considered a pollutant.

In general, suspended load may be considered a pollutant when it exceeds natural concentrations and has a detrimental effect on water quality in its biologic and esthetic sense. In these respects an increase in turbidity can be considered a type of pollution in that it affects the biotic balance. A few examples may illustrate.

The upper Pecos and a small tributary, the Holy Ghost, in New Mexico were once considered prime trout streams that were always clear, even after a rain. In the 1940's a road was built farther upstream than any that had existed previously, the upper reaches having been accessible only by foot or on horse. The road was the only obvious change, but since its construction the streams flow muddy or murky after nearly every rain. Studies of the effect of logging roads sustain the inference that the post-storm turbidity can be attributed more to erosion on roads than to any other source.

An increase of silt or fine-sediment concentration in trout streams has several biologic effects. Trout lay eggs in shallow depressions swept out of the gravel by strong side-wise action of the fish's tail. The eggs do not wash out of the depression because at low flow, water at the head of a riffle moves downward into the gravel and through it in a downstream direction, emerging into the channel flow at the foot of the riffle. The downward motion of water into the gravel is utilized by trout and salmon not merely to keep eggs in the shallow nest but to move the eggs downward into the interstices between the rocks where the alevin or developing minnow can grow using the nutrients of the egg as food. The development from egg to fry utilizes oxygen of the water that enters the gill structure of the organism, a structure easily clogged by silt. Thus, the increase in water turbidity and the clogging of interstices in stream-bed gravel by silt has a detrimental effect on the reproduction cycle of salmon and trout (Stuart 1953).

Sediment concentration is determined from a water sample obtained with a DH-48 or D-49 described in the reports of the U.S. Federal Interagency

Sedimentation Project (1941, 1957). The usual procedure is to mix the samples taken in no less than 3 and usually 5 to 10 verticals in a given river cross section. Samplers are fabricated for, and may be purchased through, the Office of the Chief Hydrologist, Water Resources Division, U.S. Geological Survey, Reston. Virginia 22092.

It is well for the environmental planner to have in mind the range of values of suspended-load concentration. It is usually given in milligrams per liter, but in some publications concentration is given in parts per million, the two being essentially equivalent. The largest value known to us was recorded on the Rio Puerco at Puerco Station, New Mexico, a notoriously muddy stream, at 500,000 ppm, that is, the sample was half sediment by weight. Values of concentration tend to increase with discharge and increase as the sampler gets closer to the stream bed.

In the Rio Grande at Bernalillo, New Mexico, a stream with a high suspended load, Nordin and Dempster (1963) measured concentrations of 1000 to 6000 mg/liter and downstream below the mouth of the Rio Puerco, 6000 to 160,000 mg/liter. In contrast, in the clear mountain streams of the Salmon River basin, Idaho, Emmett (1975) measured concentrations between one and 2000 mg/liter.

Turbidity is an expression of the optical property of water that scatters light. The scattering increases with suspended particulate matter, which may be organic or inorganic. The presence of fine materials of colloidal, clay, or even silt size gives water a cloudy or opaque appearance. Values of turbidity are often expressed in Jackson Turbidity Units. Standard methods for the collection and analysis of water samples are described by Rainwater and Thatcher (1960).

Turbidity increases with, but not as fast as, suspended-load concentration. Typical relations between them are given by Emmett (1975) as

$$t = cG_c^n \tag{19-1}$$

in which t is turbidity in Jackson Turbidity Units, G_c is suspended-sediment concentration, c is a coefficient varying between 0.7 and 1.3, and n is an exponent between 0.6 and 0.7.

Turbidity reduces the depth to which sunlight penetrates and thus alters the rate of photosynthesis. When a formerly turbid or muddy stream is made relatively clear as it usually is for some distance downstream from a reservoir, the greater clarity promotes the growth of algae. This effect is highly visible in the Colorado River below Glen Canyon Dam; where the uncontrolled river was muddy it is now clear and green. Whereas this increase in photosynthesis increases the dissolved oxygen during sunlight hours, the large amount of organic material tends to reduce greatly the available dissolved oxygen during hours of darkness. Hence it can be seen that merely reducing turbidity does not necessarily solve problems, but may create new ones. It is the change from natural conditions that usually causes difficulties, often unforeseen.

Waste Heat

The temperature of rivers and lakes varies daily and seasonally in response to the energy budget described in Chapter 4. Temperatures generally lie in the range 0° to 35°C, and aquatic ecosystems are adjusted to the seasonal pattern and to the constraints the temperatures impose. Because of the increasing interest in water temperatures as an important parameter of water quality, networks of recording thermometers are now operated, and the data are published in a variety of tabulated and summary forms (Woodward 1971, Williams 1971, Stevens et al. 1975).

The normal temperature patterns of streams and lakes are being disrupted, however, as larger and larger amounts of water are used as coolant in the condensers of steam turbines in thermal and nuclear power plants. In the United States, for example, where power consumption has reached its most insane level, the generating capacity of power plants is expected to exceed 570,000 megawatts by 1980. The cooling-water requirements of these plants will exceed 750 million cubic meters per day, or about one-sixth of the average daily runoff from the 48 conterminous states. Cooling water can, of course, be re-used many times, so the problem is not as bad as it seems at first glance. On the other hand, these requirements represent about one-third of the average dry-season runoff of the country and an even larger proportion of the low flow of streams in subhumid areas. In some areas, then, we have already come up against hydrologic constraints on the supply of cooling water.

Heated effluents also reach rivers and lakes from industrial plants, where water is used for such purposes as the quenching of steel. Municipal wastewaters also become slightly warmer. Power plants, however, create a special problem because of their growing number and size, and because of their inefficiency. The average efficiency of new North American generating stations lies in the range of 33–40 percent; that is, for every unit of electrical energy generated, two more units of energy must be wasted and dissipated to the environment as heat. Nuclear plants tend to be less efficient than thermal stations; in nuclear plants this excess heat is transferred to the cooling water as the latter passes through the condenser of the steam turbine (see Figure 19-1). In thermal plants, 85 percent of the waste heat is dissipated in this way while the remainder is lost from the stack.

The transfer of vast amounts of heat to receiving waters in rivers and lakes causes large rises of temperature (up to 15°C) immediately below the outfall from some large plants. The magnitude of this rise depends on the size of the plant, its method of cooling, and the size of the river. Such a rise is mainly caused by a large increase in the advection term (Q_v) in Equation 4-2. A review of the energy budget in Chapter 4 will convince the reader that the increase of temperature will accelerate the dissipation of heat from the water body, mainly by increasing the outflow of longwave radiation, sensible heat, and the latent heat required for evaporation. As the water moves away from the outfall of the power plant, therefore, it will cool.

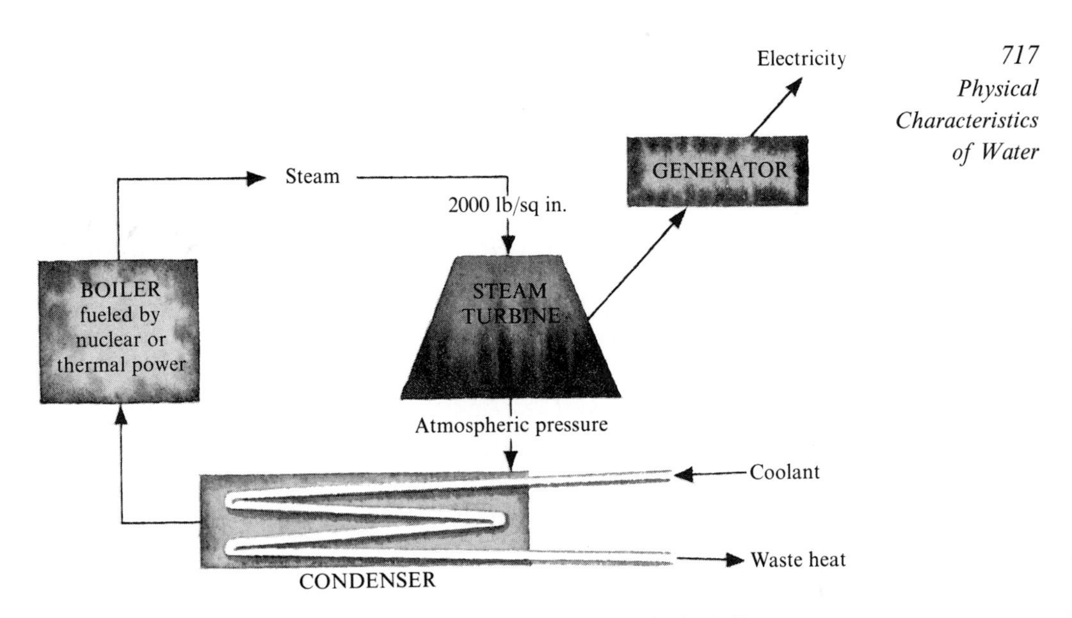

Figure 19-1 Schematic diagram of a power plant. Steam produced in the boiler enters the turbine under very high pressure and passes through at a high velocity produced by the pressure drop through the turbine. The low pressure at the outlet is produced by condensing the steam in the condenser with the aid of a stream of coolant.

The rate of cooling will depend on atmospheric conditions and characteristics of the water body, as indicated in the earlier discussion of the energy budget. Cooling will also take place by mixing with unheated water in the receiving body, the energy budget of which will also be altered. If such mixing takes place the temperature of the effluent will be lowered, but the heat will not be dissipated as quickly as if it remains concentrated in the effluent at high temperatures.

There is a zone, therefore, in which the water temperature is elevated above its natural level. The situation can be aggravated if the water is re-used several times along a river, as shown in Figure 19-2. If the elevation of temperature has deleterious effects upon the suitability of the water for aquatic ecosystems or for various uses, the situation is referred to as *thermal pollution.* Not all temperature rises are deleterious, but along most heavily used rivers a definite problem is arising.

The problem can be illustrated with the following figures. In 1945 the Tennessee Valley Authority installed its first thermoelectric power plant. It had a generating capacity of 240 megawatts, and with a water circulation of 18 m^3/sec, the water temperature was raised by 6°C at full load. A 2600-mw fossil-fuel plant installed in 1970 needed a flow of 100 m^3/sec to limit the temperature increase to 7°C. A 3450-mw nuclear plant is now planned which will need 140 m^3/sec to limit the increase to 14°C. Such plants must be located on large rivers, lakes, or seacoasts, where these large

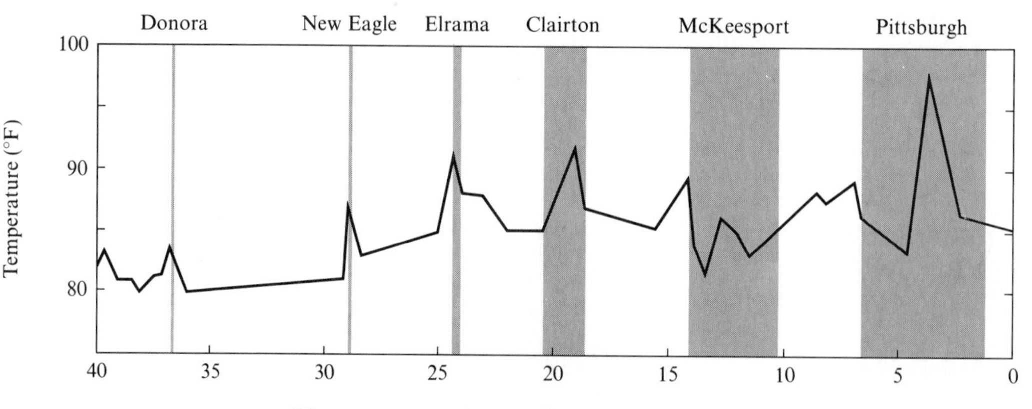

Figure 19-2 Pattern of water temperature along the Monongahela River. High temperature peaks are superimposed upon a general increase. (From Clark 1969.)

volumes of heated effluent can be mixed with even larger volumes of cold water. The low flow is the important constraint on a stream's ability to assimilate such large volumes of hot water, and few streams have low flows of the magnitudes referred to above. The tendency to concentrate several plants along important rivers in industrial regions aggravates this problem.

Smaller alterations of temperature can be changed by impounding water behind a dam. Large lakes smooth out the seasonal temperature fluctuations of rivers as well as their variations of flow. By drawing off water from different depths behind a large dam, the summer temperature of a stream can be depressed to maintain conditions suitable for fish or for more efficient cooling in power plants downstream. Williams (1971) has documented the effects of reservoir releases on the summertime temperature of the upper Delaware River. The temperature was depressed by 13°C 13 km below the reservoir and by 1°C 90 km downstream. Large numbers of unshaded, small impoundments in a basin, however, can cause an elevation of water temperatures that is usually deleterious.

Changes of temperatures along small streams can result from alterations of land use. The best documented cases of this are from urban and forest regions. Pluhowski (1970) compared the energy balance of a natural, shaded stream on Long Island with that of four other streams whose banks had been cleared of vegetation during urbanization. In addition to this alteration of the radiation balance of the stream, urbanization had reduced the inflow of groundwater and increased the inputs of stormflow, which were warmer than baseflow in summer and colder in winter. An increase of ponds and lakes in the urbanized basin further altered the radiation balance and increased the storage time of surface water in the basin. The result was an average water temperature increase of up to 8°C in the urbanized basins during the summer, and a lowering of average winter temperatures by 1.5°–3°C.

In forested areas, removal of trees from the banks of small streams during clearcutting causes a dramatic increase in the solar radiation input to small streams. Brown and Krygier (1970) reported increases of annual maximum stream temperature after clearcutting ranging up to 15°C and increases of average monthly maxima ranging up to 8°C. Because many forested streams are important spawning grounds for cold-water game fish such as salmon and trout, there is great concern about the continued impact of clearcutting upon stream temperatures (Lantz 1971). In some logging operations a buffer strip of trees is left standing along riverbanks. This has other benefits such as a reduction of organic waste in the stream, less damming of the stream by log jams, and less damage to stream banks.

Effects of Raising Water Temperatures

Some of the results of raising water temperature are relatively simple. Lowering of the viscosity of the water, for example, causes faster settling of solid particles in the settling ponds of sewage-treatment plants. Other effects are more complex. An increase of temperature causes a decrease in the solubility of oxygen, which is needed for oxidation of biodegradable wastes. At the same time, the rate of oxidation is accelerated, imposing a faster oxygen demand on the smaller supply and thereby depleting the oxygen content of the water still further (see Chapter 20).

Temperature also affects the lower organisms in the aquatic food chain, such as plankton and crustaceans. In general, the higher the temperature, the less desirable are the types of algae in water. In cooler waters diatoms are the predominant phytoplankton in water that is not heavily eutrophic. With the same nutrient levels, green algae begin to become dominant at higher temperatures and diatoms decline, while at the highest water temperatures, blue-green algae thrive and often develop into heavy blooms. Many pathogenic bacteria thrive when the temperatures of some streams are slightly increased, and their abundance can be very harmful to fish (Brett 1956).

In extreme cases, fish have been killed by lethal high temperatures or rapid temperature changes below power plants. More commonly, however, many desirable species of fish suffer in such locations because of imposed changes in their feeding or spawning behavior, their susceptibility to disease, or some other aspect of their existence that becomes unfavorable to the population even if it is not fatal to individuals. At high temperatures fish metabolism is accelerated greatly while the efficiency of their oxygen use decreases.

Some fish can survive these conditions, and if the temperature rise is not excessive they can even thrive and concentrate in zones of heated effluent. Most of the best game fish, however, do not do this; they are the ones with the highest oxygen requirements and the lowest optimal and lethal temperatures. The spawning and migration of many fish are triggered by temperature and this behavior can be disrupted by thermal pollution. Some species will not spawn at the temperatures imposed along developed rivers, and in other locations eggs will no longer survive the new temperature regime. There is great concern about this problem with respect to the chinook salmon, for example, in the Pacific Northwest, where the fish is an important game and commercial species. While the diverse reactions of different species of aquatic organisms make generalization difficult, many aquatic biologists view the projected growth of power generation with great concern, in spite of the fact that places can be found where no significant disruption of stream ecology has occurred, or even where a favorable change has been recognized.

Valuable reviews of the problem and case studies are given by Mount (1970), Clark (1969), Gibbons and Sharitz (1974), Krenkel and Parker (1969). A contrary view is presented by Wurtz (1967).

Strategies for Reducing Thermal Pollution

Because the problem of thermal pollution is already great in some rivers and lakes and because the problem is likely to grow at an increasing rate, the planner should be aware of various strategies for reducing the adverse effects of waste heat discharges to water bodies. This area of investigation will grow as experience is gained with established techniques and as new methods are devised. It is important that planners and citizens generally remain informed about new developments and are able to ask engineers the right questions about what is feasible. The technical and public relations literature and the records of court hearings are full of "brave new world" statements about how waste heat can be used. So far, however, most such schemes have encountered problems, and there is no panacea yet available for this troublesome issue.

A partial solution, of course, can (and will in fact have to) come from a reduction in the rate of growth of energy consumption. This can be accomplished by a reduction of use, and by more effective utilization (Berg 1973). In this chapter, however, we will concentrate upon the hydrologic issues concerning dissipation of the waste heat itself.

Several methods are available for cooling water once it has been used. They differ as to the gross use and the consumptive use (see Chapter 12) of water, the capital investment and operating costs, and their effectiveness in maintaining natural water temperatures. In order to assess the impact of various dissipation methods upon water temperatures, the plant designer must solve the energy budget in Equation 4-3. The task is complicated by hydrodynamic problems such as predicting the effects of differences in water density and river velocity upon the stratification of water and the extent of mixing zones. These problems have to be solved for each individual site, but a simplified energy-budget computation is useful for a preliminary exploration of probable temperature rises and consumptive use in the planning stage. The subject is clearly reviewed by Sefchovich (1970), where the reader will find an interesting application of the concepts introduced in Chapter 4. It is hoped that the planner can become acquainted with the principles sufficiently to understand the technical reports outlining proposals for waste heat disposal and to ask the appropriate questions.

The most straightforward cooling method is the *single-pass system* in which water is pumped from a river or lake through the condenser and back to the river. The waste heat is then dissipated gradually and at a distance from the plant by energy exchange and mixing as it flows through a lake or along a river channel. The single-pass method is cheap and effective if the water body used is large, cold, and not overtaxed by re-use in several closely spaced plants. In such a system a 1000-mw plant requires a through-flow of 31 m^3/sec if the temperature rise of the coolant is to be limited to 11°C. The gross use of water then is very large. In current plants the temperature rise usually lies between 5° and 17°C. Some of the heat may be dissipated from the water in a canal on its way back to the river. The effluent then mixes with the receiving water in an ill-defined mixing zone whose lateral and vertical extent will vary with weather conditions and the flow of the river. The most rapid dissipation of heat to the atmosphere occurs when the effluent has a high temperature and is allowed to spread over the surface of the receiving body so that longwave radiation and sensible and latent heat losses are high. If the effluent becomes thoroughly mixed with cooler water, the temperature is quickly depressed, but the heat is stored in the larger water body for a longer time.

If the heated effluent is thoroughly mixed with the stream into which it is discharged, the amount of heat, K, that is dissipated to the atmosphere over a channel reach of length, L, can be calculated very approximately from a formula developed by LeBosquet (1946):

$$K = 8{,}323{,}515Q \frac{\log_{10}\left(\dfrac{T_O + 17.8}{T_L + 17.8}\right)}{wL} \tag{19-2}$$

where K is in cal/m^2/hr/°C excess of water temperature over air temperature, Q is discharge of the stream after mixing (m^3/sec), T_O and T_L are the excess water temperatures over air temperature at the outfall and at distance L downstream (°C), w is the width of the stream (m), and L is the distance downstream (km). Values of K are usually obtained first from a set of temperature readings, and the equation can then be used to compute expected temperatures at various distances downstream under new conditions, such as increased discharge. Values of K usually vary between about 5000 cal/m^2/°C/hr for sluggish streams and 15,000 cal/m^2/°C/hr for swift, wide streams. Another use of the formula involves calculating the discharge, Q, required to keep T_L below some critical value.

The consumptive use of single-pass cooling is generally about one percent of the gross use. The disadvantages of the method include the large gross-use requirements, and high temperatures and temperature gradients in the receiving water body. Because of the high gross use, single-pass systems are often built at coastal sites, particularly in countries such as England, which have only relatively small streams.

If the available water body is not large enough to accommodate a single-pass system, the heated effluent may be discharged to a *cooling pond.* Here, the heat dissipation can be rapid and localized if the pond temperature is allowed to rise to high levels as the effluent is spread over its surface. After cooling, the water can be recycled through the plant with the addition of only a relatively small amount of outside water to compensate for the consumptive use involved in evaporative losses from the pond. Depending on the depth of water, the meteorological conditions, and the allowed temperature limit, most ponds have an area of 0.4–0.8 hectares per megawatt of generating capacity, or 4 to 8 square kilometers for a 1000-megawatt plant.

If the cooling pond is integrated into a regional water-management plan, as large ponds should be, fish, wildlife, and other ecological and recreational values have to be considered in setting temperature limits, both for the whole lake and for the mixing zone. These values may sometimes conflict with the most efficient use of the lake for cooling. From an efficiency point of view, the best cooling strategy is to remove cold water from the lower zone of the lake (hypolimnion) and to discharge effluent into the warmer surface layers (epilimnion) for rapid dissipation of heat to the atmosphere. This method causes a thickening of and increased biologic productivity in the epilimnion. Upon death the remains of the plants and animals sink into the oxygen-rich, colder hypolimnion to decay. The increased oxygen demand imposed on the shrunken hypolimnion can cause a reduction of its oxygen content with adverse effects upon aquatic organisms. Another problem with the management of cooling ponds is that they frequently contain noxious chemicals used in the cleaning of the cooling system. Such chemicals should be diverted to a treatment plant but often are not. There have been some fine examples of cooling-pond management, but their development costs money and usually needs pressure from the government and especially from informed planners and citizens groups. There are many other examples of cooling ponds that have a very poor aquatic population of plants and animals, and are saline, smelly, and unsightly.

If land is too valuable for the use of a cooling pond, waste heat can be dissipated on-site by means of a *spray pond* or a *wet cooling tower* (as shown in Figure 19-3). Both of these methods work on the same principle and depend heavily on evaporative cooling. Consequently their consumptive use is high (about 1.1 m^3/sec per 1000 mw for a cooling tower). They can be integrated with a small cooling pond or the water can be immediately recycled through the plant. Chemicals must be added to the water to retard the growth of algae on the wooden lattice and to inhibit corrosion of metal in pumps. These must be periodically flushed from the system and should be treated before reaching outside water bodies.

The moist air rising from ponds or from cooling towers can sometimes produce fogging or icing problems if local meteorologic and topographic conditions are adverse. Wet cooling towers cannot be used with saline or brackish water because the fallout of salt injures surrounding vegetation. Another drawback is the cost of the towers, which was roughly $8 to $13

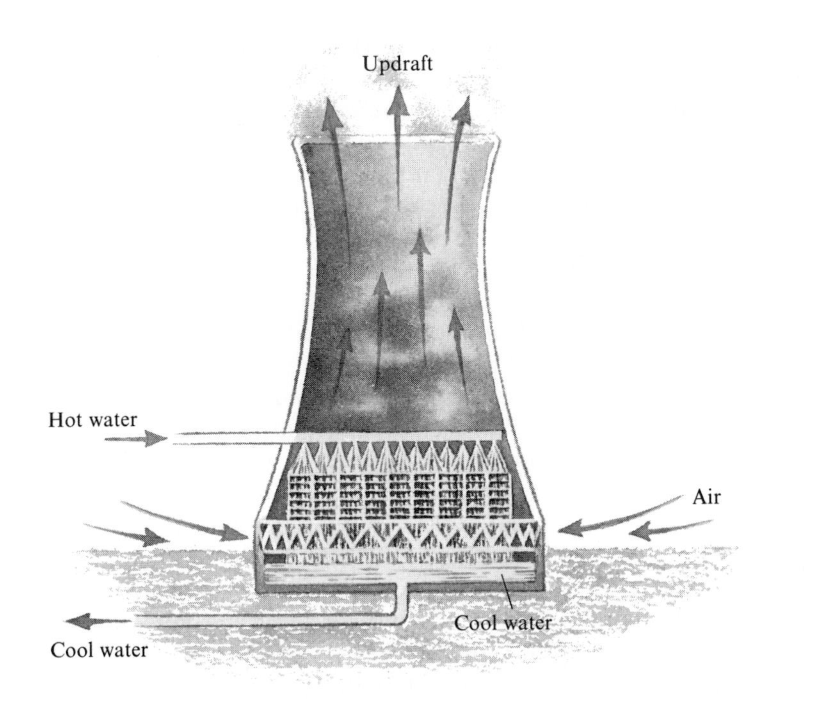

Figure 19-3 Wet cooling tower. The hot water is sprayed over a wooden lattice and cools by evaporation during its fall. The warmed, moist air rises and draws in cool air at the base. Evaporative cooling towers are most efficient when the atmospheric relative humidity is low.

per kw at 1970 prices in the United States. In Britain, however, where rivers are small, land for cooling ponds is rare and where cooling demands are heavy, wet cooling towers of this kind are used extensively.

Where the addition of moisture to the atmosphere would cause excessive fog or icing, or where consumptive use must be limited, the designer may have to install a *dry cooling tower,* which works like a giant automobile radiator. Water is circulated through the tower in a pipe. Air blown up through the tower by large fans cools the water by heat transfer through the pipe walls. There is no evaporative cooling, and consequently more vigorous air movement is necessary; also the system has a lower efficiency than the wet cooling tower. Installation, maintenance, and operating costs are very high, the towers must be even larger than the wet type, and the fans are very noisy. Even though they essentially eliminate consumptive use, it is questionable whether they will ever become generally used, except in special situations.

Because of the cost and residual environmental impact of waste heat disposal, several uses of heated effluent are being explored. Major difficulties are involved with almost all of them, and in spite of a good deal of public relations literature, most water managers and biologists are unconvinced about their general applicability.

On a small scale, hot water and steam from power plants have been used for municipal heating. However, most effluent is not hot enough for this use. Large-scale planning of communities around power plants would be necessary, along with an expensive distribution system and improved insulation.

Warm-water irrigation has been proposed as a way of using partly cooled effluent. The argument has been proposed that sprinkling of fruit trees with warm water can offset killing frosts and that crops mature earlier when irrigated with warm water. There are many drawbacks with this idea, however. The cost of distributing the water to farms is high and the proposed uses would not accommodate the 30 m^3/sec discharged by a 1000-mw, single-pass system. Irrigation is needed for only a few weeks or months of the year, and not at all in some regions; killing frosts occur on only a few days of the year, or not at all in many regions. Other analysts have questioned whether the application of warm water or the circulation of warm water in buried pipes to lengthen growing seasons and increase yields are as effective as has been claimed by the utilities, and whether such practices might favor plant diseases (Carter 1969).

In coastal locations there is some optimism about the use of heated effluents to favor the growth of shellfish in aquaculture experiments. Shrimp have been farmed in this way in Florida, and oysters in Long Island Sound. Other proposals have been made for using waste heat to speed up the digestion of sewage and some industrial wastes. Whether such uses can accommodate the amounts of waste heat predicted to be generated in the next 5 to 10 years remains doubtful.

Temperature Standards

Various approaches to the problem of setting temperature standards for streams are being tried at present. There is as yet no consensus about which method works best. In some rivers no further heat discharge is allowed that would increase the water temperature beyond the maximum permissible limit for the local fish species. This limit may relate to spawning, egg survival, or general growth conditions. In other river basins, limitations are only set to avoid raising the temperature of water in the spawning grounds of cold-water fish. Elsewhere, rather arbitrary rules are being tried. The limit for heating any stream in the most favorable season may be set at, say, 2°C outside a mixing zone of a certain size. In other places the limits of temperature increase are set for the least favorable, hot, low-flow season. A difficulty arises because the deleterious effects of a temperature rise on fish are synergistic with the effects of sewage, toxic chemicals, and perhaps other pollutants. We do not yet have a method of taking account of these interactions when setting discharge limits for heat or chemicals.

Bibliography

BERG, C. A. (1973) Energy conservation through effective utilization; *Science,* vol. 181, pp. 128–139.

BRETT, J. R. (1956) Some principles in the thermal requirements of fishes; *Quarterly Review of Biology,* vol. 31, pp. 75–81.

BROWN, G. W. AND KRYGIER, J. T. (1970) Effects of clearcutting on stream temperature; *Water Resources Research,* vol. 6, pp. 1133–1140.

CARTER, L. J. (1969) Warm-water irrigation: an answer to thermal pollution? *Science,* vol. 165, pp. 478–480.

CLARK, J. R. (1969) Thermal pollution and aquatic life; *Scientific American,* vol. 220, no. 3, pp. 19–27.

DELAY, W. H. AND SEADERS, J. (1966) Predicting temperature in rivers and reservoirs; *Journal of Sanitary Engineering Division, American Society of Civil Engineers,* vol. 92, no. SA1, pp. 115–134.

EMMETT, W. W. (1975) The channels and waters of the Upper Salmon River area, Idaho; *U.S. Geological Survey Professional Paper 870-A,* 115 pp.

GIBBONS, J. W. AND SHARITZ, R. R. (1974) Thermal alteration of aquatic ecosystems; *American Scientist,* vol. 62, pp. 660–670.

KRENKEL, P. A. AND PARKER, F. L. (1969) *Thermal pollution,* vol. I, *Biological aspects;* Vanderbilt University Press, 407 pp.

LANTZ, R. L. (1971) Influence of water temperature on fish survival, growth, and behavior; in *Proceedings of symposium on forest land use and the stream environment,* Oregon State University, School of Forestry, pp. 182–193.

LEBOSQUET, M. (1946) Cooling water benefits from increased river flows; *New England Waterworks Association Journal,* vol. 60, pp. 111–116.

MOUNT, D. I. (1970) Environmental effects of thermal discharges: ecological elements; in *Environmental effects of thermal discharges,* American Society of Mechanical Engineers, New York, pp. 7–9.

NORDIN, C. F., JR., AND DEMPSTER, G. R., JR. (1963) Vertical distribution of velocity and suspended sediment Middle Rio Grande, New Mexico; *U.S. Geological Survey Professional Paper 462-B,* 20 pp.

PARKER, F. L. AND KRENKEL, P. A. (1969) *Thermal pollution,* vol. II, *Engineering aspects;* Vanderbilt University Press, 351 pp.

PLUHOWSKI, E. J. (1970) Urbanization and its effects on the temperature of the streams on Long Island, N.Y.; *U.S. Geological Survey Professional Paper 627-D.*

RAINWATER, F. H. AND THATCHER, L. L. (1960) Methods for collection and analysis of water samples; *U.S. Geological Survey Water Supply Paper 1454,* 301 pp.

RAPHAEL, J. M. (1962) Prediction of temperatures in rivers and reservoirs; *Journal of Power Division, American Society of Civil Engineers,* vol. 88, no. P02, pp. 157–181.

SEFCHOVICH, E. (1970) The preliminary thermal analysis of a body of water in power plant siting; in *Environmental effects of thermal discharges,* American Society of Mechanical Engineers, New York, pp. 19–25.

STEVENS, H. H., FICKE, J. F. AND SMOOT, G. F. (1975) Water temperature–influential factors, field measurement and data presentation; *U.S. Geological Survey Techniques of Water Resources Investigations,* book 1, chap. D1.

STUART, T. A. (1953) *Spawning, migration, reproduction, and young stages of loch trout;* Freshwater and Salmon Fisheries Research Station, H. M. Stationary Office, Edinburgh, 39 pp.

U.S. FEDERAL INTER-AGENCY SEDIMENTATION PROJECT (1941) Methods of analyzing sediment samples; Report No. 4, St. Paul, MN, U.S. Dept. Army.

U.S. Federal Inter-Agency Sedimentation Project (1957) Measurement and analysis of sediment loads in streams; Report No. 12, Washington, DC, U.S. Government Printing Office.

Williams, O. O. (1971) Analysis of stream temperature variations in the upper Delaware River basin, N.Y.; *U.S. Geological Survey Water Supply Paper 1999-K.*

Woodward, T. H. (1971) Summary of data on temperature of streams in North Carolina; *U.S. Geological Survey Water Supply Paper 1895-A.*

Wurtz, C. B. (1967) Thermal pollution: the effect of the problem; in *Environmental problems* (ed. B. R. Wilson), Lippincott, Philadelphia, pp. 131–145.

Typical Problem

19-1 Analysis of the Effects of Waste Heat Discharge on the Temperature of a River

Find a small river that receives heated discharges from an industrial plant or other source. At a station immediately above the plant, measure the water temperature, air temperature, and the river discharge using a current meter or float, as described in Chapter 16. Estimate the discharge and temperature of the effluent from the plant. Then choose several stations downstream of the plant and measure the stream width and the water temperature at each one. Pay particular attention to temperature changes along and across the channel in the immediate vicinity of the plant. If the discharge changes significantly along the reach, as, for example, below a tributary, measure the discharge again.

Write a short report to describe and explain qualitatively the pattern of stream temperature.

Use your field data to calculate from Equation 19-2 values of K, the amount of heat that can be dissipated over a reach of channel for the measured width and air temperature. Evaluate the usefulness of Equation 19-2 for predicting the streamwater temperature at various distances from the plant. If it gives answers that fit your field data well, use it to predict the effects on the elevation of water temperature at some distance L as a result of

1. doubling the volume of heated effluent,
2. doubling the volume of effluent and increasing its temperature by 10°C,
3. discharging the present volume of effluent, increasing its temperature by 20°C, but doubling the discharge of the stream by flow regulation in a reservoir upstream. Assume that the change of stream discharge will make a negligible difference to the channel width.

Repeat the calculations for an air temperature and an initial stream temperature at another season of the year.

20

Chemical Characteristics of Water

The chemical constituents of water affect its use for drinking, recreation, industry, and irrigation, and its suitability for aquatic organisms. Optimal concentrations of many of the common chemicals such as calcium carbonate or sodium for a variety of uses are well known. There is less certainty about the significance of minor elements such as lead or cadmium, although some attention is now being focused on them. Since World War II the great expansion of chemical and pharmaceutical industries has led to the release of vast quantities of chemicals into surface and ground waters. Many of these substances are toxic in the biosphere, and the effects of many others are completely unknown. The recent development of the radioisotope industry has also begun to spread toxic materials into waterways and could become a significant health hazard as the industry grows.

The present volume cannot cover the whole subject of what is often referred to as "chemical water quality," but many excellent textbooks are available (e.g., Camp 1962, Klein 1962, Nemerow 1974). Instead, the planner will be given a general introduction to the sources of the dissolved substances listed below. We will briefly discuss the processes by which these materials enter the hydrologic cycle, how they interact, and how they limit the suitability of water for various uses. Our brief review will be structured as follows: Major solutes in natural waters, Major ions affected by human activity, Trace metals, Biodegradable wastes and the oxygen balance, Major plant nutrients: nitrogen and phosphorus, and Pesticides.

Major Solutes in Natural Waters

Hem (1970) provides a comprehensive and very useful review of the chemical characteristics of natural waters. He describes the collection of data and various methods for expressing the results of water analyses. He also reviews the origins, controls, and significance of each property and constituent reported in chemical analyses of water.

A part of the solutes in groundwater and surface water is supplied by rainwater, but most of the dissolved substances are mobilized by the chemical weathering of rocks. The weathering processes generate a solid residue, which accumulates as a soil mantle, as described in Chapter 15, and dissolved substances that are flushed slowly into streams by groundwater. As a typical example, consider the weathering of a common rock-forming mineral, orthoclase feldspar, which is a potassium alumino-silicate compound and a major constituent of granite. In the presence of an acid, the feldspar is decomposed in the following reaction:

$$\underset{\text{orthoclase feldspar}}{2KAlSi_3O_8} + \underset{\text{acid}}{2H^+} + \underset{\text{water}}{9H_2O} \rightarrow \underset{\text{clay mineral (kaolinite)}}{Al_2Si_2O_5(OH)_4} + \underset{\text{dissolved silica}}{4H_4SiO_4} + \underset{\text{dissolved potassium}}{2K^+} \qquad (20\text{-}1)$$

The main source of the acid is the solution of carbon dioxide to form carbonic acid. The carbon dioxide comes from the atmosphere and especially from the soil atmosphere where it is produced by bacterial decay of plant debris and by root respiration. The decay of plants also produces organic acids, which can participate in reactions like the one described above and can also form soluble complexes with metals, such as zinc, which are otherwise almost insoluble.

The clay mineral indicated in Equation 20-1 accumulates as a soil constituent. The dissolved products, in this case silica and potassium, are leached into the groundwater and thence to streams. Other minerals weather to yield dissolved sodium, calcium, fluorine, and all the other naturally occurring solutes, including minor amounts of metals, such as iron, lead, and zinc. A typical set of chemical analyses of groundwater is given in Table 20-1.

Some rocks contain minerals that are very soluble; others are resistant to chemical weathering. In limestone regions, for example, the groundwater usually contains high concentrations of calcium bicarbonate. In rocks, such as basalt, rich in iron and magnesium minerals, these elements are abundant in groundwater. On the other hand, in chemically resistant rocks, such as quartzite, concentrations of solutes in the groundwater are generally low.

Climate and hydrology are other controls. Deep, stagnant groundwater bodies develop high concentrations, especially in arid areas where recharge rates are low. In wet regions, groundwater is recharged and flushed out at faster rates and for the same geologic conditions tends to be less concentrated. The geologic, climatic, and hydrologic environment, therefore, controls the amount and the type of minerals in solution. Some examples of

Table 20-1 Chemical analyses of natural waters. (From Ministry of Water Development, Nairobi; and Miller 1961.)

	SOURCE								
CONSTITUENT (MG/LITER)	GROUND-WATER IN MARINE SEDIMENTS IN AN ARID CLIMATE	GROUND-WATER IN BASALT IN A HOT, WET CLIMATE	GROUND-WATER IN PHONOLITE LAVA IN A COOL, WET CLIMATE	GROUND-WATER IN CORAL LIMESTONE NEAR A TROPICAL COAST	STREAM DRAINING PHONOLITE LAVA AND GRANITE IN A SEMI-ARID CLIMATE	STREAM DRAINING BASALT IN A SEMI-ARID CLIMATE	STREAM DRAINING METAMORPHIC ROCKS IN A HOT, WET CLIMATE	STREAM DRAINING GRANITE IN COLD MOUNTAINS OF NEW MEXICO	STREAM DRAINING CALCAREOUS SANDSTONE IN COLD MOUNTAINS OF NEW MEXICO
pH*	8.9	7.1	5.7	8.1	7.5	8.3	6.8	7.0	8.0
Total dissolved solids	9135	85	51	468	295	740	31	15	90
H_4SiO_4	40	64	32	24	88	40	16	5.6	7.7
Na	3150	12	2	62	72	128	2	0.9	0.8
K	—	3	4	0.4	9	22	1	0.4	0.4
Ca	27	4	4	104	22	50	3	2.8	31
Mg	121	Trace	0.7	30	9	38	0	0.2	0.7
Fe	0	Trace	0.3	0	0.2	0.5	0.2	0.1	0.02
Carbonate alkalinity	260	0	0	—	0	52	0	—	—
Bicarbonate alkalinity	660	42	22	—	158	392	—	—	—
Carbonate (temporary) hardness	570	12	13	220	90	286	8	—	—
Noncarbonate (permanent) hardness	—	0	0	160	0	0	0	—	—
Cl	3900	8	9	158	74	33	—	0.2	—
F	3.8	0.2	1.0	0.4	1.3	3.3	0.3	—	—
SO_4	965	Trace	0	—	20	128	2	1.9	7.3
NO_3	0	0.09	0	0	0.1	0.2	0.05	0.2	0.1
HCO_3								10	91

*pH is measured in logarithmic units of hydrogen ion concentration.

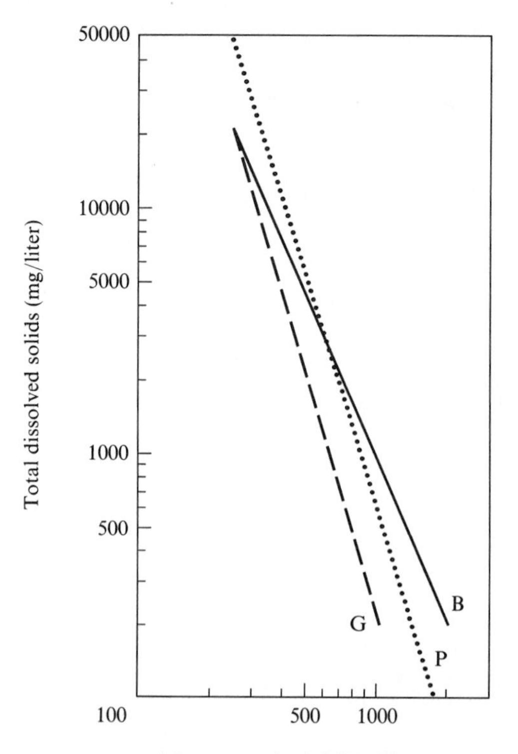

Figure 20-1 Graph showing variation of total dissolved solids in groundwater with mean annual rainfall for various rock types in Kenya. G indicates granitic rocks, P phonolite lavas, and B basalt.

these relations are illustrated in Figure 20-1, although much remains to be done in refining such predictions.

There are also strong associations between geology and the concentration of minor, but important, dissolved species such as fluoride. In Kenya, for example, groundwater in trachyte lavas and the lake sediments derived from them generally contain high concentrations of fluoride, with median values in the range 3–4 mg/liter and maximum values exceeding 100 mg/liter, which is well into the range of concentrations that promote bone disorders in human beings. In other Kenyan rocks, groundwater concentrations of fluoride are less than 4 mg/liter and have median values in the range 0.5–1.5 mg/liter. In a few marine sedimentary rocks the concentrations are less than 1.0 mg/liter and lie mainly in the range that favors dental caries. A geologic map, therefore, is a useful indicator of this one aspect of water chemistry.

Stream-water chemistry is more complex than groundwater chemistry because runoff consists of a variable mixture of waters that have reached the channel by various routes, as outlined in Chapter 9. Groundwater remains in contact with weathering rock and soil minerals for periods of time ranging from a few days to hundreds of years. It generally has a higher concentration of dissolved solids than water that reached the stream by some surface route, or after only a brief residence in the soil. During dry weather, when the baseflow of the stream is dominated by groundwater drainage, therefore, the concentration of solutes in stream water tends to be high. At times of high flow, the concentration tends to be diluted by surface runoff and shallow subsurface flow. The result for most streams is an inverse correlation between discharge and the concentration of total dissolved solids, as shown in Figure 20-2. The situation is not always so simple (Gunnerson 1967, Hendrickson and Krieger 1960), and for some particular solute the correlation may even be positive, but the situation portrayed in Figure 20-2 is very common. Because of this variation of chemical composition with discharge, it is during periods of low flow that concentrations of major anions and cations become limiting for uses such as irrigation and drinking. This is also true if human activity increases the supply of solutes into the stream because at low flow little dilution can take place. On the other hand, the storage of water in large reservoirs can mix stormwater and baseflow and can even out the fluctuations of water chemistry along with fluctuations of discharge.

The hydrologist has only limited ability to predict the variation of stream-water chemistry. It varies with the geology of the drainage basin, as might be expected (see Table 20-2), but the mixture of water from tributary basins

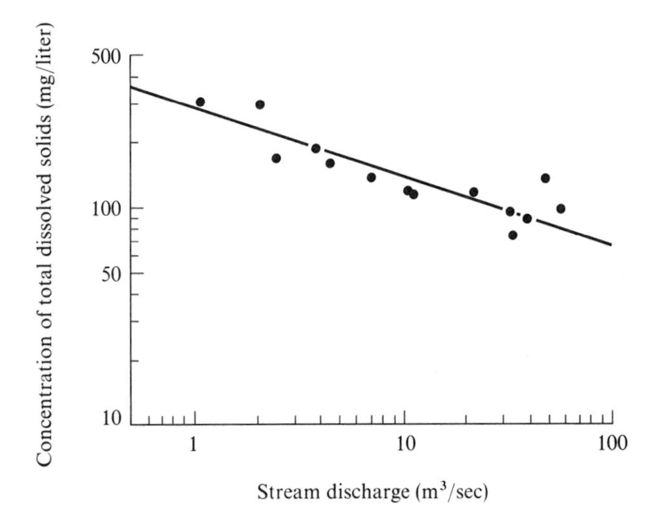

Figure 20-2 Variation of total dissolved solids concentration with stream discharge for the Athi River at Ol Donyo Sabuk, Kenya.

with different lithologies complicates the picture. There is also a rough correlation of total dissolved solids concentration (often called simple "salinity") with climate and hydrology. Streams in arid areas tend to have high concentrations, though their total annual solute transport (tonnes of salt per square kilometer per year) is low because of their low runoff. The situation is reversed in humid regions where, for a similar rock type, concentrations tend to be lower and total annual yields of salt higher than in drier climates (Figure 20-3).

Beyond such broad generalizations it is necessary to sample the stream at one station many times to construct some empirical relationship such as that shown in Figure 20-2, which can then be used in conjunction with a frequency analysis of streamflow (either extreme low flows or flow-duration

Table 20-2 Average concentrations of total dissolved solids in dry-weather flow from small drainage basins on single rock types in the Sangre de Cristo Mountains, New Mexico. (From Miller 1961.)

ROCK TYPE	AVERAGE CONCENTRATION OF TOTAL DISSOLVED SOLIDS (MG/LITER)
Calcareous sandstone	115
Granite	35–38
Quartzite	18

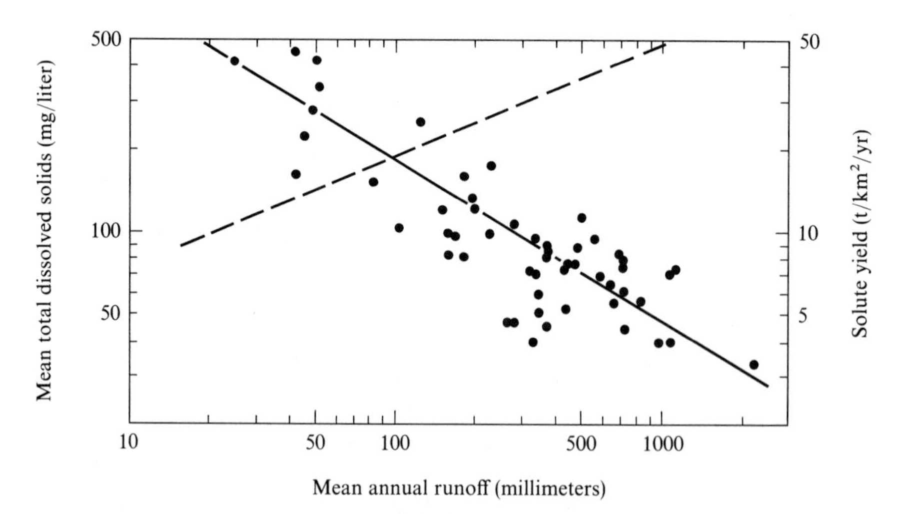

Figure 20-3 Variation of dissolved-solids concentration (solid line and circles) and of annual solute transport (dashed line) for streams in Kenya. Mean annual runoff represents the effect of climate.

analysis) to predict the concentration of chemicals at flows of various frequencies. A regional pattern of water chemistry can then sometimes be defined by mapping concentrations of various solutes at flows of chosen recurrence interval (Emmett 1975).

Major Ions Affected by Human Activity

In the United States 150 billion to 200 billion cubic meters of water are used consumptively each year for the irrigation of crops and pasture. Approximately 25 billion cubic meters are used consumptively in cooling plants and industrial processes. In the typical diversion for irrigation, 50–70 percent of the water applied is evaporated. This causes a twofold or threefold concentration of the dissolved salts. Some of these salts remain in the soil and can accumulate to damaging levels (Richards 1954). Most of the concentrated salts, however, are returned to the river. If the stream water is used several times over for irrigation, it will undergo severe concentration. On its way back to the stream, the water can dissolve even more salts from the subsoil and rocks, or flush out old, saline groundwater. Figure 20-4 shows the change in concentration along the course of the Sevier River in Utah, where seven complete stream diversions cause a twenty-fold increase in salinity over a 320-km reach of river. The ratio of sodium salts to calcium and magnesium salts also increases, and this together with the overall salinization alters the river chemistry from a state that is suitable for irrigation to one that is totally unsuitable.

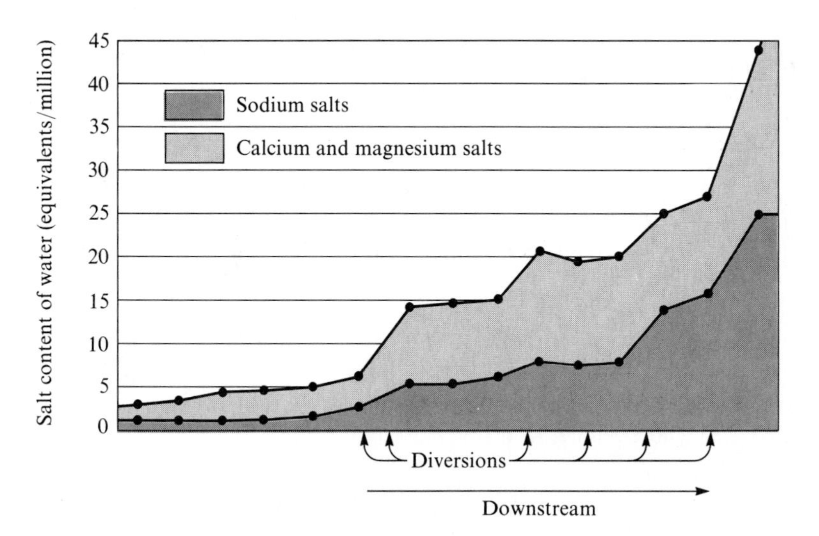

Figure 20-4 Changes in salt content of the Sevier River, Utah, as a result of repeated diversion for irrigation. (From Thorne and Peterson 1967. Copyright 1967 by the American Association for the Advancement of Science.)

Urban and industrial use also increases the concentration of salt. Treated municipal sewage adds to the receiving water about 35 kg of inorganic salts per year for each person served. Municipalities without major industries generally increase the salinity of their water supplies by 200–300 mg/liter. Industrial cities commonly cause increases of 300–500 mg/liter between intake and outfall because the water picks up large quantities of solutes when used in iron and steel manufacture, cement making and other industries. The salinity of water passing through the Los Angeles County water system increases by 1240 mg/liter. Of this 640 mg/liter comes from brines from the underlying oil fields; 350 mg/liter comes from industrial processes; 225 mg/liter is added by municipal use; and 25 mg/liter from water treatment (quoted by Thorne and Peterson 1967). Comparison of these figures with the water quality standards of most states and countries indicates that the water is useful for little else after disposal unless it is greatly diluted with clean water. The cumulative effect of industrial and municipal waste disposal upon the total dissolved solids content of a stream is well shown by the data in Figure 20-5 for the Passaic River in New Jersey. During the period 1947–1964 the drainage basin of the river underwent massive industrial and residential development.

On a smaller scale, the effects of activities such as salting of roads in cold regions can raise the sodium chloride content in groundwater and streams to nuisance levels. Kunkle (1971), for example, measured chloride concentrations exceeding 200 mg/liter in seeps polluted by drainage from a road in a rural area of Vermont where the chloride content of streams not in-

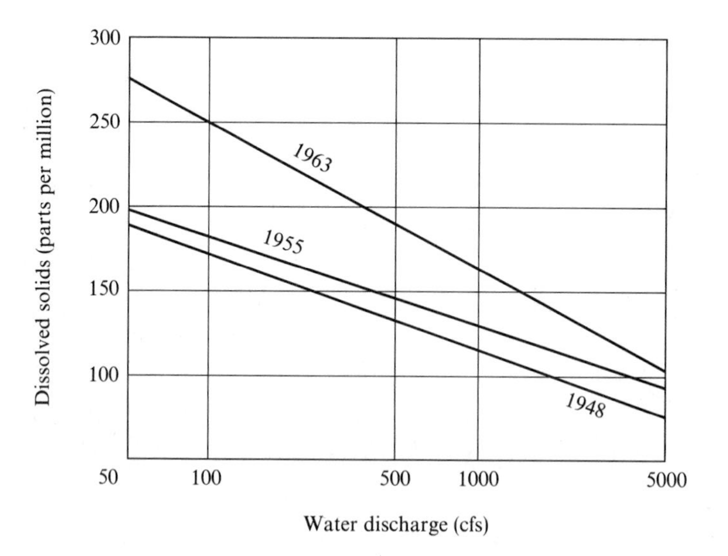

Figure 20-5 Relation between water discharge and concentration of dissolved solids in the Passaic River at Little Falls, New Jersey, for three periods during the industrial and municipal development of the basin. (From Anderson and Faust 1965.)

fluenced by highways averaged 3–7 mg/liter. Salt solutions dissolve metals into streams in an urban environment, causing further problems. An even more severe regional problem occurs in oil fields or near potash mines, where brines with concentrations of salt ranging between 10,000 and 150,000 mg/liter must be disposed of. Most of these solutions are re-injected into deep rocks, but leakage into aquifers and streams often occurs (Leonard 1972).

The net effect of these and many other additions of major ions has been to raise the salinity of most rivers in developed areas. Ackermann et al. (1970) compiled data extending back to the late nineteenth century for Lake Michigan and three large rivers in a heavily populated region of the midwestern United States. Although the record was fragmentary and difficult to interpret, they concluded that total dissolved solids concentrations have increased by 18 to 50 percent through the period of study. The increases were much larger for some constituents, such as chloride and sulfate.

The Colorado River has been greatly affected by human activities. Domestic and industrial uses, which consume 28 million cubic meters of water, add 33,000 tonnes of solutes to the river above Lee's Ferry, Arizona (1 tonne = 1000 kg). Irrigation, which consumes 2183 million cubic meters of water, adds 3.4 million tonnes of salt. But for human activity the average salt concentration at Lee's Ferry would be 263 mg/liter; it is now 501 mg/liter and rising by 32.8 mg/liter for every 100,000 hectares of newly irrigated land (Iorns et al. 1965). Increased mining activity in the basin is likely to increase this figure.

Trace Metals

Metals, such as iron, zinc, lead, molybdenum, and copper, are released in very small quantities by rock weathering (Dethier 1975, Emmett 1975). They have low solubilities and are often mobilized by forming a soluble complex with organic molecules or by becoming attached to clay particles. They participate in the biosphere as "trace elements" and are necessary for plant and animal growth. During the past few years increasing attention has been focused on their significance for human health (Cannon and Hopps 1971, Maugh 1973, U.S. Environmental Protection Agency 1973). When these metals are released into the environment in larger than "natural" concentrations, however, they can be highly toxic and can cause major disruptions of aquatic ecosystems and a general lowering of the suitability of water for industrial and domestic use.

Klein (1962) reviews the toxic effects of heavy metals in the biosphere, including a case in which industrial effluent produced a 1–2 mg/liter concentration of copper in an English river and exterminated all animal life for 16 km downstream. The effect could be observed for 30 km below the outfall, even after dilution to 0.1 mg/liter of copper. Various plant and animal species, of course, have different levels of susceptibility, but many metals cause damage at extremely low levels. Chromium at concentrations as low as 0.01 mg/liter can kill or arrest the development of many organisms, including the microorganisms that break down organic wastes in rivers and sewage treatment plants. The effects of several metals can be synergistic, and their effects can be aggravated by other ions in solution. Fish, for example, are more susceptible to toxic metals in soft water than in hard water.

Loading of metals into the hydrologic cycle can occur by direct dumping of industrial wastes or by some activity such as mining, which inadvertently encourages the mobilization of the metals. Heavy metals are frequently discharged in the company of other toxic compounds such as acids or brines, which further pollute the waterway and suppress the biologic processes that tend to purify the river. The difficulties are particularly great where the entire low flow of a stream is used by an industrial plant and discharged without dilution.

In the example shown in Figure 20-6, wastes from a plating factory have been discharged into the groundwater through a permeable disposal basin. The cadmium and chromium in the effluent have been traced as a plume within the groundwater system 1300 m long, up to 300 m wide, and up to 20 m deep. The plume now flows at a speed of 170 m per year into and under a nearby small stream. Chromium concentrations in the plume range up to 40 mg/liter and up to 3 mg/liter in the headwaters of the stream. Cadmium ranges from 10 mg/liter in the groundwater to 0.1 mg/liter in the

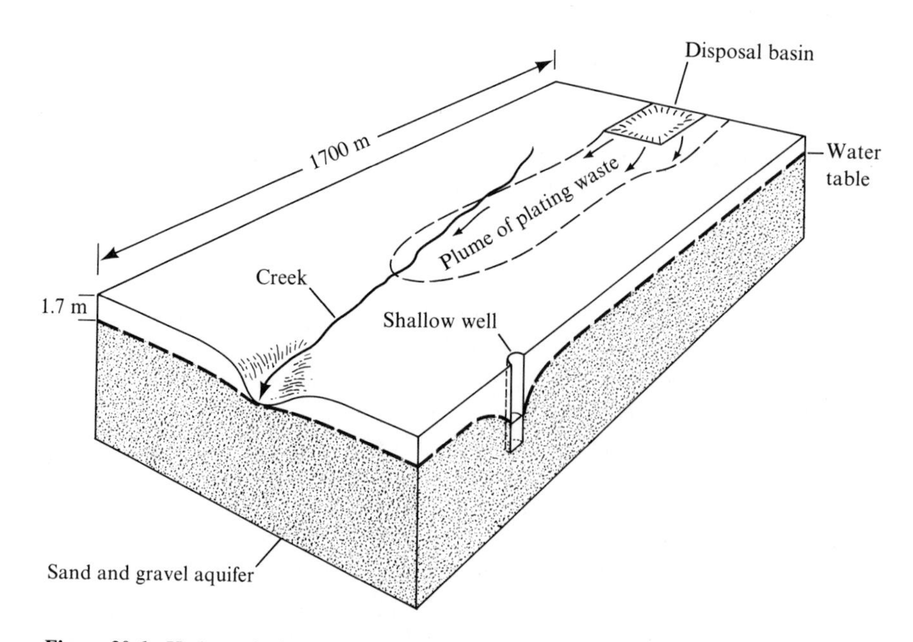

Figure 20-6 Hydrogeologic conditions into which cadmium and chromium plating wastes are discharged in Nassau County, Long Island, New York. (From Perlmutter and Lieber 1970.)

stream. With increasing dilution downstream, both metals become untraceable. This is only a small example of heavy metal pollution, but the concentrations are locally far in excess of drinking-water or food-processing water standards, and the example indicates what will occur on a much larger scale unless planners and other citizens remain concerned and vigilant about waste disposal of toxic materials. Because of the slow rate of movement of contaminated groundwater, pollutants may not reach a well or stream for years after their injection. Even if the source is then cleaned up immediately (which it usually is not), many years may elapse before the pollutant is flushed from the aquifer.

Solid wastes in municipal dumps also contain metals, and if these are placed in the wrong hydrogeologic environment, groundwater can leach out the metals, along with a variety of other pollutants. In some localities a careful site analysis is made before dumping is initiated, but at most dumps this is not the case. A few years ago, for example, a municipal dump was initiated in a locality known to us. The chosen site was a convenient, small unchannelled swale in Pleistocene sediments. No hydrogeologic site analysis was undertaken; no one asked why such a swale might exist in the first place; no one remembered having seen water seeping at many different levels from the sediments that cover most of the region. The valley was rapidly filled with garbage, and then groundwater began seeping through it, dissolving metals, mobilizing other wastes, and spreading pollution and a foul odor. A consulting firm hired to propose a solution suggested pumping groundwater out of the sediments before it could enter the waste. The initial installation cost would be $70,000; and since a simple geologic reconnaissance had still not been done at the time this proposal was made, no one really knew where the water was coming from!

An example of dumping of toxic effluent from a single plant is documented by Goolsby (1969) in Florida. A small creek within a drainage area of 11.4 square kilometers is almost entirely used in an industrial plant, which requires a throughflow of about 0.01 m^3/sec. Upon its return to the river this water carries a load of 1.5 tonnes per day of acid and 0.4 tonnes per day of dissolved ferrous iron (Fe^{2+}). The pH of the stream averages 2.0 (for comparison, the pH of table vinegar is about 4.0), and the iron concentration averages 500 mg/liter, zinc averages 5.5 mg/liter, manganese 0.3 mg/liter, and copper 0.14 mg/liter. As the water flows downstream, the acid is neutralized (mainly by effluent from another chemical plant that intermittently releases slugs of alkaline waste), and the ferrous iron is oxidized to ferric iron (Fe^{3+}) and is precipitated as ferric oxide (Fe_2O_3) and ferric hydroxide ($Fe(OH)_3$). These precipitates accumulate as a fine reddish mud on the stream bottom at the rate of about 340 kg/day along a 3.2-km stretch of river. They are very damaging to aquatic life, precipitating on fish gills causing suffocation, smothering eggs and food sources. The oxygen required for the oxidation of the ferrous compounds to ferric compounds comes out of solution in the stream water, and the dissolved oxygen content is depressed almost to zero within 3 km of the plant.

In many mining operations, the ore or the waste contains sulfides of metals such as iron, copper, lead, and zinc. When buried deep in rock strata, they are normally under reducing conditions and the sulfides are stable. Upon exposure to the atmosphere and to oxygen-rich water the sulfides are oxidized to metal sulfates and sulfuric acid. In streams and pools the rate of oxidation is accelerated several hundredfold by sulfur- and iron-metabolizing bacteria, although in groundwater systems the process occurs more slowly without biologic catalysis. The oxidation of ferrous sulfide, for example, produces a solution of sulfate ions (which make the water hard), ferric ions, and acid. The chemical composition of some representative coal-mine waters is given in Table 20-3. As described above, the ferrous ion is oxidized to ferric hydroxide, which is precipitated on the stream bed with damaging results. The oxidation process depletes the oxygen content of the stream and, together with the high acidity (low pH), high metal content, and high content of dissolved solids, produces water that is unsuitable for aquatic life, drinking, or industrial use.

Table 20-3 Chemical analyses of mine waters from Lancashire, England. (From Klein 1962.)

pH	3.2–6.7
Total dissolved solids	1,683–20,220 mg/liter
Ferrous iron	0–2,305 mg/liter
Ferric iron	0–238 mg/liter
Sulfate	605–3,390 mg/liter
Total hardness (as calcium carbonate)	860–1,560 mg/liter

The problem is best known from the coalfields of Appalachia in the United States and of Britain, where ferrous sulfide is associated with the coal seams. Not all coalfields have acid mine drainage problems, only those in which metal sulfide ores occur. Acid mine water may drain from waste piles, opencast mines, or deep mines for decades after mining has ceased. The drainage may be continuous or an intermittent slug of very acidic water during wet periods. The west branch of the Susquehanna River in Pennsylvania, for example, experienced 20 major fish kills between 1948 and 1962 as a result of such flushing. More than 8000 km of major streams are seriously affected in Appalachia alone. Besides being toxic to aquatic organisms, mine drainage requires extensive treatment before use and corrodes machinery and structures along the rivers.

Because of the intensity and extent of damage to water resources caused by acid mine drainage, a great deal of attention has been focused on halting the drainage, but without much success. One reclamation technique involves sealing deep mines to prevent the ingress of oxygen. Some demonstration projects on individual mines have decreased the drainage of acid and iron, but the number of mines, and the presence of joints and bedding planes, which allow the percolation of oxygen-rich water, suggest that it would be prohibitively expensive to make a significant regional impact by this method. Pumping out of groundwater before it reaches the sulfide-bearing zones enhances the success of the method. Other projects have demonstrated neutralization of the effluent with crushed limestone and settling out of the iron precipitate. Expense again is the major constraint. Because the iron sulfide oxidizes rather slowly, it will be necessary to treat the effluent for decades. There seems to be a greater chance of success in sealing off from groundwater circulation the more recently exploited open-pit mines.

The long-term nature of the problem is well illustrated by the results of heavy metal mining in northern Wales (Klein 1962, Elderfield et al. 1971). Lead and zinc mines, depleted early this century, continue to lose heavy metals in solution and in the sediments eroded from spoil heaps. During the mining era, the numbers of individuals, species, and phyla of aquatic organisms were severely depleted in the streams of the region. Although there has been an increase in the numbers and diversity of organisms since mining ceased, the streams have not fully recovered in 50 years. Fish, molluscs, and crustaceans remain particularly depleted, and a new coastal shellfish farming experiment seems to be suffering. Even where the concentrations of dissolved heavy metals have declined, the ecosystems continue to suffer from the channel-bed instability caused by the transport of polluted sediment from the mine tailings.

Biodegradable Wastes and the Oxygen Balance

Though oxygen is consumed by inorganic reactions described in the preceding section, the greatest consumption of oxygen in streams occurs through the aerobic chemical and microbial breakdown of long-chained organic molecules into simpler, stable end-products such as carbon dioxide, water phosphate, and nitrate. General forms of these reactions can be represented by the following examples:

Carbohydrate → Carbon dioxide
Carbohydrate → Water

Proteins → Amino acids → Ammonia → Nitrite → Nitrate
Proteins → Amino acids → Sulfate
Proteins → Amino acids → Phosphate

Each of these steps consumes oxygen dissolved in the stream water, which in turn is replenished from the atmosphere. If the dissolved oxygen is exhausted, aerobic decomposition ceases and further breakdown must be accomplished by anaerobic bacteria, which can obtain energy from oxygen bound into other substances such as sulfate compounds. The products of anaerobic decomposition are generally noxious, and the process is much slower than aerobic digestion.

The relationship of biodegradable wastes to the amount of dissolved oxygen in stream water, therefore, is fundamental to the maintenance of environmental quality along streams that are used for waste disposal. A moderately high dissolved oxygen content is necessary for the maintenance of healthy aquatic ecosystems, and particularly for the most prized game fish such as salmon and trout.

The breakdown of organic compounds occurs even in a rural drainage basin as leaves, bark, and similar materials are decomposed. A problem arises, however, when large amounts of biodegradable materials such as sewage, animal wastes, wastes from the processing of dairy products, crops, or wood pulp enter the hydrologic cycle. Their breakdown requires large amounts of oxygen, and this demand lowers the dissolved oxygen content of the water body. The ability of the water system to replenish its oxygen is then vital to maintaining good water quality.

The oxygen demand imposed by a biodegradable pollutant will depend on the amount of the pollutant and its chemistry, or more particularly the mass of oxygen required to oxidize a unit mass of the substance to a stable state. The pollutional strength of such wastes is measured by an index called the *Biochemical Oxygen Demand* (BOD). The BOD of a waste is the amount of oxygen consumed by living organisms (mainly bacteria) while utilizing the organic matter in the waste.

If a sample of waste is isolated with a fixed supply of oxygen, it is possible to measure the amount of oxygen consumed over time. The rate at which the demand is exerted varies with the temperature, but the general pattern of the demand can be represented as shown in Figure 20-7. The first smooth rise of each curve represents the consumption of oxygen by bacteria, which are digesting the carbonaceous matter. After this demand has slowed, nitrifying bacteria (which convert ammonia to nitrate) begin to multiply rapidly and exert an appreciable oxygen demand of their own. This second process cannot yet be predicted well enough for planning purposes, and so most analyses of the oxygen budget of streams concentrate on the more tractable first stage of decomposition when the largest oxygen demand is being exerted most rapidly.

Because of the complexity of the processes involved, the biochemical oxygen demand is usually defined by measurement under arbitrary conditions in the laboratory. The sample of wastewater is confined with a fixed store of oxygen for 5 days at 20°C, and the depletion of the oxygen is measured. The rate of depletion indicates how rapidly the substance is breaking down (i.e., how rapidly the oxygen demand is being exerted). For domestic sewage and many industrial wastes, about 70–80 percent of the total BOD is exerted within 5 days, but for more resistant chemicals, the 5-day BOD underestimates the full pollution load. In some situations, such as predicting the impact of pulp and paper wastes rich in chemically resistant lignin, a 20-day BOD is used. The details of the test are described by Sawyer (1960). An index known as the *Chemical Oxygen Demand* (COD) can be used to

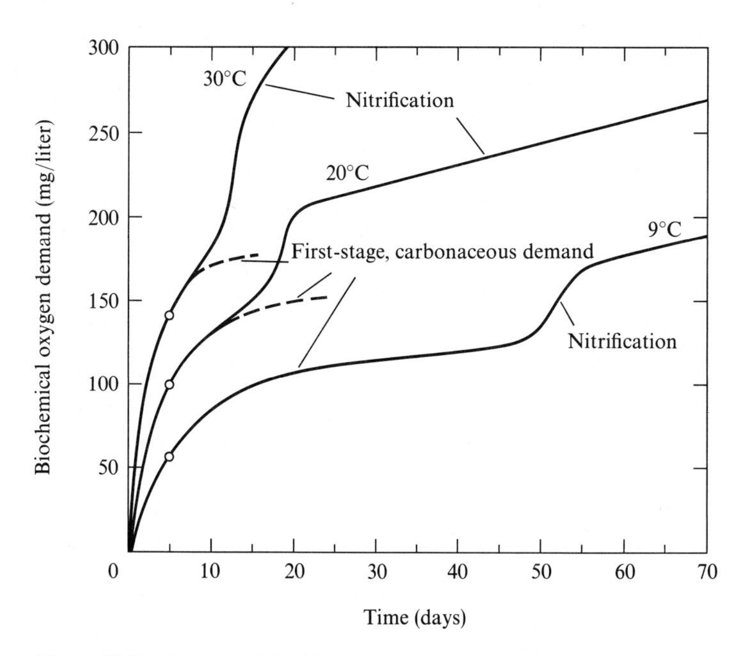

Figure 20-7 Progress of the biochemical oxygen demand at various temperatures. The circled point on each curve represents the 5-day BOD. (From Theriault 1927.)

obtain a rough estimate of the BOD because it can be measured in a few hours. Both of these indices are expressed as concentrations (milligrams of oxygen required to support oxidation of the compounds in one liter of the liquid waste).

It is then possible to compare the pollutional strengths of various wastes according to their 5-day BOD, as shown in Table 20-4. Going one step further, it is possible to rank the output of waste from a single factory or from an entire industry on the basis of *population equivalents.* The BOD of untreated human sewage is 100–300 mg/liter. From the average daily volume of waste it is possible to calculate that 0.076 kg of oxygen per day is required to assimilate the waste of one person; this is one population equivalent. Table 20-5 shows the BOD of some agricultural processing industries in terms of the potential population equivalents of their untreated wastes. The growing magnitude of the industrial waste problem in the United States is shown by Figure 20-8.

Results of laboratory measurements of BOD also show differences between wastes in the rate at which the demand is exerted. Many agricultural processes release wastes that can be rapidly oxidized by microorganisms. Dairy wastes, high in proteins, and coffee-factory wastes, rich in sugars, are examples. They exert their demand rapidly, and if the oxygen supply is not sufficient, a stream receiving such wastes can become anaerobic. In

Table 20-4 Five-day biochemical oxygen demand of some common effluents. (From various reports.)

EFFLUENT SOURCE	BOD_5 AT 20°C (MG/LITER)
Distilling	10,000–30,000
Pulp and paper processing	20–20,000
Wool scouring	200–10,000
Canning industry	400–4,000
Cattle barns and dairies	200–4,000
Meat packing plants	600–2,000
Feedlots	400–2,000
Sugar-beet mills	400–2,000
Dairy processing plants	200–2,000
Cotton mills	50–1,750
Breweries	500–1,250
Untreated domestic sewage	100–400
Combined sewer systems	50–400
Urban storm runoff	>10

Table 20-5 Pollution potential of untreated wastes from major industries in the United States. One population equivalent of BOD is 0.076 kg of oxygen per day. (From Hoover and Jasewicz 1967. Copyright 1967 by the American Association for the Advancement of Science.)

INDUSTRY	BOD_5 (MILLIONS OF POPULATION EQUIVALENTS)
Paper and pulp	216.
Meat products	13.
Dairy	11.9
Canning	8.
Cotton and woolen textiles	5.7
Sugar	4.8
Poultry processing	1.3
Single, large potato processing plant	0.3

treated sewage effluent, on the other hand, the more easily oxidized compounds have been digested in the sewage treatment plant, and the remaining oxygen demand is exerted more slowly. Pulp wastes are assimilated even more slowly.

The impact of an oxygen-demanding waste on the oxygen budget of a stream can best be demonstrated by a simplified analysis of the type used

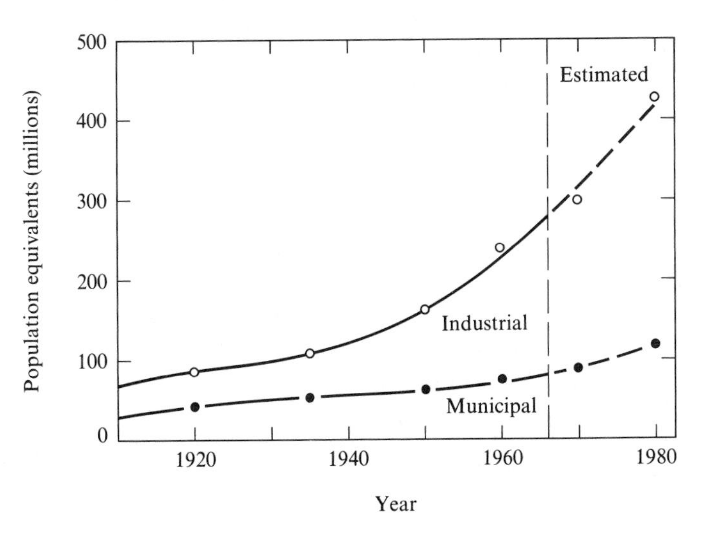

Figure 20-8 Growth of the total U.S. output of biodegradable wastes from industrial and municipal sources before treatment. (From Hoover and Jasewicz 1967. Copyright 1967 by the American Association for the Advancement of Science.)

in most design problems. The technique originated during early studies by the U.S. Public Health Service in the heavily industrialized Ohio River basin (Streeter and Phelps 1925). Since that time, many modifications have been proposed, and the latest developments are summarized by O'Conner (1967) and Nemerow (1974). The reader who needs a fuller discussion of the oxygen budget of rivers, lakes, and marine waters subjected to waste disposal should consult these sources or various textbooks on sanitary engineering. On large rivers receiving heavy discharges of pollutant, some of which may temporarily settle on the stream bed, field studies are usually made.

When a waste enters a stream, it becomes diluted, and after a short distance of travel the BOD of the stream becomes the discharge-weighted average BOD of the effluent and of the stream above the outfall. The mixture may also have an initial oxygen deficit, which means a dissolved oxygen content lower than the maximum concentration possible for water at that temperature (see Table 20-6) for the solubility of oxygen at various temperatures). The oxygen demand of the mixture will begin to exert itself at a rate proportional to the concentration of oxidizable material in the water. As this material is assimilated, therefore, the magnitude of the oxygen deficit will increase, but the *rate* of deoxygenation of the water will decline, as shown in Figure 20-9(a).

As oxygen is being consumed in this way, it can be replenished by photosynthesis of aquatic plants and by incorporation and solution of oxygen from the atmosphere. The former source is usually ignored in small-scale design calculations. The rate of reaeration from the atmosphere depends

Table 20-6 Solubility of oxygen in water at various temperatures.

TEMPERATURE (°C)	SOLUBILITY OF OXYGEN (MG/LITER)
0	
5	12.8
10	11.3
15	10.0
20	9.0
25	8.2
30	7.4

mainly on the oxygen deficit, the width of the stream, the degree of turbulence, the water temperature, and the nature of the waste in the water. For fixed stream conditions the rate of reaeration will depend on the oxygen deficit (D) and therefore as the water is deoxygenated, reaeration will accelerate as shown in Figure 20-9(b). However, the rapid initial reaeration would partially offset the original oxygen demand. The subtraction of reaeration from the oxygen demand yields the change in the oxygen deficit from its initial concentration (D_i). This is shown in Figure 20-9(c). The reaeration rate is now proportional to a new oxygen deficit. If the oxygen deficit of the stream increases during this first time period, the rate of incorporation of oxygen at the stream surface increases. Eventually the accelerating rate of reaeration will overtake the declining rate of deoxygenation, and the dissolved oxygen content of the stream will start to increase again, as shown in Figure 20-9(c). It may increase beyond the original value for the mixture and approach the solubility of oxygen for the ambient temperature.

The profile in Figure 20-9(c), which results from the interplay of deoxygenation and reaeration, is referred to as the *dissolved-oxygen sag.* The equation used to calculate it is

$$D = \frac{kL}{r - k}(e^{-kt} - e^{-rt}) + D_i e^{-rt} \tag{20-2}$$

where D = the oxygen deficit of the stream at any time (mg/liter)
D_i = the initial oxygen deficit of the mixture (mg/liter)
k = the deoxygenation constant, or rate of exertion of the BOD (per day)
L = the initial BOD of the mixture (mg/liter)
r = the reaeration rate (per day)
t = time from initial mixing (days)
e = base of natural logarithms (2.7183).

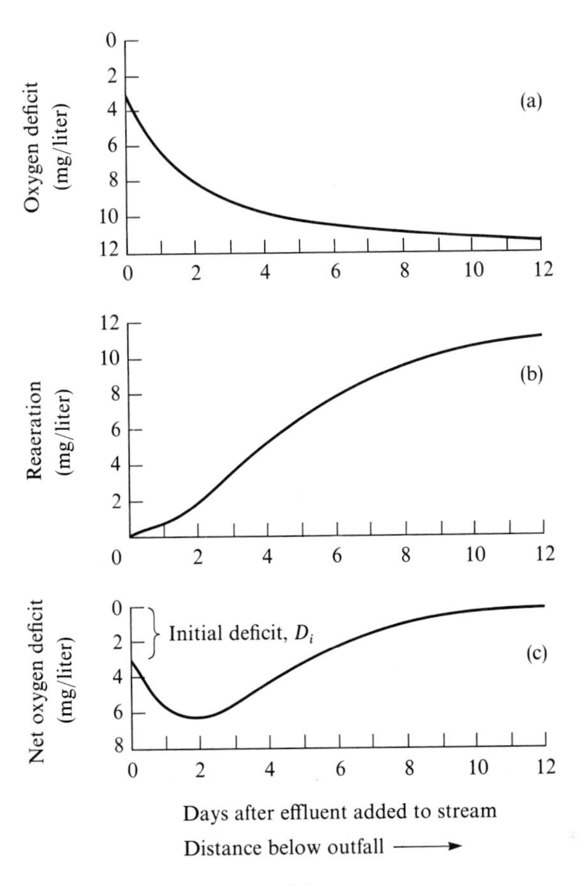

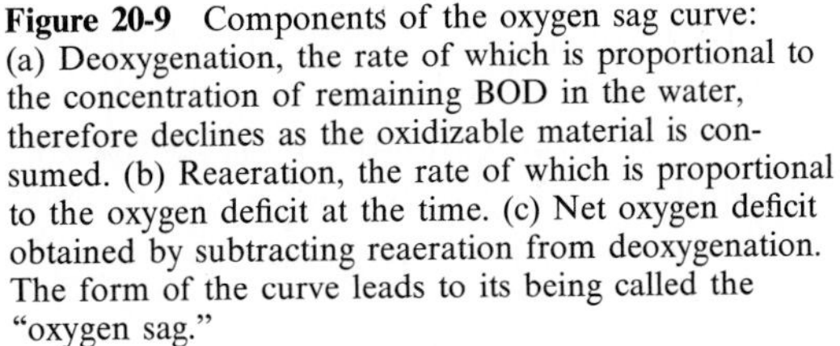

Figure 20-9 Components of the oxygen sag curve: (a) Deoxygenation, the rate of which is proportional to the concentration of remaining BOD in the water, therefore declines as the oxidizable material is consumed. (b) Reaeration, the rate of which is proportional to the oxygen deficit at the time. (c) Net oxygen deficit obtained by subtracting reaeration from deoxygenation. The form of the curve leads to its being called the "oxygen sag."

Equation 20-2 can be used to calculate the dissolved oxygen content of a stream at any time and (if the velocity of the stream can be estimated) at any distance below an outfall discharging a biodegradable pollutant (see Typical Problem 20-1 and Figure 20-10).

The parameters L and k have already been discussed briefly. Table 20-4 lists some representative values of L for wastes before dilution with stream water. The deoxygenation constant, k, or the rate of exertion of the BOD varies with the chemistry of the waste. Typical values for k are 0.23 for sewage effluent and 0.005 for chemically resistant humic materials. The

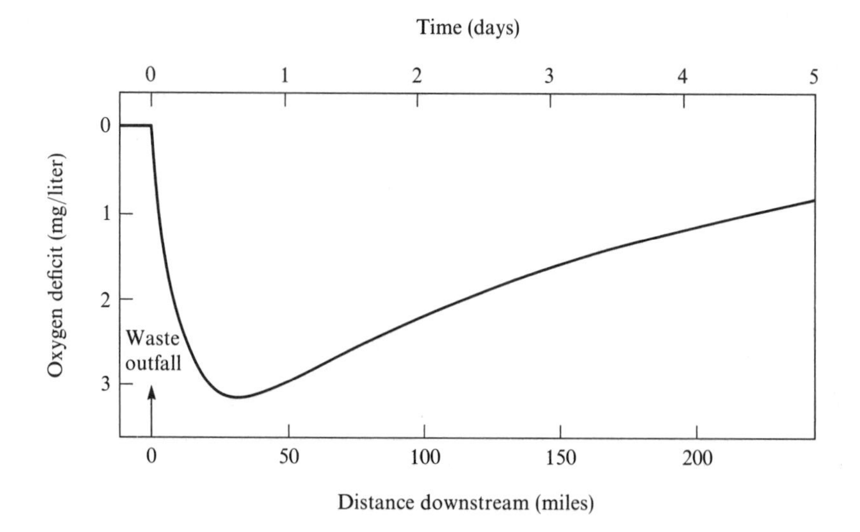

Figure 20-10 Dissolved oxygen sag calculated from Equation 20-2 and the data given in Typical Problem 20-1.

deoxygenation rate also varies with temperature, as shown in Figure 20-7. This effect can be estimated from the equation

$$k_T = k_{20} 1.05^{(T-20)} \tag{20-3}$$

where k_T and k_{20} are the deoxygenation constants at the temperature (T) of the stream and at the temperature of the standard BOD test (20°C), respectively.

The reaeration constant, r, is more difficult to predict. It varies with the degree of turbulent mixing of the stream water, and therefore with the velocity and depth of flow. Several equations have been used to define this relationship, for example,

$$r_{20} = \frac{3.3u}{d^{1.33}} \tag{20-4}$$

where r_{20} = the reaeration coefficient at 20°C (per day)
u = mean stream velocity (ft/sec)
d = mean water depth (ft).

Langbein and Durum (1967) have shown that since velocity and depth vary with discharge both at a station and downstream, the reaeration coefficient will also vary in a manner that can be calculated approximately from data on the hydraulic geometry (see Figure 20-11).

Temperature also affects the reaeration coefficient, approximately according to the relationship

$$r_T = r_{20} 1.02^{(T-20)} \tag{20-5}$$

Some pollutants, particularly some detergents, can slow the rate of reaeration, and an ice cover can reduce it virtually to zero, creating very serious deoxygenation problems in rivers and lakes in cold regions.

Figure 20-12 portrays oxygen sags resulting from the variation of some of the controlling factors. The critical feature of each curve is the magnitude and location of the minimum dissolved oxygen concentration. Without calculating the whole curve, this minimum can be located by using the following equations for the maximum dissolved oxygen deficit (D_c) and its timing (t_c):

$$t_c = \frac{1}{r-k} \ln\left[\frac{r}{k}\left(1 - \frac{D_i(r-k)}{Lk}\right)\right] \tag{20-6}$$

$$D_c = \frac{kL}{r} e^{-kt_c} \tag{20-7}$$

Equation 20-2 is not a very accurate predictor of the pattern of dissolved oxygen in a river. There are too many uncertainties about the rate at which the waste is digested, the onset of appreciable nitrification, the factors affecting reaeration, the importance of photosynthesis in supplying oxygen, as well as various hydrologic uncertainties about inputs of groundwater and surface water along the watercourse, the variation of stream temperature, and inputs of stormwater and other pollutants downstream. With a great deal of care and with field checking, this form of analysis is useful in evaluating the assimilative capacity of a stream at various discharges and for calculating the approximate pattern of dissolved oxygen along the stream

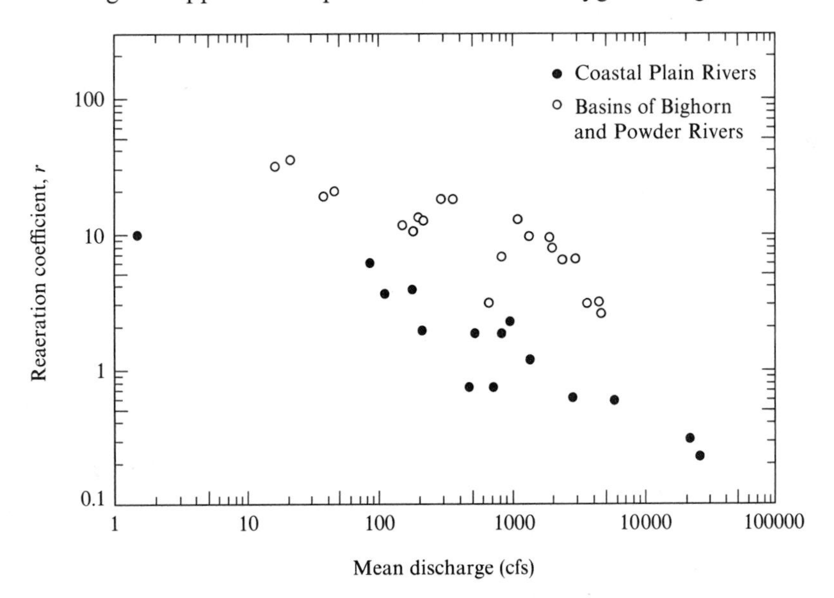

Figure 20-11 Regional contrast in the reaeration coefficient for rivers at their mean discharge. Rivers in the northern Rocky Mountains (open circles) have steeper slopes and higher reaeration coefficients than lowland rivers of the same size. (From Langbein and Durum 1967.)

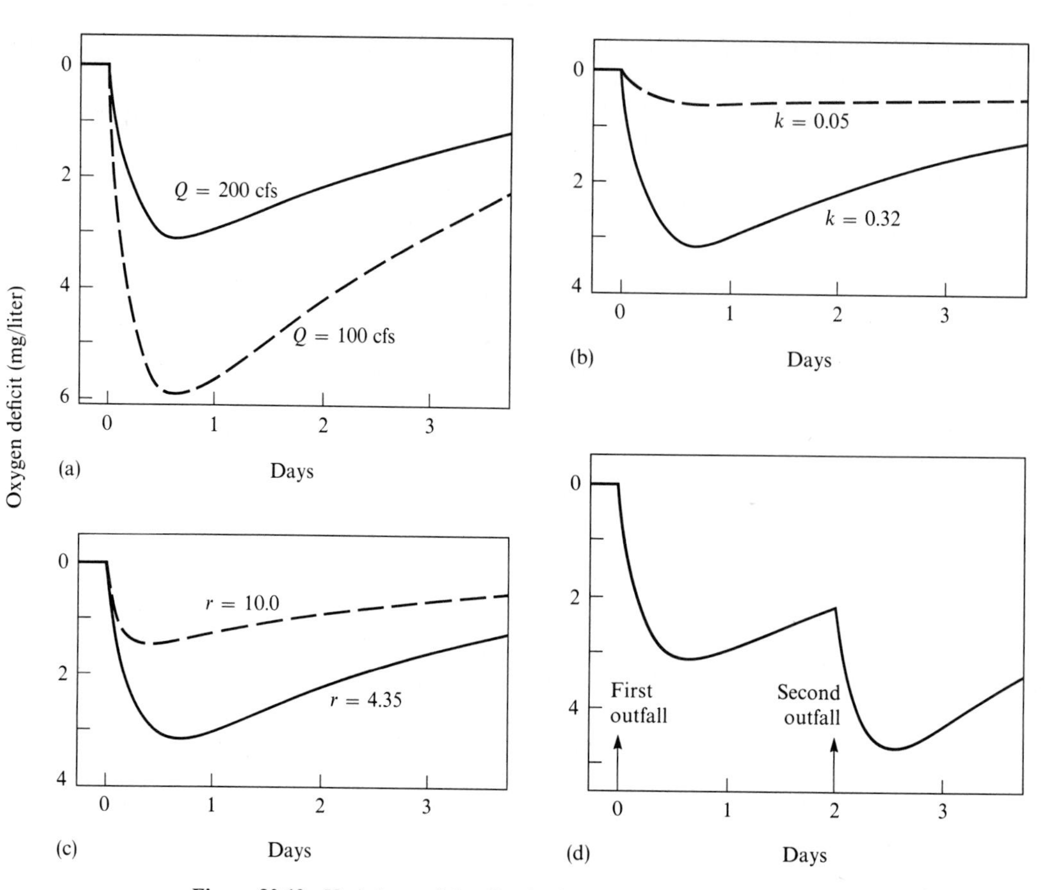

Figure 20-12 Variations of the dissolved oxygen sag. The solid curve is copied from Figure 20-10 and represents conditions described in Typical Problem 20-1. In each graph, (a), (b), and (c), various parameters in Equation 20-2 were changed one at a time, while all other values are as in the original problem. The dashed line in each graph indicates the effect of each change. In graph (d), the effect of a second outfall is shown as described in Typical Problem 20-1.

and indicating where major problems are likely to arise. It can also be used to predict the probable effect of various kinds of treatment to reduce BOD, or the result of various other changes that would alter the stream temperature, discharge, depth, velocity, ice cover, and other characteristics of the stream. We have introduced the matter here in moderate detail so that the planner can appreciate quantitatively what is involved in managing the oxygen balance of a stream. For more detailed analysis the planner would need to contact a sanitary engineer.

Below each source of oxygen-demanding pollution, therefore, a roughly definable spatial pattern of dissolved oxygen exists and will vary with the flow of the river, as well as the other variables discussed above. The oxygen sag becomes particularly critical at low flow during warm weather or under an ice cover. An example of a sag below a point source of inorganic oxygen-demanding, iron-rich effluent is shown in Figure 20-13. The situation becomes more complicated and more serious on large rivers receiving large quantities of heavy industrial and municipal wastes. Figure 20-14 shows the summer and winter oxygen sags for the Mississippi River below Minneapolis–St. Paul. The winter values are lower because an ice cover suppresses reaeration.

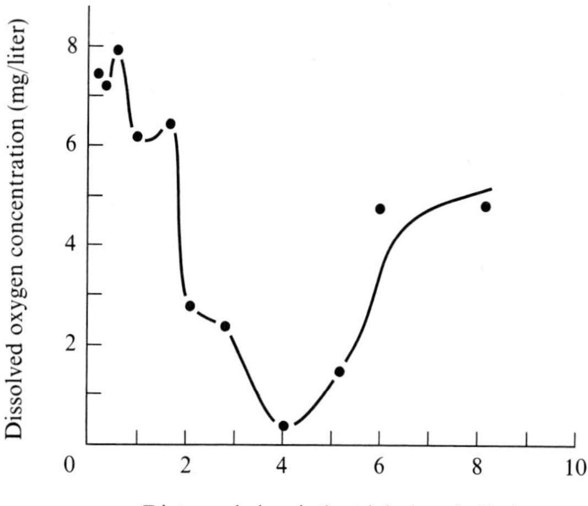

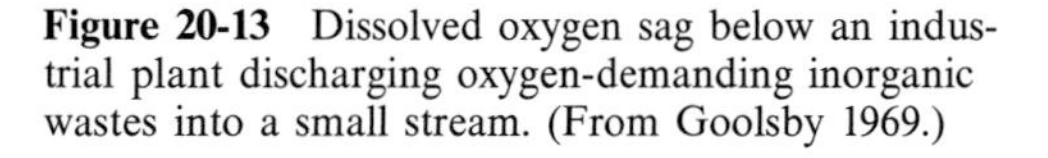

Figure 20-13 Dissolved oxygen sag below an industrial plant discharging oxygen-demanding inorganic wastes into a small stream. (From Goolsby 1969.)

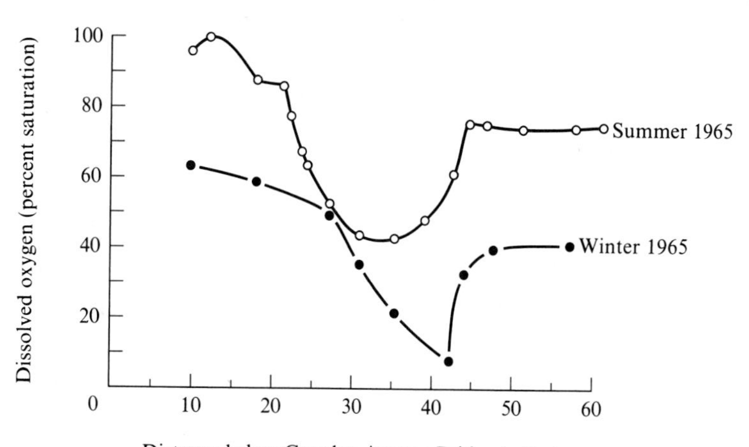

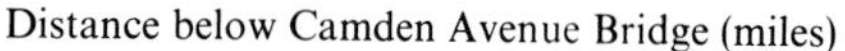

Figure 20-14 Dissolved oxygen levels in the Mississippi River below Minneapolis–St. Paul. (From M. G. Wolman, *Science,* vol. 174, pp. 905–918, 26 November 1971. Copyright 1971 by the American Association for the Advancement of Science.)

In many rivers, increased waste production has exceeded the growth in treatment facilities, and the dissolved oxygen content of receiving streams has declined for many years (see Figure 20-15). On an optimistic note, however, Wolman (1971) has demonstrated that on some large rivers in industrial regions of the United States, dissolved oxygen levels either have remained about constant or have improved since the 1930's (see, for example, Figure 20-16). These welcome changes, which have generally been ignored in the popular literature, are the results of large expenditures on treatment facilities. Similar advances have been recorded during the past few years for the Thames in England and several large European rivers. A continuing problem, however, is the geographic spread of urbanization, light industry, and agro-industrial facilities such as feedlots into rural areas. These activities often release oxygen-demanding wastes from poorly operating septic tanks, waste lagoons, storm sewers, and small, inadequate treatment plants so that many small streams and lakes in rural areas are beginning to suffer oxygen depletion.

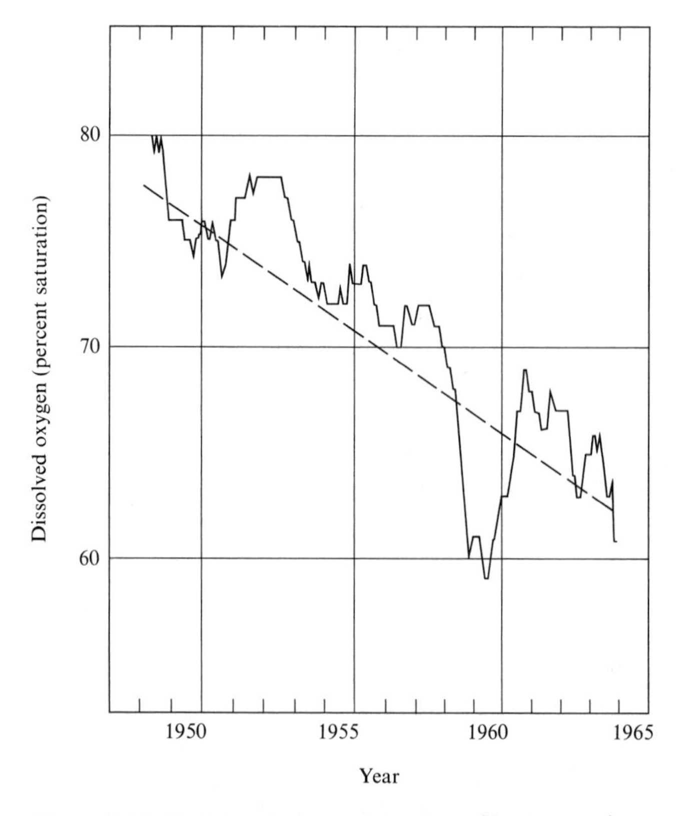

Figure 20-15 Two-year moving average of monthly concentrations of dissolved oxygen in water in the Passaic River at Little Falls, New Jersey. Dashed line indicates the general trend in oxygen content. (From Anderson and Faust 1965.)

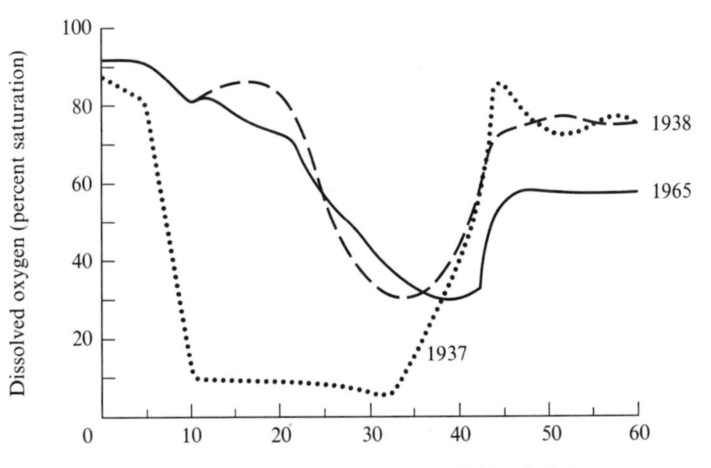

Figure 20-16 Average annual concentrations of dissolved oxygen in the Mississippi River below Minneapolis–St. Paul. The improvement between 1937 and 1938 resulted from the construction of major waste treatment facilities. Since 1938, oxygen levels have been maintained in spite of industrial and residential growth by the construction of additional treatment plants. (From Wolman 1971.)

Strategies for Reducing Dissolved-Oxygen Problems in Rivers

Dissolved-oxygen conditions have attracted the concern of sanitary engineers and river managers for many years, and great advances have been made in conserving the oxygen reserves of streams. The cost of treatment, however, has been high and is escalating. At the same time, environmental concerns have been increasing as the public becomes more aware of the value of healthy aquatic ecosystems. Yet sources of pollution are spreading along small streams and lakes. The growth of suburban housing and industry is responsible for this in some regions. Many tropical countries are developing agricultural processing industries that must discharge their effluents into small, warm streams where they endanger meager but important fisheries. For these and many other reasons, the planner needs to be aware of some of the strategies for reducing the impact of oxygen-demanding wastes on rivers.

Organic wastes such as domestic sewage, farm wastes, and biodegradable industrial wastes can be treated to reduce their oxygen demand by a set of processes that are often referred to as primary, secondary, and tertiary treatment.

Primary treatment involves the settling of suspended mineral and organic solids. The process may be accelerated by the addition of chemicals to flocculate or precipitate solids. Such treatment removes 40 to 90 percent of the suspended solids and 25 to 85 percent of the 5-day BOD (Fair et al. 1968). The settled waste must then be disposed of by dumping in a large body of water or in a landfill, and the liquor it contains must be removed, disposed of, or treated separately. The rest of the effluent passes directly to the receiving water or undergoes further treatment.

Secondary treatment involves one of two biologic forms of processing. In a *trickling filter,* wastewaters are sprayed onto columns of crushed stone and flow in thin films over biologic growths covering this substrate. The organisms, which include bacteria, fungi, and protozoa and insects which graze on them, absorb and break down the dissolved organic substances in the wastewater. Some of the breakdown products such as carbon dioxide escape to the atmosphere; others such as nitrate remain in solution, while others are absorbed into cells of the biologic growths, which eventually slough off to be carried to settling tanks by the wastewater stream. The other kind of secondary treatment is the *activated sludge* process in which flocs of bacteria, fungi, and protozoa are stirred in the wastewater. The result is about the same as during trickling filtration. Both processes remove between 50 and 95 percent of the remaining settleable solids and 50 to 95 percent of the BOD. The efficiency of secondary treatment is lowered if the plant is overloaded so that the effluent must be passed through too quickly. Toxic industrial chemicals can also retard the biologic processes. Many municipal sewer systems accommodate both sanitary sewage and storm runoff. During rainstorms these combined systems carry far too much water to be handled in a treatment plant, and large volumes of polluted runoff bypass the plant and enter receiving waters. Separation of the sewer systems, which is expensive, reduces this problem.

As the demand for cleaner water grows, *tertiary treatment* of wastewaters is being recommended and accepted in a few places. The technology is still developing and includes such practices as using a strong oxidizing agent like ozone to remove the final vestiges of BOD, odor, and taste, and the addition of alum to precipitate phosphates which are not removed efficiently by the other treatment processes.

Another method of accomplishing a final cleansing step after secondary treatment is to spray chlorinated effluent onto cropland, forests, or mine tailings. The method has several advantages over direct discharge of effluent into receiving waters. First, the remaining BOD is removed by biologic digestion in the soil. Second, phosphorus and nitrogen are absorbed by the soil and plants; if crops or timber are harvested from the site, large quantities of nutrients can be disposed of over long periods of time. Third, controlled experiments have shown that crop yields, and to a lesser extent timber yields, can be enhanced slightly by irrigation with nutrient-rich water in many areas. Finally, the groundwater is recharged with clean water, which is then available for re-use or for augmenting dry-weather streamflow (Kardos 1970). Considerable research is being carried out with this method and is being extended to the disposal of sludge liquor and sludge on forestland. There is still some concern about economics, technological details of application, the fate of pathogenic viruses, and the long-term ability of soil profiles to protect the quality of groundwater. Several studies have shown great promise, however, and indicate that the treated effluent from cities of 100,000 people could be disposed of on 6.4 square kilometers of productive land.

In many rural and suburban areas, where population densities are less than about 1000 per square kilometer, it has been difficult to justify the cost of a sewer system and treatment plant. In many places small, mechanically aerated digestion tanks called *package plants* are now being introduced to handle the wastes from villages, shopping centers, and apartments, but many problems remain with the design and operation of these systems, and their effluent should still be disposed of into soil.

Elsewhere, individual households dispose of their wastes in buried *septic tanks,* which are small sedimentation basins and anaerobic sludge digestion tanks. The effluent moves from the tank through a tile field or seepage pit, which spreads it into the surrounding soil where aerobic biologic breakdown of dissolved and solid organic compounds takes place, the nutrients are mainly trapped, and clean water recharges the groundwater. The soil depth and permeability must be such that the effluent can move freely into the soil, but its movement must be slow enough to allow the necessary biologic and chemical reactions to occur before the effluent reaches the groundwater. The physical conditions required for good septic-tank performance were discussed in Chapter 6. Unfortunately many systems do not work well because they are sited and designed poorly. Many rapidly growing suburbs rely upon septic tanks, and the oxygen balance and nutrient levels of nearby streams and lakes have suffered. Many of the homeowners have also suffered when their inefficient septic tanks have to be pumped out each month at a cost of $30.

Industrial and agricultural wastes present a more complex and growing range of problems because of their volume, timing of release, and special problems. Some are toxic or have a BOD too high for treatment in a conventional municipal treatment plant. Some factory-type agricultural operations, such as feedlots, have the same pollution potential as cities with populations of over 100,000. Yet their wastes are given only limited treatment or none at all. It is little wonder then that many large fish kills in the United States are traced to the oxygen depletion of streams below feedlots.

Many wastes from agriculture and agricultural processing industries can be treated separately by the primary and secondary methods described above. Alternatively, wastes can be processed in *stabilization ponds* in which aerobic microbial organisms digest the organic substances as described above, but the pond is also colonized by heavy growths of green and blue-green algae, which release oxygen to the wastewater during photosynthesis. Solids settle as sludge and decompose anaerobically on the bed. Small volumes of highly seasonal wastes with high BOD can be spread or sprayed onto soils, as described previously.

Some pollution problems associated with food production and processing could be ameliorated by reducing waste. Hoover (1974) has catalogued recent advances in the technology of food processing that are attractive to manufacturers because they improve product yield. Similar opportunities exist in other industries, and advances are being made in this area (Besselievre 1969). Many wastes, such as whey or the peelings from fruit and vegetables, can be used as animal feed, but there are problems of distribution costs, and usually only very large farms in advantageous locations can use most of these wastes.

The treatment of industrial wastes is a very complicated field because of the variety of the effluents. Increased pressure by government agencies and environmental groups during recent years has persuaded many industrialists to treat their wastes, but many wastes are still discharged untreated. Some of these are diverted to the local sanitary sewer system under agreement with local sanitation authorities. Many biodegradeable industrial wastes are treated separately by the primary, secondary, or stabilization pond processes described above.

In some cases wastes from different plants can be advantageously mixed before treatment. If the planner needs to have more specific information on the treatment of industrial wastes, he or she should consult one of the many textbooks in that field (e.g., Besselievre 1969) and be able to frame a set of very specific questions about the situation of interest. The list may include: What will be the volume of effluent? What is the timing and peak rate of discharge? What is the nature of any treatment process to be undertaken? What are the chemical characteristics of the effluent, including the minor toxic constituents and any constituents whose effects are not well understood? What are the probable effects of the effluent on a sewage system into which it will be discharged? What are the operating principles of the treatment facilities and the technical expertise of the personnel? What are the characteristics of the receiving water, especially at critical periods of low flow, low oxygen, or ice cover? What are the present and probable future quality of the receiving water above the waste disposal site? What are the present and planned uses of the receiving water and the ultimate disposal method for any solid wastes or concentrated fluids produced during the treatment process? Other questions may become obvious in a particular situation.

Because the topic of waste treatment is so complicated, there is a natural tendency for planners, legislators, and other concerned citizens to feel overwhelmed by the technical information in design proposals. This does not have to be the case if one keeps in mind some simple principles of hydrology and water chemistry and asks with persistence some very specific questions.

Major Plant Nutrients: Nitrogen and Phosphorus

The breakdown of organic compounds described in the last section does not remove all pollution problems from streams. The compounds decay eventually to stable phosphates, nitrates, sulfates, carbon dioxide, water, and minor constituents. A typical set of changes in river chemistry below

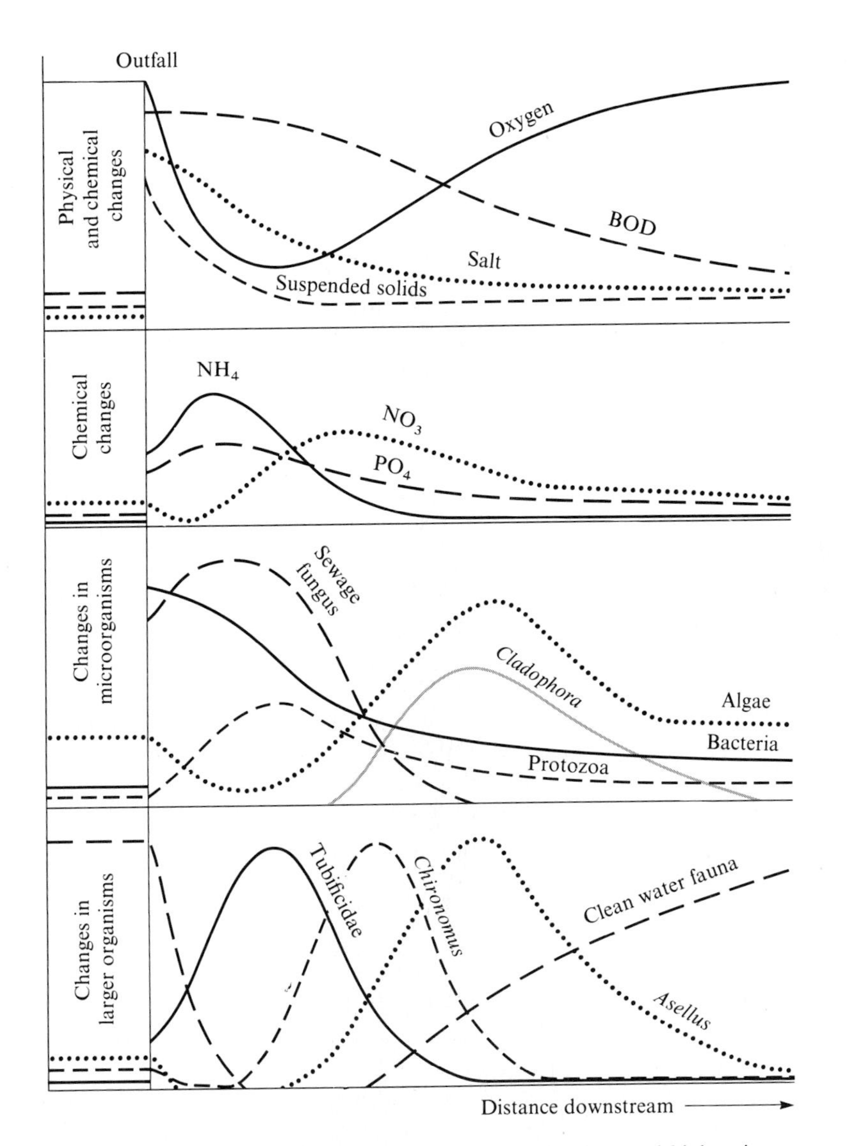

Figure 20-17 Diagram of typical changes in water chemistry and biology in a river below a source of organic pollution. (From Hynes 1963.)

a source of organic pollution is shown in Figure 20-17. The concentrations of nitrate and phosphate decrease downstream because of dilution as the river discharge increases.

Under the right conditions, one or more of these stable oxidation products can act as a nutrient for plants growing in the receiving stream or lake (see Figure 20-17). If the enrichment with nutrients causes a significant increase in the rate of plant growth, the process is known as eutrophication (Stewart and Rohlich 1967).

The plants may be phytoplankton (suspended or floating) or periphyton (attached to submerged surfaces), and those that cause the greatest concern are algae, which can grow prolifically to cause a number of problems n some polluted waters. Algal blooms are often unsightly; they can produce discoloration and a bad taste in drinking water; they may increase the costs of water treatment; and upon their death and decay they impose an oxygen demand on the water that can deoxygenate a river or lake causing fish kills and other problems referred to in the last section.

Many types of algae are found in natural water bodies, and some will be described in Chapter 21. Their identification can help the planner to assess the chemical and biologic state of the water body.

There is a great deal of controversy about what triggers algal blooms in rivers and lakes. Some writers put all of the blame on phosphorus and call for the abolition of phosphorus-containing detergents. A chemist for one of the large companies that produce detergents has called phosphorus an "innocent pawn in pollution politics" and has suggested that carbon dioxide produced by bacterial breakdown of organic wastes is the dominant factor. The situation can probably be summarized best as follows:

Primary production, or the weight of plant material produced by photosynthesis, can be limited by one of many factors. These controls include light, temperature, or the presence of some dissolved nutrient such as carbon, phosphorus, nitrogen, or a trace element. It is sometimes difficult to define which control is the limiting factor because many genera of plants are involved, and because some of them can compensate to an extent for the low availability of one nutrient, so long as other conditions are favorable. Also, as various algal species bloom and then decay, one nutrient may replace another as the limiting factor. If the abundance of the limiting factor at any time is increased by some outside agency, however, primary production will be increased until some other factor becomes scarce enough to be the limit.

The nutrient that most frequently limits algal production in most waters seems to be phosphorus (Schindler 1977); hence the importance attached to this element in current discussions of water pollution. The next most common control seems to be nitrogen, and if both phosphorus and nitrogen are relatively abundant, some other scarce element may be the control on algal growth. Even if the nutrients are abundant, the water may be too cold to allow rapid plant growth. A dense forest cover over a small stream or high sediment concentration in a large deep river may reduce the penetration of light enough to limit photosynthesis. In rivers or lakes with nutrient concentrations suitable for copious production of algae, the rapid flushing of water through the system prevents the onset of conditions favoring the growth of the most productive and obnoxious phytoplankton, the blue-green algae. Rickert et al. (1977) discuss such conditions with reference to the Willamette River in Oregon. If all other conditions are equal, nuisance growths of aquatic plants develop in lakes at lower nutrient concentrations than those that become critical in streams.

Small quantities of phosphorus are weathered from rock minerals and are held in soils. The soil water leaches low concentrations of phosphorus into the groundwater and eventually to streams. Natural waters from undisturbed forest catchments usually have dissolved phosphorus concentrations (as orthophosphate) in the range 0.005 to 0.05 mg/liter. Concentrations of 0.01 mg/liter will support algal growth, but a level of 0.08 to 0.10 mg/liter is usually necessary to trigger blooms. The element also occurs in solid form in streams from undisturbed catchments. Some orthophosphate is adsorbed on eroded soil particles. Other phosphorus constitutes a portion of any organic detritus carried by the river. These two solid forms may exceed the soluble phosphorus content by tenfold or more. They are deposited with the sediment and whether they later become available as dissolved plant nutrients depends on the chemistry of the sediments, the depth of the water body, and many other factors. The total phosphorus concentration is therefore another important characteristic. Most uncontaminated streams have less than 0.03 mg/liter of total phosphorus, although streams with less than 0.10 mg/liter of total phosphorus may not be obviously polluted.

In catchments altered by human activity, phosphorus enters natural waters from a large number of sources. Orthophosphates applied to agricultural land as fertilizers are held tightly by the soil. Small amounts enter the stream in solution, however, and larger amounts are carried by eroded soil particles. In heavily fertilized regions, such as the San Joaquin Valley of California and the Yakima River of Washington, the return flow from irrigated fields carries 0.1 to 0.4 mg/liter of soluble phosphorus, while the total concentration of the element may be twice as high as this value. Verduin (1967) and Biggar and Corey (1969) review agricultural sources of plant nutrients in aquatic ecosystems.

Animal wastes are rich in phosphorus, but if the animals graze on open ranges or well-drained pastures, the loss of phosphorus to the stream is only slightly greater than on undisturbed land. The major losses are associated with eroded soil particles or organic detritus. If the animals are confined on cattle feedlots, poultry farms, or similar facilities, the runoff contains very high concentrations of phosphorus, nitrogen, and oxygen-demanding wastes. Loehr (1968) reports dissolved phosphate concentrations as high as 30 mg/liter in stormwater from a cattle feedlot. Unless they are removed from the water such concentrations have disastrous local effects on aquatic ecosystems.

Domestic sewage is another important contributor of phosphorus to streams, because primary and secondary treatment removes only about 20 to 30 percent (about 2 mg/liter) of the element from sewage. More than one-half of the total phosphorus in domestic wastewater is contributed by detergents and is in a form known as condensed phosphate, which may be in solution or in solid form. Stormwater runoff from urban areas can also wash high concentrations of organic wastes into streams. Sylvester (1961) gives values of up to 1.4 mg/liter for total phosphorus (average of 0.2 m/liter) and up to 0.7 mg/liter for soluble phosphorus (average of 0.08

mg/liter) for urban storm runoff. Weibel (1968) also discusses this source of pollution.

Industrial sources of phosphorus effluents are highly variable, depending on the form of the enterprise. Phosphate mining and milling and the processing of agricultural and animal products are heavy contributors, as are some chemical industries including the detergent industry.

Inputs of nitrogen generally originate from the same sources as phosphorus. There is also a variety of forms of nitrogen ranging from that incorporated into organic detritus, compounds of ammonia that are in the process of being degraded, and nitrates, which are the fully oxidized, stable nitrogen compounds in most surface waters. Streams from forested catchments have nitrate concentrations of about 0.1 mg/liter and organic nitrogen contents of the same magnitude. Agricultural regions, heavily fertilized with ammonium salts, sometimes yield runoff with nitrate concentrations exceeding 1.0 mg/liter, and smaller amounts of organic nitrogen. Nitrates are held only loosely in soils, and are easily leached to the groundwater. Beneath barnyards, feedlots, and heavily fertilized croplands, the nitrate concentration of groundwaters can exceed 100 mg/liter, and have been linked to methemoglobinemia ("blue-baby syndrome") among infants drinking the water (Bosch 1950). Case studies of this problem have been reported by Bosch (1950) and Gillham and Webber (1969).

Domestic sewage averages about 20 mg/liter of total nitrogen and urban stormwater averages more than 1.0 mg/liter of total nitrogen. Industrial sources of nitrogen are at least as variable as those of phosphorus, with food processing and chemical industries, such as the manufacture of fertilizers and explosives being among the worst offenders.

The approximate annual amounts of nitrogen and phosphorus supplied to the waterways of the United States have been estimated by a committee of the American Water Works Association as shown in Table 20-7. The data indicate that agricultural runoff is the greatest single contributor of nitrogen and phosphorus to waterways, although concentrations are usually greatest in domestic and industrial wastes.

Several strategies have been proposed for controlling the influx of nitrogen and phosphorus to aquatic ecosystems. The planner should be aware of the range of techniques available for limiting both "point" sources and "non-point" sources of the pollutants. The former are easier to control.

In some regions, phosphate detergents have been banned or restricted as a first step toward the control of eutrophication. This action is not seen as a universal remedy but as one of the easiest, cheapest, and quickest measures that can be taken in the near future. Schindler (1974) has shown that the reduction of the phosphorus input to lakes has rapid effects in reducing algal production, and Murphy (1973) has shown that legislation reducing the phosphorus content of detergents in New York State was followed by a rapid decrease in the concentration of total organic phosphate from 1.74 to 0.74 mg/liter in the upper layers of Onondaga Lake in the urban area near Syracuse, New York. The decrease was accompanied by the elimination of blue-green algae, which had formed most of the unsightly scum on the lake.

Table 20-7 Estimated nutrient contributions to the waterways of the United States. (Reprinted from *Journal of the American Water Works Association*, Volume 59, by permission of the Association. Copyrighted 1967 by the American Water Works Association, Inc., 6666 West Quincy Avenue, Denver, Colorado 80235.)

SOURCE	NITROGEN LOAD (THOUSANDS OF TONNES/YEAR)	NITROGEN USUAL CONCENTRATIONS IN DISCHARGE (MG/LITER)	PHOSPHORUS LOAD (THOUSANDS OF TONNES/YEAR)	PHOSPHORUS USUAL CONCENTRATIONS IN DISCHARGE (MG/LITER)
Domestic waste	500–725	18–20	90–230	3.5–9
Industrial waste	>450	0–10,000	*	*
Rural runoff:				
Agricultural land	680–6800	1–70	54–540	0.05–1.1
Nonagricultural land	180–860	0.1–0.5	68–340	0.04–0.2
Farm animal waste	>450	*	*	*
Urban runoff	50–500	1–10	5–77	0.1–1.5
Rainfall onto water surfaces	14–270	0.1–2.0	1.4–4.1	0.01–0.03

*Insufficient data available to make an estimate.

Diversion of sewage effluent from a eutrophic stream or lake is another way of reducing the concentrations of plant nutrients below levels that trigger nuisance blooms of phytoplankton. In the 1950's, limnological studies by W. T. Edmondson of the University of Washington laid the foundation for a plan to divert the phosphorus-rich effluent from sewage treatment plants around Lake Washington. The effluent was piped out into Puget Sound where it was greatly diluted, and the abundance of nuisance algae in the lake declined drastically within a few years. In many other cases, however, there is no convenient large water body that can be used for dilution of nutrient concentrations. It may then be necessary to remove phosphorus and nitrogen from the sewage effluent.

Nutrient removal can be accomplished by adding a tertiary step to the sequence of sewage treatment practices described in the previous section or by modifying the primary and secondary steps. Phosphates can be precipitated from the effluent after secondary treatment by the addition of alum, lime, or an iron salt. Nitrogen can be removed by ion exchange. The technology was reviewed by Rohlich (1969) but is developing rapidly. Kardos (1970), among others, has reported on successful disposal of treated sewage effluent by spraying onto soil. By the time the water percolates to the groundwater its concentration of 6-7 mg/liter of both nitrogen and phosphorus had decreased to less than 3 mg/liter of nitrogen and less than 0.04

mg/liter of phosphorus. The nutrients were stored in the soil and recycled through crops of timber and agricultural plants.

Agricultural sources of nutrients are usually of the non-point variety and thus are difficult to define and to control. Their measurement usually involves monitoring the total load of nitrogen and phosphorus in a stream and the load imposed from point sources. The non-point load can then be calculated by subtraction. In large North American river basins, point sources contribute 70–90 percent of the total orthophosphorus and 50–90 percent of the total nitrogen (Rickert et al. 1977). In agricultural regions, however, their contribution is smaller.

It is difficult to locate and to design abatement strategies for non-point pollution. The most effective control is usually improved soil and water conservation and better fertilization practice. If, for example, most of the nitrogen is reaching streams by leaching from heavily fertilized croplands, a study of the soil-water balance and a lowering of application rates may be called for. Subsurface water does not usually carry high concentrations of phosphorus to streams; the element is strongly bound to soil particles. Both Horton overland flow and saturation overland flow can carry phosphorus into streams in dissolved and solid form. The majority of this transport seems to occur in the form of phosphate fertilizer adsorbed onto eroded soil particles or as organically bound phosphorus in manure and other organic detritus washed from the land surface. The phosphorus can later become available to aquatic plants in a variety of ways. These inputs can be reduced by the methods of soil conservation reviewed in Chapter 15. If the land manager also uses some of the principles outlined in Chapters 6 and 9 for recognizing zones of a catchment that are most likely to generate overland flow, other practices can be designed to reduce the loss of nutrients from these zones.

Point sources of agricultural pollution include the return flow from irrigated fields and the runoff from feedlots. Treatment of such effluent is possible but is generally too expensive to be widespread at present.

This brief review does not adequately cover all the methods of controlling the nutrient influx to rivers and lakes. It indicates only the existence of a range of strategies. Planners need to keep abreast of new developments in this field.

Pesticides

Since 1945 there has been a rapid growth in the use of complex organic compounds as insecticides and herbicides. Sheets (1967) has estimated that approximately 5 percent of the land area and about 20 percent of the cropland in the conterminous United States are sprayed with pesticides in an average year. Forests and rangelands are sprayed occasionally and rights-of-way along power lines and similar access routes are sprayed heavily and frequently with herbicides. In some parts of the tropical world, insecticides are heavily used for controlling the diseases that affect humans and livestock. Many of the substances are dispersed by wind and water and are widely distributed in wildlands, coastal marshes, and ocean sediments. A planner may become involved in decisions about pesticide application, and so we will review some aspects of the problem related to soil and water. Woodwell et al. (1971) present some data on the fate of one pesticide, while Odum (1971) and Woodwell (1967) review related ecological questions and suggest further reading.

Insecticides have been the focus of most concern during recent years. The most widely used family of insecticides are the organochlorines, which include DDT, heptachlor, aldrin, dieldrin, and BHC. These compounds were originally valued because they are lethal to a large range of pests and because, being resistant to chemical breakdown, they remain in the soil for years. Few organisms can metabolize them, and the first few stages of breakdown often produce compounds that are also toxic. DDT, for instance, is metabolized to toxic DDE and DDD. The half-life of these substances in soil is about 15 years.

It has since been realized that these characteristics, along with the high solubility of organochlorines in fats, cause damage to animal populations. This concern has led to the banning or restricted use of several organochlorine insecticides in the United States during recent years, although some are still authorized for use. In some other parts of the world, however, pest problems are so acute that on balance the continued use of dangerous insecticides is judged to be worthwhile for combating diseases such as malaria. A second family of pesticides, the organophosphates, includes such compounds as parathion and malathion. They are less resistant to breakdown than are the organochlorines but are more toxic to mammals, including humans.

Most of the information about insecticides in hydrologic systems concerns organochlorines, which enter aquatic ecosystems by direct fallout, are then adsorbed onto solid particles of organic material or soil eroded from the catchment, or go into solution. These compounds are not very soluble in water but will dissolve easily in the fatty substances in organic material. Their concentrations in stream water, therefore, are very low, but the river transports the solids in which the pesticides are carried.

Because of their solubility in fatty substances, organochlorines are concentrated within phytoplankton in the water. As the detritus or phytoplankton are eaten by zooplankton or fishes, the pesticide is dissolved by fatty substances in the gut and is stored within the body. As these organisms in turn are eaten by larger organisms, the concentration of the pesticide in the body of the predator increases because the organochlorine dissolves in fats rather than being excreted to the water.

Woodwell et al. (1967) have illustrated the effects of this process with the data in Table 20-8. Marshes on Long Island, New York, were sprayed with DDT to control mosquitoes. The concentrations were kept low to avoid poisoning fish or birds. Several years later the spraying had to be stopped

Table 20-8 Concentrations of toxic DDT, DDE, and DDE residues in samples from an estuary on Long Island, New York. The values are in parts per million wet weight of the whole organism. (Abstracted from a table by G. M. Woodwell et al., *Science*, vol. 156, pp. 821–824, 12 May 1967. Copyright 1967 by the American Association for the Advancement of Science.)

	CONCENTRATION (PPM)
Water	0.00005
Plankton, mostly zooplankton	0.04
Blanket weed algae	0.08
Shrimp	0.16
Atlantic silverside minnow	0.23
Chain pickerel (predator fish)	1.33
Green heron (predator bird)	3.54
Herring gull (scavenger)	6.50
Merganser duck (predator)	22.8
Cormorant (preys on large fish)	26.4
Ring-billed gull (preys on large fish)	75.5

because it was realized that the pesticide was being concentrated as it moved up the food chain. In the Central Valley of California, where agricultural spraying is particularly heavy, concentrations of insecticides in some streams exceed that listed in Table 20-8 by tenfold. In forestlands, where applications are lighter, concentrations are generally less than half of that shown in the table. Far greater concentrations can occur locally below manufacturing plants or municipal sewage outfalls. Spills of insecticides are responsible for a large number of fish kills and some human deaths each year.

The toxic effects of DDT and similar compounds are not yet fully known. The compounds cause the thinning of egg shells and therefore breakage and low reproduction in some birds. Odum (1971) points out that this situation is an example of how pesticide levels that are not lethal to an individual may be lethal for a population. This distinction is not made by many advocates of pesticide use. Reproduction rates are also lowered in some fishes because DDT concentrated in their egg yolks poisons the embryos. Similar bad effects on the reproduction of other wildlife have been recognized. The effects of these substances on humans are not well understood, though they are known to be accumulative in our bodies.

Herbicides are not as resistant as organochlorine insecticides to breakdown in the soil; in a forest floor they have half-lives of 10 to 300 days. Nor do they accumulate in mammals, although repeated spraying can cause a buildup to toxic concentrations. Careful application results in only minor increases in the herbicide concentrations in streams (Norris 1971), although runoff from organic-rich soils in marshy areas carries much higher concentrations for days or weeks after spraying.

Short-term effects of careful herbicide spraying are not usually deleterious, and this has encouraged managers of land and water resources to use these pesticides widely on crops and forestland. Their enthusiasm, however, seems premature in view of the limited amount of data available on the effects of long-term, low-level exposure. Some herbicides or their contaminants are suspected of causing birth defects and other diseases among humans. Ecologists have found that blanket aerial spraying of large areas is particularly detrimental to the structure of ecosystems, is uneconomic, and is commonly ineffective.

Any plans for pesticide application should be reviewed with the utmost care, and the planner should be prepared to ask pointed, specific questions and to seek a range of expert advice about this field, which is both emotional and rapidly changing.

Bibliography

Ackermann, W. C., Harmeson, R. H. and Sinclair, R. A. (1970) Long-term trends in water quality of rivers and lakes; *EOS, American Geophysical Union Transactions,* vol. 51, pp. 516–522.

American Water Works Association (1967) Sources of nitrogen and phosphorus in water supplies; *Journal of the American Water Works Association,* vol. 59, pp. 344–366.

Anderson, P. W. and Faust, S. D. (1965) Changes in quality of water in the Passaic River at Little Falls, New Jersey, as shown by long-term data; *U.S. Geological Survey Professional Paper 525-D,* pp. D214–D218.

Bell, R. C. (1950) Fertilization of natural lakes in Michigan; *Transactions of the American Fisheries Society,* vol. 78, pp. 145–155.

Besselievre, E. B. (1969) *The treatment of industrial wastes,* McGraw-Hill, New York, 403 pp.

Biesecker, J. E. and George, J. R. (1966) Stream quality in Appalachia as related to coal-mine drainage, 1965; *U.S. Geological Survey Circular 526.*

Biesecker, J. E. and Leifeste, D. K. (1975) Water quality of hydrologic bench marks—an indicator of water quality in the natural environment; *U.S. Geological Survey Circular 460-E.*

Bosch, H. M. (1950) Methemoglobinemia and Minnesota well supplies; *Journal of the American Water Works Association,* vol. 42, pp. 161–170.

Britton, L. J., Averett, R. C. and Ferreira, R. F. (1975) An introduction to the processes, problems, and management of urban lakes; *U.S. Geological Survey Circular 601-K.*

Camp, T. R. (1962) *Water and its impurities,* Reinholdt, New York, 355 pp.

Cannon, H. E. and Hopps, H. C. (1971) Environmental geochemistry in health and disease; *Geological Society of America Memoir 123,* 230 pp.

Dethier, D. P. (1975) Trace metals in a subalpine Cascade catchment; *Geological Society of America, Programs with Abstracts,* vol. 7, p. 1050.

Edmondson, W. T. (1969) Eutrophication in North America; in *Eutrophication: causes, consequences, correctives,* National Academy of Sciences, Washington, DC, pp. 124–149.

ELDERFIELD, H., THORNTON, L. AND WEBB, J. S. (1971) Heavy metals and oyster culture in Wales; *Marine Pollution Bulletin*, vol. 2 (NS), no. 3, pp. 44–47.

EMMETT, W. W. (1975) The channels and waters of the Upper Salmon River Area, Idaho; *U.S. Geological Survey Professional Paper 870-A*.

FAIR, G. M., GEYER, J. C. AND OKUN, D. A. (1968) *Water and wastewater engineering*, vol. 2, *Water purification and wastewater treatment and disposal*, John Wiley & Sons, New York.

GILLHAM, R. W. AND WEBBER, L. R. (1969) Nitrogen contamination of groundwater by barnyard leachates; *Journal of the Water Pollution Control Federation*, vol. 41, pp. 1752–1762.

GLOYNA, E. F. AND ECKENFELDER, W. W., eds. (1968) *Advances in water quality improvement;* 3 vols., University of Texas Press, Austin.

GOOLSBY, D. A. (1969) Effect of industrial effluent on water quality of Little Six-mile Creek near Jacksonville, Florida; *U.S. Geological Survey Professional Paper 650-D*, pp. D240–D243.

GUNNERSON, C. G. (1967) Streamflow and quality in the Columbia River basin; *Proceedings of the American Society of Civil Engineers, Journal of the Sanitary Engineering Division*, vol. 39, HY1, pp. 1–16.

HEM, J. D. (1970) Study and interpretation of the chemical characteristics of natural water; *U.S. Geological Survey Water Supply Paper 1473*.

HENDRICKSON, G. E. AND KRIEGER, R. A. (1960) Relationship of chemical quality of water to stream discharge in Kentucky; *Proceedings of the 21st International Geological Congress*, Part I, pp. 66–75.

HENDRICKSON, G. E. AND KRIEGER, R. A. (1964) Geochemistry of natural waters of the Blue Grass Region, Kentucky; *U.S. Geological Survey Water Supply Paper 1700*.

HINES, W. G., RICKERT, D. A., MCKENZIE, S. W., and BENNETT, J. P. (1975) Formulation and use of practical models for river quality assessment; *U.S. Geological Survey Circular 715-B*.

HOOVER, S. R. (1974) Prevention of food-processing wastes; *Science*, vol. 183, pp. 824–828.

HOOVER, S. R. AND JASEWICZ, L. B. (1967) Agricultural processing wastes: magnitude of the problem; in *Agriculture and the quality of our environment* (ed. N. C. Brady), American Association for the Advancement of Science Publication 85, Washington, DC, pp. 187–204.

HYNES, H. B. N. (1963) *The biology of polluted waters;* Liverpool University Press, 202 pp.

IORNS, W. V., HEMBREE, C. H. AND OAKLAND, G. L. (1965) Water resources of the Upper Colorado River basin: technical report; *U.S. Geological Survey Professional Paper 441*.

KARDOS, L. T. (1970) A new prospect: preventing eutrophication of our lakes and streams; *Environment*, vol. 12, no. 2, pp. 10–27.

KLEIN, L. (1962) *River pollution*, vol. II, *Causes and effects*, Butterworths, London, 456 pp.

KUNKLE, S. H. (1971) Effects of road salt on a Vermont stream: *Journal of the American Waterworks Association*, vol. 64, pp. 290–295.

LANGBEIN, W. B. AND DURUM, W. H. (1967) The aeration capacity of streams; *U.S. Geological Survey Circular 542*.

LEONARD, R. B. (1972) Chemical quality of water in the Walnut River basin, south central Kansas; *U.S. Geological Survey Water Supply Paper 1982*.

LOEHR, R. C. (1968) *Pollution implications of animal wastes: a forward oriented review;* Federal Water Pollution Control Administration, Ada, Oklahoma, 175 pp.

MACKENTHUN, K. M. (1968) The phosphorus problem; *Journal of the American Water Works Association*, vol. 60, pp. 1047–1054.

MATHER, J. R. (1953) The disposal of industrial effuent by woods irrigation; *EOS, American Geophysical Union Transactions*, vol. 34, pp. 227–239.

MAUGH, T. H. (1973) Trace elements: a growing appreciation of their effects on man; *Science*, vol. 181, pp. 253–254.

MILLER, J. P. (1961) Solutes in small streams draining single rock types, Sangre de Cristo Range, New Mexico; *U.S. Geological Survey Water Supply Paper 1535-F*.

Ministry of Housing and Local Government (1961) *Pollution of the tidal Thames;* Her Majesty's Stationery Office, London, 68 pp.

Murphy, C. B. (1973) Effect of restricted use of phosphate-based detergents on Onondaga Lake; *Science,* vol. 182, pp. 379–381.

Nemerow, N. L. (1974) *Scientific stream pollution analysis;* McGraw-Hill, New York, 358 pp.

Norris, L. A. (1971) Chemical brush control: assessing the hazard; *Journal of Forestry,* vol. 69, pp. 715–720.

O'Connor, D. J. (1967) The temporal and spatial distribution of dissolved oxygen in streams; *Water Resources Research,* vol. 3, pp. 65–79.

Odum, E. P. (1971) *Fundamentals of ecology;* Saunders, Philadelphia, 574 pp.

Perlmutter, N. M. and Lieber, M. (1970) Disposal of plating wastes and sewage contaminants in groundwater and surface water, South Farmingdale-Massapequa area, Nassau County, N.Y.; *U.S. Geological Survey Water Supply Paper 1879-G.*

Richards, L. A., ed. (1954) Diagnosis and improvement of saline and alkali soils; *U.S. Department of Agriculture Handbook No. 60,* 160 pp.

Rickert, D. A., Petersen, R. R., McKenzie, S. W., Hines, W. G., and Wille, S. A. (1977) Algal conditions and the potential for future algal problems in the Willamette River, Oregon; *U.S. Geological Survey Circular 715-G.*

Rohlich, G. A. (1969) Engineering aspects of nutrient removal; in *Eutrophication; causes, consequences, correctives,* National Academy of Sciences, Washington, DC, pp. 371–382.

Rohlich, G. A., ed. (1970) *Eutrophication: causes, consequences, correctives;* National Academy of Sciences, Washington, DC, 661 pp.

Sawyer, C. N. (1960) *Chemistry for sanitary engineers;* McGraw-Hill, New York, 367 pp.

Schindler, D. W. (1974) Eutrophication and recovery in experimental lakes: implications for lake management; *Science,* vol. 184, pp. 897–899.

Schindler, D. W. (1977) Evolution of phosphorus limitation in lakes; *Science,* vol. 195, pp. 260–262.

Sheets, T. J. (1967) The existence and seriousness of pesticide buildup in soils; in *Agriculture and the quality of our environment* (ed. N. C. Brady), American Association for the Advancement of Science, pp. 311–330.

Stewart, K. M. and Rohlich, G. A. (1967) *Eutrophication: a review;* Report No. 34, State Water Quality Control Board, Sacramento, CA, 188 pp.

Streeter, H. and Phelps, E. (1925) A study of the purification of the Ohio River; *U.S. Public Health Service Bulletin No. 146.*

Stumm, W. (1973) The acceleration of hydrogeochemical cycling of phosphorus; *Water Resources Research,* vol. 7, pp. 131–144.

Sylvester, R. O. (1961) Nutrient content of drainage water from forested, urban, and agricultural areas; *Algae and metropolitan wastes, U.S. Public Health Service Report SEC TR W61-3,* pp. 80–87.

Sylvester, R. O. and Anderson, G. C. (1964) A lake's response to its environment; *American Society of Civil Engineers, Journal of the Sanitary Engineering Division,* vol. 90, no. SA1, pp. 1–22.

Theriault, E. J. (1927) The oxygen demand of polluted waters; *U.S. Public Health Service Bulletin 173.*

Thorne, W. and Peterson, H. B. (1967) Salinity in United States waters; in *Agriculture and the quality of our environment* (ed. N. C. Brady), American Association for the Advancement of Science Publication 85, Washington, DC, pp. 221–240.

U.S. Environmental Protection Agency (1973) *Cycling and control of metals;* National Environmental Research Center, Cincinnati, OH, 187 pp.

Verduin, J. (1967) Eutrophication and agriculture in the United States; in *Agriculture and the quality of our environment* (ed. N. C. Brady), American Association for the Advancement of Science, Washington, DC, pp. 163–172.

WALL, G. J. AND WEBBER, L. R. (1970) Soil characteristics and subsurface sewage disposal; *Canadian Journal of Public Health,* vol. 61, pp. 47–54.

WEIBEL, S. R. (1969) Urban drainage as a factor in eutrophication; in *Eutrophication: causes, consequences, correctives,* National Academy of Sciences, Washington, DC, pp. 383–403.

WOLMAN, M. G. (1971) The nation's rivers; *Science,* vol. 174, pp. 905–918.

WOODWELL, G. M. (1967) Toxic substances and ecological cycles; *Scientific American,* vol. 216, no. 3, pp. 24–31.

WOODWELL, G. M., CRAIG, P. P. AND JOHNSON, H. A. (1971) DDT in the biosphere: where does it go? *Science,* vol. 174, pp. 1101–1107.

WOODWELL, G. M., WURSTER, C. F. AND ISAACSON, P. A. (1967) DDT residues in an east coast estuary: a case of biological concentration of a persistent insecticide; *Science,* vol. 156, pp. 821–824.

Typical Problem

20-1 Calculation of the Dissolved-Oxygen Sag for a Polluted River

Suppose that a waste with a discharge of 5000 gallons per minute and a BOD of 1000 mg/liter enters a river with a low flow discharge of 200 cfs, a mean velocity of 3 ft/sec, and a depth of 2 ft. The temperature of the mixture is 25°C, and the stream is initially saturated with oxygen. The waste originates in a food-processing plant, and a deoxygenation constant of 0.25 per day is thought to be appropriate at 20°C.

Calculate the dissolved oxygen sag for the stream water as it flows downriver.

Solution

Calculate first the BOD of the mixture of waste and river water. The total discharge is 211.1 cfs or 5980 liters/sec.

The input rate of the BOD is 315 liters/sec × 1000 mg/liter = 315.000 mg/sec.

The BOD concentration of the mixture, therefore, is 315,000 mg/sec ÷ 5980 liters/sec = 52.68 mg/liter, which is the value of L in Equation 20-2.

The deoxygenation constant of the waste at 20°C put into Equation 20-3 indicates that at 25°C the appropriate value for k_{25} is 0.32 per day.

From Equation 20-4, $r_{20} = 3.94$ per day, and with the aid of Equation 20-5, $r_{25} = 4.35$ per day. Figure 20-11 indicates that this value is a reasonable one.

If these values are inserted into Equation 20-2, with an initial oxygen deficit of zero, values of the oxygen deficit can be calculated for a selection of times and distances downstream. The results are plotted in Figure 20-10. Note that the calculations are simplified in the sense that constant discharge is assumed in the downstream direction. To make the results more realistic, tributary discharges should be added along the stream and a new value of L derived below each junction.

The concentration of waste remaining unoxidized after t days is Le^{-kt}. As an exercise calculate the effect on the oxygen sag of discharging a second waste, identical in magnitude and strength to the first one, into the river two days downstream. Assume constant discharge and values of k and r in Equation 20-2. A new value of L must be computed below the second outfall, and the value of D_i in Equation 20-2 is the value of D two days below the first outfall. The results are shown in Figure 20-12(d).

21

Stream Biota and Biologic Health

River water supports a whole world of its own, much of which cannot be seen. The microorganisms alone comprise a surprising variety and number of forms. The biotic health of a stream is indeed indicated by the variety and the composition of the population, visible and microscopic.

The environmental planner has a stake in the health of the watercourse because it affects the level of amenities, the potential for recreation, and public health. Because it is possible to make a rough assessment of the biologic condition of stream water with only a modicum of training, the present discussion has some practicality for the planner. Although a person untrained in biology cannot do a sophisticated evaluation, with a little knowledge of the observable forms, he or she can make a usable estimate of stream health by making field observations. The present discussion is directed to planners and environmentalists who, with little or no previous training, desire to learn a technique for making a basically sound, though admittedly rudimentary, evaluation of stream biota based on field observation alone without the help of laboratory or microscopic facilities.

A healthy stream has a large variety of organisms and a moderate population of most taxa. Diversity is thus a key attribute. High diversity is indicated by many forms or species, but not a predominance of any one form. As diversity decreases, one of the indications of loss of health, there are fewer taxa, and one or several will have a large population at the expense of other forms. When the population as a whole becomes small and only a few forms are present, either heavy organic pollution or toxic poisoning is indicated.

The second principal expression of health, or lack of it, is the mix of particular genera. As health deteriorates, some forms will become rare or absent and others more common. Therefore, the evaluation is based on the combination of families or genera present and the relative abundance of these.

It is somewhat easier to list some key genera that can be identified than to describe unequivocally the abundance. The latter may be approached by specifying the amount of time required to make a stream survey. Patrick and Grant (1971) recommended that half an hour be spent working in a stream less than 75 feet wide, but an hour is needed to examine a stream 75–150 feet wide. The time is used examining rocks picked up from the bed, looking at invertebrates on these rocks, examining floating or submerged debris such as deadwood, and sweeping deep water at stream margins with a net or porous sweep. During the time of examination, biotic forms may be observed on the rock with a hand lens or taken off the rock by tweezers and examined. On a rock of fist size one may find 3 to 15 crawling insects and a few to dozens of larval cases. Such numbers would signify very common or common species, but when 1 or 2 forms per rock are found, they are considered less common. If few insects are found when several rocks are examined, they are considered rare.

In the course of working in both pool and riffle, if any fish are present some will be seen even if they evade the net. If none are seen at all, they might be considered rare.

Following is a brief description of the macroscopic character of the key genera. The description should be compared to the sketches in Figures 21-1 and 21-2. More details of field procedure are given in Slack et al. (1973) and of identification in Needham and Needham (1962).

Diatoms–Algae Slime

These microscopic members of the plant kingdom display beautiful designs under the microscope, their siliceous or organic structure taking on a great variety of rodlike, ovoid, or streamlined shapes; they proceed quickly across the field of view. But because the field observer cannot see these forms, the macroscopic character must be observed.

One common form of diatom bloom is a brownish slime covering a rock. It often has the color of iced tea. Under water the slime coating may appear to have a palpable thickness, like a piece of velvet with each hair thin and waving.

Another common form looks initially like an oil slick, brownish or yellow-brown in color and iridescent with blue or violet hues. It is often seen in still water or an unmoving backwater.

Algae may be green or brown, the former taking on its color from the

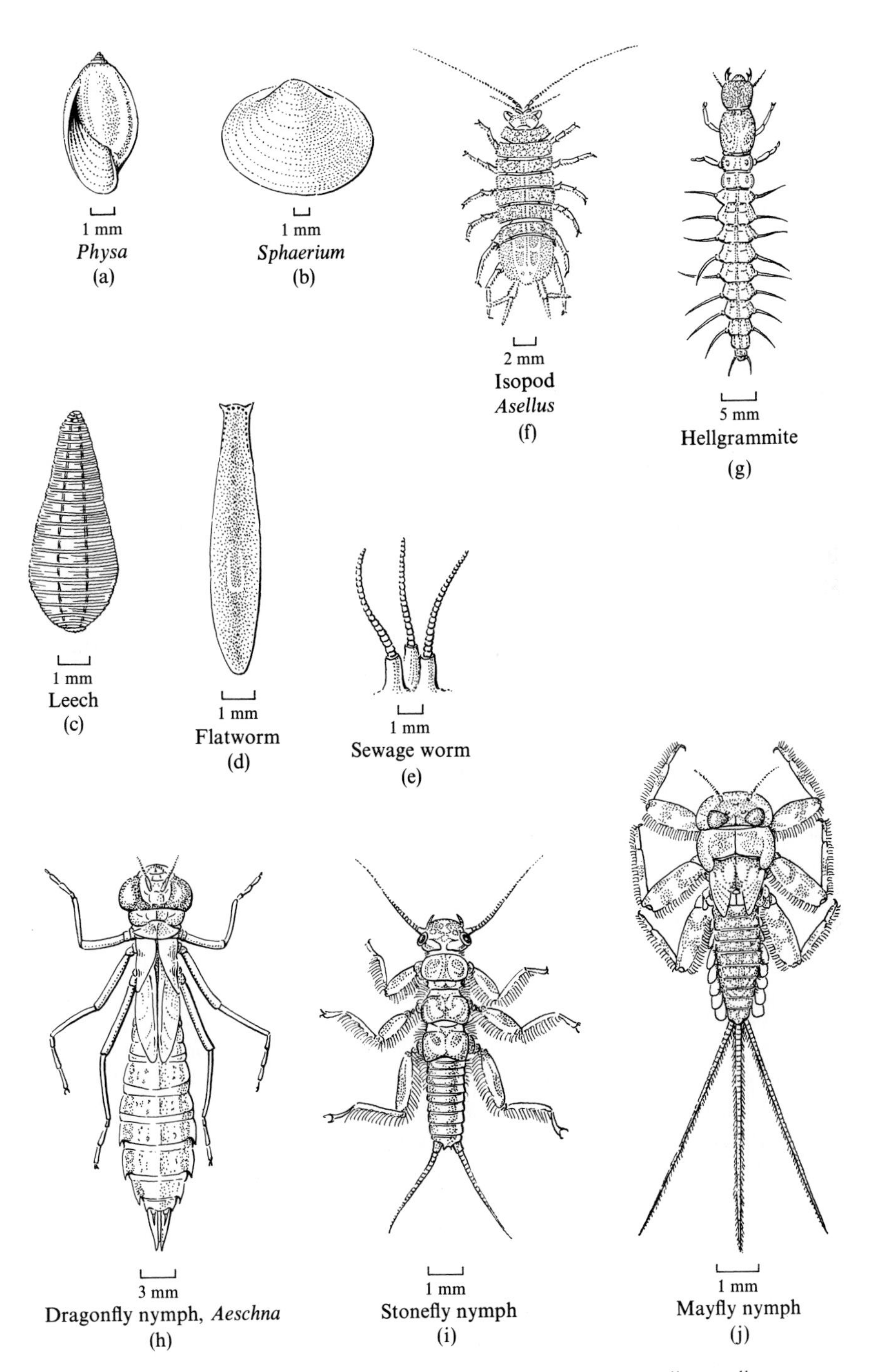

Figure 21-1 Some biologic forms observable in streams: (a) *Physa* snail, a mollusc; (b) *Sphaerium*, fingernail clam; (c) leech; (d) flatworm; (e) sewage worms, tubificids; (f) *Asellus*, an isopod crustacean; (g) hellgrammite; (h) dragonfly nymph; (i) stonefly nymph; (j) mayfly nymph.

chlorophyll it contains. Most diatoms appear more brown than green when viewed microscopically, but they may still give water a bright green or yellow-green opaqueness, as in lake water experiencing a bloom.

Filamentous or Matted Algae

Light green or yellow-green, waving, thin filaments several inches to more than a foot long trailing downstream from banks or debris would probably be *Spirogyra*, an unbranched filamentous green algae. Macroscopically, *Cladophora*, *Oedegonium*, and *Stigeoclonium* may appear similar. Some of these may hang as a thick fringe from a wall or a stick in fast flowing water. Each will be a light green color and feel slimy.

Blue-green algae are mostly matted and appear as a clump or mass of velvety unwaving material, like a piece of sod covered with moss that has been submerged. But the color is diagnostic. Blue-green algae are hunter green or dark green, in contrast with the light or yellow tint of the green algae.

The blue-green *Nostoc* produces colonies that resemble jelly-balls about the size of a marble, each ball having a gelatinous coating. Another mat-forming blue-green algae is *Phormidium,* which produces a tough, compact, brownish, gelatinous mass fixed to the substrate.

Of taxa ordinarily occurring in lakes, but sometimes found in the still water of a stream, two may be mentioned. A dense gelatine mass floating at the surface in still water having a moldy appearance may be the blue-green *Anabaena.* Greenish strings resembling lawn clippings floating in water may be colonies of *Aphanizomen.*

Invertebrates

There are several macroscopic invertebrates that are in a general way diagnostic of stream health. The ordinary flatworms, *Dugesia* or *Planaria* (Figure 21-1(d)) are shaped like an elongated arrowhead, 5–8 mm in length. They live in the mud or in decomposing organic debris on stream margins. When abundant, they indicate organic enrichment.

The tubificids or sewage worms closely resemble ordinary earthworms, even in the reddish color (see Figure 21-1(e)). They generally are 20–40 mm in length. Like flatworms, tubificids occur in mud or organic debris. The tubificids, especially *Tubifex,* are tolerant of organic pollution and if very common (and other invertebrates scarce) indicate pollution.

Leeches of the order *Oligochaetes* (Figure 21-1(c)) are well known to youngsters in the old swimming hole. They are dark brown, leathery, with a fast-hold at one end that can attach to the skin and suck blood, although many are also invertebrate predators. When abundant, leeches indicate organic pollution.

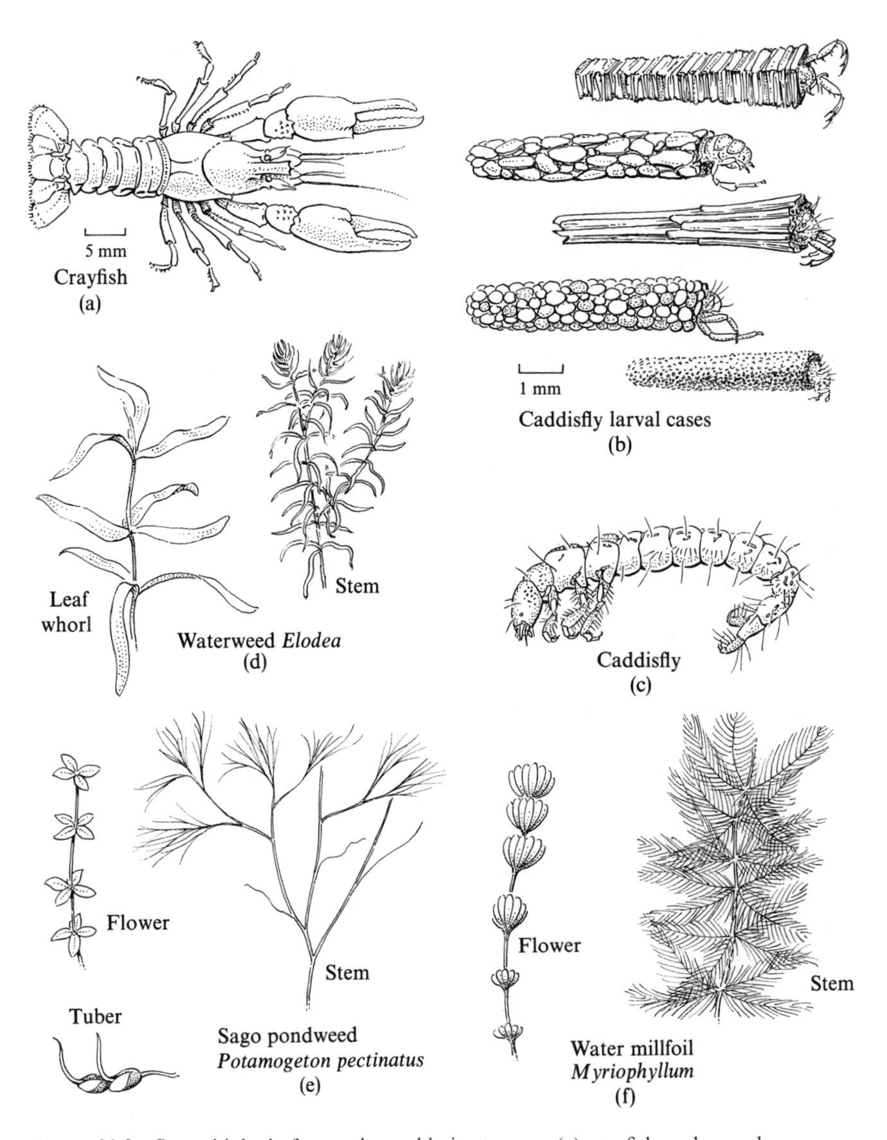

Figure 21-2 Some biologic forms observable in streams: (a) crayfish, a decapod crustacean; (b) five types of larval case of caddisfly; (c) a caddisfly larva; (d) *Elodea,* water weed—a whorl of leaves shown enlarged; (e) Sago pondweed, *Potamogeton*—whorls of flowers and root tuber shown enlarged; (f) water millfoil, *Myriophyllum*—flower whorls shown enlarged.

Molluscs

The mollusc shells of fresh water are commonly of two types: a spiral built by a snail and a bivalve of the clam. Both common and diagnostic is the snail, *Physa,* having a left-hand spiral shell 2–7 mm in length (Figure

21-1(a)). When common, *Physa* indicates organic enrichment or organic pollution. They are, however, sensitive to toxic poisoning and disappear in toxic pollution conditions.

A small clam of the size of a fingernail with a rather thick shell, less than 20 mm in size, is *Sphaerium* (see Figure 21-1(b)). When common, the conditions are organically enriched. It becomes very common when the water is polluted with organic material.

A source of difficulty in irrigation canals in the western states is the freshwater clam or mussel, *Corbicula,* whose bivalve shells range from a fraction of an inch to several inches in length. These clams have a distinctive concentric sculpture of the shell. They burrow in mud or sand in the stream bed. When colonies die, they can create a large dense mass of shells.

Crustaceans

In the class Crustacea there are two forms, relatively easy to identify macroscopically, that have diagnostic value. *Asellus* is one of the isopods resembling somewhat a sowbug with six legs on each side and antennae as long as the longest hind legs (see Figure 21-1(f)). When this form is common it is an indicator of enriched or polluted conditions.

The crayfish, a decapod crustacean, resembles a diminutive lobster with a hinged abdomen that curls the tail under the body. Two long and two shorter antennae project forward over claws that may be long like a lobster's or short (see Figure 21-2(a)). A large population suggests organic enrichment, but there are streams such as those in Yellowstone Park and in the Lake Tahoe region where crayfish are common and do not have diagnostic value.

As in the case of the invertebrates and molluscs, the crustaceans are found on the stream margins and in organic debris.

Insects

The insects in a stream constitute a major part of the population on rocks of the stream bed. Insects have two typical life histories. One, illustrated by a fly, begins when an adult deposits its eggs into a stream. The eggs hatch under the water and produce wormlike larvae that attach themselves to the solid substrate, such as a rock, or move over its surface. The larval stage may last several weeks or a year, at the end of which a second immature form, a pupa, develops. The pupa may look like a mummified version of the adult and lasts typically for a short period before the winged adult cracks out of the back of the pupal skin. It then moves out of the water to the atmosphere.

The second type of life history proceeds from egg to nymph (or naiad), a wingless version of the adult and equipped with gills in stream varieties.

The nymph grows in size for up to a year or more before emerging as an adult and crawls out of the water onto nearby vegetation or rocks.

The eggs deposited by the adult insect begin the cycle again.

Identification of insects on the rocks of a stream bed usually concerns either the larval stage or the nymphal stage. Though the free-flying insect or swimming adults such as beetles may be observed, the population can best be estimated by the immature stages–larvae, pupae, or nymphs.

Stonefly Nymph (Plecoptera)

The stonefly nymph is characterized by two tails in contrast with the three of the mayfly. Its six legs are usually fringed with hairs and have a double claw. The abdomen tends to be of the same order of length as the thorax (see Figure 21-1(i)) and shows no gills except near each end. The nymphs vary in length from 6 to 20 mm. This creature is fairly sensitive to pollution and to silt in the water, and will become rare as nutrient enrichment becomes great.

Mayfly Nymph (Ephemeroptera)

The mayfly nymph usually has three tails but always has a single claw per leg. The tails may be feathered with hairs or may have short rough spines (see Figure 21-1(j)). Many mayfly nymphs are fairly sensitive to pollution.

Dragonfly Nymphs (Odonata)

The dragonflies, Anisoptera, tend to have a long abdomen and usually only a short spine for a tail. They have no feathery gills (Figure 21-1(h)). The dragonfly nymph tolerates a moderate degree of enrichment.

Caddisflies (Trichoptera)

The presence of caddisflies can be shown either by the larva or by the larval cases. Some caddisflies do not build cases, but spin nets or are free-living, that is, move about the substrate. Figure 21-2(c) illustrates one of the many caddisfly larvae. But even when the larva is not observed, cases may remain as evidence of the former presence of the case builder

The cases are of several types (see Figure 21-2(b)) and may consist of small grains of sand or mineral grains cemented by an organic web (or may resemble cutin) built of organic material.

The caddisfly larva is tolerant of moderate to severe organic load. Mayflies and caddisflies in high population may characterize a healthy, enriched condition that stoneflies will not tolerate.

Hellgrammites (Megaloptera)

These forms have the usual three pairs of legs, as do all insects, but often have several or many paired appendages that resemble legs along their abdomen (see Figure 21-1(g)). The hellgrammites will tolerate a heavy organic load.

Bacteria

An Actinomycete, named *Sphaerotilus,* is an indicator of severe pollution, especially by large amounts of carbohydrates. It occurs, therefore, in streams receiving wastewaters from such industrial processes as pulp and paper manufacturing.

It appears as a gray, gray-brown, or brown mass of waving hairs trailing down from fixed objects in the stream. The mass may be a foot or more in length and has the appearance of moldy growth often seen on rotting food.

Rooted Aquatics or Aquatic Weeds

The submersed aquatics may be used as one indicator of stream health. They are scarce in oligotrophic waters, which are low in nutrients. As the nutrient level increases, so does the population of these plants, which flourish in enriched water. But the combination of factors governing the abundance of these plants is complicated, so their presence is not diagnostic of nutrient level. The presence of favorable factors for their growth—slow water, silty bars on the streambed, and optimum sunlight—affect the abundance. The three common and diagnostic genera are waterweed, water millfoil, and pondweed.

Waterweed, *Elodea,* is a submersed plant with distinct ovoid leaves occurring in whorls of three of four at nodes along the stem (Figure 21-2(d)). The main stems branch in paired forks. The floating plant gives the impression of a cylindrical uniform lei or garland.

Water millfoil, *Myriophyllum,* is mostly submerged, but flowering spikes or a few top branches may stick out of the water (Figure 21-2(f)). The stems are solid and thick, in simple not compound branching form. Leaves occur in whorls of three of four at nodes along the stem. In some species the whole plant has a brownish-purple color. The general appearance of millfoil is feathery because of a fine division of leaflets.

Pondweed, *Potamogeton,* has a large number of species. This genus will withstand more severe organic pollution than will waterweed or millfoil. Some pondweeds are locally known as sago weed. The leaves of the sago type are very long and narrow, often triangular in cross section, so that the difference between leaf and stem is not as marked as in most plants (Figure 21-2(e)). Nutlike tubers occur on underground stems and are a preferred food for puddle ducks. Flowers and nuts or fruits occur at the end of a long slender terminal stem that floats on or at the water surface.

Use of the Key

There are many ways of organizing data in a matrix of interrelationships, among which is the widely used key of taxonomy. Unfortunately, the presence or absence of various biologic forms and their numbers in a stream do not lead inexorably to a single classification of a degree of health. Further, the mix of the population changes with season. Indeed, the rise and fall of populations during seasonal changes in weather and streamflow constitute one of the ways the biotic community maintains its diversity.

Thus it must be understood that the use of the enclosed key gives only an indication of stream health. A trained observer would produce a more assured evaluation by closer identification, especially of microscopic forms. Using the same principles, the trained observer's experience would provide far greater sophistication, for there is no formula by which the different indicators are weighted.

In the case of the key, by far the most tenuous step is the evaluation of the population and composition of the algae. This can be done properly only by examination under a microscope. However, a lack of assurance of this first choice in the use of the key is not critical. Perhaps, in a particular case the observer does not see the diatomaceous or algal slime as described here. The abundance of algae is greatly influenced also by the amount of sunlight, because those forms that are photosynthetic require a sunny location for optimum growth. The observer should then move on to the next step in the key and look at the insect population and look for the rooted aquatics. If there are no submerged plants, but there are many insects and a diversity of forms, it would be logical to conclude that the condition is in the healthy, upper part of the key and not in the polluted or toxic condition.

Experience in teaching others the generalized identification of observed forms and the use of the key for deriving an estimate of the conditions of health give confidence that with a little practice in observation, reasonable estimates of biotic health can be made even by persons quite untrained in the details of biotic taxonomy.

The reasoning used in the following key was developed by Dr. Ruth Patrick. We have reorganized her observations into the present form, but any errors of fact or wording must be attributed to us.

Key to Biotic Health of Stream

V = very common *C* = common *L* = less common *R* = rare

Population low but diverse–many forms, but few of any single form.
Insects *C*, including mayflies, stoneflies, caddisflies.
Snails, crustacea present but *R*; no blue-green algae.
No rooted aquatics, no green algae.
Healthy, low nutirent.
Green algae *R*, rooted aquatics *R* or *L*.
Healthy, low nutrient.

Population high and diverse–many forms and high population of some forms.
Diatoms *V* or *C*.
Insects *V* or *C*, includes mayflies, caddisflies, and stoneflies *C*.
Blue-green algae absent or *R*; fish *C* to *L*.
Green algae *R*, *Spirogyra R*.
Rooted aquatics, *Oedegonium* or *Cladophora R* or *L*.
Healthy, small amount of nutrients present.
Spirogyra C, rooted aquatics *C*, crayfish, fingernail clams *C*, *Physa* snails *C* or *L*.
Healthy, but enriched.
Diatoms *L* or *R*
Insects *C* to *L*, green algae, *Spirogyra C*, rooted aquatics *C*.
Blue-green algae *C* or *L*.
Nutrient-enriched.
Insects *L*; rooted aquatics including *Elodea, Potamogeton, V* to *C*; *Spirogyra V* to *C*.
Blue-green algae *V* to *C*; mayflies *R* to absent.
Physa snails, tubificids, flatworms *C*. *Asellus C*.
Sphaerotilus R to absent.
Organically polluted.
Sphaerotilus V to *C*.
Heavily polluted.

Population low; not diverse, not many forms observed.
Blue-green algae or *Sphaerotilus* present.
Physa snails, tubificids, flatworms, *Asellus C* to *L*.
Heavily polluted.
Physa snails, tubificids, flatworms *R* to absent.
Toxic pollutants present.

Bibliography

NEEDHAM, J. G. AND NEEDHAM, R. P. (1962) *A guide to the study of freshwater biology;* Holden-Day, San Francisco, 108 pp.

PATRICK, RUTH AND GRANT, R. R., JR. (1971) Observations of stream fauna, in Hydrology of two small river basins in Pennsylvania before urbanization; *U.S. Geological Survey Professional Paper 701-A,* pp. 24–39.

PRESCOTT. G. W. (1964) *How to know the fresh-water algae;* W. C. Brown, Dubuque, IA, 272 pp.

SLACK, K. V., AVERETT, R. C., GREESON, P. E. AND LIPSCOMB, R. G. (1973) Methods for the collection and analysis of aquatic biological and microbiological samples; *U.S. Geological Survey Techniques Water Resources Investigation,* book 5, chapter A4, 165 pp.

USINGER, R. L., ed. (1956) *Aquatic insects of California;* University of California Press, Berkeley, 508 pp.

22

Evaluation of Esthetics

A principal goal of environmental planning is to maintain or enhance the amenities of landscape. Too often, engineering plans for development—whether they be for urban construction, flood control, transportation facilities, or other—are concentrated on the structural, architectural, or layout aspects, with but scant attention to the visual, ecologic, or hydrologic harmony of the result. Many of the elements that may be classified as amenities are more subjective and less clearly of economic value than the engineering or construction aspects of development work. But it has become increasingly clear that people are sufficiently concerned about the nonmonetary values of the environment that they are willing to forgo some opportunities for gain or are even willing to pay directly for the maintenance of the amenities of the landscape. These values consist of two parts: how the landscape looks and how the natural processes operate. That is, there are values to society gained from the maintenance of beauty and from the knowledge that the ecosystem continues to operate without undue degradation.

It is logical then to include in a discussion of land hydrology in environmental planning a survey of the techniques available for assessing, weighing, and comparing such amenities. Landscape architects or planners will be called upon more often to evaluate the hydrologic aspects of a proposed development than to carry to a final stage the hydrologic design of an engineering work. But they will often be in a position to assess the environmental impact on the nonmonetary or esthetic values of a hydrologic system or its components. That is, a professional hydrologist or hydraulic engineer may make the final design for a proposed flood-control measure, but the evaluation of the scenic or esthetic impacts of that proposed development may fall clearly within the sphere of the environmental planner or landscape architect.

Developments of a wide variety impinge on hydrologic features or functions. Furthermore, landscape amenities are often concentrated in areas where water is present, along rivers, along lakeshores, or at least on valley floors created and maintained by the channel system. Therefore, the esthetics of the riverscape is an essential part of modern hydrology as far as the environmental planner is concerned.

The purposes of esthetic evaluation are generally to make one or more of the following judgments:

1. How will a proposed development alter the visual or nonmonetary values of a landscape?

2. What would be the relative impacts of alternative plans to achieve a particular goal?

3. Are the esthetic values of a particular site sufficient to demand no development, preservation, or limited use?

The first of these is the most common type of problem. An evaluation of esthetic components may lead to some alteration of a proposed development to reduce the impact. Less often, alternatives are proposed in sufficiently equal detail to allow an esthetic evaluation to influence the choice among them. Still fewer in number are the instances in which an esthetic assessment is likely to lead to purchase, designation as park or protected land, or protective regulation. Yet, federal and increasingly state legeslation concerning environmental quality and the requirement of environmental impact statements have increased the importance of techniques for esthetic evaluation.

Fabos (1971) and Melhorn et al. (1975) provide reviews of the various methodologies proposed for measurement and ranking systems. The latter classifies techniques into four types: graphic, interviews, viewing, and matrix.

Graphic methods have in common the designation of land cells or units of various sizes in which each cell has a combination of factors that may or may not be weighted. The unit cells may vary from resource regions of large size to small acreages selected as optimal for particular uses. The units may be derived from a computer sorting of factors, and computer maps may also be produced.

Interview and viewing techniques connote, respectively, determination of preference among alternative sites or scenes by a sample of individuals either interviewed or exposed to a series of photographs for ranking. These have in common two possible goals of investigation: (1) the psychological attributes that determine individual preferences or (2) the pragmatic ranking of scenes in order of preference for the purpose of obtaining a quality ranking of the scenes presented. For the most part, the first goal has dominated the use of these procedures.

The matrix approach has much in common with graphic methods. Key elements chosen to describe the characteristics of a site or scene are rated for their relative presence or absence.

The literature on studies of perception and assessment methodology is now large, and there are good reviews that compare methods. The bibliography at the end of the present discussion will guide the interested reader into a sampling of this large and growing field.

Our purpose here is to present examples that illustrate some of the questions that arise in environmental planning and evaluation schemes that have been used to answer the questions. There is no single scheme universally applicable to all questions. After presentation of the examples, we summarize the elements that seem generally involved in all schemes.

All the evaluation schemes that have been applied to practical problems rather than to a study of psychological determinants of viewer preference consist basically of two steps: first, the determination of the degree to which certain chosen attributes are present or absent in each areal unit of landscape. This first step is thus the collection and organization of basic data. The second step is the application to those basic data of a procedure for ranking or weighting for the purpose of integrating or summarizing, the final result of which is to place each areal unit into a rank order or a classification that will reflect the use, capability, or relative value for chosen purposes.

This two-step approach allows an important requirement to be met, namely, the separation of facts and characteristics from value judgments or assignment of rank based on preferences of the evaluator. Recognizing that the assignment into some category representing the degree to which a particular characteristic is present or absent is usually an estimate or a subjective judgment, this assignment can and should be uninfluenced by personal preference. Experience has shown that despite the necessity for estimating, different observers will categorize the characteristics of a sample with similar and reproducible results if the definitions of the categories have been written with care. The application of preferences should properly be made only in the second step so that the basic data may be considered existential facts.

Maintaining this separation of facts from value judgments seems obviously important. It is useful to know where and how personal or administrative preference enters the assessment scheme. If the evaluator shows what he or she prefers and where those preferences enter, another observer can understand better how the earlier judgment was reached and why differences in assessment exist.

As in other problems to which an analytical technique is to be applied, there must first be given a statement of the goal. In the examples reviewed here the goals are different, and the goal or desired end product determines in part the list of factors or characteristics for which basic data will be collected. Regardless of goal, however, the initial stage in step 1 is a checklist of characteristics to be observed.

A checklist of observed characteristics might best be demonstrated by example. This one is drawn from a published paper (Leopold 1969) that includes more detail than will be presented here. The goal or statement of the problem might be formulated as follows: Hells Canyon of the Snake River in northwestern Idaho is a spectacular gorge extending about 100 miles below several already constructed reservoirs for power production between Boise and the mouth of the Salmon River near Lewiston, Idaho. This reach of the Snake River has now been designated a National Recreation Area. How does it compare in esthetic character with some other river reaches in central Idaho?

Having stated the question, two aspects of step 1 must be considered, the universe within which the comparison is to be made, and the list of factors or characteristics that will be evaluated for each sample within the universe. The universe chosen consisted of 12 localities on mountain rivers all presently used to some extent for recreation, most within national forests. Because the Snake River is the largest in the state, none of the other 11 sites are on such a large river, and this alone makes Hells Canyon somewhat different from the others. This fact also illuminates the importance of the choice of the universe from which the samples are to be drawn. In the present example, the universe excluded desert landscapes and other scenic environments of character different from mountain rivers.

For this particular analysis the checklist of factors chosen included those characteristics that influenced the character of the valley and the character of the river. The factors were seven in number, and five categories describe the relative presence or magnitude of each factor, as shown in Table 22-1.

Table 22-1 Checklist of descriptive factors with five categories of presence or absence of each.

	EVALUATION NUMBER FOR DESCRIPTIVE CATEGORIES				
	1	2	3	4	5
PHYSICAL FACTORS					
River width (ft)	< 3	3–10	10–30	30–100	> 100
River depth (ft)	< .5	.5–1	1–2	3–5	> 5
Valley width (ft)	<100	100–300	300–500	500–1000	>1000
Height of hills (ft)	<300	500	1000	2000	>3000
HUMAN-INTEREST FACTORS					
Urbanization	No buildings → Many buildings				
Vistas	Views of far places → View confined				
Pools and rapids	Smooth → Rapids and falls				

This example demonstrates that the descriptive factors can often be categorized as physical, biologic, and human use or interest. The five used here are of the physical and human-interest types. The example also shows that some factors can be quantitatively determined, or at least estimated in usual measurement units of feet, miles, or percent. River width, even when estimated, is unequivocal. The presence or absence of scenic views or vistas, however, is of the more subjective type and can best be expressed relatively, in this case in one of five categories denoting frequency.

Valley character was determined as follows. Width of valley was plotted against height of nearby hills, reasoning that where high mountains exist next to narrow valleys, the scenery is of large scale. In contrast, low hills adjacent to a wide valley lack the grandeur of large landscape scale. To describe this, the bottom graph of Figure 22-1 was prepared. Values of valley width and height of hills, estimated in the field and expressed in units of feet, are plotted against one another on logarithmic scales. Each of the 12 numbered points plotted represents a site, that is, a reach of river valley chosen as a sample to be compared with the other river sites.

To reduce the combination of height of hills and width of valley to a single parameter, here called landscape scale, the projection of the position of each plotted point on a 45° diagonal line was carried out. This projection is indicated by the dashed lines leading from each point at a 45° angle upward to the right in the lower graph.

Note that at this place in the analysis we have entered step 2, in which the weighting of factors begins. The projection to a 45° line is an assignment of equal weight to ordinate and abscissa values. Had some other angle been used, the ordinate and abscissa values would be given different weights.

The position of each point projected onto the diagonal line is a quantitative assessment of position on the yardstick of landscape scale, which can be used for construction of a new graph, shown in the middle of Figure 22-1. Landscape scale is plotted against the degree of view confinement or the presence of vistas. Where distant vistas are available in large-scale landscape, one has the impression of spectacular scenery. In contrast, confinement of view by heavy cover or where views are blocked by adjacent hills, especially when combined with subdued landscape, the result is esthetically ordinary from the scenic point of view.

In the center diagram, the position of each point can again be projected to a 45° line, providing a new yardstick of landscape interest, the factors scenic outlook and landscape scale being given equal weight. Landscape interest may be considered to vary between the spectacular and the ordinary.

In the upper diagram in Figure 22-1, landscape interest is plotted against the degree of urbanization. The combination of spectacular scenery with a low degree of urbanization falls in the lower left portion of the diagram along a scale called valley character, and might be called spectacular and wild. The combination of ordinary landscape interest and much urbanization falls in the upper right portion of the graph, where the combination yields a landscape character that may be called ordinary and urban.

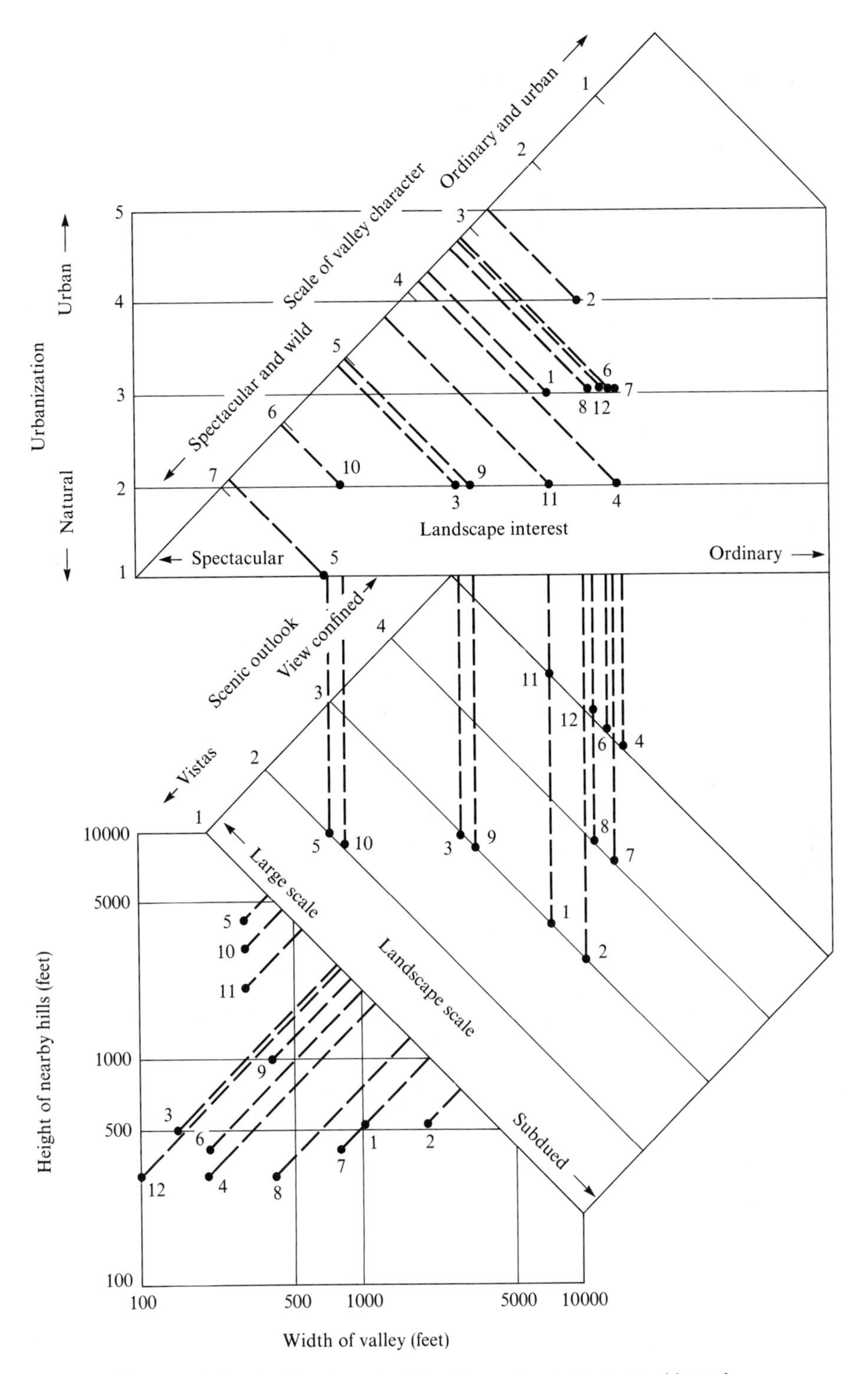

Figure 22-1 Analysis of valley character. The site number is labeled beside each plotted point. (From Leopold 1969.)

In this manner equal weights are given progressively to valley width, height of hills, scenic outlook, and urbanization to give a quantitative relative value on a scale of valley character.

Similarly, the position of each of the 12 sites on another scale called river character is shown in Figure 22-2, constructed from combining three factors–river width, river depth, and the prevalence of rapids. Again, the factors were given equal weights by projection on to 45° diagonal lines.

Now for the 12 sites, relative values on scales of valley character and river character are available and can be plotted against each other as shown in Figure 22-3. Site 5 has the highest value on each scale and relative to the

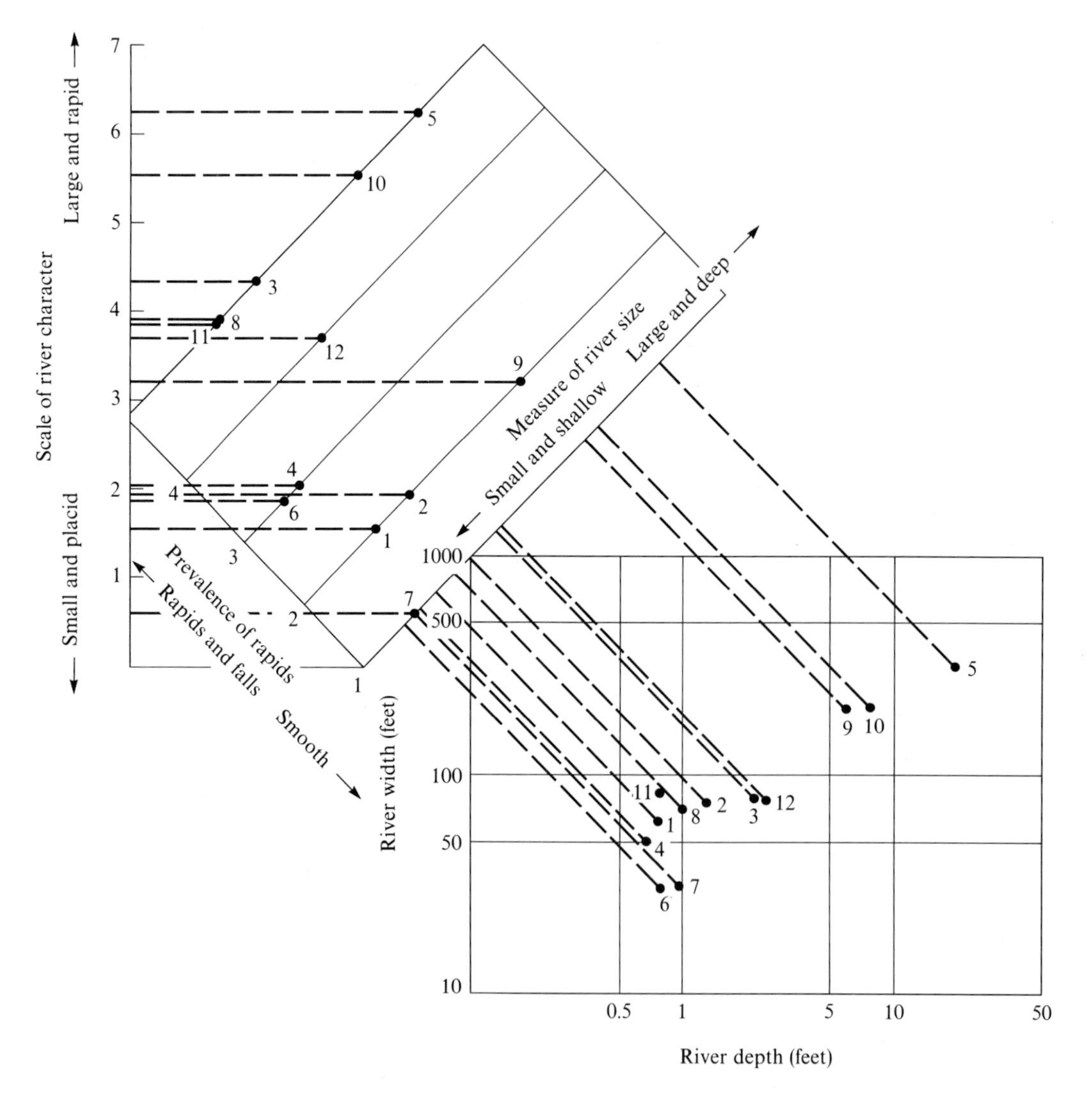

Figure 22-2 Analysis of river character. The site number is labeled beside each plotted point. (From Leopold 1969.)

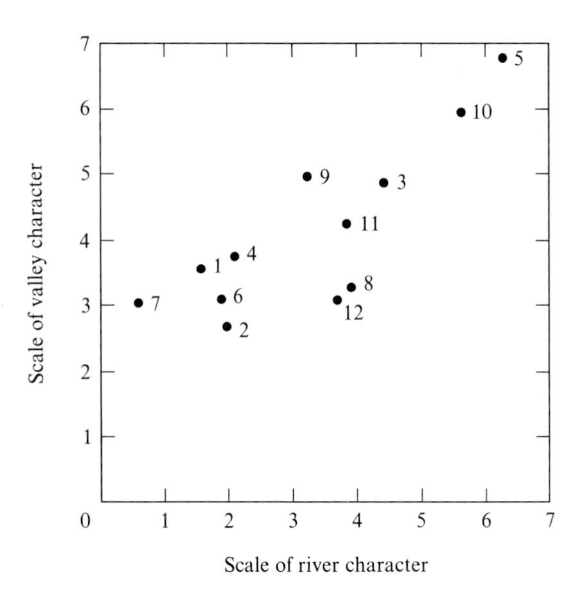

Figure 22-3 Relation of valley to river character for surveyed sites in Idaho. (From Leopold 1969.)

other sites is the most spectacular and wild, and the river is the largest and most rapid, a combination of highest esthetic appeal according to the assumptions used. In contrast, site 7 has a river that is smallest and most shallow in a valley most ordinary and urban. The graph in Figure 22-3 then presents a quantitative estimate of the answer to the question posed, how does Hells Canyon of the Snake River, site 5, compare with some other river reaches in central Idaho?

To illustrate how this technique of analysis may be applied to quite a different type of question, an example is presented in brief form drawn from a report by Michael Miernik, entitled "Multiple flood-plain use study, Fairfax County, Virginia" (1974). To make this example comparable to the one in Idaho just described, we chose from Miernik's report the following statement, which sets the goal or presents the question to be answered by the analysis. The study determines ". . . the relative recreational suitability of the large amount of open space adjacent to and within . . . [the] flood plains [of Fairfax County]."

The universe chosen consisted of 330 segments or reaches of channel in the county, a typical segment being 1000 feet wide and 6000 feet long, consisting of a strip of valley land containing the stream channel.

Six factors were chosen to provide a measure of the compatibility of various forms of recreation to the physical features of the stream segment. The factors were

Development pattern (eight categories ranging from Intermediate City Pattern to Forest-Wildland Pattern)

Width of valley flat

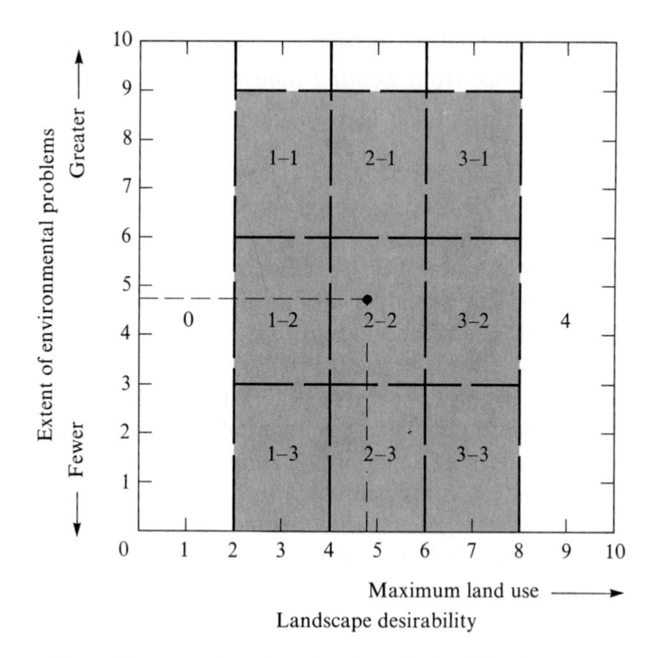

Figure 22-6 Construction of a chart of suitability for different forms of recreation. The zones in lower left are most suitable for nature trails and wildland use and those in upper right for golf courses, campgrounds, and mass uses. (From Miernik 1974.)

other uses of wild land. Rectangles toward the upper right, such as 3–1, include stream segments most suitable for golf courses, campgrounds, trailer accommodations, or other uses requiring easy access and ancillary facilities.

The graphical method of combining several observed or basic data factors into a composite estimate of hierarchical rank is only one of many ways to proceed. One of the simplest is the mere addition of evaluation numbers to produce a score or total. An example of such a method is in Knudson's (1976) evaluation of scenic rivers. His goal or question was to develop an "objective evaluation of rivers for classification in state 'scenic river' programs" (p. 281). The universe of examples consisted for 30 rivers in Indiana or segments thereof. To be included, each segment had to meet six conditions among which were a free-flowing segment of at least 10 miles and no more than 5 miles of paralleling roads within 1000 feet of the stream in any 10-mile reach.

Eight physical factors were evaluated, each having four categories or degrees representing the presence or absence of the factor. The factors were

Naturalness of bank vegetation

Vegetation depth-length index

Physical modification of stream course

Human development of floodplains, slopes
Special natural features
Esthetic quality of water
Paralleling roads
Crossings.

These factors were evaluated on a four-point scale by measurement from topographic maps, aerial photographs, and field reconnaissance. This constituted the basic data. Step 2 included the assignment of weights to the four categories, an example of which follows, using the factor "crossings," the last in the above list:

0 points	Disqualified–10 or more road, railroad, or overhead utility line crossings per 10 miles of stream
1 point	Six to 10 crossings per 10 miles
2 points	Four or 5 crossings per 10 miles
3 points	Zero to 3 crossings per 3 miles

Each factor was evaluated for each segment or river in the analysis. For the most part, the maximum number of points accrued for a factor was three, as in the example for "crossings" shown above. The total score or evaluation index for a given reach was merely the addition of all points accrued; the maximum possible score for the most desirable river from the scenic point of view would be three points for each of the eight factors or 24 points, and the minimum would be zero.

The measure of the degree to which a river segment is appropriate for scenic river classification is thus the total number of points or score accrued in the evaluation process described. Note that, in general, the weights assigned or points scored give equal importance to each of the eight factors. Under other conditions unequal weights might be assigned.

Another approach involves a more complicated and less rigorous selection of the universe or samples that must be outlined on a map before data collection begins. In such a scheme the units to be evaluated in step 2 differ in size and shape, so the selection or mapping of the units becomes a more complicated matter than in the previous examples, consisting of the designation of characteristics and then the delineation on a map of areal units that fit the combination of characteristics. An example of this type of mapping of areal units is drawn from the "Report on Visual Analysis–Teton County, Wyoming," prepared by R. Dunblazier of the U.S. Forest Service and published by the firm of Livingston and Associates (1976).

To follow again the pattern of discussion, a statement of goal or purpose is drawn from the summary of the Livingston and Associates report: "The purpose of the analysis was to evaluate visual resources in order to prevent development which would be discordant with Jackson Hole's magnificent natural setting" (p. 35).

The delineation of areal units in this study was based on outlining areas on a map considered to have characteristic combinations of slope, topography, land form, and vegetation. This resulted in six map units having irregular boundaries and various sizes. They are described as

Hillslopes with hardwood or coniferous cover

Hillslopes with dense coniferous cover

Exposed, steeply sloping side hills

Open slopes

River terraces and floodplains

Unvegetated fans and gently sloping areas.

Each of these units or types was then subjected to a rating system as in previously described examples. In this study the factors used were slope, landscape diversity, vegetative and land-form screening, revegetation potential, and soil color contrast. Each was rated on every map unit on a scale of one to three, for which appropriate definitions were developed. The reasoning behind the choice of these factors flows from the statement of the goal—the determination of how readily each landscape type can absorb visual intrusions. "Slope affects visual absorption capability because the steeper the slope, the more likely that land disturbances and developments can be seen at greater distances. Landscape diversity is an important consideration because disruptions are more visible on a homogeneous surface. . . . Screening [is] the opportunity to block the view with natural vegetation. . . . Revegetation potential was studied so that development of areas known to have low potential can be avoided" (p. 35). Soil color contrast affects how visible are disruptions of land by clearing or grading.

The rating numbers or scores were merely added to give a total score representing an estimate of the ability of each map unit to absorb visual intrusion by development, construction, or disruption of soil and vegetation. On the basis of this rating, recommendations for types or intensities of use could be made and coordinated with other constraints or opportunities dictated by hydrology, seismic and soil-movement hazards, water quality, and other considerations.

Another approach differs from those previously presented in a key element. It avoids the assignment of preferences by adopting the philosophy that a landscape that is unique—that is, different from others or uncommon—has more significance to society than that which is common. The unique qualities that enhance its value to society are those with some esthetic, scenic, or human-interest connotation. Some unique qualities may present a human-interest connotation for a reason of a different sort—it may be uniquely bad. But the principle remains the same. Human concern attaches to both.

The computation of a uniqueness index or ratio was first developed by Leopold and Marchand (1968), extended by Leopold (1969), and refined and programmed for computer by Melhorn et al. (1975). Further explanation of the philosophical premise is stated by the latter authors as follows:

> The evaluation of a landscape initially requires an evaluation of its "relative" uniqueness. . . . For example if we define an "aesthetic river" as those rivers or sections of rivers which are clear running, unpolluted and unlittered, then a stream which is polluted and cluttered with garbage dumps along the banks would be ranked very low as an aesthetic river even though it may be a relatively unique river for the region. (p. 35)

The goal or question posed is to develop a scheme by which the physical, biologic or water-quality, and human-use or human-interest factors will define the relative scarcity, commonness, or uniqueness of river reaches.

A checklist of factors under the three headings listed in the goal above was prepared, consisting of 46 in the Leopold scheme and 31 in the improved Melhorn analysis. Experience showed that many of the factors originally included were either too difficult to assess, nondiagnostic, or otherwise inappropriate. The degree of presence of each was expressed in five categories that might be called, "evaluation numbers of descriptive categories." The factors used by Melhorn, most of which were included in Leopold's original list, are as follows:

Physical

- Channel width
- Low-flow discharge
- Average discharge
- Basin area
- Channel pattern
- Ratio of valley width to height
- Bed material
- Bed slope
- Width of valley flat
- Erosion of banks
- Valley slope
- Sinuosity
- Number of tributaries

Biologic and water-quality

- Water color
- Floating material
- Algae
- Land plants–floodplain
- Land plants–hillslope
- Water plants

Human-use and human-interest

- Trash–number per 100 feet
- Variability of trash
- Artificial control
- Utilities, bridges, roads
- Urbanization
- Historical features
- Local scene
- View confinement
- Rapids and falls
- Land use
- Misfits

The uniqueness ratio is defined as the reciprocal of the number of sites sharing the same evaluation number. For example, if a factor is present to the same degree in 12 sites, then a site shares this characteristic with 11 others. The uniqueness ratio for each of the 12 sites is 1 ÷ 12, or .08, the reciprocal of the number of sites having the same characteristic. If, however, only a single site has a given characteristic, its uniqueness ratio for that characteristic is 1 ÷ 1, for it shares that feature with no others. Thus, uniqueness is defined on a scale of 0 to 1.0. A uniqueness ratio is computed for each of the 31 factors for each site, and the ratios for a given site may be averaged and these averages compared among sites.

The Melhorn "uniqueness index" is similar but defined differently. It is the percentage of the total possible uniqueness. The authors began by computing a uniqueness ratio for each factor for each site, just as described above. These ratio values were added for a given site and expressed as a ratio to the maximum possible sum of ratios. For example, if a given site shares with one other site the same evaluation number of each of 10 factors, it has 10 values of a uniqueness ratio of 1 ÷ 2, or 0.5, the sum of which is $10 \times 0.5 = 5$. But that site had the possibility of not sharing with any other site its evaluation number for 10 factors. Its maximum possible score would then be $1 \div 1 = 1.0$ for the 10 factors, or $10 \times 1 = 10$. The uniqueness index is the percentage of actual ratio sums to the maximum possible, or 50 percent in the instance cited above.

To place all percentages on a common base regardless of the number of factors used, the index is expressed as parts per 1000, so the example above would have a final index of 500.

The uniqueness ratio totals or index values can be computed separately for physical, biologic, and human-use factors if desired. Leopold (1969) plotted a graph of the total uniqueness ratio for human-interest factors against the total for biologic factors as one way of comparing sites.

Far more sophisticated than the methods outlined above are those developed by Grant R. Jones, probably the most advanced practitioner in the field of quantitative esthetic evaluation. That he is a landscape architect–planner emphasizes the need for and practicality of evaluation techniques, the application of which constitutes one of the skills needed in modern professional practice. Advanced and often time-consuming techniques are justified where large and expensive projects are being planned or when large populations experience the esthetic impact of the project on the landscape. We apply the word sophisticated to techniques that involve the identification of levels of landscape attributes that are abstract or involve the acquisition of data through the use of a sample population of viewers, or both. Only one (1975) among several techniques developed by Jones will be discussed here, and that merely to illustrate the principles rather than a complete exposition of procedure. Other techniques have also been used (Jones 1973, 1975).

Jones states the problem: "to predictively evaluate any changes in the visual quality of a landscape resulting from the physical introduction of a . . . [structure or development]." The problem was broken into the following needs: (1) to derive a workable definition of visual impact and (2) to invent a reliable procedure for measuring visual quality. The definition of visual quality stemmed from an earlier analysis pioneered by R. Burton Litton, who identified three components of landscape that influence the nature and degree of esthetic experience: variety, vividness, and unity. Though Litton et al. (1974) described and profusely illustrated various degrees of these components, Jones provided a method of assigning numerical ratings to them by introducing for each component a checklist of descriptive terms to be chosen by the evaluator when viewing a sample landscape. Note the similarity to the procedures described earlier.

Jones changed slightly the names of the three components and his discriptive terms for differing degrees of their presence or absence further the understanding of the definitions. Intactness, vividness, and unity are used in place of Litton's similar components:

Intactness of a viewscape is a measure of its apparent degree of natural condition.

1. very highly natural, no apparent man-made development
4. moderately natural, moderate degree of development
7. naturalness nearly absent, very high degree of man-made development

Vividness is the memorability of the visual impression received from the viewscape or its elements.

1. boundary very highly distinct/legible
4. boundary moderately distinct/legible
7. boundary very highly indistinct/illegible

Unity is a measure of the degree to which individual elements in the viewscape join together to form a single, coherent, and harmonious visual unit.

1. very high overall unity between viewscape elements
4. moderate overall unity
7. very low overall unity

The above summary does not purport to give sufficient detail to allow direct application of Jones' method, for each of the components are further subdivided and illustrated with pictorial sketches in the original paper.

These definitions of level are applied to rate a series of twenty photographs taken of the site of the proposed facility, representing the view from different compass directions and different distances from the site. Consideration is given both to distance of the observer from the site and to his vertical location relative to the proposed facility. Having obtained the photographs and chosen among them eight or twelve, two copies of each are made in color, 8 × 10 inches in size. On one a representation of the facility is dubbed in, with due consideration to distance and angle of view.

The paired photographs are submitted to a jury of persons "of the design profession, whose backgrounds include the practice of architecture, urban design, and landscape architecture." The visual quality of each set is rated using the criteria briefly discussed above. The quality before construction as compared with after construction is expressed as a percentage change. This visual-impact-change ratio, varying from .33 to 1.50 in the example given, is multiplied by the estimated "population viewer contacts per year" to yield a total visual impact.

The population factor is the "number of probable visual contacts per year that the facility will receive by some population from a given viewpoint. For a resident population visual contact from a representative viewpoint located in a settlement or community, an average of 75% of that community might see the facility from that viewpoint at some time during the course of a day, year round" (p. 688). Because in the example given by Jones the population visual contacts per year number from 400,000 to 10,000,000, this large number had to be reduced to a figure commensurate with the visual-impact ratio so the seventh root of the large number was taken yielding a "population contact factor" varying from 6.3 to 10. Thus the product, giving the final value of visual impact, varied in the example from 2.1 to 14.6, a range sufficient to convey a definite impression of the total change in landscape quality caused by construction of the facility.

The examples cited illustrate only some of the possible schemes for evaluation of esthetic quality, but they have certain common elements that may be used as a guide for the environmental planner who wishes to make an assessment. These commonalities may be expressed as an outline of procedure:

1. State the question or goal. The purpose of an assessment is to rank sites, to compare landscapes, or categorize rivers. The question or goal must be clearly and carefully stated because it controls the choice of factors to be evaluated.

2. Select samples and the definition of the universe. Often the universe is an areal unit such as a state, drainage basin, or private land within a county. The sample to be evaluated may include all the area of the population universe divided into areal units, or selected samples of areal units or selected sites. Sampling should be thought of as a form of statistical problem even though statistical methods may not be applied. That is, selection of samples for study should represent the universe.

3. Choose factors concerning which basic data are collected. The factors, of course, relate to the goal or question. The number of factors may be small as in the first example, or extensive as in the last one discussed.

4. Specify levels of presence or absence of each factor. The number of such categories must be three or more, but a number larger than five usually is impractical because such fine distinctions cannot be clearly discerned.

Definitions for each level or category must be written carefully, avoiding overlap to the extent possible. However, for many factors only a gradation or relative degree is possible to specify.

5. Inspect the samples or areal units with checklist in hand, and consider, evaluate, and record each applicable factor. As this process begins, the evaluator always has difficulty. He will revise definitions and even change the list of factors. But after completing the evaluation of a few sites, the experience gained increases both ease and speed. It is usual to go back to the first few sites after the definitions and choice of factors have been revised through actual trial.

6. The steps above are basic data collection. They should not involve preference or bias, but should be carried out uniformly and objectively. Having the data in hand, preference, weighting, and desirability are introduced. Care should be taken to recognize when and how weighting is introduced so that the reader or consumer will understand the reason for and degree of weight assignment.

7. Summarization of evaluation numbers, indices–weighted or assigned equal weight–is usually necessary. The summary value or index for each site or areal unit should express the heirarchical position, rank, or score for each site or areal unit in such a form that it is responsive to the question posed or goal desired.

Management, land-use decisions, development plans, or operational procedures will, we hope, be influenced by the results of the esthetic assessment, but the evaluation process is separate from the action program. To the extent that the question posed or goal sought is framed in a manner applicable to or necessary for a determination of policy or action, its potential influence will be enhanced. That is, the logic behind the assessment process must be appropriate to the types of action or policy possible.

As in all other techniques described in this book, there is no substitute for actual trial. It is hoped that the interested reader will not be content with reading about a technique, but will try applying it to a field problem.

Bibliography

DUMBLAZIER, R. (1976) Report on visual analysis: Teton County, Wyoming; in *Teton County, growth and development alternatives;* multilith, Livingston and Associates, San Francisco, pp. K1–13.

FABOS, J. G. (1971) An analysis of environmental quality ranking systems; in *Recreation Symposium Proceedings,* U.S. Department of Agriculture, Northeastern Forest Experimental Station, Upper Darby, PA, pp. 40–45.

JONES, GRANT R. ET AL. (1973) *The Nooksack Plan;* Jones and Jones, 105 S. Main St., Seattle, WA 98104.

JONES, GRANT R. ET AL. (1975) A method for the quantification of esthetic values for environmental decision making; *Nuclear Technology,* vol. 25, pp. 682–713.

JONES, GRANT R. ET AL. (1975) *An inventory and evaluation of the environmental, esthetic, and recreational resources of the Upper Susitna River, Alaska;* Jones and Jones, 105 S. Main St., Seattle, WA 98104.

KNUDSON, D. M. (1976) A system for evaluating scenic rivers; *Water Resources Bulletin,* vol. 12, no. 2, pp. 281–289.

LEOPOLD, L. B. (1969) Quantitative comparison of some esthetic factors among rivers; *U.S. Geological Survey Circular 620,* 16 pp.

LEOPOLD, L. B. AND MARCHAND, M. O. (1968) On the quantitative inventory of the riverscape; *Water Resources Research,* vol. 4, no. 4, pp. 709–717.

LITTON, R. B., TETLOW, R. J., SORENSEN, J. AND BEATTY, R. A. (1974) Water and landscape; Water Information Center, Port Washington, NY, 314 pp.

MELHORN, W. M., KELLER, E. A. AND MCBANE, R. A. (1975) Landscape esthetics numerically defined (land system): Application to fluvial environments; Purdue University, Water Resources Research Center, West Lafayette, IN, Studies in fluvial geomorphology, no. 1, 101 pp.

MIERNIK, M. (1974) Multiple flood plain use study, Master plan for flood control and drainage, Fairfax County, Virginia; multilith, Parsons, Brinckerhof. Quade & Douglas, Denver, CO.

MURTHA, P. A. AND GRECO, M. (1975) Appraisal of forest esthetic values: an annotated bibliography; *Forest Management Institute Information Report,* FMR-X-79, Canadian Forest Service, Department of the Environment, 56 pp.

Conversions and Equivalents

Length

1 mm	= 0.1 cm
	= 0.001 m
	= 0.0394 in
1 cm	= 0.01 m
	= 0.394 in
	= 0.0328 ft
1 m	= 100 cm
	= 3.281 ft
	= 39.37 in
1 km	= 100,000 cm
	= 1000 cm
	= 3281 ft
	= 0.6214 mi

1 in	= 25.4 mm
	= 2.54 cm
1 ft	= 12 in
	= 30.48 cm
	= 0.3048 m
1 yd	= 36 in
	= 91.44 cm
	= 0.9144 m
1 mi	= 5280 ft
	= 1609 m
	= 1.609 km

Author Index

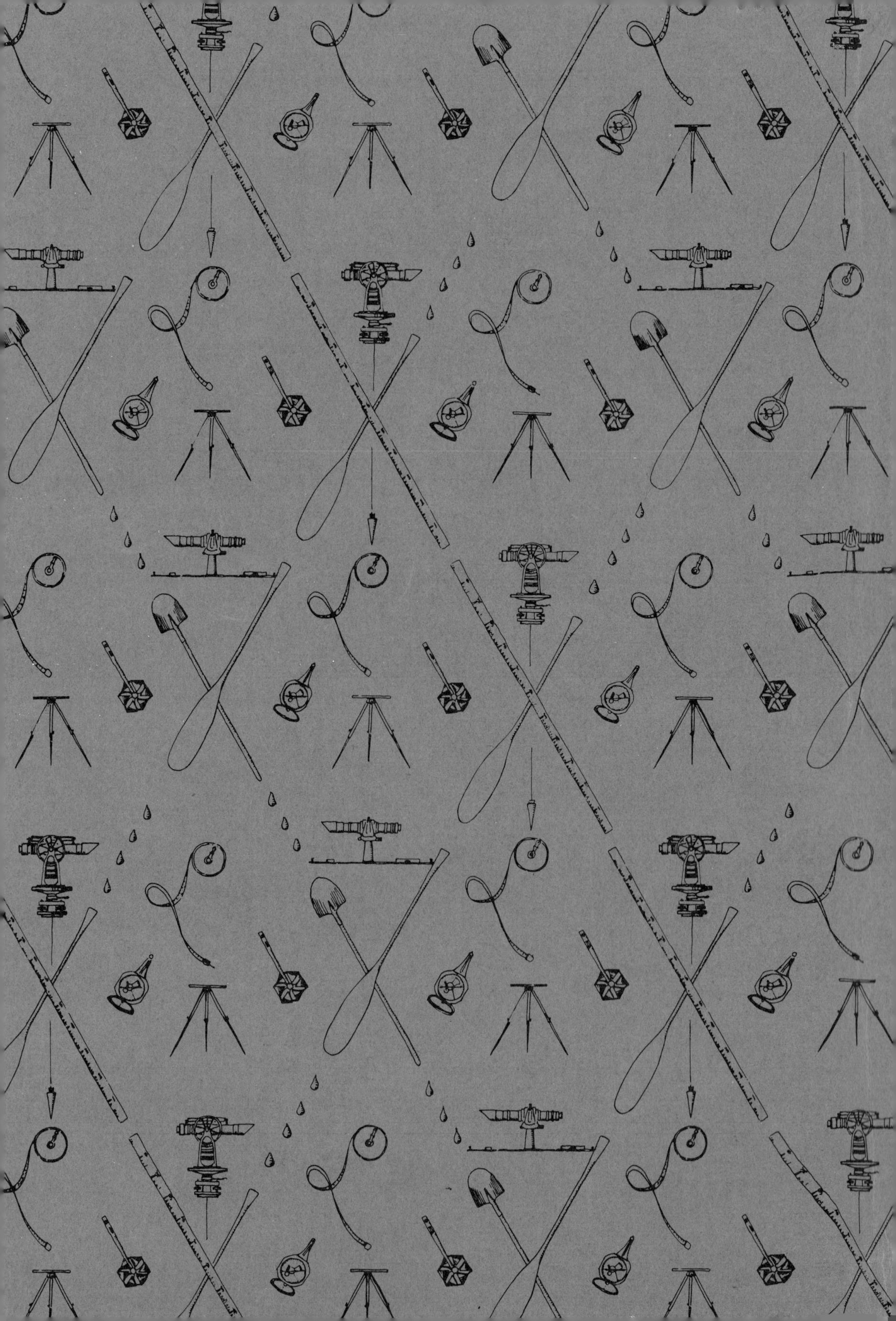